Advances in agri-food robotics

It is widely recognised that agriculture is a significant contributor to global warming and climate change. Agriculture needs to reduce its environmental impact and adapt to current climate change whilst still feeding a growing population, i.e. become more 'climate-smart'. Burleigh Dodds Science Publishing is playing its part in achieving this by bringing together key research on making the production of the world's most important crops and livestock products more sustainable.

Based on extensive research, our publications specifically target the challenge of climate-smart agriculture. In this way we are using 'smart publishing' to help achieve climate-smart agriculture.

Burleigh Dodds Science Publishing is an independent and innovative publisher delivering high quality customer-focused agricultural science content in both print and online formats for the academic and research communities. Our aim is to build a foundation of knowledge on which researchers can build to meet the challenge of climate-smart agriculture.

For more information about Burleigh Dodds Science Publishing simply call us on +44 (0) 1223 839365, email info@bdspublishing.com or alternatively please visit our website at www.bdspublishing.com.

Related titles:

Precision agriculture for sustainability
Print (ISBN 978-1-78676-204-7); Online (ISBN 978-1-78676-206-1, 978-1-78676-207-8)

Robotics and automation for improving agriculture
Print (ISBN 978-1-78676-272-6); Online (ISBN 978-1-78676-274-0, 978-1-78676-275-7)

Advances in precision livestock farming
Print (ISBN 978-1-80146-321-8); Online (ISBN 978-1-80146-322-5, 978-1-80146-323-2)

Advances in plant phenotyping for sustainable crop production
Print (ISBN 978-1-78676-856-8); Online (ISBN 978-1-78676-858-2, 978-1-78676-859-9)

Advances in sensor technology for sustainable crop production
Print (ISBN 978-1-80146-098-9); Online (ISBN 978-1-80146-100-9, 978-1-80146-101-6)

Chapters are available individually from our online bookshop:
https://shop.bdspublishing.com

climate•SMART•publishing
IN AGRICULTURAL SCIENCE

BURLEIGH DODDS SERIES IN AGRICULTURAL SCIENCE
NUMBER 139

Advances in agri-food robotics

Edited by Professor Eldert van Henten, Wageningen University, The Netherlands and Professor Yael Edan, Ben-Gurion University of the Negev, Israel

burleigh dodds
SCIENCE PUBLISHING

Published by Burleigh Dodds Science Publishing Limited
82 High Street, Sawston, Cambridge CB22 3HJ, UK
www.bdspublishing.com

Burleigh Dodds Science Publishing, 1518 Walnut Street, Suite 900, Philadelphia, PA 19102-3406, USA

First published 2024 by Burleigh Dodds Science Publishing Limited

Library of Congress Control Number: 2023952297

British Library Cataloguing in Publication Data
A catalogue record for this book is available from the British Library

ISBN 978-1-80146-277-8 (print)
ISBN 978-1-80146-279-2 (PDF)
ISBN 978-1-80146-278-5 (ePub)
ISSN 2059-6936 (print)
ISSN 2059-6944 (online)

DOI: 10.19103/AS.2023.0124

Typeset by Deanta Global Publishing Services, Dublin, Ireland

Contents

Series list xiii
Introduction xxiii
Acknowledgement xxx

Part 1 Technologies: sensing and perception

1 Advances in visual perception for agricultural robotics 3
Gert Kootstra, Wageningen University & Research, The Netherlands

1 Introduction 3
2 A short introduction to visual perception in agriculture 4
3 Challenges in visual perception for agricultural robotics 9
4 Dealing with the challenge of variation 11
5 Dealing with the challenge of incomplete information 18
6 Directions for future research 21
7 Conclusion 25
8 References 26

2 Advances in world modeling for agri-food robotics 37
Jordy Senden, Elena Torta and René van de Molengraft, Eindhoven University of Technology, The Netherlands; and Herman Bruyninckx, Eindhoven University of Technology, The Netherlands, Katholieke Universiteit Leuven and Flanders Make, Belgium

1 Introduction 37
2 The concept of a world model 38
3 Invariant-based world models 40
4 Topological maps with guarded motion 42
5 Task redundancy through multi-modal sensing 48
6 Explainable world models for skill configuration 53
7 Multi-robot cooperation through shared world models 59
8 Future perspective: world models for reasoning 61
9 Conclusion 63
10 Glossary 64

11 Where to look for further information 65
12 References 66

3 Advances in local perception for orchard robotics 75
Jose Blasco, Centro de Agroingeniería - Instituto Valenciano de Investigaciones Agrarias (IVIA), Spain; and Francisco Rovira-Más, Agricultural Robotics Laboratory (ARL) - Universitat Politècnica de València, Spain

1 Introduction 75
2 Dealing with orchard environments 79
3 Sensor and data analysis technologies to address orchard environments 80
4 Advances in multidimensional sensors for navigating orchard environments 84
5 Sensor fusion for augmented perception 88
6 Using perception technologies in agricultural operations: harvesting, crop protection and other applications 91
7 Conclusion and future trends 96
8 Where to look for further information 97
9 Acknowledgements 99
10 References 99

4 Advances in machine learning for agricultural robots 103
Polina Kurtser, Örebro University and Umeå University, Sweden; Stephanie Lowry, Örebro University, Sweden; and Ola Ringdahl, Umeå University, Sweden

1 Introduction 103
2 Applications of machine learning in agri-robotics 107
3 Challenges 112
4 Integration and field-testing use-cases 121
5 Conclusion 124
6 Where to look for further information 125
7 References 125

Part 2 Technologies: operational aspects

5 Autonomous navigation and path planning for agricultural robots 137
John F. Reid, University of Illinois at Urbana-Champaign, USA

1 Introduction 137
2 Guidance and navigation in agricultural machine systems: current practice 141
3 Mission planning for agricultural machines 145
4 Path planning for agricultural machines 147
5 Search and optimization methods 150
6 Navigation and motion control 153

7 Case studies 156
8 Conclusion and future trends 163
9 Where to look for further information 165
10 References 165

6 Advances in human-robot collaboration in agricultural robotics 169
George Adamides, Agricultural Research Institute, Cyprus; and Yael Edan, Ben-Gurion University of the Negev, Israel

1 Introduction 169
2 Scope of human-robot collaboration 171
3 Functionality, design and taxonomy 172
4 Interfaces/interaction design and usability 176
5 Conclusion and future trends 178
6 Where to look for further information 180
7 References 180

7 Implementing a digital twin for flexible operation of agricultural robotics 185
Frits K. van Evert, Trim Bresilla, Ard Nieuwenhuizen, Bram Veldhuisen and Jochen Hemming, Wageningen University & Research, The Netherlands; Giannis Avgoustakis, Spyros Fountas, Michael Koutsiaras and Nikos Mylonas, Agricultural University of Athens, Greece; Suzanne Baron, Agricultural Robotics and Automatization, France; Thanos Dritsopoulos and Markos Legas, Pegasus, Greece; Han Hilbrands and Jeroen Wolters, Smart Agri Technology BV, The Netherlands; Klaas Jan Hommes, Abemec BV, The Netherlands; Panagiotis Karagiannis, Sotiris Makris and George Michalos, University of Patras, Greece; Dimitrios Paraforos and Galibjon Sharipov, University of Hohenheim, Germany; Søren Marcus Pedersen, University of Copenhagen, Denmark; Bertrand Pinel, Terrena Societe Cooperative Agricole, France; Mladen Radisic, Foodscale Hub, Serbia; Juan F. Rascón and Carlos Rizzo, Eurecat, Centre Tecnològic de Catalunya, Spain; Raul Sanchez, Serrater, Spain; Alea Scovill, AgroIntelli, Denmark; Oriol Serra, Giropoma, Spain; and Josep Vidal, Teyme, Spain

1 Introduction 185
2 Agricultural system under study 190
3 Agricultural robotic system 193
4 Case study results 214
5 Discussion 215
6 Conclusion 217
7 Where to look for further information 218
8 Acknowledgements 218
9 References 218

8 Advances in connectivity and distributed intelligence in agricultural robotics 223
Liisa Pesonen, Natural Resources Institute Finland (Luke), Finland; Daniel Calvo Alonso, Atos, Spain; Juha Backman and Jere Kaivosoja, Natural Resources Institute Finland (Luke), Finland; Jarmi Recio Martinez, Atos, Spain; and Juha-Pekka Soininen, VTT Technical Research Centre of Finland, Finland

1 Introduction 223
2 Robotic system architecture 227
3 Communication networks 229
4 Internet of things to link robots 231
5 Distributed intelligence 233
6 Mission control platforms 238
7 Service framework 241
8 Data spaces serving robots 243
9 Case study: FlexiGroBots concept 247
10 Conclusion 248
11 Future trends in research 250
12 Acknowledgements 252
13 Where to look for further information 252
14 References 253

9 Improving fault detection and isolation in agricultural robotics 263
Nicolas Tricot, Mahmoud Almasri and Roland Lenain, Université Clermont Auvergne, France

1 Introduction 263
2 Principles of fault diagnosis 264
3 Methods for fault diagnosis 266
4 Designing a robust fault detection and isolation system 272
5 Hybrid diagnostic method 274
6 Human/robot interaction module 279
7 Implementing the proposed fault detection and isolation system: key components 287
8 Modelling robot movement and potential faults 291
9 Testing fault diagnosis using Kalman filters 295
10 Testing fault diagnosis using deep learning 298
11 Building the hybrid diagnostic method and human/robot interaction module 307
12 Conclusion and future trends 314
13 Abbreviations 315
14 Acknowledgements 316
15 Where to look for further information 316
16 References 316

Part 3 Technologies: actuation

10 Advances in mobility platforms for agricultural robots 323
Renato Vidoni and Giovanni Carabin, Free University of Bozen-Bolzano, Italy; Giuseppe Quaglia and Andrea Botta, Politecnico di Torino, Italy; and Giulio Reina and Rocco Galati, Politecnico di Bari, Italy

1 Introduction 323
2 Agricultural unmanned ground vehicles: mobility classification 326
3 Wheeled mobile robots 332
4 Tracked mobile robots 339
5 Legged mobile robots 344
6 Hybrid and other unconventional unmanned ground vehicles mobility platforms for agriculture applications 346
7 Conclusion and future trends 347
8 Where to look for further information 348
9 References 349

11 Advances in grasping techniques in agricultural robots 355
George Kantor and Francisco Yandun, Carnegie Mellon University, USA

1 Introduction 355
2 Robotic manipulators 356
3 Mechanics of grasping 358
4 Gripper design 363
5 Systems integration 367
6 Real-world applications 370
7 Conclusion 378
8 Where to look for further information 379
9 References 380

12 Advances in soft grasping in agriculture 387
Ali Leylavi Shoushtari, Wageningen University & Research and Wageningen Robotics, The Netherlands

1 Introduction 387
2 Challenges of action in agricultural robotics 389
3 How soft robotics can address challenges of action 391
4 Developing an assessment matrix for soft grippers 394
5 Performance analysis of current soft agri-food grippers 396
6 Performance analysis: dealing with object variations 399
7 Designing the ideal versatile gripper 405
8 Case study: the use of soft robotics in intercropping 406
9 Conclusion 408
10 Where to look for further information 409
11 References 410

13 Advances in agricultural unmanned aerial vehicles: focus on sensing applications 415
Tarin Paz-Kagan, Ben-Gurion University of the Negev, Israel

1 Introduction 415
2 Unmanned aerial vehicle remote sensing sensors 417
3 Platforms for precision agriculture 424
4 Flight planning and pre-processing 426
5 Application in precision agriculture 429
6 Conclusion and future trends 443
7 Where to look for further information 444
8 References 445

Part 4 Social, ethical and economic aspects

14 Regulatory frameworks and standards for agricultural robotics in the European Union 461
Andrea Bertolini and Rocco Limongelli, Scuola Superiore di Studi Universitari e di Perfezionamento Sant'Anna - Istituto Dirpolis, Italy

1 Introduction 461
2 Product safety regulation and standardization 463
3 The European Union directive on liability for defective products 468
4 Data privacy and data ownership 471
5 The European Union's Artificial Intelligence Act 474
6 Conclusion 476
7 References 477

15 Economics of agricultural robotics 481
James Lowenberg-DeBoer, Harper Adams University, UK

1 Introduction 481
2 History of adoption of agricultural robotics 484
3 Expected benefits of agricultural robotics 488
4 Potential agricultural robotics in low- and middle-income countries 494
5 Broader implications of agricultural robotics for the farm sector 497
6 Social impact 499
7 Policy, regulatory and institutional issues 501
8 Conclusion 504
9 Where to look for further information 505
10 Acknowledgements 506
11 References 506

16 Social and ethical considerations for agricultural robotics 513
Kirsten Ayris, University of Reading, UK; and David Christian Rose, Cranfield University, UK

1 Introduction 513
2 Responsible robotics: anticipating impacts 514
3 Inclusion for responsible agricultural robotics 517
4 Responsible robotics: reflexivity and responsiveness 523
5 Conclusion 525
6 Where to look for further information 525
7 Acknowledgements 526
8 References 526

Part 5 Applications

17 Advances in the use of robotics in crop phenotyping 533
M. Wattad, Technion – Israel Institute of Technology, Israel; V. Alchanatis, Volcani Institute – Agricultural Research Organization, Israel; Y. Edan, Ben-Gurion University of the Negev, Israel; S. Shriki and T. Sandovsky, Volcani Institute – Agricultural Research Organization and Ben-Gurion University of the Negev, Israel; and S. Filin, Technion – Israel Institute of Technology, Israel

1 Introduction 533
2 The role of robotics in phenotyping 535
3 Two-dimensional plant imaging and analysis for phenotyping 537
4 Three-dimensional plant imaging and analysis for phenotyping 541
5 Four-dimensional plant imaging and analysis for phenotyping 543
6 Case study: single view 3D reconstruction and analysis for phenotyping 545
7 Case study: abiotic stress 547
8 Case study: biotic stress 549
9 Conclusion and future perspectives 554
10 Where to look for further information 555
11 References 555

18 Advances in the use of robots in field crop cultivation 567
Avital Bechar, Agricultural Research Organization, Israel; and Dionysis Bochtis, Institute for Bio-Economy and Agri-Technology (iBO), Greece

1 Introduction 567
2 Requirements for robots involved in field cultivation 569
3 Enabling technologies for in-field robots 571
4 Applications: transplanting 574
5 Applications: monitoring and control of weeds, plant diseases and pests 575
6 Applications: harvesting 578

7 Conclusion and future trends 578
8 References 579

19 Advances in the use of robotics in orchard operations 587
Manoj Karkee and Qin Zhang, Washington State University, USA; Uddhav Bhattarai, University of California Davis, USA; and Xin Zhang, Mississippi State University, USA

1 Introduction 587
2 Ground platforms and in-orchard navigation 591
3 Situation awareness and operation management 595
4 Actuation 606
5 Use cases of robotic orchard operations 615
6 Challenges and opportunities 620
7 Conclusion 624
8 Where to look for further information 624
9 References 626

20 Advances in the use of robotics in greenhouse cultivation 635
Jochen Hemming and Jos Balendonck, Wageningen University & Research, The Netherlands

1 Introduction 635
2 Key challenges in using robotics in protected cultivation 636
3 State of the art and challenges: selective harvesting 638
4 State of the art and challenges: crop management 647
5 State of the art and challenges: plant propagation, grafting and planting 649
6 State of the art and challenges: crop scouting/disease and insect detection and control 650
7 State of the art and challenges: transportation and logistics 652
8 Conclusion and future research trends 653
9 Where to look for further information 655
10 References 656

21 Advances in the use of robotics in livestock production 661
Kees Lokhorst, Wageningen Livestock Research, The Netherlands; and Tomas Norton, Katholieke Universiteit Leuven, Belgium

1 Introduction and main drivers for livestock robotics 661
2 The role of robotics in precision livestock farming 663
3 Examples of robotic development: poultry 668
4 Examples of robotic development: pig production 670
5 Examples of robotic development: dairy 673
6 Challenges for research 677
7 Conclusion 685
8 References 686

Index 689

Series list

Title	Series number
Achieving sustainable cultivation of maize - Vol 1 From improved varieties to local applications Edited by: Dr Dave Watson, CGIAR Maize Research Program Manager, CIMMYT, Mexico	001
Achieving sustainable cultivation of maize - Vol 2 Cultivation techniques, pest and disease control Edited by: Dr Dave Watson, CGIAR Maize Research Program Manager, CIMMYT, Mexico	002
Achieving sustainable cultivation of rice - Vol 1 Breeding for higher yield and quality Edited by: Prof. Takuji Sasaki, Tokyo University of Agriculture, Japan	003
Achieving sustainable cultivation of rice - Vol 2 Cultivation, pest and disease management Edited by: Prof. Takuji Sasaki, Tokyo University of Agriculture, Japan	004
Achieving sustainable cultivation of wheat - Vol 1 Breeding, quality traits, pests and diseases Edited by: Prof. Peter Langridge, The University of Adelaide, Australia	005
Achieving sustainable cultivation of wheat - Vol 2 Cultivation techniques Edited by: Prof. Peter Langridge, The University of Adelaide, Australia	006
Achieving sustainable cultivation of tomatoes Edited by: Dr Autar Mattoo, USDA-ARS, USA and Prof. Avtar Handa, Purdue University, USA	007
Achieving sustainable production of milk - Vol 1 Milk composition, genetics and breeding Edited by: Dr Nico van Belzen, International Dairy Federation (IDF), Belgium	008
Achieving sustainable production of milk - Vol 2 Safety, quality and sustainability Edited by: Dr Nico van Belzen, International Dairy Federation (IDF), Belgium	009
Achieving sustainable production of milk - Vol 3 Dairy herd management and welfare Edited by: Prof. John Webster, University of Bristol, UK	010
Ensuring safety and quality in the production of beef - Vol 1 Safety Edited by: Prof. Gary Acuff, Texas A&M University, USA and Prof. James Dickson, Iowa State University, USA	011
Ensuring safety and quality in the production of beef - Vol 2 Quality Edited by: Prof. Michael Dikeman, Kansas State University, USA	012
Achieving sustainable production of poultry meat - Vol 1 Safety, quality and sustainability Edited by: Prof. Steven C. Ricke, University of Arkansas, USA	013
Achieving sustainable production of poultry meat - Vol 2 Breeding and nutrition Edited by: Prof. Todd Applegate, University of Georgia, USA	014

Achieving sustainable production of poultry meat - Vol 3 015
Health and welfare
Edited by: Prof. Todd Applegate, University of Georgia, USA

Achieving sustainable production of eggs - Vol 1 016
Safety and quality
Edited by: Prof. Julie Roberts, University of New England, Australia

Achieving sustainable production of eggs - Vol 2 017
Animal welfare and sustainability
Edited by: Prof. Julie Roberts, University of New England, Australia

Achieving sustainable cultivation of apples 018
Edited by: Dr Kate Evans, Washington State University, USA

Integrated disease management of wheat and barley 019
Edited by: Prof. Richard Oliver, Curtin University, Australia

Achieving sustainable cultivation of cassava - Vol 1 020
Cultivation techniques
Edited by: Dr Clair Hershey, formerly International Center for Tropical Agriculture (CIAT), Colombia

Achieving sustainable cultivation of cassava - Vol 2 021
Genetics, breeding, pests and diseases
Edited by: Dr Clair Hershey, formerly International Center for Tropical Agriculture (CIAT), Colombia

Achieving sustainable production of sheep 022
Edited by: Prof. Johan Greyling, University of the Free State, South Africa

Achieving sustainable production of pig meat - Vol 1 023
Safety, quality and sustainability
Edited by: Prof. Alan Mathew, Purdue University, USA

Achieving sustainable production of pig meat - Vol 2 024
Animal breeding and nutrition
Edited by: Prof. Julian Wiseman, University of Nottingham, UK

Achieving sustainable production of pig meat - Vol 3 025
Animal health and welfare
Edited by: Prof. Julian Wiseman, University of Nottingham, UK

Achieving sustainable cultivation of potatoes - Vol 1 026
Breeding improved varieties
Edited by: Prof. Gefu Wang-Pruski, Dalhousie University, Canada

Achieving sustainable cultivation of oil palm - Vol 1 027
Introduction, breeding and cultivation techniques
Edited by: Prof. Alain Rival, Center for International Cooperation in Agricultural Research for Development (CIRAD), France

Achieving sustainable cultivation of oil palm - Vol 2 028
Diseases, pests, quality and sustainability
Edited by: Prof. Alain Rival, Center for International Cooperation in Agricultural Research for Development (CIRAD), France

Achieving sustainable cultivation of soybeans - Vol 1 029
Breeding and cultivation techniques
Edited by: Prof. Henry T. Nguyen, University of Missouri, USA

Achieving sustainable cultivation of soybeans - Vol 2 030
Diseases, pests, food and non-food uses
Edited by: Prof. Henry T. Nguyen, University of Missouri, USA

Achieving sustainable cultivation of sorghum - Vol 1 031
Genetics, breeding and production techniques
Edited by: Prof. William Rooney, Texas A&M University, USA

Achieving sustainable cultivation of sorghum - Vol 2 032
Sorghum utilization around the world
Edited by: Prof. William Rooney, Texas A&M University, USA

Achieving sustainable cultivation of potatoes - Vol 2 033
Production, storage and crop protection
Edited by: Dr Stuart Wale, Potato Dynamics Ltd, UK

Achieving sustainable cultivation of mangoes 034
Edited by: Prof. Víctor Galán Saúco, Instituto Canario de Investigaciones Agrarias (ICIA), Spain and Dr Ping Lu, Charles Darwin University, Australia

Achieving sustainable cultivation of grain legumes - Vol 1 035
Advances in breeding and cultivation techniques
Edited by: Dr Shoba Sivasankar et al., formerly International Crops Research Institute for the Semi-Arid Tropics (ICRISAT), India

Achieving sustainable cultivation of grain legumes - Vol 2 036
Improving cultivation of particular grain legumes
Edited by: Dr Shoba Sivasankar et al., formerly International Crops Research Institute for the Semi-Arid Tropics (ICRISAT), India

Achieving sustainable cultivation of sugarcane - Vol 1 037
Cultivation techniques, quality and sustainability
Edited by: Prof. Philippe Rott, University of Florida, USA

Achieving sustainable cultivation of sugarcane - Vol 2 038
Breeding, pests and diseases
Edited by: Prof. Philippe Rott, University of Florida, USA

Achieving sustainable cultivation of coffee 039
Edited by: Dr Philippe Lashermes, Institut de Recherche pour le Développement (IRD), France

Achieving sustainable cultivation of bananas - Vol 1 040
Cultivation techniques
Edited by: Prof. Gert H. J. Kema, Wageningen University and Research, The Netherlands and Prof. André Drenth, University of Queensland, Australia

Global Tea Science 041
Current status and future needs
Edited by: Dr V. S. Sharma, formerly UPASI Tea Research Institute, India and Dr M. T. Kumudini Gunasekare, Coordinating Secretariat for Science Technology and Innovation (COSTI), Sri Lanka

Integrated weed management 042
Edited by: Prof. Emeritus Robert L. Zimdahl, Colorado State University, USA

Achieving sustainable cultivation of cocoa 043
Edited by: Prof. Pathmanathan Umaharan, Cocoa Research Centre – The University of the West Indies, Trinidad and Tobago

Robotics and automation for improving agriculture 044
Edited by: Prof. John Billingsley, University of Southern Queensland, Australia

Water management for sustainable agriculture 045
Edited by: Prof. Theib Oweis, ICARDA, Jordan

Improving organic animal farming 046
Edited by: Dr Mette Vaarst, Aarhus University, Denmark and Dr Stephen Roderick, Duchy College, UK

Improving organic crop cultivation 047
Edited by: Prof. Ulrich Köpke, University of Bonn, Germany

Managing soil health for sustainable agriculture - Vol 1 048
Fundamentals
Edited by: Dr Don Reicosky, Soil Scientist Emeritus USDA-ARS and University of Minnesota, USA

Managing soil health for sustainable agriculture - Vol 2 049
Monitoring and management
Edited by: Dr Don Reicosky, Soil Scientist Emeritus USDA-ARS and University of Minnesota, USA

Rice insect pests and their management 050
E. A. Heinrichs, Francis E. Nwilene, Michael J. Stout, Buyung A. R. Hadi and Thais Freitas

Improving grassland and pasture management in temperate agriculture 051
Edited by: Prof. Athole Marshall and Dr Rosemary Collins, IBERS, Aberystwyth University, UK

Precision agriculture for sustainability 052
Edited by: Dr John Stafford, Silsoe Solutions, UK

Achieving sustainable cultivation of temperate zone tree fruit and berries – Vol 1 053
Physiology, genetics and cultivation
Edited by: Prof. Gregory A. Lang, Michigan State University, USA

Achieving sustainable cultivation of temperate zone tree fruit and berries – Vol 2 054
Case studies
Edited by: Prof. Gregory A. Lang, Michigan State University, USA

Agroforestry for sustainable agriculture 055
Edited by: Prof. María Rosa Mosquera-Losada, Universidade de Santiago de Compostela, Spain and Dr Ravi Prabhu, World Agroforestry Centre (ICRAF), Kenya

Achieving sustainable cultivation of tree nuts 056
Edited by: Prof. Ümit Serdar, Ondokuz Mayis University, Turkey and Emeritus Prof. Dennis Fulbright, Michigan State University, USA

Assessing the environmental impact of agriculture 057
Edited by: Prof. Bo P. Weidema, Aalborg University, Denmark

Critical issues in plant health: 50 years of research in African agriculture 058
Edited by: Dr Peter Neuenschwander and Dr Manuele Tamò, IITA, Benin

Achieving sustainable cultivation of vegetables 059
Edited by: Emeritus Prof. George Hochmuth, University of Florida, USA

Advances in breeding techniques for cereal crops 060
Edited by: Prof. Frank Ordon, Julius Kuhn Institute (JKI), Germany and Prof. Wolfgang Friedt, Justus-Liebig University of Giessen, Germany

Advances in Conservation Agriculture – Vol 1 061
Systems and Science
Edited by: Prof. Amir Kassam, University of Reading, UK and Moderator, Global Conservation Agriculture Community of Practice (CA-CoP), FAO, Rome, Italy

Advances in Conservation Agriculture – Vol 2 062
Practice and Benefits
Edited by: Prof. Amir Kassam, University of Reading, UK and Moderator, Global Conservation Agriculture Community of Practice (CA-CoP), FAO, Rome, Italy

Achieving sustainable greenhouse cultivation 063
Edited by: Prof. Leo Marcelis and Dr Ep Heuvelink, Wageningen University, The Netherlands

Achieving carbon-negative bioenergy systems from plant materials 064
Edited by: Dr Chris Saffron, Michigan State University, USA

Achieving sustainable cultivation of tropical fruits 065
Edited by: Prof. Elhadi M. Yahia, Universidad Autónoma de Querétaro, Mexico

Advances in postharvest management of horticultural produce 066
Edited by: Prof. Chris Watkins, Cornell University, USA

Pesticides and agriculture 067
Profit, politics and policy
Dave Watson

Integrated management of diseases and insect pests of tree fruit 068
Edited by: Prof. Xiangming Xu and Dr Michelle Fountain, NIAB-EMR, UK

Integrated management of insect pests 069
Current and future developments
Edited by: Emeritus Prof. Marcos Kogan, Oregon State University, USA and Emeritus Prof. E. A. Heinrichs, University of Nebraska-Lincoln, USA

Preventing food losses and waste to achieve food security and sustainability 070
Edited by: Prof. Elhadi M. Yahia, Universidad Autónoma de Querétaro, Mexico

Achieving sustainable management of boreal and temperate forests 071
Edited by: Dr John Stanturf, Estonian University of Life Sciences , Estonia

Advances in breeding of dairy cattle 072
Edited by: Prof. Julius van der Werf, University of New England, Australia and Prof. Jennie Pryce, Agriculture Victoria and La Trobe University, Australia

Improving gut health in poultry 073
Edited by: Prof. Steven C. Ricke, University of Arkansas, USA

Achieving sustainable cultivation of barley 074
Edited by: Prof. Glen Fox, University of California-Davis, USA and The University of Queensland, Australia and Prof. Chengdao Li, Murdoch University, Australia

Advances in crop modelling for a sustainable agriculture 075
Edited by: Emeritus Prof. Kenneth Boote, University of Florida, USA

Achieving sustainable crop nutrition 076
Edited by: Prof. Zed Rengel, University of Western Australia, Australia

Achieving sustainable urban agriculture 077
Edited by: Prof. Johannes S. C. Wiskerke, Wageningen University, The Netherlands

Climate change and agriculture 078
Edited by Dr Delphine Deryng, NewClimate Institute/Integrative Research Institute on Transformations of Human-Environment Systems (IRI THESys), Humboldt-Universität zu Berlin, Germany

Advances in poultry genetics and genomics 079
Edited by: Prof. Samuel E. Aggrey, University of Georgia, USA, Prof. Huaijun Zhou, University of California-Davis, USA, Dr Michèle Tixier-Boichard, INRAE, France and Prof. Douglas D. Rhoads, University of Arkansas, USA

Achieving sustainable management of tropical forests 080
Edited by: Prof. Jürgen Blaser, Bern University of Life Sciences, Switzerland and Patrick D. Hardcastle, Forestry Development Specialist, UK

Improving the nutritional and nutraceutical properties of wheat and other cereals 081
Edited by: Prof. Trust Beta, University of Manitoba, Canada

Achieving sustainable cultivation of ornamental plants 082
Edited by: Emeritus Prof. Michael Reid, University of California-Davis, USA

Improving rumen function 083
Edited by: Dr C. S. McSweeney, CSIRO, Australia and Prof. R. I. Mackie, University of Illinois, USA

Biostimulants for sustainable crop production 084
Edited by: Youssef Rouphael, Patrick du Jardin, Patrick Brown, Stefania De Pascale and Giuseppe Colla

Improving data management and decision support systems in agriculture 085
Edited by: Dr Leisa Armstrong, Edith Cowan University, Australia

Achieving sustainable cultivation of bananas - Volume 2 086
Germplasm and genetic improvement
Edited by: Prof. Gert H. J. Kema, Wageningen University, The Netherlands and Prof. Andrè Drenth, The University of Queensland, Australia

Reconciling agricultural production with biodiversity conservation 087
Edited by: Prof. Paolo Bàrberi and Dr Anna-Camilla Moonen, Institute of Life Sciences - Scuola Superiore Sant'Anna, Pisa, Italy

Advances in postharvest management of cereals and grains 088
Edited by: Prof. Dirk E. Maier, Iowa State University, USA

Biopesticides for sustainable agriculture 089
Edited by: Prof. Nick Birch, formerly The James Hutton Institute, UK and Prof. Travis Glare, Lincoln University, New Zealand

Understanding and improving crop root function 090
Edited by: Emeritus Prof. Peter J. Gregory, University of Reading, UK

Understanding the behaviour and improving the welfare of chickens 091
Edited by: Prof. Christine Nicol, Royal Veterinary College - University of London, UK

Advances in measuring soil health 092
Edited by: Prof. Wilfred Otten, Cranfield University, UK

The sustainable intensification of smallholder farming systems 093
Edited by: Dr Dominik Klauser and Dr Michael Robinson, Syngenta Foundation for Sustainable Agriculture, Switzerland

Advances in horticultural soilless culture 094
Edited by: Prof. Nazim S. Gruda, University of Bonn, Germany

Reducing greenhouse gas emissions from livestock production 095
Edited by: Dr Richard Baines, Royal Agricultural University, UK

Understanding the behaviour and improving the welfare of pigs 096
Edited by: Emerita Prof. Sandra Edwards, Newcastle University, UK

Genome editing for precision crop breeding 097
Edited by: Dr Matthew R. Willmann, Cornell University, USA

Understanding the behaviour and improving the welfare of dairy cattle 098
Edited by: Dr Marcia Endres, University of Minnesota, USA

Developing sustainable food systems 099
Dave Watson

Plant genetic resources 100
A review of current research and future needs
Edited by: Dr M. Ehsan Dulloo, Bioversity International, Italy

Developing animal feed products 101
Edited by: Dr Navaratnam Partheeban, formerly Royal Agricultural University, UK

Improving dairy herd health 102
Edited by: Prof. Émile Bouchard, University of Montreal, Canada

Understanding gut microbiomes as targets for improving pig gut health 103
Edited by: Prof. Mick Bailey and Emeritus Prof. Chris Stokes, University of Bristol, UK

Advances in Conservation Agriculture – Vol 3 104
Adoption and Spread
Edited by: Professor Amir Kassam, University of Reading, UK and Moderator, Global Conservation Agriculture Community of Practice (CA-CoP), FAO, Rome, Italy

Advances in precision livestock farming 105
Edited by: Prof. Daniel Berckmans, Katholieke University of Leuven, Belgium

Achieving durable disease resistance in cereals 106
Edited by: Prof. Richard Oliver, formerly Curtin University, Australia

Seaweed and microalgae as alternative sources of protein 107
Edited by: Prof. Xin Gen Lei, Cornell University, USA

Microbial bioprotectants for plant disease management 108
Edited by: Dr Jürgen Köhl, Wageningen University & Research, The Netherlands and Dr Willem Ravensberg, Koppert Biological Systems, The Netherlands

Improving soil health 109
Edited by: Prof. William R. Horwath, University of California-Davis, USA

Improving integrated pest management in horticulture 110
Edited by: Prof. Rosemary Collier, Warwick University, UK

Climate-smart production of coffee 111
Improving social and environmental sustainability
Edited by: Prof. Reinhold Muschler, CATIE, Costa Rica

Developing smart agri-food supply chains 112
Using technology to improve safety and quality
Edited by: Prof. Louise Manning, Royal Agricultural University, UK

Advances in integrated weed management 113
Edited by: Prof. Per Kudsk, Aarhus University, Denmark

Understanding and improving the functional and nutritional properties of milk 114
Edited by: Prof. Thom Huppertz, Wageningen University & Research, The Netherlands and Prof. Todor Vasiljevic, Victoria University, Australia

Energy-smart farming 115
Efficiency, renewable energy and sustainability
Edited by: Emeritus Prof. Ralph Sims, Massey University, New Zealand

Understanding and optimising the nutraceutical properties of fruit and vegetables 116
Edited by: Prof. Victor R. Preedy, King's College London, UK and Dr Vinood B. Patel, University of Westminster, UK

Advances in plant phenotyping for more sustainable crop production 117
Edited by: Prof. Achim Walter, ETH Zurich, Switzerland

Optimising pig herd health and production 118
Edited by: Prof. Dominiek Maes, Ghent University, Belgium and Prof. Joaquim Segalés, Universitat Autònoma de Barcelona and IRTA-CReSA, Spain

Optimising poultry flock health 119
Edited by: Prof. Sjaak de Wit, Royal GD and University of Utrecht, The Netherlands

Advances in seed science and technology for more sustainable crop production 120
Edited by: Dr Julia Buitink, INRAE, France and Prof. Olivier Leprince, L'Institut Agro Rennes Angers, France

Understanding and fostering soil carbon sequestration 121
Edited by: Dr Cornelia Rumpel, CNRS, Sorbonne University, Institute of Ecology and Environmental Sciences Paris, France

Advances in sensor technology for sustainable crop production 122
Edited by: Dr Craig Lobsey, University of Southern Queensland, Australia and Prof. Asim Biswas, University of Guelph, Canada

Achieving sustainable cultivation of bananas - Vol 3 123
Diseases and pests
Edited by: Prof. André Drenth, The University of Queensland, Australia and Prof. Gert H. J. Kema, Wageningen University, The Netherlands

Developing drought-resistant cereals 124
Edited by: Prof. Roberto Tuberosa, University of Bologna, Italy

Achieving sustainable turfgrass management 125
Edited by: Prof. Michael Fidanza, Pennsylvania State University, USA

Promoting pollination and pollinators in farming 126
Edited by: Emeritus Prof. Peter Kevan and Dr D. Susan Willis Chan, University of Guelph, Canada

Improving poultry meat quality 127
Edited by: Prof. Massimiliano Petracci, Alma Mater Studiorum - Università di Bologna, Italy and Dr Mario Estévez, Universidad de Extremadura, Spain

Advances in monitoring of native and invasive insect pests of crops 128
Edited by: Dr Michelle Fountain, NIAB-EMR, UK and Dr Tom Pope, Harper Adams University, UK

Advances in understanding insect pests affecting wheat and other cereals 129
Edited by: Prof. Sanford Eigenbrode and Dr Arash Rashed, University of Idaho, USA

Understanding and improving crop photosynthesis 130
Edited by: Dr Robert Sharwood, Western Sydney University, Australia

Modelling climate change impacts on agricultural systems 131
Edited by: Prof. Claas Nendel, Leibniz Centre for Agricultural Landscape Research (ZALF), Germany

Understanding and minimising fungicide resistance 132
Edited by: Dr Francisco J. Lopez-Ruiz, Curtin University, Australia

Advances in sustainable dairy cattle nutrition 133
Edited by: Prof. Alexander N. Hristov, The Pennsylvania State University, USA

Embryo development and hatchery practice in poultry production 134
Edited by: Dr Nick French

Developing circular agricultural production systems 135
Edited by: Prof. (UZ) Dr Barbara Amon, University of Zielona Góra, Poland and Leibniz Institute for Agricultural Engineering and Bioeconomy (ATB), Germany

Advances in ensuring the microbiological safety of fresh produce 136
Edited by: Prof. Karl R. Matthews, Rutgers University, USA

Frontiers in agri-food supply chains 137
Frameworks and case studies
Edited by: Prof. Sander de Leeuw, Dr Renzo Akkerman and Dr Rodrigo Romero Silva, Wageningen University, The Netherlands

Improving water management in agriculture 138
Irrigation and food production
Edited by: Prof. Jerry Knox, Cranfield University, UK

Advances in agri-food robotics 139
Edited by: Prof. Eldert van Henten, Wageningen University, The Netherlands and Prof. Yael Edan, Ben-Gurion University of the Negev, Israel

Key issues in agricultural ethics 140
Edited by: Prof. Emeritus Robert L. Zimdahl, Colorado State University, USA

Advances in plant factories 141
New technologies in indoor vertical farming
Edited by: Toyoki Kozai and Eri Hayashi

Improving the quality of apples 142
Edited by: Prof. Fabrizio Costa, University of Trento, Italy

Protecting natural capital and biodiversity in the agri-food sector 143
Edited by: Prof. Jill Atkins, Cardiff University, UK

Consumers and food 144
Understanding and shaping consumer behaviour
Edited by: Professor Marian Garcia Martinez, The University of Kent, UK

Advances in cultured meat technology 145
Edited by: Prof. Mark Post, Maastricht University, The Netherlands, Prof. Che Connon, Newcastle University, UK and Dr Chris Bryant, University of Bath and Bryant Research, UK

Understanding and preventing soil erosion 146
Edited by: Dr Karl Manuel Seeger, University of Trier, Germany

Smart farms 147
Improving data-driven decision making in agriculture
Edited by: Prof. Claus Sørensen, Aarhus University, Denmark

Improving standards and certification in agri-food supply chains 148
Ensuring safety, sustainability and social responsibility
Edited by: Prof. Louise Manning, University of Lincoln, UK

Managing biodiversity in agricultural landscapes 149
Conservation, restoration and rewilding
Edited by: Emeritus Prof. Nick Reid and Dr Rhiannon Smith, University of New England, Australia

Improving nitrogen use efficiency in crop production 150
Edited by: Adjunct Prof. Jagdish Kumar Ladha, University of California-Davis, USA

Understanding and utilising soil microbiomes for a more sustainable agriculture 151
Edited by: Prof. Kari Dunfield, University of Guelph, Canada

Advances in pig breeding and reproduction 152
Edited by: Prof. Jason Ross, Iowa State University, USA

Advances in organic dairy cattle farming 153
Edited by: Dr Mette Vaarst, Aarhus University, Denmark, Dr Stephen Roderick, Duchy College, UK and Dr Lindsay Whistance, Organic Research Centre, UK

Insects as alternative sources of protein for food and feed 154
Edited by: Ms Adriana Casillas, Tebrio, Spain

Advances in pig nutrition 155
Edited by: Prof. Julian Wiseman, University of Nottingham, UK

Advances in temperate agroforestry 156
Edited by: Prof. Maria Rosa Mosquera-Losada, Universidade de Santiago de Compostela, Spain, Dr Ladislau Martin, Embrapa, Brazil, Prof. Anastasia Pantera, Agricultural University of Athens, Greece and Dr Allison Chatrchyan, Cornell University, USA

Sustainable production and postharvest handling of avocado 157
Edited by: Emeritus Prof. Elhadi M. Yahia, Autonomous University of Querétaro, Mexico

Advances in bioprotection against plant diseases 158
Edited by: Prof. Shashi Sharma, Murdoch University, Australia and Dr Minshad Ansari, Bionema UK

Advances in poultry nutrition 159
Edited by: Prof. Todd Applegate, University of Georgia, USA

Weed management in Conservation Agriculture systems 160
Edited by: Prof. Gottlieb Basch, Prof. Emilio González-Sánchez, Mr John Geraghty, Dr Seyed Vahid Eslami, Prof. Sjoerd Willem Duiker, Dr Saidi Mkomwa and Dr Marie Bartz

Improving poultry meat safety and sustainability 161
Edited by: Prof. Steven C. Ricke, University of Wisconsin-Madison, USA

Advances in sheep production 162
Edited by: Ms Lesley Stubbings OBE, Consultant, UK and Ms Kate Phillips, Independent Livestock Consultant, UK

Women and smallholder farming: Addressing global inequities in agriculture 163
Edited by: Emerita Professor Carolyn E. Sachs, Pennsylvania State University, USA and Dr Paige Castellanos, Oxfam America (OUS), USA

Introduction

The global population is expected to reach 9 billion by 2050. Feeding this growing population more sustainably is a huge challenge facing agriculture. Developing agricultural robotics is seen as one potential solution to tackling this challenge.

This volume reviews the utilisation of agricultural robots to deal with increasing labour shortages in agriculture whilst bringing greater precision and efficiency into farming operations. Chapters are split into five parts: Part 1 chapters focus on sensing and perception technologies, which is then followed by Part 2 chapters that examine operational aspects in agricultural robotics. Chapters in Part 3 draw attention to actuation technologies, which is then followed by an overview of the social, ethical and economic aspects of agricultural robotics in Part 4. Chapters in Part 5 focus on particular applications of robotics in agriculture, such as crop phenotyping, field crop cultivation, orchard cultivation, greenhouse cultivation and livestock production.

Part 1 Technologies: sensing and perception

The first chapter of the book discusses advances in visual perception for agricultural robotics, specifically to deal with the challenges of variation and incomplete information. Chapter 1 begins with a short introduction to visual perception which includes a description of different sensors and sensor systems with their advantages and disadvantages as well as a set of typical perception tasks. The chapter then goes on to describe the main challenges for visual perception in the agricultural domain. Approaches to deal with these challenges presented in literature are then discussed in subsequent sections. The chapter also discusses methods to deal with variation with a focus on the use of deep neural networks, then goes on to review methods to deal with incomplete information, motivating the need for active perception. A section proposing future research directions is also included.

Chapter 2 focuses on advances in world modelling for agri-food robotics. The chapter provides an introduction to the concept of a world model, then goes on to describe invariant-based world model and topological maps with guarded motion. In each section, the chapter focuses on current state of the art technology, the problems that can arise and also a suggested approach for their use. For the topological maps with guarded motion discussion, a case study is also included. Moving on from this, the chapter goes on to describe task redundancy through multi-modal sensing and also provides an overview of current state of the art, problems, suggested approach for their use and a case study. Explainable world models for skill configuration, multi-robot cooperation through shared world models and the use of world models for reasoning are described in the same way.

The next chapter of Part 1 reviews advances in local perception for orchard robotics. Chapter 3 begins by highlighting the challenges of automating orchard crop management. The chapter moves on to examine sensor and data analysis technologies to address orchard environments, which is then followed by an overview of advances in multidimensional sensors for navigating orchard environments. A section on sensor fusion for augmented perception is also included. The chapter also examines using perception technologies in agricultural operations, focusing specifically on detecting fruit for harvesting, plant monitoring, detection of pests, diseases and stress, the detection of weeds and also canopy detection for automated spraying of crop protection products. The chapter concludes by highlighting the importance of dealing with challenges currently facing automated orchard crop management.

The final chapter of Part 1 examines advances in machine learning for agricultural robots. Chapter 4 begins by providing a definition of artificial intelligence, machine learning and deep learning. It goes on to describe the sensors currently in use for machine learning in agri-food robotics, which is followed by an overview of the challenges in employing artificial intelligence in various applications, including agri-food robotics. After this overview, the chapter goes on to describe applications of machine learning in agri-food robotics, focusing specifically on using machine learning for perception and robotic planning. This is then followed by a more in-depth analysis of the challenges facing the industry in implementing machine learning, such as environment complexity and variability, availability of data and long cycle times. The chapter also includes a section on integration and field-testing use cases, before concluding with an overview of why machine learning could be a crucial component within the agri-food robotics sector.

Part 2 Technologies: operational aspects

Part 2 of the book opens with a chapter that focuses on autonomous navigation and path planning for agricultural robots. Chapter 5 highlights the current practices in precision agriculture involving guidance and navigation in agricultural machine systems. It also reviews mission planning for agricultural machines and also describes path planning for agricultural machines. The chapter moves on to provide an overview of search and optimisation methods, such as heuristic-based algorithms, Rapidly-exploring Random Trees (RRT) and genetic algorithms. A section on navigation and motion control is also provided, which is then followed by two case studies that illustrate applications of navigation and planning in agricultural robots, the first reviews basic orchard mowing and the second focuses on multi-vehicle mission planning for citrus grove operations.

Chapter 6 discusses advances in human-robot collaboration in agricultural robotics. The chapter begins by first focusing on the scope of human-robot collaboration in agriculture. It then goes on to examine the functionality, design and taxonomy of this collaboration, focusing specifically on concepts and examples within the agricultural sector. A section on interfaces/interaction design and usability is also included, followed by a review of major developments through the inclusion of two case studies, using harvesting and spraying as a basis for discussion. Future challenges that can generate a positive trend in performance are also summarised, before the chapter concludes with an overview of why it is important to develop human-robot collaborative systems for future use in the agricultural sector.

The subject of Chapter 7 is implementing a digital twin for flexible operation of agricultural robotics. The chapter first provides a definition of what a digital twin is, then goes into more detail as to how a digital twin can be used in agriculture. The chapter is based on a four-year EU-funded project called ROBS4CROPS to provide a specific example of digital twin technology in an agricultural system, focusing on the specific scenarios ROBS4CROPS uses: France, Greece, Spain and The Netherlands are the four particular countries used. The chapter moves on to describe the major agricultural robot elements involved in the project, such as vehicles, smart implements, communication and Farming Controller. A section providing an overview of the case study results is also included, which is followed by a review of how the proposed digital twin robotic system could work outside of the ROBS4CROPS project.

Chapter 8 draws attention to advances in connectivity and distributed intelligence in agricultural robotics. The chapter begins by first providing an overview of robotic system architecture, then moves on to discuss the challenges of communication networks and interfacing methods of robots and their support systems. The Internet of Things concept is reviewed in terms of how it can be used in robotics, followed by a review of the use of distributed intelligence such as Artificial Intelligence (AI), edge computing and distributed AI and also smart agriculture solutions and applications. A section on mission control platforms is also included by first highlighting the need for mission control, which is followed by sections that provide state-of-the-art examples of ground robot control systems and open-source drone platforms currently in development. The use of a service framework on farms is also discussed and a review of data spaces serving robots is also included. The chapter also provides a case study that focuses on the FlexiGroBots concept.

The final chapter of Part 2 discusses improving fault detection and isolation (FDI) in agricultural robotics. Chapter 9 provides an overview of the principles of fault diagnosis in robotics, then goes on to highlight various diagnostic methods by dividing them into four main categories: model-based methods, knowledge-based methods, data-based methods and hardware redundancy. After these descriptions, the chapter includes a discussion comparing the diagnostic methods and their advantages and disadvantages. The chapter also considers the Robot Fault Diagnosis Supervision System (S2D2R), which is composed of a Hybrid Diagnostic Method (MHD) and a human robot interaction module (MIHR). Key components for implementing an FDI system are also described, followed by a review of modelling robot movement and potential faults and how fault diagnosis can be tested using Kalman Filters and deep learning. A section on building the MHD and MIHR is also provided.

Part 3 Technologies: actuation

The first chapter of Part 3 reviews the use of (unmanned) ground robots in agriculture, focusing on the main different solutions in terms of mobility and kinematic architecture. Chapter 10 discusses ways of classifying mobility platforms based on criteria such as size, type of locomotion, suspension system or chassis/frame design and coupling. It then reviews wheeled, tracked and legged mobile robots as well as hybrid solutions. In each case the chapter assesses different solutions and their pros and cons, Two case studies demonstrate design and operational challenges. The chapter concludes by outlining future trends in researching mobility solutions in agricultural robotics.

Chapter 11 provides an overview of the state of the art for grasping and manipulation in agricultural settings. It begins with a review of the robotic mechanisms commonly used for manipulation and grasping. The discussion then addresses issues associated with the integration of different technologies required to create fieldable manipulation systems, namely perception and control. Finally, a review of some specific application areas being addressed is provided, including harvest, pruning, and food handling. The chapter is intended to serve as a useful starting point for researchers and practitioners interested in learning more about the challenges and associated approaches being used for grasping and manipulation for agricultural applications.

The next chapter focuses on advances in soft grasping technologies in agriculture. Chapter 12 begins by highlighting the challenges of the physical interaction between robot and agricultural product which leads into the main discussion of the chapter that draws attention to how soft robotics can address these challenges. The chapter then goes on to developing an assessment matrix for soft grippers, which is then followed by a performance analysis of current soft agri-food grippers. Dealing with object variations through performance analysis is also described, focusing specifically on variations such as side pose, object softness, flat objects and concave objects. A section on designing the ideal versatile gripper is also included, which is followed by a case study that highlights the use of soft robotics in intercropping.

Chapter 13 draws attention to advances in agricultural unmanned aerial vehicles (UAVs), with a particular focus on sensing applications. The chapter begins by providing an overview of current UAV remote sensing sensors, then goes on to discuss the platforms for UAVs in precision agriculture. This is followed by a review of the importance of flight planning and pre-processing for UAVs, drawing attention to processes such as image pre-processing. The chapter then goes on to provide examples of UAV application in precision agriculture, such as abiotic and biotic stress detection, precision irrigation of water, nutrient precision fertilisation, precision weed management, insect and pathogen detection, phenotyping, growth vigour and yield assessment and finally precision livestock farming. The chapter concludes by providing an overview of why successful implementation of UAV technology in precision agriculture is important in terms of making agricultural processes more efficient.

Part 4 Social, ethical and economic aspects

Part 4 opens with a chapter that focuses on current regulatory frameworks and standards for agricultural robotics in the European Union (EU). Chapter 14 begins by describing product safety regulation and standardisation according to the EU, which is followed by a review of the EU directive on liability for defective products. A section on data privacy and data ownership is also included, focusing on how important it is for data yielded via farming to be protected to prevent farmer exploitation. The EU's Artificial Intelligence Act is also described, drawing attention to the three categories AI can be classified into: low risk, high risk and banned. The chapter concludes by focusing on how legislation should be considered an integral part of the innovation process.

Chapter 15 draws attention to the economics of agricultural robotics. The chapter reviews the adoption history of robotic technology in agriculture, such as in livestock production, the use of autonomous crop machines and also UAVs. This is then followed by an overview of the expected benefits of agricultural robotics, focusing specifically on two technologies and their economic benefits: milking robots and autonomous crop machines. A section on the potential of agricultural robotics in low- and middle-income countries is also included, drawing attention to the use of UAVs in smallholder farms and low-cost autonomous crop machines. The chapter also discusses the broader implications of agricultural robotics for the farm sector, as well as the social impact agricultural robotics can have. A section on policy, regulatory and institutional issues is also included.

Part 4 closes with a chapter that draw attention to the social and ethical considerations for agricultural robotics. Chapter 16 uses a responsible innovation framework to anticipate the possible challenges involved in the scaling of agricultural robotics, as well as how to include a wide range of stakeholder views. The chapter also discusses which stakeholders should be included in setting trajectories for agri-robotics, as well as how to engage harder to reach voices in a meaningful way. The chapter then turns to how these stakeholder views can be reflexively incorporated into responsive practices, such as standards and codes of practice, to mitigate against some of the potential negative impacts of robotics.

Part 5 Applications

Part 5 of the book opens with a chapter that reviews advances in the use of robotics in crop phenotyping. Chapter 17 first highlights the role of robotics in phenotyping, then moves on to discuss three forms of dimensional imaging and analysis: two-dimensional (2D), three-dimensional (3D) and four-dimensional (4D). A section is dedicated to each form. The chapter also provides two case studies, the first focuses on models to detect abiotic stress in corn plants using spectral reflectance and high spectrum images of plant leaves. The second case study draws attention to biotic stress and compares leaf point spectra and whole plant spectral images in the early detection of Fusarium infection in corn plants.

Chapter 18 reviews the use of robots in field crop cultivation. The chapter begins by providing an overview of current requirements for robots involved in field cultivation, focusing primarily on requirements relating to the environment. The chapter then goes on to describe enabling technologies for in-field robots. The chapter also provides several examples of in field crop cultivation robot application, such as for transplanting, monitoring and control of weeds, plant diseases and pests and also harvesting. The chapter concludes by highlighting how developing sensing technologies for use in field crops will significantly improve field crop cultivation in the future.

Moving on from Chapter 18, Chapter 19 draws attention to the use of robotics in orchard operations. The chapter reviews the use of ground platforms and in-orchard navigation, focusing specifically on wheel-type platforms and their steering mechanism, vision-based guidance and simultaneous localisation and mapping. The is followed by an overview of situation awareness and operation management, drawing specific attention to the latest advancements in sensing and machine vision systems used in robotic orchard operations whilst also addressing the robot-canopy interactions and strategies that can improve robotic operations in orchards. A section on actuation in robotic orchard operations is also included, focusing on robotic manipulation and end-effector technologies. The chapter then provides use cases of robotic orchard operations, such as robotic pruning, thinning and pollination. It concludes by highlighting current challenges and opportunities for orchard robotics.

Chapter 20 discusses the role of robotics in greenhouse cultivation, starting with key challenges. Current technology is then presented, grouped by the main tasks executed by the robots. The selective harvesting of major greenhouse crops (tomato, cucumber, sweet pepper, strawberries and flowers) is explored, followed by a description of crop maintenance operations, including automatic leaf removal. Plant propagation operations, including grafting and autonomous planting of cuttings, are then analysed. Attention is then given to crop scouting and the detection of disease and insects and control tasks using robots and drones. A section on autonomous transport and logistics in the greenhouse concludes this section, before future trends in greenhouse robotics research are discussed. Despite considerable technical advances, success rates, accuracy and speed of most developed systems remain insufficient. Future research is needed in most areas, including enhancement of grippers, sensors and manipulators. Using AI for sensing and control of greenhouse robots is expected to intensify in the future.

The final chapter of the book looks at advances in the use of robotics in livestock production. Chapter 21 discusses developments in robotics in the poultry, pig and dairy sectors and explores the main drivers for these innovative solutions. The chapter begins by highlighting the overall role of robotics in precision livestock farming, then goes on to provide specific examples of advances in poultry, pig and dairy production. For poultry, the chapter highlights key developments in dead bird detection, litter quality monitoring and abnormal behaviour of broiler chickens. In the pig section, the chapter draws attention to technologies used for bedding distribution, cleaning and precision feeding. Examples of dairy robotic developments focus on feeding, monitoring and management, cleaning and milk processing. Predictions from the EU-Robotics topic group and the agROBOfood network are also supplied in order to achieve a better understanding of the challenges facing the livestock sector.

Acknowledgements

Cover image copyright:

- Cover image copyright: A sweet pepper harvesting robot developed in the EU project CROPS (Clever Robots for Crops). 2014. © Wageningen University & Research.

We wish to acknowledge the following for their help in reviewing particular chapters:

- Chapter 1: Dr Francisco Rovira-Más, Polytechnic University of Valencia, Spain, Dr Andrea Botta, Polytechnic of Turin, Italy and Professor Armando Jorge Sousa, INESCTEC, Portugal
- Chapter 2: Dr Giovanni Colucci, Politecnico di Torino, Italy
- Chapter 3: Dr Cheryl McCarthy, University of Southern Queensland, Australia
- Chapter 4: Dr Ali Moghimi, Univ. of California-Davis, USA
- Chapter 6: Professor Marc Hanheide, University of Lincoln, UK and Dr George Kantor, Carnegie Mellon University, USA
- Chapter 9: Professor Pilar Barreiro Elorza, Universidad Politecnica de Madrid, Spain
- Chapter 10: Professor Alessandro Gasparetto, Università degli Studi di Udine, Italy
- Chapter 11: Dr Giuseppe Carbone, University of Calabria, Italy and Professor Francisco Rodriguez Diaz, University of Almeria, Spain
- Chapter 12: Professor Steve Davis, University of Birmingham, UK
- Chapter 14: Dr Rohan Rainbow, Managing Director – Crop Protection Australia, Australia
- Chapter 15: Dr Dimitrios Paparas and Dr Karl Behrendt, Harper Adams University, UK
- Chapter 16: Dr Thomas Daum, Institute of Agricultural Sciences in the Tropics (Hans-Ruthenberg-Institute), University of Hohenheim, Germany
- Chapter 20: Dr Giovanni Matranga, STEMS-CNR, Italy
- Chapter 21: Professor Daniel Berckmans, Katholieke Universiteit Leuven, Belgium, Professor James Salfer, University of Minnesota, USA and Dr Bernadette O'Brien, Teagasc, Ireland

We'd also like to thank Dr Sjaak Wolfert, Wageningen Economic Research, Wageningen University & Research, The Netherlands for their help during the publishing process.

Part 1

Technologies: sensing and perception

Chapter 1

Advances in visual perception for agricultural robotics

Gert Kootstra, Wageningen University & Research, The Netherlands

1 Introduction
2 A short introduction to visual perception in agriculture
3 Challenges in visual perception for agricultural robotics
4 Dealing with the challenge of variation
5 Dealing with the challenge of incomplete information
6 Directions for future research
7 Conclusion
8 References

1 Introduction

Like any autonomous system, an agricultural robot is in continuous interaction with the environment. There are two aspects to this interaction: the robot senses the environment and it acts on the environment. To decide on appropriate actions to fulfill its task, the robot needs to understand the state of the environment. That is, it needs to know about the relevant aspects of the environment that influence the decision-making process. Some of this knowledge might be expert knowledge describing facts or regularities that are generally true. However, as the agricultural environment is uncertain, variable, cluttered, and constantly changing, a fundamental part of the knowledge about the state of the environment needs to be extracted from the sensor data. The process to extract relevant information from sensor data is called perception.

Modern robots are equipped with numerous sensors, which can be divided into proprioceptive and exteroceptive sensors (Siegwart et al., 2011). Proprioceptive sensors measure quantities related to the internal state of the robot, such as the temperature of its electronics, battery charge, motor speed, and joint angles. Exteroceptive sensors measure quantities of the robot's environment, such as light intensity, distance, sound amplitude, and the earth's magnetic field. Although all these sensors are relevant to develop an autonomous agricultural robot, this chapter focuses on the use of vision-based

http://dx.doi.org/10.19103/AS.2023.0124.01

sensors and range sensors providing two-dimensional (2D) and three-dimensional (3D) images of the environment, as these are the predominant exteroceptive sensors used in current robotics. Using such sensors, notably vision cameras and LiDAR sensors, a wealth of information about the spatial arrangement and spectral properties of the environment can be extracted through the process of visual perception.

Robotic perception in agriculture faces two major challenges: the challenge of dealing with variations in the environment and the challenge of operating with incomplete information. This chapter discusses advances in visual perception for agricultural robotics with respect to these two challenges and it provides an outlook on future research to better deal with the challenges. The aim of the chapter is to give a broad and comprehensive overview of current academic literature on this topic, to provide insights into the main research directions, and to propose directions for future research. Examples given in this chapter are mainly for arable, greenhouse, orchard, and livestock farming, as the author has personal expertise in these areas, but it must be noted that forest robotics is facing similar challenges. For an overview of forest robotics, the reader can refer to Billingsley et al. (2008) and Oliveira et al. (2021).

In Section 2, a short introduction to visual perception will be given, including a description of different sensors and sensor systems with their advantages and disadvantages and a set of typical perception tasks. Section 3 describes the main challenges for visual perception in the agricultural domain. Approaches to deal with these challenges presented in the literature are then discussed in subsequent sections. Section 4 discusses methods to deal with variation with a focus on the use of deep neural networks, and Section 5 discusses methods to deal with incomplete information, motivating the need for active perception. Section 6 proposes future research directions and the chapter is concluded in Section 7.

2 A short introduction to visual perception in agriculture

Visual perception is defined as the process to extract relevant information about the environment for the operation of the robot based on data from imaging sensors. This section covers different imaging sensors often used in agriculture (Section 2.1), image acquisition on robotic platforms (Section 2.2), and the visual information extracted from the sensors (Section 2.3).

2.1 Imaging sensors

Imaging sensors detect electromagnetic radiation and form an image to represent that information. An image is a 2D array with pixels representing the intensity of the radiation. Imaging sensors exist for different parts of the electromagnetic spectrum, such as radar, X-ray scanners, and Teraherz cameras. However, for agrirobotic applications, typically, cameras are used for wavelengths in the visual and (near) infrared (NIR) spectrum to get information about natural objects in the robot's surroundings under natural or artificial lighting. Hence, the remainder of this chapter discusses these camera systems. The camera measures the light reflected by the objects.

Red-green-blue (RGB) color cameras are popular for robotic systems. The quality, price, size, and energy efficiency of these sensors have improved drastically in the past decades with the introduction of mobile phones. They allow to get color information about the surroundings of the robot in high spatial resolution. This provides sufficient information to detect objects in the agricultural environment and to get geometrical and color information about these objects needed to make decisions for navigation and operation. In environments with natural illumination, such as arable fields or greenhouse, there can be high differences in the intensity in reflectance of sun-lit and shadow parts of the scene, which can cause over- or under-exposure of camera images. To deal with the large intensity difference between sun and shadow, cameras with a high dynamic range can be used, for instance, to segment vegetation and soil in the images (Suh et al., 2018a). Such high-intensity differences can also be prevented by using covers and artificial lighting (e.g. Arad et al., 2019).

To get more information about the status of the crop or its produce, spectral imaging systems (also called imaging spectroscopy) can be used. Where RGB cameras use three spectral bands (corresponding to red, green, and blue), spectral cameras use multiple bands in the electromagnetic spectrum, in some cases ranging up to hundreds of spectral bands (Polder and Gowen, 2021). In the past, this technique would be referred to as multi- or hyper-spectral imaging, but nowadays the term 'spectral imaging' is preferred (Polder and Gowen, 2021). The cameras provide high spectral resolution, combining the benefits of imaging systems with spectroscopy, allowing chemometrics to assess the chemical composition. Information about pigments in plants or fruits, such as chlorophyll or carotenoids, can be obtained from the visible part (400–700 nm) and information about, for instance, sugars, proteins, and water can be obtained from the NIR part (700–2500 nm). This can, for instance, be used for disease detection (Mishra et al., 2020; Polder et al., 2019) to establish food safety and quality (Qin et al., 2013) or to quantify variations in fish feeding behavior (Zhou et al., 2017). The downside of spectral cameras is the higher price and the lower spatial resolution.

For robots to be able to operate in the agricultural environment, it is important that they get 3D information about objects in the environment. To this end, depth cameras or light-detection-and-ranging (LiDAR) sensors are used (Horaud et al., 2016). Depth cameras, often in combination with color (so-called RGB-D cameras) get depth information typically from triangulation using stereo vision (comparing left and right camera images) or using project light (comparing an emitted pattern with the reflected pattern observed in a camera image). LiDAR is based on the time-of-flight principle, where the time it takes for an emitted laser beam to reflect back in the sensor is measured to get depth. Although a LiDAR is strictly speaking not an imaging sensor, it can be used as such in combination with an RGB camera. Both RGB-D cameras and LiDAR can provide 3D point clouds, which are a set of points with their 3D position coordinates. Active 3D camera systems, like LiDAR and projected-light cameras, emit light onto the environment. These systems sometimes have trouble in outdoor situations with strong sun light interfering with the projected light or rain and snow disturbing the distance measurements, although the quality of these sensors is rapidly improving, making outdoor applications possible. Stereo-vision cameras are passive sensors that have no trouble dealing with outdoor situations. However, the downside of stereo vision is the lack of 3D information on non- or low-textured surfaces. Using 3D information allows, for instance, fruit detection and localization in orchards (Fu et al., 2020) or 3D localization and size estimation for robotic harvesting of broccoli heads in arable fields (Blok et al., 2021).

2.2 Image-acquisition systems

The acquisition of imaging data can be done with different robotic devices that can generally be divided into mobile robots – unmanned aerial vehicles (UAVs) and unmanned ground vehicles (UGVs) – and robotic arms (Corke, 2017). Unmanned aerial vehicles are often used for monitoring and phenotyping applications on arable fields (Tsouros et al., 2019; Xie and Yang, 2020; Chlingaryan et al., 2018). They have the advantage that they can easily cover large areas in a short time. Despite the high flying altitudes, modern high-resolution cameras still allow sub-centimeter pixel resolution, sufficient to monitor individual plant growth of, for instance, cabbage and pumpkin (Jamil et al., 2022). The downside of UAVs, however, is that they need relatively stable weather conditions and have a limited flight time. Furthermore, due to limited payload and the disturbance of the crop when flying at low altitudes, UAVs are not very suitable to perform operations.

Unmanned ground vehicles are able to get much closer to the crops, providing higher resolution imaging data and allowing robotic operations. Autonomous weeding robots, for instance, typically classify plants and detect their location based on high-resolution color and/or depth data to determine spraying actions (Ruigrok et al., 2020) or mechanical weeding actions (Pretto et al., 2021). Other examples are mobile robots in orchards for yield prediction (Bargoti and Underwood, 2017), and autonomous harvesting (Silwal et al., 2017; Williams et al., 2019). Another advantage of a UGV for image acquisition is the potential to use artificial light and a cover to shield off ambient light. The resulting controlled illumination conditions improve perception, for instance, to detect the subtle visual changes caused by potato diseases (Polder et al., 2019). In other work, a strong short-range flashlight is used to minimize the effect of ambient light, for instance, Arad et al. (2019).

Eye-in-hand systems, where a camera is mounted on a robot arm, make it possible to easily reposition the camera to observe a target object from different viewpoints, providing more complete information and dealing with occlusions (Hemming et al., 2014). In Barth et al. (2016), an eye-in-hand setup was developed to actively search for bell peppers to harvest in a greenhouse. Using visual servoing, the robot could subsequently find an optimal pose for harvesting (Arad et al., 2020). Using an array of nine cameras in the robot's hand, the next viewpoint that provided the most unoccluded view on the object of interest could be determined, which iteratively resulted in the best local view on the object (Lehnert et al., 2019). The topics of multi-view perception and active vision are discussed in more detail in Section 5.

Besides the use of the above-mentioned mobile robots, fixed image-acquisition systems are often used for plant phenotyping or monitoring tasks in confined spaces. Examples are Gantry systems that transport cameras and other sensors over the crop, a so-called sensor-to-plant system, for instance, Chaudhury et al. (2019), sometimes used in combination with a mobile robot (Nicolas et al., 2016), and plant-to-sensor systems, where plants are transported to a sensor station for detailed scanning, for instance, (Golbach et al., 2016; York, 2019). It must also be mentioned that the robotic imaging systems do not necessarily need to be fully autonomous, but can, for instance, also be mounted on implements that are still navigated with human assistance.

2.3 Visual information

The purpose of the image-acquisition system is to provide relevant information to the robot about the objects in its environment. On the one hand, this concerns information that is relevant to control the robot's actions, for instance, the detection and locations of plants or trees to navigate through a row (Bonadies and Gadsden, 2019; Blok et al., 2019), or the location of a ripe fruit to perform harvesting actions (Zhao et al., 2016). On the other hand, it concerns information that needs to be provided to the farmer or a management system, such as the count of apple blossoms to determine thinning actions and predict future yield (Bulanon et al., 2020) or the plumage condition of chickens to prevent welfare issues (Lamping et al., 2022).

Three types of information are distinguished that are relevant to extract from image data: (1) spatial information about objects, (2) information about object properties, and (3) temporal information about objects. These types of information are discussed in the next paragraphs.

Object pose is an example of spatial information. Typical, object pose is defined as the 3D position and 3D orientation (six-dimensional pose) of an object. Methods have been developed, for instance, to predict the pose of fruits to provide information for grasping (Wagner et al., 2021; Guo et al., 2020). When there is rotational symmetry in the fruits, the dimensionality of the pose can be reduced to, for instance, 5D for apples (Kang et al., 2020). For animals, often another definition of pose is used. Animal pose is typically defined by a set of key points representing different body parts, providing information about the posture, for instance, Mathis et al. (2018) and Russello et al. (2022). Also, the object shape can be relevant, for instance, to grade kiwifruits (Fu et al., 2016), to determine the size and dimensions of orchard trees for variable-rate spraying (Zhang et al., 2018; Abbas et al., 2020), to estimate heifer height and body mass of young cows (Nir et al., 2018), or to estimate feed intake of individual cows (Saar et al., 2022). Furthermore, the spatial arrangement of objects in the environment is important information for robot operation, for instance, information about the location of trees for navigation in orchards (Blok et al., 2019), information about crop rows for localization and navigation in arable fields (Winterhalter et al., 2021), and the spatial arrangements of fruits to plan harvesting actions (Kurtser and Edan, 2020).

Object properties are visual and geometrical properties that provide information about the objects themselves. Some of these properties are low-level features that can directly be observed, such as color properties and texture, and others are more high-level features that can be derived from low-level features, such as ripeness and the detection of diseases. In Halstead et al. (2018), for instance, the ripeness of bell peppers on the plant is estimated from color images. Other examples are the detection and classification of defects of broccoli heads in the field based on color images (Blok et al., 2022), the detection of the potato virus Y on potato plants based on spectral images (Polder et al., 2019), and the automatic scoring of the body condition of dairy cows (Spoliansky et al., 2016).

Temporal information is crucial when analyzing the gait of animals. Based on video frames, the animal pose can be determined, for instance, for cows (Russello et al., 2022) and poultry (Doornweerd et al., 2021). By tracking the animal over time, the changes of the pose provide information about the gait, which can be used as a welfare indicator (Nasiri et al., 2022). Also for robot perception, it can be important to track objects over time, for instance, to prevent double counts when assessing fruit yield in a greenhouse (Smitt et al., 2021; Halstead et al., 2018; Halstead et al., 2021).

3 Challenges in visual perception for agricultural robotics

In the previous section, many examples of agricultural robotic systems have been discussed. However, despite the active research field, there are not many commercially successful agricultural robots on the market yet. This is a sharp contrast with, for instance, the automotive or electronics industry, where many robots are operational in the production line to date. The main difference between the agricultural and automotive environment is the level of control over the environment. The environment for robots working in an automobile factory is completely under control, with controlled lighting, clean rooms, identical objects, and full knowledge about the location and shape of all the objects. This provides the robot with complete knowledge of the environment, which eliminates any uncertainties, allowing even pre-programmed behaviors. The agricultural environment, on the other hand, is far less structured, giving rise to uncertainty about the state of the environment. These uncertainties stem from variations in the objects and environmental conditions, and from incomplete information about the environment (Kootstra et al., 2020; Kootstra et al., 2021). In Kurtser and Edan (2018), a statistical model was developed to model the detectability of bell peppers depending on factors like the viewing distance and viewing angle. Their analyses clearly illustrate both the challenge of incomplete information and the variation over different greenhouses and cultivars of bell-pepper plants. Both challenges will be discussed in the next subsections.

3.1 The challenge of variation

The agricultural task environment contains a lot of variation. Four different types of variation are distinguished (Kootstra et al., 2020; Kootstra et al., 2021): (1) object variation, (2) environmental variation, (3) variation in cultivation systems, and (4) task variation.

1. The object variation is a consequence of natural variation. Every instance of a plant, fruit, vegetable, or animal is unique in appearance, geometry, mechanical properties, and behavior. Already within one species, the variation can be substantial and it is even more apparent between species. Also, non-living matter, such as soil comes in many different variations.
2. The environmental variations are, for instance, differences in weather conditions, causing differences in illumination (brightness and color) and humidity, and also deliberate changes made to the physical environment throughout production cycles, such as plowing and irrigation.
3. Different farmers use different cultivation systems. Apple trees, for instance, can be grown in orchards using the tall spindle system, v-trellis

system, or espalier support system, and animals can be raised outdoors in a meadow or indoors in a barn.

4 Task variation is caused by the need for farmers to use a robot not for one single task, but for a variety of different tasks. A robot that can be used for harvesting, deleafing, and rehanging of tomato plants, for instance, has more economical value to a grower than a robot that is specialized in harvesting only.

All these different sources of variation cause a lot of uncertainty for the robots operating in agricultural environments. Specifically, for visual perception, it results in a lot of variation in the appearance of the objects in the images and point-cloud data. This requires that perception methods can deal with this variation and that they can generalize to new environments. In Ruigrok et al. (2023), the generalization of a deep neural network in the agricultural context was tested using imaging data from many different sugar-beet fields, containing, among others, variation in cultivar, soil type, growth stage, and weed pressure. It was shown that even state-of-the-art deep neural networks suffer from a loss in performance when tested on a new field, the so-called generalization gap. Such a gap was also found when applying a trained neural network on a new variety of broccoli (Blok et al., 2020). In Section 4, approaches to deal with variation are discussed.

3.2 The challenge of incomplete information

An agricultural robot has only partial information about the environment due to (1) sensor limitations and (2) occlusions.

1 Apart from the limits in which sensors operate, such as the spectral wave lengths and field of view, imaging sensors occasionally fail in agricultural environments, especially under uncontrolled illumination. Examples are overexposure due to direct sun light, under exposure due to strong shadows, and lens flares. Image data are also often noisy, which is particularly apparent in depth data, which is incomplete at untextured surfaces and near object boundaries.

2 When objects are occluded, they are hidden from the camera view by other objects or by the object itself (self-occlusion). Especially in cluttered environments, such as orchards and high-wire greenhouses, occlusions form a substantial challenge for robots to perform their tasks. A study of the visibility of bell peppers in a greenhouse showed that even with a requirement of only 50% visibility, at best 69% of the fruits were visible in the images (Hemming et al., 2014). Combining five viewpoints could raise this to a 90% detection rate. Similarly, in Boogaard et al. (2020), 36

viewpoints per plant were needed to reliably observe the leaf and fruit nodes in cucumber plants. Bulanon et al. (2009) analyzed the visibility of fruits on orange trees and reported that for a single viewpoint, only 19-47% of the fruits were visible, while combining 9 viewpoints resulted in 91% visibility.

In Section 5, approaches to deal with incomplete information, specifically with occlusion, are discussed.

4 Dealing with the challenge of variation

Deep neural networks boosted the ability of agricultural robots to deal with variation, as will be discussed in Section 4.1. However, even with the use of deep learning, variation in agricultural environments remains a problem as the generalization of perception methods is not sufficient. Section 4.2 discusses different approaches in the literature to minimize the generalization gap.

4.1 The era of deep learning

Generally speaking, a machine vision algorithm consists of two parts; a feature-detection part to extract relevant image features from the camera images and a decision-making part, where the features are used to perform perception tasks, such as segmentation, classification, or object detection. For many decades, developments in both parts were focused around hand-crafted algorithms. Many powerful feature extractors were developed, such as the scale-invariant feature transform (SIFT) (Lowe, 2004) and pyramids of histograms-of-gradients (Bosch et al., 2007), which could be used, for instance, to match the current camera image with features of objects stored in a database. Later, the decision-making part was replaced with machine learning algorithms. In (Suh et al., 2018b), for instance, a support-vector machine, random forest and neural network were used to classify sugar-beet and potato plants based on a bag-of-visual-words model using SIFT features. A downside of these methods, however, is that the quality of the machine vision algorithm is limited by the quality of the image features. And it turned out to be very challenging to develop feature algorithms that are invariant to various variations in the appearance of objects in agricultural environments.

With the success of AlexNet (Krizhevsky et al., 2012) in the ImageNet competition (Russakovsky et al., 2015), a new class of methods started to dominate the field of machine vision; those based on deep learning, specifically on convolutional neural networks (CNNs). With a combination of convolutional layers, pooling layers, and fully collected layers, these networks can jointly learn feature extraction and decision-making. With end-to-end learning, machine vision tasks can be optimized based on a large training set with images and corresponding desired output values (Goodfellow et al., 2016).

These techniques also revolutionized machine vision for robotic applications in agriculture. In a survey of 40 studies on the use of deep learning for machine vision tasks in agriculture and food production, it was concluded that deep neural networks outperform other existing techniques (Kamilaris and Prenafeta-Boldu´, 2018). Zhu et al. (2018) provided an overview of deep learning for the use of smart agriculture. Deep neural networks were shown to be able to detect seven different types of fruits in Sa et al. (2016) using a multi-modal version of Faster R-CNN based on RGB and NIR images. Also when dealing with 3D point clouds, deep neural networks have shown their benefits (Boogaard et al., 2021; Shi et al., 2019) although in the 3D domain, more classical feature-based methods are still competitive (Dutagaci et al., 2020).

However, even with the use of deep neural networks, a generalization gap is often observed, limiting the extent to which agricultural robots can deal with variations in the environment. The difference in the distribution of the images in the training set and that of the new images during execution is causing a drop in performance. This was, for instance, observed when dealing with a new cultivar of broccoli (Blok et al., 2020) and when confronted with a new sugar-beet field with, among others, a variation in cultivar, soil types, growth stages, and illumination (Ruigrok et al., 2023). This limitation to deal with variation is a showstopper for commercial applications, as these need a guaranteed performance in any situation occurring in practice. Falling back on non-deep-learning approaches with limited but more predictable behavior can then be preferred. This calls for methods to minimize the generalization gap.

4.2 Minimizing the generalization gap

This subsection discusses a few approaches to minimize the generalization gap.

4.2.1 Transfer learning and image augmentation

To allow deep neural networks to generalize better to new situations - that is, to let them deal better with the variation present in the agricultural environment - it is important to apply regularization techniques, especially when there is limited available training data. Transfer learning, the process of first training a neural network based on a large general image dataset, such as ImageNet, followed by fine-tuning on a specific dataset in the relevant domain, was shown to also be beneficial for agriculture tasks (Chen et al., 2020; Thenmozhi and Reddy, 2019). Another common method is to apply image augmentation to increase the variation of the training set. This was shown, for instance, to increase the level of generalization to a new species of broccoli in Blok et al. (2020) and improve the identification of sheep (Hitelman et al., 2022). Smart sampling techniques during learning can improve performance in the presence of class imbalance (Boogaard et al., 2022).

4.2.2 Composition of the training set

The generalization gap arises when the distribution of images in the training set is not covering the distribution of images that are encountered during execution (or in the test set). Figure 1 illustrates a two-class classification problem with the total distribution covering all possible variations in the shaded green and red areas. On the left, you see a typical situation, where training data is taken from only a limited set of different conditions (fields, cultivars, etc.). In such situations, there is a high change that the trained perception method is confronted with new situations that cannot be handled. One way to improve the quality of the training set is to take more training samples from the training conditions (bottom of Fig. 1). This does cover a somewhat larger part of the total distribution. However, large parts are still not covered. Getting training data from more different conditions is more effective to cover a larger area of the total distribution (top of Fig. 1). However, both are needed to properly deal with all variations that can be encountered during execution. This illustration matches with the results found in Ruigrok et al. (2023), where data from 25 different fields with sugarbeet and potato plants were collected, with variation in cultivars, spoil types, growth stages, and weed pressure, to train a deep neural network for plant detection. Keeping the number of training images fixed to 500, the performance on the five independent test sets was shown to increase significantly when training data were used from more different fields.

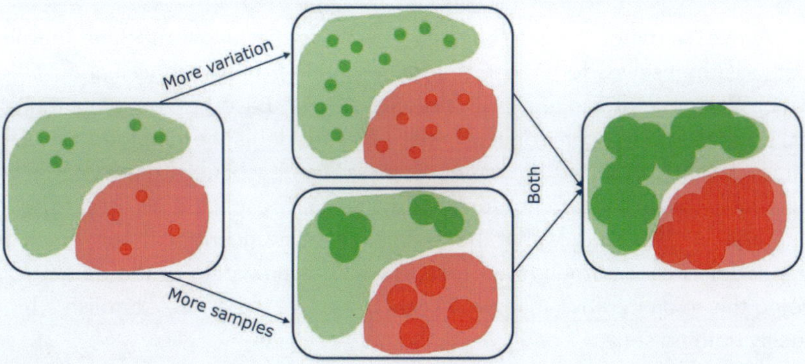

Figure 1 An illustration of the two ways to improve the quality of the training set. The shaded green and red areas symbolize the total distribution of all possible images of plants A and B. The darker areas illustrate the distribution of the training data. The training data can better cover the total distribution by adding more samples of the same fields, but more importantly, by adding more variation by adding data from new fields.

The performance was further improved by including more training images. This shows the importance to have sufficient variation and samples in the training set in order to minimize the generalization gap.

4.2.3 Incremental learning

However, even when using all training data from 20 different fields, a significant generalization gap was shown for some of the test fields that still contained something unique not found in the training fields (Ruigrok et al., 2023). This generalization gap could effectively be mitigated using incremental learning, where the neural network trained on the large and varied data set was fine-tuning on only a small number of training images from the specific field. The same was observed in Blok et al. (2020), where only a handful of images were needed to successfully fine-tune an instance-segmentation network to detect broccoli heads of a new cultivar.

4.2.4 Active learning

To obtain a high-quality dataset, it is important to add the right samples to the training data, that is, to add training samples that better cover the variation in possible future encounters. Randomly adding new training samples runs the risk of adding redundant samples containing information that was already covered in the training set. Instead, a different strategy is needed to select the most relevant new training samples, allowing the perception methods to achieve greater accuracy with fewer labeled training samples. This is important, especially in the field of agricultural robotics, where there is plenty of unlabeled data, but where it is difficult and time-consuming to get high-quality annotated data.

Active learning refers to a set of methods that allow machine learning methods themselves to choose the training data from which to learn (Settles, 2012). The general concept is illustrated in Fig. 2. Based on an initial training set, an active-learning model is trained. This model is then applied to all the data in a large pool of unlabeled images. For each unlabeled image, the method provides a prediction-uncertainty value. These are then used to select a number of new images that the method is most uncertain about, which are then labeled by a human (the oracle) to add as new data to the training set. Doing this in an iterative manner, allows the active learner to establish a high-quality training set.

The key in active learning is to estimate the prediction uncertainty of the model, also called the epistemic uncertainty (Kendall and Gal, 2017). In Bayesian neural networks, the epistemic uncertainty is modeled by including a distribution over the weights of the network, basically capturing the set

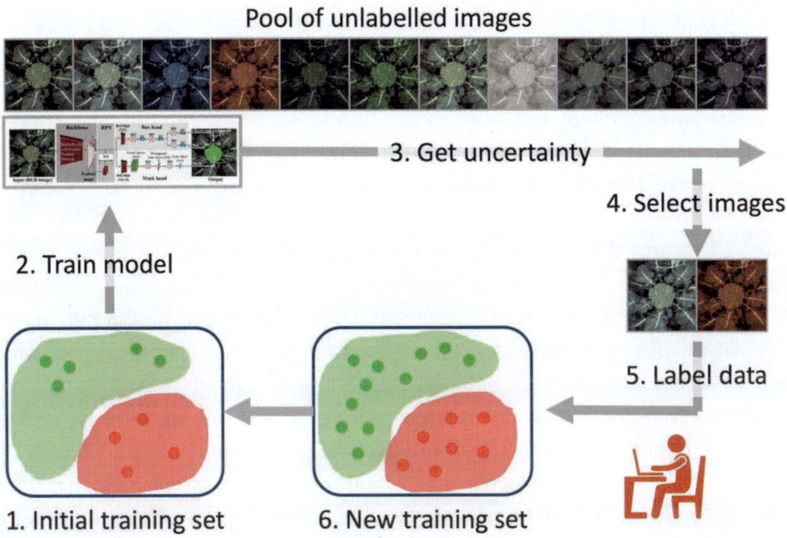

Figure 2 An illustration of the active-learning cycle for a classification task (green/red). The initial training set covers only part of the total distribution that can occur in reality (shaded green and red parts). Starting from this initial training set, the perception model is trained. This model is then applied to all available unlabeled images or during runtime. The images that the model is most uncertain about are then selected for human annotation to create an improved new training set. This process is continuously repeated.

of all plausible models given the training data. With more training data, the estimation of the model weights typically becomes more certain. However, to keep things tractable, usually an approximation of this Bayesian formulation is used by applying Monte-Carlo dropout (Gal and Ghahramani, 2016). In this procedure, multiple stochastic forward passes are applied to the same input. In each inference run, random dropout is applied, which basically disables part of the network connections at random. Prediction uncertainty can then be obtained by measuring the variation in the predictions made in the multiple stochastic forward passes, for instance, using the variance of the predictions for regression or using entropy for classification. Building upon Morrison et al. (2019), in Blok et al. (2022) Monte-Carlo dropout was used to estimate model uncertainty for an instance segmentation task of the detection of broccoli heads and the classification of different defects. Here, dropout was applied solely in the different heads of the Mask R-CNN network, not in the feature extraction part, speeding up computation time. The overall epistemic uncertainty consisted of semantic uncertainty (class), spatial uncertainty (bounding-box and mask), and uncertainty in object presence.

Using the epistemic uncertainty for active learning resulted in significantly better performance with the same amount of training data compared to the random selection of training data (Blok et al., 2022), indicating a better selection of training data using epistemic uncertainty. This was further supported by the fact that the method selected more images containing underrepresented classes, resulting in a more balanced training set. A similar active-learning approach was taken in Chandra et al. (2020) to improve the performance of a deep neural network for panicle detection in cereal crops. Older work applied active learning to classical machine learning methods to classify crops and weeds using spectral imaging data (Pantazi et al., 2016). However, it must be noted that active learning is not always successful. In Rawat et al. (2022), active learning outperformed random sampling in only one of three semantic segmentation tasks for fruit detection. This might be caused by the composition of the datasets and the specific uncertainty metric used. More research is needed to ensure effective active learning in all cases.

Active learning is a human-in-the-loop machine learning approach (Monarch, 2021). By combining the expertise of human annotators with the ability of the machine to process large amounts of data, this hybrid approach allows the efficient selection of training data. Moreover, it allows live-long or continuous learning, where the agricultural robot asks for human assistance whenever it is uncertain about its own perception, becoming more and more capable to deal with the variation in agricultural environments.

4.2.5 Unsupervised and self-supervised

Where labeled data is costly to acquire, the costs of getting unlabeled data are typically quite low, for instance, by using UAVs to acquire images of fields. Unsupervised learning methods can exploit large unlabeled datasets by finding patterns in the data in order to learn good representations of the data. In dos Santos Ferreira et al. (2019), for instance, unsupervised deep clustering methods were used on two large image datasets with weed plants, which allowed the unsupervised training of visual features that were representative of the data. This was then combined in a semi-supervised approach, where human annotators annotated the clusters, greatly reducing the amount of human annotations needed, while achieving high classification results.

Another use of unsupervised methods is dimensionality reduction. Auto encoders are a popular way to reduce the dimensionality of image data and to capture the most important patterns in the data. This was applied to disease detection on peach leaves in Bedi and Gole (2021). They used a convolutional auto-encoder to learn a latent/embedded representation of the images, which was then used as input for a simple CNN to classify images as healthy or diseased, combining the power of unsupervised and supervised learning.

Self-supervised learning is another brand of unsupervised methods that gaining a lot of momentum and was promoted by LeCun and Misra (2022). Like supervised learning, these methods also learn from input–output pairs, but different from supervised learning, these outputs are not provided by human annotation, but collected by the system itself from the unlabeled data. This can be framed as predictive learning, for instance, by learning to predict one part of the image based on another part (image completion), or by predicting the future based on the past using temporal data.

Contrastive learning is a self-supervised learning approach where a network is trained to learn an embedding that minimizes the distance between images from the same domain (positive samples) while maximizing the distance between images from different domains (negative samples). These positive and negative samples can be taken, for instance, from datasets of different crops, or a positive pair can be made with an image and a heavily distorted/augmented version of that image, while the negative pair are two different images. This way, contrastive learning results in a meaningful feature representation of the images capturing essential patterns in the unlabeled data. This principle was used in the agricultural domain, for instance, in Güldenring and Nalpantidis (2021), such an approach was applied for unsupervised pre-training of a CNN, which was then fine-tuned for a downstream task (plant classification) on a small amount of labeled data. This was shown to outperform conventional methods that use pre-training based on ImageNet, especially when limited amounts of labeled data are available.

4.2.6 Synthetic data and generative learning

The lack of training data to get robust perception models that generalize well can also be solved by enriching the training set using synthetic data. In Turgut et al. (2022), 3D synthetic rosebush models were successfully used to pre-train 3D point-based deep neural networks for plant-part segmentation. Using realistic 3D structural plant models, Barth et al. (2018) and Barth et al. (2019) rendered a large set of synthetic images to bootstrap training of a CNN for the segmentation of plant parts for a bell-pepper harvesting robot. They further increased the realism of these images using a generative adversarial network (GAN), more specifically CycleGAN (Barth et al., 2020). A conditional GAN was used in (Abbas et al., 2021) to generate additional synthetic images of tomato leaves, boosting the performance of a CNN for disease classification. A similar approach was taken in Arsenovic et al. (2019) using StyleGAN to generate additional synthetic images. Although several successes of the use of GANs to generate additional training data are reported in the literature, van Marrewijk et al. (2022) noted that in some studies, the evaluation was not sound as the GANs themselves were trained on data outside of the original training set, giving an unfair advantage over training on only the original set. They also concluded that the advantage of using synthetic data from a GAN is limited and time spent on setting up the GAN and optimizing the hyperparameters might be better spent on labeling some additional training data. Hence, the effectiveness of using generative learning to improve the generalization of neural networks needs to be further investigated.

5 Dealing with the challenge of incomplete information

Section 5.1 discusses some passive approaches to deal with incomplete information. However, as discussed in Section 5.2, multi-view and active-vision approaches have been proposed that allow agricultural robots to actively gather more information about the environment to minimize the information gap.

5.1 Passive visual perception

Despite the fact that robots have the ability to actively gather new information, most of the work on machine vision for agricultural applications have focused on passive perception, that is, trying to get as much information out of a single image. Whenever the objects of interest are well visible in the image, that approach is perfectly valid. However, when the objects are not well visible in the image, for instance, due to sensor noise, flares, or because the objects are partially or completely occluded, a passive approach is inadequate.

To some extent, deep neural networks can deal with partial occlusions using a single image by developing and training them to be robust to occlusion and even to predict the occluded parts. In Li et al. (2022), for instance, a method was developed for the localization of partially occluded apples. Blok et al. (2021) trained a neural network to predict the complete 2D mask of partially occluded broccoli heads, improving the size estimation. Similarly, the full 3D point cloud, including the non-observable parts, could be predicted from monocular images (Zeng et al., 2018). Others used a conditional GAN to learn to generate non-occluded versions of images with occluded grape bunches (Kierdorf et al., 2022).

Although these methods gain some improvement in performance, they can only predict the occluded parts based on patterns observed in the training data. If there is sufficient regularity or symmetry in the shape of the objects, this can work well. However, many agricultural objects are quite irregular, making it much more challenging to accurately predict the occluded parts, which is often needed for robotic operation. Furthermore, in agricultural production, anomalies and deviations from the norm are often relevant features, which cannot be predicted from partial information at all. Moreover, in cluttered environments, many objects are heavily or even fully occluded. In a bell-pepper greenhouse, for instance, in 31% of the cases, less than half of the fruit surface was visible in the images (Hemming et al., 2014), many were even completely hidden from view. To properly deal with such situations, the information gap needs to be minimized by allowing robots to gather more information.

5.2 Minimizing the information gap

This subsection discusses a few approaches to minimize the information gap.

5.2.1 Multi-view visual perception

Hemming et al. (2014) showed that in a cluttered environment, the detection of bell peppers was significantly improved when combining multiple viewpoints with an eye-in-hand system; going from 69% for a single viewpoint to 90% for five viewpoints. In Sa et al. (2017), a fixed scanning path was applied to acquire a more complete 3D point-cloud reconstruction of a bell pepper from multiple viewpoints to be used to guide the harvesting operation. A similar approach was taken in Barth et al. (2016), where apart from a 3D reconstruction, the scan was used to find the best view on the to-be-harvested bell pepper.

In (Halstead et al., 2018), a vision system was developed to estimate the yield in a bell-pepper greenhouse. They developed a multi-view approach with the robot taking images while traversing the crop row. To prevent double counting and to deal with fruits that are (partially) occluded in some frames, a multi-object tracker was implemented, based on the overlap in object detections in consecutive frames. The tracking method was improved in Smitt et al. (2021) using the reprojection of the object mask in the current frame to the expected location in the next frame using depth data and wheel odometer. This method was further refined and applied in two different use-cases in Halstead et al. (2021). Kurtser and Edan (2018) analyzed the detectability of bell pepper depending on viewing distance and angle, which can be used to select more profitable sets of viewpoints to increase detectability. Not only can multiple views increase the chance of having the objects of interest in view, they can also be used to improve the detection. Observing plants from ten different camera viewpoints, for instance, did not only result in more complete 3D reconstructions but it also improved the plant-part segmentation (Shi et al., 2019). Van Essen et al. (2022) showed that fruit maturity classification could be improved by dynamically selecting additional viewpoints. In van Essen et al. (2021), the species classification of fish on a conveyor belt was improved by tracking the fish over multiple frames and combining the predictions of a neural network probabilistically. Russello et al. (2021) showed that the pose estimation of partially occluded walking cows could be improved with a temporal network including multiple video frames.

It is clear that multi-view perception has many advantages over passive perception. Still, the viewpoints in the above-mentioned studies are set in advance by the developer and not tuned to the current situation, which does not guaranty that the required information is acquired. To further improve perception, agricultural robots would benefit from actively viewing the world.

5.2.2 Active visual perception

Using a matrix of nine cameras connected to the end effector of a robot, Lehnert et al. (2019) developed a visual servoing method that determines the next viewpoint to get a less occluded view of a bell pepper. This system was used to train a deep neural network to predict the next viewpoint from just one camera image (Zapotezny-Anderson and Lehnert, 2019).

The downside of visual-servoing approaches is that they require a non-occluded starting viewpoint and can get stuck in local optima. Next-best-view (NBV) planners based on occupancy grids can deal with this, as they are based on a calculation of information gain that guides the robot to explore unseen parts of space. As illustrated in Fig. 3, based on a current reconstruction of the scene, the robot evaluates a few candidate viewpoints to then choose the viewpoint with the highest expected information gain. In Zaenker et al. (2021), this was, for instance, used to detect fruits on a number of plants in a small simulated greenhouse. It showed beneficial to combine the global NBV

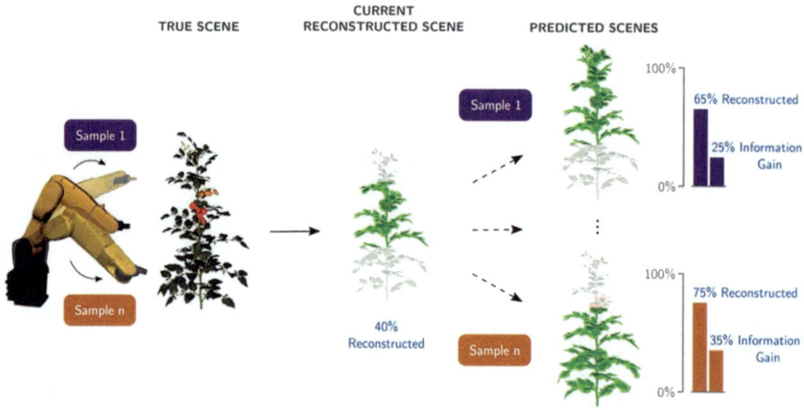

Figure 3 An illustration of the active-perception next-best-view paradigm. The robot has a current reconstruction of the world (a tomato plant in this case), which represents the seen and unseen parts of space. Based on this, a prediction can be made of the expected information gain from different new viewpoints, in order to select the best next view.

planning with local view planning based on visual servoing (Zaenker et al., 2021). Burusa et al. (under review) added an attention mechanism to efficiently reconstruct specific plant parts and showed that it outperformed random and fixed multi-view approaches.

The NBV planning could be further optimized and accelerated using deep reinforcement learning, providing additional coverage of unexplored space and new information about the relevant parts as a reward to train the system (Zeng et al., 2022). Alternatively, in Han et al. (2022), the information gain for new viewpoints was learned by a network in a supervised manner, acquiring the training data by comparing the reconstruction to a known 3D point cloud of the object.

5.2.3 Sensor fusion

Instead of redirecting the same sensor, the challenge of incomplete information can also be approached by combining different complementary sensor modalities. The detection of immature citrus fruit, for instance, was boosted by combining color and thermal images (Gan et al., 2018). Semantic segmentation of apple trees (Kang and Wang, 2023) and cucumber plants (Boogaard et al., 2021) was improved with a fusion of 3D and color data. For the automatic identification of plant diseases using UAVs, the fusion of thermal and spectral imaging was often successful (Neupane and Baysal-Gurel, 2021).

6 Directions for future research

Sections 4 and 5 discuss approaches to deal with the challenges of variation and incomplete information. However, the challenges are far from being resolved and future research needs to further progress along these lines. There are some suggestions for future research on the short term to better deal with the challenges, and some suggestions for future research on the long term to get to a higher level of autonomy in agricultural robotics.

6.1 On the short term: dealing better with variation and incomplete information

From this review, it is clear that the perception for agricultural robotics has greatly improved in the past decade. Building upon the progress in the field of deep learning, the performance of machine vision algorithms improved significantly and methods became more robust to variations present in the agricultural environment. However, as discussed in Section 3, the challenge of variation is not yet solved. Similarly, progress in the general field of robotics provided methods for multi-view and active perception, yet, the challenge of incomplete information remains for robots to effectively deal with cluttered agricultural environments. In this subsection, a few specific directions for future research are being discussed.

6.1.1 Quality of training set

It was shown in Ruigrok et al. (2023) that the generalization of deep neural networks to new situations greatly benefits from a rich and varied training set. This calls for better sharing of labeled datasets. Even research institutes are cautious with sharing their data, but especially companies consider data the new gold. Although this is possibly true, keeping the data private hampers progress in the field. Moreover, many of the small startups that currently enter the field are too small to collect rich-enough datasets to deal with all the variance present in agricultural environments. Alternatively, datasets can be enriched using image augmentation. Currently, mainly general augmentation methods are applied, which are suboptimal or sometimes even harmful in a specific domain (Balestriero et al., 2022). Instead, domain-specific augmentation can improve the generalization of the perception methods (e.g. Su et al., 2021; Ratner et al., 2017).

6.1.2 Active learning and uncertainty

Perception models need to become better aware of the uncertainty in the prediction. This is important to improve the safety and reliability of agricultural robots. When uncertain, actions can be postponed, new observations can be acquired or human assistance can be called to prevent errors. The latter might sound like a weak option, and fully autonomous systems are indeed the aim, but for the coming decade, robots need the assistance of humans to operate over long periods of time and in many different environments. More research on human-in-the-loop systems therefore needs to be done. Not only on the topic of active learning but also on effective interfaces to minimize the human effort to assist the robot (Monarch, 2021).

6.1.3 Unsupervised learning

In his cake analogy, Yann LeCun views machine intelligence as a cake, with reinforcement learning as the cherry on top, supervised learning as the icing, and unsupervised learning as the bulk of the cake (LeCun, 2016). In other words, unsupervised learning is essential to create intelligent systems. It allows to exploit large sets of unlabeled data, typically available also in agriculture, to learn powerful low-dimensional representations of the data, which allows efficient supervised learning of downstream tasks based on small labeled datasets. To unlock the power of unsupervised learning, more research is needed in the development and application of unsupervised techniques like predictive learning, contrastive learning, deep clustering, domain adaptation, and self-supervised learning.

6.1.4 Use of domain knowledge

The completely data-driven deep-learning approaches ignore the available domain knowledge to solve agricultural tasks. Combining data-based and model-based methods would allow to put constraints on the learning process to achieve better performance especially when training data are limited. This can be done, for instance, by adding domain knowledge to create synthetic data based on models, for instance, using plant models (Turgut et al., 2022; Barth et al., 2019), to include specific image features as additional input (Liu et al., 2021; Milioto et al., 2018), or to use domain knowledge in the loss function to enforce consistency of the network predictions, such as geometrical and temporal constraints for pose detection (Dabral et al., 2018). Domain knowledge can also be used to deal with missing information, for instance, by using leaf templates to reconstruct plant leaves in occluded scenarios (Marks et al., 2022).

6.1.5 Active perception

Robots have the ability to control their own sensory input. This fact is hardly explored in current agricultural-robotic research, which instead focuses predominantly on the processing of single images. As discussed in Section 5, multi-view and active NBV planning improves the performance of perception methods, especially in occluded environments. Further research is needed to make these methods even more effective, for instance using attention mechanisms (Burusa et al., under review), adding prior knowledge for effective object search, or allowing robots to learn optimal view planning from their own experience (Zeng et al., 2022; Han et al., 2022). The robot's active capabilities can also be used in an active learning setting, where the robot actively plans a path to collect useful training data to be labeled by a human (Rückin et al., 2022). The next step is to develop interactive object learning, where a robot actively explores objects in the environment to learn and improve its perception models without or with minimal human intervention (Lyubova et al., 2016).

6.2 On the long term: autonomous agricultural robotics

With the research directions described earlier, in the next decade, the field can develop perception methods for agricultural robots to robustly perform specific tasks, such as harvesting fruits or removing weeds. These robots will probably work semi-autonomously, needing occasional input from a human operator. On the long term, however, robots are needed that can perform multiple tasks and operate fully autonomously. This requires the next level of artificial intelligence, going beyond the understanding of a single domain to mastering multiple domains. Some first steps in generalizing over different tasks have recently been taken with Gato, a generalist agent (Reed et al., 2022). However, it also requires a true understanding of the world, going beyond patterns extracted from training data to grounding the external world in the robot's own sensorimotor system (Harnad, 1990). This research challenge is enormous, and I definitely do not want to claim to know the solution. However, there are a few principles that I believe are important to reach this goal.

Most importantly, perception and action need to be tighter integrated to achieve sensorimotor coordination. When a robot takes an action, it immediately results in a change of the sensory observations. When the robot changes its position, it gets a new view of the world, and when it interacts with objects, it changes the state of the objects. This applies not only to the sense of vision but to all senses. When a robot interacts with an object, for instance, it receives visual feedback about its actions, but also tactile and auditory. For a robot to truly understand the world, it needs to be able to predict the sensory consequences of its actions, the so-called sensorimotor contingencies (O'Regan and Noë, 2001). This requires that the robot has a model of the world containing knowledge about the state and how this will change as a consequence of an action, including object affordances containing the knowledge of the actions that can be applied to objects and the resulting effects of these actions (Gibson, 2014; Krüger et al., 2011). This world model needs to be continuously updated to be kept in sync with the physical world based on new sensory observations and experiences of the robot.

The predominant sense-think-act cycle in robotics then needs to change to an act-predict-sense-compare cycle, where the predictions of the consequences of actions are compared to the new observations after applying the action. When the predictions are correct, the world model is correct and the next planned action can be executed. When the predictions do not match the observations, the world changes differently than expected, requiring attention to update the model. This is in line with current theories on human cognition (see, e.g. Clark, 2015).

The consequences of actions applied to an object can be learned by the robot autonomously by interacting with that object, thereby grounding the object in its sensory-motor experience (Harnad, 1990; Pfeifer and Bongard, 2006). This process of playful manipulation of the world allows the robot to learn causality (Pearl, 2019). Furthermore, during execution, the robot continuously observes the sensory consequences of actions, allowing self-supervised and life-long online learning of the prediction models. Importantly, the robot should not only be equipped with visual sensors but also with other sensory modalities.

Inspiration for robot learning can furthermore be taken from developmental psychology on how humans learn to perceive and interact with the world (Cangelosi and Schlesinger, 2015; Lungarella et al., 2003). An important aspect is that learning is incremental, starting from learning basic skills, like moving your hand to an object in space and grasping that object, which allows learning about objects in the world and on which more complex skills can be built, like harvesting all ripe apples in an orchard. Initially, learning is more constraint due to limitations in the sensorimotor system and through scaffolding by caretakers, simplifying the learning process, which gradually becomes less constraint to learn more complex relations. Another aspect is that of curiosity. When a certain skill is learned and predictions of the sensory consequence can be made successfully, it is time to explore the world for new challenges.

With this sensorimotor learning, robots can start to understand the world from their own perspective in an unsupervised way. This builds the foundation for further learning. To connect to the human world and perform a variety of agricultural tasks, future robots will no longer be programmed, but learn from human demonstrations (Ravichandar et al., 2020). Similar to how infants learn, imitation learning will provide a common taxonomy of the world and bootstrap learning of tasks. With this in place, robots can learn to fine-tune the execution of tasks with their own body morphology using reinforcement learning (Kober et al., 2013; Arulkumaran et al., 2017). These three levels of learning (unsupervised, supervised, and reinforced) correspond to the three ingredients of LeCun's cake analogy.

7 Conclusion

To conclude this survey of the advances of visual perception for agricultural robotics, it can be observed that current robots can operate in environments that are relatively structured and controlled. The challenges of variation and incomplete information, however, hamper the application of agricultural robots on a larger scale. If the amount of variation increases, for instance, by going to a variety of fields, the generalization of current perception methods is not yet good enough to be commercially applied. At the same time, when the amount of occlusion increases in cluttered environments, such as high-wire greenhouses and orchards, current robots are not able to locate all relevant objects. To improve future agricultural robots, a number of promising research directions to deal with the challenges of variation and incomplete information have been provided. This will allow the field to develop robots that can successfully perform specific tasks in agricultural environments in the coming decade. To achieve more versatile robots that can perform any task on a farm, steps need to be taken to get closer to artificial general intelligence in the coming decades. As an important principle, robot perception should no longer be studied in isolation, but in tight integration with the robot's actions within the framework of sensorimotor systems. Furthermore, for effective learning, robots should be able to explore the world and learn from their own experiences.

8 References

Abbas, I., Liu, J., Faheem, M., Noor, R. S., Shaikh, S. A., Solangi, K. A. and Raza, S. M. Different sensor based intelligent spraying systems in agriculture, *Sensors and Actuators. Part A* 316, 112265 2020, issn: 0924-4247. doi: 10.1016/j.sna.2020.112265. [Online]. Available at: https://www.sciencedirect.com/science/article/ pii/S0924424720309274.

Abbas, A., Jain, S., Gour, M. and Vankudothu, S. Tomato plant disease detection using transfer learning with c-gan synthetic images, *Computers and Electronics in Agriculture* 187, 106279 2021.

Arad, B., Kurtser, P., Barnea, E., Harel, B., Edan, Y. and Ben-Shahar, O. Controlled lighting and illumination-independent target detection for real-time cost-efficient applications. The case study of sweet pepper robotic harvesting, *Sensors* 19(6) 2019, issn: 1424-8220. doi: 10.3390/s19061390. [Online]. Available at: https://www.mdpi.com/1424-8220/19/6/1390.

Arad, B., Balendonck, J., Barth, R., Ben-Shahar, O., Edan, Y., Hellström, T., Hemming, J., Kurtser, P., Ringdahl, O., Tielen, T. and van Tuijl, B. Development of a sweet pepper harvesting robot, *Journal of Field Robotics* 37(6), 1027–1039 2020. doi: 10.1002/rob.21937.

Arsenovic, M., Karanovic, M., Sladojevic, S., Anderla, A. and Stefanovic, D. Solving current limitations of deep learning based approaches for plant disease detection, *Symmetry* 11(7), 939 2019.

Arulkumaran, K., Deisenroth, M. P., Brundage, M. and Bharath, A. A. Deep reinforcement learning: A brief survey, *IEEE Signal Processing Magazine* 34(6), 26–38 2017.

Balestriero, R., Bottou, L. and LeCun, Y. *"The Effects of Regularization and Data Augmentation Are Class Dependent," arXiv Preprint ArXiv:2204.03632* 2022.

Bargoti, S. and Underwood, J. P. Image segmentation for fruit detection and yield estimation in apple orchards, *Journal of Field Robotics* 34(6), 1039–1060 2017. doi: 10.1002/rob.21699. eprint. [Online]. Available at: https://onlinelibrary.wiley.com/doi/abs/10.1002/rob.21699.

Barth, R., Hemming, J. and van Henten, E. J. Design of an eye-in-hand sensing and servo control framework for harvesting robotics in dense vegetation, *Biosystems Engineering* 146, 71–84 2016. doi: 10.1016/j.biosystemseng.2015.12.001.

Barth, R., IJsselmuiden, J., Hemming, J. and Van Henten, E. J. Data synthesis methods for semantic segmentation in agriculture: A capsicum annuum dataset, *Computers and Electronics in Agriculture* 144, 284–296 2018. doi: 10.1016/j.compag. 2017.12.001. [Online]. Available at: https://www.sciencedirect.com/science/article/pii/S0168169917305689.

Barth, R., IJsselmuiden, J., Hemming, J. and Van Henten, E. J. Synthetic bootstrapping of convolutional neural networks for semantic plant part segmentation, *Computers and Electronics in Agriculture*. Hemming 161, 291–304 2019. doi: 10.1016/j.compag.2017.11.040. [Online]. Available at: http://www.sciencedirect.com/science/article/pii/S0168169917307664.

Barth, R., Hemming, J. and Van Henten, E. J. Optimising realism of synthetic images using cycle generative adversarial networks for improved part segmentation, *Computers and Electronics in Agriculture* 173, 105378 2020, issn: 0168-1699. doi: 10.1016/j.compag.2020.105378. [Online]. Available at: https://www.sciencedirect.com/science/article/pii/ S0168169919320794.

Bedi, P. and Gole, P. Plant disease detection using hybrid model based on convolutional autoencoder and convolutional neural network, *Artificial Intelligence in Agriculture* 5, 90–101 2021.

Billingsley, J., Visala, A. and Dunn, M. Robotics in agriculture and forestry. In: *Springer Handbook of Robotics* 2008. Springer, Berlin, Heidelberg Siciliano, B. and Khatib, O. (Eds). doi: 10.1007/978-3-540-30301-5_47.

Blok, P. M., van Evert, F. K., Tielen, A. P. M., van Henten, E. J. and Kootstra, G. The effect of data augmentation and network simplification on the image-based detection of broccoli heads with Mask R-CNN, *Journal of Field Robotics*, 1–20 2020. doi: 10.1002/rob.21975. eprint. [Online]. Available at: https://onlinelibrary.wiley.com/doi/abs/10.1002/rob.21975.

Blok, P. M., van Henten, E. J., van Evert, F. K. and Kootstra, G. Image-based size estimation of broccoli heads under varying degrees of occlusion, *Biosystems Engineering* 208, 213–233 2021.

Blok, P. M., Kootstra, G., Elghor, H. E., Diallo, B., van Evert, F. K. and van Henten, E. J. Active learning with maskal reduces annotation effort for training mask r-cnn on a broccoli dataset with visually similar classes, *Computers and Electronics in Agriculture* 197, 106917 2022, issn: 0168-1699. doi: 10.1016/j.compag.2022.106917. [Online]. Available at: https: //www.sciencedirect.com/science/article/pii/S0168169922002344.

Blok, P. M., van Boheemen, K. and van Evert, F. K., iJsselmuiden, J., and Kim, G.-H., Robot navigation in orchards with localization based on particle filter andabelln filter, *Computers and Electronics in Agriculture* 157, 261–269 n.d. doi: 10.1016/j.compag.2018.12.046. [Online]. Available at: https://www.sciencedirect.com/science/article/ pii/S0168169918315230.

Bonadies, S. and Gadsden, S. A. An overview of autonomous crop row navigation strategies for unmanned ground vehicles, *Engineering in Agriculture, Environment and Food* 12(1), 24–31 2019, issn: 1881-8366. doi: 10.1016/j.eaef.2018.09.001. [Online]. Available at: https://www.sciencedirect.com/science/article/pii/S188183661730188X.

Boogaard, F. P., Rongen, K. S. A. H. and Kootstra, G. W. Robust node detection and tracking in fruit-vegetable crops using deep learning and multi-view imaging, *Biosystems Engineering* 192, 117–132 2020, issn: 1537-5110. doi: 10.1016/j.biosystemseng.2020.01.023. [Online].

Boogaard, F. P., van Henten, E. J. and Kootstra, G. Boosting plant-part segmentation of cucumber plants by enriching incomplete 3D point clouds with spectral data, *Biosystems Engineering* 211, 167–182 2021. doi: 10.1016/j.biosystemseng.2021.09.004.

Boogaard, F. P., van Henten, E. J. and Kootstra, G. Improved point-cloud segmentation for plant phenotyping through class-dependent sampling of training data to battle class imbalance, *Frontiers in Plant Science* 13, 838190–838190 2022.

Bosch, A., Zisserman, A. and Munoz, X. Representing shape with a spatial pyramid kernel. In: Proceedings of the 6th ACM International Conference on Image and Video Retrieval 2007, pp. 401–408.

Bulanon, D. M., Burks, T. F. and Alchanatis, V. Fruit visibility analysis for robotic citrus harvesting, *Transactions of the ASABE* 52(1), 277–283 2009.

Bulanon, D. M., Hestand, T., Nogales, C., Allen, B. and Colwell, J. Machine vision system for orchard management. In: *Machine Vision and Navigation* Sergiyenko, O., Flores-Fuentes, W. and Mercorelli, P. (Eds). Springer, 2020.

Burusa, A., van Henten, E. and Kootstra, G. *"Attention-driven active vision for efficient reconstruction of plants and targeted plant parts," arXiv Preprint ArXiv:2206.10274* 2022. doi: 10.48550/arXiv.2206.10274.

Cangelosi, A. and Schlesinger, M. *Developmental Robotics: From Babies to Robots*. MIT Press 2015.

Caron, M., Misra, I., Mairal, J., Goyal, P., Bojanowski, P. and Joulin, A. Unsupervised learning of visual features by contrasting cluster assignments, *Advances in Neural Information Processing Systems* 33, 9912–9924 2020.

Chandra, A. L., Desai, S. V., Balasubramanian, V. N., Ninomiya, S. and Guo, W. Active learning with point supervision for cost-effective panicle detection in cereal crops, *Plant Methods* 16(1), 34 2020.

Chaudhury, A., Ward, C., Talasaz, A., Ivanov, A. G., Brophy, M., Grodzinski, B., Huner, N. P. A., Patel, R. V. and Barron, J. L. Machine vision system for 3d plant phenotyping, *IEEE/ACM Transactions on Computational Biology and Bioinformatics* 16(6), 2009–2022 2019. doi: 10.1109/TCBB.2018.2824814.

Chen, J., Chen, J., Zhang, D., Sun, Y. and Nanehkaran, Y. A. Using deep transfer learning for imagebased plant disease identification, *Computers and Electronics in Agriculture* 173, 105393 2020.

Chlingaryan, A., Sukkarieh, S. and Whelan, B. Machine learning approaches for crop yield prediction and nitrogen status estimation in precision agriculture: a review, *Computers and Electronics in Agriculture* 151, 61–69 2018, issn: 0168-1699. doi: 10.1016/j.compag.2018.05.012. [Online]. Available at: https://www.sciencedirect.com/science/article/pii/ S0168169917314710.

Clark, A. *Surfing Uncertainty: Prediction, Action, and the Embodied Mind*. Oxford University Press 2015.

Corke, P. *Robotics, Vision and Control, Ser., Springer Tracts in Advanced Robotics*. Springer, Cham 2017.

Dabral, R., Mundhada, A., Kusupati, U., Afaque, S., Sharma, A. and Jain, A. Learning 3d human pose from structure and motion. In: Proceedings of the European Conference on Computer Vision (ECCV) 2018, pp. 679–696.

Doornweerd, J. E., Kootstra, G., Veerkamp, R. F., Ellen, E. D., van der Eijk, J. A. J., van de Straat, T. and Bouwman, A. C. Across-species pose estimation in poultry based on images using deep learning, *Frontiers in Animal Science* 2 2021. doi: 10.3389/fanim.2021.791290. [Online].

dos Santos Ferreira, A., Freitas, D. M., da Silva, G. G., Pistori, H. and Folhes, M. T. Unsupervised deep learning and semi-automatic dataabellingg in weed discrimination, *Computers and Electronics in Agriculture* 165, 104963 2019.

Dutagaci, H., Rasti, P., Galopin, G. and Rousseau, D. Rose-x: an annotated data set for evaluation of 3d plant organ segmentation methods, *Plant Methods* 16(1), 28 2020.

Fu, L., Sun, S., Li, R. and Wang, S. Classification of kiwifruit grades based on fruit shape using a single camera, *Sensors* 16(7) 2016, issn: 1424-8220. doi: 10.3390/s16071012. [Online]. Available at: https://www.mdpi.com/1424-8220/16/7/1012.

Fu, L., Gao, F., Wu, J., Li, R., Karkee, M. and Zhang, Q. Application of consumer rgb-d cameras for fruit detection and localization in field: A critical review, *Computers and Electronics in Agriculture* 177, 105687 2020, issn: 0168-1699. doi: 10.1016/j.compag.2020.105687. [Online]. Available at: https://www.sciencedirect.com/science/article/pii/ S0168169920319530.

Gal, Y. and Ghahramani, Z. Dropout as aabellinn approximation: representing model uncertainty in deep learning. In: International Conference *on* Machine Learning, PMLR, 2016 pp. 1050–1059.

Gan, H., Lee, W. S., Alchanatis, V., Ehsani, R. and Schueller, J. K. Immature green citrus fruit detection using color and thermal images, *Computers and Electronics in Agriculture* 152, 117–125 2018.

Gibson, J. J. *The Ecological Approach to Visual Perception: Classic Edition*. Psychology Press 2014.

Golbach, F., Kootstra, G., Damjanovic, S., Otten, G. and van de Zedde, R. Validation of plant part measurements using a 3d reconstruction method suitable for high-throughput seedling phenotyping, *Machine Vision and Applications* 27(5), 663–680 2016.

Goodfellow, I., Bengio, Y. and Courville, A. *Deep Learning*. MIT Press 2016. Available at: http://www.deeplearningbook.org.

Güldenring, R. and Nalpantidis, L. Self-supervised contrastive learning on agricultural images, *Computers and Electronics in Agriculture* 191, 106510 2021.

Guo, N., Zhang, B., Zhou, J., Zhan, K. and Lai, S. Pose estimation and adaptable grasp configuration with point cloud registration and geometry understanding for fruit grasp planning, *Computers and Electronics in Agriculture* 179, 105818 2020, issn: 0168-1699. doi: 10.1016/j.compag.2020.105818. [Online]. Available at: https://www.sciencedirect.com/science/ article/pii/S0168169920314046.

Halstead, M., McCool, C., Denman, S., Perez, T. and Fookes, C. Fruit quantity and ripeness estimation using a robotic vision system, *IEEE Robotics and Automation Letters* 3(4), 2995–3002 2018. doi: 10.1109/LRA.2018.2849514.

Halstead, M., Ahmadi, A., Smitt, C., Schmittmann, O. and McCool, C. Crop agnostic monitoring driven by deep learning, *Frontiers in Plant Science* 12, 786702 2021, issn: 1664-462X. doi: 10.3389/fpls.2021.786702. [Online]. Available at: https://www .frontiersin.org/articles/10.3389/fpls.

Han, Y., Zhan, I. H., Zhao, W. and Liu, Y.-J. A double branch next-best-view network and novel robot system for active object reconstruction. In: IEEE Publications International Conference on Robotics and Automation (ICRA) (vol. 2022), 2022, pp. 7306–7312.

Harnad, S. The symbol grounding problem, *Physica D: Nonlinear Phenomena* 42(1–3), 335–346 1990.

Hemming, J., Ruizendaal, J., Hofstee, J. W. and van Henten, E. J. Fruit detectability analysis for different camera positions in sweet-pepper, *Sensors* 14(4), 6032–6044 2014. doi: 10.3390/s140406032.

Hitelman, A., Edan, Y., Godo, A., Berenstein, R., Lepar, J. and Halachmi, I. Biometric identification of sheep via a machine-vision system, *Computers and Electronics in Agriculture* 194, 106713 2022.

Horaud, R., Hansard, M., Evangelidis, G. and Ménier, C. An overview of depth cameras and range scanners based on time-of-flight technologies, *Machine Vision and Applications* 27(7), 1005–1020 2016. doi: 10.1007/s00138-016-0784-4.

Jamil, N., Kootstra, G. and Kooistra, L. Evaluation of individual plant growth estimation in an intercropping field with UAV imagery, *Agriculture* 12(1), 10 2022, issn: 2077-0472. doi: 10.3390/agriculture12010102. [Online]. Available at: https://www.mdpi.com/2077-0472/12/1/102.

Kamilaris, A. and Prenafeta-Boldú, F. X. Deep learning in agriculture: A survey, *Computers and Electronics in Agriculture* 147, 70-90 2018, issn: 0168-1699. doi: 10.1016/j.compag.2018.02.016. [Online]. Available at: http://www.sciencedirect.com/science/ article/pii/S0168169917308803.

Kang, H. and Wang, X. Semantic segmentation of fruits on multi-sensor fused data in natural orchards, *Computers and Electronics in Agriculture* 204, 107569 2023.

Kang, H., Zhou, H., Wang, X. and Chen, C. Real-time fruit recognition and grasping estimation for robotic apple harvesting, *Sensors* 20(19) 2020, issn: 1424-8220. doi: 10.3390/s20195670. [Online]. Available at: https://www.mdpi.com/1424-8220/20/19/5670.

Kendall, A. and Gal, Y. What uncertainties do we need inabellinn deep learning for computer vision?, *Advances in Neural Information Processing Systems* 30 2017.

Kierdorf, J., Weber, I., Kicherer, A., Zabawa, L., Drees, L. and Roscher, R. Behind the leaves: estimation of occluded grapevine berries with conditional generative adversarial networks, *Frontiers in Artificial Intelligence* 5, 830026 2022.

Kober, J., Bagnell, J. A. and Peters, J. Reinforcement learning in robotics: A survey, *The International Journal of Robotics Research* 32(11), 1238–1274 2013.

Kootstra, G., Bender, A., Perez, T. and van Henten, E. J. 2020. Robotics in agriculture. In *Encyclopedia of Robotics*, pp. 1–19. Springer, isbn: 978-3-642-41610-1. doi: 10.1007/9783-642-41610-1_43-1. [Online]. doi: 10.1007/978-3-642-41610-1_43-1.

Kootstra, G., Wang, X., Blok, P. M., Hemming, J. and van Henten, E. Selective harvesting robotics: current research, trends, and future directions, *Current Robotics Reports* 2(1), 95–104 2021, issn: 2662-4087. doi: 10.1007/s43154-020-00034-1. [Online].

Krizhevsky, A., Sutskever, I. and Hinton, G. E. 2012. Imagenet classification with deep convolutional neural networks. In *Advances in Neural Information Processing Systems* Pereira, F., Burges, C. J. C., Bottou, L. and Weinberger, K. Q. (Eds) (vol. 25), pp. 1097–1105. Curran Associates, Inc. [Online]. Available at: https://proceedings.neurips.cc/paper/2012/file/c399862d3b9d6b76c8436e924a68c45bPaper.pdf.

Krüger, N., Geib, C., Piater, J., Petrick, R., Steedman, M., Wörgötter, F., Ude, A., Asfour, T., Kraft, D., Omrčen, D., Agostini, A. and Dillmann, R. Object-action complexes: grounded abstractions of sensory-motor processes, *Robotics and Autonomous Systems* 59(10), 740–757 2011.

Kurtser, P. and Edan, Y. Statistical models for fruit detectability: spatial and temporal analyses of sweet peppers, *Biosystems Engineering* 171, 272–289 2018.

Kurtser, P. and Edan, Y. Planning the sequence of tasks for harvesting robots, *Robotics and Autonomous Systems* 131, 103591 2020, issn: 0921-8890. doi: 10.1016/j.robot.2020.103591. [Online]. Available at: https://www.sciencedirect.com/science/article/ pii/S0921889020304310.

Lamping, C., Derks, M., Groot Koerkamp, P. and Kootstra, G. ChickenNet-- an end-to-end approach for plumage condition assessment of laying hens in commercial farms using computer vision, *Computers and Electronics in Agriculture* 194, 106695

2022, issn: 0168-1699. doi: 10.1016/j.compag.2022.106695. [Online]. Available at: https://www.sciencedirect. com/science/article/pii/S0168169922000126.

LeCun, Y. [Online], *"Predictive Learning, Keynote at NIPS2016."* 2016. Available at: https://youtu.be/Ount2Y4qxQo (visited on 24/07/2022).

LeCun, Y. and Misra, I. Self-supervised learning: the dark matter of intelligence, Facebook 2022, [Online] Available at: https://ai.facebook.com/blog/self-supervised-learning-the-dark-matter-ofintelligence (visited on 15/07/2022).

Lehnert, C., Tsai, D., Eriksson, A. and McCool, C. 3d move to see: multi-perspective visual servoing towards the next best view within unstructured and occluded environments 2019. In: IEEE Publications International, R. S. J. (Ed.) Conference on Intelligent Robots and Systems (IROS) (vol. 2019), pp. 3890–3897. doi: 10.1109/IROS40897.2019.8967918.

Li, T., Feng, Q., Qiu, Q., Xie, F. and Zhao, C. Occluded apple fruit detection and localization with a frustum-based point-cloud-processing approach for robotic harvesting, *Remote Sensing* 14(3) 2022, issn: 2072-4292. doi: 10.3390/rs14030482. [Online]. Available at: https://www.mdpi.com/20724292/14/3/482.

Liu, X., Yu, S. Y., Flierman, N. A., Loyola, S., Kamermans, M., Hoogland, T. M. and De Zeeuw, C. I. Optiflex: multi-frame animal pose estimation combining deep learning with optical flow, *Frontiers in Cellular Neuroscience* 15, 621252 2021.

Lowe, D. G. Distinctive image features from scale-invariant keypoints, *International Journal of Computer Vision* 60(2), 91–110 2004.

Lungarella, M., Metta, G., Pfeifer, R. and Sandini, G. Developmental robotics: A survey, *Connection Science* 15(4), 151–190 2003.

Lyubova, N., Ivaldi, S. and Filliat, D. From passive to interactive object learning and recognition through self-identification on a humanoid robot, *Autonomous Robots* 40(1), 33–57 2016.

Marks, E., Magistri, F. and Stachniss, C. Precise 3d reconstruction of plants from UAV imagery combining bundle adjustment and template matching. In: IEEE Publications International Conference on Robotics and Automation (ICRA) (vol. 2022) 2022, pp. 2259–2265.

Mathis, A., Mamidanna, P., Cury, K. M., Abe, T., Murthy, V. N., Mathis, M. W. and Bethge, M. Deeplabcut: markerless pose estimation of user-defined body parts with deep learning, *Nature Neuroscience* 21(9), 1281–1289 2018. doi: 10.1038/s41593-018-0209-y.

Milioto, A., Lottes, P. and Stachniss, C. Real-time semantic segmentation of crop and weed for precision agriculture robots leveraging background knowledge in cnns. In: IEEE Publications IEEE International Conference on Robotics and Automation (ICRA) (vol. 2018) 2018, pp. 2229–2235.

Mishra, P., Polder, G. and Vilfan, N. Close range spectral imaging for disease detection in plants using autonomous platforms: a review on recent studies, *Current Robotics Reports* 1(2), 43–48 2020. doi: 10.1007/s43154-020-00004-7.

Monarch, R. M. *Human-in-the-Loop Machine Learning: Active Learning and Annotation for Humancentered AI*. Simon and Schuster 2021.

Morrison, D., Milan, A. and Antonakos, E. Uncertainty-aware instance segmentation using dropout sampling. In: Proceedings of the Robotic Vision Probabilistic Object Detection Challenge (CVPR 2019 Workshop), Long Beach, CA 2019, pp. 16–20.

Nasiri, A., Yoder, J., Zhao, Y., Hawkins, S., Prado, M. and Gan, H. Pose estimation-based lameness recognition in broiler using cnn-LSTM network, *Computers and*

Electronics in Agriculture 197, 106931 2022, issn: 0168-1699. doi: 10.1016/j.compag.2022.106931. [Online]. Available at: https://www.sciencedirect.com/science/article/pii/S0168169922002484.

Neupane, K. and Baysal-Gurel, F. Automatic identification and monitoring of plant diseases using unmanned aerial vehicles: a review, *Remote Sensing* 13(19), 3841 2021.

Nir, O., Parmet, Y., Werner, D., Adin, G. and Halachmi, I. 3D Computer-vision system for automatically estimating heifer height and body mass, *Biosystems Engineering* 173, 4–10 2018.

O'Regan, J. K. and Noë, A. A sensorimotor account of vision and visual consciousness, *Behavioral and Brain Sciences* 24(5), 939–73; discussion 973 2001.

Oliveira, L. F. P., Moreira, A. P. and Silva, M. F. Advances in forest robotics: A state-of-the-art survey, *Robotics* 10(2), 53 2021. doi: 10.3390/robotics10020053.

Pantazi, X.-E., Moshou, D. and Bravo, C. Active learning system for weed species recognition based on hyperspectral sensing, *Biosystems Engineering* 146, 193–202 2016.

Pearl, J. The seven tools of causal inference, with reflections on machine learning, *Communications of the ACM* 62(3), 54–60 2019.

Pfeifer, R. and Bongard, J. *How the Body Shapes the Way We Think: a New View of Intelligence*. MIT Press 2006.

Polder, G. and Gowen, A. The hype in Spectral Imaging, *Spectroscopy Europe* 33(3), 12–14 2021.

Polder, G., Blok, P. M., de Villiers, H. A. C., van der Wolf, J. M. and Kamp, J. Potato virus Y detection in seed potatoes using deep learning on hyperspectral images, *Frontiers in Plant Science* 10, 209 2019. doi: 10.3389/fpls.2019.00209. [Online]. Available at: https://www.frontiersin.org/ article/10.3389/fpls.2019.00209.

Pretto, A., Aravecchia, S., Burgard, W., Chebrolu, N., Dornhege, C., Falck, T., Fleckenstein, F., Fontenla, A., Imperoli, M., Khanna, R., Liebisch, F., Lottes, P., Milioto, A., Nardi, D., Nardi, S., Pfeifer, J., Popovic, M., Potena, C., Pradalier, C., Rothacker-Feder, E., Sa, I., Schaefer, A., Siegwart, R., Stachniss, C., Walter, A., Winterhalter, W., Wu, X. and Nieto, J. Building an aerial & ground robotics system for precision farming: an adaptable solution, *IEEE Robotics and Automation Magazine* 28(3), 29–49 2021. doi: 10.1109/MRA.2020.3012492.

Qin, J., Chao, K., Kim, M. S., Lu, R. and Burks, T. F. Hyperspectral and multispectral imaging for evaluating food safety and quality, *Journal of Food Engineering* 118(2), 157–171 2013.

Ratner, A. J., Ehrenberg, H., Hussain, Z., Dunnmon, J. and Ré, C., Learning to compose domain-specific transformations for data augmentation, *Advances in Neural Information Processing Systems* 30 2017.

Ravichandar, H., Polydoros, A. S., Chernova, S. and Billard, A. Recent advances in robot learning from demonstration, *Annual Review of Control, Robotics, and Autonomous Systems* 3(1). ARTICLE, 297–330 2020.

Rawat, S., Chandra, A. L., Desai, S. V., Balasubramanian, V. N., Ninomiya, S. and Guo, W. How useful is image-based active learning for plant organ segmentation?, *Plant Phenomics* 2022, 9795275 2022.

Reed, S., et al. *"A generalist agent," arXiv Preprint ArXiv:2205.06175* 2022.

Rückin, J., Jin, L., Magistri, F., Stachniss, C. and Popović, M. Informative path planning for active learning in aerial semantic mapping. In: IEEE Publications International Conference on Robotics and Automation (ICRA) 2022.

Ruigrok, T., van Henten, E., Booij, J., van Boheemen, K. and Kootstra, G. Application-specific evaluation of a weed-detection algorithm for plant-specific spraying, *Sensors* 20(24), 7262 2020. doi: 10.3390/s20247262.

Ruigrok, T., van Henten, E., Dirks, J. P. and Kootstra, G. Generalization of deep neural networks for weed detection, *Computers and Electronics in Agriculture* 204, 107554 2023. doi: 10.1016/j.compag.2022.107554. Available: https://doi.org/10.1016/j.compag.2022.107554.

Russakovsky, O., Deng, J., Su, H., Krause, J., Satheesh, S., Ma, S., Huang, Z., Karpathy, A., Khosla, A., Bernstein, M., Berg, A. C. and Fei-Fei, L. Imagenet large scale visual recognition challenge, *International Journal of Computer Vision* 115(3), 211–252 2015.

Russello, H., van der Tol, R. and Kootstra, G. T-leap: occlusion-robust pose estimation of walking cows using temporal information, *Computers and Electronics in Agriculture* 192, no. 106559 2022.

Sa, I., Ge, Z., Dayoub, F., Upcroft, B., Perez, T. and McCool, C. Deepfruits: A fruit detection system using deep neural networks, *Sensors* 16(8), 1222 2016, issn: 1424-8220. doi: 10.3390/s16081222. [Online]. Available at: https://doi.org/10.3390/s16081222.

Sa, I., Lehnert, C., English, A., McCool, C., Dayoub, F., Upcroft, B. and Perez, T. Peduncle detection of sweet pepper for autonomous crop harvesting–combined color and 3-d information, *IEEE Robotics and Automation Letters* 2(2), 765–772 2017.

Saar, M., Edan, Y., Godo, A., Lepar, J., Parmet, Y. and Halachmi, I. A machine vision system to predict individual cow feed intake of different feeds in a cowshed, *Animal* 16(1), 100432 2022.

Settles, B. Active learning, *Synthesis Lectures on Artificial Intelligence and Machine Learning* 6(1), 1–114 2012.

Shi, W., van de Zedde, R., Jiang, H. and Kootstra, G. Plant-part segmentation using deep learning and multi-view vision, *Biosystems Engineering* 187, 81–95 2019.

Siegwart, R., Nourbakhsh, I. and Scaramuzza, D. *Introduction to Autonomous Mobile Robots*. The MIT Press 2011.

Silwal, A., Davidson, J. R., Karkee, M., Mo, C., Zhang, Q. and Lewis, K. Design, integration, and field evaluation of a robotic apple harvester, *Journal of Field Robotics* 34(6), 1140–1159 2017. doi: 10.1002/rob.21715. [Online]. Available at: https://onlinelibrary. Available at: wiley.com/doi/abs/10.1002/rob.21715.

Smitt, C., Halstead, M., Zaenker, T., Bennewitz, M. and McCool, C. Pathobot: A robot for glasshouse crop phenotyping and intervention. In: IEEE International Conference on Robotics and Automation (ICRA) (vol. 2021), 2021, pp. 2324–2330. doi: 10.1109/ICRA48506.2021.9562047.

Spoliansky, R., Edan, Y., Parmet, Y. and Halachmi, I. Development of automatic body condition scoring using a low-cost 3-dimensional Kinect camera, *Journal of Dairy Science* 99(9), 7714–7725 2016.

Su, D., Kong, H., Qiao, Y. and Sukkarieh, S. Data augmentation for deep learning based semantic segmentation and crop-weed classification in agricultural robotics, *Computers and Electronics in Agriculture* 190, 106418 2021, issn: 0168-1699. doi: 10.1016/j.compag.2021.106418. [Online]. Available at: https://www.sciencedirect.com/science/article/pii/ S016816992100435X.

Suh, H. K., Hofstee, J. W. and van Henten, E. J. Improved vegetation segmentation with ground shadow removal using an hdr camera, *Precision Agriculture* 19(2), 218–237 2018a. doi: 10.1007/s11119-017-9511-z.

Suh, H. K., Hofstee, J. W., IJsselmuiden, J. and van Henten, E. J. Sugar beet and volunteer potato classification using bag-of-visual-words model, scale-invariant feature transform, or speeded up robust feature descriptors and crop row information, *Biosystems Engineering* 166, 210–226 2018b, issn: 1537-5110. doi: 10.1016/j.biosystemseng.2017.11.015. [Online]. Available at: https://www.sciencedirect.com/science/article/pii/S1537511017301629.

Thenmozhi, K. and Reddy, U. S. Crop pest classification based on deep convolutional neural network and transfer learning, *Computers and Electronics in Agriculture* 164, 104906 2019.

Tsouros, D. C., Bibi, S. and Sarigiannidis, P. G. A review on UAV-based applications for precision agriculture, *Information* 10(11) 2019, issn: 2078-2489. doi: 10.3390/info10110349. [Online]. Available at: https://www.mdpi.com/2078-2489/10/11/349.

Turgut, K., Dutagaci, H., Galopin, G. and Rousseau, D. Segmentation of structural parts of rosebush plants with 3d point-based deep learning methods, *Plant Methods* 18(1), 20 2022.

van Essen, R., Nguyen, L., van Helmond, E., Batsleer, J., Poos, J.-J. and Kootstra, G. doi: 10.1093/icesjms/fsab233. [Online]. Automatic bycatch registration using deep learning and object tracking, *International Journal of Marine Science* 78(10), 3834–3846 2021. doi: 10.1093/icesjms/fsab233.

van Essen, R., Harel, B., Kootstra, G. and Edan, Y. Dynamic viewpoint selection for sweet pepper maturity classification using online economic decisions, *Applied Sciences* 12(9), 4414 2022.

van Marrewijk, B., Polder, G. and Kootstra, G. Investigation of the added value of cyclegan on the plant pathology dataset. In Proceedings of the 7th IFAC Conference on Sensing, Control and Automation Technologies for Agriculture. Munich, Germany 2022.

Virlet, N., Sabermanesh, K., Sadeghi-Tehran, P. and Hawkesford, M. J. Field scanalyzer: an automated robotic field phenotyping platform for detailed crop monitoring, *Functional Plant Biology* 44(1), 143–153 2016.

Wagner, N., Kirk, R., Hanheide, M. and Cielniak, G. Efficient and robust orientation estimation of strawberries for fruit picking applications. In: IEEE International Conference on Robotics and Automation (ICRA) (vol. 2021), 2021, pp. 13857–13863. doi: 10.1109/ICRA48506.2021.9561848.

Williams, H. A. M., Jones, M. H., Nejati, M., Seabright, M. J., Bell, J., Penhall, N. D., Barnett, J. J., Duke, M. D., Scarfe, A. J., Ahn, H. S., Lim, J. and MacDonald, B. A. Robotic kiwifruit harvesting using machine vision, convolutional neural networks, and robotic arms, *Biosystems Engineering* 181, 140–156 2019, issn: 1537-5110. doi: 10.1016/j.biosystemseng.2019.03.007. [Online]. Available at: https: //www.sciencedirect.com/science/article/pii/S153751101830638X.

Winterhalter, W., Fleckenstein, F., Dornhege, C. and Burgard, W. Localization for precision navigation in agricultural fields—beyond crop row following, *Journal of Field Robotics* 38(3), 429–451 2021. doi: 10.1002/rob.21995. eprint. [Online]. Available at: https://onlinelibrary.wiley/abs/10.1002/rob.21995.

Xie, C. and Yang, C. A review on plant high-throughput phenotyping traits using UAV-based sensors, *Computers and Electronics in Agriculture* 178, 105731 2020, issn: 0168-1699. doi: 10.1016/j.compag.2020.105731. [Online]. Available at: https://www.sciencedirect. com/science/article/pii/S0168169919320046.

York, L. M. Functional phenomics: an emerging field integrating high-throughput phenotyping, physiology, and bioinformatics, *Journal of Experimental Botany* 70(2), 379–386 2019, issn: 0022-0957. doi: 10.1093/jxb/ery379. eprint. Available at: https://academic.oup.com/jxb/article-pdf/70/2/379/27435247/ery379.pdf. [Online].

Zaenker, T., Smitt, C., McCool, C. and Bennewitz, M. Viewpoint planning for fruit size and position estimation. In: IEEE Publications International, R. S. J. (Ed.) Conference on Intelligent Robots and Systems (IROS) (vol. 2021), 2021 pp. 3271–3277.

Zaenker, T., Lehnert, C., McCool, C. and Bennewitz, M. Combining local and global viewpoint planning for fruit coverage. In: IEEE Publications European Conference on Mobile Robots (ECMR) (vol. 2021), 2021 pp. 1–7.

Zapotezny-Anderson, P. and Lehnert, C. Towards active robotic vision in agriculture: A deep learning approach to visual servoing in occluded and unstructured protected cropping environments, *IFAC-PapersOnLine* 52(30), 120–125 2019, 6th IFAC Conference on Sensing, Control and Automation Technologies for Agriculture AGRICONTROL, issn: 2405-8963. doi: 10.1016/j.ifacol.2019.12.508. [Online]. Available at: https://www.sciencedirect.com/science/article/pii/S2405896319324243.

Zeng, W., Karaoglu, S. and Gevers, T. 2018. Inferring point clouds from single monocular images by depth intermediation. doi: 10.48550/ARXIV.1812.01402. [Online] Available at: https://arxiv.org/ abs/1812.01402.

Zeng, X., Zaenker, T. and Bennewitz, M. Deep reinforcement learning for next-best-view planning in agricultural applications. In: IEEE Publications International Conference on Robotics and Automation (ICRA) (vol. 2022), 2022, pp. 2323–2329.

Zhang, Z., Wang, X., Lai, Q. and Zhang, Z. Review of variable-rate sprayer applications based on realtime sensor technologies, *Automation in Agriculture-Securing Food Supplies for Future Generations*, 53–79 2018.

Zhao, Y., Gong, L., Huang, Y. and Liu, C. A review of key techniques of vision-based control for harvesting robot, *Computers and Electronics in Agriculture* 127, 311–323 2016, issn: 01681699. doi: 10.1016/j.compag.2016.06.022. [Online]. Available at: https://www.sciencedirect.com/science/article/pii/S0168169916304227.

Zhou, C., Zhang, B., Lin, K., Xu, D., Chen, C., Yang, X. and Sun, C. Near-infrared imaging to quantify the feeding behavior of fish in aquaculture, *Computers and Electronics in Agriculture* 135, 233–241 2017.

Zhu, N., Liu, X., Liu, Z., Hu, K., Wang, Y., Tan, J., Huang, M., Zhu, Q., Ji, X., Jiang, Y. and Guo, Y. Deep learning for smart agriculture: concepts, tools, applications, and opportunities, *International Journal of Agricultural and Biological Engineering* 11(4), 21–28 2018.

Chapter 2

Advances in world modeling for agri-food robotics

Jordy Senden[*], Elena Torta and René van de Molengraft, Eindhoven University of Technology, The Netherlands; and Herman Bruyninckx, Eindhoven University of Technology, The Netherlands, Katholieke Universiteit Leuven and Flanders Make, Belgium

1 Introduction
2 The concept of a world model
3 Invariant-based world models
4 Topological maps with guarded motion
5 Task redundancy through multi-modal sensing
6 Explainable world models for skill configuration
7 Multi-robot cooperation through shared world models
8 Future perspective: world models for reasoning
9 Conclusion
10 Glossary
11 Where to look for further information
12 References

1 Introduction

Robots have been primarily used as motion-controlled positioning systems in controlled structured environments, such as manufacturing facilities, to perform repetitive tasks as fast and precise as possible. In contrast, the agricultural environment contains many different objects or actors, like humans, animals or plants, which can exhibit variations in both their appearance and behavior. These organic environments are complex, unstructured, heterogeneous and dynamic over time. Although in some 'semi-structured' environments, like a greenhouse, environmental conditions are controlled as much as possible, the natural objects (e.g. plants or fruit) have uncontrollable variations in properties like shape, reflectance or mass, both in space and time.

* EMAIL: j.p.f.senden@tue.nl.

http://dx.doi.org/10.19103/AS.2023.0124.02

Robots operating in such environments must deal with an unbounded set of possible situations, caused by the variations. Since it is impossible to predict all situations at design-time, the traditional sense–plan–act paradigm (Srivastava, 2019) is insufficient. Static discrete controllers, e.g. finite state machines, rely on the assumptions that (1) all necessary information to perform a task can be derived from sensor data and (2) all possible situations encountered by a robot are known and can be accounted for in advance. Currently, information is mostly derived from the raw sensor data, but purely data-driven approaches lack the flexibility to adapt to the variations in unstructured agricultural and other environments (Yuille and Liu, 2021). Instead, it is important to understand the context in which robots are deployed. Some key information cannot be derived from sensor data alone, like an object's fragility or flexibility, or the relations between objects (e.g. plants aligned in a row or tomatoes in a truss). Such non-measurable information comes from our knowledge of the world, captured in a so-called 'world model' (WM).

In this chapter, we review the potential role of WMs to address the challenges in agri-food robotics. We will outline important aspects of systems that can robustly perform tasks in unstructured environments (Bruyninckx, 2022). The chapter uses examples of harvesting, specifically tomatoes, to illustrate how WMs can be deployed in practice.

2 The concept of a world model

To perform a given task robustly when faced with variations in the environment, robots need a number of deliberation functions like planning, acting, observing, monitoring, goal reasoning and learning. A simplified overview of these functions is shown in Fig. 1 (adapted from Ingrand and Ghallab, 2017), where the WM has a central role.

A WM involves a model-based representation of the robot and its environment. It may include knowledge on how plants grow over time or how to manipulate plant parts to deal with sensor occlusions. A WM allows the robot to make decisions about which motion and sensing actions to execute, considering a much larger horizon in space and time than what can be provided by the robot's sensors alone.

Robots need to have different ways to obtain the required information or to execute the required manipulation, which are called 'skills' in this chapter. To be robust against 'disturbances', redundant skills are necessary, i.e. having other alternatives in case the first one fails. A WM should hold the necessary information to determine which skills and resources are needed for a task and it holds the relevant parameters to configure the skills.

Since there are multiple action and detection skills that share the same information, it is obvious that a WM needs to get a central role in the whole

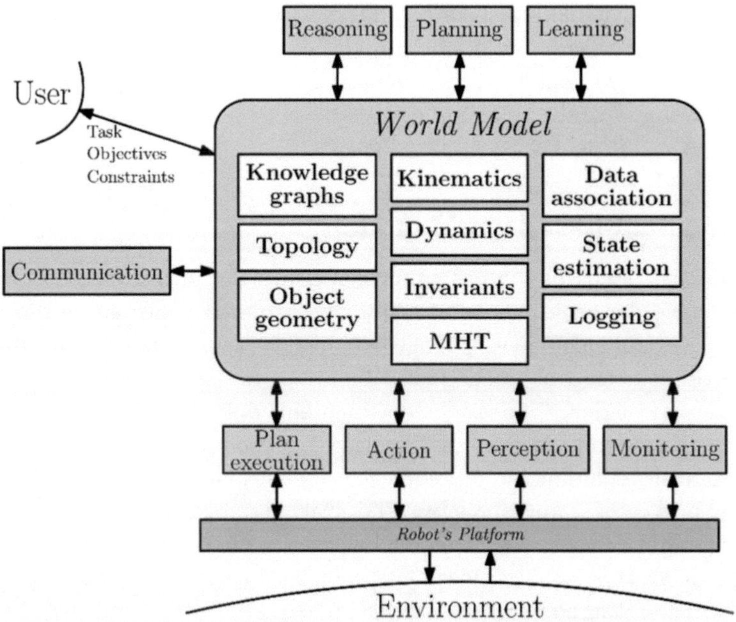

Figure 1 Simplified overview of robotic functions. This is an extension of the work on deliberation functions by Ingrand and Ghallab (2017), with an emphasis on the different functionalities of the world model.

robot application. This allows for a flexible architecture in which individual sub-systems such as sensors, data fusion and path planning communicate via the WM. Another advantage of an explicit WM as a central component in the software architecture is the ability to share knowledge and intentions between robots and to host deterministic and explainable rules via which coordination problems between several robots can be solved. A shared WM makes it easier for robots to collaborate with human operators (Sakai and Nagai, 2022) and with each other to work on the same objective based on distributed shared task. In the future, this may be necessary to enable fleets of robots to share and perform tasks in parallel.

A WM needs to be a multi-layered, multi-scale composition of features. It should contain not only geometrical representations of objects but also semantic information like affordances (Ardón et al., 2020) and invariants. Affordances model the cause and effect of an action that an actor (e.g. a robot) can perform on an object (Andries et al., 2018). An invariant is a mathematical constraint that remains unchanged under a particular transformation in a system (Senden et al., 2022). Describing the world on the abstraction level of invariants allows robots to deal with variations and anomalies in the environments and objects they encounter. For example, an invariant feature of tomatoes in a single truss is that their sizes (diameters) are relatively equal. This holds for most truss tomatoes regardless of their actual diameter, which can depend on the type (e.g. cherry tomatoes) and ripeness stage. The next sections discuss how task robustness could be increased in several of these functions using WMs based on invariants.

3 Invariant-based world models

This section discusses the use of invariants to be robust against geometric variations in the environment. In many environments, especially in the agro-food context, objects can vary a lot when measuring exact parameters like size and weight. Instead of creating skills that have to handle these variations in parameters, the description of the environment could be chosen in such a way that these variations do not exist anymore. For example, each tomato plant is geometrically different, but topologically all plants have a main stem to which trusses and leaves are connected. The exact distance between consecutive trusses might vary between species, or per season, but this distance is kept equal for each plant. Such invariants contained in the WM can be used to configure the skills and monitor the system (Fig. 2).

3.1 Current state of the art

Classical robot localization is often based on precise metric parameterization using a grid map. In open-field applications (Gasparino et al., 2020; Blok

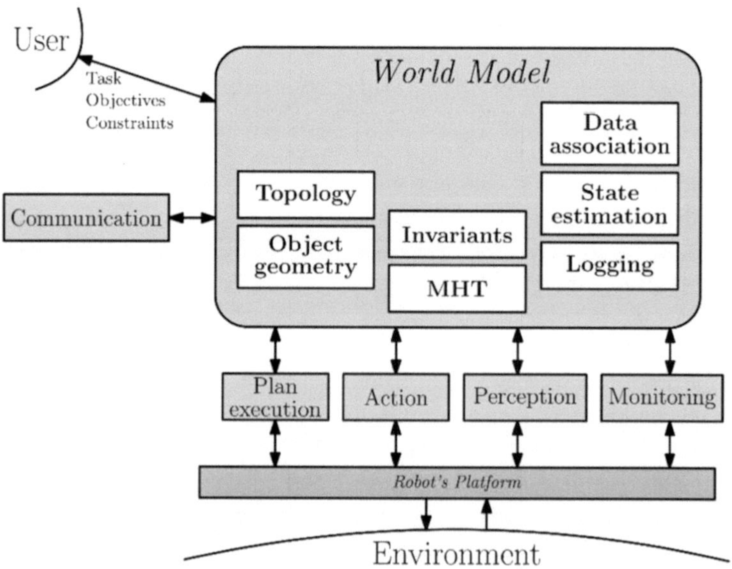

Figure 2 Overview of functionalities of the world model that are discussed in Section 3.

et al., 2019; Cobbenhagen et al., 2022), these grid maps can be combined with a Global Navigation Satellite System (GNSS) to determine its absolute coordinates (Cheein et al., 2016; Auat Cheein et al., 2017; Bawden et al., 2017). A review on navigation systems for autonomous agricultural vehicles is given in Mousazadeh (2013) and localization methods are reviewed in Panigrahi and Bisoy (2021).

When a map is not provided, the robot can use its onboard sensors to create its own map and localize itself therein, which is called SLAM (simultaneous localization and mapping) (Rosen et al., 2021). In the agro-food context, this approach is used to create a 3D map of an orchard (e.g. Underwood et al., 2016) or to characterize tree canopies (Auat Cheein et al., 2015; Yandún Narváez et al., 2016; Pfeiffer et al., 2018).

3.2 Problem

Geometric maps that model the environment in Euclidean space with respect to some origin have the benefit of mathematical descriptions of, e.g. optimality and physics. This often requires calibration steps to map sensor values to world coordinates. Skills that are dependent on absolute parameters or distances in the world are prone to variations in these parameters.

3.3 Our approach

Instead of describing the world with exact geometric properties, the expected variations should be embedded in the semantics of the WM. When designing a skill, it should be based on properties that will not change and are thus invariant, e.g. inherent features like all tomatoes having a spherical shape or relative features like all tomatoes belonging to one truss having a similar diameter. When exact measures of dimension and distance become irrelevant to solve a task, the WM can be described in affine space. An affine space is a geometric structure that captures notions of parallel lines and ratios which are independent of distances and angles. The WM in affine space is size agnostic and relies solely on ratios captured by the invariants. Although the configuration of skills should not be dependent on exact sizes and parameters, estimations thereof can be a result of executing a skill. A more in-depth explanation and example can be found in Senden et al. (2022), where an affine WM of a semi-automated soccer table is created to be robust against the movement of the vision system that is used to track the ball. The sections in this chapter refer to different kinds of invariants and how these are used to perform robotic skills.

4 Topological maps with guarded motion

This section discusses how a robot can deal with an environment that has geometrical variation but is topologically invariant, e.g. a tomato plant growing on a highwire system. Figure 3 summarizes the functionalities discussed in this section.

4.1 Current state of the art

For any robot, it is important to be aware of its surroundings. It needs to know which objects are in its environment (object detection), where they are (object localization) and where they are located in the world (localization). This is often done by measuring or estimating distances to surrounding objects. Geometric maps describe the shape of objects and the distance between objects in the world, often with respect to a chosen fixed reference frame. A popular type of geometric map is the occupancy grid map (OGM) (Tsardoulias et al., 2016), which divides the environment into an evenly spaced grid. The cells of a 2D grid map are called pixels, while the cells of a 3D grid map are called voxels. The size of the cells in this grid is dependent on the level of detail necessary for the task. Each cell can either be occupied (e.g. containing a part of an object) or be free, which means there is nothing detected at that position.

Some of the main sensors for object detection and localization capture their data in a discrete, pixelated format e.g. RGB (red, green, blue) cameras, depth

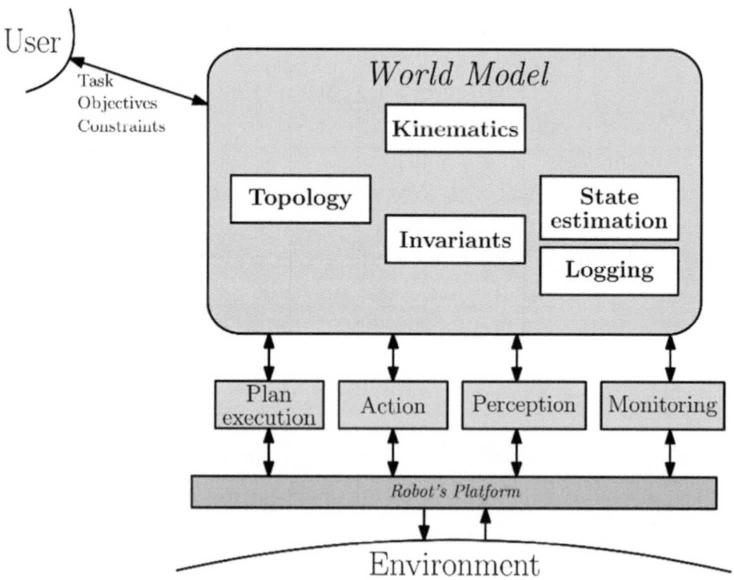

Figure 3 Overview of functionalities of the world model that are discussed in Section 4.

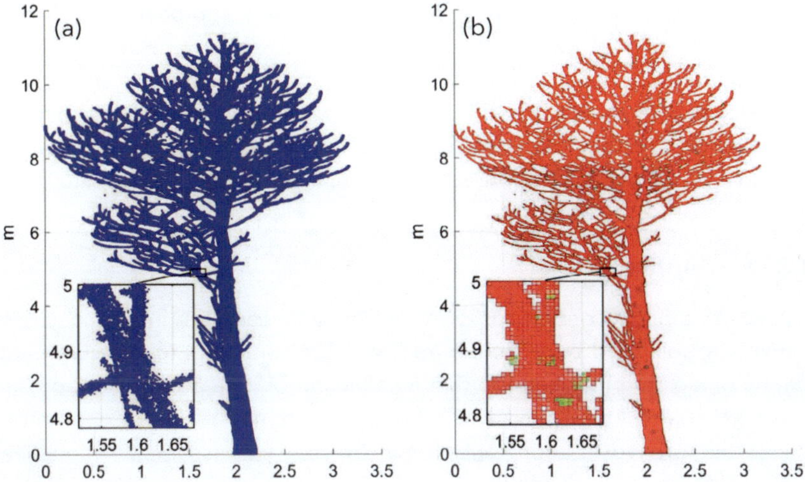

Figure 4 An occupancy grid map of a cherry tree (Prunus avium), where (a) shows the raw point cloud data from the LiDAR sensor and (b) shows the occupied voxels in the OGM. This map is used for canopy porosity and crown surface estimation. Source: Adapted from: Pfeiffer et al. (2018).

cameras, LiDAR (light detection and ranging) and LRF (laser range finder). Using this pixelated data directly, such as a 3D point cloud, an OGM of the environment can be created. Figure 4 shows an example (Pfeiffer et al., 2018) of how an OGM of a cherry tree is created, based on point cloud data retrieved from a LiDAR sensor. Detected objects are often modeled having six degrees of freedom (6DOF) pose in the grid map, in order to plan a geometric path toward, e.g. the strawberry harvester described by Ge et al. (2019). The path itself is often constructed to be optimal in time, energy or control effort required (Wang et al., 2018), as well as to avoid collisions (Boryga et al., 2015; Bac et al., 2016; Wei and Ren, 2018).

4.2 Problem

Determining an optimal path is dependent on the starting and end pose of the end effector (EE), where the latter is dependent on the pose of the object. Detecting this pose from vision sensors is difficult, especially in semi-structured, dynamic environments where each instance of an object is geometrically different, and there is a lot of clutter. A slight deviation in the detected pose might lead to a completely different optimal path. Moreover, it is hard to find an optimal path that avoids potential obstacles and collisions.

A robot therefore needs to be reactive to a changing environment. Scoccia et al. (2021) suggest an approach that first generates an off-line trajectory on the assumption of a static world but then adapts to a dynamically changing world through local replanning in order to avoid collisions with moving objects.

Marturi et al. (2019) propose a dynamic switching approach that decides whether to locally adapt the trajectory to a moving object or to calculate an entirely new plan. These approaches rely on a known geometric state, position and velocity of all the objects in the environment to make corrections. In reality, however, it is difficult to measure exact values for the states of dynamic objects, making it hard to define an optimal geometric trajectory for the robot to follow.

4.3 Our approach

Instead of a purely geometric OGM, which only captures free or occupied space, a robot could benefit from adding more semantics to measurements, e.g. by being able to recognize that a collection of data points can represent a type of object. Adding this kind of higher-level information to geometrical maps ('this set of voxels represent a tomato truss') can shift some functionalities, like localization and planning, beyond simple low-level instructions such as 'drive 5 meters straight, then turn 90 degrees to your right and drive 2 meters again' to more high-level and semantically rich guidance such as 'drive along a row of plants until you reach a plant with ripe trusses, then harvest all trusses of that plant'.

Topological maps can capture knowledge of the relation between different features, objects or areas, which allows for a semantic description to calculate routes (Marinakis and Dudek, 2010). Metric and topological maps can be combined, as shown in Fig. 5, to create hybrid metric–topological SLAM suitable for navigating complex environments such as agricultural settings (Blanco et al., 2007; Kostavelis and Gasteratos, 2015). Prior knowledge about the environment is necessary to translate low-level sensor data into topological maps (Liu and von Wichert, 2014; Liu et al., 2016). The higher level of semantics makes these maps robust against measurement errors and suitable to share between different robots (Ravankar et al., 2020; Taniguchi et al., 2019).

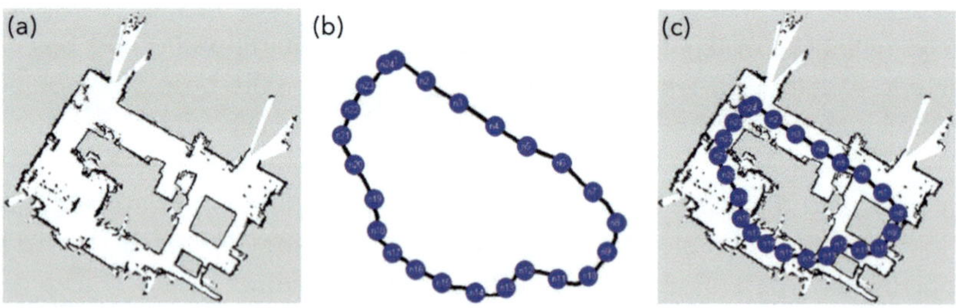

Figure 5 (a) Example of a 2D metric map of an indoor environment, (b) the topological map of that environment, (c) the hybrid topometric map, where the topological map is registered with a spatial specific region of the occupancy grid. Source: Adapted from: Kostavelis and Gasteratos (2015).

Though plants are geometrically different, their topology abides by certain rules. This led to the development of the L-system to model plant morphology and growth (Lindenmayer, 1987) and used for visualization (Prusinkiewicz et al., 1997, 2018). A topological structure is generated by a set of rules and added geometric properties are used to render a 3D model. This model initiated the main-stem paradigm (Soetens, 2015), where the stem is used as a path for the robot ('follow the stem upwards until a side branch is reached') using a form of guarded motion (Hill and Sword, 1973; Will and Grossman, 1975; Mason, 1981).

4.4 Case study

This approach relies on physical contact with the stem (Eizentals, 2016), while estimating the contact forces. Figure 6 shows how a robot arm, with a two-finger EF, makes contact with a plant stem and follows it up until it feels a peduncle with a connected tomato truss. Contact forces are estimated using torque sensors in the arm joints in order to monitor and direct movement. This approach is based on a topological model of the tomato plant (Fig. 7). It integrates the closed-loop inverse kinematic (CLIK) method and adaptive gain adaptive bias (ABAG) controllers to achieve robust, reactive control (Fig. 8). A finite state machine with guard monitors is used as discrete control plan (Fig. 9). Small contact forces and singularities in arm configuration can limit full wrench estimation from all joint torques. Therefore, the robot uses situation

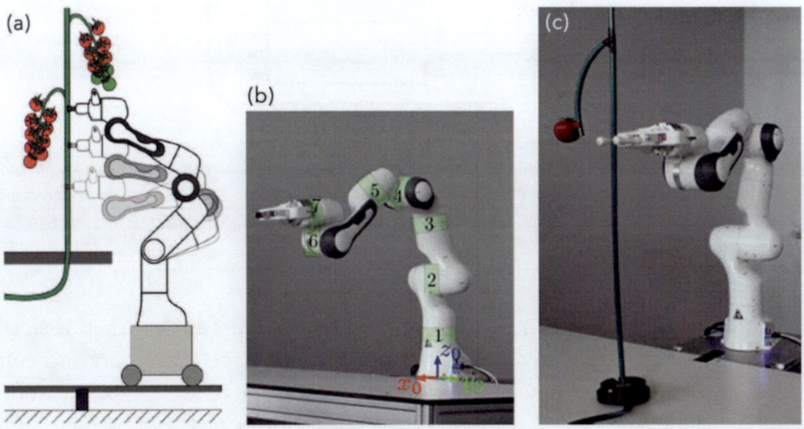

Figure 6 (a) Visualization of the main-stem paradigm; a robotic arm follows the stem until it 'feels' the peduncle. (b) Franka Emika Panda cobot with joint numbering. (c) Experimental set-up with the Panda cobot following a dummy plant with a mock tomato. The standard fingers are replaced with passively rolling fingers.

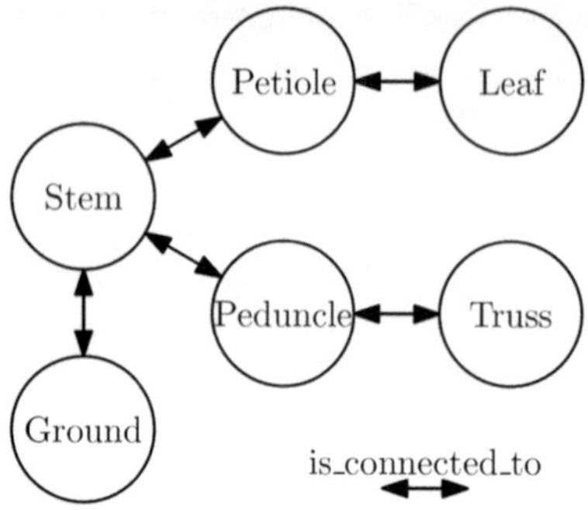

Figure 7 Topological world model of a tomato plant. This model captures the connections between key features of a tomato plant while omitting scale- or distance information.

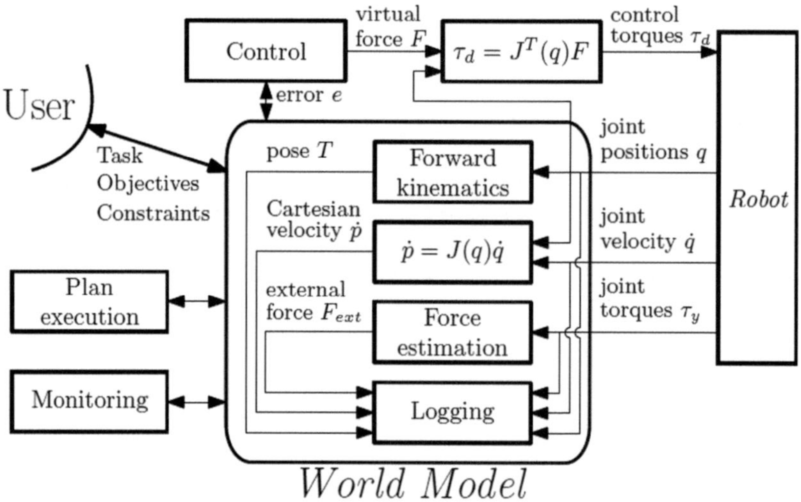

Figure 8 CLIK scheme to compute control inputs τ_d . The 'world model' block contains all the information required for computing the tracking error and for the monitoring to inform the higher level discrete control (plan execution) when to switch (part of) the task specification.

aware active force sensing (SAAFS) to estimate expected forces in a single direction from a single joint. In this example, joints 1 and 6 are perpendicular to the expected forces in y and z respectively. Figure 10 shows experimental results and a video of a successful completion of this task can be viewed by scanning the QR code in Fig. 11.

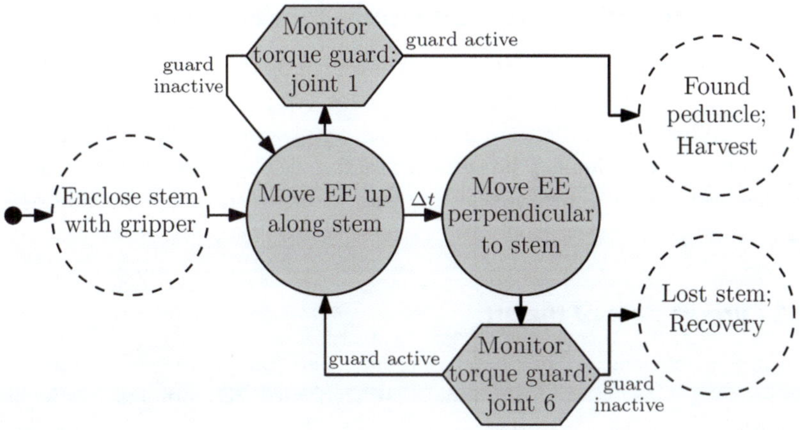

Figure 9 Discrete control plan, with monitors, for detecting the stem and peduncle using torque feedback on the joints of the robot.

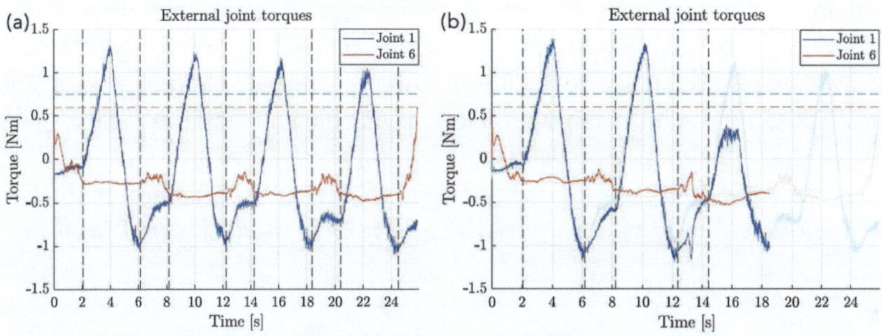

Figure 10 (a) Results of the FollowStem task, where the task is terminated after the tomato is reached. (b) Results of the FollowStem task, showing a prematurely terminated task. The stem is pulled away, so the robot cannot feel it anymore with its current sensors. State transitions are indicated by the vertical black dashed lines. The torque guards on respectively joints 1 and 6 are indicated by the horizontal blue and orange dashed lines.

Figure 11 QR code that links to a video of a successfully completed experiment (https://youtu.be/XilY-1XNK-c).

5 Task redundancy through multi-modal sensing

Object detection in robotics often relies on computer vision systems like cameras or depth sensors. However, humans use more senses, e.g. touch, to find an object at the bottom of a backpack or navigate through a house when the lights are off by feeling around. This section discusses how different sensor modalities to perform similar skills can lead to more robust task execution (Fig. 12).

5.1 Current state of the art

Solving robotic tasks, like object detection and localization, often rely on a single sensor modality. Commonly used types of 2D cameras in machine vision/image analysis in agri-food applications include mono-chromatic, RGB-color, multispectral, hyper-spectral and thermal cameras. The review on fruit harvesting techniques (Li et al., 2011) spends a large portion on machine vision systems and image analysis methods. Kapach et al. (2012) also discusses the commonly used vision systems and image analysis algorithms and extends this with a review on visual clues from the fruits and challenges ahead (Patel et al., 2012) gives an extensive review of machine vision systems for quality inspection of different food products.

From mid-2010, depth sensors like LiDAR, LRF or time-of-flight (ToF) technologies started to gain popularity as vision systems for fruit localization

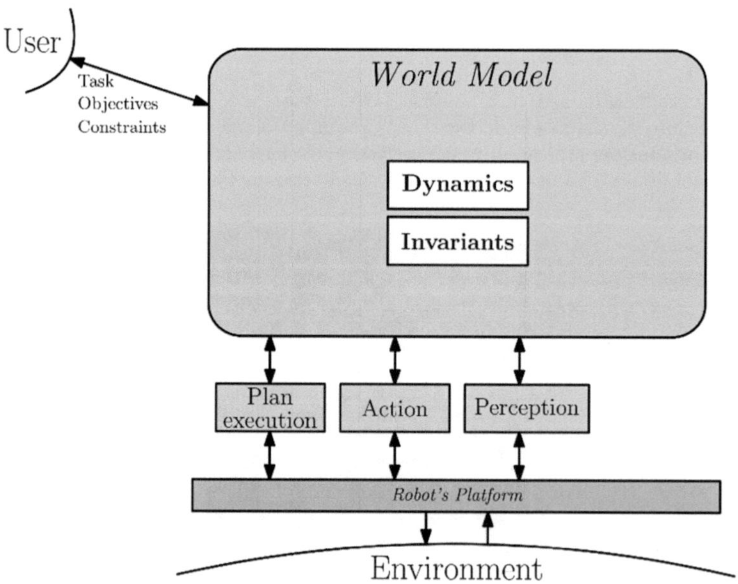

Figure 12 Overview of functionalities of the world model that are discussed in Section 5.

(Gongal et al., 2015;Lin, 2015; Zujevs et al., 2015) and (Zhao et al., 2016). A more recent development in image analysis is the use of artificial neural networks to detect and classify objects (Koirala et al., 2019; Botta et al., 2022).

5.2 Problem

The downside of using vision-based sensors is that they require a free line of sight toward the objects they need to detect. In natural and dense environments, such as a greenhouse, fruits are typically surrounded by dense foliage and other fruits, increasing the chance of occlusion.

One solution is active perception: moving the sensor to better viewpoints to detect the object (Bajcsy et al., 2018; Zaky et al., 2020). However, an object can still remain occluded and undetectable, e.g. when light conditions are poor or the fruit is fully covered by leaves. Another challenge is the skills required for a robot to adopt and discriminate between different viewpoints (Magalhães et al., 2022), which requires knowledge of the environment.

Another approach is to use other sensor technologies such as tactile sensors, proximity sensors and chemical sensors (Li et al., 2011; Zujevs et al., 2015; Lehnert et al., 2017). Examples used in the agri-food sector include ultrasonic sensors to estimate the canopy density of apple trees (Palleja and Landers, 2015), electronic noses to identify fruits and assess ripeness and quality (Baietto and Wilson, 2015), dielectric elastomer actuators to estimate apple firmness (Hiruta et al., 2021) and acoustic techniques to determine fruit quality (Ding et al., 2021). These examples are focused on quality inspection, rather than localization.

5.3 Our approach

This section proposes a method to extend the tactile truss localization approaches discussed in Section 4.1. The aim is to identify trusses and estimate the progression made toward them while traveling up the stem before contact is made with the peduncles. This is done by understanding and utilizing the dynamic response of the tomato plant to an excitation force applied to the stem. Tracking this change makes the approach invariant to the exact physical properties of the environment, which will vary between plants. A review of how plant dynamics on various scales can be modeled is given in (de Langre, 2019). A more in-depth explanation about this section can be found in (Senden et al., 2022).

5.4 Case study

A 1DOF actuator (Fig. 13) was used for excitation and force sensing. The 1DOF actuator was manually placed at different heights on the stem (h_z). The system

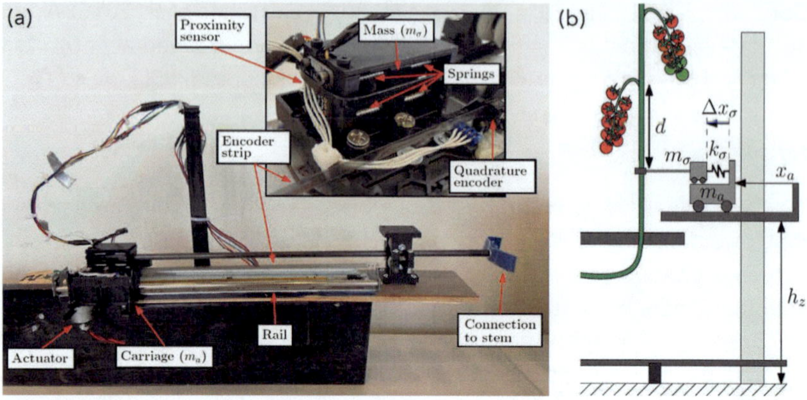

Figure 13 (a) Picture of the 1DOF setup, with a more detailed view of the carriage and sensors. (b) Schematic overview of the test setup, where the actuator is connected to the stem at a distance d to the peduncle.

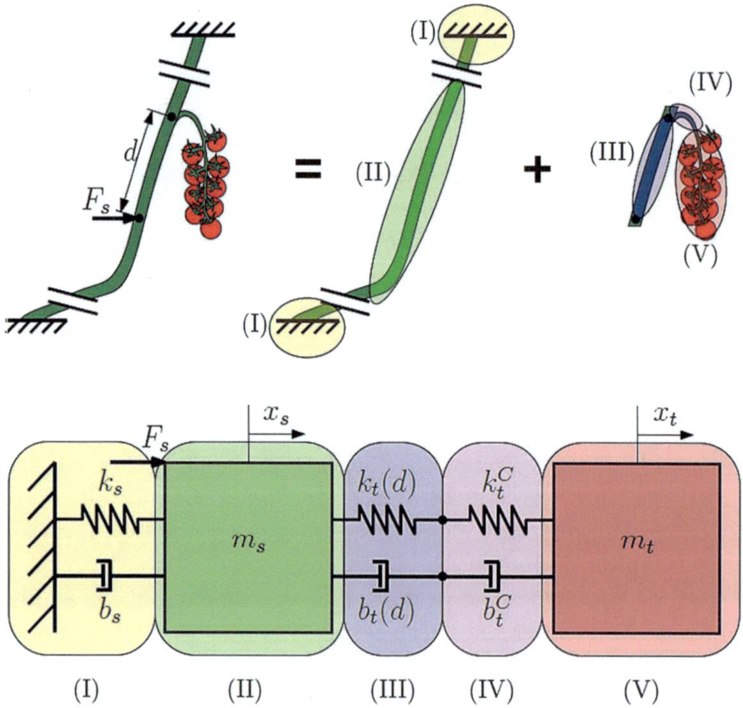

Figure 14 World model of a plant with only one truss (I) is the connection of the plant to the fixed world, pulling it back to its equilibrium position, (II) represents the mass of the stem, (III) represents the stiffness and damping of the part of the stem between the actuator and the peduncle, (IV) represents the stiffness and damping of the peduncle, and (V) represents the mass of the tomato truss.

is connected to the plant by tightly wrapping a Velcro strap around the stem, emulating the grasp of an EF.

The carriage (m_a) can move linearly in a range of 30 cm, while the position of mass (x_a) and distance (Δx_σ) relative to the stem are measured. Mass (m_σ) is assumed to be negligible with respect to the mass of the stem and tomatoes. Knowing the stiffness of the springs ($k_\sigma = 0.72NM$) and the distance Δx_σ, the force between the stem and the actuator can be estimated using equation (1). The position of the connection point is calculated using equation (2). Measuring the displacement x as a result of the applied force F_a gives the dynamic response of the plant.

$$F_a = k_\sigma \Delta x_\sigma \tag{1}$$

$$x_s = x_a + \Delta x_\sigma \tag{2}$$

The method was first tested by exciting a dummy plant, showing that it behaves like a fourth-order mass spring damper system, schematically shown in Fig. 14. The data in Fig. 15 show that the antiresonance frequency, where the phase crosses the −90 degree line upward, increases when the distance to the peduncle decreases.

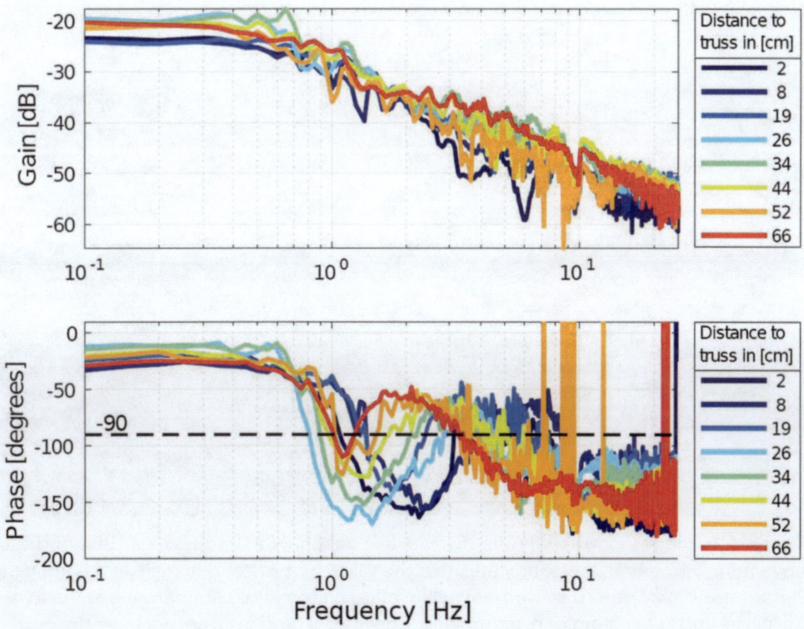

Figure 15 Frequency response of the dummy plant with a truss, connected at a height of 150 cm from the floor. The stem is excited at different distances below the truss, measured in centimeters. This plot shows the average response of multiple measurements at each given height. It can be seen that the antiresonance frequency, where the phase crosses the −90 degree line upward, increases when the distance to the peduncle decreases.

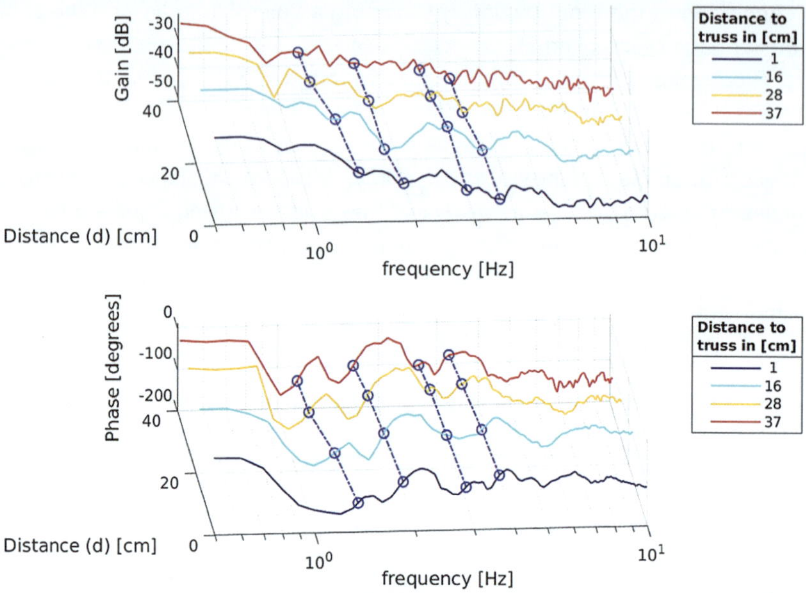

Figure 16 Anti-resonance frequencies of the four trusses at different distances to the first truss.

Figure 17 Images corresponding to the five resonance frequencies. Blur indicates movement, while sharpness indicates that the plant part is standing still. In the image (a) all parts are blurry due to movement, while at (e) all tomatoes are hanging still and only the leaves and lower stem part are moving. Between (a) and (e), a consecutive decoupling of the trusses is observed.

These tests were repeated on a real plant with multiple trusses. The average plots for multiple measurements at four different heights are shown in Fig. 16. The anti-resonance frequencies felt on the real plant happen at the eigen frequencies of each truss. These frequencies are distinguishable for the first three trusses, where the truss with the highest eigen frequency is closest to the actuator. Starting from the truss farthest away from the actuator, the motion of the trusses decouples from the motion of the stem, which is shown in Fig. 17. Approaching the trusses cause the eigen-frequencies of trusses to increase by approximately 0.7Hz/m. This method works without the need of exact values for masses, springs and dampers in the model. Instead of identifying the exact parameters in the model, the resonance frequencies can be determined at different actuation positions. Tracking if these frequencies increase or decrease is sufficient to know whether we are approaching a tomato or moving further away from it.

6 Explainable world models for skill configuration

This section shows how a tomato detection skill is configured by a WM. Detecting red and green tomatoes based on color will fail in a mainly green environment. Instead, the reflection difference between a tomato and the environment is used as an invariant feature. Since the amount of tomatoes is unknown, multiple possible detections are tracked (Fig. 18).

6.1 Current state of the art

Harvesting requires the ability to detect a fruit and its location. Identification is typically based on establishing precise parameters such as color (reflection),

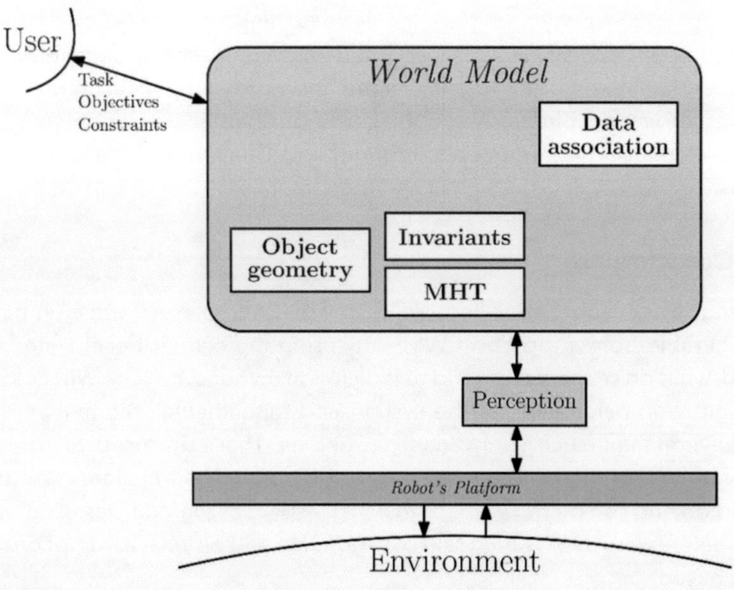

Figure 18 Overview of functionalities of the world model that are discussed in Section 6.

shape, size or impedance properties (e.g. to determine firmness). An example is a strawberry harvesting system in (Xiong et al., 2019a,b) that uses an RGB-d camera for strawberry detection. These cameras allow affordable 3D sensing solutions to detect fruit e.g. using a support vector machine algorithm (Barnea et al., 2016) or a neural network to interpret image data (Gené-Mola et al., 2019; Koirala et al., 2019). Images are typically filtered using a color threshold followed by noise filtering, which is prone to illumination issues. To circumvent these issues, Lee et al. (2019) use a laser to aid color-based detection and Wei et al. (2018) developed a Retinex algorithm to enhance image colors during the night-time harvesting of apples. LiDAR-type sensors can also be used to locate horticultural produce (Eizentals and Oka, 2016). The scanned point cloud of the fruit surface is fitted to a model of the fruit using a coherent point drift (CPD) algorithm or an Iterative Closest Point algorithm (Myronenko and Song, 2010).

6.2 Problem

Vision-based detection is vulnerable to occlusions, lighting condition and variations in object properties like shapes, colors and the amount of expected target objects. Neural networks try to circumvent this by including a large set of parameter variations in the training dataset. Problems with this approach include the need for all potential variations to be captured beforehand and problems of overfitting models toward a variation that might only appear sporadically. More fundamentally, neural networks are black box models that prohibit explainability and deliberative reasoning. In complex and dynamic environments, robots need to adapt to new unforeseen situations. They need explainable models to understand and justify their measurements and actions and reversible models of the sensing capabilities to understand how reality is mapped into percepts (Ingrand and Ghallab, 2017; Yuille and Liu, 2021).

6.3 Our approach

We suggest an approach where skills provide bottom-up information but are configurable from a top-down WM, e.g. 'using the current belief state of the world, what do I expect the world to look like in my sensors?' The WM holds the configuration parameters of the system and fundamental information about the environment, such as invariant properties. These are used to configure action and detection skills and interpret bottom-up information. Actions are based on and can be traced back to which aspect of the WM was used, which parameters were used during skill configuration and on what assumptions they were based.

6.4 Case study

This case study aims to develop a tomato truss detection skill to detect ripe and unripe tomatoes with a 2D RGB camera. The skill should be configurable using information from the WM and be explainable. To achieve this, the number of parameters used in the detection skill is kept to a minimum and is physically meaningful. Because all tomatoes are more or less spherical, regardless of size, the shape is an (absolute) invariant that is used for detection. Since unripe tomatoes reflect similar light wavelengths as the surrounding foliage, color-based identification may not work. However, there is a difference in how light is reflected isotropically from the smooth, curved surface of tomatoes compared to the rougher surfaces of leaves and stems (Fig. 19). This reflective difference is invariant to color and holds for both ripe and unripe tomatoes (Pothen and Nuske, 2016).

The two main reflection mechanisms are diffuse reflection (or body reflection) and specular reflection (or surface reflection) which often occur simultaneously. The reflectance of an object can be expressed by the bidirectional reflectance distribution function (BRDF) (Naya et al., 1991a). The BRDF of a surface can be modeled in several ways (Naya et al., 1991b; Duvenhage et al., 2013). In this case, the aim is not to estimate reflection precisely but to use reflection properties as the differentiation invariant to separate the tomatoes from their environment, making it possible to estimate the position and size of a tomato.

The color image from an RGB camera is used to create an intensity image by converting it to HSV (hue, saturation, value) image space and looking at the V channel (Fig. 20). The intensity profile of neighboring pixels that represent a tomato is smoother and more predictable than that of its surroundings. Scanlines are used to find these smooth profiles in the intensity image (see Fig. 20b showing horizontal and vertical scanlines). In a fully automated setting, these positions can be based on visual cues, also called seeds, in the image (e.g. reflection spots). After choosing the lines, the intensity values of the pixels

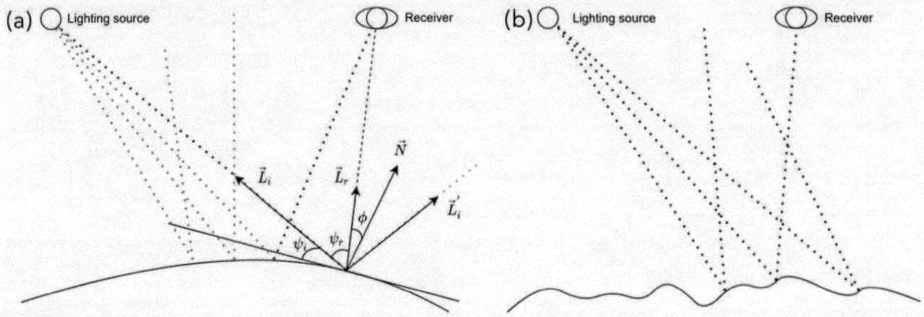

Figure 19 Schematic representation of light reflection from a smooth surface (a) versus light scattering from a rough surface (b).

on these lines are evaluated. The normalized intensity values for each pixel on the horizontal scanline in Fig. 20b are plotted in Fig. 21.

The intensity values on the scanline show that the reflection on the tomatoes results in a smooth curve with respect to the noisy background. The curves resemble semi-circles due to the cylindrical shape of the tomatoes and a uniform light distribution. The aim is to find semicircles that fit these smooth regions. These semicircles are modeled in Equation (3), where a_i represents the radius and b_i the center position of the circle i on the scanline. The reflection distribution, normally modeled as a semantically meaningful BRDF,

Figure 20 (a) Example of an RGB image of a tomato truss. (b) V-channel of the image after converting to HSV image space. The green lines are the 'scanlines', used for tomato detection.

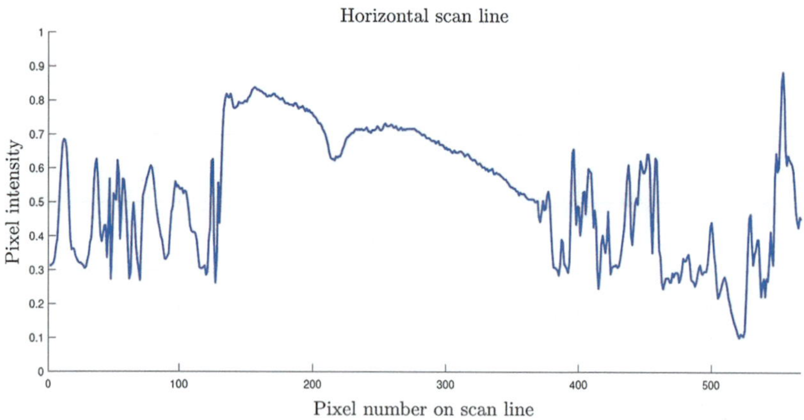

Figure 21 Normalized intensity values for each pixel on the horizontal scanline in Figure 20b.

Figure 22 (a) Image of a tomato truss inside a greenhouse. (b) Intensity map of the image with green scanlines. The red circle is drawn around the tomato that is found by fitting model (3) to the data on the scanlines. An example is given for the vertical scanline in Fig. 23.

is simplified where c_i captures the light distribution and d_i represents the mean intensity:

$$G(\theta_i)\begin{cases} x_f = a_i \cos(\phi) + b_i \\ y_f = c_i \sin(\phi) + d_i \end{cases}$$

$$\theta_i = (a_i, b_i, c_i, d_i)$$

$$\phi = \left[-\frac{\pi}{2}, \frac{\pi}{2}\right] \tag{3}$$

To fit Equation (3) to the scanline, an initial value for the parameters in θ_i is chosen and a cost function, which calculates the difference between the model and the data, is evaluated. The parameters a_i and b_i are adapted sequentially, until a (local) minimum of the cost function is found. In Fig. 22, an example is shown where three tomatoes ($i = 3$) are found on the vertical scanline on the truss in Fig. 23b. Doing this for both the vertical and horizontal scanlines results in an estimation of the center and radius of tomatoes in the 2D image visualized in Fig. 23b.

To initialize detection, seeds and initial parameters need to be chosen. One way to choose seeds is to use a neural network approach to suggest pixels in the image that might be trussed. Another approach is to add an active lighting system to the RGB camera. Figure 24 shows how a blue light, placed behind the camera, causes clear specular reflection spots on the tomatoes.

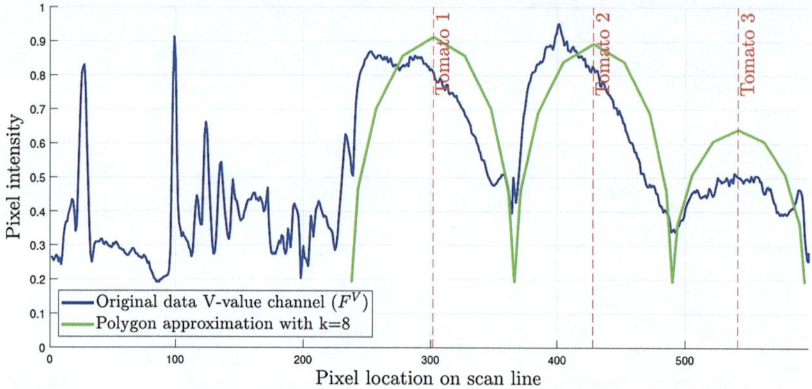

Figure 23 Normalized intensity values for each pixel on the vertical scanline in Fig. 22b. The green lines represent three semi-circles that are fitted to the data.

Figure 24 (a) RGB image with only natural lighting. (b) RGB image where the environment is illuminated with a blue monochromatic light source behind the camera. (c) Result of subtracting image (a) from (b) showing the spectral reflection spot caused by the blue light.

By subtracting an image without the blue light (Fig. 24a), from the image with the light (Fig. 24b), the difference in intensity clearly shows these spots (Fig. 24c). These spots can act as seeds and a first estimate of the center points of the tomatoes. The parameters a_i and b_i in model (3) can be configured with estimated properties of the environment and the robot. Using an estimation of the distance to the trusses, the size of a tomato, and the camera parameters, the radius and center point of the tomatoes in the image can be estimated with a pinhole camera model (Zhang, 2000).

7 Multi-robot cooperation through shared world models

This section discusses the role of sharing a WM in a multi-robot system that needs to work together. Without a centralized global controller, robots need to share knowledge to create a more complete view of the environment than they can individually and come up with a shared plan to perform a task. Combining the perception of a space or object, and communicating information between team members allows cooperative action such as passing the ball in a football match (Fig. 25) or allocating tasks in a fleet of harvesting robots (Emmi et al., 2014; Gonzalez-de-Santos et al., 2017; Mapes and Xu, 2022).

7.1 Current state of the art

A standard approach when automating a task is to design a single robotic system with a centralized controller. Skills required to perform (sub)tasks rely only on their own local representation of the world and are created as pieces of stand-alone code. This approach often follows the sense-plan-act paradigm where sensor data (often from a single) has to be analyzed before an action can be planned or performed. This single-solution approach has resulted in a mass pursuit of higher detection rates, e.g. from neural networks, to try to

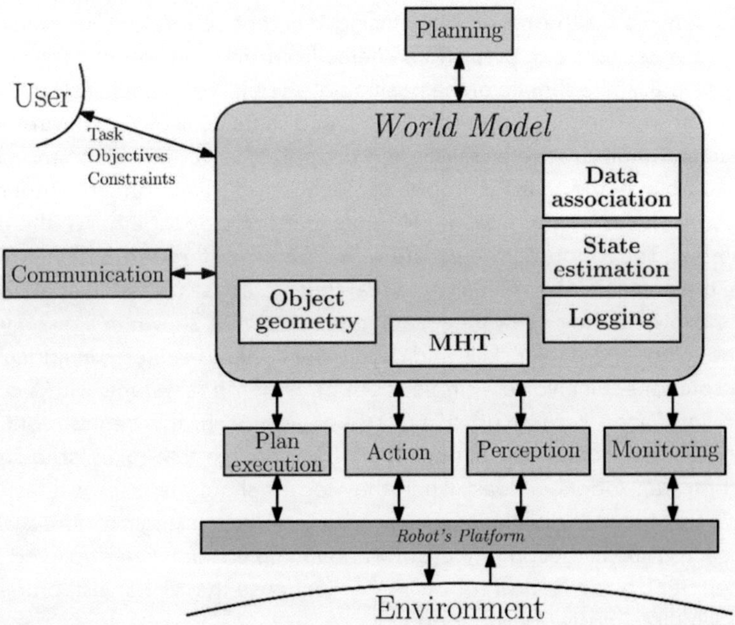

Figure 25 Overview of functionalities of the world model that are discussed in Section 7.

offset the variability and unpredictability of unstructured environments such as agricultural settings.

7.2 Problem

Designing skills as stand-alone applications can lead to information inconsistency. Separate skills hold their own belief about the same information, which can result in problems such as 'magic numbers' throughout the software. Ideally, the information that each skill retrieves should be centrally stored in an explicit WM, creating a consistent world view. This ensures that it can be enriched from a range of sources, it can be used by a multitude of skills, and that it can be shared with other robots. This kind of centralized data collection requires fusion and data association (Elfring et al., 2011, 2013, 2015; Houtman et al., 2021).

7.3 Our approach

We suggest sharing local information with a WM that has a central part within the information architecture (Waibel et al., 2011). This way, when the necessary information is already in the WM, a sensing step can potentially be skipped. Moreover, sharing information enriches detection. This approach of sharing explicit WMs and using them as a central part of the information architecture can be used for multi-robot systems. The shared WM acts as an extra sensor to the sensors on the robot itself, fusing different sources of information to improve detection, e.g. by using cameras from different robots to achieve a more accurate 3D estimate of the position of an object (Kuijpers et al., 2017). By combining sensor data, robots can create and share a full picture environment, understand their relative positions and cooperate to achieve a common goal. An example is the TURTLE robot of TechUnited (Tech United Eindhoven, 2022), which consists of a team of robots that can play together in the game of football. The robots communicate with each other, through WIFI via UDP, to share their local information using a real-time database (Almeida et al., 2004). At the center of their software architecture is the WM, shown in Fig. 26, which is used to configure their skills and determine a shared strategy. In agriculture, most solutions still involve a single robot or multi-robot systems with a central controller. When scaling up a multi-robot approach, the central controller can become a bottleneck. A decentralized approach that relies on a shared WM between robots, at least when they locally share a space or a task, can relieve some responsibilities of the central controller, making it a more scalable approach. This will become relevant when multiple robots are deployed in an environment, e.g. a fleet of robots in a greenhouse that all harvest, prune and trellis simultaneously or even together.

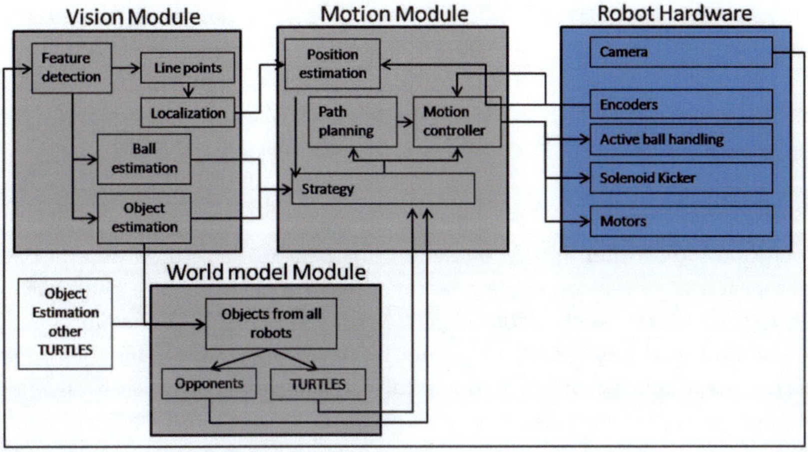

Figure 26 A schematic representation of the hardware and software modules of the TechUnited soccer robots (de Best et al., 2010). Each module has its own skills to fulfill and runs in parallel at a different rate. The world model is filled with object estimation from its own sensors and from peer robots, through Wi-Fi communication.

8 Future perspective: world models for reasoning

8.1 Current state of the art

Sharing a WM between robots requires the information to adhere to a known meta-model and be standardized. Well-known examples of such standardization include open street map (OSM) for navigation (Haklay and Weber, 2008) and the building information model (BIM) for representing buildings in 3D (Eastman et al., 2011). The high level of semantics included in a BIM allows Hendrikx et al. (2021) to do semantically rich feature-based localization, rather than a pure data-based approach. A tool specific for navigation is ED, which provides a 3D geometric, object-based world representation system for robots (Environment Descriptor, 2022).

Geometric maps for mobile robotic tasks are being combined with semantic information (ontologies) for robot navigation (Crespo et al., 2020). Semantic maps model knowledge that does not include geometry, like mereological or topological relations between objects (Kostavelis and Gasteratos, 2015). The semantic mapping framework SEMAP has been used to plan harvesting and other agricultural operations (Deeken et al., 2018, 2019). More broadly, there have been initiatives to develop general ontologies of key concepts and relations which can then be deployed in service robotics (Thosar et al., 2019; Paulius and Sun, 2019; Olivares-Alarcos et al., 2019). A well-known language for knowledge representation is RoboEarth (Tenorth et al., 2013), which can be used for knowledge processing systems for robots such as KnowRob (Tenorth and Beetz, 2009, 2013, 2017). A more recent development, driven by multi-robot cooperation and robust task execution, is the use of knowledge graphs to represent high-level concepts and relations on the symbolical level (Ingrand and Ghallab, 2017; Wilcox et al., 2022).

8.2 Problem

Such ontologies often model an action that can be performed on an object as a property of that object, e.g. a cup can be picked up or a chair can be moved. However, such affordances (Ardón et al., 2020) are very much dependent on the system that executes them. Modeling affordances as object properties requires a different model for each object-system pair even where the dynamics (e.g. the process of picking up different types of object) are the same. By modeling affordances directly as object properties, the semantics of the underlying dynamics are lost, reducing the ability of a robot to adapt to unforeseen conditions.

8.3 Our approach

The world representation must be decoupled from the intended use of the information and the robot that is going to perform the task. Introducing a new task to an existing robot entails the same problems as introducing a new robot to an existing task. We call this problem the composition problem: composing existing skills to solve new tasks (van Niekerk et al., 2019). This requires the decomposition of task and robot capabilities. Decomposition is challenging since system actions are often specifically designed for a certain task. To create composable systems, the task should be modeled agnostic to the robotic system and the robot's capabilities should be modeled without a specific task in mind. The composition of the correct capability to perform a task should be done at runtime (Tenorth et al., 2013). Examples include the composition of a set of basic skills to perform several tasks (Tasse et al., 2022) and assigning different robots to a task that fits their individual capabilities (Joo et al., 2022).

The WM can be split into an environment model (EM) (incorporating objects in the environment) and a robot model (RM) (incorporating robot subsystems). Tasks should be agnostic to the system and thus be described on the EM level, where robot actions are described on the RM level. A task is described as a desired change in the EM and a resource is a possibility for change in the RM. Skills are what link robot resources to the task. They contain the functionality to make changes in the world (action skills) or to retrieve information from it (perception skills). Each skill contains a pre- and post-condition in the form of a sub-graph. By matching these graphs to the WM and sequencing the appropriate skills, possible task execution plans can be created. As an example, the system can choose how to move a truss into a box if it knows the physical connection: it can either pick it up depending on the mass of the truss or slide it over, depending on the friction between the truss and the box. Instead of modeling actions as a property of an object, the system uses interaction models to select possible actions it can undertake on an object.

Labeled property graphs (LPGs) can be used to separate tasks and actions (Angles, 2018). LPG support highly interconnected datasets. An LPG is made up of vertices (V) which represent objects (and their properties) and edges (E) representing types of connections between objects. The set of vertices and edges constitutes the graph (G(V,E)). Both vertices and edges can contain properties. An example of this approach, where the world is modeled as LPG and the task and skills are driven by the WM, is shown in Senden et al. (2023).

9 Conclusion

Robots are more than just motion-controlled positioning systems. However, many companies still build robots using this motion-control paradigm aiming for high precision and repeatability. However, when operating in unstructured environments such as agricultural settings, positional accuracy and speed of operation are of secondary importance to the ability to cope with variability, unpredictability and incomplete sensor data. This requires a paradigm shift in the way robots are designed.

In mainstream robotic systems, there is also typically one model per type of sensor, one for each level of control (joint space, Cartesian space and task space) and one for each type of monitoring. The WM should act as the central part of the information architecture to use and maintain these models. Larger scale applications require a hierarchy in the information architecture, each with a complementary level of abstraction and resolution, each with a separate WM. A central WM acts as a database to store data and as a knowledge base to provide context to this data.

To achieve robust task execution, skills must be able to deal with disturbances to their nominal execution order. If a skill does not succeed in acquiring the correct information or performing the desired action, there must be a 'Plan B' (and ideally also a 'Plan C', 'Plan D', ...) to cope with the disturbance. The major missing functionality in mainstream robotic systems is the addition of reasoning capabilities to the WM to compose an appropriate sequence of possible skill plans to fulfill a task in complex, real-world conditions. The skills should be configurable based on a WM.

Robustness in skills should rely on different, complementary sensor types. For example, if a detection skill using a camera fails, e.g. due to poor lighting conditions, force or touch sensing techniques for localization can be used.

Combining force/touch sensing and vision can increase the robustness of task execution.

Invariants are a powerful way to increase the robustness of task execution. In some cases, a task can be solved without requiring exact information, like geometric measurements, size or exact color. It is often sufficient to know only the relative properties of the object, e.g. all tomatoes in a truss should have a similar diameter. Developing skills based on invariants will be more robust in dealing with variations.

10 Glossary

World Model – A WM involves a model-based representation of the robot and its environment. The WM can be split into an environment model (EM) and a robot model (RM). The EM includes knowledge of objects and their behavior in the environment, e.g. how plants grow over time, where the RM describes the robot and its resources. A WM allows the robot to make decisions about which motion and sensing actions to execute, considering a much larger horizon in space and time than what can be provided by the robot's sensors alone.

Composition – Creating a whole out of smaller subparts. Examples are creating a sequence of sub-tasks to form a task plan or choosing which skills to perform subsequential or in parallel to form a skill plan.

Configuration – Determining the values of parameters that are necessary for a skill to function. All functionalities or algorithms that are necessary for the robot to function rely on certain parameters, which can have a specific value. Instead of hardcoding these values into the algorithm, which leads to magic numbers in the software, these should be queried from the WM. The skill should become dependent on the configured parameter. If a situation changes, causing the values of parameters in the WM to change, the skill execution adapts to this.

Task – A task is a desired change in the environment; that what is desired to happen. The description of a task should be agnostic to the system that is going to perform it (the resource or how). A task describes a change in the EM, from the current state to the desired state, adhering to constraints expressed in the EM. As an example, the task of harvesting can be formulate as 'separate all ripe trusses form a plan, by severing the peduncle'. Notice the use of the more general verbs 'separate' and 'severing' instead of 'cutting', the latter which would imply a system that possess a cutting device like a knife or scissors.

Skill – A skill is what connects a resource to a task, it is the functionality of the system. Skills describe how a task can be fulfilled and contain elements (e.g. constraints) of the EM as well as RM. Action skills use resources to perform a change in the environment, e.g. cutting a peduncle with a knife or physical interaction to move occlusions out of the way. Perception skill use resources to acquire information from the environment, e.g. using the data from the vision system to acquire the ripeness, size or position of a truss.

Resource – A resource is a possibility for change described in the RM; that what the robot can do or provide. The description of a resource should be agnostic to the task that needs to be performed. For example, a robot can drive around, move its EF up and down or capture data from a vision system.

Invariant – An invariant is a mathematical constraint that remains unchanged under a particular transformation in a system (Senden et al., 2022). It exists in many forms, like a structured topology, an absolute shape or a relative property. For example, the invariant features of tomatoes in a single truss are that they are ball-shaped and their sizes (diameters) are relatively equal to each other. This holds for most truss tomatoes regardless of their actual diameter, which can depend on the type (e.g. cherry tomatoes) and ripeness stage.

11 Where to look for further information

Wageningen University & Research (WUR) is worldwide the most experienced and innovative organization in the field of agri-food robotics research and development:

- https://www.wur.nl/en/research-results/projects-and-programmes/agro-food-robotics.htm.

To keep track of the research and development of robots in a broader context, such as the team of reigning world champions in soccer robots or care-robots, you can visit the website of the Eindhoven University of Technology:

- https://www.tue.nl/en/research/institutes/eindhoven-artificial-intelligence-systems-institute/robotics.

The main drivers behind many ideas set forth in this chapter have a starting point in the online book (work in progress) *Building blocks for complicated and situational aware robotic and cyber-physical systems*, which can be found at:

- https://robmosys.pages.gitlab.kuleuven.be/.

The origin of this book is the H2020 project RobMoSys (2017–2020), which aimed at an European ecosystem for model-driven engineering for robotics:

- https://robmosys.eu/.

Sharing knowledge for collaboration between robots was one of the key drivers behind the RoboEarth project. Some important outputs of this project are also already discussed in this chapter, such as WIRE and KnowRob:

- https://www.roboearth.ethz.ch/index.html.
- https://www.roboearth.ethz.ch/wp-content/uploads/2011/06/Waibel2011RoboEarth.pdf.

To facilitate the sharing of knowledge and online composition of skills, this chapter suggests to create a task-description that is agnostic to the system that performs the task. Decomposing the task into the 'why' and the 'how' in order to facilitate online composition has its roots in the CAFCR model for system design:

- https://www.gaudisite.nl/SABP.html.
- https://www.gaudisite.nl/BasicCAFCRPaper.pdf.

12 References

Almeida, L., Santos, F., Facchinetti, T., Pedreiras, P., Silva, V. and Seabra Lopes, L. (2004). Coordinating distributed autonomous agents with a real-time database: the CAMBADA project. International Symposium on Computer and Information Sciences (ISCIS). Available at: https://www.researchgate.net/publication/221579370.

Andries, M., Chavez-Garcia, R. O., Chatila, R., Giusti, A. and Gambardella, L. M. (2018). Affordance equivalences in robotics: A formalism. *Frontiers in Neurorobotics* 12, 26. https://doi.org/10.3389/fnbot.2018.00026.

Angles, R. (2018). The property graph database model. In: Olteanu, D. and Poblete, B. (Eds). Proceedings of the 12th Alberto Mendelzon International Workshop on Foundations of Data Management (AMW 2018). Available at: http://ceur-ws.org/Vol-2100/.

Ardón, P., Pairet, È., Lohan, K. S., Ramamoorthy, S. and Petrick, R. P. A. (2020). Affordances in robotic tasks: A survey, pp. 1–18. Available at: http://arxiv.org/abs/2004.07400.

Auat Cheein, F. A., Guivant, J., Sanz, R., Escolà, A., Yandún, F., Torres-Torriti, M. and Rosell-Polo, J. R. (2015). Real-time approaches for characterization of fully and partially scanned canopies in groves. *Computers and Electronics in Agriculture* 118, 361–371. https://doi.org/10.1016/j.compag.2015.09.017.

Auat Cheein, F., Torres-Torriti, M., Hopfenblatt, N. B., Prado, Á. J. and Calabi, D. (2017). Agricultural service unit motion planning under harvesting scheduling and terrain constraints. *Journal of Field Robotics* 34(8), 1531–1542. https://doi.org/10.1002/rob.21738.

Bac, C. W., Roorda, T., Reshef, R., Berman, S., Hemming, J. and van Henten, E. J. (2016). Analysis of a motion planning problem for sweet-pepper harvesting in a dense obstacle environment. *Biosystems Engineering* 146, 85–97. https://doi.org/10.1016/j.biosystemseng.2015.07.004.

Baietto, M. and Wilson, A. D. (2015). Electronic-nose applications for fruit identification, ripeness and quality grading. *Sensors* 15(1), 899–931. https://doi.org/10.3390/s150100899.

Bajcsy, R., Aloimonos, Y. and Tsotsos, J. K. (2018). Revisiting active perception. *Autonomous Robots* 42(2), 177–196. https://doi.org/10.1007/s10514-017-9615-3.

Barnea, E., Mairon, R. and Ben-shahar, O. (2016). Colour-agnostic shape-based 3D fruit detection for crop harvesting robots. *Biosystems Engineering* 146, 57–70. https://doi.org/10.1016/j.biosystemseng.2016.01.013.

Bawden, O., Kulk, J., Russell, R., McCool, C., English, A., Dayoub, F., Lehnert, C. and Perez, T. (2017). Robot for weed species plant-specific management. *Journal of Field Robotics* 34(6), 1179–1199. https://doi.org/10.1002/rob.21727.

Blanco, J.-L., Fernandez-Madrigal, J.-A. and Gonzalez, J. (2007). A new approach for large-scale localization and mapping: hybrid metric-topological SLAM. Proceedings 2007 IEEE International Conference on Robotics and Automation, 2061–2067 (vol. 2007). https://doi.org/10.1109/ROBOT.2007.363625.

Blok, P. M., van Boheemen, K., van Evert, F. K., IJsselmuiden, J. and Kim, G. (2019). Robot navigation in orchards with localization based on Particle filter and Kalman filter. *Computers and Electronics in Agriculture* 157, 261–269. https://doi.org/10.1016/j.compag.2018.12.046.

Boryga, M., Graboś, A., Kołodziej, P., Gołacki, K. and Stropek, Z. (2015). Trajectory planning with obstacles on the example of tomato harvest. *Agriculture and Agricultural Science Procedia* 7, 27–34. https://doi.org/10.1016/j.aaspro.2015.12.026.

Botta, A., Cavallone, P., Baglieri, L., Colucci, G., Tagliavini, L. and Quaglia, G. (2022). A review of robots, perception, and tasks in precision agriculture. *Applied Mechanics* 3(3), 830–854. https://doi.org/10.3390/applmech3030049.

Bruyninckx, H. (2022). *Building Blocks for Complicated and Situational Aware Robotic and Cyber-physical Systems Design of Complicated Systems (Online Book). Work in Progress: Vol. Version: 32427a*. Available at: https://robmosys.pages.gitlab.kuleuven.be/.

Cheein, F. A. A., Scaglia, G., Torres-Torriti, M., Guivant, J., Prado, A. J., Arnò, J., Escolà, A. and Rosell-Polo, J. R. (2016). Algebraic path tracking to aid the manual harvesting of olives using an automated service unit. *Biosystems Engineering* 142, 117–132. https://doi.org/10.1016/j.biosystemseng.2015.12.006.

Cobbenhagen, A. T. J. R., Schoonen, L. P. A., van de Molengraft, M. J. G. and Heemels, W. P. M. H. (2022). Optimal Irrigation Allocation for Large-Scale Arable Farming in *IEEE Transactions on Control Systems Technology* 30(4), 1484–1493. https://doi: 10.1109/TCST.2021.3118296.

Crespo, J., Castillo, J. C., Mozos, O. M. and Barber, R. (2020). Semantic information for robot navigation: A survey. *Applied Sciences* 10(2). https://doi.org/10.3390/app10020497.

de Best, J. J. T. H., Bruijnen, D. J. H., Hoogendijk, R., Janssen, R. J. M., Meessen, K. J., Merry, R. J. E., van de Molengraft, M. J. G., Naus, G. J. L. and Ronde, M. J. C. (2010). *Tech United Eindhoven Team Description 2010*. Available at: http://www.techunited.nl.

de Langre, E. (2019). Plant vibrations at all scales: a review. *Journal of Experimental Botany* 70(14), 3521–3531. https://doi.org/10.1093/jxb/erz209.

Deeken, H., Wiemann, T. and Hertzberg, J. (2018). Grounding semantic maps in spatial databases. *Robotics and Autonomous Systems* 105, 146–165. https://doi.org/10.1016/j.robot.2018.03.011.

Deeken, H., Wiemann, T. and Hertzberg, J. (2019). A spatio-semantic approach to reasoning about agricultural processes. *Applied Intelligence* 49(11), 3821–3833. https://doi.org/10.1007/s10489-019-01451-2.

Ding, C., Feng, Z., Wang, D., Cui, D. and Li, W. (2021). Acoustic vibration technology: toward a promising fruit quality detection method. *Comprehensive Reviews in Food Science and Food Safety* 20(2), 1655–1680. https://doi.org/10.1111/1541-4337.12722.

Duvenhage, B., Bouatouch, K. and Kourie, D. G. (2013). Numerical verification of bidirectional reflectance distribution functions for physical plausibility. *ACM International Conference Proceeding Series*, 200–208. https://doi.org/10.1145/2513456.2513499.

Eastman, C., Teicholz, P., Sacks, R. and Liston, K. (2011). *BIM Handbook: A Guide to Building Information Modeling for Owners, Managers, Designers, Engineers and Contractors* (2nd edn.). John Wiley & Sons.

ED: Environment Descriptor. (2022). [Online] Accessed on 29/04/2022. Available at: https://github.com/tue-robotics/ed.

Eizentals, P. (2016). *Picking System for Automatic Harvesting of Sweet Pepper*. Kochi University of Technology, Kochi, Japan.

Eizentals, P. and Oka, K. (2016). 3D pose estimation of green pepper fruit for automated harvesting. *Computers and Electronics in Agriculture* 128, 127–140. https://doi.org/10.1016/j.compag.2016.08.024.

Elfring, J., van de Molengraft, M. J. G., Janssen, R. J. M. and Steinbuch, M. (2011). Two level world modeling for cooperating robots using a multiple hypotheses filter, pp. 815–820. In: Proceedings of the - IEEE International Conference on Robotics and Automation. https://doi.org/10.1109/ICRA.2011.5980219.

Elfring, J., van den Dries, S., van de Molengraft, M. J. G. and Steinbuch, M. (2013). Semantic world modeling using probabilistic multiple hypothesis anchoring. *Robotics and Autonomous Systems* 61(2), 95–105. https://doi.org/10.1016/j.robot.2012.11.005.

Elfring, J., van de Molengraft, R. and Steinbuch, M. (2015). Semi-task-dependent and uncertainty-driven world model maintenance. *Autonomous Robots* 38(1), 1–15. https://doi.org/10.1007/s10514-014-9393-0.

Emmi, L., Gonzalez-De-Soto, M., Pajares, G. and Gonzalez-De-Santos, P. (2014). New trends in robotics for agriculture: integration and assessment of a real fleet of robots. *The Scientific World Journal* 2014, 404059. https://doi.org/10.1155/2014/404059.

Gasparino, M. V. v, Higuti, V. A. H., Velasquez, A. E. B. and Becker, M. (2020). Improved localization in a corn crop row using a rotated laser rangefinder for three-dimensional data acquisition. In: Proceedings of the 14th International Conference on Precision Agriculture.

Ge, Y., Xiong, Y., Tenorio, G. L. and From, P. J. (2019). Fruit Localization and Environment Perception for Strawberry Harvesting Robots. *IEEE Access* 7, 147642–147652. https://doi.org/10.1109/ACCESS.2019.2946369.

Gené-Mola, J., Vilaplana, V., Rosell-Polo, J. R., Morros, J.-R., Ruiz-Hidalgo, J. and Gregorio, E. (2019). Multi-modal deep learning for Fuji apple detection using RGB-D cameras and their radiometric capabilities. *Computers and Electronics in Agriculture* 162(May), 689–698. https://doi.org/10.1016/j.compag.2019.05.016.

Gongal, A., Amatya, S., Karkee, M., Zhang, Q. and Lewis, K. (2015). Sensors and systems for fruit detection and localization: a review. In: *Computers and Electronics in Agriculture* 116. Elsevier B.V. https://doi.org/10.1016/j.compag.2015.05.021.

Gonzalez-de-Santos, P., Ribeiro, A., Fernandez-Quintanilla, C., Lopez-Granados, F., Brandstoetter, M., Tomic, S., Pedrazzi, S., Peruzzi, A., Pajares, G., Kaplanis, G., Perez-Ruiz, M., Valero, C., del Cerro, J., Vieri, M., Rabatel, G. and Debilde, B. (2017). Fleets of robots for environmentally-safe pest control in agriculture. In: *Precision Agriculture*. Springer 18(4), 574–614. https://doi.org/10.1007/s11119-016-9476-3.

Haklay, M. and Weber, P. (2008). openstreetMap: user-Generated street maps. *IEEE Pervasive Computing* 7(4), 12–18. Available at: www.openstreetmap

Hendrikx, R. W. M., Pauwels, P., Torta, E., Bruyninckx, H. P. J. and van de Molengraft, M. J. G. (2021). Connecting semantic building information models and robotics: an application to 2D LiDAR-based localization. In: IEEE International Conference on Robotics and Automation (ICRA), Icra, 11654–11660 (vol. 2021). https://doi.org/10.1109/ICRA48506.2021.9561129.

Hill, J. W. and Sword, A. J. (1973). Manipulation based on sensor-directed control: an integrated end effector and touch sensing system. In: 17th Annual Human Factors Society Convention.

Hiruta, T., Hosoya, N., Maeda, S. and Kajiwara, I. (2021). Experimental evaluation of frequency response and firmness of apples based on an excitation technique using a dielectric elastomer actuator. *Sensors and Actuators. Part A* 330, 112830. https://doi.org/10.1016/j.sna.2021.112830.

Houtman, W., Siagkris-Lekkos, A., Bos, D. J. M., van den Heuvel, B., den Boer, R., Elfring, J. and van de Molengraft, M. J. G. (2021). Automated flower counting from partial detections: Multiple hypothesis tracking with a connected-flower plant model. *Computers and Electronics in Agriculture* 188, 106346. https://doi.org.10.1016/j.compag.2021.106346.

Ingrand, F. and Ghallab, M. (2017). Deliberation for autonomous robots: A survey. *Artificial Intelligence* 247, 10–44. https://doi.org/10.1016/j.artint.2014.11.003.

Joo, S., Bae, S., Choi, J., Park, H., Lee, S., You, S., Uhm, T., Moon, J. and Kuc, T. (2022). A flexible semantic ontological model framework and its application to robotic navigation in large dynamic environments. *Electronics* 11(15). https://doi.org/10.3390/electronics11152420.

Kapach, K., Barnea, E., Mairon, R., Edan, Y. and Ben-Shahar, O. B. (2012). Computer vision for fruit harvesting robots: state of the art and challenges ahead. *International Journal of Computational Vision and Robotics* 3(1/2), 4–34.

Koirala, A., Walsh, K. B., Wang, Z. and McCarthy, C. (2019). Deep learning: method overview and review of use for fruit detection and yield estimation. *Computers and Electronics in Agriculture* 162(April), 219–234. https://doi.org/10.1016/j.compag.2019.04.017.

Kostavelis, I. and Gasteratos, A. (2015). Semantic mapping for mobile robotics tasks: A survey. *Robotics and Autonomous Systems* 66, 86–103. https://doi.org/10.1016/j.robot.2014.12.006.

Kuijpers, W., Neves, A. J. R. and van de Molengraft, R. (2017). Cooperative sensing for 3D ball positioning in the RoboCup middle size league, pp. 268–278. In: RoboCup 2016: Robot World Cup XX (vol. 2016). https://doi.org/10.1007/978-3-319-68792-6_22.

Lee, B., Kam, D., Min, B., Hwa, J. and Oh, S. (2019). A vision servo system for automated harvest of sweet pepper in Korean greenhouse environment. *Applied Sciences* 9(12), 2395.

Lehnert, C., English, A., Mccool, C., Tow, A. W. and Perez, T. (2017). Autonomous sweet pepper harvesting for protected cropping systems. *IEEE Robotics and Automation Letters* 2(2), 872–879.

Li, P., Lee, S. and Hsu, H. (2011). Review on fruit harvesting method for potential use of automatic fruit harvesting systems. *Procedia Engineering* 23, 351–366. https://doi.org/10.1016/j.proeng.2011.11.2514.

Li, Z., Li, P. and Liu, J. (2011). Physical and mechanical properties of tomato fruits as related to robot's harvesting. *Journal of Food Engineering* 103(2), 170–178. https://doi.org/10.1016/j.jfoodeng.2010.10.013.

Lin, Y. (2015). LiDAR: an important tool for next-generation phenotyping technology of high potential for plant phenomics? In: *Computers and Electronics in Agriculture* 119. Elsevier B.V. https://doi.org/10.1016/j.compag.2015.10.011.

Lindenmayer, A. (1987). *Models for Multicellular Development: Characterization, Inference and Complexity of L-Systems*, pp. 138–168. https://doi.org/10.1007/3540185356_36.

Liu, Q., Li, R., Hu, H. and Gu, D. (2016). Extracting semantic information from visual data: A survey. *Robotics* 5(1). https://doi.org/10.3390/robotics5010008.

Liu, Z. and von Wichert, G. (2014). Extracting semantic indoor maps from occupancy grids. *Robotics and Autonomous Systems* 62(5), 663–674. https://doi.org/10.1016/j.robot.2012.10.004.

Magalhães, S. A., Moreira, A. P., Santos, F. N. D. and Dias, J. (2022). Active perception fruit harvesting robots: A systematic review. *Journal of Intelligent and Robotic Systems* 105(1). https://doi.org/10.1007/s10846-022-01595-3.

Mapes, M. and Xu, Y. (2022). Row allocation negotiation for a fleet of strawberry harvesting robots. *ASME Letters in Dynamic Systems and Control* 2(3), 1–7. https://doi.org/10.1115/1.4054644.

Marinakis, D. and Dudek, G. (2010). Pure topological mapping in mobile robotics. *IEEE Transactions on Robotics* 26(6), 1051–1064. https://doi.org/10.1109/TRO.2010.2081410.

Marturi, N., Kopicki, M., Rastegarpanah, A., Rajasekaran, V., Adjigble, M., Stolkin, R., Leonardis, A. and Bekiroglu, Y. (2019). Dynamic grasp and trajectory planning for moving objects. *Autonomous Robots* 43(5), 1241–1256. https://doi.org/10.1007/s10514-018-9799-1.

Mason, M. T. (1981). Compliance and force control for computer controlled manipulators. *IEEE Transactions on Systems, Man, and Cybernetics* 11(6), 418–432. https://doi.org/10.1109/TSMC.1981.4308708.

Mousazadeh, H. (2013). A technical review on navigation systems of agricultural autonomous off-road vehicles. *Journal of Terramechanics* 50(3), 211–232. https://doi.org/10.1016/j.jterra.2013.03.004.

Myronenko, A. and Song, X. (2010). Point set registration: coherent point drift. *IEEE Transactions on Pattern Analysis and Machine Intelligence* 32(12), 2262–2275. https://doi.org/10.1109/TPAMI.2010.46.

Naya, S. K., Ikeuch, K. and Kanade, T. (1991a). Surface reflection: physical and geometrical perspectives. *IEEE Transactions on Pattern Analysis and Machine Intelligence* 13(7), 611–634. https://doi.org/10.1109/34.85654.

Naya, S. K., Ikeuch, K. and Kanade, T. (1991b). Surface reflection: physical and geometrical perspectives. *IEEE Transactions on Pattern Analysis and Machine Intelligence* 13(7), 611–634. https://doi.org/10.1109/34.85654.

Olivares-Alarcos, A., Beßler, D., Khamis, A., Goncalves, P., Habib, M. K., Bermejo-Alonso, J., Barreto, M., Diab, M., Rosell, J., Quintas, J., Olszewska, J., Nakawala, H., Pignaton, E., Gyrard, A., Borgo, S., Alenyà, G., Beetz, M. and Li, H. (2019). A review and comparison of ontology-based approaches to robot autonomy. *Knowledge Engineering Review* 34, 1-29. https://doi.org/10.1017/S0269888919000237.

Palleja, T. and Landers, A. J. (2015). Real time canopy density estimation using ultrasonic envelope signals in the orchard and vineyard. *Computers and Electronics in Agriculture* 115, 108–117. https://doi.org/10.1016/j.compag.2015.05.014.

Panigrahi, P. K. and Bisoy, S. K. (2021). Localization strategies for autonomous mobile robots: a review. In: *Journal of King Saud University - Computer and Information Sciences*. King Saud Bin Abdulaziz University. https://doi.org/10.1016/j.jksuci.2021.02.015.

Patel, K. K., Kar, A., Jha, S. N. and Khan, M. A. (2012). Machine vision system: a tool for quality inspection of food and agricultural products. *Journal of Food Science and Technology* 49(2), 123–141. https://doi.org/10.1007/s13197-011-0321-4.

Paulius, D. and Sun, Y. (2019). A survey of knowledge representation in service robotics. *Robotics and Autonomous Systems* 118, 13–30. https://doi.org/10.1016/j.robot.2019.03.005.

Pfeiffer, S. A., Guevara, J., Cheein, F. A. and Sanz, R. (2018). Mechatronic terrestrial LiDAR for canopy porosity and crown surface estimation. *Computers and Electronics in Agriculture* 146(January), 104–113. https://doi.org/10.1016/j.compag.2018.01.022.

Pothen, Z. S. and Nuske, S. (2016). Texture-based fruit detection via images using the smooth patterns on the fruit. In: International Conference on Robotics and Automation, pp. 5171–5176.

Prusinkiewicz, P., Cieslak, M., Ferraro, P. and Hanan, J. (2018). Modeling plant development with L-systems. In: *Mathematical Modelling in Plant Biology*. Springer International Publishing, pp. 139–169. https://doi.org/10.1007/978-3-319-99070-5_8.

Prusinkiewicz, P., Hanan, J., Hammel, M. and Mech, R. (1997). L-systems: from the theory to visual models of plants. In: Michalewicz, M. T. (Ed). CSIRO, pp. 1–27. *Plants to Ecosystems. Advances in Computational Life Sciences*.

Ravankar, A. A., Ravankar, A., Emaru, T. and Kobayashi, Y. (2020). *Multi-robot Mapping and Navigation Using Topological Features* (vol. 68). https://doi.org/10.3390/ecsa-6-06580.

Rosen, D. M., Doherty, K. J., Terán Espinoza, A. and Leonard, J. J. (2021). Advances in inference and representation for simultaneous localization and mapping. *Annual Review of Control, Robotics, and Autonomous Systems* 4(1), 215–242. https://doi.org/10.1146/annurev-control-072720-082553.

Sakai, T. and Nagai, T. (2022). Explainable autonomous robots: a survey and perspective. *Advanced Robotics* 36(5–6), 219–238. https://doi.org/10.1080/01691864.2022.2029720.

Scoccia, C., Palmieri, G., Palpacelli, M. C. and Callegari, M. (2021). A collision avoidance strategy for redundant manipulators in dynamically variable environments: on-line perturbations of off-line generated trajectories. *Machines* 9(2), 1–16. https://doi.org/10.3390/machines9020030.

Senden, J., Janssen, L., van der Kruk, R., Bruyninckx, H. and van de Molengraft, R. (2022). Exploiting plant dynamics in robotic fruit localization. *Computers and Electronics in Agriculture* 196. https://doi.org/10.1016/j.compag.2022.106860.

Senden, J., Jebbink, K., Bruyninckx, H. and van de Molengraft, R. (2022). Invariant-based world models for robust robotic systems demonstrated on an autonomous football table. *IEEE Robotics and Automation Letters* 7(3), 8542–8549. https://doi.org/10.1109/LRA.2022.3185767.

Senden, J., Hollands, K., Rapado-Rincon, D., Burusa, A. K., Herremans, B., Bruyninckx, H. and van de Molengraft, R. (2023). Multi-hypothesis tracking in a graph-based world model for knowledge-driven active perception. *IEEE Robotics and Automation Letters* 8(9), 5934–5941.

Soetens, R. (2015). *Towards Automatic Harvesting of Truss Tomato in a High-Wire Cultivation System. Master thesis, TU/e.*

Srivastava, A. (2019). *Sense-Plan-Act in Robotic Applications*. https://doi.org/10.13140/RG.2.2.21308.36481.

Taniguchi, T., Piater, J., Worgotter, F., Ugur, E., Hoffmann, M., Jamone, L., Nagai, T., Rosman, B., Matsuka, T., Iwahashi, N. and Oztop, E. (2019). Symbol emergence in cognitive developmental systems: A survey. *IEEE Transactions on Cognitive and Developmental Systems* 11(4), 494–516. https://doi.org/10.1109/TCDS.2018.2867772.

Tasse, G. N., Jarvis, D., James, S. and Rosman, B. (2022). Skill machines: temporal logic composition in reinforcement learning. Available at: http://arxiv.org/abs/2205.12532.

Tech United Eindhoven. (2022). [Online] Accessed on 19/04/2022. Available at: https://www.techunited.nl/.

Tenorth, M. and Beetz, M. (2009). KNOWROB — knowledge processing for autonomous personal robots. 2009 IEEE/RSJ International Conference on Intelligent Robots and Systems, pp. 4261–4266. https://doi.org/10.1109/IROS.2009.5354602.

Tenorth, M. and Beetz, M. (2013). KnowRob: A knowledge processing infrastructure for cognition-enabled robots. *International Journal of Robotics Research* 32(5), 566–590. https://doi.org/10.1177/0278364913481635.

Tenorth, M. and Beetz, M. (2017). Representations for robot knowledge in the KnowRob framework. *Artificial Intelligence* 247, 151–169. https://doi.org/10.1016/j.artint.2015.05.010.

Tenorth, M., Perzylo, A., Lafrenz, R. and Beetz, M. (2013). The RoboEarth language: representing and exchanging knowledge about actions, objects, and environments. IJCAI International Joint Conference on Artificial Intelligence (vols. 3091–3095). https://doi.org/10.1109/ICRA.2012.6224812.

Thosar, M., Zug, S., Skaria, A. M. and Jain, A. (2019). A review of knowledge bases for service robots in household environments. *CEUR Workshop Proceedings* 2418(October), 98–110.

Tsardoulias, E. G., Iliakopoulou, A., Kargakos, A. and Petrou, L. (2016). A review of global path planning methods for occupancy grid maps regardless of obstacle density. *Journal of Intelligent and Robotic Systems* 84(1–4), 829–858. https://doi.org/10.1007/s10846-016-0362-z.

Underwood, J. P., Hung, C., Whelan, B. and Sukkarieh, S. (2016). Mapping almond orchard canopy volume, flowers, fruit and yield using lidar and vision sensors. *Computers and Electronics in Agriculture* 130, 83–96. https://doi.org/10.1016/j.compag.2016.09.014.

van Niekerk, B., James, S., Earle, A. and Rosman, B. (2019). Composing value functions in reinforcement learning. In: Chaudhuri, K. and Salakhutdinov, R. (Eds). Proceedings of the 36th International Conference on Machine Learning. PMLR, pp. 6401–6409. Available at: https://proceedings.mlr.press/v97/van-niekerk19a.html.

Waibel, B. M., Beetz, M., Civera, J., Andrea, R. D., Waibel, M., Beetz, M., Civera, J., D'Andrea, R., Elfring, J., Galvez-Lopez, D., Haussermann, K., Janssen, R., Montiel, J. M. M., Perzylo, A., Schiessle, B., Tenorth, M., Zweigle, O. and van de Molengraft, R. (2011). A World Wide Web for robots. *IEEE Robotics and Automation Magazine* 18(2), 69–82. https://doi.org/10.1109/MRA.2011.941632.

Wang, Y., Ma, J., Zhang, Z., Zhao, H., Teng, J., He, Y. and Xu, H. (2018). Harvesting sequence planning of a harvesting robot based on the principle of minimum energy consumption. *The Journal of Engineering* 2018(16), 1447–1456. https://doi.org/10.1049/joe.2018.8325.

Wei, J., Zhijie, Q., Bo, X. and Dean, Z. (2018). A nighttime image enhancement method based on Retinex and guided filter for object recognition of apple harvesting robot. *International Journal of Advanced Robotic Systems* 15(1), 1–12. https://doi.org/10.1177/1729881417753871.

Wei, K. and Ren, B. (2018). A method on dynamic path planning for robotic manipulator autonomous obstacle avoidance based on an improved RRT algorithm. *Sensors* 18(2), 571. https://doi.org/10.3390/s18020571.

Wilcock, G. and Jokinen, K. (2022). Conversational AI and knowledge graphs for social robot interaction, pp. 1090–1094 17th ACM/IEEE International Conference on Human-Robot Interaction (HRI), Sapporo, Japan (vol. 2022), https://doi.org/10.1109/HRI53351.2022.9889583.

Will, P. M. and Grossman, D. D. (1975). An experimental system for computer controlled mechanical assembly. *IEEE Transactions on Computers* C–24(9), 879–888. https://doi.org/10.1109/T-C.1975.224333.

Xiong, Y., Grimstad, L., Yuanyue, G., and From, P. J. (2019a). An autonomous strawberry-harvesting robot: Design, development, integration, and field evaluation. *Journal of Field Robotics, March*, 1–23. https://doi.org/10.1002/rob.21889.

Xiong, Y., Peng, C., Grimstad, L., From, P. J. and Isler, V. (2019b). Development and field evaluation of a strawberry harvesting robot with a cable-driven gripper. *Computers and Electronics in Agriculture* 157, 392–402. https://doi.org/10.1016/j.compag.2019.01.009.

Yandún Narváez, F. J., Salvo del Pedregal, J., Prieto, P. A., Torres-Torriti, M. and Cheein, F. A. A. (2016). LiDAR and thermal images fusion for ground-based 3D characterisation of fruit trees. *Biosystems Engineering* 151, 479–494. https://doi.org/10.1016/j.biosystemseng.2016.10.012.

Yuille, A. L. and Liu, C. (2021). Deep nets: what have they ever done for vision? *International Journal of Computer Vision* 129(3), 781–802. https://doi.org/10.1007/s11263-020-01405-z.

Zaky, Y., Paruthi, G., Tripp, B. and Bergstra, J. (2020). Active perception and representation for robotic manipulation. Available at: http://arxiv.org/abs/2003.06734.

Zhang, Z. (2000). A flexible new technique for camera calibration. *IEEE Transactions on Pattern Analysis and Machine Intelligence* 22(11), 1330–1334.

Zhao, Y., Gong, L., Huang, Y. and Liu, C. (2016). A review of key techniques of vision-based control for harvesting robot. *Computers and Electronics in Agriculture* 127, 311–323. https://doi.org/10.1016/j.compag.2016.06.022.

Zujevs, A., Osadcuks, V. and Ahrendt, P. (2015). Trends in robotic sensor technologies for fruit harvesting : 2010–2015. *Procedia Computer Science* 77, 227–233. https://doi.org/10.1016/j.procs.2015.12.378.

Chapter 3

Advances in local perception for orchard robotics

Jose Blasco, Centro de Agroingeniería – Instituto Valenciano de Investigaciones Agrarias (IVIA), Spain; and Francisco Rovira-Más, Agricultural Robotics Laboratory (ARL) – Universitat Politècnica de València, Spain

1 Introduction
2 Dealing with orchard environments
3 Sensor and data analysis technologies to address orchard environments
4 Advances in multidimensional sensors for navigating orchard environments
5 Sensor fusion for augmented perception
6 Using perception technologies in agricultural operations: harvesting, crop protection and other applications
7 Conclusion and future trends
8 Where to look for further information
9 Acknowledgements
10 References

1 Introduction

Speciality crops continue to be an important part of agriculture in regions such as Europe. This can be seen, e.g. in production levels for 2020 (Agricultural Outlook 2020 Report):

- Olive oil (2 233 000 tons);
- Wine (3.2 million ha and 159 million hl);
- Apples (516 million ha and 11 527 000 tons);
- Peaches and nectarines (3 120 000 tons and 170 000 ha); and
- Oranges (6 505 000 tons and 273 million ha).

Figure 1 shows the distribution of fruit production in the European Union, totalling 3.4 million ha of land used for fruit production, 25.7 million tons of

http://dx.doi.org/10.19103/AS.2023.0124.03

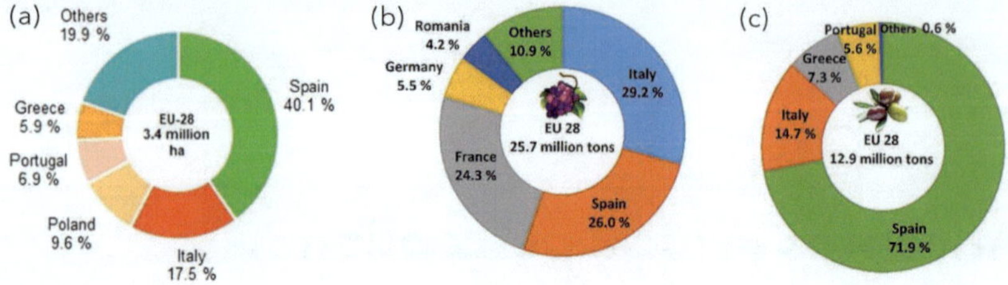

Figure 1 Fruit production in Europe: (a) land, (b) grapes for wine and (c) olives for oil. Source: Eurostat, 2018.

grapes for wine and 12.9 million tons of olives for oil. Speciality crops encompass a wide variety of plants that are distinct from commonly grown staple crops. These crops are typically produced in smaller quantities and require specific growing conditions, husbandry and expertise.

Orchard crops are particularly important speciality crops and primarily consist of fruit-bearing orchards, bushes and vineyards. Orchards require long-term investment and management, as they are permanent crops that can provide long-term yields. Consumer demand for seasonal fresh produce allows growers to command higher prices than commodity crops, potentially leading to greater profitability. Globally consumers often favour speciality fruits such as apples, cherries, stone fruits, avocados or citrus fruits, presenting opportunities for growers to tap into export markets and expand revenue streams. Other crops, such as olive trees and grapevines, have significant economic potential as ingredients in high added-value products such as olive oil and wine. Orchard crops allow consistent and sustained production over the crop's lifespan, allowing growers to invest in resources, infrastructure and technology.

Two opposing forces have driven the introduction of automation and robotics in agricultural sectors such as speciality crops. Firstly, the growing availability of technological solutions such as sensors, computer processors and algorithms; secondly, the major sustainability challenges faced by agriculture. Digitalisation is steadily reaching rural areas but, no matter how accessible novel technology becomes, it will only be adopted if it solves problems in a way that improves on current practices and deals with the challenges that agriculture faces. It is crucial therefore to identify the current challenges faced by the agri-food sector and analyse to what extent they can be mitigated by introducing automation and robotics (Jin et al., 2021). Three key challenges are noteworthy: the first impacts developing countries, and the others mainly affect industrialised agriculture:

1. Increase in world population;
2. Ageing of professional farmers; and
3. Unsustainable production costs.

These are discussed in more detail later.

According to Foley (2014), the world population will increase by 35% by 2050 to 9.5 billion, with a greater impact in Asia and Africa. This rise in population requires an increase of 100%, in food production which implies doubling current production by 2050. The use of bigger machines to increase production creates problems of its own, such as increased emissions and soil compaction which may decrease fertility and compromise long-term sustainability. In doubling food production whilst avoiding further expansion of cultivated land, farm equipment must increase efficiency. Automation, sensors, robotics and artificial intelligence (AI) appear promising solutions to resolve these problems. The integration of farm robotics is particularly significant in orchard production, with an estimated global fruit yield of 909 million tons in 2020, cultivated within a land area of 64.86 hectares (FAO, 2022).

In general, labour participation in agriculture declines as the economic prosperity of a country grows. Industrialised countries face a severe problem due to a lack of young professionals in agriculture to ensure a seamless generational transition. The average age of the American farmer in 2017, e.g. was 57.5 years (USDA, 2017) with a similar situation in Europe. Introducing robotics and automation may both reduce the need for labour and attract younger generations to agriculture since those born in the digital era find computers and electronics familiar and attractive.

Labour costs for the production of fruits in the USA often range between 40% and 60% of total production costs (Burks and Schmoldt, 2008), with a similar situation in Europe. In addition to labour, farmers require fuel for machinery, certified seeds, fertilisers, irrigation water and crop protection products, with costs constantly increasing due to increasing environmental regulations. These production costs have been growing over the last few decades whilst the profits of farmers have remained stagnant or even decreased (Giller et al., 2021). Robotics and automation offer a potential way to reduce labour and other costs and optimise input use.

Innovation and modernisation through the automation of agricultural tasks, digitisation and the introduction of robotic technologies can contribute to the optimal management of resources, reduce production costs, prevent productive land from being abandoned, promote the involvement of young professionals, especially women, in agriculture, allowing a generational shift capable of growing more competitive crops (Botta et al., 2022).

As a result, interest in farm robots has grown exponentially over the last decade, although their distribution is not uniform. Figure 2a shows the global spread of the agricultural robots market for 2017 and market projection for 2025. Overall, as shown in the world map in Fig. 2, North America dominates the global agricultural robots market, driven by growing automation in the

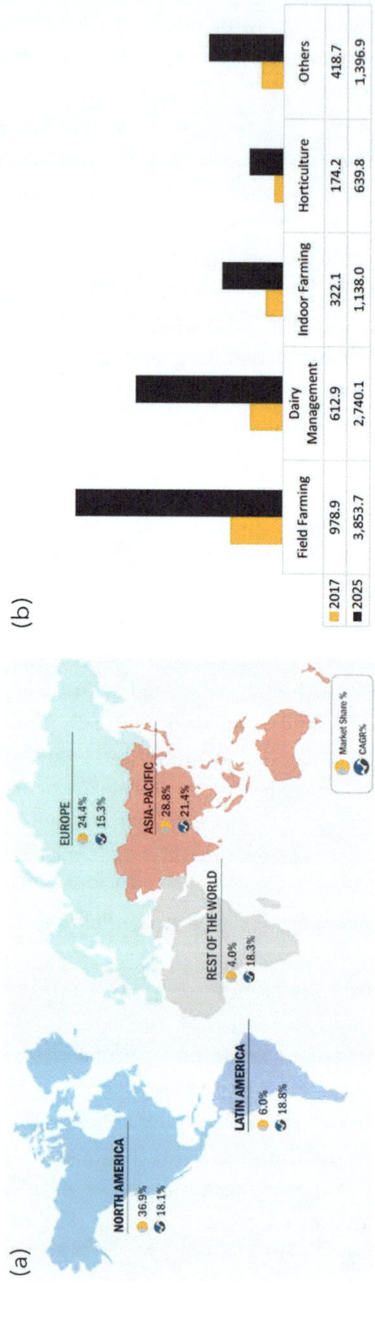

Figure 2 Global market of agricultural robots. Source: VMI, 2018.

agriculture sector in the USA and Canada. Europe, considered a mature market, is projected to show relatively slower growth from 2018 to 2025 (VMI, 2018).

A map showing the distribution of farm robots in the world might be misleading if most of the robots considered were static milking robots, as used to be the case at the turn of the twenty-first century. However, this is no longer the case, partially due to the advances made in perception systems. Figure 2b shows that field robots are currently more numerous than dairy management robots, and the largest growth within farm robots is expected to occur in the area of field robots. Field farming robots in Europe represent US$238.9 million out of the US$611.7 million estimated for 2017, with the projected market for field robots in Europe in 2025 being US$755.3 million (VMI, 2018). However, despite these figures, overall adoption of robots in farms is relatively low, which can be attributed to technical factors related to the degree of autonomous operation agricultural robots are currently capable of, as well as financial pressures manufacturers face during the development and commercialisation of their products and services (Gil et al., 2023).

2 Dealing with orchard environments

This chapter focuses on the challenges of automating the management of orchard crops through robots that can navigate, sense and act in outdoor environments. Unlike sensors in industrial settings, where environmental conditions are controlled, manufactured objects are identical, and tasks are repetitive, agricultural open fields are subject to unpredictable and changing environmental conditions, heterogeneous objects (such as fruit) and diverse applications.

Robots are valuable potential tools for precision crop management in orchards. However, orchard crops present a series of specific challenges for robotic platforms with autonomous navigation capabilities, since the environment is particularly complex and unstructured when compared, e.g. with field crops. The great diversity of orchard conditions and crops makes navigating and perceiving the environment a major challenge, hindering the complete adoption of these technologies (Marinoudi et al., 2019).

Robotic applications in outdoor agriculture must first address dynamic, complex and unstructured environments for different crops that display high variability and rapid changes. The irregular characteristics of terrain and vegetation profile can significantly vary from one point of the plot to another. Ambient light conditions and visibility also tend to change continuously, which makes feature identification imprecise and uncertain. In addition, working conditions in agriculture are challenging; wind, dust, rain, extreme temperatures, humidity and vibrations are unpredictable. Sensors must be able to cope with these harsh operating conditions. The structure of crops in vertical cultivation

Figure 3 Crop structures favourable to orchard automation: (a) apples, (b) cherries and (c) olives.

systems such as vineyards or orchards also complicates autonomous navigation and Global Navigation Satellite System (GNSS) reception (Ruangurai et al., 2022).

In addition, agricultural robots must undertake a wide range of tasks which affects the type of sensor to use and how it should be deployed. Picking robots, e.g. need sensors capable of measuring the distance to the target fruit (Gené-Mola et al., 2019). Applications include the identification of fruits for monitoring or harvesting, the inspection of fruit or trees to detect diseases, pests, nutritional deficiencies or stresses, the detection and elimination of weeds, husbandry operations such as pruning as well as picking and sorting fruit at harvest.

Before the widespread adoption of tractors and other machinery, orchards were designed with narrow pathways to allow movement by farmers and horses within the field. The growing use of mechanisation has resulted in a fundamental redesign of modern orchards in a regular pattern with enough row spacing to ensure the safe movement of machinery. In addition, canopies have been reshaped into wall-like trellises, where canopies are more regular in shape and fruits are more accessible. This is the case for vineyards, which have been transformed from traditional goblet cropping to a modern vertical shoot position formation amenable to mechanical harvesting. The success of vertical canopies in vineyards is now being adopted in other high-value crops such as apples (Fig. 3a), cherries (Fig. 3b) or olive groves (Fig. 3c). However, trimming vigorous trees into regular flat hedges often results in tall narrow canopies that pose significant challenges to the real-time reception of GNSS signals with sufficient quality and integrity. This highlights the continuing challenges that even these more standardised environments still pose. The development of more precise perception systems is therefore a priority, using sensors and algorithms for processing data from the environment, including novel AI algorithms and robotic and sensing technology.

3 Sensor and data analysis technologies to address orchard environments

Sensing in agriculture refers to using advanced technologies to collect and analyse data about the agricultural ecosystem. Sensing, via sensors, helps farmers to make informed decisions regarding crop management, such as determining when to plant, irrigate, fertilise or harvest. There are various types of sensors used in agriculture, including the following:

- Weather sensors to monitor temperature, humidity, rainfall, wind speed and other weather-related data;
- Soil sensors to measure soil compactness, moisture, nutrient levels, salinity and pH;
- Crop sensors to analyse crop health, biomass, yield, pests and diseases and nutrient status; and
- Remote sensing to capture images of crops and land from satellites, drones or other airborne devices.

This chapter centres on sensors capable of perceiving, extracting and interpreting useful information about their environment. Perception is achieved by sensors collecting data and software that processes the data and extracts the relevant information. Perception can be divided into active, i.e. the sensors interact with the scene by emitting signals or pulses of energy, or passive sensors that collect the data from nature or detect and measure the natural energy radiated from the target (Magelhães et al., 2022).

Optical (Cubero et al., 2016), spectral (Barbedo, 2023) and thermal (Zhou et al., 2021) sensors are the most popular sensing techniques to collect crop data from different platforms. Colour-based cameras capture optical information from their surroundings by combining the three primary colours, red, green and blue (RGB). Image processing techniques can discriminate objects by shape or colour (e.g. vegetative structures) or determine the presence of pigments related to particular features, such as the maturity stage of fruits (Lee and Blasco, 2021, Blasco and Aleixos 2023).

Spectral sensors collect information by measuring the radiation emitted, reflected or transmitted from the vegetation. The sensors used to obtain this information, such as colour cameras, spectral imaging systems, point spectrometers or thermographic cameras, measure the electromagnetic energy reflected or emitted by the vegetation in different spectral ranges or at particular wavelengths (Walsh et al., 2020). In contrast, multispectral and hyperspectral sensors are not limited to these three visible bands and can collect complete spectral information beyond the visible part of the electromagnetic spectrum, typically in the near-infrared (NIR) and short-wave infrared (IR) regions (Lorente et al., 2012). The spectral response of vegetation and fruits is affected by several structural and biochemical components such as leaf area, porosity, chlorophyll and water content or nutrient concentration, which typically change and may result in stress potentially detectable by spectral devices. Healthy, stressed or diseased plants have distinct spectral signatures, which allow the detection of problems before the appearance of external symptoms.

The stress or appearance of vegetative problems can influence the temperature of the leaves, as stomatal activity is an essential physiological trait for plant growth and development. As a result, leaf temperature can be correlated with stomatal conductance and further related to yield and stress tolerance. Thermal imaging provides images in which each pixel represents a temperature value; hence, thermal cameras can determine the physiological status of plants. This technology is a valuable stress detection method before the appearance of symptoms, as it operates in the IR band, out of the visual range of the spectrum. It can even be used for determining water stress and irrigation needs. However, careful consideration must be taken when interpreting thermographs, as the plant surface temperature is significantly affected by environmental factors such as air temperature and relative humidity. In addition, plant features, such as leaf size, biomass or canopy structure, can further influence the plant surface temperature, and environmental factors such as wind speed and light intensity can also affect canopy temperatures.

Essential information can also be obtained from the analysis of the three-dimensional (3D) structure of plants, which can be affected by biotic and abiotic stresses. This information can be collected with devices capable of estimating the plant volume or structure by measuring the distance between the device and the plant, creating point clouds to render a scene, such as stereo cameras, time-of-flight (ToF) cameras (Moreno et al., 2020a) or laser scanners as light detection and ranging (LiDAR) devices (Moreno et al., 2020b). Binocular stereo cameras consist of two identical cameras aligned to produce a stereoscopic effect. This effect mimics how human vision works by capturing two images of the same scene, separated by approximately 7 cm to produce similar but not equal images. By comparing the relative position of the pixels of the same objects in both images, the distance from the camera to the objects can be calculated.

The ToF cameras, by contrast, use IR light to determine depth information. The sensor emits a light signal, which hits the objects in the scene and returns to the sensor. The time it takes to bounce and return is then measured, obtaining a depth map. LiDAR is another sensor commonly used to measure the distance to the nearest objects by emitting a laser beam in a given direction; the time elapsed between the emission and reception of the signal is then used to calculate the distance to the target. With these sensors, biometric parameters of interest, such as leaf area index, height, width, volume and other structural parameters of the canopy, can be estimated and related to vegetative growth. This information is also useful for guiding robots in autonomous navigation applications, detecting and picking fruits, and adjusting the spraying rates to the canopy for variable rate spraying or fertilising, as the plant's nutritional needs strongly depend on its biomass.

The data captured by sensors must be efficiently analysed to obtain useful information for growers. Unlike controlled industrial environments with predictable conditions, agricultural fields are open, uneven and constantly evolving over time. Their characteristics depend on multiple factors such as crop, variety, management, soil and weather, among other factors. As a result, there is a need to collect a significant volume of data from a range of sources in order to account for and make sense of this complex, unstructured environment. Data analysis techniques must effectively combine and process large amounts of data to rapidly adapt to changing conditions and make predictions capable of informing operational decisions.

The AI algorithms can make decisions by learning from the data they process (Fountas et al., 2022). This learning capacity is generally achieved using several techniques, particularly by neural network approaches known as machine learning and deep learning (DL) techniques. These algorithms can identify complex patterns among large amounts of data by inferring their own rules to detect similar patterns in new data sets. They can learn to predict behaviours, automatically detect anomalies or recommend optimal decisions. In the case of agricultural robots, AI-based algorithms are currently being applied for autonomous navigation and the prediction of crop variables for intelligent decision-making user support.

Deep multilayer networks like deep convolutional neural networks (CNN) provide a promising way to deal with massive amounts of data (Li et al., 2021). In particular, they have shown exceptional performance in image classification. However, the acquisition and labelling of the massive number of images required to train DL models is a tedious and costly process. Moreover, training data need to encapsulate the range of conditions expected to be encountered, which can be highly variable and seasonal in a field environment. Recently, several attempts have been made to improve CNN performance in classification tasks for agriculture. Specifically, various techniques have been developed for the segmentation of hyperspectral images, such as region-based CNN (R-CNN), fast R-CNN, faster R-CNN and more recently, the 'You Only Look Once' (YOLO) algorithm in its different versions (currently YOLOv8), which are particularly promising in the recognition of fruits in complex orchard environments (Saleem et al., 2021; Sekharamantry et al., 2023).

As noted earlier, the most common sensors currently in use are based on photonics, such as standard and non-standard artificial vision cameras, laser or spectral sensors (Wang et al., 2022), but there are other technologies, such as ultrasound, that can be used to detect the presence of vegetation. However, optical sensors, especially colour cameras, are highly susceptible to changes in ambient light. Natural lighting varies throughout the day, but it can present strong variations over short periods due to the position of the sun or the passage of clouds. Direct sunlight in agricultural settings usually causes high contrasts and darkens the colours of target objects, making it much more difficult to identify objects and their features. Image sensors are particularly affected by these changes and are often set to specific exposure times to optimise the dynamic range of the images and make the acquisition conditions comparable.

Many image segmentation algorithms are based on applying thresholds within regions of the image based on colour differences or brightness levels. However, the segmentation process can prove ineffectual when lighting conditions cannot be maintained. This requires segmentation techniques that adapt to illumination changes and consider relative brightness differences instead of predefined absolute thresholds. Minimum error thresholding, the measure of fuzziness, the K-means algorithm, entropic thresholding, the moment-preserving algorithm, ISODATA or the Otsu method rank among the most popular automatic segmentation algorithms (Russ, 2011).

In the case of mobile robots, the movement and vibrations caused by traversing uneven terrain can influence onboard vision systems and, consequently, the quality of the captured images (e.g. causing blurring). In the case of fruit-picking robots, unexpected movements (e.g. caused by wind) may make it challenging to obtain the necessary references to precisely target fruit with end-effectors. Sudden changes in lighting can introduce changes in light intensity, brightness or shadows that result in false positive or negative detection of fruits. Moving at higher speeds (to optimise efficiency, e.g. in harvesting) requires short exposure times to avoid blurry images, which aggravates lighting problems given short exposures. Additionally, there is a need to capture and process images in real time to harvest fruit quickly and efficiently. Further problems arise from a lack of GNSS signal in the open field or fruit deep within the canopy that cannot be seen in the line of sight (occlusions). This means that, despite impressive technological advances, it is still challenging to implement onboard perception systems for robots capable of adapting to any crop and set of environmental conditions.

4 Advances in multidimensional sensors for navigating orchard environments

The systematic awareness of its surroundings is essential to allow an autonomous robot to move in complex environments. Some sensors only provide a one-dimensional straight distance assessment between the sensor and the sensed object, also known as the range, whilst others work in a plane and provide two dimensions (2D). A complete scene rendering can be achieved with a sensor providing 3D points in clouds, i.e. clusters of points with their three coordinates x, y and z relative to the sensor. Several off-the-shelf sensor solutions exist, but four particular technologies have yielded the most reliable results (Vázquez-Arellano et al., 2016; Bai et al., 2023):

- Binocular stereoscopic vision;
- ToF 3D sensors;
- Multi-beam LiDAR; and
- Sonar networks.

These are discussed later.

Binocular stereo cameras are widely used for autonomous navigation. The comparison of left and right images is based on texture and carried out pixel by pixel, with each offset (horizontal displacement or shift between corresponding pixels in the left and right images) called a disparity. As a result, an image with 640 × 480 pixels has 307 200 disparities. The algorithm must identify and locate the pixel in each image that points at the same real point and compute its disparity. There are various algorithms that can perform the matching process in real time and calculate the disparity image. However, regardless of the algorithm used, mismatches often appear, leading to erroneous points that must be detected and corrected. The two main advantages of stereo cameras that make them ideal for outdoor perception are as follows:

- The richness of information available at high frequencies, as 3D point clouds of several hundred thousand points can result from moderate resolution images taken at 25–30 Hz.
- Because perception is based upon a comparison between two images taken simultaneously, sudden illumination changes affect both images equally, offsetting the distorting effect of lighting changes.

The robustness of perception under unpredictable illumination is a major advantage for outdoor agricultural applications. There are, however, two disadvantages in using a stereo camera. The richness of information carries a high computational cost for coordinate calculation and point cloud handling, which may affect selection for some real-time applications, lowering requirements by decreasing the image resolution and sampling rate. The second disadvantage when comparing stereo vision with other active perception sensors is the need for ambient light to generate the images. Consequently, when light is weak, the texture is poor, and pixel matching becomes inefficient and prone to error, as near pixels are not distinguishable. Fortunately, orchard scenes and tree canopies are rich in texture, and pixel matching tends to be effective. Currently, the principal applications of stereo vision in agriculture are in-field navigation and obstacle detection (Wei et al., 2005). The use of stereo vision

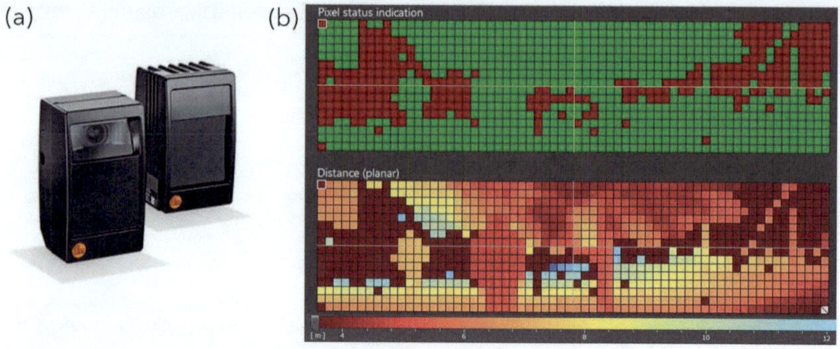

Figure 4 Time-of-flight three-dimensional sensing: (a) sensor IFM O3M and (b) obstacle maps.

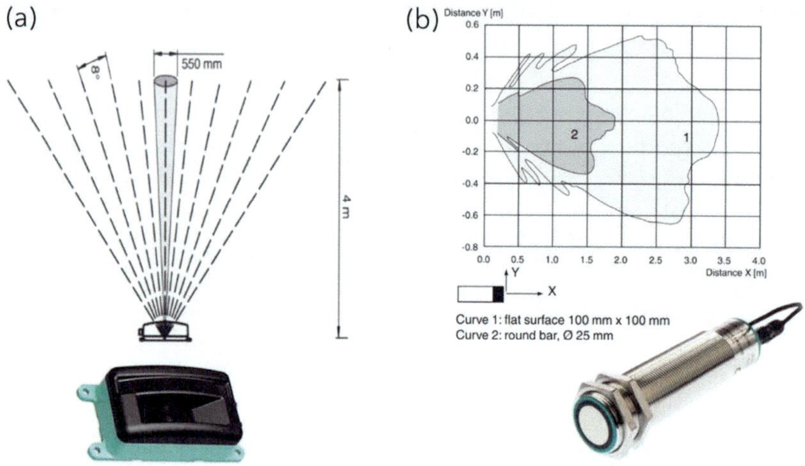

Figure 5 Two-dimensional and one-dimensional sensing: (a) multi-beam LED LiDAR and (b) ultrasonic sensor. Source: Courtesy of Pepperl+fuchs. Note: LiDAR, light detection and ranging.

for the navigation of a scouting robot in vineyards has been reported by Rovira-Más et al. (2015).

A second option, and a competing technology to stereo vision in 3D rendering for navigation, is the use of ToF sensors (Gimenez et al., 2022). Although low-cost sensors inherited from video games (Kinect 3D sensor, Microsoft) became popular in the last decade, heavy-duty IP-67 sensors of high reliability, as shown in Fig. 4a, offer a valid alternative for outdoor sensing. The main advantage of ToF sensors is independence from ambient illumination, with minor interference from sunlight and the ability to work well at night. Its primary disadvantage is low resolution and a reduced field of view compared with stereoscopic cameras. Figure 4b shows the range map of a ToF sensor

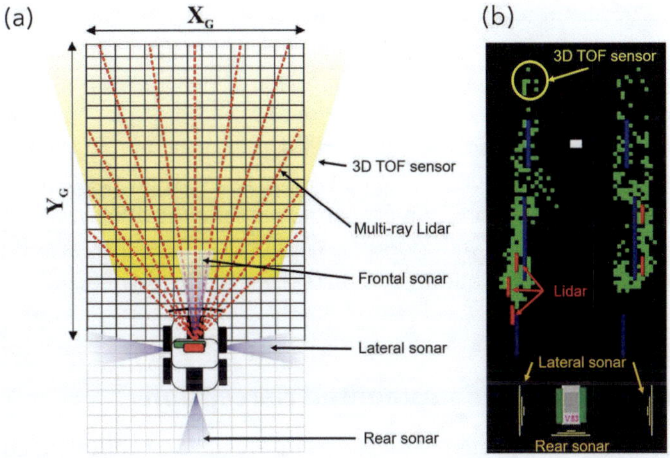

Figure 6 Augmented perception approach: (a) space discretisation and (b) multi-sensor obstacle map.

with two obstacles in an olive grove. Note that the resolution of the map is 64 × 16, yielding 1024 potential pixels, not all of them valid. The top map of Fig. 4b segments the pixels carrying valid information (green) from those discarded (red), either because ranges are too far or because they have a high likelihood of being wrong and have been filtered out by the matching algorithm. The scale at the bottom of the figure provides the specific range for each pixel in the bottom map. The multi-sensor perception map shown in Fig. 6 includes the information from the ToF sensor in Fig. 4.

A third alternative used extensively in navigation to perceive the environment is rotating LiDAR. This sensor is highly accurate (with accuracy of 1 mm in many cases) but prone to errors when a vehicle traverses rough off-road terrain. A practical option for a cost-efficient solution is provided by multi-beam LiDARs, which provide precise laser ranging but drop one dimension in order to generate 2D maps. Figure 5a depicts a LiDAR featuring 11 beams with an angular separation of 8° between beams to fully cover a perception angle of 88° (Pepperl+fuchs, Mannheim, Germany). This sensor is certified to work outdoors in agricultural environments. The use of LiDAR for obstacle detection has been used effectively as a combination of several rotating heads, such as the Grand Challenge winning robot Stanley, or as a combination of diverse technologies, as in the case of VineScout (Rovira-Más et al., 2021a). The perception map shown in Fig. 6 integrates the outputs of the LiDAR of Fig. 6a for short ranges (under 4 m) ahead of an autonomous vehicle.

The last technology reviewed for local perception is based on ultrasonics and consists of estimating ranges by timing the round trip made by sound waves. This sensor only provides one dimension and has been used in agriculture due to its robustness as the rangefinder sonar (Pepperl+fuchs, Mannheim, Germany), as shown in Fig. 5b. The main disadvantage is that sound waves are transmitted in irregular cones (Fig. 5b), and several sensors constituting a network are typically necessary for a sonar-based practical application. The behaviour of the waves is dependent on the reflective nature of the bouncing obstacles, and thus the multiple orientations of leaves within tree canopies may result in inconsistent readings unless sensors are of high quality. Sonar sensors are usually cost-effective with good perception for short ranges and typically used for safety and navigation, as in the sensor fusion approach described in the following section.

5 Sensor fusion for augmented perception

A multipurpose sensor capable of capturing everything needed for consistent awareness of the immediate surroundings of a vehicle is not generally available for practical outdoor applications. A robust solution requires redundancy in case a sensor fails or the combination of different sources of information. Combining several perception sensors with diverse technologies offers a more reliable solution for outdoor applications. In particular, combining the perception sensors described in the previous section provides redundancy as they are based on 3D imaging, laser ranging and ultrasonics, all using different physical principles for sensing. The question is how to fuse information from such diverse technologies in real time.

An effective means of achieving this is by dividing the surrounding space into a regular grid; cells in the grid may be filled with discrete data from any perception sensor, depending on accuracy, reliability and confidence rules. Figure 4b features a discretisation of space with only one (3D) sensor. The following step involves augmentation of the grid with complementary perception sensors. This can be seen in the case of a multi-beam LiDAR sensor (Fig. 5a) and four sonar rangefinders (Fig. 5b) according to the approach shown in Fig. 6a. The mathematical foundation of this data fusion approach can be followed in Rovira-Más et al. (2021a). The unified obstacle map resulting from this fusion algorithm is depicted in Fig. 6b and was used to steer a scouting robot automatically. A detailed description of its features follows later.

The perception universe represented in Fig. 6a is divided into two regions:

1 The straight guidance area comprising a rectangular grid of dimensions $X_G \times Y_G$
2 A close vicinity perimeter 2 m around the robot

The former consists of square cells filled by a 3D ToF sensor and an 11-beam LiDAR to identify the canopy, as combined in Fig. 6b. The latter includes four ultrasonic sensors perceiving up to 2 m from the robot. For the scouting robot, shown in Fig. 8, monitoring vineyards and olive groves (Rovira-Más et al., 2021b), X_G was 6 m and Y_G was 8 m. The white square in Fig. 6b indicates the target point where the robot is directed in order to stay centred in the middle of the lane between adjacent rows. Note that the LiDAR provides redundant perception in the first 4 m from the robot, whereas the ToF sensor provides a look-ahead distance of over 5 m, complementing each other in estimating the target point. Unlike the frontal sonar, which is not shown in the obstacle map of Fig. 6b, the lateral and rear sonars plot the map, although none are used to calculate the target point. The frontal sonar is used for safety purposes, stopping the robot when an obstacle appears, and activating the rear manoeuvre for headland turning if the turn becomes too tight. The lateral sonars provide safety corrections if the robot drifts and gets too close to the canopy. Finally, the rear sonar is also for safety, halting the robot when it moves backwards. The network of ultrasonic rangefinders plays a crucial role in the headland turning algorithm (i.e. to determine turning the robot when reaches the end of a row), as this complex manoeuvre requires reliable and prompt corrections. Headland turning is a key requirement for farm vehicle automation. This complex manoeuvre usually requires as much attention as regular inside-row steering for a complete solution.

It is also essential to devise a systematic procedure to evaluate the performance of a navigation algorithm for orchard automation, ensuring vehicle safety and optimising performance. The comparison of steering commands with actual wheel angles (Fig. 7a) provides a measure of how the control system reacts in the field when following a succession of commands but cannot provide a complete assessment of how the vehicle has traversed a row in relation to the referencing canopies on both sides. Fortunately, perception sensors (esp. sonar sensors) may help provide such information. The perception system configured in Fig. 6 uses lateral sonar rangefinders to track the distance - left and right - of the vehicle in relation to row-enclosing canopies; the more equal both distances are, the more centred the robot is in the lane. Figure 7b plots the left and right vehicle-row distances for a trajectory along a vineyard row with a spacing of 2 m. Both distances can be combined in a unique index to assess guidance stability: the offset ratio ρ. This offset ratio can be defined as $\rho = |1 - R_S/L_S|$, where R_S is the right range and L_S is the left range. Stability increases as ρ approaches zero. Figure 8 illustrates how R_S and L_S were measured in a commercial vineyard for the monitoring robot VineScout (Rovira-Más et al., 2021b).

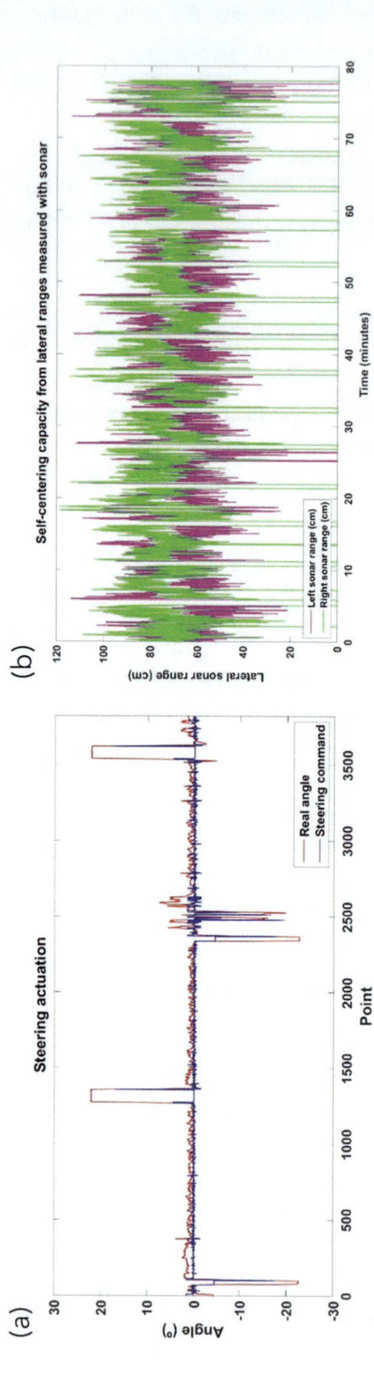

Figure 7 Evaluation of automated guidance: (a) steering actuation in an olive grove and (b) self-centring capacity of a robot estimated with the lateral ranges given by sonar sensors.

Figure 8 Evaluation of the self-centring ability of monitoring robot VineScout VS-3. Source: Rovira-Más et al., 2021b.

6 Using perception technologies in agricultural operations: harvesting, crop protection and other applications

6.1 Detecting fruit for harvesting

Harvesting robots must detect fruit, which may be partially or totally hidden by branches, leaves or other fruits in complex agricultural environments. The problem is exacerbated by uneven light distribution through gaps in the canopy, and the need to detect fruits with a colour similar to leaves. These robots must operate in real time as they move through crop lanes, including detecting potential obstacles affecting the fruit-picking behaviour of the robot. Computer vision, stereo vision, ToF cameras, LiDAR and spectral sensing are the most commonly employed technologies to detect fruits on trees. The 2D sensors capture colour images but do not provide information on the distance to target objects ahead, limiting their use in calculating the position of objects and guiding end-effectors. However, stereo cameras provide information to calculate the position of fruits and their distance from the robot. The 3D sensors can map the 3D geometry of trees and identify objects with specific shapes, such as fruits or branches (Gené-Mola et al., 2019).

When the goal is to detect and approach fruits similar in colour to leaves, the information based on geometry or the study of specific reflection patterns may be advantageous. Multispectral cameras can be convenient alternatives to differentiate fruits from other organs of the tree, such as leaves or branches that present different spectral reflections, but they do not provide direct information on the distance and position of the fruits either. Other technologies, such as hyperspectral cameras, help identify wavebands of interest. However, they are slower, more expensive and significantly more complex than standard colour cameras for outdoor use by robots that must capture images, process them, identify the fruits and harvest them in a very short time.

Depending on the imaging technology employed, different algorithms can be used to detect fruits in crop images. Both the technology used to capture the images and the algorithms set to analyse them depend on each specific problem. When the colour of fruits is very different from the leaves, as in the case of red or orange fruits, image segmentation algorithms based on adaptive thresholding can be applied to binarise the images. However, once the fruit has been detected, the problem of finding individual fruits in clusters or detecting those that are partially occluded or hidden remains. There are various methods to extract the contours of the fruits from binary images, such as Sobel, Roberts, Prewitt and Canny operators. Once the contours have been extracted, other algorithms, such as the Hough transform, can be applied to identify the circular shapes of the fruits (Russ, 2011) and to detect individual fruits despite bunches or occlusions.

Recently, AI algorithms like those based on R-CNN or YOLO have become popular in addressing fruit recognition problems. YOLO uses DL and CNN to detect objects in images and differs from other ANN-based algorithms in that it analyses each image only once, allowing the algorithm to work quickly but at the expense of accuracy (Jiang et al., 2022). This speed allows easy detection of objects in real time. A set of training images must be used to train the algorithm. Each image must be accompanied by a text file indicating the coordinates and the class to which the target objects belong. The algorithm first divides the image into a grid of size S × S to detect these particular target objects. In each of the windows, it finds n possible 'bounding boxes' (possible objects) and, for each of them, it calculates the level of certainty (or probability) of being a searched object. Most of them obtain a very low certainty level and are discarded. A non-max suppression step is applied to the remaining boxes, eliminating possible duplicate objects and obtaining the most exact match. In the case of harvesting robots, the training set is made up with images of the crop together with files that indicate the position of the fruits identified in each image. The algorithm must be trained with thousands of captured images, including the widest possible lighting situations, occlusions, diverse colours and the presence of leaves and branches, to achieve highly successful recognition rates. Practical implementations of YOLO algorithm have been used for the recognition of sweet peppers (Ning et al., 2022), green peppers (Li et al., 2021a), table grapes (Li et al., 2021b), mango (Zheng et al., 2021), apples (Wu et al., 2021) or kiwifruit (Suo et al., 2021), among others. To date, the latest version of the algorithm is YOLO v8.

6.2 Plant monitoring

Comprehensive knowledge of plant characteristics is essential for precision farming applications such as selective harvesting, precise spraying, fertilisation, smart weeding and effective crop management. However, the diverse nature of phenotypic traits makes the potential variety of imaging devices and analysis algorithms very high. A primary condition for selecting one technology over another is the plant structure and range of wavelengths required in the case of images. Computer vision systems are cost-effective and convenient for detecting structures of different colours and shapes, but most cameras do not obtain volume information. In contrast, advances in acquisition and computing technology have resulted in other optical sensors that allow high-speed 3D phenotyping, such as LiDAR, ToF cameras or photogrammetry techniques (Gené-Mola et al., 2021). The LiDARs produce 3D models of plants and thus their structure. The RGB cameras separate targeted objects by colour or through geometric analysis of the objects in the image. For example, distinguishing leaves from woody parts of a tree is essential for accurately estimating biomass. One of the key challenges of imaging technologies is the presence of shadow areas. This can be addressed by using many laser beams to track various points, creating an envelope of the object. Currently, multi-beam LiDARs that simultaneously emit 8, 11, 16, 32 or 128 beams with different angles are increasing in popularity. Structures are reconstructed with greater accuracy detail as the number of beams increases. However, the more information collected, the more computation is needed to process the data. The scouting robot in Fig. 9 features an 11-beam horizontal LiDAR to aid in orchard navigation, an IR radiometer to track canopy heat, and two compact spectral sensors to register plant properties through the Normalised Difference Vegetation Index (NDVI) and Photochemical Reflectance Index.

6.3 Detection of pests, diseases and stress

Plants close their stomata in situations of stress, e.g. due to lack of water, or when attacked by diseases or pests. This results in reduced evapotranspiration and an increased plant temperature. Stomata closure produces changes in photosynthetic activity which can be detected with sensors such as thermal or

Figure 9 Monitoring robot VineScout for speciality crops: (a) VS-2 in vineyards and (b) VS-3 in an olive grove. Source: Rovira-Más et al., 2021b.

hyperspectral cameras. Thermal cameras can be used to detect water stress non-invasively, as water stress is strongly related to an increase in the temperature of leaves. To minimise the influence of solar radiation, it is essential to take the measurements at the same time of day, for instance at solar noon. Qiu et al. (2021) fused thermal images with ToF cameras on a mobile platform to obtain 3D plant characteristics of maize, as a first step to studying the temperature distribution of the plant structure and the influence of height or leaf exposure. In this case, the challenge was combining two types of images captured with different cameras and resolutions to obtain consistent information about the same plant.

To better understand the effects of water stress on wine grape properties, Fernández-Novales et al. (2021) developed the VineScout robot that, among other sensing devices, incorporates an IR radiometer and an environmental probe capable of creating water status maps in vineyards. To generate hydric status models, the robot measures air temperature, relative humidity, atmospheric pressure and CO_2 concentration in real time, together with two spectral sensors that introduce illumination corrections to calculate the NDVI of the vegetation at high resolution.

However, although an increase in temperature is usually associated with water stress, an increase in canopy temperature can also occur when a plant is diseased or infested by a pest. As a result, it is sometimes necessary to combine thermal and spectral sensors to create high-resolution maps indicating the distribution and severity of the damage caused by pests or diseases. Rey et al. (2019) developed a small field robot (XF-ROVIM) for the early detection of *Xylella fastidiosa* at the leaf level in olive orchards (Fig. 10a). The robot was equipped with different proximal sensors and a GNSS receiver. It was able to capture thermal, spectral and structural information from a thermal camera, a multispectral camera and LiDAR. A similar robot (RobHortic) was developed by Cubero et al. (2020), including the same set of sensors to detect infections caused by *Candidatus liberibacter* solanacearum in horticultural crops (Fig. 10b). More recently, CropScout has been developed, using visible, thermal and spectral information to detect disease and estimate biomass in citrus, persimmon and avocado (Fig. 10c).

6.4 Detection of weeds

The automatic discrimination between weeds and crops using optical methods is usually achieved by analysing spectral reflectance differences or image analysis techniques, with the first robotic weeders using these technologies dating from the late 1990s. Since then, automatic weed detection, identification and removal technology has come a long way. In addition to detecting weeds, the problem of discriminating between weeds and crops must be efficiently

Figure 10 Robots for crop scouting and disease detection developed at IVIA: (a) XF-ROVIM, a robot for orchard monitoring (Rey et al., 2019), (b) RobHortic, a platform for horticultural crops monitoring (Cubero et al., 2020) and (c) CropScout, self-navigating robot for orchard monitoring.

solved. This may be relatively simple in the case of trees since any plant on the ground between the rows of trees can be considered a weed. Unfortunately, in the case of commodity or horticultural crops, the problem is more complex because the crop can be easily confused with weeds. Colour cameras are typically used to detect weeds. The plant is segmented from the soil through thresholding algorithms, and the weed is differentiated from the crop through geometric recognition algorithms or pattern recognition techniques that identify the specific shapes of the leaves for weeds and crops. In horticultural crops, imaging technology usually works well during the early growing season, before canopy closure, when the leaves of weeds and plants do not overlap and the density is relatively low.

An alternative to RGB cameras is the use of cameras with customised spectral bands, as reported by Machleb et al. (2021). These authors used a bispectral camera sensitive to red and NIR spectral bands to detect and remove weeds in sugar beet cultivation. The differential image calculated by comparing the NIR and red images pixel by pixel resulted in binary images distinguishing vegetation from the background. A database of images containing weeds and sugar beet plans helped to apply morphological operators to distinguish the weeds harmful to the crop from other species that can be beneficial, for example, for biological pest control. Eliminating weeds can be undertaken with different techniques such as mechanical cutting, using laser beams whose energy is strongly absorbed by the plant tissues leading to lethal damage, or high-voltage electrical discharges (Blasco et al., 2002). Currently, the two main methods for weeding are spraying and mechanical weeding.

6.5 Canopy detection for automated spraying of crop protection products

Technologies for the precise application of crop protection products have improved over the last two decades to achieve greater efficiency. Excessive application causes drift with resulting pollution of the neighbouring environment. As a result, it is necessary to spray only where it is needed and using the correct amount, thus reducing contamination and conserving product. This requires estimating canopy volume and leaf density to calculate the optimal dose for each tree. The first sensors used to detect tree vegetation were based on ultrasound. Chueca et al. (2008) developed a system to estimate the shape of vegetation using ultrasonic rangefinders to adjust for the most appropriate spraying pressure and droplet size. Sensors like LiDAR and ToF cameras are now used to map tree canopies with higher precision. Whilst unmanned aerial vehicles (UAV) have advantages such as speed of application or the possibility of spraying more inaccessible areas, their relatively low battery life and limited payload make ground robotics a more realistic alternative. Partel et al. (2021) mounted different sensors to map vegetation, calculate the volume and height of the trees and even count fruits in citrus orchards. Sensors included LiDAR, cameras and flow sensors, with information processed using a CNN. A hybrid approach was proposed by Ortí et al. (2022), where a UAV-generated prescription map based on canopy vigour can be uploaded to a smart blast sprayer capable of applying variable rates of crop protection products by controlling an arc of 16 pulse width modulated solenoid valves from a ruggedized field computer equipped with a GNSS receiver.

7 Conclusion and future trends

Considerable recent progress has been made in the ability of robots to perceive the environment, navigate autonomously through complex environments such as agricultural fields, and obtain the information needed to carry out particular tasks. However, many challenges still remain in the widespread implementation of robotics in agriculture, some related to perception systems and real-time analysis of the resulting data. Among these, the following are worth mentioning:

Challenges in outdoor autonomous navigation: Safety remains a key issue. Large manufacturers have been reluctant to release robotic solutions to the global market to avoid jeopardising their reputation, despite multiple prototypes being successfully demonstrated over the last decade. However, commercialisation is now a realistic prospect. The reliability of sensors and actuators is continuously improving, as are ways of developing intelligent machines able to operate autonomously in dealing safely with complex, changing and unpredictable outdoor environments.

Fleets of robots and co-operative work: Sometimes, a single robot cannot do all the necessary work, either due to the size of the farm or because operations are complex. Co-operation between ground and aerial robots is still challenging, requiring accurate communication and synchronisation between all vehicles involved in operations. A drone can see more widely and can guide ground robots to specific targets, e.g. by warning about obstacles or detecting areas of interest, such as patches of weeds. In other cases, cooperation is necessary to complete complex tasks. The challenge lies in achieving real-time communication and accurately registering the position of each element of the fleet. Successful business cases have been reported in the last 5 years by GUSS in California and Jacto in Brazil for orchard spraying, and Vitirover in France for vineyard weeding.

Communication and cloud processing: Atmospheric events can reduce global positioning accuracy or communication coverage. Large areas with poor coverage also remain. Some technologies, such as LoRaWAN, or operators such as Sigfox, can be used to alleviate this problem by transmitting data in areas with little or no coverage. Although the transmission rate is currently limited, they can be used for positioning in case of an emergency, trajectory replanning or essential communications, but they cannot be used to send data to be processed in the cloud for feedback in real time to execute agricultural operations on-site.

AI and agriculture 4.0: Precision agriculture involves using data to detect crop variability and perform precise operations. Agriculture 4.0 goes further and is related to digital technologies and intelligent systems. Concepts such as Agriculture 4.0, the Internet of Things or big data are expected to be widely implemented in the coming years. These technologies make it possible to use data (including data from sensors on robots) to optimise both overall farm operations and, indeed, the entire production chain from the field to the table.

Trends and new market solutions: The difficult challenges posed by the reliability and safety of autonomous machines have led to intermediate solutions involving robotics and automation in regular farming operations. Three approaches, in particular, are attractive to technology-based entrepreneurs: (1) reduction of the size, weight and power of farm equipment, which allows the dual advantage of reducing the risk of causing damage and introducing electrical power as a clean energy source; (2) specialisation of each machine to focus on particular groups of tasks, such as spraying or harvesting, which lowers complexity in the decision-making algorithms; and (3) the advent of new business models, such as service providers of robot fleets, which releases farmers from the duty of maintaining complex machines full of unfamiliar electronics and software. The combined effect of these three strategies is already bearing fruit, and the appearance of robotic solutions in trade shows and equipment exhibitions is growing daily.

8 Where to look for further information

The following references provide a good overview of the topics addressed in this chapter:

- Zhang, Q. (2023). *Encyclopedia of Digital Agricultural Technologies*. Springer Nature Switzerland AG, Cham, Switzerland. ISBN: 978-3-031-24860-3.
- Rovira-Más, F., Zhang, Q. and Hansen A. C. (2011). *Mechatronics and Intelligent Systems for Off-road Vehicles*. Springer-Verlag, London Limited 2011. ISBN: 978-1-84996-467-8.
- Manoj Karkee, M. and Zhang, Q. (2021). *Fundamentals of Agricultural and Field Robotics*. Springer International Publishing, Cham, Switzerland. ISBN: 978-3-030-70399-8.
- Blasco, J., González González, M. G., Chueca, P., Cubero, S. and Aleixos, N. (2019). Advances in automated in-field grading of harvested crops. Chapter 10. In: Billingsley, J. (Ed). *Robotics and Automation for Improving Agriculture*, pp. 215–232. Burleigh Dodds Science Publishing, Sawston, Cambridge. ISBN: 978-1786762726.
- Fountas, S., Mylonas, N., Malounas, I., Rodias, E., Hellmann Santos, C. and Pekkeriet, E. (2020). Agricultural robotics for field operations. *Sensors (Basel)* 20(9), 2672. DOI: 10.3390/s20092672.
- Fountas, S., Malounas, I., Athanasakos, L., Avgoustakis, I. and Espejo-Garcia, B. (2022). AI-assisted vision for agricultural robots. *AgriEngineering* 4, 674–694. DOI: 10.3390/ agriengineering4030043.
- Botta, A., Cavallone, P., Baglieri, L., Colucci, G., Tagliavini, L. and Quaglia, G. (2022). A review of robots, perception, and tasks in precision agriculture. *Applied Mechanics* 3, 830–854. DOI: 10.3390/applmech3030049.
- Magalhães, S. A., Moreira, A. P., Santos, F. N. D. and Dias, J. (2022). Active perception fruit harvesting robots – a systematic review. *Journal of Intelligent & Robotic Systems* 105, 14. https://doi-org.ezproxy.library.wur.nl/10.1007/s10846-022-01595-3.
- Xie, D., Chen, L., Liu, L., Chen, L. and Wang, H. (2022). Actuators and sensors for application in agricultural robots: A review. *Machines* 10(10), art. no. 913. DOI: 10.3390/machines10100913.

Sources of information include the following:

- GOFAR: Global Organization for Agricultural Robotics. https://www.agricultural-robotics.com/.
- FIRA: International Forum of Agricultural Robotics. https://www.fira-agtech.com/c/fira.
- AgriExpo, Farm Robots. https://www.agriexpo.online.
- Future Farming. https://www.futurefarming.com/.
- ATLAS Webpage. https://www.atlas-h2020.eu/.

Centres of expertise include:

- UK Agri-Tech Centres. https://www.agritechcentres.com/.
- Lincoln Agri-Robotics (LAR). https://lar.lincoln.ac.uk/.

- Australian Centre of Field Robotics (ACFR). https://www.sydney.edu.au/.
- Intelligent Robotics LAB. https://smartrobabcbgu.wixsite.com/iemirl.
- AgroFood Robotics. https://www.wur.nl/en/research-results/projects-and-programmes/agro-food-robotics.htm.

Laboratories in which the authors are involved are as follows:

- Centro de Agroingeniería, Instituto Valenciano de Investigaciones Agrarias (IVIA), Spain.
- Agricultural Robotics Laboratory (ARL), Universitat Politècnica de València, Spain.

9 Acknowledgements

This work is co-funded through projects GVA-IVIA 52204 and the European Union through the European Regional Development Fund of the Generalitat Valenciana 2014-2020, MICIN - AEI (TED2021-130117B-C31) Smart autonomous electrical robot for a digital and sustainable agriculture in the Valencian Community (AgriSmartRobot), and the European Union's Horizon 2020 research and innovation program with grant agreement number 737669.

10 References

Bai, Y., Zhang, B., Xu, N., Zhou, J., Shi, J. and Diao, Z. (2023). Vision-based navigation and guidance for agricultural autonomous vehicles and robots: a review. *Computers and Electronics in Agriculture* 205, art. no. 107584. DOI: 10.1016/j.compag.2022.107584.

Barbedo, J. G. A. (2023). A review on the combination of deep learning techniques with proximal hyperspectral images in agriculture. *Computers and Electronics in Agriculture* 210, art. no. 107920. DOI: 10.1016/j.compag.2023.107920.

Blasco, J. and Aleixos, N. (2023). Computer Vision in Agriculture In: Zhang, Q. (eds). *Encyclopedia of Smart Agriculture Technologies* 165–176. Springer Nature Switzerland AG. Cham, Switzerland. DOI: 10.1007/978-3-030-89123-7_45-1.

Blasco, J., Aleixos, N., Roger, J. M., Rabatel, G. and Moltó, E. (2002). Robotic weed control using machine vision. *Biosystems Engineering* 83(2), 149. DOI: 10.1016/S1537-5110(02)00155-1.

Burks, T. F. and Schmoldt, D. L. (2008). U. S. specialty crops at a crossroad. *Hi-Tech or Else? Resource: September*. ASABE, St. Joseph, MI.

Chueca, P., Garcera, C., Molto, E. and Gutierrez, A. (2008). Development of a sensor-controlled sprayer for applying low-volume bait treatments. *Crop Protection* 27(10), 1373. DOI: 10.1016/j.cropro.2008.05.004.

Cubero, S., Lee, W. S., Aleixos, N., Albert, F. and Blasco, J. (2016). Automated systems based on machine vision for inspecting citrus fruits from the field to postharvest: a review. *Food and Bioprocess Technology* 9(10), 1623. DOI: 10.1007/s11947-016-1767-1.

Cubero, S., Marco-Noales, E., Aleixos, N., Barbé, S. and Blasco, J. (2020). RobHortic: a field robot to detect pests and diseases in horticultural crops by proximal sensing. *Agriculture*. Special Issue "Enhancing Surveillance and Detection of Invasive Harmful Plant Pathogens and Pests" 10(7), 276. DOI: 10.3390/agriculture10070276.

Eurostat (2018). Agriculture, forestry, and fishery statistics, 2108 edition. *European Union 2018*. https://ec.europa.eu/eurostat/product?code=ks-fk-18-001.

FAO (2022). *World Food and Agriculture – Statistical Yearbook 2022*. Rome. DOI: 10.4060/cc2211en.

Fernández-Novales, J., Saiz-Rubio, V., Barrio, I., Rovira-Más, F., Cuenca-Cuenca, A., Santos Alves, F., Valente, J., Tardaguila, J. and Diago, M. P. (2021). Monitoring and mapping vineyard water status using non-invasive technologies by a ground robot. *Remote Sensing*. Diago, M.P. 13(14), 2830. DOI: 10.3390/rs13142830.

Foley, J. (2014). A five-step plan to feed the world. *National Geographic* May.

Fountas, S., Malounas, I., Athanasakos, L., Avgoustakis, I. and Espejo-Garcia, B. (2022). AI-assisted vision for agricultural robots. *AgriEngineering* 4(3), 674. DOI: 10.3390/agriengineering4030043.

Gené-Mola, J., Gregorio, E., Guevara, J., Auat, F., Sanz-Cortiella, R., Escolà, A., Llorens, J., Morros, J. R., Ruiz-Hidalgo, J., Vilaplana, V. and Rosell-Polo, J. R. (2019). Fruit detection in an apple orchard using a mobile terrestrial laser scanner. *Biosystems Engineering* 187, 171. DOI: 10.1016/j.biosystemseng.2019.08.017.

Gené-Mola, J., Sanz-Cortiella, R., Rosell-Polo, J. R., Escolà, A. and Gregorio, E. (2021). In-field apple size estimation using photogrammetry-derived 3D point clouds: comparison of 4 different methods considering fruit occlusions. *Computers and Electronics in Agriculture* 188, 106343. DOI: 10.1016/j.compag.2021.106343.

Gil, G., Casagrande, D. E., Pérez Cortés, L. P. and Verschae, R. (2023). Why the low adoption of robotics in the farms? Challenges for the establishment of commercial agricultural robots. *Smart Agricultural Technology* 3, 100069. DOI: 10.1016/j.atech.2022.100069.

Giller, K. E., Delaune, T., Silva, J. V., Descheemaeker, K., van de Ven, G., Schut, A. G. T., van Wijk, M., Hammond, J., Hochman, Z., Taulya, G., Chikowo, R., Narayanan, S., Kishore, A., Bresciani, F., Teixeira, H. M., Andersson, J. A. and van Ittersum, M. K. (2021). The future of farming: who will produce our food? *Food Security* 13(5), 1073. DOI: 10.1007/s12571-021-01184-6.

Gimenez, J., Sansoni, S., Tosetti, S., Capraro, F. and Carelli, R. (2022). Trunk detection in tree crops using RGB-D images for structure-based ICM-SLAM. *Computers and Electronics in Agriculture* 199, art. 107099. DOI: 10.1016/j.compag.2022.107099.

Jiang, P., Ergu, D., Liu, F., Cai, Y. and Ma, B. (2022). A review of Yolo algorithm developments. *Procedia Computer Science* 199, 1066. DOI: 10.1016/j.procs.2022.01.135.

Jin, Y., Liu, J., Xu, Z., Yuan, S., Li, P. and Wang, J. (2021). Development status and trend of agricultural robot technology. *International Journal of Agricultural and Biological Engineering* 14(3), 1. DOI: 10.25165/j.ijabe.20211404.6821.

Lee, W. S. and Blasco, J. (2021). Sensors I: Color imaging and basics of image processing. In: Karkee, M. and Zhang, Q. (Eds). *Fundamentals of Agricultural and Field Robotics*, pp. 13-37. Springer International Publishing, Cham, Switzerland. ISBN: 978-3-030-70399-8.

Li, H., Li, C., Li, G. and Chen, L. (2021a). A real-time table grape detection method based on improved YOLOv4-tiny network in complex background. *Biosystems Engineering* 212, 347. DOI: 10.1016/j.biosystemseng.2021.11.011.

Li, X., Pan, J., Xie, F., Zeng, J., Li, Q., Huang, X., Liu, D. and Wang, X. (2021b). Fast and accurate green pepper detection in complex backgrounds via an improved

YOLOv4-tiny model. *Computers and Electronics in Agriculture* 191, 106503. DOI: 10.1016/j.compag.2021.106503.

Li, Y., Li, M., Qi, J., Zhou, D., Zou, Z. and Liu, K. (2021c). Detection of typical obstacles in orchards based on deep convolutional neural network. *Computers and Electronics in Agriculture* 181, 105932.

Lorente, D., Aleixos, N., Gómez-Sanchis, J., Cubero, S., García-Navarrete, O. L. and Blasco, J. (2012). Recent advances and applications of hyperspectral imaging for fruit and vegetable quality assessment. *Food and Bioprocess Technology* 5(4), 1121. DOI: 10.1007/s11947-011-0725-1.

Machleb, J., Peteinatos, G. G., Sökefeld, M. and Gerhards, R. (2021). Sensor-Based Intrarow Mechanical Weed Control in Sugar Beets with Motorized Finger Weeders. *Agronomy* 11, 1517. https://doi.org/10.3390/agronomy11081517.

Magalhães, S. A., Moreira, A. P., Santos, F. N. d. and Dias, J. (2022). Active perception fruit harvesting robots: a systematic review. *Journal of Intelligent and Robotic Systems* 105(1), 14. Available at: https://doi-org.ezproxy.library.wur.nl/10.1007/s10846-022-01595-3.

Marinoudi, V., Sørensen, C. G., Pearson, S. and Bochtis, D. (2019). Robotics and labour in agriculture. A context consideration. *Biosystems Engineering* 184, 111. DOI: 10.1016/j.biosystemseng.2019.06.013.

Moreno, H., Rueda-Ayala, V., Ribeiro, A. and Bengochea-Guevara, J. M., Lopez, J. M., Peteinatos, G., Valero, C., Andújar, D. (2020a). Evaluation of vineyard cropping systems using onboard RGB-depth perception. *Sensors*, 20, 6912. DOI: 10.3390/s20236912.

Moreno, H., Valero, C., Bengochea-Guevara, J. M., Ribeiro, Á., Garrido-Izard, M. and Andújar, D. (2020b). On-ground vineyard reconstruction using a LiDAR-based automated system. *Sensors* 20(4), 1102. DOI: 10.3390/s20041102.

Ning, Z., Luo, L., Ding, X. M., Dong, Z., Yang, B., Cai, J., Chen, W. and Lu, Q. (2022). Recognition of sweet peppers and planning the robotic picking sequence in high-density orchards. *Computers and Electronics in Agriculture* 196, 106878. DOI: 10.1016/j.compag.2022.106878.

Ortí, E., Cuenca, A., Pérez, M., Torregrosa, A., Ortiz, C. and Rovira-Más, F. (2022). Preliminary evaluation of a blast sprayer controlled by pulse width modulated nozzles. *Sensors* 22(13), 4924. DOI: 10.3390/s22134924.

Partel, V., Costa, L. and Ampatzidis, Y. (2021). Smart tree crop sprayer utilising sensor fusion and artificial intelligence. *Computers and Electronics in Agriculture* 191, 106556.

Qiu, R., Miao, Y., Zhang, M. and Li, H. (2021). Detection of the 3D temperature characteristics of maise under water stress using thermal and RGB-D cameras. *Computers and Electronics in Agriculture* 191, 106551. DOI: 10.1016/j.compag.2021.106551.

Rey, B., Aleixos, N., Cubero, S. and Blasco, J. (2019). XF-ROVIM. A field robot to detect olive trees infected by Xylella fastidiosa using proximal sensing. *Remote Sensing*. Special Issue Advances in Remote Sensing Applications for the Detection of Biological Invasions 11(3), 221. DOI: 10.3390/rs11030221.

Rovira-Más, F., Millot, C. and Saiz-Rubio, V. (2015). *Navigation Strategies for a Vineyard Robot*. 2015 ASABE Annual International Meeting, *ASABE Paper n° 152189750*.DOI: 10.13031/aim.20152189750

Rovira-Más, F., Saiz-Rubio, V. and Cuenca-Cuenca, A. (2021a). Augmented perception for agricultural robots navigation. *IEEE Sensors Journal* 21(10), 11712.

Rovira-Más, F., Saiz-Rubio, V. and Cuenca-Cuenca, A. (2021b). Sensing architecture for terrestrial crop monitoring: harvesting data as an asset. *Sensors* 21(9), 3114. DOI: 10.3390/s21093114.

Ruangurai, P., Dailey, M. N., Ekpanyapong, M. and Soni, P. (2022). Optimal vision-based guidance row locating for autonomous agricultural machines. *Precision Agriculture* 23(4), 1205. DOI: 10.1007/s11119-022-09881-8.

Russ, J. C. (2011). *The Image Processing Handbook* (6th edn.). ISBN: 978-1439840450. CRC Press, Boca Raton.

Saleem, M. H., Potgieter, J. and Arif, K. M. (2021). Automation in agriculture by machine and deep learning techniques: a review of recent developments. *Precision Agriculture* 22(6), 2053. DOI: 10.1007/s11119-021-09806-x).

Sekharamantry, P. K., Melgani, F. and Malacarne, J. (2023). Deep learning-based apple detection with attention module and improved loss function in YOLO. *Remote Sensing* 15(6), 1516. DOI: 10.3390/rs15061516.

Suo, R., Gao, F., Zhou, Z., Fu, L., Song, Z., Dhupia, J., Li, R. and Cui, Y. (2021). Improved multi-classes kiwifruit detection in orchard to avoid collisions during robotic picking. *Computers and Electronics in Agriculture* 182, 106052. DOI: 10.1016/j.compag.2021.106052.

USDA (2017). Census of agriculture. Last Visit 3 March 2023. Available at: https://www.nass.usda.gov/Publications/Highlights/2019/2017Census_Farm_Producers.pdf.

Vázquez-Arellano, M., Griepentrog, H. W., Reiser, D. and Paraforos, D. S. (2016). 3-D imaging systems for agricultural applications: a review. *Sensors* 16(5), 618. DOI: 10.3390/s16050618.

Verified Market Intelligence (VMI) (2018). *Global Agricultural Robots: Market Size, Status and Forecast to 2025*. Boonton.

Walsh, K. B., Blasco, J., Zude-Sasse, M. and Sun, X. (2020). Visible-NIR 'point' spectroscopy in postharvest fruit and vegetable assessment: the science behind three decades of commercial use. *Postharvest Biology and Technology* 168, 111246. DOI: 10.1016/j.postharvbio.2020.111246.

Wang, T., Chen, B., Zhang, Z., Li, H. and Zhang, M. (2022). Applications of machine vision in agricultural robot navigation: a review. *Computers and Electronics in Agriculture* 198, art. no. 107085. DOI: 10.1016/j.compag.2022.107085.

Wei, J., Rovira-Más, F., Reid, J. F. and Han, S. (2005). Obstacle detection using stereo vision to enhance safety of autonomous machines. *Transactions of the ASAE* 48(6), 2389.

Wu, L., Ma, J., Zhao, Y. and Liu, H. (2021). Apple detection in complex scene using the improved YOLOv4 model. *Agronomy* 11(3), 476. DOI: 10.3390/agronomy11030476.

Xie, D., Chen, L., Liu, L., Chen, L. and Wang, H. (2022). Actuators and sensors for application in agricultural robots: a review. *Machines* 10(10), art. no. 913. DOI: 10.3390/machines10100913.

Zheng, C., Chen, P., Pang, J., Yang, X., Chen, C., Tu, S. and Xue, Y. (2021). A mango picking vision algorithm on instance segmentation and key point detection from RGB images in an open orchard. *Biosystems Engineering* 206, 32. DOI: 10.1016/j.biosystemseng.2021.03.012.

Zhou, Z., Majeed, Y., Diverres Naranjo, G. and Gambacorta, E. M. T. (2021). Assessment for crop water stress with infrared thermal imagery in precision agriculture: a review and future prospects for deep learning applications. *Computers and Electronics in Agriculture* 182, art. no. 106019. DOI: 10.1016/j.compag.2021.106019.

Chapter 4

Advances in machine learning for agricultural robots

Polina Kurtser, Örebro University and Umeå University, Sweden; Stephanie Lowry, Örebro University, Sweden; and Ola Ringdahl, Umeå University, Sweden

1 Introduction
2 Applications of machine learning in agri-robotics
3 Challenges
4 Integration and field-testing use-cases
5 Conclusion
6 Where to look for further information
7 References

1 Introduction

Agricultural robots operate in and interact with complex and highly variable physical environments. Within these complex environments, agri-robots must navigate and localize themselves, recognize and detect crops, and perform complex tasks that require interacting with and manipulating the environment, such as weeding or harvesting. Furthermore, agricultural robots must operate in a wide range of different types of environments: fields, greenhouses, and orchards. There can be significant changes in operational conditions between seasons, time of the day, varieties of the same crop, growing conditions, and locations of the targets.

The complexity and variability of the agricultural environments in which agri-robots must operate mean that it is difficult to develop analytical representations and hand-tailored feature extraction methods to the level that is required for robust generalized robotic operation. Instead, agricultural robots increasingly rely on machine learning (ML): i.e. algorithms capable of learning a model from previous experiences. In recent years, ML algorithms have outperformed all other methods on a large number of perception and planning tasks for robots and have set a new standard in terms of performance in many domains [1]. It is therefore unsurprising that ML algorithms are included in all aspects of robust autonomous or semi- autonomous agricultural operations. These include examples like perception operations [2–4] for the

http://dx.doi.org/10.19103/AS.2023.0124.04

detection of crops, weeds, and diseases and evaluating ripeness and plant stress levels.

Navigation through the fields and vineyards relies on ML [5] for operations like path planning and obstacle avoidance. Machine learning is also used in motion planning for accurate interaction with the crop [6–8] while harvesting, pruning, and spraying.

This chapter presents an overview of the usage of ML for agri-robotics, with a focus on the use of ML for data analysis and decision-making for perception and navigation. The chapter is structured as follows. First, we briefly define some relevant terms in ML and artificial intelligence (AI) (Section 1.1), summarize the robotic sensors available for use in ML (Section 1.2), and introduce the grand challenges for AI and ML in agri-robotics (Section 1.3). Section 2 discusses the applications of ML in agri-robotics, looking at both robot perception (Section 2.1) and robot planning (Section 2.2). Section 3 presents some of the major challenges for the widespread deployment of ML in agricultural operations: the environmental complexity (Section 3.1), the need for widespread availability of appropriate training data (Section 3.2), and the challenges of real-time operation in agri-robotic interactions (Section 3.3). It also presents the ongoing work to address these challenges, through the production of open-source annotated datasets (Section 3.2.1) and the adoption of new techniques for coping with minimal data (Section 3.2.2). Section 4 reviews the current state-of-the-art in fully integrated agri-robotic systems and the specific ML algorithms that these systems use. The chapter closes with a discussion of the rapid advances in ML in the last decade and the ongoing open research challenges (Section 5).

1.1 Definition of artificial intelligence, machine learning and deep learning

Machine learning provides computer systems with the ability to automatically improve their performance through experience. Complex physical environments are often too variable and complex to be fully defined analytically, and thus empirical and statistical approaches, heavily based on algorithms capable to learn a model from previous experiences, are at the cornerstone of robust and intelligent robotic operation.

Machine learning can be divided into supervised learning where an agent observes some example input–output pairs and learns a function that maps from input to output, unsupervised learning where the agent learns patterns in the input even though no explicit feedback is supplied, and reinforcement learning (RL) where the agent learns from a series of reinforcements – rewards or punishments [9].

Currently, most of the focus of supervised ML is on end-to-end learning approaches, such as deep learning (DL). Previously, the ML methods in agricultural operations relied heavily on hand-crafted features [2], followed by classical ML techniques like decision trees, support vector machines (SVMs), or nearest neighbor search. Deep learning is a class of ML algorithms that uses artificial neural networks with multiple layers to progressively extract higher-level features from the raw input [10]. The complexity of the DL architecture allows representations of data with multiple levels of abstraction [1]. For example, in image processing, lower layers may identify edges, while higher layers may identify the concepts relevant to a human such as digits, letters, or faces. Therefore, DL is extremely effective for learning complex sensor data such as color images from cameras or dense three-dimensional (3D) sensors. For this reason, DL has had a dramatic impact on the capabilities of many perception algorithms, including object recognition and detection, which make it a natural choice for perception problems in the complex environments agri-robots must operate.

Reinforcement learning is a field of ML where the agent, often a robot, learns how to map situations to actions through a series of trial-and-error agent-environment interactions, to maximize a reward signal [11]. The application of RL to robotics constitutes complex interactions between a mechanical system and its environment. Due to the trial-and-error nature of the interactions, the overall modeling of the environment required can be much less meticulous compared to alternative supervised learning algorithms, as it is directly learned from the robot–environment interactions. As a result, the implementation of RL in robotics applications is gaining popularity as robots are expected to engage with highly complex, variable, and poorly modeled environments in an intelligent way [12].

Machine learning is itself a subfield of artificial intelligence (AI). AI is a very broad field with many different domains such as intelligent robotics, planning, perception, natural language processing, and ML, to mention a few. Figure 1 illustrates the relation between AI, ML, and DL.

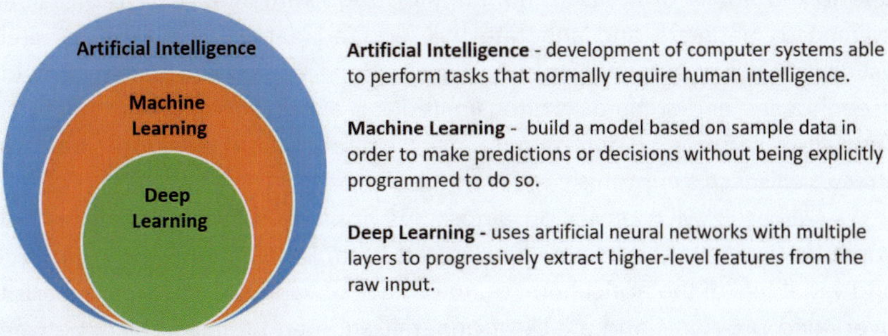

Figure 1 How artificial intelligence, machine learning, and deep learning are related.

1.2 Sensors for machine learning in agri-robotics

Machine learning algorithms depend on data from the external environment as well as the internal state of the robot to learn and develop a model. The algorithms are highly dependent on the nature of the available sensor data; therefore, a critical part of the ML design is the sensors providing the data. Specifically, in agri-robotics the sensors providing most of the rich data required for model training are visual sensors such as near-infrared (NIR) or infrared (IR), X-ray, fluorescence, acoustic, two dimensional (2D), and 3D vision as well as positioning sensors such as laser, radar, sonar, or stereo cameras. The data are used for perceiving the environment for safe and accurate operation.

But these sensors are not the only ones robots are equipped with to ensure safe and robust field operations. Typically [13], an agricultural robot is also equipped with motion sensors (e.g. odometry and inertial sensors), positioning sensors (e.g. GPS), and a range of mechanical sensors (e.g. tactile contact forces and torques sensors).

All these sensors accumulate a large amount of data in regular robotic operations, mostly for perception and navigation. Analyzing this data is the key to intelligent autonomous or semi-autonomous operations.

1.3 Introduction to challenges in employing artificial intelligence

Despite the grand success and penetration of AI and ML into various applications, including agri-robotics, the field faces several major challenges. Some challenges are very general and apply to most application fields of ML, while some are more specific and unique to the agri-robotics field.

First, the accuracy of ML algorithms relies heavily on the quality of the data used to train ML models. As discussed further in this chapter, some fields of ML research developed rapidly thanks to large open-access datasets collected through major crowd-sourcing efforts. These efforts were made possible largely due to the ability of a vast community of users with almost no professional training to generate and annotate the required data. This is, unfortunately, usually not the case for agri-robotics applications. Access to agricultural fields, greenhouses, and orchards is often limited and annotation efforts often require specialists capable of determining plants' phenotypical traits, maturity, and stress or disease symptoms.

As a result, ML perception algorithms applied in the agri-robotics field continue to be limited by smaller datasets, unable to capture the complexity and variability of the agricultural domain. Lack of variable data in turn leads to overfitting problems and a lack of generalization and transferability between similar applications. We discuss examples and implications of these challenges in detail in Section 3.1.

Additionally, most ML models, particularly DL models, suffer from a lack of explainability (i.e. behaving like a 'black box'). This trait means it is hard to interpret what the model learned or which features were most important for the result. As a result, determining if the model has an overfitting issue is often challenging and will generalize poorly in a new setting. Moreover, the explainability of the algorithms is also instrumental in the adoption of the technology. Lack of insight and understanding by the users is often considered to be a key bottleneck in technology adoption. Explainable AI is a research field within AI that provides methods for explaining the learned models such that humans can understand them [14]. This helps to determine what features are important for the learning and to see if the result is transferable, i.e. if it will work well in another similar application.

2 Applications of machine learning in agri-robotics

Many ML algorithms are used in precision agriculture [15, 16], and ML plays a particularly important role in agri-robotics. A robot needs to perceive and interpret the environment around it, as well as navigate and interact within this environment [17], and ML algorithms contribute to both these aspects of robotics: robot perception and robot navigation and manipulation. For robot perception, ML provides techniques to interpret incoming visual sensor data. For robot navigation and manipulation, ML is used to develop planning algorithms to enable efficient and collision-free motions.

This section presents these two major application areas and identifies the ongoing challenges in the deployment of ML algorithms for use with robotics in the agricultural domain. It also presents some of the ongoing research directions in ML for agri-robotics, which at the present time has a particular focus on DL. The development of DL has been one of the most significant advances in ML in the last decade [1]. Deep learning systems use neural network architectures with many layers and have had a dramatic impact on the performance of ML systems. Deep learning architectures are used for both perception tasks (with both supervised and unsupervised learning) and for planning tasks (with RL) and are responsible for the remarkable success of ML algorithms in many application domains in recent years.

2.1 Machine learning for perception

Perception is an essential task for any agri-robot, particularly an autonomous one: the robot must be able to see the environment and use that incoming sensor data to learn about the world. This enables the robot to (1) perform its designated tasks, (2) evaluate the operational result, and (3) operate safely with regard to detected obstacles in the environment. Agricultural tasks that depend on perception include detecting objects such as fruit [18, 19], weeds [20], or visible signs of diseases [21]. A robot that can detect and label relevant agricultural elements in the environment can then perform other quantitative tasks such as yield estimation [22, 23] and canopy cover measurements [24, 25]. Perception is also a precursor to many actions and interventions that agri-robots perform, such as weed spraying or harvesting: the robot must first detect the weed or the fruit for harvest before taking any further action.

Most ML-based perception algorithms are based on supervised learning. In the supervised learning pipeline, the system is first trained using labeled sensor data. The ML system learns a model that, when provided with previously unseen sensor data, can then correctly label the unseen data. Supervised learning systems have demonstrated great effectiveness at performing recognition, detection, and segmentation tasks, especially in recent years with the development of deep neural network architectures that can learn very sophisticated models [1].

Supervised perception algorithms can in theory be performed using any sensor that receives data from the environment. Most commonly it is applied to visual data, such as images received from RGB cameras [26]. However, more recently 3D sensors such as RGB-D cameras and 3D LiDARs have been used for learning and perception. This section will discuss these two sensor modalities and the role of supervised learning in developing perception algorithms for agri-robot applications.

2.1.1 Learning using two-dimensional perception data

Visual sensing is central to the automation of agricultural tasks. Integrated autonomous in-field robots use color and 3D range data in operations such as harvesting [6, 27], weeding [28, 29], and pruning [130]. Using these sensors, a set of phenotypic traits are often collected such as objects size [23], count [31], and color [32] measures. What differentiates agri-robotics from autonomous high-throughput phenotyping platforms [33], is the translation of the phenotypic measures collected into decisions and actions in real time. For example, ML algorithms that rely on the collection of color measures of sweet peppers for the estimation of their maturity levels [34] are used by fully integrated autonomous harvesting robots [27] to avoid harvesting unripe peppers. Weed detection algorithms that rely on ML [35] are translated into object 3D localization and size estimation algorithms for successful mechanical weeding operations [28, 29, 36]. In-field volume estimation of grape vines, often used for vineyard yield estimation [23, 24], is translated into precision selective pesticide spraying operations [37].

Computer vision has been revolutionized by DL models. Perception for agri-robotics is no exception with a strong dominance of DL models integrated into the perception system to perform detection and segmentation of crops, weeds, and pests [3]. Over the last decade, the available DL architectures have evolved into lighter, less computationally complex, and more efficient architectures. Well-known architectures are AlexNet [38], CaffeNet [39], VGG [40], GoogleNet [41], and ResNet[42]. More recently, MobileNet [43] and EfficientNet [44] have continued the trend of powerful image recognition techniques with increasing computational efficiency.

These architectures focus on image classification: identifying and labeling elements in an image. For many domains, including agri-robotics, it is useful to also localize elements within the image using a bounding box or segmentation mask: these techniques are known as object detection and semantic segmentation. Well-known architectures include Faster-RCNN [45], Mask-RCNN [46], and YOLO [47–49].

An important aspect of DL is that the architectures can be pre-trained. Since training complex DL models can be extremely demanding in terms of available data and computational requirements, the availability of open-source pre-trained models makes these techniques widely accessible. These architectures are pre-trained on large image datasets such as ImageNet [50], a dataset consisting of more than 14 million images. Once the architecture is pre-trained, it only requires adjustment of the weights to the specific detection problem at hand, using additional more relevant data. This technique is known as transfer learning.

Deep learning networks are widely employed in the agri-robotics domain. Examples include YOLO architecture being employed for pest detection [51–54] and fruit detection and localization [55, 56]. Similarly, VGG models are employed for similar tasks [57–59], and many other examples can be found [16]. However, despite the popularity of DL, other less data-demanding algorithms are still widely used in the agricultural domain. Even when transfer learning is applied, DL models still need sufficient data for fine-tuning. Therefore, classical ML algorithms such as SVMs [60] are still commonly used supervised learning classification methods in the agri-robotic domain. For example, SVMs are used for weed detection [61], fruit classification [62], and multiple perception steps in a grapevine pruning robot [63].

All of these algorithms are designed for and applied to 2D visual data: i.e. data from RGB cameras. However, recently there has been an increase in the availability of 3D sensors for a wider range of agricultural perception tasks. The wider availability of these sensors, combined with increased computational power for ML tasks, and more sophisticated ML algorithms, makes the use of ML for 3D data an exciting new advancement in agri-robotics. This topic is discussed in the following section.

2.1.2 Learning using three-dimensional perception data

The application of advanced ML techniques, particularly DL, to 3D sensor data is a relatively recent development. While agri-robots have frequently utilized 3D sensors (such as LiDARs, radars, and stereo cameras) in field conditions for many years, the 3D data were almost solely applied to navigation and obstacle avoidance. However, there have been significant advances in available 3D sensors: both laser technology and commercial-grade NIR stereo cameras have become cheaper and more robust to outdoor lighting conditions [64]. This technological development has great potential for agricultural applications, as 3D data contains valuable information about size and depth that a 2D image cannot provide. This additional geometric information is critical for size estimation of fruit [65] and other crops [66]. The use of 3D data is also vital for robots that perform manipulation tasks, such as harvesting [6].

The benefits of 3D data are not only restricted to ML scenarios. For example, agri-robots that are only equipped with 2D RGB cameras must currently perform target localization using time-consuming operations such as visual servoing [67], which could be replaced by more efficient algorithms that exploit the geometric information intrinsic in the 3D data. Similarly, algorithms for the collection of phenotypes – including physical size estimation such as stem length or leaf area – that were originally developed for indoor controlled conditions can now be integrated into field operations [68].

While many of these systems depend on ML algorithms, many of them only use the RGB part of the data and not the depth information, thereby reverting the sensor data to that captured by a 2D camera [69]. This approach means the added value of the 3D information is lost but allows the use of widely available 2D ML detection algorithms such as YOLO [47] and Faster R-CNN [45]. Other systems do use the depth information combined with a lightweight ML algorithm such as SVM [60]. Support vector machines have been used with 3D data for detection tasks, e.g. on sweet peppers [70, 71] and citrus fruits [72]. Unsupervised clustering algorithms can also be applied to 3D data. Examples in outdoor conditions include clustering for detection and size estimation of grape bunches [23] and apple segmentation [73].

Furthermore, the computational power required to apply DL algorithms to rich 3D data has become more widely available, and there has been a resulting flourishing in DL algorithms for 3D data [74–77]. PointNet [74] and its successor PointNet++ [78] use 3D point-cloud data to classify objects and scenes and are widely considered pioneering algorithms in the field of DL for point clouds. Furthermore, PointNet has been successfully applied to agricultural scenarios such as segmenting and labeling grape clusters [79].

There is no doubt that 3D detection and segmentation approaches are more computationally intensive than 2D frameworks, but an exciting trend in recent years is that 3D algorithms are becoming both more efficient and higher performing. An algorithm such as LatticeNet [77] can achieve better semantic segmentation than PointNet and much faster, demonstrating how rapidly the field of DL for 3D point clouds has developed within the last 5 years. It is likely that this trend will continue, as 3D algorithms experience the same rapid developments that 2D algorithms achieved in the last decade since the watershed year in DL of 2012 [38].

However, a major challenge in adapting and developing algorithms and models for ML using 3D data is the lack of high-quality training data specifically for agricultural operations. While high-quality models such as Pointnet++ [74] and LatticeNet [77] exist, transfer learning is necessary for applying these algorithms in agri-robotics. To date, open-access datasets, collected in real-world conditions for site-specific operations, are focused almost entirely on RGB data [80]. This is an obstacle in employing advanced ML algorithms for 3D data in the agricultural domain and will need to be addressed for these approaches to reach their full utility in the future.

2.2 Machine learning for robotic planning

Navigation, path planning, and area coverage algorithms are all at the cornerstone of autonomous or semi-autonomous operation of an agricultural robot. Agri-robots are most often implemented in the form of a mobile platform (UGV) or a drone [unmanned aerial vehicles (UAV)] equipped with a multitude of sensors and manipulators to perform the designated monitoring or intervention task. As a result, the basic requirement of such a robot is the ability to navigate the platform through the environment or move a robotic arm toward the selected target safely and accurately.

Traditionally, a lot of the navigation and path planning performed by autonomous robots does not require a learning component [5, 81]. In fact, a recent review of path planning algorithms [81] shows that ML is not widely used for navigation in agri-robotics applications. However, this approach is now slowly changing, and in recent years, RL algorithms (particularly deep RL) have been used for agri-robotic path planning.

Reinforcement learning directly maps inputs (such as a camera image captured by the robot) to the robot's actions. Training for RL consists of the system exploring the policy space and receiving rewards and, often, penalties for its actions. Over time, the system learns successful behaviors due to the rewards it receives. Deep RL extends the RL concept by including deep neural network architectures. This allows the deep RL system to learn from complex input, such as a camera image. Such complex ML architectures are necessary for agri-robotics where the environments are complex and the tasks the robots must achieve are challenging, and the recent development of deep RL methods means RL can scale to problems with high-dimensional state and action spaces [82], which has the potential to radically increase the application of RL to agricultural robots. Very few RL systems could be found in agri-robots before the very last few years (although a few exceptions exist [83, 84]). However, in recent years, RL methods have been deployed for path planning based on area coverage [85, 86], vineyard in-row navigation [87], and collision-free path planning [88].

Reinforcement learning requires a large number of trials, many of which end in failure. It is clearly not feasible nor safe for a robot to perform RL itself. Instead, the system should be trained in a simulation environment, until a successful policy has been trained. Then, it can be deployed on a robot within a real-world environment [88].

It is important to note, however, that not all RL systems for agri-robotics are deployed and tested in a real environment, with Santos et al. [81] observing that fewer than half of the reviewed papers presented tests in a real environment. Similarly, both Din et al. [85] and Martini et al. [87] only performed simulation experiments for their deep RL systems. However, Wang et al. [86] evaluated a deep RL path planning algorithm in a real-world kiwifruit orchard and concluded it achieved area coverage more efficiently than a standard path planning approach, while Lin et al. [88] demonstrated a deep RL planning algorithm performing collision-free guava picking more effectively than a traditional path planner.

While RL for agri-robotics is still in its infancy, the development of deep RL algorithms that depend on complex input to directly map to control policies seems likely to have a dramatic effect on the directions of research for agri-robotics in the near future. Since agricultural environments are incredibly complex, hand-crafted path planning algorithms can be brittle and prone to failure, and deep RL provides a more flexible and reactive approach to the interactions of the robot and the environment The challenge will be to develop sufficiently realistic simulation environments so that RL systems can be safely and effectively trained and then directly deployed on real-world robots.

3 Challenges

3.1 Environment complexity and variability

The agricultural environment, especially commercial fields, orchards vineyards, and greenhouses, is inherently complex for robotic operations and constitutes a number of challenges. First, the sensors that agricultural robots are typically equipped with are often prone to errors in agricultural conditions. Dust, soil, and other particles often generated by the presence of agricultural machinery and the robotic platform itself have an adversary effect on time-of-flight laser and light-based range sensors such as LiDARs [89]. Direct sunlight, often aggregated purposely for enhanced growing conditions in environments, such as greenhouses, or naturally present in environments, such as open sky fields and orchards, undermines the performance of structured light-based RGB-D cameras [80]. Commercial orchards and vineyards are often sufficiently dense to cause significant disruptions to GPS signals [90] causing significant errors in GPS-based localization and autonomous navigation.

This adversary effect of outdoor conditions on accuracy has led to some robotic platforms being equipped with several sensors with overlapping functions, coupled with possible sensor-fusion or sensor selection algorithms. For example, in the study by Mengoli et al. [91], the authors propose an orchard management platform equipped with LiDAR, inertial measurement unit, and GPS, all for navigational purposes to perform under different environmental conditions. Some authors opt to directly simplify the agricultural setting. For example, the autonomous field robot LadyBird [92] is equipped with wings (Fig. 2) carrying solar panels doubling as an imaging canopy, blocking direct sunlight. Under the installed wings, the authors placed the thermal and stereo imaging cameras used for perception. A similar solution is inherent to most weeding robots operating in a top-down viewpoint [28, 29], where the robotic body occludes most of the ambient light. Other robotic harvesters opt for operation at night time with the presence of artificial lighting [93]. Artificial lighting is also used to combat ambient light in daytime conditions through techniques such as Flash-NoFlash [94]. To reduce scene complexity due to vegetation resulting in high levels of occlusions, alterations to the crop-growing protocols have been proposed [95].

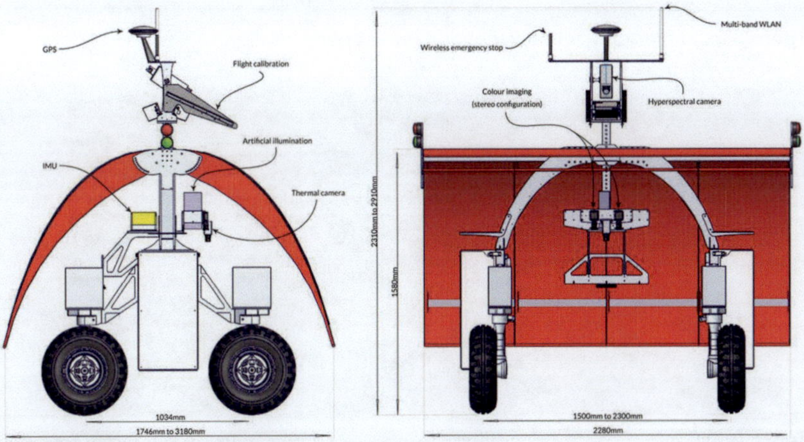

Figure 2 LadyBird field robot equipped with a range of imaging sensors covered by solar panels doubling as an imaging canopy. Source: [92].

Additionally, the same crop possesses a large variety of visually differing attributes in different varieties and maturity levels. Figure 3 illustrates the problem by presenting an example of the visual differences in appearance of 13 sweet pepper varieties discussed in the study by Boyhan et al. [96]. An ML algorithm trained to detect 'sweet peppers' to be implemented, e.g. in a harvesting robot, using a dataset consisting of a single variety of sweet peppers will most likely not be able to generalize well to detect a significantly different variety. Given that commercial greenhouses typically grow a low number of different varieties of the same crop, at the same time, a data acquisition session performed in a commercial greenhouse would most likely consist of a rather limited number of different varieties and therefore would lead to low variability in the training data. Therefore, researchers aiming to acquire variable datasets must do so in many different physical locations in collaboration with a wide number of partners.

Figure 3 Examples of different varieties of sweet peppers. Source: [96].

Figure 4 Different growing conditions for the same crop can have a large impact on the difficulties of detection and robotic harvesting. The left image shows hanging strawberries while in the right image, they grow on the ground.

Moreover, often the same crop of the same variety can be grown in a flora of different growing conditions. Figure 4 shows an example of two different growing methodologies employed for growing the same crop - strawberry. The classical method of on-ground cropping possesses several challenges such as high occlusion levels and limitations on robot reachability due to a high number of obstacles (leaves). The top-down cropping method is aimed to simplify localization and harvesting operations for human workers and benefits a robotic harvester. It is clear that sensory data collected for training a perception algorithm in one of these conditions is likely not to generalize well to the other.

The example earlier presents a rather radical case of growing conditions variability, and one could suggest that entirely different robotic systems are to be designed for such variations. However, more subtle changes in growing conditions still impose significant challenges on the generalizability and scalability of ML algorithms in the agricultural setting. For example, one might expect significant changes in operational conditions in the same field between seasons. In the beginning of the season, seasonal plants tend to be smaller. This can make it harder to perceive them accurately due to resolution limitations in the hardware. On the other hand, as plant density is typically lower there are less occlusions, making the detection problem less complex [97]. In the attempt to address these issues, researchers opt for the collection of longitudinal datasets, which monitor the plants for several weeks or an entire season [20, 92]. These datasets are still rare as the collection requires often significant resources.

Finally, the sensory data acquired in field conditions possess significant variability when acquired even during different times of the day. As mentioned earlier, some researchers opt for sensing in night conditions with artificial light to avoid the adversary effect of ambient light on the imagery data. Perception algorithms trained on these data acquisitions will most likely not generalize well to daytime acquisitions both due to the significant change in illumination and also due to morphological changes in plant structure during different times of the day, especially night, a phenomenon described in the literature as early as in 1755 by Carl von Linné [98].

3.2 Availability of data for machine learning in agri-robotics

Machine learning algorithms rely on previous experience to learn. Therefore, the amount of data available for training and its quality have a significant impact on the performance of the algorithms. Some algorithms require less data than others, but still, the need for data has been a major bottleneck in the agri-robotics domain where datasets are scarce and generalize poorly [3]. Open-access datasets often provide low variation among the different classes [99], and the collection of a variable accurate dataset can often be inaccessible to the research group. When a group does have access to the hardware and funding required for the acquisition of large variable datasets, the need for manual annotation is often an additional obstacle. Lack of or false annotations are major sources of error and a limiting factor in producing large quantities of relevant data [3]. Some annotations require experts to be involved (e.g. disease detection [100]), and some annotations are simply harder and more prone to human errors (e.g. weed detection [101] or annotation of IR [102] and multi-spectral images [103]).

Nevertheless, with the recent developments in ML that we describe further in this section, the amount of data required to obtain robust performance and accuracy is not as high as it used to be. This can be partially attributed to the ability to transfer knowledge from other domains by fine-tuning pre-trained models and partially to the developments in unsupervised and single-class training. Some algorithms rely on less common data types, such as RGB-D or multi-spectral. Pre-trained models for these datasets are scarce, and therefore, performance is expected to suffer for these algorithms.

3.2.1 Available datasets

Despite the mentioned challenges, large annotated datasets are being collected in the agricultural application domain [104, 105] in an attempt to learn about agricultural ecosystems. A great deal of these datasets are acquired using remote sensing from satellites, airplanes, and static towers [106, 107], but some are collected using UAV with data being more directly translatable to robotic in-field operations [108, 109]. A few of the notable in-field datasets that were collected in the last few years include LadyBird [92] and SugarBeets2016 [20].

The LadyBird dataset [92] includes weekly scans of cauliflower and broccoli fields. It consists of stereo color, thermal, and hyperspectral data, as well as ground truth measures (greenness, morphological size, and fresh and dry weight), and environmental measures (weather and soil sensors). The uniqueness of the dataset for the agri-robotics community lies in the capturing of the dataset using the Ladybird robot platform that autonomously followed GPS waypoints in a farm map and localized using RTK-GPS. This resulted in a dataset acquisition protocol in highly realistic conditions for future operations of similar robots. The authors also demonstrate examples of how the dataset can be used for semantic segmentation and object detection.

Similarly, the SugarBeets2016 dataset [20] consists of scans of sugar beets growing in field conditions, captured using the BoniRob field robot, which was manually controlled. The dataset includes RGB and depth information from various sensors mounted on the platform (RGB, RGB-D, LiDAR, and laser range sensors). The dataset contains ground truth in the form of labeled segmentation masks.

Datasets generated by manual acquisition of images in agricultural and controlled conditions have been reviewed in several major review papers including [2, 3, 26]. Of the datasets reviewed, worth mentioning here are PlantVillage [110] and iCassava2019 [111]. The PlantVillage dataset [110] contains more than 54000 images of healthy and disease-affected leaves of different varieties on homogeneous backgrounds. iCassava2019 [111] consists of 9436 labeled images covering healthy cassava leaves and 4 common diseases.

Most commonly ground truth annotations are divided into two types of collected information. The first is the annotation of the acquired images. This can include a bounding box or pixel-level mask annotation of the acquired image. The second type is manual collection of physical traits in the scene of operation. These traits include the physical size, count of targets, or physical location of the target. Both types of annotations are typically manually gathered, prone to human error, and highly labor-intensive.

3.2.2 Developments in coping with lack of data

Several approaches in ML algorithm development have been applied to cope with the lack of available high-quality variable data. First, data augmentation [112], where collected images are subject to artificial manipulations for the generation of new slightly different images while retaining the original annotation. Examples of such augmentations include flipping the image, adjusting its colorspace, adding noise, cropping, rotating, and translating. Data augmentation not only directly increases the number of images available for training but also helps prevent overfitting of the models by introducing noisy data.

Second, the generation of synthetic datasets by bootstrapping real-world data helps generate large quantities of quality annotated data. Examples of such open-access datasets include a barely seeds dataset in a controlled environment [113], soybean pods dataset in a controlled environment [114], and sweet peppers in greenhouse conditions [115]. Synthetic images help overcome the annotated data bottleneck, but the generated images have to remain realistically resembling empirical data to be useful in real-world scenarios [116].

Third, unsupervised learning algorithms, specifically clustering and classification such as K-means [117] and K-NN [118] are widely used in the domain of agri-robotics, as we discuss in the use-cases section of this chapter. Despite being proven to under-perform compared to the supervised learning algorithms, especially the DL-based ones, the reduced need for data remains a driving force behind their development. Specifically, a direction gaining popularity in the field is single-class training for anomaly detection [119]. This group of algorithms fits a model to data for which annotations are available only for a single class (e.g. healthy leaf). Once the model is fed with a new image with a class unseen before (e.g. plant diseases), the model is expected to generate a high error that can be detected and identified as a possible anomaly. Testing these kinds of algorithms on plant disease datasets such as PlantVillage [110] showed promising results [120, 121], but the field is rather new and therefore requires further exploration.

Fourth, transfer learning has been shown to reduce the amount of application-specific data for training ML algorithms. In this method, a model is trained on a large dataset from a different application and then fine-tuned by training on a smaller dataset more relevant to the specific application. Bosilj et al. [122] showed that by training a segmentation model (SegNet-Basic) on the openly available SugarBeets dataset discussed earlier [20], the authors were successful in segmenting carrot and onion plants from RGB and IR images. The authors used 283 images from the sugar beets dataset and only had to collect 20 images of carrots and onions for fine-tuning. The authors also show that the effect of the annotation precision in the fine-tuned dataset was minimal. Similar attempts are reported in the studies by Narvekar and Rao [123] and Nguyen et al. [124] who trained a flower classifier and an insect classifier previously pre-trained on the ImageNet dataset [50]. These results suggest that pre-trained networks should be considered when evaluating a possible classification network, but the amount and variability of the application-specific data are to be evaluated for the specific use case.

Fifth, similarly to transfer learning in DL architectures, some research work attempts to develop ML algorithms that specifically rely on smaller datasets. These algorithms typically aim to learn a small number of parameters tailor-made for the specific application and environmental conditions as an input for a clustering or other morphological operation. These algorithms often rely on domain knowledge and can potentially generalize poorly to other domains. For example, Zemmour et al. [125] suggested an algorithm for learning adaptive thresholds using very small datasets (less than 20 images) for fruit detection in vineyard, orchard, and greenhouse conditions. Similarly, in the study by Ostovar et al. [126], an adaptive segmentation of fruits based on color using RL was developed. Both works show promising results for one type of crop, but generalization suffers when transferring the results to other crops.

Finally, some work exists in the literature attempting to automate data collection and ground truth acquisition using robots to increase the amount of annotated data collected. Kurtser et al. [127] used an industrial robotic manipulator to automate the collection of a visual data set using RGB and depth data (Fig. 5). For each scene, the robotic manipulator collected up to 12 images from different viewpoints. As a result, the collection of ground truth for a single scene resulted in 12 fully annotated images. The idea of using

Figure 5 Automated data collection for sweet pepper harvesting. A robotic industrial manipulator equipped with a sensory rig acquires images from a set of pre-defined directions to create an annotated dataset. Source: http://www.sweeper-robot.eu/.

geometric reasoning to translate the location of objects scanned from one viewpoint to the next has also been used in the study by Kurtser et al. [79]. A mobile platform was equipped with an RGB-D camera for detection and size estimation of grape clusters. That work employed the fact that the motion of the robot along the lane is linear and showed that annotating only 20% of the images is sufficient to accurately infer the annotations for sequential images. The annotated images were used to successfully train a PointNet [74] DL network for grapes segmentation from the acquired point clouds. Recently, the *Journal of Field Robotics*, a major publication venue for quality agri-robotics research papers, introduced a special issue[1] on building intelligent data collectors in agri-robotics, where we expect additional exciting work in this domain to be published.

3.3 Long cycle times

Very recent reviews [128] continue to stress that long cycle times remain a major challenge preventing the commercialization of agricultural robots in general and harvesting robots in particular. Given that previous reviews published over a decade ago [6] have outlined the same challenge, it seems that cycle times and real-time remain a major pain point unsolved by recent advancements in hardware and software.

As agricultural robots are expected to interact with a dynamic environment, real-time operations are important for safe and timely responses. As a result,

[1] https://onlinelibrary.wiley.com/pb-assets/assets/15564967/Agricultural%20Robots%20for%20Ag%204.0%20-%20Building%20intelligent%20data-collectors%20CfP%20SI%203.0-1630061792947.pdf - last accessed Aug 2023.

navigation and obstacle avoidance algorithms are expected to react quickly to dynamic obstacles, and swift motion re-planning is required. Currently proposed navigation algorithms tested in field conditions seem to address the real-time requirements. For example, in the study by Lin et al. [88], the authors report calculating collision-free paths within a few fractions of a second. But navigation is not the only aspect of agricultural robotics requiring time optimization and quick performance.

Machine learning perception algorithms, specifically detection and localization, can achieve real-time performance given the robot is equipped with capable hardware. Training a DL model can be done offline and on more powerful hardware, e.g. a PC or even in the cloud. However, when running the trained model, one needs to consider that some robots rely on a resource-constrained embedded system. It is important to choose a network that produces a model that fits the processor used. One example of this is MobileNets [43], a family of mobile-first computer vision models for TensorFlow. They are designed to maximize accuracy while working on the restricted resources for an on-device or embedded application. If possible, it is also good to choose hardware (particularly CPU) on the robot that better fits the requirements of the chosen DL model. There are several examples of successful use of MobileNets in agricultural robotics. Chechliński et al. [129] developed a system using MobileNets and some other techniques to identify weeds and crops using a Rasberry Pi 3 in 10 FPS. In the study by Gunturu et al. [130], the authors used MobileNets for wheat detection and obstacle avoidance. They tested the system on several embedded computing platforms suitable for integration with a robotic platform. The trained model and camera could achieve 18–23 frames per second, enabling the robot to process its surroundings once every 5–7 cm of driving.

It is important to note that many robotic systems have no immediate need to be so resource-constrained. As modern hardware advances with more and more resources available on smaller and cheaper devices, many agricultural robots relying on UGV platforms opt for a PC [27, 88, 131] with even a GPU card on board [27]. It is reasonable to assume that in the near future, the problem of real-time operation sourcing from hardware limitations is to become less and less acute in most robotic applications.

An additional source of increased cycle times has been inaccurate or missing depth information in field conditions. This led to the development of accuracy compensating algorithms such as visual servoing [67] aiming at slow robotic arm manipulation and constant re-detection of the target as the plant manipulation is performed. With advancements in sensor development, target localization accuracy is expected to be improved. As a result, time-consuming algorithms aiming to compensate for the sensor's poor performance are expected to become redundant in currently tested environments. Nevertheless, the robots are expected to operate in more and more challenging environments, often undermining the sensors (e.g. challenging weather and environmental conditions), and therefore, some of these algorithms might remain needed till more accurate sensors are developed.

Finally, to achieve robust performance in competitive cycle times, some researchers opt for task planning algorithms and scheduling algorithms [132]. These algorithms further optimize the overall operations of the robotic system to achieve better cycle times. Questions like the order of target manipulation or planning and scheduling of sensing operations can possibly be addressed more efficiently with the help of ML.

4 Integration and field-testing use-cases

Reviewing the literature of the past 10 years (since 2012), we were interested to find which of the ML algorithms developed and reported on in the literature were employed and reported on in fully integrated agri-robotic systems. Quite an impressive number of robotic systems have been reported in the literature in the last decade as can be attested by the recent review by Oliveira et al. [133]. However, only a few were extensively tested in operation in field conditions. Here we elaborate on a few representative examples of reported robots for sweet pepper harvesting [27, 134], kiwifruit picking [86], guava harvesting [88], weed control [29, 135], apple harvesting [131], strawberry harvesting [136], and spraying [137]. In this section, we outline the algorithms used for perception and path planning in these setups. We specifically focus on the ML algorithms, based on the authors' reports. For additional similar examples, the reader is advised to read the reviews for pest control [8] and harvesting [6, 7, 128].

Wang et al. [86] developed a deep RL-based path planning method for kiwifruit picking robot coverage called re-DQN (an improved DQN algorithm). A LiDAR is used to construct a 2D grid map from the environment. The location of the fruits is determined using a YOLOv4 DL network and a real-sense camera. Based on this and the effective picking area of the robot, the grid is divided into regions and the re-DQN algorithm plans an efficient path through each region. The results show that the novel algorithm produced 32% shorter paths and the overall navigation time was reduced by 35%.

Another deep RL-based path-planning algorithm was proposed by Lin et al. [88] for generating collision-free paths for a guava-harvesting robot arm. A recurrent neural network was adopted to remember and exploit the past states observed by the robot. Then a deep deterministic policy gradient algorithm (DDPG) used the states to predict a collision-free path. To enable recurrent DDPG to generalize to real-world scenarios, a simulation environment with randomized parameters was used during the training phase. For fruit and obstacle detection, a Kinect V2 RGB-D camera was used together with a convolutional network to obtain a fruit and branch binary map, which was then converted into a 3D point cloud. Field tests show that recurrent DDPG can increase grasp, detachment, and harvest success rates by 19.43%, 9.11%, and 10.97%, respectively, compared to cases where no collision-free path-planning algorithm is implemented. The authors report that the algorithm took just 29 ms to plan a collision-free path with a success rate of 90.90%.

Silwal et al. [131] reported on the testing of a fully integrated robot for apple harvesting. Instead of employing ML in their detection, they instead created favorable sensing conditions by using a black curtain background and sensing in multiple exposure times. Their detection algorithms were Circular Hough Transformation and blob detection, which employ domain knowledge of the apple shape. For the localization of the apple, they employed a statistical outlier removal technique (modified Thompson rule) and center of mass calculation. They used ML for path planning by employing k-NN and solving the traveling salesman problem for travel optimization. The authors report a success rate of 84% with an average picking time of 6.0 s per fruit.

Arad et al. [27] reported on a sweet pepper harvesting robot in greenhouse conditions. The authors placed an RGB-D camera on the harvesting end-effector. A shape and color-based detection algorithm was used to detect fruits, using both RGB and the point cloud. Holes in the detected region were removed using morphological operations. The DL model based on a VGG-16 architecture trained on a synthetic dataset [138] was employed to detect and localize the stem. The angle between the fruit and stem was determined to enable picking the fruit from in front of the stem. The lazy PRM* motion planner was used to plan a path for the robotic arm between fruits. The authors report harvest success rates of 61% in modified crop conditions and 18% in current crop conditions.

A similar setup for sweet pepper harvesting in greenhouse conditions was reported by Lehnert et al. [134] with an RGB-D camera placed in an eye-in-hand position at the end of the robotic manipulator. For perception and sweet pepper detection, the authors used an unsupervised autoencoder (Sparse Auto-Encoder) to generate a probability map used as an input to a simpler blob detection algorithm. The detected targets are then used in an Euclidean clustering algorithm employing the depth data collected for localization and grasp pose calculation. For robotic arm path planning, the TRAC-IK algorithm was used. The authors report achieving a 46% success rate for an unmodified crop, and 58% for a modified crop.

Wu et al. [29] developed a weed control robot. Three RGB and NIR cameras were employed for weed detection, using the normalized difference vegetation index, which utilizes both color (RGB) and NIR information. Detection was performed using convolutional neural networks for crop and weed detection coupled with a Naive Bayesian classifier with a biased probabilistic model to filter out falsely classified weeds and crops.

Several fully integrated robots reported on in the literature manage to get competitive performance without using ML. Xiong et al. [136] reported on the development of a strawberry harvesting robot. An RGB-D camera was mounted directly on the mobile base of the robot. Detection of the berries is done through thresholding of the RGB component, followed by morphological operations. Localization of the berry is done through the identification of the corresponding point cloud and filtering of it, based on the depth component. The guidance of the robotic arm is done through visual servoing based on an IR sensor mounted on the robotic arm. The authors report that in farm settings, the average picking success rate is 53.6%, mostly due to occlusion and clustering of fruits. Another robot was developed by Bawden et al. [135] for weed management. The authors employed only RGB sensors, motivating the choice with the high price range of NIR sensors. For target perception, they used local binary patterns and covariance features. The authors report a significant reduction in weed coverage after treatment with the robot and above 95% weed detection and classification rates in various outdoor conditions. Cantelli et al. [137] developed a spraying robot. The spraying was performed continuously and therefore all sensing was performed for localization and navigation and not for target detection. For navigation purposes, a stereo camera, an RTK-DGPS GNSS receiver, an attitude and heading reference system, a laser scanner rangefinder, ultrasonic sensors, and motor encoders were used. For navigation, a potential field method was used. The robot managed to keep within 0.23 m of the path with most errors occurring during turning at the end of the lane. While in agri-robotics, these major advances have been investigated and many ML algorithms for perception in the agricultural domain exist [139]. However, there are still relatively few fully integrated ML perception systems tested in field conditions. Furthermore, integrated systems tend to avoid using sophisticated perception algorithms. The fully integrated experiments may use a simplified scene (e.g. a scene where occlusions have been removed [27]). As a result, somewhat simpler case studies have been successful enough to be proposed and evaluated as fully integrated robots in real conditions. The prevalence of the harvesting and weeding robots in semi-controlled conditions and the choice of crops with high contrast to the background (e.g. yellow pepper or red strawberries and green vegetation background) are vivid examples. More complex detection algorithms, such as for pest control and disease detection, are still tested mostly in laboratory or very controlled conditions.

We can see that the fully integrated systems seem to use less advanced ML methods than what is reported in individual experiments in the field. A recent review of robotic harvesting systems [128] shows that the cycle time for harvesting varies between 6 s and 64 s and the harvesting success between ca.

60–90%. This indicates that one reason for using less advanced ML methods could be that when developing a fully integrated system that combines many different aspects, there are many other problems to solve (i.e. ML is not the bottleneck) so most researchers opt for simpler solutions instead of advanced ML algorithms.

5 Conclusion

There have been rapid advances in ML in the last decade. These improvements have exciting implications for many domains, including agri-robotics. There is a wide range of publications and systems investigating and evaluating ML algorithms, in particular for perception [139] and, more recently, path planning for both navigation and manipulation [85–88].

However, there are still relatively few fully integrated ML systems tested in field conditions. Agricultural robots must operate in and interact with very challenging and complex environments. Furthermore, the tasks that these ML algorithms are designed for are often very challenging. Fruit detection must be robust to leaf occlusion. Weed and disease detection must be highly accurate. Pruning and harvesting must be achieved accurately and safely in complex and occluded environments. Therefore, although the potential for ML in agri-robotics is still extremely valuable, at present, the available ML algorithms are not yet sufficiently robust to be widely deployed in fully integrated autonomous systems.

Machine learning algorithms are critically dependent on the incoming sensor data, and therefore, depend on the development not just of software and algorithms but also novel and improved hardware. This is evident, as an example, in the rapid adaptation of commercial-grade RGB-D cameras. These sensors started showing adequate sensing quality in outdoor conditions at a low price range only in the last few years [64, 80], but are already now integrated into many agri-robotic systems. The wide availability of these sensors allows agri-robots to now have access to 3D data as well as 2D imagery in outdoor field conditions as well as controlled post-harvest environments. The 3D data simplifies many algorithms, including segmentation, target localization, obstacle detection and avoidance, and collection of physical morphological phenotype traits. Having access to high-quality and relevant sensor data will undoubtedly allow more reliable ML algorithms that can have a critical impact on the ability of robots to navigate and operate autonomously and safely in complex agricultural environments.

A key ongoing challenge is the acquisition of representative data. The pre-trained DL networks available today provide the developer of agri-robotics systems with the possibility to adjust the model to the specific environment in which the robot operates using a relatively low amount of additional data, compared to training the model 'from scratch'. Nevertheless, for this transfer learning to be effective, it still requires environment-specific data that accounts for the majority of possible variability in environmental conditions in which the robot is to operate. Attempt to automate this data collection [79, 127] is a possible direction to address this need, as is the release of collected data into the public domain.

At present, it is not the ML algorithms that are the limiting factor in the wide deployment of fully integrated agri-robotic systems. A fully integrated system is dependent on all of its components: sensors, actuators, and algorithms. A robot needs to convert perception and planning into action, and the challenging nature of the agricultural domain means there are also many mechanical challenges to be addressed for tasks such as harvesting, weeding, and pruning. The progress that ML algorithms have been making to develop robust perception and planning solutions is part of the ongoing wider development in agri-robotics.

6 Where to look for further information

The topic of machine learning is both vast and rapidly expanding. For more information on deep learning systems, the book *Deep Learning* by Ian Goodfellow, Yoshua Bengio, and Aaron Courville contains detailed information on a broad range of topics in deep learning, covering both mathematical and conceptual background, deep learning techniques used in industry, and research perspectives. Similarly, for readers interested in learning more about reinforcement learning, the book *Reinforcement Learning* by Richard S. Sutton and Andrew G. Barto presents the key ideas and algorithms in the field of reinforcement learning. To know more about the latest developments of machine learning applied in agri-robotics, state-of-the-art techniques can be found in the major journals in the field, such as *Journal of Field Robotics* (JFR) and *Computers and Electronics in Agriculture* (COMPAG).

7 References

1. LeCun, Y., Bengio, Y. and Hinton, G. Deep learning. *Nature* 521(7553):436–444 2015.
2. Kapach, K., Barnea, E., Mairon, R., Edan, Y. and Ben-Shahar, O. B. Computer vision for fruit harvesting robots–state of the art and challenges ahead. *International Journal of Computational Vision and Robotics* 3(1/2):4–34 2012.
3. Kamilaris, A. and Prenafeta-Boldú, F. X. Deep learning in agriculture: a survey. *Computers and Electronics in Agriculture* 147:70–90 2018.
4. Tang, Y., Chen, M., Wang, C., Luo, L., Li, J., Lian, G. and Zou, X. Recognition and localization methods for vision-based fruit picking robots: a review. *Frontiers in Plant Science* 11:510 2020.

5. Mousazadeh, H. A technical review on navigation systems of agricultural autonomous off-road vehicles. *Journal of Terramechanics* 50(3):211-232 2013.
6. Bac, C. W., van Henten, E. J., Hemming, J. and Edan, Y. Harvesting robots for high-value crops: state-of-the-art review and challenges ahead. *Journal of Field Robotics* 31(6):888–911 2014.
7. He, L. and Schupp, J. Sensing and automation in pruning of apple trees: a review. *Agronomy* 8(10):211 2018.
8. Meshram, A. T., Vanalkar, A. V., Kalambe, K. B. and Badar, A. M. Pesticide spraying robot for precision agriculture: a categorical literature review and future trends. *Journal of Field Robotics* 39(2):153–171 2022.
9. Russell, S. J. and Norvig, P. *Artificial Intelligence: a Modern Approach* (3rd edn.). Pearson 2009.
10. Deng, L., Dong, Y. Deep learning: methods and applications. *Foundations and Trends in Signal Processing* 7(3–4):197–387 2014. ISSN 1932-8346. doi: 10.1561/2000000039.
11. Sutton, R. S. and Barto, A. G. *Reinforcement Learning: an Introduction*. MIT Press 2018.
12. Polydoros, A. S. and Nalpantidis, L. Survey of model-based reinforcement learning: applications on robotics. *Journal of Intelligent and Robotic Systems* 86(2):153–173 2017.
13. Bechar, A. and Vigneault, C. Agricultural robots for field operations: concepts and components. *Biosystems Engineering* 149:94–111 2016.
14. Vilone, G. and Longo, L. Notions of explainability and evaluation approaches for explainable artificial intelligence. *Information Fusion* 76:89–106 2021. ISSN 1566-2535. doi: 10.1016/j.inffus.2021.05.009. Available at: https://www.sciencedirect/science/article/pii/S1566253521001093.
15. Liakos, K. G., Busato, P., Moshou, D., Pearson, S. and Bochtis, D. Machine learning in agriculture: a review. *Sensors* 18(8):2674 2018.
16. Benos, L., Tagarakis, A. C., Dolias, G., Berruto, R., Kateris, D. and Bochtis, D. Machine learning in agriculture: a comprehensive updated review. *Sensors* 21(11):3758 2021.
17. Siciliano, B. and Khatib, O. *Introduction*. Springer 2008, pages 1–4. ISBN 978-3-540-30301-5. doi: 10.1007/978-3-540-30301-5_1.
18. Gené-Mola, J., Vilaplana, V., Rosell-Polo, J. R., Morros, J.-R., Ruiz-Hidalgo, J. and Gregorio, E. Multi-modal deep learning for fuji apple detection using rgb-d cameras and their radiometric capabilities. *Computers and Electronics in Agriculture* 162:689–698 2019. ISSN 0168-1699. doi: 10.1016/j.compag.2019.05.016. Available at: https://www.sciencedirect.com/science/article/pii/ S0168169919301413.
19. Gené-Mola, J., Vilaplana, V., Rosell-Polo, J. R., Morros, J.-R., Ruiz-Hidalgo, J. and Gregorio, E. KFuji RGB-DS database: Fuji apple multi-modal images for fruit detection with color, depth and range-corrected IR data. *Data in Brief* 25:104289 2019. ISSN 2352-3409. doi: 10.1016/j.dib.2019.104289. Available at: https://www.sciencedirect.com/science/article/pii/S2352340919306432.
20. Chebrolu, N., Lottes, P., Schaefer, A., Winterhalter, W., Burgard, W. and Stachniss, C. Agricultural robot dataset for plant classification, localization and mapping on sugar beet fields. *The International Journal of Robotics Research* 36(10):1045–1052 2017.
21. Tyr Wiesner-Hanks, E. L. Image set for deep learning: field images of maize annotated with disease symptoms. *BMC Research Notes*. Nicholas Kaczmar, Stewart.

Hod Lipson, Nelson 11(1):440 Jul 2018. ISSN 1756-0500. doi: 10.1186/s13104-018-3548-6. Available at: https://doi.org/10.1186/ s13104-018-3548-6.

22. Gené-Mola, J., Gregorio, E., Cheein, F. A., Guevara, J., Llorens, J., Sanz-Cortiella, R., Escolà, A. and Rosell-Polo, J. R. Fruit detection, yield prediction and canopy geometric characterization using lidar with forced air flow. *Computers and Electronics in Agriculture* 168:105121 2020. ISSN 0168-1699. doi: 10.1016/j.compag.2019.105121. Available at: https://www.sciencedirect.com/science/article/pii/S0168169919313390.
23. Kurtser, P., Ringdahl, O., Rotstein, Nati, Berenstein, R. and Edan, Y. In-field grape cluster size assessment for vine yield estimation using a mobile robot and a consumer level rgb-d camera. *IEEE Robotics and Automation Letters* 5(2):2031–2038 2020b.
24. Milella, A., Marani, R., Petitti, A. and Reina, G. In-field high throughput grapevine phenotyping with a consumer-grade depth camera. *Computers and Electronics in Agriculture* 156:293–306 2019.
25. Marani, R., Milella, A., Petitti, A. and Reina, G. Deep neural networks for grape bunch segmentation in natural images from a consumer-grade camera. *Precision Agriculture* 22(2):387–413 2021.
26. Lu, Y. and Young, S. A survey of public datasets for computer vision tasks in precision agriculture. *Computers and Electronics in Agriculture* 178:105760 2020.
27. Arad, B., Balendonck, J., Barth, R., Ben-Shahar, O., Edan, Y., Hellström, T., Hemming, J., Kurtser, P., Ringdahl, O., Tielen, T. and van Tuijl, B. Development of a sweet pepper harvesting robot. *Journal of Field Robotics* 37(6):1027–1039 2020.
28. Harders, L. O., Czymmek, V., Knoll, F. J. and Hussmann, S. Area yield performance evaluation of a nonchemical weeding robot in organic farming. In: IEEE International Instrumentation and Measurement Technology Conference (I2MTC) 2021, pages 1–6. IEEE Publications.
29. Wu, X., Aravecchia, S., Lottes, P., Stachniss, C. and Pradalier, C. Robotic weed control using automated weed and crop classification. *Journal of Field Robotics* 37(2):322-340 2020.
30. Zahid, A., Mahmud, Md. S., He, L., Heinemann, P., Choi, D. and Schupp, J. Technological advancements towards developing a robotic pruner for apple trees: a review. *Computers and Electronics in Agriculture* 189:106383 2021.
31. Khoroshevsky, F., Khoroshevsky, S. and Bar-Hillel, A. Parts-per-object count in agricultural images: solving phenotyping problems via a single deep neural network. *Remote Sensing* 13(13):2496 2021.
32. Kicherer, A., Herzog, K., Pflanz, M., Wieland, M., Rüger, P., Kecke, S., Kuhlmann, H. and Töpfer, R. An automated field phenotyping pipeline for application in grapevine research. *Sensors* 15(3):4823–4836 2015.
33. Araus, J. L. and Cairns, J. E. Field high-throughput phenotyping: the new crop breeding frontier. *Trends in Plant Science* 19(1):52–61 2014.
34. Harel, B., Parmet, Y. and Edan, Y. Maturity classification of sweet peppers using image datasets acquired in different times. *Computers in Industry* 121:103274 2020.
35. Wang, Aichen, Zhang, W. and Wei, X. A review on weed detection using ground-based machine vision and image processing techniques. *Computers and Electronics in Agriculture* 158:226–240 2019.

36. Bakker, T., van Asselt, K., Bontsema, J., Müller, J. and van Straten, G. An autonomous weeding robot for organic farming. In: *Field and Service Robotics* 2006, pages 579-590. Springer.
37. Berenstein, R., Ben Shahar, O. B., Shapiro, A. and Edan, Y. Grape clusters and foliage detection algorithms for autonomous selective vineyard sprayer. *Intelligent Service Robotics* 3(4): 233–243 2010.
38. Krizhevsky, A., Sutskever, I. and Hinton, G. E. Imagenet classification with deep convolutional neural networks. *Advances in Neural Information Processing Systems* 25: 1097–1105 2012.
39. Jia, Y., Shelhamer, E., Donahue, J., Karayev, S., Long, J., Girshick, R., Guadarrama, S. and Darrell, T. Caffe: convolutional architecture for fast feature embedding. In: Proceedings of the 22nd ACM International Conference on Multimedia 2014, pages 675–678.
40. Simonyan, K. and Zisserman, A. Very deep convolutional networks for large-scale image recognition. *arXiv Preprint ArXiv:1409.1556* 2014, pages 4. https://doi.org/10.48550/arXiv.1409.1556.
41. Szegedy, C., Liu, W., Jia, Y., Sermanet, P., Reed, S., Anguelov, D., Erhan, D., Vanhoucke, V. and Rabinovich, A. Going deeper with convolutions. In: Proceedings of the IEEE Conference on Computer Vision and Pattern Recognition 2015, pages 1–9.
42. He, K., Zhang, X., Ren, S. and Sun, J. Deep residual learning for image recognition. In: IEEE Conference on Computer Vision and Pattern Recognition (CVPR) 2016, (vol. 2016), pages 770–778. doi: 10.1109/CVPR.2016.90.
43. Howard, A. G., Zhu, M., Chen, B., Kalenichenko, D., Wang, W., Weyand, T., Andreetto, M. and Adam, H. Mobilenets: efficient convolutional neural networks for mobile vision applications. *arXiv Preprint ArXiv:1704.04861* 2017, pages 9. https://doi.org/10.48550/arXiv.1704.04861.
44. Mingxing, T. and Le, Q. V. Efficientnet: rethinking model scaling for convolutional neural networks. *Proceedings of the 36th International Conference on Machine Learning*, in Proceedings of Machine Learning Research 97:6105-6114. Available from https://proceedings.mlr.press/v97/tan19a.html. 2020.
45. Ren, S., He, K., Girshick, R., Sun, J. and Faster, R.-C. N. N. Towards realtime object detection with region proposal networks. In: *Advances in Neural Information Processing Systems* (vol. 28) Cortes, C., Lawrence, N., Lee, D., Sugiyama, M. and Garnett, R. (Eds). Curran Associates, Inc. 2015. Available at: https://proceedings.neurips, page cc/paper/2015/file/14bfa6bb14875e45bba028a21ed38046-Paper.pdf.
46. He, K., Gkioxari, G., Dollár, P., Girshick, R. and Mask, R.-C. N. N. IEEE International Conference on Computer Vision (ICCV) 2017, (vol. 2017), pages 2980–2988. doi: 10.1109/ICCV.2017.322.
47. Redmon, J., Divvala, S., Girshick, R. and Farhadi, A. You only look once: unified, real-time object detection. In: Proceedings of the IEEE Conference on Computer Vision and Pattern Recognition 2016, pages 779–788.
48. Redmon, J. and Farhadi, A. Yolo9000: better, faster, stronger. In: Proceedings of the IEEE Conference on Computer Vision and Pattern Recognition (CVPR) 2017:6517–6525.
49. Redmon, J. and Farhadi, A. Yolov3: an incremental improvement. 2018. Available at: http://arxiv.org/abs/1804.02767.

50. Deng, J., Dong, W., Socher, R., Li, L., Li, K. and Fei-Fei, L. Imagenet: a large-scale hierarchical image database. In: IEEE Conference on Computer Vision and Pattern Recognition 2009, pages 248–255. IEEE Publications.
51. Lippi, M., Bonucci, N., Carpio, R. F., Contarini, M., Speranza, S. and Gasparri, A. A yolo-based pest detection system for precision agriculture 2021. In: 29th Mediterranean Conference on Control. and Automation (MED), pages 342-347. IEEE Publications.
52. Tang, Z., Chen, Z., Qi, F., Zhang, L. and Chen, S. Pest-yolo: deep image mining and multi-feature fusion for real-time agriculture pest detection. In: IEEE International Conference on Data Mining (ICDM) 2021, pages 1348–1353. IEEE Publications.
53. Liu, J. and Wang, X. Tomato diseases and pests detection based on improved yolo v3 convolutional neural network. *Frontiers in Plant Science* 11:898 2020.
54. Legaspi, K. R. B., Sison, N. W. S. and Villaverde, J. F. Detection and classification of whiteflies and fruit flies using yolo. In 13th International Conference on Computer and Automation Engineering (ICCAE) 2021, pages 1–4. IEEE Publications.
55. Liu, G., Nouaze, J. C., Mbouembe, P. L. T. and Kim, J. H. Yolo-tomato: a robust algorithm for tomato detection based on yolov3. *Sensors* 20(7):2145 2020.
56. Tian, Y., Yang, G., Wang, Z., Wang, H., Li, E. and Liang, Zize. Apple detection during different growth stages in orchards using the improved yolo-v3 model. *Computers and Electronics in Agriculture* 157:417–426 2019.
57. Li, X. and Rai, L. Apple leaf disease identification and classification using resnet models. In: IEEE 3rd International Conference on Electronic Information and Communication Technology (ICEICT) 2020, pages 738–742. IEEE Publications.
58. Nagaraju, Y., Swetha, S., Stalin, S. and Stalin, S. Apple and grape leaf diseases classification using transfer learning via fine-tuned classifier. In: IEEE International Conference on Machine Learning and Applied Network Technologies (ICMLANT) 2020, pages 1–6. IEEE Publications.
59. Parraga-Alava, J., Alcivar-Cevallos, R., Riascos, J. A. and Becerra, M. A. Aphids detection on lemons leaf image using convolutional neural networks. In: International Conference on Systems and Information Sciences 2020, pages 16–27. Springer.
60. Cortes, C. and Vapnik, V. Support-vector networks. *Machine Learning* 20(3):273–297 1995.
61. Murawwat, S., Qureshi, A., Ahmad, S. and Shahid, Y. Weed detection using svms. *Engineering, Technology and Applied Science Research* 8(1):2412–2416 2018.
62. Nosseir, A. and Ahmed, S. E. A. Automatic classification for fruits' types and identification of rotten ones using k-nn and svm. *International Journal of Online and Biomedical Engineering* 15(3) 2019.
63. Botterill, T., Paulin, S., Green, R., Williams, S., Lin, J., Saxton, V., Mills, S., Chen, X. Q. and Corbett-Davies, S. A robot system for pruning grape vines. *Journal of Field Robotics* 34(6): 1100–1122 2017.
64. Vit, A. and Shani, G. Comparing RGB-D sensors for close range outdoor agricultural phenotyping. *Sensors* 18(12): 4413 2018.
65. Gené-Mola, J., Sanz-Cortiella, R., Rosell-Polo, J. R., Escolà, A. and Gregorio, E. PFuji-size dataset: a collection of images and photogrammetry-derived 3d point clouds with ground truth annotations for fuji apple detection and size estimation in field conditions. *Data in Brief* 39:107629 2021. ISSN 2352-3409. doi: 10.1016/j.dib.2021.107629. Available at: https://www.sciencedirect.com/science/ article/pii/S2352340921009045.

66. Blok, P. M., van Henten, E. J., van Evert, F. K. and Kootstra, G. Image- based size estimation of broccoli heads under varying degrees of occlusion. *Biosystems Engineering* 208:213–233 2021. ISSN 1537-5110. doi: 10.1016/j.biosystemseng.2021.06.001. Available at: https://www.sciencedirect.com/science/article/pii/S1537511021001203.
67. Barth, R., Hemming, J. and van Henten, E. J. Design of an eye-in-hand sensing and servo control framework for harvesting robotics in dense vegetation. *Biosystems Engineering* 146:71–84 2016.
68. Fu, L., Gao, F., Wu, J., Li, R., Karkee, M. and Zhang, Q. Application of consumer RGB-D cameras for fruit detection and localization in field: a critical review. *Computers and Electronics in Agriculture* 177:105687 2020.
69. Halstead, M., Denman, S., Fookes, C. and McCool, C. Fruit detection in the wild: the impact of varying conditions and cultivar. In: *Digital Image Computing (DICTA)* 2020:1–8 2020. doi: 10.1109/DICTA51227.2020.9363407.
70. Barnea, E., Mairon, R. and Ben-Shahar, O. Colour-agnostic shape-based 3d fruit detection for crop harvesting robots. *Biosystems Engineering* 146:57–70 2016. ISSN 1537-5110. doi: 10.1016/j.biosystemseng.2016.01.013. Available at: https://www.sciencedirect.com/science/article/pii/S1537511016000131. Special Issue: Advances in Robotic Agriculture for Crops.
71. Sa, I., Lehnert, C., English, A., McCool, C., Dayoub, F., Upcroft, B. and Perez, T. Peduncle detection of sweet pepper for autonomous crop harvesting–combined color and 3-d information. *IEEE Robotics and Automation Letters* 2(2):765–772 2017b. doi: 10.1109/LRA.2017.2651952.
72. Lin, G., Tang, Y., Zou, X., Li, J. and Xiong, J. In-field citrus detection and localisation based on rgb-d image analysis. *Biosystems Engineering* 186:34–44 2019. ISSN 1537-5110. doi: 10.1016/j.biosystemseng.2019.06.019. Available at: https://www.sciencedirect.com/science/article/pii/S1537511018312765.
73. Nguyen, T. T., Vandevoorde, K., Wouters, N., Kayacan, E., De Baerdemaeker, J. G. and Saeys, W. Detection of red and bicoloured apples on tree with an rgb-d camera. *Biosystems Engineering* 146:33–44 2016.
74. Qi, C. R., Su, H., Mo, K. and Guibas, L. J. Pointnet: deep learning on point sets for 3d classification and segmentation. In: Proceedings of the IEEE Conference on Computer Vision and Pattern Recognition 2017a, pages 652–660.
75. He, Y., Chiu, W.-C., Keuper, M. and Fritz, M. Std2p: RGBD semantic segmentation using spatio-temporal data-driven pooling. In: Proceedings of the IEEE Conference on Computer Vision and Pattern Recognition 2017, pages 4837–4846.
76. Hu, Yaosi, Chen, Z. and Lin, W. RGB-D semantic segmentation: a review. In: IEEE International Conference on Multimedia & Expo Workshops (ICMEW) 2018, pages 1-6. IEEE Publications.
77. Rosu, Radu A., Schütt, P., Quenzel, J. and Behnke, S. Latticenet: fast spatio-temporal point cloud segmentation using permutohedral lattices. *Autonomous Robots* 46(1):45–60 2022.
78. Qi, C. R., Yi, L., Su, H. and Guibas, L. J. Pointnet++: deep hierarchical feature learning on point sets in a metric space. *Advances in Neural Information Processing Systems* 30:5099–5108 2017b.
79. Kurtser, P., Ringdahl, O., Rotstein, Nati and Andreasson, H. Pointnet and geometric reasoning for detection of grape vines from single frame RGB-D data in outdoor

conditions. In Proceedings of the Northern Lights Deep Learning Workshop 3rd Northern Lights Deep Learning Workshop, Tromsö, Norway 1:1–6 2020a.

80. Kurtser, P. and Lowry, S. RGB-D perception for automation of site specific agricultural operations - survey of public datasets. *Computers and Electronics in Agriculture* 212 2023.
81. Santos, L. C., Santos, F. N., Solteiro Pires, E. J., Valente, A., Costa, P. and Magalhães, S. Path planning for ground robots in agriculture: a short review. In: IEEE International Conference on Autonomous Robot Systems and Competitions (ICARSC) 2020, pages 61–66. IEEE Publications.
82. Arulkumaran, K., Deisenroth, M. P., Brundage, M. and Bharath, A. A. Deep reinforcement learning: a brief survey. *IEEE Signal Processing Magazine* 34(6):26–38 2017.
83. Makino, T., Yokoi, H. and Kakazu, Y. Development of a motion planning system for an agricultural mobile robot. In: *Sice'99*. Proceedings of the 38th SICE Annual Conference. International Session Papers 1999, pages 959–962. IEEE Publications (*IEEE. Cat. No. 99TH8456*).
84. Patino, H. D., Tosetti, S. and Capraro, F. Adaptive critic designs-based autonomous unmanned vehicles navigation: application to robotic farm vehicles. In: IEEE Symposium on Adaptive Dynamic Programming and Reinforcement Learning 2009, pages 233–237. IEEE Publications.
85. Din, A., Ismail, M. Y., Shah, B., Babar, M., Ali, F. and Baig, S. U. A deep reinforcement learning-based multi-agent area coverage control for smart agriculture. *Computers and Electrical Engineering* 101:108089 2022.
86. Wang, Y., He, Z., Cao, D., Ma, L., Li, K., Jia, L. and Cui, Y. Coverage path planning for kiwifruit picking robots based on deep reinforcement learning. *Computers and Electronics in Agriculture* 205:107593 2023.
87. Martini, M., Cerrato, S., Salvetti, F., Angarano, S. and Chiaberge, M. Position-agnostic autonomous navigation in vineyards with deep reinforcement learning 2022. In: 18th International Conference on Automation Science and Engineering (CASE), pages 477–484. IEEE Publications. IEEE Publications.
88. Lin, G., Zhu, L., Li, J., Zou, X. and Tang, Y. Collision-free path planning for a guava-harvesting robot based on recurrent deep reinforcement learning. *Computers and Electronics in Agriculture* 188:106350 2021.
89. Mona, L., Liu, Z., Müller, D., Omar, A., Papayannis, A., Pappalardo, G., Sugimoto, N. and Vaughan, M. Lidar measurements for desert dust characterization: an overview. *Advances in Meteorology* 2012:1–36 2012.
90. Costley, A. and Christensen, R. Landmark aided gps-denied navigation for orchards and vineyards. In: Location and Navigation Symposium (PLANS) 2020, pages 987–995. IEEE Publications.
91. Mengoli, D., Tazzari, R. and Marconi, L. Autonomous robotic platform for precision orchard management: architecture and software perspective 2020. In: IEEE International Workshop on Metrology for Agriculture and Forestry (MetroAgriFor), pages 303–308. IEEE Publications.
92. Bender, A., Whelan, B. and Sukkarieh, S. A high-resolution, multimodal data set for agricultural robotics: a ladybird's-eye view of brassica. *Journal of Field Robotics* 37(1):73–96 2020. doi: 10.1002/rob.21877. Available at: https://onlinelibrary. Available at: wiley.com/doi/abs/10.1002/rob.21877.

93. Weikuan, J., Dean, Z., Chengzhi, R., Xiaoyang, L., Yu, C. and Wei, J. Combined method for night vision image denoising based on wavelet transform and ICA. *Nongye Jixie Xuebao/Transactions of the Chinese Society of Agricultural Machinery* 46(9): 9–17 2015.
94. Arad, B., Kurtser, P., Barnea, E., Harel, B., Edan, Y. and Ben-Shahar, O. Controlled lighting and illumination-independent target detection for real-time cost-efficient applications. The case study of sweet pepper robotic harvesting. *Sensors* 19(6):1390 2019.
95. van Herck, L., Kurtser, P., Wittemans, L. and Edan, Y. Crop design for improved robotic harvesting: a case study of sweet pepper harvesting. *Biosystems Engineering* 192:294–308 2020.
96. Boyhan, G. E., McGregor, C., O'Connell, S., Biang, J. and Berle, D. A comparison of 13 sweet pepper varieties under an organic farming system. *HortTechnology* 30(1):135–143 2020. doi: 10.21273/HORTTECH04455-19. Available at: https://journals.ashs.org/horttech/view/journals/ horttech/30/1/article-p135.xml.
97. Kurtser, P. and Edan, Y. Statistical models for fruit detectability: spatial and temporal analyses of sweet peppers. *Biosystems Engineering* 171:272–289 2018.
98. von Linné, C. *Somnus plantarum* in dissertatione academica propositus, consens. experient. Facult. Med. in Reg. acad. Upsaliensi, præside... Carolo Linnæo... a Petro Bremer, Helsingo. In audit. Gustav. d. X. Decembr. anni MDCCLV.1755.
99. Yalcin, H. Plant phenology recognition using deep learning: deep-pheno. In: 6th International Conference on Agro-Geoinformatics 2017, pages 1–5. IEEE Publications.
100. Amara, J., Bouaziz, B. and Algergawy, A. A deep learning-based approach for banana leaf diseases classification. *Datenbanksysteme für Business, Technologie und Web (BTW 2017)-Workshopband* Bonn: Gesellschaft für Informatik e.V. 2017 pages 79–88..
101. Sørensen, R. A., Rasmussen, J., Nielsen, J. and Jørgensen, R. N. Thistle detection using convolutional neural networks. In: EFITA WCCA 2017 Conference 2017, pages 2–6. Montpellier Supagro, Montpellier, France.
102. Wachs, J., Stern, H. I., Burks, T., Alchanatis, V. and Bet-Dagan, I. Apple detection in natural tree canopies from multimodal images. In: Proceedings of the 7th European Conference on Precision Agriculture, Wageningen, the Netherlands (vol. 68) 2009, page 293–302.
103. Sa, I., Chen, Zetao, Popović, M., Khanna, R., Liebisch, F., Nieto, J. and Siegwart, R. WeedNet: dense semantic weed classification using multispectral images and mav for smart farming. *IEEE Robotics and Automation Letters* 3(1):588–595 2017a.
104. Kamilaris, A., Kartakoullis, A. and Prenafeta-Boldú, F. X. A review on the practice of big data analysis in agriculture. *Computers and Electronics in Agriculture* 143:23–37 2017.
105. Chi, M., Plaza, A., Benediktsson, J. A., Sun, Z., Shen, J. and Zhu, Y. Big data for remote sensing: challenges and opportunities. *Proceedings of the IEEE* 104(11):2207–2219 2016.
106. Shanmugapriya, P., Rathika, S., Ramesh, T. and Janaki, P. Applications of remote sensing in agriculture-a review. *International Journal of Current Microbiology and Applied Sciences* 8(1):2270–2283 2019.
107. Weiss, M., Jacob, F. and Duveiller, G. Remote sensing for agricultural applications: a meta-review. *Remote Sensing of Environment* 236:111402 2020.

108. Zhang, H., Wang, L., Tian, T. and Yin, J. A review of unmanned aerial vehicle low-altitude remote sensing (uav-lars) use in agricultural monitoring in china. *Remote Sensing* 13(6) 1221.
109. Muchiri, G. N. and Kimathi, S. A review of applications and potential applications of uav. In: Proceedings of the Sustainable Research and Innovation Conference 2022, pages 280–283.
110. Hughes, D., Salathé, M. An open access repository of images on plant health to enable the development of mobile disease diagnostics. *arXiv Preprint ArXiv:1511.08060* 2015.
111. Mwebaze, E., Gebru, T., Frome, A., Nsumba, S. and Tusubira, J. 2019. iCassava 2019 fine-grained visual categorization challenge. Available at: https://arxiv.org/abs/1908.02900.
112. Shorten, C. and Khoshgoftaar, T. M. A survey on image data augmentation for deep learning. *Journal of Big Data* 6(1):1–48 2019.
113. Toda, Y., Okura, F., Ito, J., Okada, S., Kinoshita, T., Tsuji, H. and Saisho, D. Training instance segmentation neural network with synthetic datasets for crop seed phenotyping. *Communications Biology* 3(1):173 2020.
114. Yang, S., Zheng, L., Yang, H., Zhang, M., Wu, T., Sun, S., Tomasetto, F. and Wang, M. A synthetic datasets based instance segmentation network for high-throughput soybean pods phenotype investigation. *Expert Systems with Applications* 192:116403 2022.
115. Barth, R., IJsselmuiden, J., Hemming, J. and Henten, E. J. V. Data synthesis methods for semantic segmentation in agriculture: a Capsicum annuum dataset. *Computers and Electronics in Agriculture* 144:284–296 2018.
116. Barth, R., Hemming, J. and Van Henten, E. J. Optimising realism of synthetic images using cycle generative adversarial networks for improved part segmentation. *Computers and Electronics in Agriculture* 173:105378 2020.
117. Ahmed, M., Seraj, R. and Islam, S. M. S. The k-means algorithm: a comprehensive survey and performance evaluation. *Electronics* 9(8) 1295.
118. Guo, G., Wang, H., Bell, D., Bi, Y. and Greer, K. Knn model-based approach in classification. In: OTM Confederated International Conferences 2003, pages 986–996. Springer.
119. Wang, J. and Cherian, A. Gods: generalized one-class discriminative subspaces for anomaly detection. In: Proceedings of the IEEE/CVF International Conference on Computer Vision 2019, pages 8200–8210.
120. Katafuchi, R. and Tokunaga, T. Image-based plant disease diagnosis with unsupervised anomaly detection based on reconstructability of colors. *arXiv Preprint ArXiv:2011.14306* 2020. pages 13 https://doi.org/10.48550/arXiv.2011.14306.
121. Ivanovska, M. and Štruc, V. Y-gan: learning dual data representations for efficient anomaly detection. *arXiv Preprint ArXiv:2109.14020* 2021.
122. Bosilj, P., Aptoula, Erchan, Duckett, T. and Cielniak, G. Transfer learning between crop types for semantic segmentation of crops versus weeds in precision agriculture. *Journal of Field Robotics* 37(1):7–19 2020.
123. Narvekar, C. and Rao, M. Flower classification using cnn and transfer learning in cnn-agriculture perspective. In: 3rd International Conference on Intelligent Sustainable Systems (ICISS) 2020, pages 660–664. IEEE Publications.
124. Nguyen, T. T., Vien, Q.-T. and Sellahewa, H. An efficient pest classification in smart agriculture using transfer learning. *EAI Endorsed Transactions on Industrial Networks and Intelligent Systems* 8(26):1–8 2021.

125. Zemmour, E., Kurtser, P. and Edan, Y. Automatic parameter tuning for adaptive thresholding in fruit detection. *Sensors* 19(9):2130 2019.
126. Ostovar, A., Ringdahl, O. and Hellström, T. Adaptive image thresholding of yellow peppers for a harvesting robot. *Robotics* 7(1) 2018. ISSN 2218-6581. doi: 10.3390/robotics7010011. Available at: https://www.mdpi.com/2218-6581/7/1/11.
127. Kurtser, P., Arad, B., Ben-Shahar, O., van Bree, M., Moonen, J., van Tujil, B. and Edan, Y. Robotic data acquisition of sweet pepper images for research and development. In: The 5th Israeli Conference on Robotics, Air Force Conference Center Hertzilya 2016, Israel, 13–14 April, 2016 (vol. 2016).
128. Droukas, L., Doulgeri, Z., Tsakiridis, N. L., Triantafyllou, D., Kleitsiotis, I., Mariolis, I., Giakoumis, D., Tzovaras, D., Kateris, D. and Bochtis, D. A survey of robotic harvesting systems and enabling technologies. *Journal of Intelligent and Robotic Systems* 107(2):21 2023. doi: 10.1007/s10846-022-01793-z.
129. Chechliński, Ł., Siemiątkowska, B. and Majewski, M. A system for weeds and crops identification - reaching over 10 fps on raspberry pi with the usage of mobilenets, densenet and custom modifications. *Sensors* 19(17) 2019. ISSN 1424-8220. doi: 10.3390/s19173787. Available at: https://www.mdpi.com/1424-8220/19/17/3787.
130. Gunturu, S., Munir, A., Ullah, H., Welch, S. and Flippo, D. A spatial ai-based agricultural robotic platform for wheat detection and collision avoidance. *AI* 3(3):719–738 2022. ISSN 2673-2688. doi: 10.3390/ai3030042. Available at: https://www.mdpi.com/2673-2688/3/3/42.
131. Silwal, A., Davidson, J. R., Karkee, M., Mo, Changki, Zhang, Q. and Lewis, K. Design, integration, and field evaluation of a robotic apple harvester. *Journal of Field Robotics* 34(6):1140–1159 2017. doi: 10.1002/rob.21715. Available at: https://onlinelibrary.wiley.com/doi/abs/10.1002/rob.21715.
132. Kurtser, P. and Edan, Y. Planning the sequence of tasks for harvesting robots. *Robotics and Autonomous Systems* 131:103591 2020.
133. Oliveira, L. F. P., Moreira, A. P. and Silva, M. F. Advances in agriculture robotics: a state-of-the-art review and challenges ahead. *Robotics* 10(2):52 2021.
134. Lehnert, C., English, A., McCool, C., Tow, A. W. and Perez, T. Autonomous sweet pepper harvesting for protected cropping systems. *IEEE Robotics and Automation Letters* 2(2):872–879 2017.
135. Bawden, O., Kulk, J., Russell, R., McCool, C., English, A., Dayoub, F., Lehnert, C. and Perez, T. Robot for weed species plant-specific management. *Journal of Field Robotics* 34(6):1179–1199 2017.
136. Xiong, Y., Peng, C., Grimstad, L., From, P. J. and Isler, V. Development and field evaluation of a strawberry harvesting robot with a cable-driven gripper. *Computers and Electronics in Agriculture* 157:392–402 2019. ISSN 0168-1699. doi: 10.1016/j.compag.2019.01.009. Available at: https://www.sciencedirect.com/science/article/pii/S0168169918312456.
137. Cantelli, L., Bonaccorso, F., Longo, D., Melita, C. D., Schillaci, G. and Muscato, G. A small versatile electrical robot for autonomous spraying in agriculture. *AgriEngineering* 1(3):391–402 2019.
138. Barth, R., IJsselmuiden, J., Hemming, J. and Van Henten, E. J. Synthetic bootstrapping of convolutional neural networks for semantic plant part segmentation. *Computers and Electronics in Agriculture* 161: 291–304 2019.
139. Tian, H., Wang, Tianhai, Liu, Y., Qiao, X. and Li, Y. Computer vision technology in agricultural automation: a review. *Information Processing in Agriculture* 7(1):1–19 2020.

Part 2

Technologies: operational aspects

Chapter 5

Autonomous navigation and path planning for agricultural robots

John F. Reid, University of Illinois at Urbana-Champaign, USA

1 Introduction
2 Guidance and navigation in agricultural machine systems: current practice
3 Mission planning for agricultural machines
4 Path planning for agricultural machines
5 Search and optimization methods
6 Navigation and motion control
7 Case studies
8 Conclusion and future trends
9 Where to look for further information
10 References

1 Introduction

From the time of the draft animal and plow through the mechanization of agricultural production, path plans have been a part of performing work tasks on fields. In planting, spraying, and harvesting crops, farmers have used their best judgment to self-determine routes to cover fields efficiently and completely. The choice of path plans has varied depending on the production task, the type and size of the equipment used, and the field layout. Historically, operations have been executed in parallel passes or 'swaths' where the next pass is adjacent to the current swath, continuing this pattern to cover the area of a field.

Mechanization over the last 100 years has driven increased productivity through innovations that have resulted in wider machine systems. However, until precision agriculture, the patterns for covering fields remained similar. Tillage was an operation where plowed ground provided visible evidence of the last pass. In planting crops, marker devices were integrated into larger width implements to mark the location for the tractor to follow in an adjacent pass. Wide-boomed sprayers have required the need to mark the swath by

http://dx.doi.org/10.19103/AS.2023.0124.05

dropping foam as a visual aid to inform application coverage. Despite marking systems for planting, manual execution of navigation tasks was prone to error and resulted in path errors between swaths. Subsequent production steps in the same fields incur timeliness losses, as the machine operator needs to search and find specific locations in the headland that are aligned with the original swaths. In harvesting cereal grains, the combine harvester operator could follow the cut-edge of the harvested crop and travel in circular or spiral patterns around the field.

The era of precision agriculture has introduced new tools that have changed navigation and planning practices. Precision agriculture introduced digital technologies that have contributed to agricultural productivity through machine automation, data management, and smart-connected machines (Schueller and Reid, 2022). The resulting innovations have enabled increased productivity and convenience in performing all production job steps. Precision agriculture has provided a toolbox of capabilities that can be leveraged to increase the sustainability of agricultural production systems and prepare for the integration of agricultural robots into the landscape. Some of the critical capabilities required for both are navigation and path planning technologies.

Today's increasingly automated agricultural production systems depend on the collection, transfer, and management of information by information and computer technologies to drive increased productivity. What was once a highly mechanical system is becoming a dynamic cyber-physical system (CPS) that combines the digital domain with the physical domain (Reid, 2011). Precision agriculture solutions are the starting point in a CPS. Agricultural robots have the potential to become collaborative systems that intelligently work with the farmer to create highly productive outcomes but face many challenges in being integrated into production tasks executed by farmers. Bechar and Vigneault (2016) survey these challenges and the many crop types that have been explored in agri-robotics use-inspired research. Gao et al. (2018) provide a detailed review of wheeled mobile robots designed to address the complexity of the agricultural environment. Considering this as a continuum, the learnings from precision agriculture can provide insights that lead to highly effective commercial agricultural robot systems. This chapter describes navigation and path planning for agriculture machine systems building on the premise that precision agriculture is on a critical path that leads to smart-connected autonomous robots.

It is worth noting that many of these technologies and approaches described in this chapter also apply to robots operating in controlled environments (e.g. greenhouses or vertical farms) and for aerial platforms operating in the air space above the field. These solutions increase the degrees of freedom society has for achieving sustainable productivity to meet future food demand.

1.1 Terminology and taxonomy

Figure 1 represents the application of robotics and automation to agricultural production systems. Crop production outcomes are delivered through machine systems enabled by automation and control. Ecosystem integration makes it easier to manage automated/autonomous machine system operations through production knowledge (field terrain, boundaries, hazards, crop information, etc.) that facilitate job execution. Knowledge from the completed execution of agricultural jobs then adds to production domain knowledge.

The agricultural machine system includes a domain controller which provides supervisory control over the execution of the mission performed by the automated machine system or agricultural robot. The domain controller

- interacts with mounted/dismounted operators through a human–machine interface (HMI);
- receives inputs of production domain knowledge and situation awareness including machine state (e.g. vehicle pose and operational state information) and Local Perception of environment; and
- performs computational processes to develop a Mission Plan that provides for safe operation in the crop production application through Guidance and Navigation and Job Performance actions to achieve a defined crop production outcome.

Specific terminology that will be used throughout this chapter include the following:

- Agricultural machine system includes traditional agricultural machines and agricultural robots. This shorthand will be used to describe the collection of technologies that apply to both. Traditional agricultural machines may provide the HMI to an on-board operator/manager. Agricultural robotic systems tend to engage a dismounted or remote mission manager.
- Production job outcome is the result of an agricultural machine system successfully executing a mission to achieve specific agricultural production goals. Agricultural machine systems are often uniquely designed to achieve a specific step in the crop production process. The steps to create an outcome in the production system will be linked to the jobs-to-be-done approach of Christensen and Raynor (2003). It is likely that agri-robotics provide novel solutions that are beyond creating autonomous versions of current crop production tasks.
- Mission planning is the integration of path planning with overall objectives, goals, and tasks that an agricultural machine system accomplishes to create a job outcome. Unlike automobiles which are typically about

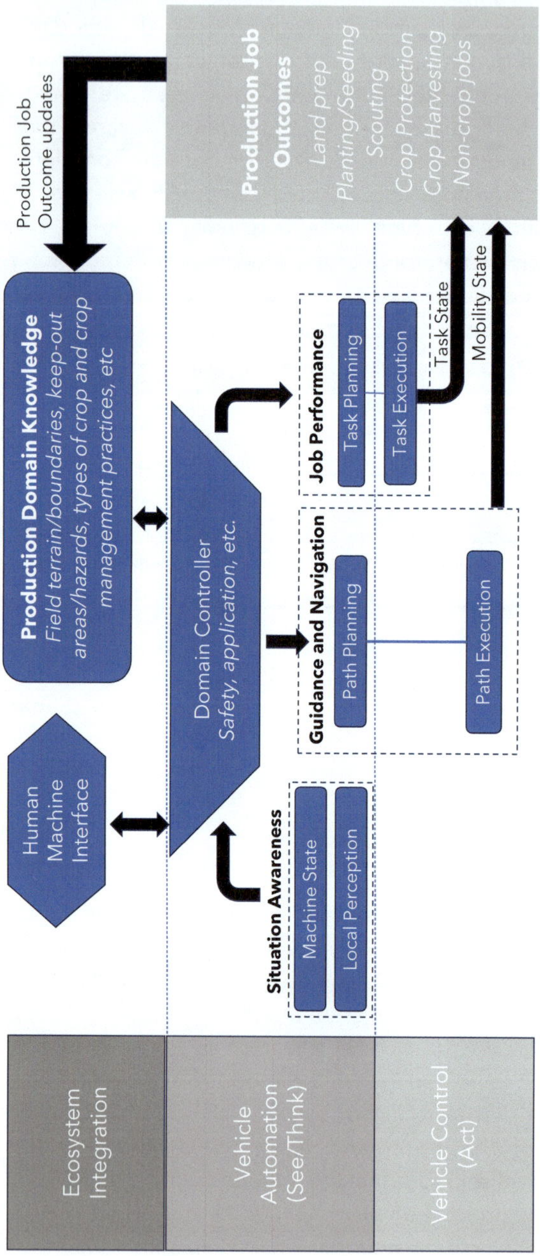

Figure 1 Key elements for navigation and planning of agricultural machine systems.

mobility, agricultural machine systems require integration of mobility and execution of the task of the agricultural machine system. Mission planning components include the planning and execution of guidance and navigation and job quality performance. This is not the same as motion planning which is an additional component that compensates for the velocities and accelerations of the agricultural machine systems.

- Path planning refers to the process of determining a sequence of actions or movements that an agricultural machine system should undertake to reach a desired goal while avoiding obstacles and adhering to path and motion constraints.
- Navigation and motion control is the process of ascertaining the position and orientation of a vehicle system and managing the agricultural machine system to follow desired trajectories with respect to a path plan.
- Pose is a machine state vector that represents the position and orientation of a machine system with respect to some reference plane. Pose provides context to the agricultural machine systems' situational awareness of the environment that can influence the path or mission plan.

2 Guidance and navigation in agricultural machine systems: current practice

Guidance and navigation of agricultural machine systems will be discussed in terms of the current practices in precision agriculture that provide many of the key elements required including highly controllable machines, precision guidance technologies, and perception sensing for situational awareness around the vehicle. In path execution, steering geometry plays a critical importance in following a path. Motion primitives (preprogrammed motion options) can be used to approximate the motion capabilities and influence the path planning process. The primary approaches that apply to common agricultural use cases will be described in terms of the contribution to global and local planning approaches.

Thomasson et al. (2019) provide an overview of modern precision agriculture technologies that have been successfully implemented by various manufacturers. Guidance and navigation control have revolutionized the way agricultural vehicles operate in the field, significantly increasing productivity and convenience. State-of-the-art precision agriculture machinery is equipped with integrated sensors that accurately estimate the vehicle's pose, including its position and orientation. This information is leveraged for the precision management of crops. Through automatic guidance, agricultural machines can effectively navigate along predefined paths or trajectories. Advanced sensors, actuators, and control algorithms work in unison to steer the vehicles, and navigation provides the necessary control commands that keep the vehicle on the intended path. The geometric relationship between the tractor and the implement is a factor in determining the guidance trajectory. In certain machine options, the implement is independently actuated to achieve the desired location in relation to the crop (i.e. active implement guidance).

Automatic guidance provides accurate navigation through electronic control of the agricultural machine system. Automatic guidance is primarily delivered via the Global Navigation Satellite System (GNSS) which uses satellite-based triangulation to provide precision position estimates. GNSS-based localization has been in commercial production for more than two decades and has evolved through several generations of change that have increased accuracy and improved performance through integration of multiple satellite constellations and signal corrections that compensate for errors induced by weather, solar, and other issues.

Figure 2 shows some of the key elements of automatic guidance and some relationships to path and mission planning. GNSS receivers mounted on the tractor enable absolute positioning of the vehicle in the field. A basic route for guidance is formed from an 'A–B' line representing a single operational plan for a swath from start goal (A) at the headland at one end of the field and a goal node (B) at the opposite end of a field. Knowing the operating width of the machine system enables the guidance system to estimate the start and goal of the next pass. This effectively creates an electronic reference line for aligning the machine system in the headland and a simple path plan for the return swath. Later these systems evolved to represent more complex A-B pathways

Field Production:

- Automatic Guidance for forming beds
- Straight A-B Guidance Lines in a Boustrophedon order (back and forth) pattern

Headland:

- Headland area to maneuver the tractor and mounted implement
- Tire tracks leave an omega-turn in the headland due to vehicle turning contraints

Figure 2 Automatic guidance enables precision management in production operations on fields. The headland area where the agricultural machine system is maneuvered for the next operational swath.

like curved tracks that parallel the shape field boundaries. Automatic guidance has been a successful innovation for precision agriculture (Schueller and Reid, 2022). Recent surveys have determined that automatic guidance through GNSS has had wide adoption by farmers (McFadden et al., 2023).

Path planning is an element of an overall mission plan that describes navigation trajectories across a field. The mission plan also includes actions to control the implement functions (e.g. raising/lowering an implement) and condition-based/autonomous behaviors along the pathway. In tree-fruit production, the execution of the mission plan may also involve path planning for one or more manipulators that perform the production work task relative to the crop.

Agricultural machine systems need to have a working area to execute transition maneuvers to access other swaths in the field. It is common in traditional farming practices for fields to have some form of headland area which is used to reorient the agricultural machine system. This area is dependent on the space required to maneuver the agricultural machine system and is typically correlated to the standard width of equipment that the farmer utilizes. The headland is typically planted with a crop as a final step in planting operations and it is often removed in the harvesting production step during the first passes. Some manufacturers of advanced agricultural machine systems offer headland management solutions to automatically perform sequences of tasks to manage an implement, automate turning, and precisely align the machine for the next pass (Fig. 3a).

Absolute positioning by GNSS does not meet the requirements for all agricultural use-cases; one example being precision guidance in standing crops (Fig. 3b). Sensor-based navigation approaches can be used to sense how the agricultural machine system needs to traverse the field with respect to the standing crop. Sensors are used to estimate the travel path or desired vehicle trajectory. A basic sensing approach is to use mechanical feelers mounted on the machine to sense the crop. More advanced systems use non-contact

(a) Advanced systems are integrating mission management and headland turns (Anonymous, 2007)

(b) Crop row orientation defining vehicle trajectory across a field (dotted line)

(c) Orchard lanes define vehicle trajectory through tree fruit (dotted line)

Figure 3 Representative global and local path plans that are commonly used in agriculture.

sensors like cameras and LIDAR to extract the guidance directrix information from the scene (Han et al., 2004).

Modern agricultural production systems use Farm Management Information Systems (FMISs) to replace the direct machine-to-operator relationship by providing process intelligence at a more sophisticated and farm operational level (Herlitzius et al., 2021). This provides farmers with management tools that can monitor and impact the quality of the job done by machine operators. FMIS is a contributor of production domain knowledge (Fig. 1) relevant to the execution of automated and autonomous missions. Setup and control values from the FMIS are automatically downloaded into specific machine systems and used in executing application maps for relevant production operations. The agricultural machine system performs its internal control loops to provide desired productivity, efficiency, and quality. Process execution data is fed back to the FMIS, where it becomes a part of the farm operation database. The concept of an FMIS is critical to success for agricultural robotics which will require significant information resources to execute a mission. For example, path planning capabilities need to be an integral part of the FMIS for autonomous agriculture, and farmers are also going to be interested in the task performance by agricultural robots. The FMIS is one approach to deploy path plans and supporting replanning. Furthermore, documentation of the completed path plans in the FMIS has value across production steps and can be shared between the various machine types for other operational steps in agricultural production.

Despite these advances, commercial guidance systems have been primarily tailored toward the large-scale production of commodity crops. In specialty crop applications, there are fewer guidance systems integrated into production machines and these solutions often require working with third parties. Orchards have other challenges including a wide variety of production practices and the need for smaller machines with sophisticated internal automation control. Additionally, a combination of plants, trees, and infrastructure constrain agricultural machine systems to traverse down specific lanes for crop management (Fig. 3c). Tree canopy vegetation, which interferes with GNSS accuracy, can impact performance. There is a continuing need to develop precision agricultural technologies to fit the diversity of production practices.

Many of the components for precision agriculture production can support and enable the commercial realization of navigation and path-planning solutions by agricultural robots. Automatic guidance of tractors and implements, machine optimization, vehicle pose, situational awareness through perception, path following across fields, and turning management in the headlands are of particular importance. Benefits of the current suite of precision agriculture technologies include

- accurate placement of crops in the field with increased plant population density;
- empowering safe and accurate operations at higher speeds;
- increasing the productivity of lower-skill operators; and
- enabling consistent productive operations over extended operating periods.

This last point includes enabling the ability to operate at night when conditions warrant it.

3 Mission planning for agricultural machines

The mission plan comprises the path plan for guidance and navigation and includes the actions required for job performance (Fig. 1) that control implement functions (e.g. raising/lowering an implement) and condition-based, autonomous behaviors along the pathway. Some machine systems have control devices to actuate the implement relative to the tractor to keep it on the proper path. In tree-fruit production, the execution of the mission plan may require path planning for one or more manipulators that perform the production task. Manipulator path planning to perform a task will not be discussed further, but some path planning methodologies like point-to-point planning can be applied to manipulator trajectory planning.

There are several approaches to integrating path planning and mission execution into agricultural machine systems. The approaches are generally related to how functions are allocated for on-vehicle vs. off-vehicle computing resources. Overall architecture design decisions will set requirements for machine system compute resources, connectivity, and edge and cloud computing. In precision agriculture, basic automatic guidance functions by GNSS-based planning or computer vision detection of crops typically include the planning and control functions for mission execution as features of on-vehicle embedded systems that communicate control commands to the low-level controllers on the vehicle. More evolved solution architectures have additional capabilities for machine-to-machine communication and cloud connectivity for advanced data management.

There are several steps in compiling and executing a production agricultural task. Using the framework of the 'jobs-to-be-done' (Christenson and Raynor, 2003), initial plans to achieve an agricultural production step can take place ahead of time with the resulting plans being distributed to the target machine system. Mission and path plans for fields can be initially preplanned as part of the steps in planning a job. In the execution phase of a mission, it is typical for local planning to be running on a domain controller (Fig. 1) comprised of embedded systems that are part of the agricultural vehicle system. There is often a need for replanning to take place when the original plan fails due to unplanned obstacles in the path or other unforeseen environmental factors. For traditional production agricultural machine systems, the operator can reposition the machine around the obstacle and reinitiate the execution of the current plan, but autonomous systems need to be designed to compensate for these events. Another common factor impacting a mission is the need to replenish the agricultural machine system with fuel or supplies. For manned precision agriculture systems, there is a need to understand where the execution of the plan was interrupted so that the machine system can return to task execution once replenished. For agricultural robots and other driverless agricultural machine systems, the need to replenish power and fuel may also be critical to successfully achieve production job outcomes. Robust mission management tools are required to remotely manage these tasks, perhaps across a fleet of smaller machines.

Autonomous missions in agricultural environments will often require a combination of global and local planning to successfully achieve mission outcomes. Global plans are descriptive of the routes required to complete a job. The job can represent the complete field area and may even include missions between fields. Global plans can be thought of as a deliberative behavior because the plan represents the requirements for the overall mission. Global planning is an important step in optimizing machine operations and takes into consideration production domain knowledge like the field size, shape, topography, known hazards, and the agronomic factors related to the production step. Modern FMISs commonly include field boundary maps and other mission-related information that the farmer can use in global planning ahead of time.

However, global plans are likely to change during mission execution because the agricultural machine system encounters situations that change that mission. As mentioned previously, covering large areas requires replenishing inputs or repowering the agricultural machine system. Replanning may also result from unexpected obstacles or from the impact of weather events that make portions of the original plan impassable. When the agricultural machine system is manned, the operator interrupts the mission, addresses the impeding challenges to allow continuation, or modifies the original plan for the remainder of the mission. Agricultural robots require an internal ability to sense, send alerts for support, or make the required adaptations to the mission plan.

Local planning functions are also important functions for agricultural machine systems to sense and react to changes in the field (e.g. obstacle detection). Reactive behaviors can be executed throughout a mission to constantly monitor and respond to local changes, obstacles, and other anomalies. Reactive planning approaches have also been used to navigate research robotics in greenhouses (Gonzalez et al., 2009) and orchards (Rovira-Más et al., 2020).

Local planning with reactive behaviors has been commercialized in precision agriculture for guiding agricultural machine systems along the direction of crop rows using computer-vision sensing (Fig. 3b) (Han et al., 2004). The structured pattern formed by planted crops provides sufficient information to define the desired pathway for short distances. Point cloud data captured by 3D sensors, such as LIDAR and stereo vision, can also be used to create a detailed representation of the local geometry around an agricultural machine system. The set of 3D points can be processed and converted into a representation suitable for local path planning. Reactive planning approaches are often part of a local planning system used for obstacle detection and avoidance.

From agricultural production missions, the quality of the execution of the mission is measured through key performance metrics including:

- level of completion of the mission plan and the effectiveness of covering an area which can also include duplication of passages. In some crop production tasks, this can include overlapping and underlapping operations during the mission;
- measuring the time engaged in the task execution (working time) vs. non-working time representing turning in the headlands or maneuvering to a new working position; and
- damage (Pour Arab et al., 2022) is another critical element due to the need to perform some jobs with respect to living plants and animal production systems.

Path planning, a critical subprocess of the overall mission plan, will be further discussed in the next section.

4 Path planning for agricultural machines

The primary planning approaches that fit the needs of agricultural machine systems are point-to-point planning and area coverage planning. Several studies have reviewed path-planning methods developed for agricultural machine systems (Santos et al., 2020; Chakraborty et al., 2022; Moysiadis et al., 2020; Yahaya et al., 2022). Computational approaches for developing path plans include occupancy grids, graphs, trees, geometric representations, and point clouds. Types of representation include information on obstacles, field boundaries, and other constraints present in the environment so that computation methods can identify pathways. Path plans have a starting point and a goal point that is specified in the coordinate system of the environment. For local plans, the starting point may be the current position of the vehicle system. Search algorithms (Section 2.4) are used to determine feasible paths from the starting point to the goal point.

Point-to-point planning involves determining a trajectory or path from a starting point to a goal point in an environment, ensuring that the agricultural machine system can navigate from one specific location to another while avoiding obstacles. Search approaches for point-to-point planning are discussed in the next section. Point-to-point planning is a fundamental building block for more complex motion planning tasks, including trajectory planning and task planning. In agricultural environments, it has significant utility, especially in the headlands of a field or in executing paths for refueling or replenishing inputs for a job step.

Area coverage path planning (Fig. 4) is of critical importance for global planning in agriculture because of the structured nature of crop production systems. Several research studies have considered optimal patterns to cover different types of field geometries (Oksanen and Visala, 2007; Pour Arab et al., 2022). Area coverage planning is influenced by various factors including the layout of the field, entry point to the field, location of known obstacles, and the crop production step to be undertaken. Field geometry, topography, and environmental factors can also influence the routes taken across agricultural fields. To maximize productive working time, the area coverage path plan may be oriented along the major axis of the field. Efficient plans improve productivity by minimizing turns (non-productive work time) and enabling longer working passes where the machine task is functioning (Oksanen and Visala, 2009). The topography of a field could be an input into planning to consider path patterns that minimize energy requirements (Anderson, 2003). For more complex fields,

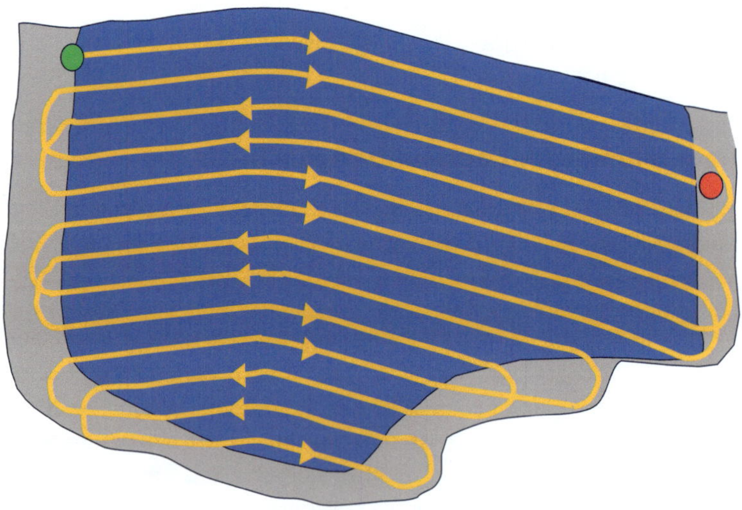

Figure 4 An area coverage plan for an agricultural field. Planning systems have defined characteristics including a starting point (green dot), a goal point (red dot). Notice that the turning patterns integrate primitives for the types of turns that the agricultural machine system can implement.

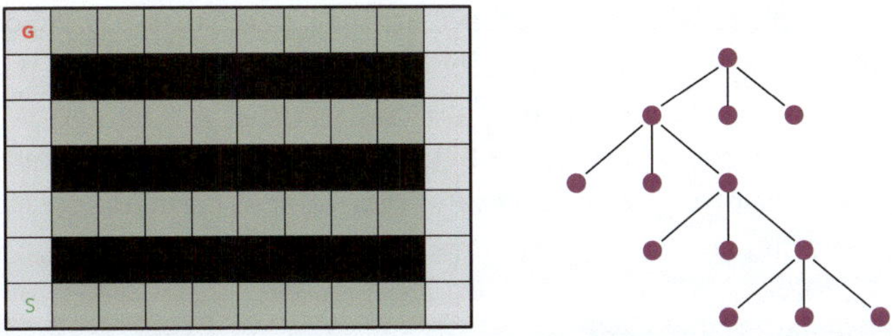

(a) Grid representation (b) Graph representation

Figure 5 Path representations for path planning. In (a) the zones of an agricultural field are represented as headland (gray), operational zones for plant management (green), and known obstacles (black). In (b) a graph representation can be used to show feasible pathways.

the path plan may need to break the field into subregions (Flann et al., 2003) with plans that connect the regions together and plan the pattern of operations within each region.

Path planning requires methods of representing the geometry of agricultural fields. Various methods aim to create digital representations of the field's shape and features, enabling efficient and accurate path planning for agricultural operations. The methods discussed here are grid-based, graph-based, and point cloud representations (Fig. 5).

In grid-based representations, a grid or lattice of cells overlays the field where each cell represents a specific area. Grid cells can be uniform in size or refined based on the field's characteristics. Grid cells that are impassable or represent obstacles can be excluded from path search algorithms. Graph-based representations use geometric primitives, such as lines, polygons, or splines, to represent the field's boundaries and features. This method provides more flexibility in capturing the exact shape and contours of the field.

Due to the parallel nature of agricultural operations, desired path trajectories can be created that have the desired parallelism or adjacency. A common approach for a path plan for a field includes the following steps:

- Characterize the perimeter of the field.
- Define the perimeters around all known obstacles and 'keep-out' areas inside of the field.
- Establish headland areas in the field for turning the agricultural machine system.
- Overlay the field with the operational paths that the vehicle must travel to accomplish the production task. These can be straight or various curved tracks based on the characteristics of the field and task.
- Connect the swaths in the headlands based on the steering geometry of the agricultural machine system to ensure complete coverage of the field.

For more complex field shapes, the field may need to be partitioned into contiguous subregions with connecting paths to move between subareas. Optimization methods can be used to connect the swaths and to connect subareas for complete coverage of the field with an agricultural machine system. There are open-source software libraries that practitioners can access to get started developing and common path plans for robotics applications (Mier et al., 2023).

Path planning functions can also be integrated into more advanced agricultural practices. Multi-vehicle plans are needed in some cases to distribute the work between two or more agricultural machine systems. There are numerous variants that can be considered. One simplified approach is to assume that machines cover separate zones or areas and develop independent plans for each zone. More complex missions may require plans for agricultural machine systems to work collaboratively in an integrated approach to achieve production objectives.

5 Search and optimization methods

Search algorithms are critical to automatically generating path plans and traversing graphs. Some approaches apply search algorithms directly to grids. Other approaches search to link tree representations of the agricultural environment. The main search and optimization methods are reviewed in this section. All the path search algorithms are based on a starting point for the robot and a goal point for the robot.

5.1 Heuristic-based algorithms

Three algorithms based on a distance-based heuristic for the cost of nodes are reviewed. Their differences are based on how the start and goal node distances factor into the heuristic, and how that choice of goal function impacts the depth of search and the optimality of the path.

In Dijkstra's algorithm, the heuristic approach for path evaluation is the shortest distance from the start node to the node being evaluated. The algorithm maintains a list of visited nodes and a list of unvisited nodes. It also maintains a cost estimate for each node, which is represented as the shortest known distance from the start node to that node. The algorithm begins by initializing the start node as the current node, setting its cost estimate to 0, and adding it to the list of unvisited nodes. It then evaluates each neighboring node of the current node and updates its cost estimates if a shorter path is found. This is done by adding the cost of traveling from the current node to the neighboring node to the current node's cost estimate and comparing it to the current cost estimate of the neighboring node. If the new cost estimate is lower, the neighboring node's cost estimate is updated, and its parent node is set to the current node. After updating the cost estimates of all neighboring nodes, the algorithm marks the current node as visited and selects the unvisited node with the lowest cost estimate as the new current node. The algorithm repeats this process until all nodes have been visited or until the goal node is reached. In most cases, Dijkstra's algorithm will find the shortest path between the start node and any other node in the graph. However, the algorithm can be slow for large graphs or graphs with many edges, since explores the graph from the starting node outward regardless of the location of the goal.

The Greedy Algorithm uses a heuristic approach that gives some value to the direction that the goal is in, without regard to the obstacles that may be in the direct line of sight. The Greedy Algorithm assesses the distance from the current location to the goal with the shortest-path distance being selected as the path step. This approach continues until the goal node is reached. This approach requires searching fewer nodes than Dijksta's algorithm but, depending on the obstacles, the resulting path can be less than optimal. A primary weakness is that the Greedy Algorithm only includes the distance from the current point on the path to the goal, irrespective of the starting point.

The A* search algorithm attempts to overcome the weaknesses of these other approaches by considering information from both the start node and goal node in the solution to find the shortest path. This is accomplished through a heuristic function to estimate the cost of traveling from the current node to the goal node and the actual cost of reaching the node from the starting node. The algorithm begins by initializing the starting node as the current node and creates an open list of nodes to explore. It then evaluates the current node and expands it to consider its neighboring nodes. For each neighboring node, the algorithm computes its cost estimate and its actual cost from the start node, using the cost function. A common cost function is a distance metric to the goal node. It then adds the current node to the open list, which is sorted by the sum of these two costs, placing the lowest-cost nodes at the front of the list. The algorithm then selects the node with the lowest cost from the open list as the new current node and repeats the process of evaluating its neighbors, computing their costs, and adding them to the open list. This process continues until the goal node is reached or until there are no more nodes to explore. A* has advantages over Dijkstra's algorithm because it does not look at unnecessary paths that do not lead to the goal. Comparative research commonly demonstrates A* outperforming other forms other heuristic search approaches (Pak et al., 2022; Vásconez et al., 2023).

There are other variants of A* that have been applied to search in path planning. Bounded A* combines depth-first branch-and-bound and A* to reduce computational time and memory requirements by implementing a cut-off to the depth of the search providing a tradeoff between optimality and computational efficiency (Flann et al., 2006). D* is an A* variant designed for dynamic environments (Stenz, 1995). D* starts with the initial path calculated by A* and includes it in a dynamic map that detects changes in obstacles. A recalculation (via A*) is triggered by significant changes in the dynamic map.

5.2 Rapidly exploring random trees

Rapidly exploring random trees (RRTs) is a sampling-based algorithm used in path planning. It is particularly effective for problems involving exploration in environments with obstacles or constraints. RRT algorithms construct a tree-like structure that rapidly explores the configuration space of the robot or system from the start point to the goal. The algorithm incrementally builds a tree by randomly sampling points in the space and connecting them to the existing tree structure. Random sampling allows the algorithm to explore different regions of the space, making it well-suited for complex and unknown environments.

The algorithm starts by initializing a tree with a single node at the starting point. Random configuration points are sampled from the space, considering any constraints or boundaries. These samples are typically generated uniformly or using a probabilistic distribution. The sampled configuration point is connected to the nearest node in the existing tree and forms an edge to a new node added to the tree. The new node is checked to confirm that the path to the sampling point is obstacle free. The sampling and expansion of the tree continue until a termination condition is met; typically, a point close to the goal with a clear pathway is reached. The path plan is the connected traversal path from the start point to the goal.

The RRT algorithm is known for its efficiency and the randomness in the sampling and the incremental tree expansion enable the algorithm to explore the space efficiently and find feasible paths. Extensions and variations of the RRT algorithm, such as RRT*, have been developed to improve optimality and convergence properties. These variations incorporate techniques like rewiring the tree, optimizing the path cost, or connecting multiple trees to enhance the planning performance.

5.3 Genetic algorithms

Genetic algorithms are a type of optimization algorithm that is inspired by the process of natural selection in biological evolution. They can be used in path-planning problems to generate optimal or near-optimal paths through complex environments (Hameed et al., 2013).

In this approach, a population of potential paths is generated, typically using random or heuristic methods. The fitness of each path in the population is evaluated by a fitness function, which measures the quality of the path based on factors such as the distance traveled, obstacles avoided, and other criteria specific to the problem. The genetic algorithm applies the principles of selection, crossover, and mutation to evolve the population toward a better set of solutions. In the selection step, paths with higher fitness values are more likely to be selected for reproduction. In the crossover step, two paths are combined to produce a new path with traits from both parents. Mutation introduces small changes in the new paths to maintain genetic diversity. After applying these steps, the population evolves over multiple generations, with each generation producing better and better solutions. Eventually, the algorithm converges on a set of near-optimal paths that reach the goal.

Genetic algorithms can find good solutions for path planning even in complex and changing environments. However, they may be computationally expensive, especially for large and complex environments, and require careful tuning of the algorithm parameters to achieve good results.

6 Navigation and motion control

The execution of a path plan is governed by navigation and motion control processes (Fig. 1) intended to efficiently allow agricultural machines to accurately follow a path plan. Overall system performance depends on powertrain architecture, the turning radius of the machine system, the effects of any trailing implement, and velocity constraints. Collectively these are known as non-holonomic constraints. Vehicle steering geometry is an overall critical factor that impacts the ability of a machine system to follow a path. The common forms of steering geometry used in agricultural vehicles are provided in Table 1.

The Ackerman steering geometry is a common form for many agricultural tractors. A typical two-wheel drive tractor with Ackermann steering geometry (Fig. 6) can move forward and backward, turn left and right, and move sideways slightly due to the steering mechanism, but it cannot move directly sideways without first turning in place. Similarly, wheeled agricultural robot with differential steering can move forward and backward and turn in place, but it cannot move sideways without executing a series of back-and-forth maneuvers. Some agriculture robots, including omnidirectional robotic systems and drones, are examples of holonomic systems that can move in any direction.

Navigation and motion control algorithms for vehicles must consider non-holonomic constraints and generate feasible paths given the vehicle's motion capabilities. One common approach is to use motion primitives, which are predefined sequences of control inputs that allow the vehicle to move in a certain

Table 1 Steering geometries used in agricultural machine systems

Steering geometry type	Description	Benefits
Ackermann Steering	A simple steering mechanism where the front wheels or tracks are linked together with a common center point, and each wheel or track follows a different turning radius.	Allows the vehicle to execute smooth turns while maintaining the desired track width.
Crab Steering	Crab steering enables a vehicle to move sideways by steering all wheels or tracks in the same direction. By moving the entire vehicle diagonally, it allows for lateral movement without changing the vehicle's orientation.	This can be useful in tight spaces or when precise positioning is required.
Differential Steering	Differential steering is commonly used in tracked vehicles and some agricultural robots. It involves controlling the speed and direction of the tracks independently.	By varying the speed of the tracks in opposite directions, the vehicle can turn on the spot, making it highly maneuverable.
Four-Wheel Steering	Four-wheel steering involves adding an additional steering mechanism to the rear wheels of a vehicle. This configuration allows the front and rear wheels to steer in the same direction, improving maneuverability and reducing the turning radius.	Four-wheel steering can enhance the agility and stability of agricultural vehicles.
Articulated Steering	Articulated steering involves dividing the vehicle into two or more segments connected by a joint. The front and rear segments can pivot relative to each other, enabling the vehicle to navigate tight turns and uneven terrain.	Articulated steering is commonly used in large agricultural machinery, such as 4WD tractors.
Skid Steering	Skid steering is a steering mechanism where the vehicle turns by selectively braking or accelerating the wheels or tracks on one side while the other side remains stationary. This results in a skidding or sliding motion, allowing the vehicle to rotate in place.	Skid steering is commonly used in small agricultural robots and unmanned ground vehicles.

way. These motion primitives can be combined to form more complex paths that satisfy the non-holonomic constraints of the vehicle. For most production agriculture vehicles, there are three primary motion primitives that are used to represent turning behaviors (Zhou et al., 2015). The forward-turn (as illustrated in the headland of Fig. 2), three-point turns, and the double-rounded corner (rounded staple-shaped turns). The forward-turn and the double-rounded corner are performed in a forward direction of motion, while three-point turns

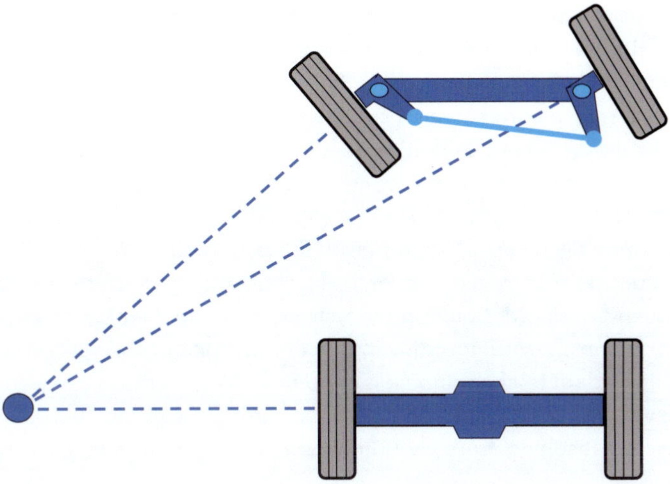

Figure 6 Ackermann steering geometry defines the types of motion an agricultural machine system can perform. The center of the turning radius is illustrated.

require forward and reverse capabilities to execute. Three-point turning can be challenging for towed implements, so they would normally use the two forward-motion turning motion primitives.

Motion control for an agricultural machine system considers the motion actuation devices on the agricultural machine systems and uses control algorithms to coordinate the movement of the systems along the desired path plan to execute the goals of a mission plan (Fig. 1).

Each type of steering geometry in Table 1 relates to a unique approach to control the overall motion of the agricultural machine system. Agricultural machine systems with Ackermann steering geometries are controlled by the steering angle of the front wheels, whereas wheeled mobile robots develop different vehicle headings by differential speeds on the drive wheels.

Situational awareness is critical for motion control and is provided by various sensors. The vehicle state can include GNSS sensors that provide the instantaneous pose of the vehicle. Local perception sensors can provide information about the relative position with respect to crops, trees, and other fiducial references in the agricultural environment that can influence navigation and motion control. Agricultural machine systems used in modern precision agriculture have benefited from the robust performance of GNSS integrated with pose sensing to account for terrain influences on navigation. Crop-referenced guidance through perception of the crop rows is another approach for high-accuracy motion control. Sensor fusion approaches (Zhang et al., 1999; Han et al., 2004; Han et al., 2001) integrate multiple sources of information together to enhance the quality of pose sensing. Applications that require localization under-canopy or in GNSS-deprived environments require the detection of environmental features to determine the relative location of the agricultural machine system (Fig. 3c).

A trajectory planner plans how the agricultural machine system moves from an arbitrary offset position to be aligned with a path plan, including the limitations that non-holonomic constraints put on the vehicle motion. The result is a local planning approach that aligns the initial state of the agricultural machine system with a goal aligned with the path plan.

Agricultural robots often operate in challenging outdoor environments with uneven terrain. Motion control systems must be capable of adapting to these conditions, including adjusting wheel or track movements to maintain stability and adapt to changing field conditions. Pose sensing can also provide valuable information on the relative motion the agricultural vehicle system is experiencing resulting from motion slips due to terrain and changing field conditions.

7 Case studies

Two examples will be used to illustrate applications of navigation and planning for agricultural robots. Both studies involve conventional agricultural equipment converted into driverless machine systems representing larger-scale agricultural robots. Modification of conventional precision agricultural machine forms to enable autonomous operation has several benefits:

- Task automation has already been significantly developed in base, thus providing an autonomous development platform with significant capability.
- Manual operation of the system can be used to overcome some of the more challenging job steps required to prepare for autonomous field operation (e.g. public road transport, moving between fields).
- Existing production domain knowledge management tools used for precision farming can have some of the necessary information required to set up autonomous operation (field boundaries, terrain characteristics, known hazards, agronomic and machine state documentation, etc.).
- System productivity is potentially at the same scale as human-operated machines providing opportunities to understand the comparative performance between manned and autonomous machines.

Some disadvantages include not pushing R&D to explore innovative form factors that might be possible through agricultural robots (Herlitzius et al., 2021). And even in a similar form-factor, refactoring an existing production system component into an autonomous system requires consideration of the additional sensing and control technologies required to replace a machine operator.

The first example will demonstrate the development and controlled execution of an orchard management mission assuming limited GNSS capabilities for absolute positioning. This application uses a combination of reactive behaviors and open loop turning behaviors to follow a predetermined path and execute missions to manage ground cover in tree-fruit operations. The second case study will demonstrate a multi-vehicle system performing orchard management under the supervision of a remote operator using a custom developed FMIS and compare the productivity to manned operations.

7.1 Basic orchard mowing mission

The motivation for this work was driven by a technology roadmap developed by the Washington State Tree Fruit industry in 2002. To compete in global markets, there was a need to look at new technologies and systems-level management approaches to address the significant challenges of labor cost and availability. Autonomous robots were of interest for addressing both production and harvest costs. This project provided an assessment of autonomous systems in the execution of a typical mission for demonstration to growers in the region.

The agricultural robot platform was a John Deere 4710 compact utility tractor (CUT) (Fig. 7) converted to driverless operation. The vehicle was modified with specialized controllers for the actuation of steering, brakes, throttle, and shifting through low-level vehicle commands on a CAN-bus. A mower unit on the CUT was used for mowing grass as the primary job to be done in the orchard.

The apple orchard had a trellised architecture with the trees trained to grow in a giant spiral trellis planar architecture. Tree height was approximately

Figure 7 (a) Experimental JD 4710 compact utility tractor (CUT) platform used for mowing operations in a vertical trellis apple tree architecture and (b) path plan for an orchard block.

4 m, spaced in rows 3 m apart, and the depth of the canopy, when trimmed, was about 0.5 m between travel lanes. Tree-fruit branches were not closely pruned, and some branches extended into the travel lanes.

The requirements for the autonomous mission were to provide safe operation including the detection of people and obstacles in the path of the vehicle. If obstacles were detected, the vehicle speed was attenuated with a full-stop at a safe distance. The 3 m spacing of rows was close to the width of the tractor platform which resulted in a single pass per lane being able to adequately mow the crop. Typical orchard blocks varied from 0.8 ha to 4.0 ha and the objective was to traverse the orchard lanes at 1.5 m/s. It was also desired to have the capability to operate day or night to understand opportunities for spraying and other production tasks.

The platform was modified with the addition of a forward-looking sensing suite that included GPS, LIDAR, RGB, and thermal cameras mounted on the rollbar of the CUT. The vehicle could be controlled in both forward and reverse directions, but there was no rearward situation awareness. The mission supervisor included an obstacle-avoidance behavior that maintained a safe speed and increased attenuated the vehicle speed based on the distance to detected obstacles, eventually stopping 2 m from an obstacle. Although GPS was available, these missions assumed that the mission was GPS-denied. Reactive planning was used to traverse the orchard lanes based on the perception sensors and visual odometry was used to estimate the distance traversed down the lanes. Detection of the end-of-lane was used to trigger a turning behavior based on dead reckoning and detection of the beginning of a lane triggered the initiation of mowing operations.

The path planning for this application was determined by a precision survey of the locations of the tree architecture and the passage lanes between trees. Although manually planned, this was able to be reused many times and would be expected to be valuable for use over the life cycle of the trellis architecture (more than 5 years).

The mission started at the topmost lane of the block and moved in a simple serpentine pattern row-by-row to the last lane of the orchard block. The mission was developed and stored in an XML file format that represented behaviors required to traverse a lane and mow grass, behaviors required to execute a turn from one lane to the next, and behaviors based on the detection of obstacles.

As part of the preparation steps for an autonomous mission, the orchard block was manually scouted to determine if there were obstacles or hazards from normal operations. Once all conditions were checked and cleared, autonomous missions were executed.

The vehicle was initiated using point-to-point planning and GNSS-based positioning to enter the first segment for orchard mowing. The remainder of the mission was executed using a reactive approach that centered the vehicle between the trellises and used odometry to estimate the distance down the row that the vehicle system had traveled. Detection of the ends of rows were used to transition into a turning behavior that transitioned the vehicle into the next orchard lane. This process was repeated until the completion of mowing along all lanes of the orchard block.

Reactive approaches require robust performance of the machine perception system. Capabilities developed for this use-case included software 'pruning' of tree branches to determine a clear path, labeling to identify trees/obstacles, detection of obstacles/human presence, and control of 'safe speed' behavior for reacting to obstacles. The system had the capability to use perception for path behaviors of following left or right tree rows, centering in the row, and detecting of end of row to initiate turning behavior. Path tracking accuracy was documented and varied from 4.4 cm to 7.6 cm from the centers of tree rows.

While functional and successful, there were many challenges related to the navigation and path planning discussed in this chapter. In general, there were challenges that could be addressed if quality GNSS could be integrated into the solution. And rearward-looking perception would have enabled safe operation in reverse that would have allowed more use of three-point turns. Finally, the highly planned XML execution files would require integration into an FMIS to simplify the systems planning and management.

Orchard growers involved in the testing and evaluation said it was difficult for them to imagine the CUT tractor as the final product, and they preferred a solution that was custom developed to meet the needs of their business. Mowing was a good representative application. One grower indicated that he has personnel involved in mowing all the daylight hours for 4 months of the year. Higher value production tasks would be for spraying of chemicals where the benefit would be to remove the operator from exposure. This was a task where it was increasingly hard to find workers that wanted to perform this operation. Growers also expressed interest in how autonomous operations could support documentation and traceability. Autonomous agricultural machine systems can provide more information on performed tasks than is typically reported by human operators. While passing through the orchard for mowing tasks, perception technologies could also document crop growth, observe the development of trees and fruit, and be of use in disease detection or changes in the canopy.

7.2 Multi-vehicle mission planning for citrus grove operations

The second use-case reports on the field deployment of multiple tractors for mission execution for citrus grove management as reported by Moorehead et al. (2012). Orchards and groves have many labor-intensive tasks which must

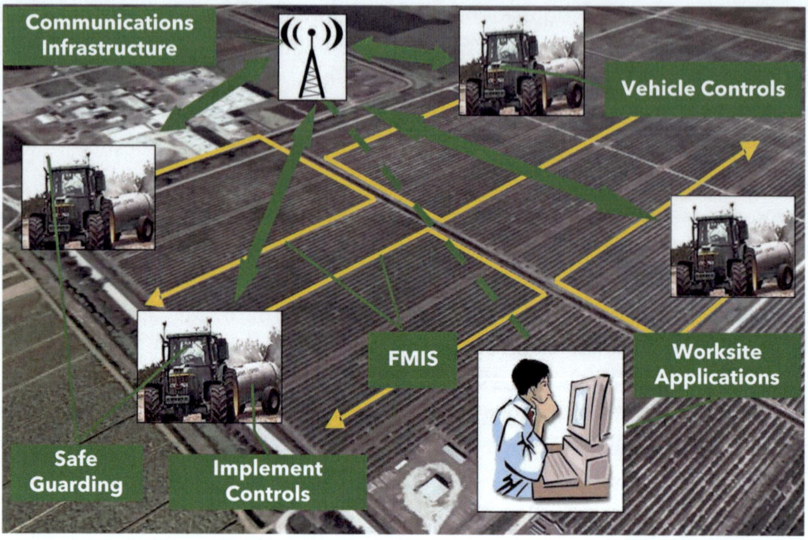

Figure 8 Representation of the multi-vehicle mission for orchard management. A remote-manager-controlled operations of multiple vehicles through a custom FMIS performing crop protection and ground maintenance. Source: Used with permission from: Moorehead et al. (2012).

be performed to help maximize yields. This work demonstrated the conversion of conventional tractors into agricultural robots supporting spraying and mowing tasks, though remotely managed by an operator (Fig. 8). The remote operator could be in the comfort of a truck near the field operations or in an office in a station where the operator could have individual user-interfaces to monitor each of the autonomous systems.

As labor costs rise, and reliable labor becomes harder to find, orchard owners are looking for methods to improve productivity and lower costs in the long run. Some labor-intensive tasks in an orchard, such as harvest, are very difficult to automate, while others, such as mowing and spraying, are easier applications for initial automation. Mowing and spraying are done with one person per tractor and some orchards have multiple tractors in the field performing these tasks simultaneously. It is possible to improve productivity by having a single operator supervise several semi-autonomous vehicles, intervening only when a tractor encounters something it cannot handle autonomously.

A modified John Deere 6430 tractor, like the traditional product used by this orchard in mowing and spray, was retrofitted for autonomous execution of mowing and spraying tasks. It has 67 PTO kW which is sufficient to power the 3.0 m and 4.6 m towed mowers as well as a blast sprayer. The base tractor included GNSS-based navigation and automatic drivetrain control that was accessed for autonomous control.

A global path planner was used to develop plans to cover an orchard block as efficiently as possible. An analysis of traditional operating practices provided evidence that planned paths would result in less distance travel in the coverage of an orchard block. A Generalized Traveling Salesman Problem (GTSP) was used with each orchard row treated as a city that can be entered from the top or the bottom. A dynamic programming approach transformed the GTSP into a normal Traveling Salesman Problem (TSP), but the TSP is combinatorial and NP-hard in the general case. The problem required optimization over approximately 40 rows of trees in an orchard block, so computing the guaranteed optimal solution is intractable. The orchard had a lot of consistent structure and in most cases neighboring rows could be traversed sequentially. In this research, the planner was initialized using a greedy solution, and then a heuristic neighborhood search was performed that swapped the order and direction of rows and blocks of rows to find a near-optimal solution to the TSP.

Figure 9 shows plans for an entire zone as generated by a greedy search algorithm (a) and the optimized planner (b). The total length of the greedy plan is 32 703 m and would take the tractor 5 h and 14 min to complete with no stops. The optimized plan has fewer duplicate rows where the tractor is driving down a row without performing work and is only 28,686 m long, which would take 4 h and 44 min to complete. This result is 9.6% time savings for a single pump zone in the orchard.

Once the path plan is generated, it was downloaded to the tractor by the remote operator for execution. The vehicle used RTK GPS to navigate the

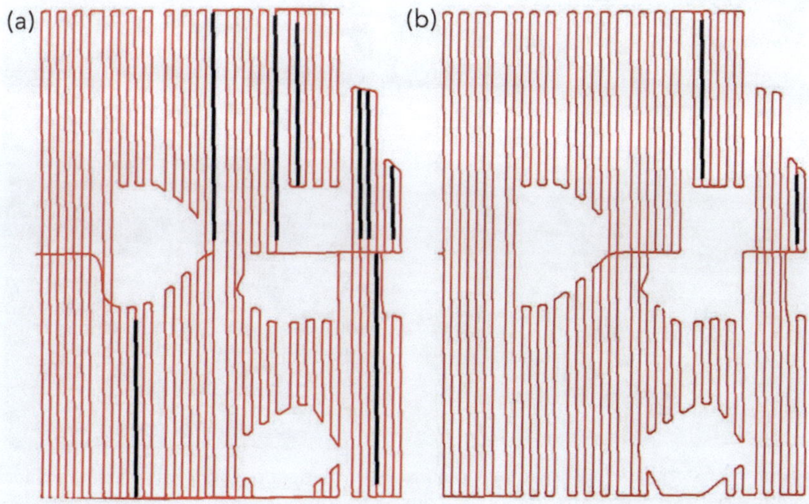

Figure 9 (a) Path plan for one zone using greedy search. (b) Path plan for the same zone using optimized planner. Rows that are traversed twice are shown in black. The optimized path is 9.6% shorter. Source: Used with permission from: Moorehead et al. (2012).

vehicle along the path plan. However, environmental variation in planting, tree growth rates and manually hedged resulted in trees not being perfectly symmetrical around the planned path. To compensate for this, a local plan was used to create a behavior for traversal down the orchard lanes. This reactive behavior was based on a tree classifier. The travel path was computed from a tree map to find the lateral offset within the row relative to the original planned path that will keep the tractor from running into trees on either side. This offset was computed continuously and applied to the original planned path to create a new path for the tractor to follow (Fig. 10).

A custom user interface was developed as a basic FMIS for use in this project. Management processes assumed the missions would be executed by a remote operator with the ability to select the desired zones to operate in and to assign missions to respective agricultural robots. Integrating the agricultural robot systems with the orchard FMIS to facilitate preparation, planning, and to supported work documentation. Tree spraying and ground maintenance missions occur over the entire orchard operation at regular frequencies all year.

Figure 10 Example of reactive path plan for row guidance. Top left is an orchard lane from the perspective of the tractor. Bottom left is a representation of the orchard lane with tree cells colored green. Right image shows a segment of the global path plan as white line and local path plan (reactive) adjusted path as pink line. Source: Used with permission from: Moorehead et al. (2012).

Daily, the remote manager planned and assigned missions to the agricultural robots and coordinated these missions with other manned systems performing similar missions.

Both manned and robotic missions were documented in this study. Manned operations did not have the advantage of a formal plan to follow, which resulted in significant inefficiency. The missions of multiple robots could be executed consistently by the remote manager. Additionally, the robotic systems could consistently maintain a higher safe travel speed than traditional machine operators would be willing to operate. Over months of operation, the remote manager controlling two agricultural robots was able to consistently achieve productivity improvements of 25–40% over manned tractors performing similar tasks.

Agricultural robot management also requires expanded skills over those required to manually drive the missions. The remote manager requires skills for planning and managing the remote mission. And, in this research, the remote manage needed to respond to challenges, like perception system detections of potential obstacles, and make decisions on the appropriate problem resolution for continuing the mission.

8 Conclusion and future trends

Navigation and path planning are critical elements of how agricultural machines create that outcomes produce food, fiber, and material. Prior to mechanization, farmers used intuitive approaches to plan how to cover fields that became standard practice as machine forms grew and productivity increased. Precision agriculture and related digital technologies have created an innovation platform that support precision agriculture and the expansion of agricultural robotics. Area coverage path planning has top significance as a deliberative planning tool for agriculture and reactive planning approaches support navigation with respect to growth crop, navigation with respect to embedded infrastructure, and obstacle detection and avoidance. Key elements for the planning and execution of path planning are part of the information in FMIS. The transition from precision agricultural machinery is likely to involve a mixture of agriculture robots and automated conventional machine systems which need to have alignment of path and mission planning practices that FMIS can provide. The continued expansion of digital technologies into agriculture and the convergence with agricultural robots provides critical resources that can be used to meet increasing food demand and enable the adoption of sustainable production practices.

This chapter has focused on how the future of autonomous navigation and path planning for agricultural robots is an extension of the existing precision agriculture platform. As commercial autonomous systems mature, it is likely that there will be mixed fleets, consisting of both conventional agriculture machines and agricultural robots. Some manufacturers have already started discussing and deploying special commercial releases of autonomous systems based on modified commercial products.

Current farming practices create some challenges for mixed fleets. For example, small agriculture robots are being developed for crop scouting and in-season cover crop applications in row crops (Rowbot; TerraSentia). In general, these systems are small enough to travel in interrow spaces. However, when headlands are planted, access to internal rows are blocked. Future field planning systems must consider planting patterns that accommodate mixed fleets and create travel paths to provide access to crop rows. High-accuracy precision planting solutions based on electric drive row units have the capacity to implement these plans providing that planning software can generate the appropriate planting patterns.

Agricultural robots heavily rely on perception sensors for situational awareness in their immediate surroundings. A review of perception technologies used in agriculture is provided by Wang et al. (2022). The robustness of sensors and their coverage of the local area plays a vital role in perception engineering. Sensors need to differentiate between areas that the robot can navigate through and impassible areas, even as crops continue to grow and potentially obstruct the path.

Integrating agricultural robots with existing FMIS will maximize their utility and effectiveness. This integration may require coordinating the efforts with human workers, as well as with other autonomous robots or machines.

Navigation and path planning can contribute to the sustainability of current agricultural practices, particularly in the context of sustainable and regenerative agriculture (Basso et al., 2021). Smaller robot systems are being developed to plant cover crops during the growing season, reducing greenhouse gas emissions through continuous ground vegetation (Rowbot; Cavendar-Bares and Lofgren, 2014; McAllister et al., 2019; TerraSentia). Full-field planning should consider the requirements of small robots and devise feasible plans that enable operation across all field areas. As regenerative agriculture emerges, with a focus on diversified crops and animals on the landscape, agricultural robots can provide flexibility in managing these systems.

The translation of research into commercial practice creates opportunities for alternative business models. Direct adoption of agricultural robot technology by farmers needs to consider ways to reduce the additional workload. Major machinery manufacturers are converting existing production machines into driverless solutions, which can be operated either with a mounted or dismounted operator, which has been discussed to provide benefits in the transition to autonomous operation. For agricultural robots to be widely adopted, the platform management systems must include well-defined user experience and understand the potential to be self-managing. Another approach is to shift the operational complexity to service providers who can offer autonomy-as-a-service. To facilitate working with patterns, FMIS need to define formats and data exchange capabilities for sharing field boundaries, obstacle maps, and plans from prior production steps with service providers.

9 Where to look for further information

Open-Source Software for robotics and path planning:

- ROS Agriculture. http://wiki.ros.org/agriculture.
- Fields2Cover. https://github.com/Fields2Cover/Fields2Cover.

Automated and Robotic Agricultural Machine Systems:

- Höffmann, M., Patel, S. and Büskens, C. (2023). Optimal Coverage Path Planning for Agricultural Vehicles with Curvature Constraints. *Agriculture* 13(11), 2112. https://doi.org/10.3390/agriculture13112112.
- Min, S., Feng, X., Pan, S., Song, X. and Jiang, L. (2023). A Collaborative Path Planning Method for Intelligent Agricultural Machinery Based on Unmanned Aerial Vehicles. *Electronics* 12(15), 3232. https://doi.org/10.3390/electronics12153232.
- Dongbo, X., Chen, L., Liu, L., Chen, L. and Wang, H. (2022). Actuators and Sensors for Application in Agricultural Robots: A Review. *Machines* 10(10), 913. https://doi.org/10.3390/machines10100913.

10 References

Anderson, N. (2003). Method and system for determining an energy-efficient path of a machine. Patent US6728607B1.

Anonymous. (2007). John Deere iTEC Pro in action. Available at: https://www.youtube.com/watch?v=8N6v5BRgAKA.

Basso, B., Jones, J. W., Antle, J., Martinez-Feria, R. A. and Verma, B. (2021). Enabling circularity in grain production systems with novel technologies and policy. *Agricultural Systems* 193, 103244. https://doi.org/10.1016/j.agsy.2021.103244.

Bechar, A. and Vigneault, C. (2016). Agricultural robots for field operations: concepts and components. *Biosystems Engineering* 149, 94–111.

Cavendar-Bares, K. and Lofgren, J. (2014). Robotic platform and method for performing multiple functions in agricultural systems. Patent US9265187B2.

Chakraborty, S., Elangovan, D., Govindarajan, P. L., ELnaggar, M. F., Alrashed, M. M. and Kamel, S. (2022). A comprehensive review of path planning for agricultural ground robots. *Sustainability* 14(15), 9156. http://doi.org/10.3390/su14159156.

Christensen, C. and Raynor, M. (2003). *The Innovator's Solution: Creating and Sustaining Successful Growth*. Cambridge, MA: Harvard University Press.

Flann, N., Hansen, S. and Gray, S. (2003). A method and system for determining an efficient path. Patent US6934615B2.

Flann, N., Hansen, S. and Gray, S. (2006). Point to point path planning. Patent US7079943B2.

Gao, X., Li, J., Fan, L., Zhou, Q., Yin, K., Wang, J., Song, C., Huang, L. and Wang, Z. (2018). Review of wheeled mobile robots' navigation problems and application prospects in agriculture. *IEEE Access* 6, 49248–49268.

Gonzalez, R., Rodríguez, F., Sánchez-Hermosilla, J. and Donaire, J. G. (2009). Navigation techniques for mobile robots in greenhouses. *Applied Engineering in Agriculture* 25(2), 153–165.

Hameed, I. A., Bochtis, D. and Sørensen, C. A. (2013). An optimized field coverage planning approach for navigation of agricultural robots in fields involving obstacle areas. *International Journal of Advanced Robotic Systems* 10(5), 231.

Han, S., Zhang, Q. and Reid, J. F. (2001). A navigation planner for automatic tractor guidance using machine vision and DGPS. *IFAC Proceedings* 34(9, July), 209–214.

Han, S., Zhang, Q., Ni, B. and Reid, J. F. (2004). A guidance directrix approach to vision-based vehicle guidance systems. *Computers and Electronics in Agriculture* 43(3), 179–195.

Herlitzius, T., Hengst, M., Grosa, A. and Fichtl, H. (2021). Fieldswarm technology for tillage and crop care: paradigm change in agricultural technology: bigger, faster, wider is going to change towards smart, connected and modular. *Automatisierungstechnik* 69(4), 316–324. https://doi.org/10.1515/auto-2020-0127.

McAllister, W., Osipychev, D., Davis, A. and Chowdhary, G. (2019). Agbots: weeding a field with a team of autonomous robots. *Computers and Electronics in Agriculture* 163, 104827, https://doi.org/10.1016/j.compag.2019.05.036.

McFadden, J., Njuki, E. and Griffin, T. (2023). Precision agriculture in the digital era: recent adoption on U.S. farms, EIB-248 USDA, Economic Research Service.

Mier, G., Valente, J. and de Bruin, S. (2023). Fields2Cover: an open-source coverage path planning library for unmanned agricultural vehicles. *IEEE Robotics and Automation Letters* 8(4), 2166–2172.

Moorehead, S., Wellington, C., Gilmore, B. and Vallespi, C. (2012). Automating orchards: a system of autonomous tractors for orchard maintenance. Proceedings of the IEEE International Conference of Intelligent Robots and Systems, Workshop on Agricultural Robotics, Vilamoura-Algarve, Portugal, 7–12.

Moysiadis, V., Tsolakis, N., Katikaridis, D., Sørensen, C. G., Pearson, S. and Bochtis, D. (2020). Mobile robotics in agricultural operations: a narrative review on planning aspects. *Applied Sciences* 10, 3453.

Oksanen, T. and Visala, A. (2007). Path planning algorithms for agricultural machines. *Agricultural Engineering International: CIGR Journal* IX, https://cigrjournal.org/index.php/Ejounral/article/view/940.

Oksanen, T. and Visala, A. (2009). Coverage path planning algorithms for agricultural field machines. *Journal of Field Robotics* 26(8), 651–668. https://doi.org/10.1002/rob.20300.

Pak, J., Kim, J., Park, Y. and Son, H. I. (2022). Field evaluation of path-planning algorithms for autonomous mobile robot in smart farms. *IEEE Access* 10, 60253–60266.

Pour Arab, D., Spisser, M. and Essert, C. (2022). Complete coverage path planning for wheeled agricultural robots. *Journal of Field Robotics* 40(6). , 1460–1503.

Reid, J. (2011). The impact of mechanization on agriculture. *The Bridge* 41(3), 22–29.

Rovira-Más, F., Saiz-Rubio, V., and Cuenca-Cuenca, A. (2020). Augmented perception for agricultural robots navigation. *IEEE Sensors Journal* 21(10), 11712–11727.

Rowbot. Rowbot Systems LLC. Available at: https://www.rowbot.com.

Santos, L. C., Santos, F. N., Solteiro Pires, E. J., Valente, A., Costa, P. and Magalhães, S. (2020). Path planning for ground robots in agriculture: a short review. In: 2020 IEEE International Conference on Autonomous Robot Systems and Competitions (ICARSC), Ponta Delgada, Portugal, pp. 61–66, https://doi.org/10.1109/ICARSC49921.2020.9096177.

Schueller, J. and Reid, J. (2022). The digitalization of production agriculture. *The Bridge* 52(2), 35–40.

Stentz, A. (1995). The focused D* algorithm for real-time replanning. In: Proceedings of the International Joint Conference on Artificial Intelligence (IJCAI), Montreal, Quebec, Canandapp 1662–1669.

Thomasson, J. A., Baillie, C. P., Antille, D. L., Lobsey, C. R. and McCarthy, C. L. (2019). Autonomous technologies in agricultural equipment: a review of the state of the art. *ASABE Distinguished Lecture* 40, 1–17.

Vásconez, J. P., Basoalto, F., Briceño, I. C., Pantoja, J. M., Larenas, R. A., Rios, J. H. and Castro, F. A. (2023). Comparison of path planning methods for robot navigation in simulated agricultural environments. *Procedia Computer Science* 220, 898–903.

Wang, T., Chen, B., Zhang, Z., Li, H. and Zhang, M. (2022). Applications of machine vision in agricultural robot navigation: a review. *Computers and Electronics in Agriculture* 198, 107085.

Yahaya, Z. A., Buyamin, S., Chiroma, H., Azimi Mahmud, M. S., Hassan, F. and Badi, A. A. H. (2022). Paths planning for agricultural robots: recent development, taxonomy, challenges, and opportunities for future research. In: 2022 IEEE 20th Student Conference on Research and Development (SCOReD), pp. 7-12. IEEE, Bangi, Malaysia. https://ieeexplore.ieee.org/document/9973836.

Zhang, Q., Reid, J. and Noguchi, N. (1999). Agricultural vehicle navigation using redundant sensors. *SAE Transactions Journal of Commercial Vehicles* 108(2), 27–31.

Zhou, K., Jensen, A. L., Bochtis, D. D. and Sørensen, C. G. (2015). Quantifying the benefits of alternative fieldwork patterns in a potato cultivation system. *Computers and Electronics in Agriculture* 119, 228–240.

Chapter 6

Advances in human–robot collaboration in agricultural robotics

George Adamides, Agricultural Research Institute, Cyprus; and Yael Edan, Ben-Gurion University of the Negev, Israel

1 Introduction
2 Scope of human–robot collaboration
3 Functionality, design and taxonomy
4 Interfaces/interaction design and usability
5 Conclusion and future trends
6 Where to look for further information
7 References

1 Introduction

The rapidly increasing global population requires significant intensification of food production while properly utilizing available resources to maintain high nutritional quality and production efficiency (Kamilaris and Prenafeta-Boldú, 2018). Smart farming has emerged as a technology-driven approach using, among others, Internet of Things, data analysis, sensors, and robotics to improve the quality and quantity of crop production (Stafford, 2000). Robotic systems can contribute to smart farming by collecting timely accurate information and applying the right treatment, in the right place at the right time (Moysiadis et al., 2020). By making timely informed localized decisions and actions based on real-time data, the overall efficiency of the farming process is improved, resulting in higher yields, improved quality, and more sustainable and efficient practices (Gebbers and Adamchuk, 2010).

Data gathered from Internet of Things stations (Moysiadis et al., 2021), drones (Panday et al., 2020), and robots (Botta et al., 2022) can improve the transparency of production, improve decision-making and resource use efficiency (e.g. water and pesticides), as well as reduce labor and input costs. The technical feasibility of agricultural robots has been proven in many agricultural tasks and greenhouse applications since the early 1980s (Edan et al., 2023), becoming an indispensable tool to reduce the workload and increase the productivity of agricultural processes. Agricultural robots have

http://dx.doi.org/10.19103/AS.2023.0124.07

been implemented for repetitive tasks, trying to reduce the farmer's workload and optimize process times and costs, such as land preparation (Oliveira et al., 2021), weeding (Slaughter et al., 2008), spraying (Berenstein and Edan, 2017), harvesting (Arad et al., 2020; Kootstra et al., 2021), monitoring and inspection (Kubota and Ozawa, 2008), and mapping (Moreno et al., 2014). Despite the many developments (Adamides and Edan, 2023; Edan et al., 2009) and intensive research and development (R&D) since the late 1980s, very few systems are used in practice (Edan et al., 2023). Most systems to date have been implemented as autonomous or semi-autonomous systems.

As opposed to industrial robots which operate in controlled environments, agricultural robots must deal with several complexities due to the non-controlled and dynamic environment they operate in (Bac et al., 2014). The operations in agricultural tasks are difficult due to the high environmental variability, complicated by the unstructured and dynamic environment and inherent variability of biological materials (Kurtser and Edan, 2018), e.g. detach a fruit crop of variable size, shape, color, and shading, and at a random unknown location in changing illumination conditions and obstructed by obstacles (branches, leaves). Robots move on unstructured and unpredictable terrain. Furthermore, the robots work under volatile and uncontrolled climatic conditions, such as wet muddy soil, strong winds, dust in the atmosphere, and different light settings depending on the sun's location or clouds. This creates a need for highly complex systems which are difficult to develop and costly to operate. Additionally, since each crop is different from the other (round large watermelons vs. elongated cucumbers vs. small cherry clustered tomatoes) and influenced by growing conditions (there are changes depending on horticulture practices, location, weather and even in the same crop along the season and between years) the variability increases the complexity. Interestingly, humans can easily fit themselves into such changing and highly variable environments due to their high perception skills. By taking advantage of the human perception and adaption skills, domain knowledge, and incorporating this with the robot's accuracy and consistency, a synergetic human-robotic system can be simplified and result in improved performance (Adamides and Edan, 2023; Edan et al., 2022; Vásconez and Cheein, 2022).

Human-robot collaboration (HRC) approaches and human-robot interaction (HRI) applications can provide a possible way to increase performance, efficiency, and effectiveness. By keeping the human-in-the-loop (Sreeram and Nof, 2021), the robotic system can be simplified, be more robust to the high variability, and enable automation of agriculture practices with economic feasibility. Human-robot collaboration Semeraro et al. (2023) is 'the approach that explores the interaction between a human and a robot, during the completion of a common objective, at the cognitive and physical level.' Human-robot interaction has been defined Feil-Seifer et al. (2020) as 'a field study dedicated to understanding, designing, and evaluating robotic systems for use by or with humans'. Collaboration means working with someone on something while interacting on/with someone else Bauer et al. (2008). Of course, both HRC and HRI aim at reaching a common goal in a collaborative human-robotic system.

Human-robot collaboration viz. synergetic systems (Aivazidou and Tsolakis, 2023) can simplify the adoption of agricultural robots in agribusiness for sustainable rural development by considering socioeconomic perceptions (i.e. social acceptability drivers and barriers). A transition to HRC under a systems thinking perspective could provide both farmers and rural communities with the necessary incentives for the digitalization of agriculture. Furthermore, it can at the same time increase in-skilled jobs and attract younger farm workers.

Human-robot collaboration links the advantages of the human system (intelligence, perception, eye-hand coordination) with the robotic system (speed, accuracy, repeatability) overcoming both limitations (human - fatigue, subjective; robot - limited perception). The outcome is an improved system that can operate in the challenging agricultural environment.

2 Scope of human-robot collaboration

Research in HRI and HRC systems (HRIC) in the agriculture domain has been very limited despite its advantages (Edan et al., 2022; Lytridis et al., 2021; Vasconez et al., 2019). Recent reviews of HRI in agriculture (Adamides and Edan, 2023; Vasconez et al., 2019) dealt with concepts related to design and taxonomy, ergonomics, and productivity from management perspectives (Benos et al., 2020).

Designing HRI systems is a challenging problem given the human complexity and the large, multiple, and complicated interactions associated with a given task. The updated taxonomy of HRI (Yanco and Drury, 2004) characterized aspects to be considered when designing HRI systems including task type, task criticality, robot morphology, composition of robot teams, ratio of people to robots, level of shared interaction, interaction roles, types of human-robot physical proximity, time/space taxonomy, and autonomy level.

Cognitive ergonomics approaches (Vasconez et al., 2019) aim to improve work efficiency, decrease human errors, and understand how humans process information when interacting or collaborating with a robot towards a task. Cognitive ergonomics support the human mental process, which is based on perception, memory, attention, mental workload (Adamides, 2016), stress and mental models, thus improving awareness (Drury et al., 2007), decision-making and problem-solving (Bouargane and Cherkaoui, 2015). In addition to these review papers, specific studies as detailed later have focused on interaction modalities and modes of collaboration.

Recently, Adamides and Edan (2023) examined three additional aspects of HRIC systems that are of critical importance, especially in agriculture: those of safety, functionality, and usability. In terms of safety, HRI cognitive ergonomics, e.g. situation awareness (SA) and attention to mental workload, are factors that support humans and contribute to the reduction of the number of accidents in agricultural fields (Vásconez and Cheein, 2022). A new safety validation approach has been proposed by defining a set of protocols (Valori et al., 2021). Functionality was investigated for cooperative and collaborative systems. In cooperative systems, robot(s) and human(s) perform distinct tasks, while in collaborative systems, robot(s) and human(s) perform the same task. Usability in HRIC systems has been directed to interface design and evaluation, modes of interaction, and optimization models for HRC. Each of these is detailed later with examples following. Then, we provide several case studies presented using these concepts.

3 Functionality, design and taxonomy

3.1 Concepts

There are four types of collaboration (Marvel et al., 2020), as illustrated in Fig. 1:

1 Separate: Human and robot tasks are kept apart; the human and the robot do not share workspaces, tools, or work pieces.
2 Sequential: Human and robot tasks are completed consecutively; workspaces, tools, and workpieces may be shared, but there is a strict serialization of the tasks such that any sharing is temporally separated.
3 Simultaneous: Human and robot tasks are executed concurrently and may involve working on different parts of the same workpiece, but the human and the robot are focused on achieving separate task goals.
4 Supportive: The human and the robot work together at the same time and with the same workpiece to complete a common task.

The functionalities involve sensing, planning/decision-making, and acting and are related to the level of automation (LOA) that can be divided into ten levels, ranging from pure teleoperation to fully autonomous (Sheridan and Verplank (1978). Each of these levels specifies a different degree to which a task is automated, meaning that automation is not all or none, but can vary across a continuum of intermediate levels, between fully manual performance and fully autonomous conditions at the two extremes (Vagia et al., 2016). Several different LOAs have been proposed by different researchers, resulting in numerous taxonomies describing the interaction between humans and machines (Vagia et al., 2016).

Figure 1 Types of collaboration. Top left: separate, top right: sequential. Bottom left: supportive, bottom right: simultaneous.

Teleoperation is the mode of operation where an operator at one location directly controls a robot to perform a task at some other location. Fong et al. (2003) explain that in a collaborative control model for teleoperation, the robot and the human work as a team to perform tasks and achieve common goals. For example, in agriculture, in the task of harvesting the human may be detaching crops from plants and a robot may be following the human carrying cases to collect the harvested crops. Such a model incorporates aspects of HRI, dialog, and switching between different levels of automation.

The LOA design can be modeled by the four stages of information processing (Parasuraman et al., 2000) denoted as the OODA loop (Brehmer, 2006) acquiring information (Observe), processing information (Orient), making decisions (Decide), and taking action (Act). Acquiring information involves collecting it before performing the task (e.g. sensing fruit locations), processing information requires generating alternatives for performing the task (e.g. finding obstacle-free paths), making decisions involves identifying which of the options to select in performing the task (e.g. selecting the shortest path to the most mature fruit), and taking action encompasses all steps associated with the decision made (e.g. moving toward the fruit and harvesting it).

In agriculture, the functionality of a robot was classified into five levels from fully manual (corresponding to Beer et al. (2014), Human Alone, in which the robot is not involved in any part of the task and the human performs all actions) - through semi-manual (corresponding to Beer et al. (2014), Human-Oriented Semi-Autonomy, in which the user must explicitly agree to suggested actions before they are performed by the robot), intermediate (combination of Human-Oriented Semi-Autonomy and Robot-Oriented Semi-Autonomy) and semi-autonomous (corresponding to Beer, Robot-Oriented Semi-Autonomy, in which the robot implements actions unless the user objects and informs the user of the implemented action following its execution) – to fully autonomous (corresponding to Beer, Robot Alone, in which the robot performs all actions without any form of human involvement) for locomotion and manipulation (Table 1, SPARKLE[1]).

3.2 Examples

Four HRC levels for target recognition tasks in unstructured environments were defined (Bechar and Edan, 2003): (a) HO – the human operator marks targets unaided; (b) HO-Rr – the human operator marks targets aided by robot recommendations; (c) HO-R – targets are identified automatically by the robot detection algorithm; the human operators' assignment is to cancel false detections and to mark the targets missed by an automatic robot detection algorithm; and (d) R – the targets are marked automatically by the system (robot). Experimental results on melon images taken in the field indicate that HO and robot collaboration increases detection and reduces the time required for detection (Bechar et al., 2009).

Seyyedhasani et al. (2020a) developed an HRC simulator to predict the efficiency of a picking crew assigned to harvest specialty crops (e.g. strawberries) and to evaluate various scheduling and dispatching policies for robot teams. Hybrid automata with stochastic parameters were used to compare the all-manual harvesting and machine aided harvesting using crop-transport robots. This latter synergy increased harvest automation by reducing the pickers' non-productive walking times, as harvest-aid robots were scheduled to dispatch empty and full trays to the picking crew. As reported in Seyyedhasani et al. (2020b), when deploying five crop-transport robots, the harvest efficiency was improved by 92% (morning) and 86.5% (afternoon), compared to 81.8% and 78.2%, respectively, for manual harvesting.

Recently, Sreeram and Nof (2021) examined two different integration modes for keeping the human-in-the-loop for agricultural robotic systems – sequential integration and shared integration. In both cases, the HRC protocol provided better performance.

[1] http://sparkle-project.eu/.

Table 1 Levels of robot autonomy (www.sparkle-project.eu)

	Robot functionality																			
	Fully dependent				Semi-dependent				Intermediate				Semi-autonomous				Fully autonomous			
Locomotion	Driven, not aware of surroundings				Driven, partly aware of surroundings				Partly self-driving, partly or not aware of surroundings				Self-driving, partly aware of surroundings				Self-driving, fully aware of surroundings			
Actions to be done: **O**bserve **O**rient (processing information) **D**ecision Taking **A**ction	O	O	D	A	O	O	D	A	O	O	D	A	O	O	D	A	O	O	D	A
Who is performing the action: **H**uman, **R**obot or a mix of both	H	H	H	H	R/H	R/H	R/H	H	R/H	R/H	R/H	R/H	R/H	R/H	R/H	R	R	R	R	R
Manipulation	Fixed actions without taking into account surroundings				Fixed actions, partly aware of surroundings				Partly independent, partly or not aware of surroundings				Autonomously selects actions, partly aware of surroundings				Autonomously selects actions, fully aware of surroundings			

4 Interfaces/interaction design and usability

4.1 Concepts

Any kind of HRC requires interaction and hence some interface (Adamides and Edan, 2022). Interface design must take into account the task, the robot, and the user. User interface usability, defined as 'a quality attribute that assesses how easy user interfaces are to use,' is an important feature of HRI (Clarkson and Arkin, 2007). Different types of input and output devices have been employed in the design of interfaces for agricultural robots as detailed later.

A developed taxonomy for agricultural robots generated using the card sorting procedure provides robot teleoperation user interface design guidelines (Adamides et al., 2015). The proposed taxonomy consists of eight distinct categories: (1) platform architecture and scalability, (2) error prevention and recovery, (3) visual design, (4) information presentation, (5) robot state awareness, (6) interaction effectiveness and efficiency, (7) robot surroundings/environment awareness, and (8) cognitive load. This taxonomy provides important guidelines for the design and evaluation of teleoperated mobile field robots to other researchers, information architects, usability experts, and developers, especially those interested in them.

4.2 Examples

An augmented reality headset system was employed to improve the SA of an operator who is on an autonomous robot in the field but at the same time is also responsible for monitoring a fleet of autonomous robots working in the same field (Huuskonen and Oksanen, 2019). The augmented reality system improved the SA of the human supervisor. However, several issues concerning the system's usability were found, especially difficulties in viewing the scene outdoors due to sunlight.

User interfaces were designed for target-specific pesticide robot sprayers (Adamides, Katsanos et al., 2017; Berenstein and Edan, 2012). In one study, different marking techniques were explored to maximize target detection while taking into account usability aspects (Berenstein and Edan, 2012).

4.3 Case studies

A review of various HRC found in agriculture (Table 2) reveals that the focus is on different applications like harvesting and spraying (Fig. 2). In agricultural harvesting operations, the harvesting cost currently exceeds the cost of crop production (Rysz et al., 2019) since manual harvesting can contribute up to 60% of the yearly operating costs per acre (Bolda et al., 2003). In addition, there is an observed shortage of labor in agriculture and farm workers (e.g. for manual

Table 2 Review of human–robot collaboration in harvesting and spraying in agriculture

Application	Collaboration type	Functionality LOA – where in OODA	Number of LOA	Interfaces – what interfaces were used	Safety – 1/0 addressed or not	Reference
Harvesting	Simultaneous	LOA fully depended **O**bserve and Taking **A**ction	2	Not reported	1	Baxter et al. (2018)
Harvesting	Supportive	LOA semi-depended, A = H All rest R/H SA **O**bserve, **O**rient (processing information)	4	Not reported	0	Huang et al. (2020)
Harvesting	Supportive	LOA intermediate all R/HAS **O**bserve, **O**rient (processing information), and Taking **A**ction	2	The GUI with 3D reconstruction, robot motion and command interface	0	Zhao et al. (2016)
Harvesting	Simultaneous	LOA semi-autonomous, A = R All rest R/H SA **O**bserve, **O**rient (processing information), **D**ecision, and Taking **A**ction	2	Non-verbal communication, hand gestures	1	Vasconez et al. (2019)
Spraying	Supportive	LOA fully depended, HHHH **O**bserve and Taking **A**ction	4	GUI and interaction modes	1	Adamides et al. (2017)
Spraying	Supportive	LOA semi-autonomous, A = R All rest R/H SA **O**bserve, **O**rient (processing information), **D**ecision, and Taking **A**ction	2	GUI	0	Berenstein and Edan (2017)

Figure 2 Left: agrirobot sprayer (Adamides et al., 2017), right: robot sprayer (Berenstein and Edan, 2017).

harvesting) because of socioeconomic factors (Christiaensen et al., 2020). Pesticides and fertilizers are widely used in agriculture for controlling pests, pathogens, and weeds, ergo are beneficial to enhance crop protection and production (i.e. reduced crop loss) (Cooper and Dobson, 2007). Howbeit, long-term exposure to pesticides (e.g. during field application) may adversely affect human health (Bonner and Alavanja, 2017). In addition, according to studies (Pimentel and Levitan, 1986; van der Werf, 1996), less than 0.3% of pesticides applied reach the target (i.e. plants) and the remaining is found elsewhere in the environment resulting in undesirable environmental effects (e.g. soil degradation). Blackmore et al. (2005) argue that 95% is the lowest barrier for the detection rate for the spraying process to be economically feasible. Therefore, human-robot synergetic spraying systems are essential and have been extensively examined as presented in the literature that follows.

5 Conclusion and future trends

Despite the technological advances, there are limited commercially available agricultural robots. This is due to the unstructured and highly variable nature of both objects and the environment making agricultural robots among the most complicated robotics systems to develop. Additionally, the developed systems must cope with harsh and unpredicted environmental conditions while keeping costs low. Hence, developing human-robot collaborative systems in which the human is retained in the loop can provide a sustainable and feasible solution. It can also simplify the overall system and thereby reduce its costs enabling the economic feasibility of agricultural robots and their penetration into market. Human-robot systems, that work together to face the challenges of the agricultural production process, can lead to improved production which is essential at this time.

Furthermore, since the farmer will want to remain involved in the farming system, the 'human-in-the-loop' set-up can provide an attractive solution, enabling easier market penetration. This set-up is also likely to attract younger generations to farming by turning it into an attractive profession with advanced technology.

Cooperative robotics in agriculture (Lytridis et al., 2021) provides strengths (e.g. reduction of task duration, improved quality) and opportunities (e.g. labor shortages of seasonal workers, improved work conditions). However, weaknesses (e.g. human–robot interfaces and action coordination) and threads (the human–robot work safety) must be addressed in order to ensure successful implementation.

Human–robot collaborations systems could advance possible solutions and serve as intermediate steps toward autonomy: combining robotic skills with human capabilities in a human–robot co-working framework would contribute to overcoming current bottlenecks in the development of fully autonomous systems and simplifying system design. For example, since not all operations can be automated (e.g. thinning, disease detection), the farmer would, in any event, continue to employ workers for many different farm tasks. Furthermore, by including a human in the loop, the robotic system can be simplified leading to its economic feasibility. In addition, it is to be expected that many farmers would not want to implement fully autonomous operations so as to ensure that they remain in the loop for exploiting their 'green thumb' talent. Therefore, systems should be developed to enable humans and robots to work together in a human–robot co-working framework (Bac et al., 2014; Kootstra et al., 2021).

With collaborative robots becoming a standard in industry and service robots penetrating the market, HRC in agriculture will become common practice. However, intensive R&D must be advanced to ensure intuitive and interactive user interfaces that can be operated by multiple users with different skill levels at different levels of collaboration. In parallel, it is important to advance the certification of systems along with standard definitions. Finally, the safe interaction of humans and agriculture machines is one of the central preconditions for the market introduction of autonomous agriculture. Like industrial robots for some applications, it may be sufficient to protect the work area and ensure safety by excluding people from the working environment. However, applications that include direct physical interaction (with people or livestock) will require higher safety levels. Safety must be embedded in the physical human–machine interface which must be able to operate in diverse conditions. Safety must also be inherently integrated into the robot system to ensure reliable and robust environmental detection. Additionally, certification of safe operation must be advanced. Adamides and Edan (2023) presented a roadmap for such a development, which would be of critical importance for the penetration of agricultural robotics into practice.

6 Where to look for further information

The following articles provide a good overview of the subject:

- Adamides, G. and Edan, Y. (2023). Human-robot collaboration systems in agricultural tasks: A review and roadmap. *Computers and Electronics in Agriculture*, 204, 107541.
- Vasconez, J. P., Kantor, G. A. and Cheein, F. A. A. (2019). Human-robot interaction in agriculture: A survey and current challenges. *Biosystems Engineering*, 179, 35–48.
- Edan, Y. (2019). Human-robot collaboration in agricultural robots. In *Robotics and Automation for Improving Agriculture* (pp. 77–100). Burleigh Dodds Science Publishing.

Key research in this area can be found at the following organizations:

- Wageningen University and Research – Agro Food Robotics (WUR) (www.wur.nl/en/research-results/projects-and-programmes/agro-food-robotics.htm).
- Ben-Gurion University of the Negev, Agricultural, Biological Cognitive Robotics Initiative (https://in.bgu.ac.il/en/robotics/Pages/default.aspx; https://smartrobabcbgu.wixsite.com/iemirl).
- Global Organization for Agricultural Robotics (FIRA) (www.agricultural-robotics.com/).
- International Institute of Tropical Africa (IITA) (www.iita.org).

7 References

Adamides, G. (2016). *User Interfaces for Human-Robot Interaction: Application on a Semi-autonomous Agricultural Robot Sprayer* (Publication Number 978-9963-695-49-2) [Doctoral Dissertation, Open University of Cyprus]. Nicosia.

Adamides, G. and Edan, Y. (2023). Human-robot collaboration systems in agricultural tasks: a review and roadmap. *Computers and Electronics in Agriculture* 204, 107541. Available at: https://doi.org/10.1016/j.compag.2022.107541.

Adamides, G., Christou, G., Katsanos, C., Xenos, M. and Hadzilacos, T. (2015). Usability guidelines for the design of robot teleoperation: A taxonomy. *IEEE Transactions on Human-Machine Systems* 45(2), 256–262. https://doi.org/10.1109/THMS.2014.2371048.

Adamides, G., Katsanos, C., Constantinou, I., Christou, G., Xenos, M., Hadzilacos, T. and Edan, Y. (2017a). Design and development of a semi-autonomous agricultural vineyard sprayer: human-robot interaction aspects. *Journal of Field Robotics* 34(8), 1407–1426. https://doi.org/10.1002/rob.21721.

Adamides, G., Katsanos, C., Parmet, Y., Christou, G., Xenos, M., Hadzilacos, T. and Edan, Y. (2017b). HRI usability evaluation of interaction modes for a teleoperated agricultural robotic sprayer. *Applied Ergonomics* 62, 237–246. Available at: http://www.sciencedirect.com/science/article/pii/S0003687017300674.

Aivazidou, E. and Tsolakis, N. (2023). Transitioning towards human-robot synergy in agriculture: A systems thinking perspective. *Systems Research and Behavioral Science* 40(3), 536–551.

Arad, B., Balendonck, J., Barth, R., Ben-Shahar, O., Edan, Y., Hellström, T., Hemming, J., Kurtser, P., Ringdahl, O., Tielen, T. and van Tuijl, B. (2020). Development of a sweet pepper harvesting robot. *Journal of Field Robotics* 37(6), 1027–1039.

Bac, C. W., Henten, E. J., Hemming, J. and Edan, Y. (2014). Harvesting Robots for High-value Crops: state of the art Review and Challenges Ahead. *Journal of Field Robotics* 31(6), 888–911.

Bauer, A., Wollherr, D. and Buss, M. (2008). Human-robot collaboration: a survey. *International Journal of Humanoid Robotics* 05(1), 47–66.

Baxter, P., Cielniak, G., Hanheide, M. and From, P. (2018). Safe human-robot interaction in agriculture. Companion of the 2018 ACM/IEEE International Conference on Human-Robot Interaction.

Bechar, A., Meyer, J. and Edan, Y. (2009). An objective function to evaluate performance of human-robot collaboration in target recognition tasks. *IEEE Transactions on Systems, Man, and Cybernetics, Part C* 39(6), 611–620.

Beer, J. M., Fisk, A. D. and Rogers, W. A. (2014). Toward a framework for levels of robot autonomy in human-robot interaction. *Journal of Human-Robot Interaction* 3(2), 74–99.

Benos, L., Bechar, A. and Bochtis, D. (2020). Safety and ergonomics in human-robot interactive agricultural operations. *Biosystems Engineering* 200, 55–72.

Berenstein, R. and Edan, Y. (2012). Evaluation of marking techniques for a human-robot selective vineyard sprayer International Conference of Agricultural Engineering, Valencia, Spain.

Berenstein, R. and Edan, Y. (2017). Human-robot collaborative site-specific sprayer. *Journal of Field Robotics* 34(8), 1519–1530. https://doi.org/10.1002/rob.21730.

Botta, A., Cavallone, P., Baglieri, L., Colucci, G., Tagliavini, L. and Quaglia, G. (2022). A review of robots, perception, and tasks in precision agriculture. *Applied Mechanics* 3(3), 830–854.

Bouargane, L. and Cherkaoui, A. (2015). Towards an explicative model of human cognitive process in a hidden hazardous situation and a cognitive ergonomics intervention in railway environment International Conference on Industrial Engineering and Systems Management (IESM) (vol. 2015).

Brehmer, B. (2006). One loop to rule them all. In: Proceedings of the 11th International Command and Control Research and Technology Symposium, Cambridge, UK.

Clarkson, E. and Arkin, R. C. (2007). Applying heuristic evaluation to human-robot interaction systems. FLAIRS Conference.

Drury, J. L., Keyes, B. and Yanco, H. A. (2007). LASSOing HRI: analyzing situation awareness in map-centric and video-centric interfaces. In: Proceedings of the ACM/IEEE International Conference on Human-Robot Interaction, 279-286.

Edan, Y., Adamides, G. and Oberti, R. (2023). Agriculture automation. In: Nof, S. Y. (Ed.). *Springer Handbook of Automation* (2nd edn.). Springer, Berlin Heidelberg, 1055–1078.

Edan, Y., Han, S. and Kondo, N. (2009). Automation in agriculture. In: Nof, S. Y. (Ed.). *Springer Handbook of Automation*. Springer, Berlin Heidelberg (pp. 1095–1128). https://doi.org/10.1007/978-3-540-78831-7_63.

Feil-Seifer, D., Haring, K. S., Rossi, S., Wagner, A. R. and Williams, T. (2020). *Where to Next? The Impact of COVID-19 on Human-Robot Interaction Research*. ACM, New York.

Fong, T., Thorpe, C. and Baur, C. (2003). Collaboration, dialogue, human-robot interaction. In: *Robotics Research*. Springer, Berlin Heidelberg (pp. 255–266).

Gebbers, R. and Adamchuk, V. I. (2010). Precision agriculture and food security. *Science* 327(5967), 828–831.

Huang, Z., Miyauchi, G., Salazar Gomez, A., Bird, R., Kalsi, A. S., Jansen, C., Liu, Z., Parsons, S. and Sklar, E. (2020). Toward robot co-labourers for intelligent farming. Companion of the 2020 ACM/IEEE International Conference on Human-Robot Interaction.

Huuskonen, J. and Oksanen, T. (2019). Augmented reality for supervising multirobot system in agricultural field operation. *IFAC-PapersOnLine* 52(30), 367–372.

Kamilaris, A. and Prenafeta-Boldú, F. X. (2018). Deep learning in agriculture: A survey. *Computers and Electronics in Agriculture* 147, 70–90.

Kootstra, G., Wang, X., Blok, P. M., Hemming, J. and Van Henten, E. (2021). Selective harvesting robotics: current research, trends, and future directions. *Current Robotics Reports* 2(1), 95–104.

Kubota, N. and Ozawa, S. (2008). Tele-operated robots for monitoring based on sensor networks. SICE Annual Conference (vol. 2008).

Kurtser, P. and Edan, Y. (2018). Statistical models for fruit detectability: spatial and temporal analyses of sweet peppers. *Biosystems Engineering* 171, 272–289.

Lytridis, C., Kaburlasos, V. G., Pachidis, T., Manios, M., Vrochidou, E., Kalampokas, T. and Chatzistamatis, S. (2021). An overview of cooperative robotics in agriculture. *Agronomy* 11(9), 1818.

Marvel, J. A., Bagchi, S., Zimmerman, M. and Antonishek, B. (2020). Towards effective interface designs for collaborative HRI in manufacturing: metrics and measures. *ACM Transactions on Human-Robot Interaction (THRI)* 9(4), 1–55.

Moreno, F.-A., Cielniak, G. and Duckett, T. (2014). Evaluation of laser range-finder mapping for agricultural spraying vehicles. In *Towards Autonomous Robotic Systems* (pp. 210–221). Springer.

Moysiadis, T., Adamides, G., Stylianou, A., Zotos, N., Giannakopoulou, M. and Alexiou, G. (2020). Use of IoT technologies for irrigation and plant protection: the case for Cypriot fruits and vegetables. In: Bochtis, D., Achillas, C., Banias, G. and Lampridi, M. (Eds). *Bio-economy and Agri-Production: Concepts and Evidence*. Academic Press.

Moysiadis, T., Adamides, G., Stylianou, A., Zotos, N., Giannakopoulou, M. and Alexiou, G. (2021). Chapter 10. Use of IoT technologies for irrigation and plant protection: the case for Cypriot fruits and vegetables. In: Bochtis, D., Achillas, C., Banias, G. and Lampridi, M. (Eds) *Bio-economy and Agri-Production*. Academic Press (pp. 175–194). https://doi.org/10.1016/B978-0-12-819774-5.00010-2.

Oliveira, L. F. P., Moreira, A. P. and Silva, M. F. (2021). Advances in agriculture robotics: A state-of-the-art review and challenges ahead. *Robotics* 10(2), 52.

Panday, U. S., Pratihast, A. K., Aryal, J. and Kayastha, R. B. (2020). A review on drone-based data solutions for cereal crops. *Drones* 4(3), 41.

Parasuraman, R., Sheridan, T. B. and Wickens, C. D. (2000). A model for types and levels of human interaction with automation. *IEEE Transactions on Systems, Man, and Cybernetics – Part A* 30(3), 286–297.

Pimentel, D. and Levitan, L. (1986). Pesticides: amounts applied and amounts reaching pests. *BioScience* 36(2), 86–91.

Rysz, M., Ganesh, P., Burks, T. F. and Mehta, S. S. (2019). Risk-averse optimization for improving harvesting efficiency of autonomous systems through human collaboration. *IFAC-PapersOnLine* 52(30), 207–212.

Semeraro, F., Griffiths, A. and Cangelosi, A. (2023). Human–robot collaboration and machine learning: A systematic review of recent research. *Robotics and Computer-Integrated Manufacturing* 79, 102432.

Seyyedhasani, H., Peng, C., Jang, W.-j. and Vougioukas, S. G. (2020a). Collaboration of human pickers and crop-transporting robots during harvesting–Part I: Model and simulator development. *Computers and Electronics in Agriculture* 172, 105324.

Seyyedhasani, H., Peng, C., Jang, W.-j. and Vougioukas, S. G. (2020b). Collaboration of human pickers and crop-transporting robots during harvesting–Part II: Simulator evaluation and robot-scheduling case-study. *Computers and Electronics in Agriculture* 172, 105323.

Sheridan, T. and Verplank, W. (1978). Human and computer control of undersea teleoperators. In: Man-Machine Systems Laboratory, Department of Mechanical Engineering: Cambridge, MA: MIT Press.

Slaughter, D. C., Giles, D. K. and Downey, D. (2008). Autonomous robotic weed control systems: a review. *Computers and Electronics in Agriculture* 61(1), 63–78.

Sreeram, M. and Nof, S. Y. (2021). Human-in-the-loop: role in cyber physical agricultural systems. *International Journal of Computers Communications and Control* 16(2), 96.

Stafford, J. V. (2000). Implementing precision agriculture in the 21st century. *Journal of Agricultural Engineering Research* 76(3), 267–275.

Vagia, M., Transeth, A. A. and Fjerdingen, S. A. (2016). A literature review on the levels of automation during the years. What are the different taxonomies that have been proposed? *Applied Ergonomics* 53(A), 190–202.

Valori, M., Scibilia, A., Fassi, I., Saenz, J., Behrens, R., Herbster, S., Bidard, C., Lucet, E., Magisson, A., Schaake, L., Bessler, J., Prange-Lasonder, G. B., Kühnrich, M., Lassen, A. B. and Nielsen, K. (2021). Validating safety in human–robot collaboration: standards and new perspectives. *Robotics* 10(2), 65.

Vasconez, J. P., Kantor, G. A. and Auat Cheein, F. A. (2019). Human–robot interaction in agriculture: A survey and current challenges. *Biosystems Engineering* 179, 35–48.

Vásconez, J. P. and Cheein, F. A. (2022). Workload and production assessment in the avocado harvesting process using human-robot collaborative strategies. *Biosystems Engineering* 223, 56–77. Available at: https://doi.org/10.1016/j.biosystemseng.2022.08.010.

Vasconez, J. P., Guevara, L. and Cheein, F. A. (2019a). Social robot navigation based on HRI non-verbal communication: a case study on avocado harvesting. In: Proceedings of the 34th ACM/SIGAPP Symposium on Applied Computing, 957–960.

Vasconez, J. P., Kantor, G. A. and Cheein, F. A. A. (2019b). Human–robot interaction in agriculture: A survey and current challenges. *Biosystems Engineering* 179, 35–48.

Yanco, H. A. and Drury, J. (2004). Classifying human-robot interaction: an updated taxonomy. IEEE International Conference on Systems, Man and Cybernetics, 10-13 Oct. 2004.

Zhao, Y., Gong, L., Liu, C. and Huang, Y. (2016). Dual-arm robot design and testing for harvesting tomato in greenhouse. *IFAC-PapersOnLine* 49(16), 161–165.

Chapter 7

Implementing a digital twin for flexible operation of agricultural robotics

Frits K. van Evert, Trim Bresilla, Ard Nieuwenhuizen, Bram Veldhuisen and Jochen Hemming, Wageningen University & Research, The Netherlands; Giannis Avgoustakis, Spyros Fountas, Michael Koutsiaras and Nikos Mylonas, Agricultural University of Athens, Greece; Suzanne Baron, Agricultural Robotics and Automatization, France; Thanos Dritsopoulos and Markos Legas, Pegasus, Greece; Han Hilbrands and Jeroen Wolters, Smart Agri Technology BV, The Netherlands; Klaas Jan Hommes, Abemec BV, The Netherlands; Panagiotis Karagiannis, Sotiris Makris and George Michalos, University of Patras, Greece; Dimitrios Paraforos and Galibjon Sharipov, University of Hohenheim, Germany; Søren Marcus Pedersen, University of Copenhagen, Denmark; Bertrand Pinel, Terrena Societe Cooperative Agricole, France; Mladen Radisic, Foodscale Hub, Serbia; Juan F. Rascón and Carlos Rizzo, Eurecat, Centre Tecnològic de Catalunya, Spain; Raul Sanchez, Serrater, Spain; Alea Scovill, AgroIntelli, Denmark; Oriol Serra, Giropoma, Spain; and Josep Vidal, Teyme, Spain

1 Introduction
2 Agricultural system under study
3 Agricultural robotic system
4 Case study results
5 Discussion
6 Conclusion
7 Where to look for further information
8 Acknowledgements
9 References

1 Introduction

One of the most important challenges in European agriculture is the cost and scarcity of labour. The cost of labour affects the profitability of farming while the scarcity of labour threatens its very existence. The latter was illustrated vividly during the COVID-19 pandemic when many EU countries were not able to welcome field workers from abroad. As a result, some crops could not be harvested and had to be left on the field.[1]

[1] https://www.nytimes.com/2020/03/27/business/coronavirus-farm-labor-europe.html.

http://dx.doi.org/11.19103/AS.2023.0124.28

The cost and shortage of labour may be addressed by adopting robotics. Robots have replaced labour in several sectors of the economy, but they have not yet reached widespread acceptance in agriculture. Robots have been proposed for all activities during the cropping cycle, from tillage to sowing, fertilization, crop protection and harvesting. In this chapter, we focus on mobile robots for spraying and mechanical weeding because these are dull, dirty, dangerous tasks that need to be done several times per season.

Agricultural robots are available commercially from vendors including Naïo,[2] Agreenculture,[3] AgroIntelli,[4] AgXeed[5] and FarmDroid.[6] Robotic research platforms include Flourish[7] and Deepfield[8]. Autonomous tractors are another class of robotic vehicles in agriculture, available commercially (Kubota,[9] only for rice, only in Japan) or as a concept (John Deere[10]). Alternatively, after-market kits are available to convert a conventional farm tractor into a robot (e.g. JCA Technologies[11] and Blue White Robotics[12]).

All these vehicles can execute basic autonomous behaviour: follow a crop row to the end of the field, make a turn and return in the next crop row. Path planning is a topic that has received ample attention, especially in the context of covering irregularly shaped fields with a minimum of overlap (Oksanen and Visala, 2009; Mier et al., 2022). It is relevant for human-operated machinery but will also be needed by autonomous vehicles. However, full autonomous behaviour is by far not realized. These vehicles must be brought to the field and pointed in the right direction. Planning and scheduling are capabilities that are not offered. Working autonomously and robustly, during a whole growing season, in a realistic environment, has not been demonstrated. Advanced route planning and dynamic re-planning were proposed 15 years ago and demonstrated in simulation (Bochtis et al., 2007); as far as we are aware, this system was never implemented. More recently, a route planning algorithm for an electrical agricultural vehicle was developed that not only takes into account hill slope and remaining battery charge but also here robust operation and dynamic response to unforeseen events are not considered (Hizatate and Noguchi, 2023).

Agricultural implements have also been equipped with autonomy. The main objective of these smart implements is to close the loop from perception to decision and action. Examples include the WEED-IT spot sprayer (Cloete, 2020) which prevents runoff of crop protection chemicals by applying them

[2] https://www.naio-technologies.com/en/.
[3] https://www.agreenculture.net/.
[4] https://agrointelli.com/.
[5] https://www.agxeed.com/.
[6] https://farmdroid.dk/.
[7] http://flourish-project.eu/ NOTE better to cite scientific papers.
[8] https://www.deepfield-robotics.com/en/ website no longer works but we can cite the scientific paper.
[9] https://www.youtube.com/watch?v=PG3vd3AsdfY.
[10] https://www.deere.co.uk/en/agriculture/future-of-farming/.
[11] https://jcatechnologies.com/.
[12] https://www.bluewhite.co/.

only where weeds are detected by sensors in real-time, the Ecorobotix ARA spot sprayer (Anken and Latsch, 2022; Anken et al., 2022) and See & Spray (ASABE Staff, 2022). The Robocrop system (Tillett and Hague, 2006) and the Steketee IC-Weeder use cameras to guide a hoe around crop plants (Hemming et al., 2018). The 'Hill Control' control system from Rauch improves the distribution accuracy by changing the drop point, disc speed and metering quantity when spreading fertilizers with a centrifugal disk spreader, especially in hilly terrain.[13] Advances in sprayers involve not only boom control systems[14] but also more advanced systems that control the full functionality of a sprayer[15] are important towards achieving an even distribution of the spray deposit. However, the information gathered by the sensors in these cases is not used to monitor the quality of the action and it is not used to alter or optimize the operation of the robotic agricultural operation within the farming system as a whole.

Agricultural robots exist but are not widely used. An important reason is that agricultural robots are not flexible. As a first example, agricultural robots do not have a mechanism that adjusts or stops the robot when the process is not going well (insufficient weed control, crop is being damaged). As a second example, agricultural robots are not flexible in dealing with unforeseen events (weather, obstacles, tank empty, working lane not being wide enough, soil trafficability, etc.). As a result, agricultural robots need a lot of supervision to check their operation and to get them going again when they get stuck. This means that the real amount of labour saved will be much less than expected.

In contrast to agriculture, robots are widely used in industrial manufacturing. Digital Twin (DT) is a concept that is important to understand how robots function in industrial manufacturing. Stark and Damerau (2019) provide the following definition of a DT:

> *A digital twin is a digital representation of an active unique product (real device, object, machine, service, or intangible asset) or unique product-service system (a system consisting of a product and a related service) that comprises its selected characteristics, properties, conditions, and behaviours by means of models, information, and data within a single or even across multiple life cycle phases.*

One of the most common applications of the DT is for the design and configuration of an assembly line (Kousi et al., 2021). The DT enables the modelling of the different levels of a production system, namely the assembly process, production station, and line level, updating its parameters in real-time synthesisng data from multiple 2D–3D sensors about the actual production process. Then an AI-based logic is deriving alternative configurations of the

[13] Rauch, 2019. HillControl: Mehr Präzision beim Düngerstreuen in Hanglagen, auf Kuppen und in Senken!
[14] Raven, 2019. AutoBoom XRT - Raven Applied Technology. https://ravenprecision.com/products/boom-controls/autoboom.
[15] Müller Electronics, 2019. ISOBUS SPRAYER-Controller MAXI. https://www.mueller-elektronik.de/en/products/spray-controller-maxi-2/.

production system, suggesting the most efficient based on a set of predefined criteria.

Another common use of DTs is to control the resources of a production cell and adapt their behaviour based on the production requirements (Kousi et al., 2019). In other words, a model of the real system is used to re-plan production in response to unforeseen events, such as a temporary unavailability of parts due to supply line issues or a breakdown; or people stepping in front of a mobile robot and impeding its motion.

Predictive maintenance addresses production alterations that occur due to unexpected events such as breakdowns and unavailability of resources. Methodologies have been developed to calculate the Remaining Useful Life (RUL) of a machine by using physics-based simulation models and the DT concept (Aivaliotis et al., 2019). In the design phase, a model of a gearbox was created and parameterized using data from the real system. This model was able to simulate the friction inside the gearbox. During the execution phase, the model was fed with data from the machine's controllers as well as from external sensors that monitor their activity. As a result, the increase in gearbox friction could be simulated in real-time and the gearbox RUL could be calculated.

A DT may be used for process control. In additive manufacturing, the quality of processes as well as the robustness and the performance are posing several challenges due to a number of uncertainties, such as the material properties. A DT was designed and implemented for uncertainty management and robust process control (Stavropoulos et al., 2021). Using Linear Matrix Inequalities, within defined uncertainty limits, the process control was made and compared to a typical control approach that uses a fine-tuned conventional proportional-integral-derivative (PID). As a result, the robust control design achieved a 68% faster response in the settling time metric, while a well-calibrated PID only achieved 38% compared to the initial model.

A physics-based model is often complex and a significant amount of time may be needed to realize simulations with it, which prevents using the DT in real-time. Creating a meta-model by using the output of a large number of simulations with the DT ('synthetic data') may address this problem. Of especial interest is the use of AI-based meta-models (Alexopoulos et al., 2020).

A DT implementation has been proposed as part of a general cyber-physical system (CPS) to enable the optimization of the planning and commissioning of human-based production processes using simulation-based approaches (Nikolakis et al., 2019). The benefit of this development is that the above activities are optimized through (1) sensor data fusion and human activity recognition in the shopfloor and (2) knowledge management mechanism for capturing the implicit knowledge of the task execution.

Finally, DTs have been used in human-robot collaboration (HRC) applications as well, for teleoperation and coordination of multiple industrial robot operations, allowing operators from different manufacturing domains to control the robot. Li et al. (2022) describe a multi-robot teleoperation system in which the DT of the real robots is visualized in augmented reality (AR) glasses. Experimental studies with a single robot and with multiple robots working on collaborative tasks have shown that the proposed system can manage the multi-robot teleoperation tasks with high efficiency.

In summary, a DT always involves a model of a real 'thing'. The model can be static or dynamic, but the defining characteristic of a DT model is that the model is parameterized for a specific real-world system.

The DT can be used in two major ways. First, it can be used to configure and optimize the system, even before the real system is created. The DT can also be used to train personnel that will operate the real system. When a DT is used in this fashion, it is sometimes called a 'digital shadow'. Second, the DT can be used as a state estimator in real-time. In this case, sensors on the real system (joint positions, proximity, temperatures, pressures, cameras, etc.) are used to obtain information about the dynamic state of the system. This information may be fused with simulation outcomes. Also future states may be simulated and control actions may be taken in the present time to ensure that undesirable future states are avoided for the real system.

We hypothesize that agriculture has requirements similar to industrial manufacturing in terms of DTs. DT modelling of agricultural fields and agricultural processes could help to optimize agricultural activities. Using a digital version of the field, optimized vehicle routes can be generated both within and outside the field. Additionally, visualization tools for the above-mentioned information could enable the farmer to better understand and adopt the new technologies, especially when these are provided through web and/or mobile UIs. On top of that, in combination with simulation libraries, the farmers could be trained with simulated fields on how they could use the DTs and generate optimized operations, before using them in real conditions. DT can be combined with machine learning and AI-based algorithms to predict farming operations before robots enter the field, as well as optimize them during their execution in the field.

ROBS4CROPS is a 4-year EU-funded project (2021–24). The project aims to develop an agricultural robotic system that uses DT technology to react to unforeseen events autonomously. The objective of our work is to create an agricultural robot system that can operate with much less supervision than current practice. At a minimum, our system will detect problems and stop operating in a safe manner, but ideally it will be able to respond to changing conditions and unforeseen events.

2 Agricultural system under study

ROBS4CROPS focuses on the needs of four specific real-world farming scenarios (Table 1). For each scenario, the project has established a large-scale pilot which serves to elicit requirements, to develop the system and to demonstrate the developed robotic system in real operating conditions. Two pilots focus on mechanical weeding: in vineyards in France, and in arable crops in the Netherlands; and two pilots focus on spraying: in vineyards in Greece, and in apple orchards in Spain.

In the pilot in France, the aim is to reduce the environmental impact of wine growing by replacing chemical weed control in vineyards with mechanical weeding. Mechanical weeding with a tractor driver is economically not attractive (five to seven passes per year, total cost €800 per ha). Also, it is difficult to attract and retain experienced tractor drivers who can operate the several different weeding implements that are used. The goal is to replace the tractor and human driver with a robot. In this pilot, the Agreenculture robot, developed specifically for vineyards, is used (Fig. 1).

The pilot in Greece focuses on spraying table grape vineyards (Fig. 2). Supermarkets and consumers impose very high-quality standards and the fruit is highly sensitive to pests and diseases. Table grapes need up to 30 applications of biocides (preventive and curative) and foliar fertilizer each growing season. Labour costs are high and access to the workforce is difficult, especially given the limited time window available for each spraying application. In this pilot, the Agreenculture robot and a retrofitted tractor (autonomous) are used to carry a lift-mounted sprayer. The amount of chemicals used is reduced by adjusting the spray rate to the density of the canopy. An extra challenge in this pilot is the accessibility of the hilly terrain and the dynamic behaviour of the robots in this pilot.

The pilot in Spain focuses on apple orchards (Fig. 3). The cultivation of apples requires many fungicide applications. Farmers typically do not consider the growing stage of the trees and the disease severity when applying chemicals (uniform application). Using a robot addresses the labour challenge, while the amount of chemicals used is reduced by adjusting the spray rate to the density of the canopy.

Table 1 Overview of the large-scale pilots

Country	Crop	Action	Vehicle
France	Vineyard (wine grapes)	Mechanical weeding	Agreenculture robot
Greece	Vineyard (table grapes)	Spraying	Retrofitted tractor and Agreenculture robot
Spain	Apple orchard	Spraying	Retrofitted tractor
The Netherlands	Sugar beet and pumpkins	Mechanical weeding	Robotti

Figure 1 The Agreenculture robot with a weeding implement attached.

Figure 2 Conventional tractor and sprayer in a table grape vineyard in Greece.

In the pilot in the Netherlands (Fig. 4), the aim is to reduce the environmental impact of arable farming by replacing chemical weed control with mechanical weeding. Sugar beets and pumpkins are both crops that grow slowly at first and therefore need several weeding passes to control the weeds. The Robotti robot, which was developed for arable farming, is used in this pilot.

The four pilots have in common that they use an autonomous vehicle with an implement attached, that field operations consist of visiting each predefined lane once, and that travel to, from and between lanes takes place in predefined

Figure 3 The EOLO sprayer towed by a New Holland tractor in an apple orchard in Spain.

Figure 4 The weeding implement used in the pilot in the Netherlands, mounted on the Robotti robot.

areas. Spraying and weeding are operations that must be performed several times during the growing season on a given field. In addition, a typical farm has several fields that need to be sprayed and/or weeded and the robot must be moved from one field to the next.

3 Agricultural robotic system

The ROBS4CROPS agricultural robotic system consists of three major elements: autonomous vehicles, smart implements and DT-based software to visualize, optimize and control the robots, which we call the Farming Controller. In addition, there is a communication protocol through which information is exchanged between vehicles, implements and the Farming Controller.

At the time of writing, the project is half-way. Consequently, some parts of the project vision have been demonstrated as proof of concept only; they will be further developed and demonstrated in the field in the two remaining years of the project.

3.1 Vehicles

3.1.1 Agreenculture robot

The robot from Agreenculture is a tracked vehicle that specifically targets vineyards. It weighs 800 kg, is powered by a 12.7 kW diesel engine and can operate continuously for 10 h on its 30 L fuel tank. The robot is driven by 48 V DC electric motors. A single electronic unit, the AGCBox, contains an RTK-GNSS receiver as well as the guiding software. The robot is equipped with bumpers for obstacle detection. A three-point linkage[16] allows the use of standard agricultural implements.

A distinguishing element of the robot is its safe fencing, a certified safety feature which ensures that no part of the robot or its tool protrudes beyond a predefined safety contour. Certification is possible because the robot makes use of a dedicated RTK base station, GNSS signals are processed by proprietary software and boundary files are stored in an encrypted format.

When the safety boundary and all lanes to be travelled have been defined, the working path for the robot (the 'mission') can be generated. The mission is validated in a simulator before the robot is deployed in the field.

In order to accommodate the smart implements, ISOBUS and TIM were implemented. Messages for GNSS-based position and velocity were added on the CAN using the ISOBUS protocol allowing the different components (of several partners) to communicate using a common framework.

The robot software was enhanced to allow it to respond to messages requesting a stop, a change of speed and a change of hitch position. This allows

[16] https://www.iso.org/standard/41233.html.

the implement to control some aspects of robot functioning. A stop request received from the weeding implement is routed to the high-level software of the robot. Responding to it does not compromise the safety system of the robot.

3.1.2 Robotti

The Robotti 150D is a field robot for power-intensive operations. It is AgroIntelli's second Robotti model and is being used in more than 15 countries. The robot weighs 3150 kg, measures 2.44 m (length) × 3.05–4.90 m (width) × 2.15 m (height). Robotti has two diesel engines with a total power of 104 kW. The wheels are driven via hydraulic power and the robot features four-wheel driving and two-wheel steering. The robot has a traditional power take-off (PTO), a three-point linkage with a lift capacity of 750 kg, and it provides one double-acting hydraulic outlet (maximum 40 L/min) with free return.

Field operations can be planned and monitored via the Robotti website. Users enter the implement's specifications, define the borders and headlands of the field and determine how the operation is to be carried out. The user can either let the system optimize the route or specify the route manually. In addition the user can configure the type of turns to be made (U-turn or zero-turn), corner paths, transit areas, and non-work areas. When the route is fully specified, it is downloaded to the robot. For safety reasons and to meet ISO standards, the robot can only be started when a person pushes a button on the computer screen on the physical robot. Once the robot has been started, its performance and the progress of the operation can be followed in real-time on the website. Footage from the front and rear surveillance cameras can also be monitored. A third camera, ImplementCam, can be put where the user wants to monitor within the 5 m cable length.

The route planning facility on the website is the only way that ROBOTTI can be programmed. This architecture is not designed to allow interaction with the Farming Controller. However, the route planning facility is accessible via an API. The desired functionality could be achieved when the Farming Controller accesses the routing API to create a new route, which is then uploaded to Robotti. This is not a step which we plan in the project.

Robotti has a CAN bus. For the project, the speed and position of the robot were made available on this bus. Additionally, the software driving the robot was modified such that it can respond to a TIM request_stop message.

3.1.3 Retrofitted tractor

One of the aims of the project is to develop a retrofit kit that can be used to convert a conventional farm tractor into an autonomous vehicle. Such an autonomous tractor can be used both in conventional farming and as part of a robotic system, thus allowing for a gradual transition to agricultural robotics. Two New Holland tractors were purchased: a T4.80F for the pilot in Greece and a T4.110F for the pilot in Spain.

In early 2022, the tractor in the pilot in Spain has been fitted with CANBUS functionality and it was used for operating the smart sprayer during the remainder of the year.

In winter 2022/23, both tractors were instrumented with actuators on the steering wheel, the accelerator and the brake pedal. An angle sensor mounted on the front axle measures the position of the front wheels. A revolution counter on the rear wheel axle measures wheel speed. Closed loop control of wheel speed and steering angle is realized through the same hardware, and similar software, that drives the Agreenculture robot. The autonomous tractor can realize a commanded linear and angular speed. It reports realized values on the CANBUS. The tractor is connected to ISOBUS-compliant smart implements through the IBBC connector.

3.1.4 Localization without relying on GNSS

In 2022, it was found that GNSS signals cannot be reliably received in two of the pilot sites. In Spain, the apple trees are sufficiently high that an RTK fix can often not be obtained. In Greece, a metal frame is used to support plastic covers to protect the grapes from rain; the GNSS signal is seriously affected by this construction. To address this issue, we have implemented a solution based on Simultaneous Localization and Mapping (SLAM).

We first build a map using a SLAM module (Grisetti et al., 2005) while teleoperating the robot, and subsequently the map is used for localization using a Monte Carlo approach (Hornung et al., 2014). This approach enables multi-waypoint navigation in a local frame (map), without relying on GNSS.

SLAM has been intensively addressed by the robotics community in the last two decades, mainly in indoor environments. In these cases, the environment is well-structured, predictable and limited in size. In contrast, in agricultural fields the environment is unstructured or semi-structured, with large extensions of land, the features evolve (e.g. the structure of the trees changes according to the seasons), and the terrain is irregular.

To address these issues, in ROBS4CROPS we have implemented an agricultural-persistent-features-based SLAM. In the case of the orchards, for example, as the structure of the trees evolve, we have focused on lower part of the trunks, which are present and visible across all seasons. To that end, we have implemented a trunk detector module, and with a graph-based approach, a persistent-features map is built, which is subsequently used for localization purposes (Fig. 5).

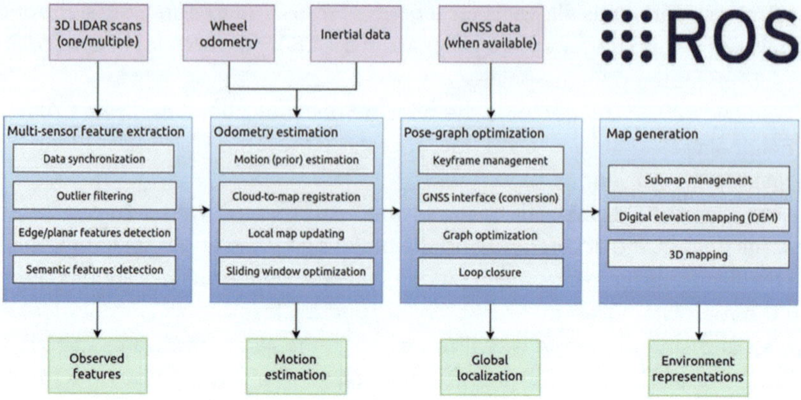

Figure 5 System diagram of the ROBS4CROPS SLAM system.

The ROBS4CROPS SLAM implementation does not rely on GNSS measurements, although these will be used if available. We use a bi-directional conversion between global coordinates (latitude, longitude and altitude) and the SLAM map frame. In this way any robot position (x, y and z) on the local map can be expressed in geodetic (latitude, longitude and altitude) coordinates and sent to the farming controller, even in the absence of GNSS data.

We rely on map-based navigation, in which the waypoints are referenced in the local frame of the map. As the localization precision of the SLAM algorithm depends on the features, we have implemented two navigation modules:

- Map-based navigation: which uses particle filtering over the map built with the SLAM module, suitable for navigation between rows or from arbitrary points of the map, where centimetric precision is not needed.
- Along row reactive navigation: as the clearance between the robot and the side trees of a row is about 0.2 m, we have developed a more robust reactive navigation module, which estimates the central path with centimetre-based precision. To this end, we extract the lines of the left and right sides given the aligned trunks using the Hough transform and derive the centre of the path. Given this information, precise linear and angular velocities corrections are executed to keep the robot centred.

3.2 Smart implements

3.2.1 Smart weeding implement for arable farming in the Netherlands

The weeding implement is a standard weeder with knives that are pulled through the soil a few centimetres below the soil surface (Fig. 4). This is a simple operation but there are several things that can reduce the efficacy of weeding. Here we focus on two common problems. First, the blades may become misaligned. This can happen when the hoe is thrown out of alignment when it hits a rock or when the fastening becomes loose. It may also be that the crop rows are not evenly spaced due to a problem during seeding. In either case, the result is that the crop plants are damaged. Second, it often happens that clods, weeds and/or crop residue collect on the vertical hoe supports and that crop plants are pulled out of the ground by this jam. A tractor driver who is weeding will frequently check the weeder to ensure that the above-mentioned problems are not happening.

To enable automatic evaluation of the performance of the weeding implement, two cameras were mounted on the Robotti, one at the front looking forward and one at the back looking rearward. The idea is to determine the quality of weeding by considering the differences between the images before and after the implement has passed. We used OAK-D Power-over-Ethernet (PoE) cameras which are designed for use in rugged environments and are rated IP67 (protect against weather and dust). Timing and speed are used to synchronize the images captured by the two cameras.

The initial algorithm to determine whether crop plants are being removed by the weeding implement relies on the assumption that only a small fraction of the ground is covered by weeds (normal situation). In this case, it is expected that the fraction of the soil surface covered by green material is substantially the same before and after weeding. Green soil cover is easily measured. First, each image is converted from RGB to HSV. Pixels with a hue between predetermined limits are considered to represent green material, resulting in a binary image in which green plants are indicated by white pixels. Only a small region of interest (ROI) close to the robot is being considered; the size of this ROI is chosen such that there is almost no overlap between the ROIs in successive images. If the fraction of green measured by the front camera is substantially larger than that measured by the rear camera, this is a strong indication that crop plants have been removed by the weeding operation. Output from the algorithm is illustrated in Fig. 6.

3.2.2 Smart weeding implement for vineyards in France

Four types of weeding implements are used in the large-scale pilot in France. All four concern weeding between the vines: serrated discs create a mound under the row (the 'ridging'), inter-vine knives cut the weeds' roots, Kress fingers scrape the soil surface and lump-breaker discs hoe the soil (Fig. 7).

The most common error encountered during the operation of these implements is that they are jammed with soil, weeds, crop residue and other debris encountered in the field. When this happens to the Kress fingers or

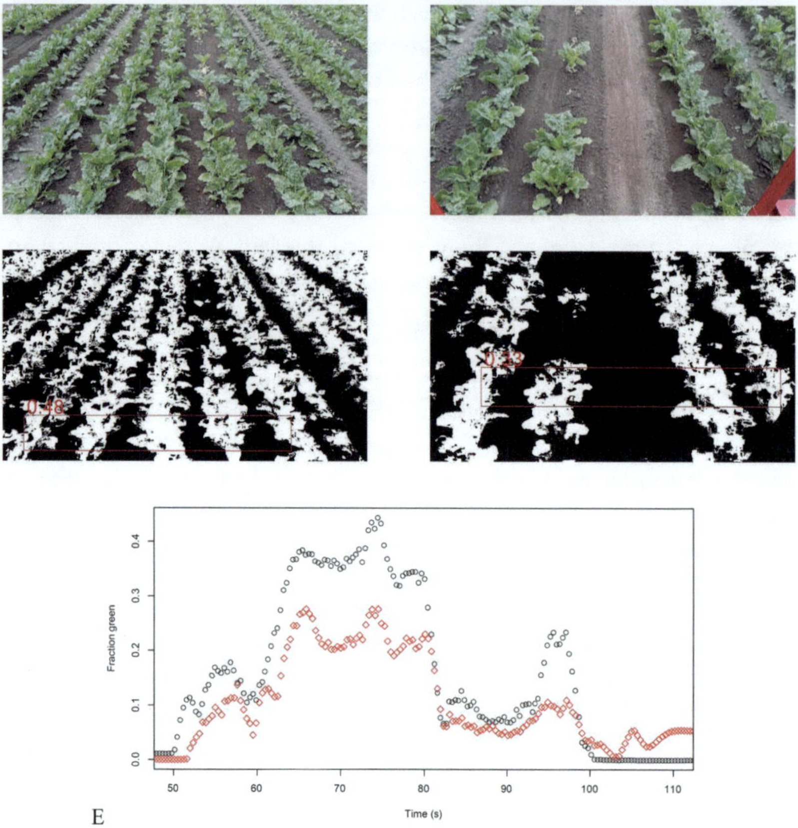

Figure 6 Smart weeder. Top left: forward-looking camera. Top right: rearward-looking camera. Middle row: binary images after transforming to HSV colour space and thresholding on hue; corresponding regions of interest in forward- and rearward-looking images are indicated, as well as the fraction of plant pixels in the ROI. Bottom: rolling median of the fraction of plant pixels in front (black circles) and rear (red diamonds) camera images. The fraction of plant pixels in the rear camera image is lower than in the front image, which is a strong indicator that crop plants have been removed.

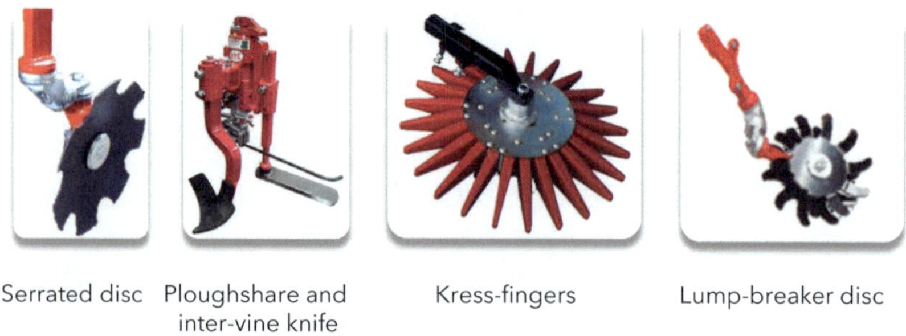

Figure 7 Weeding implements used in the pilot in France.

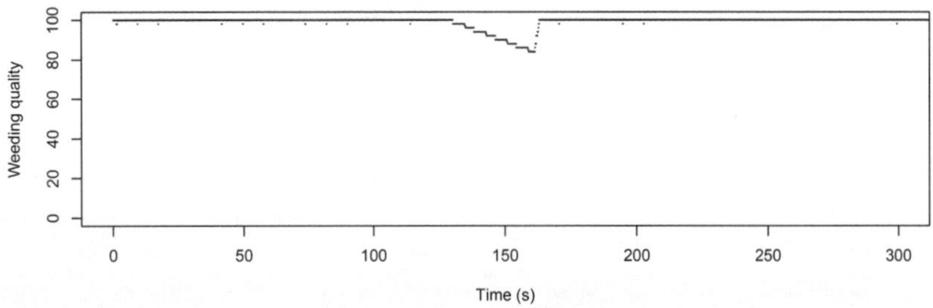

Figure 8 Weeding quality as measured by the weeding implement in the pilot in France. In this case, poor weeding quality was simulated by throwing a handful weeds into one of the Kress fingers at t = 130 s.

the discs their rotation is reduced or stopped altogether. We use revolution counters to measure the rotation of the Kress fingers and the discs. Pulses sent by the sensors are counted by a microcontroller, which sends the information to a PC mounted on the implement. Weeding quality is assumed to be 100% at the beginning. Whenever one of the Kress fingers does not send a pulse in a 1-s interval, weeding quality is decreased by 1. If weeding quality is less than 100 and all Kress fingers rotate properly, then weeding quality is increased. Example results are shown in Fig. 8.

3.2.3 Smart sprayers for vineyards in Greece and orchards in Spain

The smart sprayers consist of two elements: a variable-rate sprayer that can be controlled through ISOBUS and a perception unit (PU) that measures canopy density on-the-go and commands a spraying rate through ISOBUS.

3.2.3.1 Sprayer

Two different spraying units perform crop protection tasks in the spraying pilots. A 2000-L orchard sprayer, EOLO model, to be towed by a tractor, is used in the pilot in Spain and a 200 L lift-mounted vineyard sprayer, ASM model, is used in the pilot in Greece. Both spraying units share the same operating concepts with minor changes on sensors and actuators.

Smart sprayers gather intelligent and autonomous robotic capabilities to operate together with robotics vehicles with minimum human intervention required. Not only the spraying task but also derivative procedures are automated, as mixing phytosanitary when tank filling, tank cleaning, other maintenance and surveillance tasks as nozzle blockage control. An alarm, warnings and status dataset messages generated by the sprayer sensors are implemented on the communication bus of the FMIS-Vehicle-Sprayer architecture.

Variable rate technology (VRT) embedded in the sprayers allows an increase in efficiency and performance compared to traditional sprayers with mechanical-driven components.

The orchard (EOLO) sprayer has two horizontal sections, divided into three vertical layers (Fig. 3). The applied liquid flow rate is adjusted precisely through PWM nozzles independently for every one of the six sections according to commands received from the PU. Otherwise, if the sprayer follows commands from a prescription map, as-applied flow rate setpoint determined by the task controller for every one of the horizontal sections is applied homogeneously through the nozzles in the three vertical layers. The flow air rate assistance for canopy stirring is variable and independently controlled in the three vertical layers by hydraulic proportional valves which adjust hydraulic-driven fan speed. In the same way as liquid flow rate, air flow rate can be determined by the PU or configured through the human interface.

The vineyard (ASM) sprayer has two horizontal sections, divided in four vertical layers (Fig. 9). Liquid variable rate application operates in a similar manner as EOLO model. Flow air rate assistance for canopy stirring is adjusted according to PU commands depending on vineyard stage of vegetation, and commanded by the Sprayer ECU, which actuates the fan hydraulic proportional valve for rotating speed adjustment. Unlike EOLO, the ASM model pump is proportional hydraulic driven, providing more efficiency in adjusting rotation speed depending on liquid flow rate demand.

Figure 9 The ASM sprayer lifted by a New Holland tractor.

An ECU (electronic control unit) from EPEC (Seinäjoki, Finland) is used to control the fan and all the nozzles. ISOBUS functionality is implemented on the ECU with the Library V3 software stack (no peer-to-peer control).

3.2.3.2 Perception unit for canopy density

The amount of spraying liquid needed for effective disease control depends (among other factors) on the leaf area per unit ground surface (Duga et al., 2015). In practice, this means that early in the season, when grape vines and orchard trees carry few leaves, a low spray rate is sufficient to achieve full protection.

The PU developed in ROBS4CROPS measures the leaf area (Fig. 10). It adopts a modular camera design and mounting solution, in order to accommodate different canopy geometries and heights. This vision-based system uses depth and colour information to estimate the canopy density and generate real-time spraying recommendations. An HSV filter is used to keep only the green pixels in the depth image.

There are two main features considered when processing depth information, namely (1) the total pixel count on the depth image, which resembles to the canopy information that is present in the camera field of view (FOV) and (2i) the relative proximity index of neighbouring pixels, which corresponds to different leaf layers at different depths.

A machine learning model (multi-layer perceptron) has been trained to predict the density profile in the image, using three categorical classes based on the canopy presence: (1) no canopy, (2) sparse canopy and (3) full

Figure 10 The canopy density perception unit mounted on the hood of the tractor. The unit has two depth cameras (RealSense D435i, Intel, Santa Clara CA, USA) (one camera is facing the viewer). Also visible are two antennas for the wifi router that is part of the unit.

Figure 11 The effectiveness of the autonomous spraying system in table grapes was assessed in the field using strips of water-sensitive paper. See main text for explanation.

canopy. Annotation of depth data was performed by field experts categorizing each image in the corresponding class, using the Point Quadrant measuring technique and domain knowledge where necessary.

As a result, based on colour and depth information, the PU can modulate the default/maximum tree row spraying volume. If no canopy is detected, the spraying rate is set to 0. If a sparse canopy is detected, the spraying rate is set to 50% of the maximum rate. If a dense canopy is detected, the spraying rate is set to 100% of the maximum rate.

The PU was designed to send real-time spraying recommendations in the sprayer using the current velocity of the tractor/robot. All sending and receiving messages were based on ISOBUS and J1939 protocol communication. The software implementation uses python-can[17] for sending and receiving messages on a CAN bus line.

During the first season of deploying the autonomous spraying system in table grapes in Greece, an assessment protocol was established to measure the spraying effectiveness. Water-sensitive paper was placed on the leaves of three positions with different canopy densities. Canopy positions were selected in order to resemble grape canopies with (1) no canopy, (2) sparse canopy and (3) full canopy coverage. Each canopy was split into nine areas positioned in the upper (U) and lower (L) parts, front (F), middle (M) and back (B) (Fig. 11).

Water-sensitive papers have a specially coated yellow side that changes to blue when exposed to moisture. So, the wet area gets blue and the dry area remains yellow. We collected sprayer performance data in three contrasting settings, namely (1) a conventional sprayer, as currently used in table grape farms, (2) the ASM sprayer without PU, and (3) the ASM sprayer with PU activated (Table 2).

After the field session, a photo of each sprayed water-sensitive paper was taken in order to calculate the percentage of area covered by spray liquid on each paper, using image analysis software.

17 https://python-can.readthedocs.io/en/stable/.

Table 2 Experimental set-up for testing the sprayer. The conventional sprayer used hollow cone nozzles (Magnojet MGA 03). The variable rate sprayer used flat fan nozzles (Arag ASJ SF11004)

Treatment	Sprayer	Nozzles	Pressure (bar)	Speed (km/h)	Repetitions	Variable rate	Spraying rate (L/ha)
1	Conventional	Hollow cone	10	4	2	No	800
2	ASM	Flat fan	4	4	2	No	800
3	ASM	Flat fan	4	4	2	Yes	0, 400, 800

The PU software accurately quantified variations in foliage density in real-time. The results showed that the sprayer achieved satisfactory coverage and droplet size while saving from 10% to 50% of spray liquid compared to the conventional sprayer. It is also noteworthy that the innovative spraying operated with lower pressure and, thus, drift was reduced.

3.3 Communication

The robotic system consists of three main classes of components: autonomous vehicles, smart implements and the Farming Controller. The robotic system requires that these components are in constant communication with each other. Here we describe how this communication is effectuated.

3.3.1 Communication between implement and vehicle via ISOBUS

Communication between the implement and the vehicle is implemented with ISOBUS (ISO 11783). ISOBUS is based on the SAE J1939 communication protocol, which in turn is based on the physical CAN bus (ISO 11898).

ISOBUS is a standard for communication between agricultural vehicles and implements. However, it is in most cases not used by agricultural robots and there are still many implements without ISOBUS. Therefore, we have adopted an approach that allows the gradual introduction of ISOBUS. In this approach, ISOBUS connectivity is emulated for those vehicles and implements that do not yet offer ISOBUS connectivity. Emulation is provided by CANoe 15 SP2 software (Vector Informatik GmbH, Germany) running on an industrial PC. See Fig. 12.

The smart sprayers (Section 3.2.3) are fully ISOBUS compliant. The sprayers are designed with an ECU (EPEC, Seinäjoki, Finland) which integrates the full ISO 11783 protocol. As an input, the ECU of the sprayer receives GNSS information to geolocate the applied points and the ground-based speed to calculate the point-to-point distance from the Agreenculture robot and the autonomous tractor (AGCBox) through IBBC connector. For this, the TECU functionality of the ISOBUS is integrated into the CANoe software to enable address-claiming with the ECU of the sprayer. Another input to the ECU is the prescribed/recommended rate from the PU that is packed as a 'Process Data' (PD) message (DDI36 - App*RateVolumePerTime_Set*). The message PD is defined with the ID of *0xCCBFEFE*. The ID is usually dynamic due to the dependency on the source address (SA) and destination address (DA). The PU (Section 3.4) is partially ISOBUS compliant because of the implemented Task Controller Library Version 3.0 of the ECU Sprayer. In this library, there is no possibility to enable the 'peer-to-peer' control function of the ISOBUS. Therefore, the PU cannot directly claim an address with the Sprayer ECU as

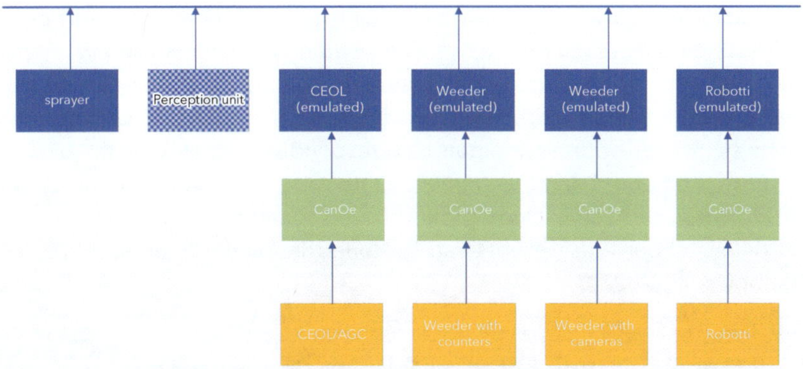

Figure 12 Schematic representation of connectivity between vehicles and implements via ISOBUS. See main text for explanation.

a fully ISOBUS compliant component. The Sprayer ECU sends the as-applied information (PD message, DDI2 – *AppRateVolumePerArea_Act*) as an output.

ISOBUS functionality for the weeding implements and the autonomous vehicles is achieved through emulation. The entire ECU of the weeding implements and ISOBUS functionality are emulated in the CANoe software. For the weeding implement in combination with the Agreenculture robot, the AGCBox emulates the GNSS information to geolocate the weeding points and the ground-based speed. The weeder in France calculates weeding quality by considering the number of revolutions of the discs or Kress fingers per metre travelled by the robot. The weeder in the Netherlands needs geolocation to correlate images taken by its forward- and rearward-looking cameras.

Tractor Implement Management (TIM) is an ISOBUS-based solution for a system where the implement can control certain tractor functions. While TIM is developed for tractor-based systems, it can be expected to be very useful for any situation in which smart implements gather information about their operation and information about the environment. For example, when a smart weeder detects that weed quality is very poor, it would be reasonable to stop the operation.

The vehicles used in our project (and indeed most tractors sold today) do not yet implement TIM. As a first step, we implemented on the vehicles used in our project the ability to respond to a single TIM command, namely the emergency stop command. The vehicles use ISOBUS message (TIM21) with an ID of *0x1C24FF21* to recognize the stop command incoming from the TIM Client node of the middleware. The ISOBUS address claiming between the TIM Client node and the implement ECU to send and receive signals is left to CANoe and defined under the 'acknowledgement' message with and ID of *0x18EEFF21*.

The weeding implement used in the pilot in France counts the number of revolutions for each of the Kress fingers or discs. A microcontroller sends this information via CAN to CANoe which claims an ISOBUS address and puts a corresponding message on the bus. The weeding implement used in the pilot in the Netherlands uses image processing software to determine the weeding quality. Here a Python script sends this information via CAN to CANoe which claims an ISOBUS address and puts a corresponding message on the bus.

Both the Robotti and the Agreenculture robot are connected via CAN to CANoe. Messages sent include ground-based speed and GNSS position data. Full details about these robots can be found in Paraforos et al. (2022) and Sharipov et al. (2023).

3.3.2 Communication between Farming Controller and implement/vehicle via Robot Operating System 2

The Robot Operating System (ROS) is a set of open-source software libraries and tools for robotics applications (Macenski et al., 2022). ROS comprises a message-based communication framework for passing information between the components of a robotic system. ROS has always enabled distributed systems but the latest version (ROS 2) greatly simplifies setting up communication between components running on different computers. All communication between the Farming Controller on the one hand and vehicles and implements on the other hand is implemented through ROS 2.

We used mobile internet and VPN to set up a Software-defined Wide Area Network (SD-WAN). This WAN is used to connect all vehicles and implements.

The Farming Controller and vehicle are on the same ROS 2 domain and therefore discover each other automatically. All topics published by all robots are always visible. We only want to log data when the robot is working, not when it is on a road or when it is in the barn. This is achieved by the Farming Controller only logging when a flag has been raised. Specifically, the robot in the pilot in Spain initiates logging by sending a message with content 'true' on the ROS 2 topic named 'lsps/trigger_flag'. When the flag is up, the Farming Controller scans the list of topics and subscribes to all topics with names starting with '/lsps/'.

3.4 Farming Controller

The Farming Controller is a DT-based software that allows the orchestration of the resources described in the previous sections (vehicles, implements and communication facilities). The Farming Controller must allow to select combinations of vehicles and implements in order to carry out different tasks. It must also allow the planning and scheduling of field operations (spraying and weeding) in a series of fields during an entire growing season; in doing so it must take into account transporting the robot from one field to the next. Once the robot is in a field, there are decisions to be made in terms of the order in which working lanes are visited (vehicle routing problem (VRP)). Once the robot is working in a field, the system must be able to detect problems with weeding and spraying, with obstacles, with fuel running out or batteries running empty and so on and replan the work if necessary.

The requirements for the Farming Controller call for the implementation of several functions of DTs as used in industrial manufacturing (Table 3). The first function, 'Design and configure', will allow farmers to generate and evaluate work schedules before any work is done in the real world. Schedules may be evaluated for the available set of fields, vehicles and implements, as well as for fields, vehicles and implements under consideration for purchasing. The second function, 'respond in real-time to changing conditions/unforeseen events', will, for example, allow the robotic system to use an alternative route if an obstacle is detected. Also, the system may respond to changes in weather: for example, if the wind increases while a spraying operation is in progress, it may be prudent to switch operations to a field where spray drift is less likely to damage an adjacent crop. The third function, 'predictive maintenance', is technically close to the second in the sense that data collected in real-time about the system is used to inform decision-making. In

Table 3 Digital twin-supported functionalities from industrial manufacturing and their application in agricultural robotics

Function	Industrial manufacturing example	ROBS4CROPS example
Design and configure (before deployment)	Optimize location and number of mobile robots and human workers in vehicle front axle assembly line (Kousi et al., 2021)	Select vehicles and implements; assign them to fields; create a work schedule to perform all necessary operations on each field within the set time frame; identify the work schedule that is optimal in some sense
Respond in real-time to changing conditions/ unforeseen events	Reconfigure assembly line when parts supply is interrupted. Re-plan robot motion path when human worker steps in front of a mobile robot (Kousi et al., 2019)	Reroute vehicle to avoid an obstacle or to accommodate unexpected field conditions
Predictive maintenance	Monitor gearbox wear and signal the need for repairs when they are needed (but not before then) (Aivaliotis et al., 2019)	Monitor battery charge (electric vehicle) or amount of fuel remaining (vehicle with combustion engine), as well as amount of biocide remaining. Reroute vehicle to filling/ charging station when needed

Published by Burleigh Dodds Science Publishing Limited, 2024.

predictive maintenance, for example, the robotic system may monitor over time how well the battery of an electric vehicle is holding up, or how much biocide is used on average in a variable-application-rate modus, and then take appropriate action.

At a high level, an example of a working schedule could be the following. The farmer would like to find the optimal path to perform the spraying and weeding activities in different fields with the following structure:

- Field 1 for spraying;
- Field 2 for spraying; and
- Field 3 for weeding.

The farmer possesses the following resources which can be used by the decision-making algorithm to generate alternative path and operations/field assignments to each resource:

- Robot 1;
- Robot 2;
- Implement 1 for weeding; and
- Implement 2 for spraying.

Firstly, a combination of the possible vehicles is generated (robot with implement), based on the compatible combinations between each robot with each implement that can be made. The suitable combinations that have been generated by the decision-making algorithm are as follows:

- Vehicle 1 (Robot 1 + Implement 1);
- Vehicle 2 (Robot 2 + Implement 1); and
- Vehicle 3 (Robot 2 + Implement 2).

The decision-making algorithm generates many alternative working schedules by allocating the above vehicles to the different fields, taking into account that some fields need spraying and others need weeding (Fig. 13). For each allocation of a vehicle to a field, the time needed for the operation is simulated (see later). The amount of time needed to complete a working schedule depends on the number of vehicles used and on how the vehicles are allocated to fields. The most appropriate alternative is selected. Since this algorithm is based in heuristics and doesn't investigate all the alternatives exhaustively, the solution is considered to be acceptable but it is not necessarily the best.

An algorithm based on the Traveling Salesman Problem (TSP) is used to optimize the manoeuvre time of a vehicle in the field, taking into consideration the turning constraints. The algorithm results in a path that visits all working

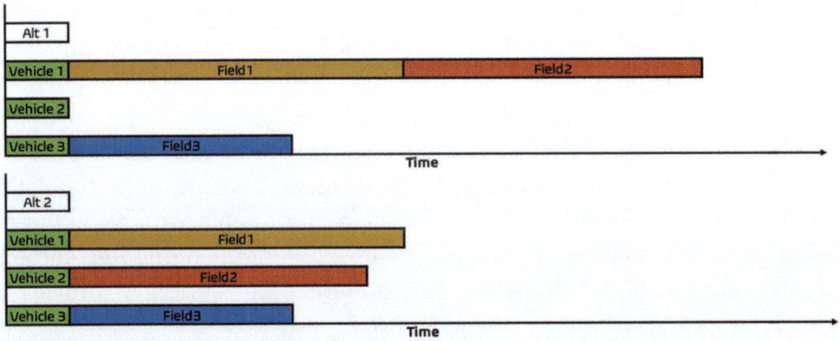

Figure 13 Gantt Chart that illustrates an example of two alternatives that are generated and evaluated by the decision-making algorithm.

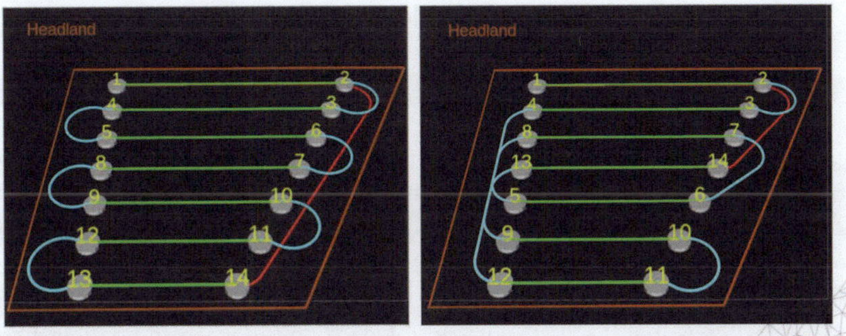

Figure 14 Two possible routes through a given field. Left: an initial route in which all working lanes are visited in sequence. Right: a route proposed by the TSP algorithm that takes into account the time needed to turn from one lane into another and minimizes the time to complete the entire route.

lanes at least once. In this algorithm, the parameters can be adapted to choose a path that is either shortest in length or shortest in travel time (Fig. 14).

The TSP algorithm gives a path that is optimal in terms of distance or travel time. Travel time is calculated by multiplying the length of each path segment by the speed with which that segment is traversed; working lanes will have a different speed than transport lanes and headland turns.

TSP cannot easily be extended to take into account spatially variable conditions (of the field), time-varying conditions (state of the vehicle and/or the field) and interactions between them. An example of a spatially variable condition of the field is that there may be zones where something is limited, such as vehicle speed, weight, width and/or height. An example of a time-varying condition is the weight of the vehicle, which may be high when a spraying operation starts (full tank) and low once the tank has been partially emptied. An example of an interaction is that a vehicle with a full tank may not be able to pass over a bridge with a low load carrying capacity, whereas the same bridge may be passable with a tank that is almost empty. Another example is that travelling uphill with a full tank requires more power and leads to more soil damage than either travelling downhill with a full tank or travelling uphill with a partially empty tank.

We use a dynamic (discrete event) simulation model to evaluate performance indicators associated with a given route. State variables of the simulation model include (for the vehicle) pose (location and orientation), battery status or fuel remaining and biocide level. The environment is monitored through indicators such as fuel consumption rate or engine power uptake (proxies for state of the soil), crop canopy size and crop disease level.

The simulated vehicle moves along the given route and activates the implement in working lanes. The speed is determined as the minimum of target speed for the operation and the maximum speed permitted by the terrain (input to the model). The model can easily be extended to include, for example, soil water content and trafficability and energy consumption as a function of vehicle weight, slope and soil condition.

An example of the model's output is given in Fig. 15 for a field which has four working lanes of 100 m length. Time and distance are simulated for two routes that start at the bottom left corner (East = North = 0). The first route starts with by going east and enters the first working lane. The second route starts by going north and then enters the second working lane. Both routes have the same length and both routes traverse twice a part of the headland. The difference is that the first route doubles up on the western headland and the second route does so on the eastern headland. In this example, speed on the western headland is restricted and therefore the first route takes longer to complete.

The full process of optimizing optimal use of resources is shown schematically in Fig. 16. Modelling takes place at three levels: Gantt chart to allocate vehicles and implements to fields during a growing season, TSP to select the sequence and direction in which working lanes are visited and physics-based simulation model to track changes in state variables as the vehicle moves along the field. Re-planning and -scheduling in response to changing conditions/unforeseen events could take place at any of the three levels, here we focus on the last two levels, i.e. responding while the vehicle is moving inside one field.

When the robot is moving in a field, a change in conditions or an unforeseen event may make it expedient to calculate a new vehicle route. A change in condition may be that the wind is picking up and that it is prudent to prioritize spraying in working lanes where there is no adjacent crop that could be damaged by spray drift. Another change in condition may happen during variable rate spraying where the application rate depends on some spatially varying characteristic of the crop (e.g. canopy density or disease infestation

Published by Burleigh Dodds Science Publishing Limited, 2024.

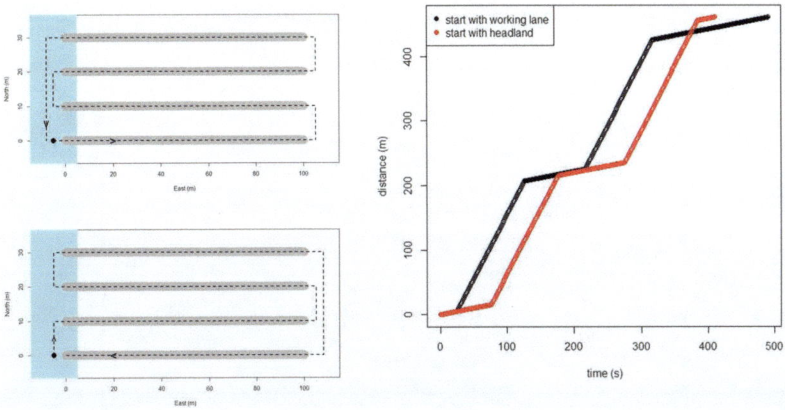

Figure 15 Example output of the agricultural simulation model, where based on working lane conditions the distance travelled and time spent can be determined during operation of the route in the field and during simulation. In this example, the terrain imposes a speed limit on the part of the field where East ≤ 5 m (indicated by the blue background). The 'start with working lane' route has a greater length in this area than the 'start with headland' route and therefore takes longer to finish even though the total length of both routes is the same.

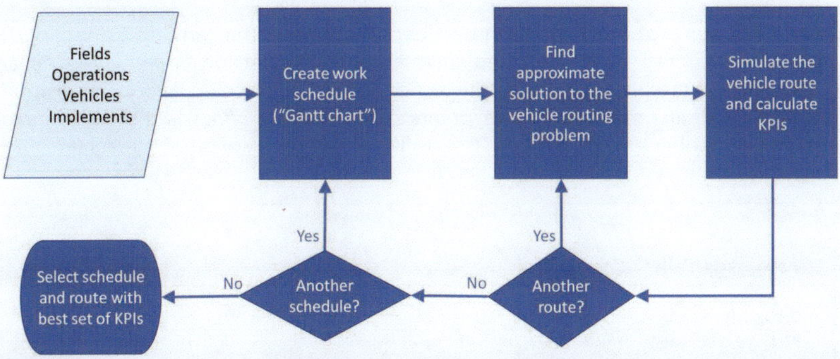

Figure 16 Flow chart of the three-level modelling process inside the Farming Controller. Given the set of fields, operations, vehicles and implements, a feasible working schedule is created. For a given working schedule, an approximate solution for the robot to move through field is created. Finally, for a given vehicle route, a dynamic simulation is performed to evaluate several key performance indicators (KPIs). Each process is repeated until a large set of alternatives is available, from which the most suitable one is selected in the last step of the flow chart.

level). In this situation, it may be that the biocide is used at a higher rate than foreseen. When the DT model is re-run with the new (observed) average application rate, a prediction can be made as to when the robot will run out of biocide and thus, at what point it must return to base for a refill.

The action communication type of ROS 2 is used to implement the robot route control mechanism (Fig. 17). The logic of this control mechanism is to

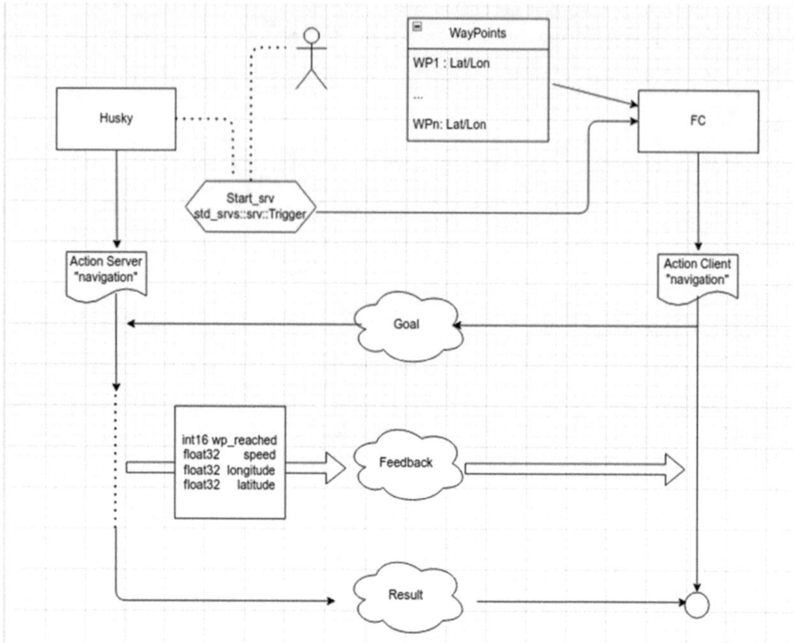

Figure 17 Schema of the interaction between a robot and the Farming Controller. The robot (here denoted as 'Husky') implements the ROS 2 Action Server interface and the Farming Controller implements the Action Client interface. In a loop, the Farming Controller sends an ordered set of waypoints to be visited (the goal) to the robot, and the robot sends feedback to the Farming Controller whenever a waypoint is reached.

```
2
# Goal
nav_msgs/Path          path
---
# Result
std_msgs/Empty         plan_result
---
#Feedback
int16                  waypoint_reached
float32                speed
float32                longitude
float32                latitude
```

Figure 18 The ROS 2 Action message that is used to send a route to the robot and receive feedback. Here 'path' is the ordered series of waypoints that define the route to be followed; 'waypoint_reached' is the sequence number of the last waypoint reached; 'speed' is the current speed of the robot (m/s); longitude and latitude indicate the current position of the robot in the WGS84 coordinate system.

get an action signal from the robot to the Farming Controller that it is ready to get navigation points and the Farming Controller is responding with the set of waypoints that the robot should visit. Once the robot confirms reaching the first point, the Farming Controller sends the next set of waypoints and this loop is iterated until the whole path is followed. In ROS 2 terms, the Farming Controller is the Action Client and the robot is an Action Server; communication is effectuated through an Action message (Fig. 18). The action message contains the list of waypoints expressed in the WGS84 coordinate system. While the robot is navigating, it sends a continuous stream of position messages (with current speed, latitude, longitude, etc.) as well as the number of each waypoint that it reaches.

When it is necessary to send the robot on a different route, a new goal (Action message) is sent to the robot. The new goal pre-empts the earlier goal and the robot starts following the new route.

The first hardware implementation of the route control mechanism was realized using a Husky (Clearpath Robotics Inc., Kitchener ON, Canada). The

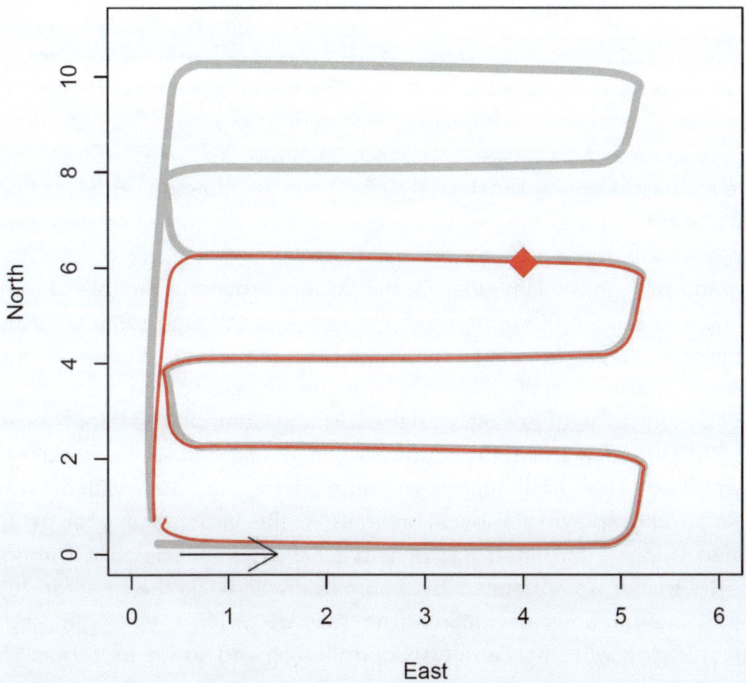

Figure 19 Results of a route control experiment. Starting near (0,0), and initially travelling in the direction indicated by the arrow, the thick grey line shows the path followed by the robot during a normal field operation. The thin red line shows the path followed by the robot during a field operation when an unexpected event occurs (indicated by the diamond symbol). The kind of the event is not specified, it could be a change in the weather, the observation that the battery is almost empty or any other non-emergency event that results in a new route being sent to the robot. When the new route is sent, the robot continues to follow the lane until it reaches the headland; from there it can continue on the alternative route.

small size of this robot makes it easier for development work than a full-sized tractor. A standard set-up using the Navigation 2 package from ROS 2 was put in place. The Action Server was implemented in Python. The experiment was conducted with the Husky located in the Netherlands and the Farming Controller running on a server in Greece. A VPN was used to establish a network connection (see Section 3.3). Results of the experiment are shown in Fig. 19. The experiment itself is trivial but it was instrumental in developing the protocol.

4 Case study results

In the context of the large-scale pilot in France, the robot operated in multiple series of weeding sessions in the field, proving its robustness and ability to perform. However, the main challenge encountered has been the frequency of stops, which posed the need for the user to intervene and restart the robot or perform troubleshooting. Thus, the user was required to always be on-site in order to supervise and act accordingly. The communication of the robot with the Farming Controller has been excellent during field trials. In 2022, nine plots with a total area of 14 ha were weeded. Each plot was weeded 6 or 7 times, with each pass taking approximately 1 h. The most common error was that the robot's sensitive front bumper was activated by a vine; in all cases, the robot could continue working after a manual override.

In the pilot, in the Netherlands, the Robotti proved to be able to perform seeding and weeding throughout the season, even though some adjustments were needed to perfectly fit all implements on the robotic platform. A human supervisor was present at all times to intervene in cases of the robot stopping. The communication of Robotti with the Farming Controller has been excellent during field trials. In 2022, the robot worked in a field of 0.8 ha; seeding, two tine weeding passes and four hoe weeding passes were done with the robot.

Regarding spraying application trails in the large-scale pilot in Spain, spraying using a retrofitted tractor and the PU was satisfactory. During the trials, the tractor was operated by a human driver, while the goal for 2023 is to apply autonomous navigation. The ISOBUS worked efficiently and the communication with the Farming Controller proved to be performing well. Sprayings took place for many hours during the entire growing season from June to September 2022 in a field of 12 ha.

As for the large-scale pilot in Greece, trials during 2022 were performed using a temporary solution of a pre-existing tractor that was modified and retrofitted. As in Spain, the tractor was driven by a human and the communication with the Farming Controller was very stable. Experimental operations lasted for 2 days in a field of 0.2 ha.

The VPN connection with the Farming Controller was very stable in all pilots. Logging data worked without fail.

5 Discussion

We have presented a vision and a preliminary implementation of an agricultural robotics system centred around a DT. This robotic system consists of smart implements and autonomous vehicles which are orchestrated by a Farming Controller. The system is developed to serve the needs of four specific farming situations and is being tested in these farms.

The smart weeder for sugar beets consists of a conventional weeding implement which is made smart by a forward-looking and a rearward-looking camera. Comparing the images captured by these cameras allows to determine the quality of the weeding operation. The algorithm developed is designed to check one specific condition, namely where weeds and/or crop residue are jammed in the hoe, leading to 'collecting' the crop plants, a situation in which the crop plants are pulled out of the ground. The algorithm is based on basic image processing techniques and is sensitive to the threshold values that are used. The algorithm may fail, for example, when dark shadows are present in the image, when a lot of weeds are present and when it is deployed on different soils. However, this initial algorithm can easily be replaced with another algorithm, possibly one based on deep learning. It is also conceivable to have a stack of algorithms running, each checking for a different error condition.

Likewise, the smart weeder for vineyards also checks for a specific (and common) error condition, namely the Kress fingers or rotating discs being jammed by weeds or soil.

The sprayers (orchard in Spain, vineyard in Greece) are smart by relying on the PU which adjusts the spraying rate based on canopy density. Limitations of the current system include step-wise instead of continuous adjustment of the spray rate and not measuring infestation level which would allow to reduce spraying rate on healthy but dense canopies. A further limitation is that the sprayer does not measure flow rate and is thus not able to detect if one or more nozzles are blocked. These shortcomings can be addressed at a later time without invalidating the robotic system itself. In the meantime, the system is already an improvement over current practices.

The robots used in this project have proven to be robust and able to work for many hours. However, working in large, open field with sugar beets, in which no obstacles are present, is easier than working in a vineyard. In the latter case, we found that the robot frequently stopped because a vine touched its safety bumper. In all cases, the robot was commanded by the supervisor to continue driving. This points to a need to be able to differentiate between vine and a real obstacle touching the bumper.

In order to transform a conventional tractor into an autonomous tractor, it is necessary to get accurate and precise feedback about the wheel speed and the angle of the front wheels. These measurements are available on the tractor's internal messaging bus, but the protocol is not documented. We have overcome this problem by installing additional sensors on the tractors.

A major, unexpected obstacle encountered was that GNSS signals were not reliable in two of our pilot sites. In the orchards in Spain, trees are sufficiently high that an RTK fix can often not be obtained. In the vineyard in Greece, we were surprised to discover that a metal frame is used to support plastic covers intended to keep rain from the grapes; naturally, the GNSS signal is seriously affected by this Faraday cage. This problem was overcome by resorting to localization based on LIDAR and a map of the orchard or vineyard.

Implementing the ability to respond to a 'stop' command has turned out to be a complex issue for the robot manufacturers in our project. The robots were designed to operate as stand-alone units and the software for navigation and for low-level control is tightly integrated. This has allowed us to provide the desired level of safety: the robots are unable to start moving without a person being physically present.

The Farming Controller has only partially been developed. However, we have delivered a proof of concept of calculating a route off-site, sending it to a robot in the field and pre-empting that route in real-time by sending an alternative route to the robot.

The smart implements, the autonomous vehicles and the Farming Controller must be in constant communication with each other for the robotic system to work. We use ISOBUS for communication between vehicles and implements. The bandwidth offered by ISOBUS is insufficient to support the delivery of a stream of images; thus we used ethernet for the cameras. We use ROS 2 over a mobile ethernet connection for communication between the Farming Controller and the vehicles. We found this to work very well, but of course this is predicated on having high-quality mobile ethernet in the field.

The work presented here will be elaborated and expanded during the remaining 2 years of the project. For implements, it is expected that camera-based monitoring of weeding will benefit from leveraging developments in deep learning to become more robust and more versatile. The bottleneck here will be collecting sufficient training data.

For the Farming Controller, it is expected that task optimization will become more important. Optimizing a task will typically be a multi-objective goal, taking into account time, fuel use, timeliness of operations and so on. The information gathered by a variety of sensors and sent to the Farming Controller can be collected in a digital twin of the farm, which gives up-to-date, precise, accurate information about the health and size of the crops; the presence of weeds, pests and diseases and trafficability of the field (Paraforos et al., 2019). Using this information, future versions of the Farming Controller may be able to initiate tasks. For example, when measurements in the field or a disease prediction model (Been et al., 2023; Kessel et al., 2018) indicate that a potato crop is at risk of being infected with late blight, the Farming Controller could start a spraying task.

But widespread adoption of agricultural robotics will require more than just technology. In this chapter, we have not addressed that a support system of training, dealers, maintenance provides, banks and insurers is needed before large-scale adoption will be reached.

In summary, the preliminary results presented here support the hypothesis that the proposed robotic system will work. Widespread adoption of an agricultural robotic system may still be some distance away but an immediately relevant contribution of our work is the data sent by the implements and the vehicles to the Farming Controller in a DT framework. This opens the way to create a rich user interface with which an operator (farmer, farm worker or contractor) can follow in real time the location of the agricultural robots in his or her fleet of robots. Such a user interface can also provide information about the status of each robot, the quality of the work being done by the robot and a live stream of the robot at work. The user interface can also present information about how long it will take to complete each task and it will offer alternative schedules when conditions change. We expect that the willingness of operators to 'trust' robots will greatly increase when they can monitor the performance of the machines on their smartphones, in real-time, wherever they are located. This will be an important step towards adopting a complete robotic system.

6 Conclusion

Our work is based on the hypothesis that digital twins will be just as useful to optimize activities in agriculture as they have proven to be helpful in industrial manufacturing. We have presented a vision of an agricultural robotics system centred around a digital twin. In addition, we have been able to provide an implementation of this system using off-the-shelf hard- and software in combination with some custom-made components. The system is being tested in four different farming systems throughout Europe. Based on initial test results, we conclude that a DT-based agricultural robot system requires less supervision than current systems and that it will be able to respond to changing conditions and unforeseen events.

7 Where to look for further information

- Field robots catalogue: https://www.futurefarming.com/dossier/field-robots/.
- Human supervision and national and EU rules and regulation is a challenge (Lowenberg-DeBoer et al., 2022; Maritan et al., 2023).
- Farmers adoption of autonomous systems are modest due to a range of factors (Gabriel and Gandorfer, 2023; Gil et al., 2023).
- Robots not only address labour shortages but can also lead to reduced CO_2 emissions (Gonzalez-de-Soto et al., 2015) and soil compaction (Duckett et al., 2018).
- Several studies indicate positive economic perspectives from using robots (Lowenberg-DeBoer et al., 2020).

7.1 Key journals/conferences

- Journal of Field Robotics. A Special Issue on agricultural robotics will appear in 2023. The 2020 Special Issue on agricultural robotics contains many relevant papers (but digital twins are not mentioned).
- FIRA (https://www.fira-agtech.com/) and SIMA (https://en.simaonline.com/) showcase many agricultural robots.
- IEEE International Conference on Robotics and Automation (ICRA) and International Conference on Intelligent Robots and Systems (IROS) are leading conferences.

8 Acknowledgements

The ROBS4CROPS project has received funding from the European Union's Horizon 2020 research and innovation programme under grant agreement No. 101016807. Wageningen University & Research has received additional funding from the Netherlands Ministry of Agriculture (project BO-67-101-023).

9 References

Aivaliotis, P., Georgoulias, K. and Chryssolouris, G. 2019. The use of digital twin for predictive maintenance in manufacturing. *International Journal of Computer Integrated Manufacturing* 32(11): 1067–1080.

Alexopoulos, K., Nikolakis, N. and Chryssolouris, G. 2020. Digital twin-driven supervised machine learning for the development of artificial intelligence applications in manufacturing. *International Journal of Computer Integrated Manufacturing* 33(5): 429–439.

Anken, T. and Latsch, A. 2022. Detection rate and spraying accuracy of Ecorobotix ARA. Lecture notes in informatics (LNI). In *Lecture Notes in Informatics (LNI) - Proceedings. Series of the Gesellschaft fur Informatik (GI).* Gesellschaft für Informatik. P-317: 45–50. ISBN 978-3-88579-711-1. Gesellschaft für Informatik.

Anken, T., Latsch, A. and Hatt, M. 2022. Spot spraying broad-leaved dock plants in meadows Performance, accuracy and sprayer test of Ecorobotix ARA. *VDI-Berichte* 2022: 279–284. https://doi.org/10.51202/9783181023952-279.

ASABE Staff. 2022. See & SprayTM select by John Deere. *Resource Magazine* 29(3): 7–8.

Been, T. H., Kempenaar, C., van Evert, F. K., Hoving, I. E., Kessel, G. J. T., Dantuma, W., Booij, J. A., Molendijk, L. P. G., Sijbrandij, F. D. and van Boheemen, K. 2023. Akkerweb and farmmaps: development of open service platforms for precision agriculture. In Cammarano, D., van Evert, F. K. and Kempenaar, C. (Eds). *Precision Agriculture: Modelling*. Springer International Publishing, Cham, pp. 269–293.

Bochtis, D., Vougioukas, S., Ampatzidis, Y. and Tsatsarelis, C. 2007. On-line co-ordination of combines and transport carts during harvesting operations. *Precision Agriculture 2007 - Papers Presented at the 6th European Conference on Precision Agriculture*, ECPA, pp. 715–721.

Cloete, S. 2020. Weed management using sensor technology. *Farmer's Weekly* 2020: 60–61.

Duckett, T., Pearson, S., Blackmore, S., Grieve, B., Chen, W.-H., Cielniak, G., et al. 2018. Agricultural robotics: the future of robotic agriculture. arXiv preprint arXiv:1806.06762.

Duga, A. T., Ruysen, K., Dekeyser, D., Nuyttens, D., Bylemans, D., Nicolai, B. M. and Verboven, P. 2015. Spray deposition profiles in pome fruit trees: effects of sprayer design, training system and tree canopy characteristics. *Crop Protection* 67: 200–213. https://doi.org/10.1016/j.cropro.2014.10.016.

Gabriel, A. and Gandorfer, M. 2023. Adoption of digital technologies in agriculture–an inventory in a European small-scale farming region. *Precision Agriculture* 24: 68–91. https://doi.org/10.1007/s11119-022-09931-1.

Gil, G., Casagrande, D. E., Cortés, L. P. and Verschae, R. 2023. Why the low adoption of robotics in the farms? Challenges for the establishment of commercial agricultural robots. *Smart Agricultural Technology* 3: 100069.

Gonzalez-de-Soto, M., Emmi, L., Garcia, I. and Gonzalez-de-Santos, P. 2015. Reducing fuel consumption in weed and pest control using robotic tractors. *Computers and Electronics in Agriculture* 114: 96–113. https://doi.org/10.1016/j.compag.2015.04.003.

Grisetti, G., Stachniss, C. and Burgard, W. 2005. Improving grid-based slam with rao-blackwellized particle filters by adaptive proposals and selective resampling. In *Proceedings of the 2005 IEEE International Conference on Robotics and Automation*. IEEE Publications, USA, pp. 2432–2437.

Hemming, J., de Jong, H., Struik, L., van Meer, J. and Van Henten, E. 2018. Field performance evaluation method for a machine vision based intra row hoeing machine. In *Proceedings of the European Conference on Agricultural Engineering AgEng2018*. Wageningen University & Research, Wageningen, The Netherlands, pp. 188–196.

Hizatate, T. and Noguchi, N. 2023. Work schedule optimization for electric agricultural robots in orchards. *Computers and Electronics in Agriculture* 210: 107889. https://doi.org/10.1016/j.compag.2023.107889.

Hornung, A., Oßwald, S., Maier, D. and Bennewitz, M. 2014. Monte Carlo localization for humanoid robot navigation in complex indoor environments. *International Journal of Humanoid Robotics* 11(2): 1441002.

Kessel, G. J. T., Mullins, E., Evenhuis, A., Stellingwerf, J., Cortes, V. O., Phelan, S., van den Bosch, T., Förch, M. G., Goedhart, P., van der Voet, H. and Lotz, L. A. P. 2018. Development and validation of IPM strategies for the cultivation of cisgenically modified late blight resistant potato. *European Journal of Agronomy* 96: 146–155. https://doi.org/10.1016/j.eja.2018.01.012.

Kousi, N., Gkournelos, C., Aivaliotis, S., Giannoulis, C., Michalos, G. and Makris, S. 2019. Digital twin for adaptation of robots' behavior in flexible robotic assembly lines. *Procedia Manufacturing* 28: 121–126.

Kousi, N., Gkournelos, C., Aivaliotis, S., Lotsaris, K., Bavelos, A. C., Baris, P., Michalos, G. and Makris, S. 2021. Digital twin for designing and reconfiguring human–robot collaborative assembly lines. *Applied Sciences* 11(10): 4620.

Li, C., Zheng, P., Li, S., Pang, Y. and Lee, C. K. M. 2022. AR-assisted digital twin-enabled robot collaborative manufacturing system with human-in-the-loop. *Robotics and Computer-Integrated Manufacturing* 76: 102321. https://doi.org/10.1016/j.rcim.2022.102321.

Lowenberg-DeBoer, J., Huang, I. Y., Grigoriadis, V. and Blackmore, S. 2020. Economics of robots and automation in field crop production. *Precision Agriculture* 21(2): 278–299.

Lowenberg-DeBoer, J., Behrendt, K., Ehlers, M. H., Dillon, C., Gabriel, A., Huang, I. Y., Kumwenda, I., Mark, T., Meyer-Aurich, A., Milics, G., Olagunju, K. O., Pedersen, S. M., Shockley, J. and Rose, D. 2022. Lessons to be learned in adoption of autonomous equipment for field crops. *Applied Economic Perspectives and Policy* 44(2): 848–864.

Macenski, S., Foote, T., Gerkey, B., Lalancette, C. and Woodall, W. 2022. Robot Operating System 2: design, architecture, and uses in the wild. *Science Robotics* 7(66): eabm6074.

Maritan, E., Lowenberg-DeBoer, J., Behrendt, K. and Franklin, K. 2023. Economically optimal farmer supervision of crop robots. *Smart Agricultural Technology* 3: 100110.

Mier, G., Valente, J. and de Bruin, S. 2022. Fields2Cover: an open-source coverage path planning library for unmanned agricultural vehicles. arXiv preprint arXiv:2210.07838.

Nikolakis, N., Alexopoulos, K., Xanthakis, E. and Chryssolouris, G. 2019. The digital twin implementation for linking the virtual representation of human-based production tasks to their physical counterpart in the factory-floor. *International Journal of Computer Integrated Manufacturing* 32(1): 1–12. https://doi.org/10.1080/0951192X.2018.1529430.

Oksanen, T. and Visala, A. 2009. Coverage path planning algorithms for agricultural field machines. *Journal of Field Robotics* 26(8): 651–668.

Paraforos, D. S., Aubé, C., Athanasakos, L., Avgoustakis, I., Baron, S., Bresilla, T., Fountas, S., Hemming, J., Karagiannis, P., Mylonas, N., Nieuwenhuizen, A. T., Garcia, F. R., Pavlenko, T., Scovill, A., Sharipov, G. M., Vidal, J. and van Evert, F. K. 2022. Connecting agricultural robots and smart implements by using ISO 11783 communication. *IFAC-PapersOnLine* 55(32): 200–205. https://doi.org/10.1016/j.ifacol.2022.11.139.

Paraforos, D. S., Sharipov, G. M. and Griepentrog, H. W. 2019. ISO 11783-compatible industrial sensor and control systems and related research: a review. *Computers and Electronics in Agriculture* 163: 104863.

Sharipov, G., Bresilla, T., Nieuwenhuizen, A., Hemming, J., Van Evert, F. K., Aubé, C., et al. 2023. *Communication between Agricultural Robot and Mechanical Weeding Machine Based on ISO 11783 Network*. IFAC, Japan.

Stark, R. and Damerau, T. 2019. Digital twin. In Chatti, S. and Tolio, T. (Eds). *CIRP Encyclopedia of Production Engineering*. Springer, Berlin Heidelberg, Berlin, Heidelberg, pp. 1–8.

Stavropoulos, P., Papacharalampopoulos, A., Michail, C. K. and Chryssolouris, G. 2021. Robust additive manufacturing performance through a control oriented digital twin. *Metals* 11(5): 708.

Tillett, N. D. and Hague, T. 2006. Increasing work rate in vision guided precision banded operations. *Biosystems Engineering* 94(4): 487–494.

Chapter 8

Advances in connectivity and distributed intelligence in agricultural robotics

Liisa Pesonen, Natural Resources Institute Finland (Luke), Finland; Daniel Calvo Alonso, Atos, Spain; Juha Backman and Jere Kaivosoja, Natural Resources Institute Finland (Luke), Finland; Jarmi Recio Martinez, Atos, Spain; and Juha-Pekka Soininen, VTT Technical Research Centre of Finland, Finland

1 Introduction
2 Robotic system architecture
3 Communication networks
4 Internet of things to link robots
5 Distributed intelligence
6 Mission control platforms
7 Service framework
8 Data spaces serving robots
9 Case study: FlexiGroBots concept
10 Conclusion
11 Future trends in research
12 Acknowledgements
13 Where to look for further information
14 References

1 Introduction

Robots can be used in several types of farms to assist or replace human work in heavy, dangerous, dull or high accuracy needing tasks, or when the availability of labour is limited (Basri et al., 2021). Automatic milking systems (AMS), also called milking robots, are widely used service robots in livestock farming, together with increasingly common feeding and cleaning robots (John et al., 2016; Stülpner et al., 2014). In horticulture, several different approaches for harvesting robots in greenhouses have been introduced (Bachche, 2015), and farm implementations are gradually gaining ground (e.g. Ridder (no date)). Weeding and pest control robots have been developed for orchards and vineyards (Horizon The EU Research and Innovation Magazine (no date)).

http://dx.doi.org/10.19103/AS.2023.0124.14

Diverse pot- and bin-handling robots for both greenhouses and orchards have been designed to assist in heavy and monotonic tasks (Ye et al., 2018). Also in arable farming, several commercial field robots have been introduced and are available in markets: drones for remote sensing and ground robots capable of harrowing, weeding, fertilising and seeding (e.g. AgroIntelli (no date); FarmDroid (no date)). However, the adoption of robotic systems in use has been slow in arable farming.

Meaningful use of robots in farms requires smart, user-friendly and the circumstance-adaptive applications. This is difficult to achieve since farms are demanding worksite environments for autonomous machines. Farm operations link with several farm tasks, heterogeneous machinery and decision support systems and people and animals (Fountas et al., 2015; Köksal and Tekinerdogan, 2019). Agricultural robots need to be well connected with other actors in the farm operations they execute as well as with supporting services (Berenstein and Edan, 2017; Vasconez et al., 2019; Anagnostis et al., 2021). Robots also need to mediate data from their operations for further use in food systems since today's data-driven food systems require data or data-based information from agriculture about its production processes, i.e. field operations and production environments to function efficiently and safely, and to improve sustainability in a documented manner (Miranda et al., 2019; Corallo et al., 2018).

For better adoption, especially smallholders benefit from robotic systems that provide well-tailored, easy-to-use system entities. Smallholders have limited economic capacity to employ workers, especially when the labour is needed only seasonally. Usually, the farmer himself executes the main part of farm tasks and takes care of farm management and businesses in parallel. To be useful working partners, the robots, as any automation system, need to be well connected to other farm systems and to the robot's maintenance systems, and the provision of the robot's services needs to fit well to often individual farm circumstances and farmer's preferences (Pesonen et al., 2008).

In this sense, operation environments in arable farming set the most challenging flexibility and configurability demands for robot systems. Arable farms vary in crop selection, production strategies and methods, farm size and technological infrastructure thereof. Arable fields concern large land areas in diverse topographic and (micro)climate environments (The World Bank (no date)). Thus, the operational environments in arable farming cannot be standardised in a way it is possible in indoor animal production or greenhouses. Farm fields locate often in rural areas and in the distance from a farm's physical operation centre, which challenges functions for monitoring, control and logistics of robot systems. There is no electricity network available in the fields, and since arable fields are operated with moving machines, the needed communication requires always wireless connections. Arable farming worksites are open environments that expose robots to varying weather conditions;

moving obstacles like people, animals and other machines; and suddenly changing working partners due to which connections to supporting services are essential (Steen et al., 2012; Kaloxylos et al., 2012; Goap et al., 2018). The ability to carry out rapid reactions and proportionate changes in plans challenges not only the connections to supporting network of services and other machines but also the intelligence of robot systems. Thus, this chapter focusses mainly on arable farming and issues encountered there.

The increased challenge is faced by the robot entrepreneur who provides RaaS for several farms. This means that the RaaS system must connect with the various machines of the customer, data sources and Farm Management Information Systems (FMIS). This usually means that the Robot entrepreneur's Management Information System (RMIS) must be able to connect with 'things' and services provided by different brands. Figure 1 illustrates the problem field of such a complex challenge. The complexity is increased when the RaaS system consists of multiple robots operating as a fleet in the same farm operation.

The RaaS system needs to connect with other systems to communicate in several operational levels of farming systems, i.e.:

- farm tasks, which include from a farm or field viewpoint what task should be done, where and when;
- fleet tasks, which indicate the allocation of farm tasks to different robots;
- robot tasks, which contain waypoints for navigation, mission parameters and main actions; and
- detailed robot tasks containing detailed instructions for the robot's actuators.

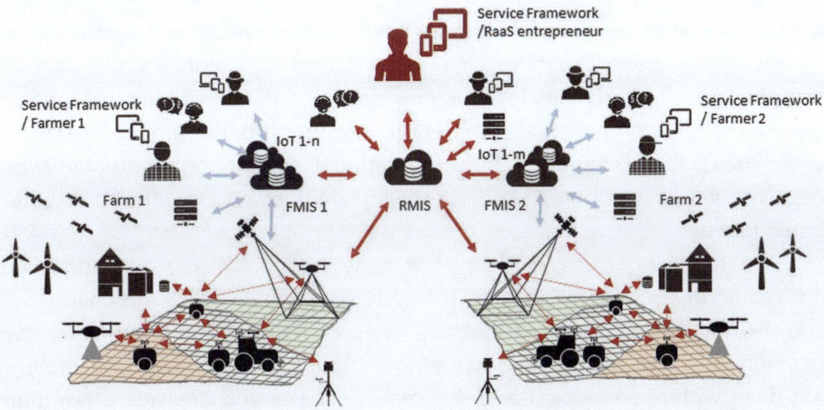

Figure 1 An illustration of a complex connectivity environment of robot systems from the viewpoint of RaaS (Robot as a Service) provider. The RaaS entrepreneur's Robot Management Information System (RMIS) must be able to communicate with several customers' Farm Management Information Systems (FMIS), Internet of Things systems and directly with robots and other machines and sensor systems in customers' fields. Data connections are depicted with arrows. Robot entrepreneur and farmers have grouped their individual sets of services as Service Frameworks.

From farmers' and robot entrepreneurs' points of view, there is a need to use and control the systems that are needed in everyday work. Internet of Things (IoT) and further the concept of Internet of Robotic Things (IoRT) (Ray, 2016) bring possibilities to include a growing number of services to farms' and entrepreneurs' toolboxes. To maintain awareness of ongoing operations and forecasts and adjust plans accordingly, there is a need for a Service Framework that offers a single access point to diverse user interfaces (UI) of the systems in use (Fig. 1).

The connectivity and intelligence of robot systems contribute to efficient, reliable and safe operation. Efficient and reliable operation requires that the robot system operates correctly even in suddenly changing operating circumstances and environments. The robot system can recover fast from disturbances, with minimal manual interference by the robot operator. Remote monitoring and notifications of deviations from plans create trust in the system. Robots must be able to operate safely as a part of various collaborating machine fleets and workforce. The functionalities must support the overall easiness to adopt and use of robot systems and economic viability of use.

The desired functionalities demand advanced connectivity of robots within a robotic system and with their support systems like robot IoT platform, FMIS, RMIS and possibly other supporting (AI) systems and data sources. The networked connections are essential to enable timely and correct autonomous decision-making through advanced distributed intelligence. Dyke Parunak (1994) explains that a system utilising distributed intelligence is a multi-agent system where 'the autonomous agent approach replaces a centralized database and control computer with a network of agents, each endowed with a local view of its environment and the ability and authority to respond locally to that environment. The overall system performance is not globally planned, but emerges through the dynamic interaction of the agents in real-time'. In agricultural environments, also the edge computing paradigm (i.e. Bierzynski et al., 2021) i.e. the sharing of the computational load among different network nodes as well as distributed problem-solving are relevant aspects of distributed intelligence.

The following sections deepen the views on how key challenges are addressed in agricultural systems. The system parts handled in this chapter are shown in Fig. 3. The system architecture is described in Section 2. The challenges of communication and interfacing methods of robots and their support systems are dealt with in Sections 3 and 4. Timely analytics and decision-making are discussed in Sections 5 and 6. Access to various farm- and business-specific data sources and services is handled in Sections 7 and 8. Section 9 gives a practical example of solving.

2 Robotic system architecture

Along with the Industry 4.0 era, robots are increasingly part of connected machinery or service systems in industrial production environments. It requires the ability of robots to function in meaningful ways, to integrate into diverse production chains and to operate efficiently and safely in demanding conditions. IoT has been one of the main technological concepts of Industry 4.0. As Ray (2016) abstracts, IoT allows a massive number of uniquely addressable 'things' to communicate with each other and to transfer data using existing internet or compatible network protocols. The concept of Cloud Robotics increases robots' processing power and data needs, i.e. by utilising distributed computing and data sources and BigData analysis (Ray, 2016). The concept of the IoRT combines these two concepts, providing means to support, control and monitor activities at deployment sites like agricultural fields (Khalid, 2021; Villa et al., 2021).

The architecture of IoRT consists of several layers, including typical hardware, network connectivity, internet connectivity, infrastructure for the robotic platform and cloud platform support, and application layers (Ray, 2016; Batth et al., 2018).

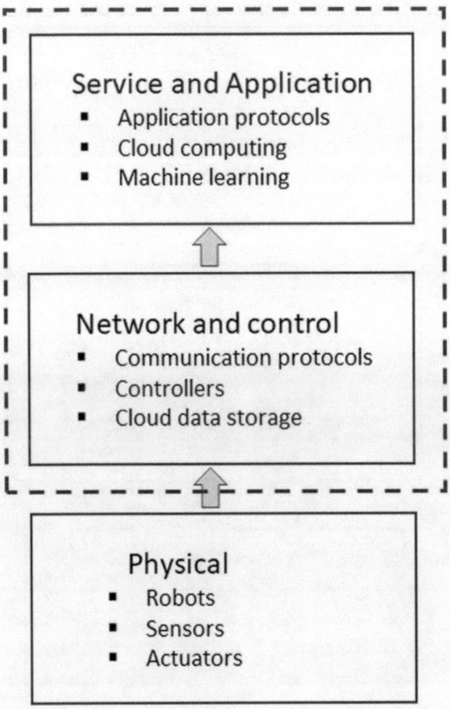

Figure 2 Internet of Robotic Things architecture overview after Villa et al. (2021). The dashed line depicts the focus of this chapter.

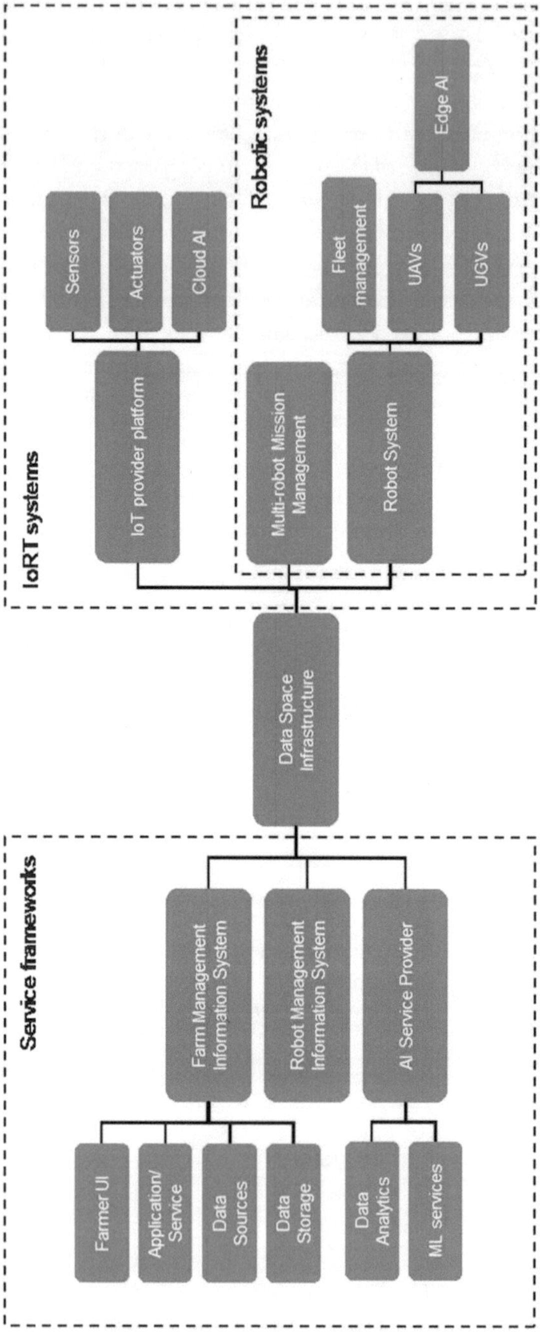

Figure 3 Advanced robotic system architecture based on connectivity and collaborative intelligence as discussed in this chapter.

The layers enable composability of the system, context awareness in operations, virtualised diversification, interoperability, dynamic and self-adaptive operations, geo-distribution of data and data processing, and ubiquitous network access. Villa et al. (2021) described the IoRT architecture overview by three main layers (Fig. 2): (1) physical layer including robots, sensors and actuators; (2) network and control layer including communication protocols, controllers and cloud data storage; and (3) service and application layer including application protocols, cloud computing and machine learning (ML).

To meet the challenges presented in Section 1, the concept of IoRT is evolving to support networked co-operation models between business entities and their heterogeneous systems (FMIS, RMIS and AI Service), as indicated in Fig. 3, in which the architecture includes distributed AI, data sovereignty and data ownership respecting data space infrastructure for agile data connections, multi-robot systems (MRS) management and service compositions as service frameworks.

3 Communication networks

In IoRT, robots are extended with functionalities in the cloud. By definition, it involves the need to be connected to the Internet and certainly, the Internet is the most important communication network of robots and the agriculture domain, too. In the case of agricultural robots, the main challenges related to the Internet have been the unpredictable delays of the network and the 'last mile connection' that needs to be wireless when agricultural robots perform their actions in fields. Recently, for example, the need to transmit large image files or video and the need to use external third-party services is creating new challenges with respect to transfer capacity, privacy and safety (Guo et al., 2020; Kaknjo et al., 2018).

Originally, consumer electronics have been the main driving force in the development of communication networks, especially cellular networks, i.e. mobile phone networks. However, in recent times, the requirements for the development of cellular networks have come from applications other than consumer applications. Industrial applications such as robotics may be more demanding than consumer applications in terms of low latency, high bandwidth, low power consumption or secure connection (i.e. Backman et al., 2019). The concept of cellular networks was developed by Bell Laboratories in 1947 (Asif, 2018). However, the first generation (1G) of commercial cellular networks was not introduced until the 1980s. The 1G networks were analogue systems and are no longer operational. However, the second-generation (2G) digital networks, which were introduced in 1990s, are still operational. Before the introduction of 2G networks, the European Telecommunications Standards Institute (ETSI) was founded in 1988 (ETSI, 2022). The Global System for Mobile

Communications standard, which is the basis for 2G networks, was the first of the standards developed and published by ETSI. To provide worldwide coverage of technology, seven telecommunications standard development organisations (ARIB, ATIS, CCSA, ETSI, TSDSI, TTA and TTC) have agreed to co-operate in Third Generation Partnership Project (3GPP, 2022). The first 3G networks were introduced in 2001 and the successor 4G in 2009. The latest generation, 5G networks, was launched in 2019 (Tang et al., 2021). All these technologies are operational at the same time. 4G was the first mobile broadband technology to enable 100 Mbps or higher data rate. The 5G systems are expected to provide a 20 Gbps peak data range and enable mission-critical applications that require ultra-high reliability and low latency (Asif, 2018).

In cellular networks, the data is transferred to public networks through the internet. In some applications, such as remote operation solutions or robot collaboration, this might be a problem if low latency, reliability and security are needed. Virtual Private Networks and other similar technologies are possible to use, but still the data is essentially transferred together with all other traffic that uses the same networks. In the future, 5G technology can be also used to construct wide-area private networks (Ericsson, 2019), but for short-range (<100 m) wireless communication, there are also other technologies developed. The Wireless LAN (WLAN) technologies are based on the IEEE 802.11 standards which contain several different protocols (IEEE 802.11). With WLAN technology, it is possible to build local closed private networks for specific use.

In agriculture, Agricultural Industry Electronics Foundation (AEF) is the organisation that develops guidelines for new communication technology standards (AEF, no date) and initiates the standardisation process in ISO. For standardisation of wireless communication between tractors and other entities, a dedicated workgroup in AEF exists. AEF has identified the use cases where wireless communication is needed and the requirements that those use cases set (AEF, 2022a). However, technological choices are still open, but the developments of road vehicles are followed. In the automotive industry, Vehicle-to-Everything (V2X) communication technologies have been developed for communication between a vehicle and other entities. The original V2X technologies were based on WLAN technologies (IEEE 802.11p-2010), but more recent development C-V2X has used cellular technologies and is expected to use 5G networks (3GPP, no date; Christopoulou et al., 2023). C-V2X technologies are interesting to employ also in agriculture since C-V2X is able to directly connect individual vehicles within the same area and also enables the development of cooperative intelligent systems using conventional mobile network (GSMA, no date; HKT, 2020), and the first use cases for agricultural machinery have been introduced by commercial actors (Autotalks, no date).

Highly accurate real-time positioning is an essential part of robotic operations. The backbone of the global level positioning is the GNSS (Global Navigation Satellite System). Single GNSS positioning is inaccurate. High-accuracy GNSS systems (Ogaja, 2022) such as real-time kinematics GNSS require communication between the satellite receiver and service-providing parameters needed to determine precise coordinates for the position. Methodologies such as PPP (precise point positioning) that require not only a single GNSS but also precise measurements of the GNSS satellites' orbits from external sources are also applied in robotics and autonomous navigation. Typical communication networks with networked transport of RTCM via internet protocol (Lenz, 2004) or separate radio links are used with the accurate GNSS. Solutions such as satellite-based augmentation system do not need two-way communication. Generally, GNSS-positioning technologies are becoming smaller, cheaper and more accurate. However, high-precision positioning relying on smartphone devices is still challenging (Zangenehnejad and Gao, 2021) but is under constant development. The current challenges include multipath errors due to polarising antennas, frequent cycle slip (signal interruption) and missing signal phase observations, and lack of phase center offset and variation information (Zangenehnejad and Gao, 2021). GNSS vulnerability (Zidan et al., 2020) remains a great challenge.

4 Internet of things to link robots

The IoT concept has been widely adopted in almost all kinds of systems, industrial and private. IoT is an important integrating technology also in smart farming (Verdouw et al., 2016). Lezoche et al. (2020) lists that IoT has positive impacts on:

- functionality (sensing, monitoring, controlling);
- economics (operational efficiency);
- environment (enhanced farming methods, resource efficiency);
- social issues (endured certification schemes in food chains, less manual labour);
- business (new business models and collaboration); and
- technology (low-power wireless sensors and machine-to-machine (M2M) connectivity).

In the same review, capital investment costs, especially for smallholders, user acceptance and current lack of technical skills, and technical issues like lack of interoperability and connectivity in rural areas, absence of data processing power, unclear data governance (data ownership), decentralisation, among others are listed as challenges.

In the case of agricultural robots, the IoT concept has two roles. First, it can provide data for the robot, and second, the robot itself can be considered as a thing connected to internet. Today's IoT solutions are typically cloud services that:

1. collect data from sensors and devices;
2. store the data both in databases meant to provide contextual information from some specific place (context brokers) and in time series databases of individual data sources; and
3. provide means to analyse the collected datasets using data analytics and artificial intelligence (AI) services.

IoT today can be considered as a platform, on which smart applications or cyber-physical systems are built.

The origins of IoT are different and consist of three main concepts. The first concept, the networked things, is to combine small digitalised or computerised physical appliances (i.e. things) into a networked system. M2M communication was the main enabler that allowed communication within the group of machines and devices, and when at least one point (machine or device) of the network was connected to the internet, the other 'things' were connected as well. The second concept was to see the connected thing as a service that could be accessed via a web browser. This approach called the Web of Things creates an application programming interface (API) to the 'things' enabling a user to access or manipulate the 'things' through it. The main difference compared to the first approach is that the 'thing' is not considered as a node in a network but as a service. The third concept was to connect things that do not have digital interfaces to internet. The unique identifiers in the form of codes and tags were used to connect the physical things to data that represented them or to the service that was linked to them. Bar codes, QR codes and Bluetooth beacons transmitting digital identifiers of things to be sensed by Bluetooth devices were the first techniques to associate objects with digital services. Later, for example, image recognition has partly removed the need for specific codes.

The Internet of Things–Architecture Reference Model (IoT-ARM) has been the baseline for the development of IoT platforms (Bauer et al., 2013). The IoT-ARM defined the five basic models of IoT system architecture: domain model, information model, functional model, communication model, and trust, security and privacy model. The IoT platforms have largely been developed based on the IoT-ARM. The FIWARE platform (FIWARE no date) is an open-source platform for smart solutions, including smart agriculture. It was based on extending the M2M communication model with virtual models of real-world entities. Its main components are an NGSI protocol that connects system modules together, a Virtual Entity model for managing the data of connected objects, and generic enablers such as a context broker, a time-series database and other functionalities that provide, for example, data analytic and visualisation services for applications. FIWARE has been applied in many agriculture solutions (Rodriguez et al., 2018; Corista et al., 2018). The smart objects for intelligent applications project extended the IoT towards smart spaces, semantic interoperability and LinkedData (LinkedData, no date). The idea of smart spaces was to create an IoT data-sharing environment for limited physical places and to enable the data-centric creation of smart services in it (Kiljander et al., 2014). Semantic interoperability is the ability to share data with unambiguous shared meaning. This is typically achieved by using common ontologies and semantic data models that are models where data elements are linked with the definitions of their meaning. When the links given in ontology models are provided between data elements, we end up with the graph model of data called LinkedData. Since these early solutions, the IoT platforms have evolved to become application domain-specific services that users can connect to with a huge variety of sensors and communication technologies and which provide advanced services for implementing cyber-physical systems (Chen, 2020; Farooq et al., 2020; Ojha et al., 2021; Ray, 2016).

The future IoT has two main roles. First, it provides means to share data of the physical world in the cyber world and enables analysis and optimisation of large-scale systems of systems. Second, it provides means to create advanced situational awareness and physical interaction in more local contexts and thereby allows developing more intelligent and autonomous robotic systems (Ploennigs et al., 2018). Digital twins are software models of real systems that are either targeted to analyse, simulate or extend the physical systems. The main use has been, for example, in maintenance, where the wearing of parts can be analysed using simulation models (Pylianidis et al., 2021). Currently, IoT feeds digital twins with real-world data, but the trend is in expanding the use of digital twins towards the exploration of alternative futures that can be a basis for better decisions today (Verdouw et al., 2021).

5 Distributed intelligence

5.1 Artificial intelligence

The evolution of humankind is permeated by revolutions that transform radically the way of life, relations and the environment. It is impossible to understand current societies, economies or philosophical and political movements without considering the impact of the cognitive, agricultural, scientific and the three industrial revolutions. Currently, the world is at the dawn of the fourth industrial revolution and many experts predict that it will bring disruption and changes of a magnitude never seen before in history (Xu et al., 2018). The convergence in the same temporal window of a great number of advances in fields like computer systems, telecommunications, biotechnology, photonics, nanoelectronics or materials engineering has enabled the development of new technologies with applications in a wide variety of verticals. Relevant examples are IoT, BigData and Blockchain but especially AI.

The basic concepts of AI were proposed in the middle of the last century. In 1943, Warren S. McCulloch and Walter Pitts designed the first computational neural model (Chakraverty et al., 2019). In 1952, Marvin Minsky built the Stochastic Neural Analog Reinforcement Calculator, one of the first implementations of neural networks, which was based on vacuum tubes. Since then, the history of AI has been a roller coaster ride, with several winters but also with golden ages, being the actual moment may be the definite one (Toosi et al., 2021). In order to explain the renewed prominence of AI, multiple factors can be mentioned:

- The advances in hardware systems and the consequent increase in processing power enable complex models and algorithms to be designed and implemented in reasonable times;
- The definite maturity and adoption of IoT, cloud computing or big data. They are basic technologies that make it possible to collect, aggregate and process huge amounts of data, which are needed for training; and
- Finally, it is also important to emphasize the release of tools that abstract the complexity of the underlying mathematical concepts.

At least, the following main areas can be identified within the AI realm (Claire, no date):

- ML comprises tools and techniques that allow systems to learn from data (Zhou, 2021). Supervised and unsupervised learning can be used depending on whether data is labelled or not (Alloghani et al., 2020). The most common forms of implementation are neural networks, deep learning (DL) and generative adversarial networks. Lately, systems based on reinforcement learning are gaining importance. They are based on algorithms that learn how to achieve a certain objective through positive or negative incentives defined by the designer (Wiering and Van Otterlo, 2012).
- Knowledge representation and reasoning is a field that focuses on how to represent information so that a system can solve complex tasks. It covers aspects such as semantic networks, ontologies and knowledge graphs to represent information (Duan et al., 2019), metamodels, fuzzy logic that allows working with uncertainty (Mittal et al., 2020), first-order logic etc.
- Search paradigms try to solve problems in which, starting from an initial state, a certain objective must be reached and to do so a sequence of

actions must be executed (Korf, 1999). Optimisation techniques are used when, in addition to finding the right sequence of actions, costs must be reduced (Mirjalili and Dong, 2020). The simplest examples are algorithms such as Depth-First Search.

- Planning enters those larger scenarios where a blind or even informed search is inefficient. In this context, planning algorithms differ in the amount of information they handle when deciding. Some relevant techniques are Stanford Research Institute Problem Solver framework (Lifschitz, 1987), planning graphs, heuristic search planners (Bonet and Geffner, 2001), inductive logic programming (Riguzzi, 2022) or decision networks.
- Multi-agent systems include concepts such as self-organisation or self-direction and multi-agent learning (Wood and DeLoach, 2000). Multi-agent systems could be framed within a larger field and distributed AI (Peteiro-Barral and Guijarro-Berdiñas, 2013), in which there is also an interesting topic such as swarm computing.
- Natural Language Processing (NLP) is with areas such as machine translation or translation from one language to another, understanding or generation of natural language (Chowdhary, 2020).
- Robotics (Craig, 2005) is where many of these techniques are applied but where we have to take into account additional factors such as physical safety issues (Haddadin, 2015), human-machine or machine–machine interaction (Yang et al., 2021), mechatronic systems (Merzouki et al., 2013), real-time requirements or physical intelligence.
- Computer vision (CV) is including aspects of recognition, identification, detection, motion analysis, scenario and image reconstruction (Forsyth and Ponce, 2011).

During the last years, ML and DL have attracted a lot of attention not only from research groups and companies but also from a more general audience (Sejnowski, 2018). In many cases, to train the models, huge datasets must be used during very long processes that consume a lot of powerful and specialised computation resources (e.g. graphics processing units, tensor processing units or even supercomputers). Thus, the catalogue of services of the main cloud providers following Infrastructure as a Service (IaaS) or Platform as a Service (PaaS) paradigms includes offerings targeting the specific needs of ML processes (Google Cloud, no date) (AWS, no date), (Azure, no date). Software as a Service (SaaS) is also starting to emerge with off-the-shelf components and models for NLP, CV or to boost the training (Hugging Face, no date).

During the industrialisation process of AI and ML technologies, it is essential to guarantee robustness, reliability and accuracy during the complete life cycle. Potential problems may have a strong and negative impact on the performance of the models and the applications or services integrating them. If the models are deployed in a production environment, it is possible that their performance may decline significantly when the input data differs slightly from the training dataset. This difference in data distribution between the training and production scenarios is referred to as 'domain shift' and is a commonly encountered issue in ML (Quinonero-Candela et al., 2008). Unsupervised domain adaptation techniques address the problem of domain shift by transferring knowledge learned from source domains, which have a large number of annotated training examples to target domains that only have unlabelled data. After deployment in production, domain drift may occur multiple times. If these changes are detected, it may be necessary to initiate a new adaptation process. There is also a need for the development and expansion of methods and tools that can monitor the error analysis of neural networks in situ and assess the robustness of neural networks to detect any shift or regression. Machine learning operations (MLOps) is a new paradigm that aims to introduce best practices to deliver efficient, repeatable and reliable end-to-end ML workflows (Alla and Adari, 2021). Tools like Kubeflow (Kubeflow, no date) or MLFlow (MLFlow, no date) are becoming more and more popular in this area.

5.2 Edge computing and distributed artificial Intelligence

While cloud-centric AI solutions are completely dominant in the consumers' applications market (e.g. personal assistants, recommendation systems), industrial or enterprise clients provide a much bigger opportunity, being the agriculture domain one of its clearest exponents. Nevertheless, some barriers must be overcome to achieve the adoption and success of AI:

- Protect data security and privacy since the information may be confidential;
- Satisfy real-time requirements for decision-making and control;
- Provide high accuracy and adaption to evolving data distributions and contexts; and
- Guarantee cost-effectiveness and scalability in scenarios with a massive amount of data sources.

In many cases, AI systems relying on cloud platforms are not able to comply with these requirements due to the need to centralise the information on a single location for the training process and the limitations of the communication protocols and network technologies during the inference phase. To mitigate these problems, edge computing offers important advantages by exploiting the increasingly powerful computing capabilities of resources like IoT devices, gateways, robots, customers' infrastructures, regional data centres, etc. (Bierzynski et al., 2021). At the software level, approaches like Tiny Machine Learning or lightweight neural network architectures enable AI to be executed over more limited edge computing resources (Ray, 2021). There are also available approaches that leverage the distributed nature of edge-based systems, being Federated Learning (FL) one of the most well-known (Yang et al., 2021). Using FL, each edge device trains its own version of the model without sharing the raw information with other devices or with a cloud platform. Then, all the partial and local models are combined by a central node resulting in a global model that is distributed to the edge devices for inference. Thus, edge computing offers important advantages for distributed intelligence enhancing both the concept of multi-agent systems and shared computational load. Distributed Artificial Intelligence or Decentralised Artificial Intelligence (DAI) aims to leverage independent and autonomous computational resources to learn, plan or make decisions (Huhns, 2012). The availability of multiple resources enables the speed-up of processes through paradigms like Parallel Programming or Distributed Training (Verbraeken et al., 2020). Multi-agent systems (MAS) have emerged also as a technology to solve problems that cannot be addressed properly by a single individual agent, having vast applicability in the control of robotics systems (Van der Hoek and Wooldridge, 2008). Multi-agent deep reinforcement learning (MARL) is being recently been proposed to implement scalable solutions in environments with partial knowledge or a high level of uncertainty (Oroojlooy and Hajinezhad, 2022).

5.3 Smart agriculture solutions and applications

A great diversity of services can be implemented powered by digitalisation and AI, contributing to addressing some of the challenges that currently limit the sustainability of food systems. Relevant examples and opportunities are being explored in farming, post-harvest operations, processing, logistics and consumer support, covering the complete farm-to-fork value chain (Marvin et al., 2022). The interplay with aerial and ground robots is also vast: on one hand, they require different types of ML models and AI algorithms for object recognition, people detection, localisation, navigation, decision-making and actuation. One example is the work released by Facebook Research with DETIC (Zhou et al., 2022), a detector that is able to identify more than 20 000 classes off-the-shelf and that can be used to analyse or process images and video streams for different tasks in which robots are involved. On the other hand, robotics systems collect high-quality and heterogeneous information from crops thanks to onboard cameras and sensors that can feed the different phases of MLOps pipelines. In precision agriculture applications, images captured by drones can be combined with satellite information to assess the crops' growth and health through the calculation of relevant indexes like NDVI (Normalised Difference Vegetation Index) (Harsh et al., 2021). Drones and ground robots with onboard cameras can be used to detect in real-time the early presence of pests and diseases, applying appropriate treatments with high accuracy, which leads to important savings in terms of costs and environmental pollution (Iost Filho et al., 2020). They can also perform a high number of actuations (e.g. soil sampling, weeding, harvesting) in an automatised way, improving the labour conditions of farmers and the sustainability of their businesses (Krishnan et al., 2020).

6 Mission control platforms

6.1 Need for mission control

Robotic systems and autonomous vehicles have become powerful instruments that complement and extend the capabilities of traditional agricultural machinery. They incorporate a variety of onboard sensors and cameras to obtain high-quality information and to perceive the environment. The integration of advanced embedded systems and connectivity, as explained in previous sections, allows processing in real-time collected data to apply AI techniques in order to make decisions or to offload part of the heavier algorithms to systems running on cloud platforms. The resulting decisions require controlling different systems (e.g. navigation and actuation) or even interacting with human actors or other devices. In the specific case of agricultural applications, an additional challenge is a need to work in highly dynamic and unstructured environments, characterised by the uncontrolled presence of people, vehicles, machinery or cattle. Even MRS, consisting of different robots with different capabilities need to be introduced to address the complexity and performance requirements (Ju et al., 2022).

The complexity of robotics missions can differ depending on the level of autonomy allowed to the robots and the possibility to dynamically update the plans according to their evolution and to the changes in the context. For instance, in the case of highly autonomous and automated robots, a mission can be seen as a sequence of locations to be visited and a set of actions to be performed, satisfying some high-level goals and constraints, e.g. mission duration and fuel consumption. In the case of simpler robots, the mission control must be closer to the actual robot control system that includes instructions on what is expected from various parts of the robots (ROS, no date; ROS2, no date). MRS can be divided into fleets of homogeneous robots, typically called robot swarms and fleets of heterogeneous robots. Robot swarm mission control typically involves spatial task allocation, formation control and task execution. Scenarios with fleets of multiple heterogeneous robots require an additional level of intelligence in the planning to determine which is part of the mission that must be assigned to each vehicle and how to orchestrate them (Wheare, 2018; Dumka et al., 2018, Rizk et al., 2019).

Mission Control is a high-level user interface for a robot. Mission Control Platforms are used to visualise the data coming from the robot as well as to give new instructions to the robot (Plura et al., 2011). Usually, the instructions are at a high level, i.e. to start or to stop operation, to change the task or to move to a certain point. A robot may have its own dedicated user interface, but a mission control platform is more general software that can be used to command multiple different robots. They can be seen as a very simple way to plan and manage the execution of the mission, relying mostly on human operator skills and knowledge.

In the following sections, we give state-of-the-art examples of agriculture ground robot control systems and open-source drone platforms that are being developed to be applicable to robot and drone fleets.

6.2 Mission control by ISOBUS and extended farm management information systems data interface

The development of the robot tractors started several decades ago (Rondelli et.al., 2022). Today, there are even commercial robot tractor applications available. However, the robot tractors are rarely in use on farms. The reason may be that the current robot control systems do not integrate well to existing farm ICT systems. Usually, those are vendor-specific closed systems. In agricultural technology, it has already been recognised that the standardisation is the key to the widespread adoption of technology. The ISO 11783 standard, market name ISOBUS, is widely accepted communication standard for agricultural machinery. The standard defines many devices that can be used in precision farming. By using the standard, the devices and the implements of different manufacturers are compatible with each other.

ISOBUS Task Controller (TC) is a standardised device that is used to document and control the precision farming operations on the field. It is based on the ISO 11783-10 standard (ISO 11783-10:2015). The data is transferred to the ISOBUS TC using the ISOBUS Task files. Despite the original purpose of the ISOBUS Task file, it can be also used to describe the robot's work on the field (Backman et al., 2022). The driving lines, implement allocations, timing of operations, description of operation etc. can be described in the Task file in a standardised way. Practically, all the information the robot needs can be included in it.

The Task files are designed in the FMIS software. Today, FMIS software is increasingly running in the cloud service, or the cloud service is part of the operation of the software. For this reason, many manufacturers have made their own solutions for transferring data to the ISOBUS TC from the cloud. There are also broker services that allow transferring of ISOBUS Task files to TCs from multiple vendors (DKE-Data GmbH & Co. KG, no date). However, these are not standards, so universal compatibility with products from all different manufacturers is not guaranteed. The universal compatibility is what the ISOBUS systems are wanted to be.

The AEF recognised the problem of incompatibility between different ISOBUS products and in 2017 set up a working group to develop vendor-independent data transfer between FMIS software and the TC. The data transfer method was named Extended FMIS Data Interface (EFDI). The AEF has published a guideline on the definition of EFDI (AEF, 2020b) and the process of ISO standardisation is ongoing (ISO/DIS 5231, 2022).

In the EFDI's development work, AEF did not develop a protocol from scratch. AEF's EFDI working group has selected appropriate existing protocols already in use on the internet and built functionality on top of them. Before the selection, the working group explored the different options and their pros and cons. The data transfer was required to be reliable and scalable to systems of different sizes, and it needed to work even with a slower network connection. Naturally, it was desired to be compatible with the definition of the existing ISOBUS Task file (ISO 11783-10:2015). As a result of the study, the Protobuf (Google, 2022) originally developed by Google was chosen as the data transfer format. The conversion from XML file to Protobuf format and back is straightforward. EFDI also defines more metadata than the original XML file, describing, for example, what should be done with the transferred Task file. The MQTT (Message Queuing Telemetry Transport, ISO/IEC 20922, 2016), which is commonly used in various IoT systems, was chosen as the data transmission. In addition to these, the HTTP protocol (IETF Trust, 2007) was chosen to log in to the system.

Using the EFDI communication, the instructions to the robot can be sent in a standardised way. The operations can be started or stopped, or the operations can be scheduled to be started at a desired time. The execution of the operation can also be monitored through the telemetry protocol that is included in the EFDI communication. The communication channel is also extensible, and it supports sending and receiving arbitrary binary files. In other words, EFDI is a suitable standardised communication channel between the robot and mission control system similar between the robot and FMIS. On the other hand, FMIS systems seem to be developing in the direction of Smart Farm Management Systems (FIWARE, 2018), capable of also managing Mission Control Systems.

6.3 QGroundControl drone control system

Drones are becoming important farm machines due to their advanced easiness of use and models that are affordable also for smallholders (Puri et al., 2017). Autopilot systems are important techniques in view of user-friendliness and safe use. Drones equipped with an autopilot are robots, which require a mission control system (Kangunde et al., 2021). QGroundControl (QGroundControl, no date) is a powerful open-source software for flight control and vehicle setup (Drone Foundation, no date). It allows mission planning for autonomous vehicles and a full configuration of PX4 vehicles or any autopilot using the MAVLink (Micro Air Vehicle Link) protocol.

PX4 is an open-source autopilot for controlling different vehicle types, including aircraft, ground vehicles and underwater vehicles (PX4 Autopilot, no date, a). It is a core part of a drone platform, along with the QGroundControl station and other hardware components, which communicate via MAVLink protocol. MAVLink is a lightweight messaging protocol for communicating with drones and between them (comparable with EFDI protocol). It follows a hybrid publish-subscribe and point-to-point design pattern where data streams are published as topics while configuration subprotocols are point-to-point with retransmission (MAVLink, no date).

QGroundControl is a multi-platform software, it runs on Windows, OS X, Linux platforms, iOS and Android devices. It provides 2D and 3D aerial maps where the user can build the mission, and which allows visualizing the location where the vehicle is during its mission. One of the most interesting features of this software is the telemetric information of the vehicle and multiple sensors that the user can view during the execution of the mission. Another key aspect of QGroundControl is the logging system, which allows downloading of binary log files from vehicles and sensors to check the status and assess specific information.

QGroundControl software provides an internal MAVLink Console to run commands and a MAVLink Inspector, where the user can inspect real-time messages of multiple parameters and create plots from this information. PX4 supports simulations on computers (SITL) and simulations using simulation firmware on a real flight controller board (HITL). Gazebo (Gazebo, no date) is one of the supported simulators for PX4, it provides a powerful 3D environment for testing object avoidance and computer vision (PX4 Autopilot, no date, b). It can be used with both simulation methods, Software in The Loop and Hardware in The Loop (Nguyen and Nguyen 2019; Hentati et al., 2018).

7 Service framework

Farms utilize increasingly digital technologies to run their farming operations and businesses. Data-driven farm management enables farms to create resource-efficient production as well as to connect digitally with other businesses and thus be part of the evolving data economy. Each farm is unique due to its location, ecological environment, size, production line, machinery, skills, business opportunities and history. Thus, the optimal set of farm technology varies from farm to farm. Typically, the machinery, equipment and management systems of a farm consist of several brands. Farms that are digitalising operations and farm management have faced problems with connectivity and data sharing between the available digital systems.

Farm, as well as machine management, in the field needs several data sources to run efficiently (Sorensen et al., 2010, Fountas et al., 2015). Cloud platforms were presented as a generic solution for the data integration problem in the 2000s and also in agriculture (Kaloxylos et al., 2012). The FMIS as a cloud platform was introduced. The FMIS platform provides the API through which diverse other services, applications, data sources and storages can be integrated into one farm management system. In addition to APIs, an FMIS platform includes data storage and analytics layers for decision support and user interface for the end user. The concept gives the farmer a value proposition that he/she and his/her robot systems can get access to all needed digital tools through the familiar FMIS.

The concept has been found challenging from the business point of view when the number of digital solutions that farms could choose from is increasing. In practice, FMIS providers cannot collaborate with all available applications or data providers that the farmers would wish to use. Therefore, Kaloxylos et al. (2012) proposed the concept of Farm Management System (FMS) which is a kind of application framework including one or more FMISs and other additional services for farmers. The FMS could be provided as a package service offered by another provider. Digesting an additional data source or application to FMIS service provision needs effort and business motivation. Thus, it was stated by Pesonen et al. (2014) that instead of gathering farm's digital tools into a single cloud platform the technology should be provided as an interlinked system of systems from which the farmers could choose the needed services and applications and utilize them through a kind of a Service Framework.

The concept of a Service Framework becomes even more important in the context of field robots. Robots require all information in digital format to function fluently and autonomously. This increases the number of data sources and analytics services that robots must be connected to when compared to manned machinery. Robots need to be connected to FMIS (i.e. using EFDI) to perform the field tasks but also to other machines in the area or linked to the task in hand. The robot also needs to connect to the services providing information about the environment like topography, obstacles, weather, soil and plant condition, roads, traffic, storage etc.

When looking at the needed management system from a robot entrepreneur's point of view, who provides RaaS services to several farms, the currently available information management systems are not adequate. The Service Framework for RaaS management is also needed. The Smart Farming Management System concept introduced by FIWARE (FIWARE, 2018) describes comprehensively the needed FMS elements and their connections. The concept

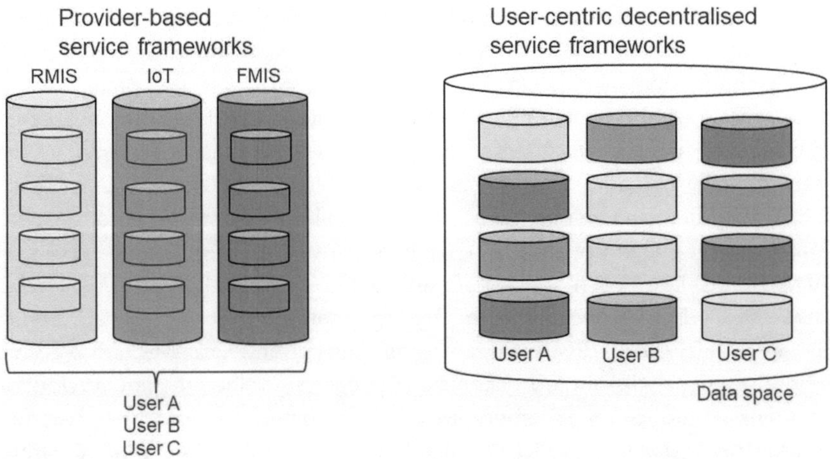

Figure 4 Illustration of a foreseen shift from provider-based service frameworks to user-centric decentralised service frameworks enabled by data space infrastructures and data sovereignty.

could be implemented as loosely coupled independent element clusters from different or selectable providers if data connections were easy to create and data management rights and obligations were clear to all parties. The concept of data space with the idea of data sovereignty provides interesting views for technical solutions (Section 8). Figure 4 illustrates how current provider-based service frameworks, i.e. created around IoT, FMIS or RMIS platforms could change to user-centric decentralised service frameworks, where users can take in only those services that fit their needs and services would be connected. The user of this kind of user-centric service framework could be a farmer as well as a RaaS entrepreneur.

8 Data spaces serving robots

8.1 Data spaces

Agricultural robots are intelligent autonomous devices that both create and consume data. Their role is changing from simple tasks to more complex operations that involve interaction with AI, analytic tasks and even with another robot. Robots are becoming parts of networks that exchange data. They deal with data that is critical to the farmer and valuable to the other actors in the food system. The data is an asset to the data owner, and it is worth the effort to invest in its ownership and governance.

Data space is defined as a decentralised infrastructure for trustworthy data sharing and exchange in data ecosystems based on commonly agreed principles (Nagel and Lycklama, 2021). Data spaces infrastructure tries to create an environment and tools where different companies and entities can share and exchange their data assets without losing control over their data. The aim is to enable both the data economy, where data is considered an asset that is traded as any real-world asset and distributed systems of systems, where data is exchanged automatically between complex digital systems (Curry et al., 2022). The first part relates to buying and selling data assets through data marketplaces (Huang et al., 2021). The main drivers are big data analytics and AI that needs large, high-quality data sets for ML with a level playing field for the participants. The second part is the implementation of cyber-physical systems (Liu et al., 2017) or industry 4.0 concepts (Alcácer and Cruz-Machado, 2019), where physically distributed systems create a single value creation network or complex control or management systems. The data space can even be seen as a distributed alternative to a service framework (Fig. 4). Examples vary from complex logistics chains to smart factories, smart city and smart farming and food production type of systems. The common feature in all of these is that the complete operation requires data from multiple data owners whose ownership must be respected.

8.2 Road towards data spaces

Even though the data spaces are meant for exchanging data between ICT systems running in cloud platforms, their roots are in the networking of systems, in the smartness of collaborative systems, and in the creation of trust between partners. Communication and networking have undergone a transition from networking computers to things and data. This has led to concepts such as IoT and semantic web or Linked Data. IoT has its roots in M2M communications that were extended with digital representations of physical things, which have further evolved into Digital Twins that are models representing physical components and systems (Vermesan and Friess, 2013). The Web of Things turned physical things into digital services providing a more software-oriented approach to real-world interaction. In IoT the main trend has been to develop IoT platforms that collect data and provide integrated and advanced services to process and visualise the data. This has led to the business model of having IoT PaaS, where data are valuable assets that has further led to the data silos and platform lock-ins making data sharing and use more challenging due to business protection reasons.

Networking based on data was started by hypertext and world-wide-web back in 1990s. The next step was to make data machine readable so that data processing could be automated. This and the open data concept-enabled web-based applications and services. The Semantic Web concept is that data elements themselves are linked with each other, and these linkages

Published by Burleigh Dodds Science Publishing Limited, 2024.

are embedded into the data. This opened the way to real data networking that means accessing data through its links to other data. Currently related technologies such as XML (extensible markup language), OWL (Web Ontology Language), RDF (Resource Description Framework), SPARQL (Simple Protocol and RDF Query Language) and JSON-LD (JavaScript Object Notation for Linked Data) are widely used in most of the ICT systems (Semantic Web, no data).

Automatic and autonomous devices and robots have been present in visions for ages. Advances in AI,ML and big data analytics have started to make them a reality. Object identification, data fusion and context modelling have been solutions that have enabled systems from autonomous driving and complex cyber-physical systems. Technology is also taking steps towards intelligent fleets of robots or smart places that involve sharing of situational awareness between systems. The data space concept provides means to combine these advanced technologies that need advanced competencies both in development and maintenance into complete systems.

Trust is a critical aspect in the data spaces (TRUSTS, no date). A specific characteristic of the data asset is that it can be easily copied and distributed making the ownership hard to supervise. The creation of trust needs both technical means to share and control the data and its use, and a legal framework that provides policies, rules and regulations for being part of the ecosystem and creating consequences in case of misbehaving. Technical means have focused on interoperability and encryption, while legal frameworks have been based on identity management and smart contracts. Bitcoin and blockchains initiated the development of distributed ledger technology (DLT) that has proven to be a powerful tool for developing transparent and trustworthy systems (Backman et al., 2017; Hummel et al., 2021).

8.3 Recent developments enabling data spaces

During the last decades, we have seen the rise of platforms that have made the owners very powerful and dominant. The security of solutions and reliability of platform providers has raised concerns that have limited digitalisation. Companies have become concerned about their ownership and control of the data. The wishes for more restricted and trustworthy internet and data infrastructure are now recognised as a necessity. Data spaces as described in EU Data Strategy are seen as an alternative to creating an environment for a data economy with more fairness and equality for the players (Anon, 2020; Anon, 2022).

The International Data Spaces Association (IDSA) was initially an industrial data-sharing initiative with a target to create a Linked Data solution that could be used by companies in sharing their confidential data. It was renamed changed as a data space initiative in 2017 and a reference architecture for data

Published by Burleigh Dodds Science Publishing Limited, 2024.

spaces was created (Otto et al., no date). The main idea of the IDSA architecture is that all users' identities are validated and that all the activities through the data space occur via open source and transparent components that have been accepted by IDSA. The data sharing includes smart contracts between partners, where usage policies are agreed. The data transfers take place through IDSA Connectors and point-to-point communication. All the traffic is encrypted using X.509 certificates. The data is not transferred to any centralised storage and the data owner has full control over its data. The IDSA architecture defines a component called Metadata Broker for publishing the metadata for the data each partner is making available in the data space. The Broker allows the data users to search available and published data assets but accessing to data must be agreed upon with the data owner.

- GAIA-X is another example of a data space and started as a German–French initiative in 2019, but it has expanded to a global initiative. The aim was to create a federation layer between the physical cloud infrastructures and the use of data (Gaia-X Architecture Document 21.12 Release, no date). The main idea of GAIA-X was to expose the physical infrastructure properties to the users so that users would know in detail where and by whom their data is stored or processed. In some cases, the trade secrets or legislation requires that data must be kept within some specific country, for example. GAIA-X also involves sovereignty preserving data sharing and it also tries to collect companies to set up data spaces in several domains. GAIA-X aims at an open-source, transparent and decentralised implementation. The aim is to create a decentralised autonomous organisation (DAO), where all the operations are defined as software.

Yet another example is the Ocean Protocol which is a crypto ecosystem using the Ethereum blockchain for sharing data. Data tokens are used for sharing URLs of data assets, access control to data services and for different access and usage policies. The network is implemented as a DAO (Ocean Protocol, no date).

iSHARE and i4Trust are examples of an approach where data spaces are built from a trust perspective. The aim of these is to create a trustworthy environment through a federated legal framework with verifiable identities and agreements. The data exchange takes place through common APIs (iSHARE, no date).

Using the data space concept in data economy or distributed and collaborative systems requires data interoperability and common operating principles. The scope of interoperability can vary from a single use case to a single global data space or to a single digital market as in the EU. The baseline for interoperability is the work that has been done in semantic data modelling and smart contracts, but various initiatives to create for example domain-specific solutions have been proposed. For example, in GAIA-X several domain working groups have been established to create common rules and policies (Tardieu, 2021). The EU has also done significant investments in setting up the data spaces in its Digital Europe programme (Anon, 2021). The purpose of these initiatives is not to create new solutions but to agree together on practical issues that are needed in creating collaboration.

8.4 Data spaces and robots for agriculture tasks

Data spaces are a key enabler for implementing robotic systems in complex and fragmented farming environments meaningfully. Complexity implies that the constellation of utilised tools, e.g. IoT solutions, robot platforms and communication channels, must be tailored individually for each farm operation, varying from farm to farm. The chosen technologies must be able to establish data connections between each other and also to other farm services when operating at different levels of farm management, i.e. farm tasks, fleet tasks, robot tasks and detailed robot tasks. Data spaces with evolving standardisation pave the way for such agile utilisation of digital solutions and data in a secure and trusted manner. The data sovereignty of both farmer and service providers (i.e. RaaS provider) enabled by common data intermediary services, consent management functionalities, service catalogues, metadata operators and user's service framework application make the planning and control of complicated MRS possible in an agile and user-friendly way. While data spaces can be seen as a distributed service framework, users like farmers and RaaS entrepreneurs can define their individual Service Framework capturing all services they need for their activities.

9 Case study: FlexiGroBots concept

As a practical example of advanced connectivity and the use of distributed intelligence in agriculture, a use case based on the research project FlexiGroBots is introduced. The use case focuses on oil seed pest management (Kaivosoja et al., 2022).

Efficient oil seed pest management can be challenging. Invasive insect pests can destroy the whole yield, and yet, the reduction of applied pesticides is encouraged. The use case studies modern possibilities for implementing drones and robotics for pest invasion scouting and for precision pesticide spraying against pollen beetles in rapeseed fields. In this case, a drone is first scouting over the rapeseed field to spot individual pollen beetles in traps and taking images for analysis of the level of invasion: how many pollen beetles per trap and in which parts of the field. The drone is operated using QGroundControl for autonomous mission planning and control. The communication uses

MAVLink protocol. The QGroundControl to receive spatial data of field maps and locations of traps from FMIS.

The images taken by drone camera are sent to an IoT cloud platform and from there further to FlexiGrobots platform for AI-assisted analysis where ML techniques are utilised to automize pollen beetle identification and counting. The AI solutions in the platform are trained by exploiting local reference data from the trap images, available from the drone-camera system's IoT platform collected to the IoT platform. After processing, the results are sent back to FMIS where they are translated into a spraying task. The spraying task can be performed by a spraying drone, small field robots, a tractor-sprayer or with different combinations of these platforms. When different types of robotic platforms can utilize similar task applications, the platform selection can be done just before the actual spraying. A spraying drone can spray hotspots rapidly without trampling the rapeseed while the tractor-sprayer can spray the entire field sufficiently.

The field operation by spraying drones follows the same task planning and mission control as in the case of scouting drone operation. When spraying takes place by a small field (ground) robot or tractor-sprayer the mission control center communicates with them and FMIS utilising the EFDI data transfer method with Protobuf format and MQTT data transmission. The service framework enables different service providers to join the use case. Such services can be an FMIS, IoT trap service, drone scouting service, service for machine vision or image classification, service for application task generation and spraying services. The data connections between participating services are arranged using data space infrastructure that follows IDS reference architecture.

10 Conclusion

Farms are demanding worksite environments for autonomous machines. The most demanding cases in agriculture where the communication and smartness of the robotic systems are challenged can be found in the context of open fields of arable farming, when MRS execute farm tasks fluently and safely. Along with the Industry 4.0 era, robots are increasingly part of connected machinery or service systems also in agricultural production environments. Robots need to be able to function in meaningful ways and integrate into diverse production chains and operate efficiently and safely in demanding conditions, especially when robots are exploited as a service (RaaS).

The concept of the IoRT defines connected robots as IoT systems which exploit cloud services to increase calculation, data storage and analytics capacities, i.e. Big Data analytics and ML. When applying the IoRT concept in agriculture, the internet is the most important communication network of robots. The main challenges related to the Internet are the unpredictable delays of the network and the 'last mile connection' that needs to be wireless when agricultural robots perform their actions in fields. For communication between vehicles and other entities, V2X communication technologies have been developed. The original V2X technologies were based on WLAN technologies, but the more recent development C-V2X uses cellular technologies and is expected to use also 5G networks.

In the case of agricultural robots, IoT has two roles: (1) it can provide data for the robot and (2) the robot itself can be considered as a thing connected to the Internet. In the future, IoT provides means to share data in physical world in the cyber world, enabling analysis and optimisation of large-scale systems of systems. It also provides means to create advanced situational awareness about the state of relevant aspects affecting the decision-making, i.e. weather, pest invasion, progress of other units' working processes and physical interaction in local contexts. Currently, IoT feeds the digital twins with real-world data, but the trend is in expanding the use of digital twins towards the exploration of alternative futures that can be a basis for better decisions today.

The data deluge captured by the heterogeneous digital systems that are currently supporting the different processes that are part of food production systems can be exploited through the application of AI techniques to create intelligent applications. ML models can be trained, deployed and productised following MLOps practices and recommendations that guarantee compliance with ethical and trust requirements. The convergence of AI techniques with the powerful capabilities of the new generation of edge computing devices facilitates addressing some of the most relevant challenges in terms of privacy preservation, scalability and real-time control which has traditionally limited the introduction of AI in the agriculture domain.

Mission Control is a high-level user interface for a robot. Mission Control Platforms are used to visualise the data coming from the robot as well as to give new instructions to the robot. Agriculture has a dedicated state-of-the-art communication protocol for ground vehicle mission control called EFDI as part of the ISOBUS standard. For drones, the QGroundControl drone control system with the MAVLink protocol has the same role.

A data space is defined as a decentralised infrastructure for trustworthy data sharing and exchange in data ecosystems based on commonly agreed principles. Data spaces infrastructure tries to create an environment and tools where different companies can share and exchange their data assets without losing their control over their data. Data spaces are a key enabler for implementing robotic systems in complex and fragmented farming environments meaningfully. Complexity implies that the constellation of utilised tools, e.g. IoT solutions, robot platforms and communication channels, must be able to tailor individually per farm operation, varying from farm to farm. While data spaces can be seen as a distributed service framework, users like farmers and RaaS entrepreneurs can define their individual Service Framework capturing all services they need for their activities.

11 Future trends in research

In agricultural robotics, the future will bring in robots that can do more complex and sensitive tasks. The capabilities developed for industrial robots will be brought to agricultural tasks and the capabilities developed for autonomous vehicles will make the field robots more maneuverable. The development of battery technologies and the need to get rid of CO_2 emissions will pave the way to electrification. This will lead to smaller robots and the need for collaborative robotics will increase (Lytridis et al., 2021). Robots will need capabilities to interact and make joint decisions more efficiently (Oliveira et al., 2021). The role and research needs of shared situational awareness of relevant information needed for decision-making among collaborating systems and humans and distributed decision-making will increase. Edge computing and real-time analysis will produce results in more and more real time, while AI and new analysing methods will offer reliability to systems. The developing connectivity makes it possible for the robots to be more purely only sensors and actuators, the computation and intelligence being separated and outsourced as edge or cloud service. This is an interesting future research topic since outsourcing can result in more precise and intelligent robot operations when several external data sources and services are connected in outsourced service to provide dedicated operational functionalities. This will also have an impact on the farming process by enabling the efficient execution of complex farming operations like mixed cropping. Also, user-centric decentralised service frameworks enabled by data space infrastructures provide new ground for ever-smarted robotic systems research.

The dependency of GNSSs on agricultural robotics and drones is critical. Methodologies for optional navigation systems such as local wireless networks and image recognition and analysis will also be future research topics in connectivity.

The research on synthetic data for training AI, especially ML systems, is topical for further research. Nowadays data is used increasingly to build Digital Twins also in agriculture. This opens possibilities to develop these virtual environments into synthetic simulation environments by utilising synthetic datasets in the environment (parallel to real-world data). These environments provide interactive simulation environments for training ML agents, i.e. reinforced learning agents (Nikolenko, 2021). Simulation carried out using

Digital Twins will also take a role in constructing reference data and providing synthetic data for measurement development (Kaivosoja, 2022).

As the degree of automation increases in agriculture, the challenge of how to make robots cost-efficient and user-friendlier will change the current practices. The possible scenario is to move towards a service model, where robots and robot fleets are owned and operated by service companies instead of farmers (Agam, 2022). This sets requirements for robots to adapt to different types of digital environments and systems, which requires research on new self-learning and cognitive capabilities of robots. User-friendliness requires higher abstraction-level user interfaces, where operators do not need to understand the details of how the robots operate. However, to enhance the operator's and end-user's trust in AI, research on explainable AI (XAI) development increasingly is important. XAI aims to explain transparently how AI performs the decision-making processes (Ryo, 2022).

The future of agricultural robotics will be based on the development of the technologies described in this chapter. Connectivity, AI and data will play a key role in developing robot systems to serve farms individually and feasibly and to assist them to comply with the requirements of food value chains, i.e. sustainability, green transitions, farm-to-fork communication and circular design. Sustainable intensification of food production requires that agricultural actors are better aware of production-related possibilities and constraints (environmental, biological, legal, ethical, economic and technical) in several levels of decision-making and information detail than today and are able to exploit the potential. On the other hand, sustainability, ethical aspects, increasing population, productivity requirements and novel production methods like vertical farming are all issues that change the ways robotics will be applied. While intensifying, production environments must foster the well-being of humans and animals. To be able to cope with all requirements, AI assistance will be embedded widely in everyday farming tools and services having diverse levels of autonomy (Duckett et al., 2018). As stated by Asseng and Asche (2019), we have the components needed to build future farms based on robots, but we need to combine the elements to fully utilize the potential. A more holistic approach than today is needed for the digitalisation of the production at farms, capturing all farming processes, as signposted by 'Hands-Free Hectare' and 'Hand-Free Farm' projects (Hands-Free Farm, no date). For example, it is needed to extend the current MRS control approaches to farm or crop mission management that covers all the phases of the crops including monitoring, analysis and various actions that are needed during growth periods. Such solutions could provide means to collect and combine the crop-related data for later phases of food production. The holistic and farm-specific approach would help to define meaningful division of roles between machines and humans while respecting farmers' individual preferences but exploiting still data-based performance optimisation. On that basis, it would be possible to construct such robot systems as an integral part of farms' workforce, where the system has the ability to react in real-time and even to anticipate the collaborators' actions as suggested by Vasconez et al. (2019). Further research on connectivity and intelligence of robot systems will have a key role in the development.

12 Acknowledgements

The chapter was supported by the European Union's Horizon 2020 research and innovation programme under grant agreement No 101017111.

13 Where to look for further information

The following articles provide a good overview on:

- Communication networks by describing technical and policy elements of telecommunication, particularly in the context of 5G: Asif (2018). 5G Mobile Communications: Concepts and Technologies. https://doi.org/10.1201/9780429466342.
- Internet of Robotic Things gives a novel perspective on the IoRT that involves communication among robotic things and humans and highlights the convergence of several technologies and interactions between different taxonomies used in the literature: Vermesan, O., Bahr, R., Ottella, M., Serrano, M., Karlsen, T., Wahlstrøm, T., Sand, H. E., Ashwathnarayan, M. and Gamba, M. T. (2020). Internet of Robotic Things Intelligent Connectivity and Platforms. Frontiers in Robotics and AI https://doi.org/10.3389/frobt.2020.00104
- Multirobot systems are used in agriculture in terms of platforms, control and application. Ju, C., Kim, J., Seol, J. and Il Son, H. I. (2022). A review on multirobot systems in agriculture. Computers and Electronics in Agriculture, 202, 107336, ISSN 0168-1699, https://doi.org/10.1016/j.compag.2022.107336.

The following initiatives and their websites are worth visiting to gain more information about:

- Data space development and standardization: Gaia-X, https://gaia-x.eu/.
- 6G research and development: Flagship 6G, https://www.oulu.fi/6g-flagship/.

14 References

3GPP. (n.d.). About 3GPP home. Available at: https://www.3gpp.org/about-3gpp/about-3gpp.

AEF. (n.d.). AEF homepage. Available at: https://www.aef-online.org/home.html (accessed 11 November 2022).

AEF. (2022a). Activity report AEF – Wireless in-field communication. Available at: https://www.aef-online.org/about-us/activities/activity-report.html (accessed 11 November 2022).

AEF. (2020b). New guideline for extended farm management information systems data interface (EFDI). Available at: https://www.aef-online.org/news/news/news/new-guideline-for-extended-farm-management-information-systems-data-interface-efdi-available.html (accessed 10 November 2022).

Agam, T. (2022). Robot-as-a-service for smart connected agriculture. Available at: https://pipelinepub.com/IoT-2022/robot-as-a-service-for-smart-connected-agriculture (accessed 7 January 2023).

AgroIntelli. (n.d.). Robotti, AgroIntelli ApS. Available at: https://agrointelli.com/robotti (accessed 11 November 2022).

Alcácer, V. and Cruz-Machado, V. (2019). Scanning the Industry 4.0: A literature review on technologies for manufacturing systems. *Engineering Science and Technology: An International Journal* 22(3), 899–919.

Alla, S. and Adari, S. K. (2021). What is mlops? In: *Beginning MLOps With MLFlow*. Apress, Berkeley, CA, pp. 79-124.

Alloghani, M., Al-Jumeily, D., Mustafina, J., Hussain, A. and Aljaaf, A. J. (2020). A systematic review on supervised and unsupervised machine learning algorithms for data science. In: *Supervised and Unsupervised Learning for Data Science*, pp. 3–21.

Amazon AWS. (n.d.). Machine learning and artificial intelligence. Available at: https://aws.amazon.com/machine-learning/.

Anagnostis, A., Benos, L., Tsaopoulos, D., Tagarakis, A., Tsolakis, N. and Bochtis, D. (2021). Human activity recognition through recurrent neural networks for human-robot interaction in agriculture. *Applied Sciences* 2021, 052188. https://doi.org/10. 3390/app11.

Anon. (2020). A European strategy for data. Communication from the commission to the European Parliament, the Council, the European Economic and Social Committee and the Committee of the Regions. *Com*. Brussels, 19.2.2020, 66 Final. Available at: https://eur-lex.europa.eu/legal-content/EN/TXT/PDF/?uri=CELEX:52020DC0066&from=EN (accessed 10 November 2022).

Anon. (2021). Regulation (EU) 2021/694 of the European Parliament and of The Council of 29 April 2021 establishing the Digital Europe Programme and repealing Decision (EU). *Official Journal of the European Union*. 11.5.2021 2015/2240. Regulations, L166. Available at: https://eur-lex.europa.eu/legal-content/EN/TXT/PDF/?uri=CELEX:32021R0694&from=EN (accessed 10 November 2022).

Anon. (2022). Commission staff working document on common European data spaces. Available at: https://ec.europa.eu/newsroom/dae/redirection/document/83562 (accessed 10 November 2022). Brussels, 23.2.2022 SWD(2022) 45 Final.

Asseng, S. and Asche, F. (2019). Future farms without farmers. *Science Robotics* 4(27), 1–2. Available at: https://robotics.sciencemag.org/content/4/27/eaaw1875 (accessed 7 January 2023).

Asif, S. (2018). *5G Mobile Communications: Concepts and Technologies*. CRC Press. https://doi.org/10.1201/9780429466342.

Autotalks. (n.d.). Vehicle to vehicle communication for heavy machinery. Available at: https://auto-talks.com/applications/heavy-machinery/ (accessed 7 January 2023).

Azure. (n.d.). Machine learning - ML as a service. *Microsoft Website*. Available at: https://azure.microsoft.com/en-us/products/machine-learning/.

Bachche, S. (2015). Deliberation on design strategies of automatic harvesting systems: A survey. *Robotics* 4(2), 194–222. https://doi.org/10.3390/robotics4020194.

Backman, J., Linkolehto, R., Koistinen, M., Nikander, J., Ronkainen, A., Kaivosoja, J., Suomi, P. and Pesonen, L. (2019). Cropinfra research data collection platform for ISO 11783 compatible and retrofit farm equipment. *Computers and Electronics in Agriculture* 166, 105008. https://doi.org/10.1016/j.compag.2019.105008.

Backman, J., Linkolehto, R., Lemsalu, M. and Kaivosoja, J. (2022). Building a robot tractor using commercial components and widely used standards. *IFAC-PapersOnLine* 55(32), 6–11.

Backman, J., Valtanen, K., Laikari, A., Vallivaara, V. and Ilomäki, J. (2017). Blockchain review: BOND project (Blockchains Boosting Finnish Industry) report. *VTT Technology*. Available at: https://publications.vtt.fi/pdf/technology/2017/T330.pdf (accessed 8 March 2022). VTT Technical Research Centre of Finland, Espoo (Vol. 330).

Basri, R., Islam, F., Binte, S. B. and Uddin, M. S. (2021). Robots and drones in agriculture–A survey. In: Uddin, M. S., Bansal, J. C. (Eds.), *Computer Vision and Machine Learning in Agriculture, Algorithms for Intelligent Systems*, pp. 9–29. https://doi.org/10.1007/978-981-33-6424-0_2.

Batth, R. S., Nayyar, A. and Nagpal, A. (2018). Internet of robotic things: driving intelligent robotics of future-concept, architecture, applications and technologies. In: *2018 4th International Conference on Computing Sciences (ICCS)*. IEEE, pp. 151–160. https://doi.org/10.1109/ICCS.2018.00033.

Bauer, M., Bui, N., de Loof, J., Magerkurth, C., Nettsträter, A., Stefa, J. and Walewski, J. W. (2013). IoT reference model. In: *Enabling Things to Talk*, pp. 113–162. https://doi.org/10.1007/978-3-642-40403-0_7.

Berenstein, R. and Edan, Y. (2017). Human-robot collaborative site-specific sprayer. *Journal of Robotics*, 21730. https://doi.org/10.1002/rob.

Bierzynski, K., Calvo Alonso, D. and Gandhi, K. (2021). AI at the edge. *EPoSS: European Technology Platform on Smart Systems Integration*. Available at: https://www.smart-systems-integration.org/system/files/document/EPoSS-White-Paper-AI.pdf (accessed 6 January 2023) (Vol. 2021). EPoSS White Paper.

Bonet, B. and Geffner, H. (2001). Planning as heuristic search. *Artificial Intelligence* 129(1–2), 5–33.

Chakraverty, S., Sahoo, D. M. and Mahato, N. R. (2019). McCulloch–Pitts neural network model. In: *Concepts of Soft Computing*. Springer, Singapore, pp. 167–173.

Chen, C. W. (2020). Internet of video things: Next-generation IoT with visual sensors. *IEEE Internet of Things Journal* 7(8), 6676–6685. https://doi.org/10.1109/JIOT.2020.3005727.

Chowdhary, K. (2020). Natural language processing. *Fundamentals of Artificial Intelligence*, 603–649.

Christopoulou, M., Barmpounakis, S., Koumaras, H. and Kaloxylos, A. (2023). Artificial Intelligence and Machine Learning as key enablers for V2X communications: A comprehensive survey. *Vehicular Communications* 39, 100569. https://doi.org/10.1016/j.vehcom.2022.100569.

Claire. (n.d.). What is AI? - claire-AI. Available at: https://claire-ai.org/what-is-ai/#RAareas (accessed 9 March 2022).

Corallo, A., Latino, M. E. and Menegoli, M. (2018). From Industry 4.0 to Agriculture 4.0: A framework to manage product data in Agri-food supply chain for voluntary traceability. *World Academy of Science, Engineering and Technology International Journal of Nutrition and Food Engineering* 12(5). Available at: https://www.researchgate.net/profile/Maria-Latino/publication/326231394_From_Industry_40_to_Agriculture_40_How_Manage_Product_Data_in_Agri-Food_Supply_Chain_for_Voluntary_Traceability_A_framework_proposed/links/5b473761a6fdccadaec1981f/From-Industry-40-to-Agriculture-40-How-Manage-Product-Data-in-Agri-Food-Supply-Chain-for-Voluntary-Traceability-A-framework-proposed.pdf (accessed 6 January 2023).

Corista, P., Ferreira, D., Giao, J., Sarraipa, J. and Goncalves, R. J. (2018). Proceedings. An IoT agriculture system using FIWARE. In: *Institute of Electrical and Electronics Engineers Inc.* IEEE International Conference on Engineering, Technology and Innovation, ICE/ITMC (Vol. 2018). https://doi.org/10.1109/ICE.2018.8436381.

Craig, J. J. (2005). *Introduction to Robotics: Mechanics and Control.* Pearson Education.

Curry, E., Scerri, S. and Tuikka, T. (2022). Data spaces: Design, deployment, and future directions. In: Curry, E., Scerri, S. and Tuikka, T. (Eds.), *Data Spaces*. Springer, Cham. https://doi.org/10.1007/978-3-030-98636-0_1.

DKE-Data GmbH & Co. KG. (n.d.). Agrirouter. Osnabrück, Available at: https://my-agrirouter.com/en/ (accessed 9 March 2022).

Drone Foundation. (n.d.). QGroundControl user guide. Available at: https://docs.qgroundcontrol.com/master/en/index.html (accessed 9 March 2022).

Duan, Y., Edwards, J. S. and Dwivedi, Y. K. (2019). Artificial intelligence for decision making in the era of Big Data–evolution, challenges and research agenda. *International Journal of Information Management* 48, 63–71.

Duckett, T., Pearson, S., Blackmore, S. and Grieve, B. (2018). Agricultural robotics: the future of robotic agriculture. UK-RAS White papers. UK-RAS robotics and automation network. ISSN 2398-4414. Available at: https://www.ukras.org.uk/publications/white-papers/agricultural-robotics/ (accessed 14 June 2023).

Dumka, S., Maheshwari, S. and Kala, R. (2018). Decentralized multi-robot mission planning using evolutionary computation. In: *IEEE Congress on Evolutionary Computation (CEC)* (Vol. 2018). IEEE Publications, pp. 1–8.

Dyke Parunak, H. V. (1994). Applications of distributed artificial intelligence in industry. *Industrial Technology Institute, ITI TR*. Available at: https://www.emse.fr/~boissier/enseignement/atelier00/jennings.pdf (accessed 12 November, 2022). USA, p. 93-03.1.

Ericsson. (2019). Critical capabilities for private 5G networks. White Paper 3/221 09-FGB 1010949, December 2019.

ETSI. (n.d.). ETSI – Standards, mission, vision, direct member participation. Available at: https://www.etsi.org/about (accessed 9 March 2022).

FarmDroid. (n.d.). FarmacoDroid - The world's first seeding and weeding robotDroid ApS. Available at: https://farmdroid.dk (accessed 11 November 2022).

Farooq, M. S., Riaz, S., Abid, A., Umer, T. and Zikria, Y. B. (2020). Role of IoT technology in agriculture: A systematic literature review. *Electronics* 9(2), 319. https://doi.org/10.3390/ELECTRONICS9020319.

FIWARE. (n.d.). FIWARE: The open source platform for our smart digital future. Available at: http://www.fiware.org/ (accessed 21 December 2022).

FIWARE. (2018). The open source platform of choice for building smart farming solutions. Available at: https://www.fiware.org/wp-content/uploads/2018/06/Smart-Agrifood_Brochure_June2018.pdf (accessed 4 January 2023).

FlexiGroBots. (n.d.). Flexible robots for intelligent automation of precision agriculture operations. *Project of the Horizon (2020–)*. European Union, p. 2.1.1. Available at: https://flexigrobots-h2020.eu/ (accessed 8 March 2022).

Forsyth, D. and Ponce, J. (2011). *Computer Vision: A Modern Approach*. Prentice Hall, Bloomington, p. 792.

Fountas, S., Sorensen, C. G., Tsiropoulos, Z., Cavalaris, C., Liakos, V. and Gemtos, T. (2015). Farm machinery management information system. *Computers and Electronics in Agriculture* 110, 131–138. http://doi.org/10.1016/j.compag.2014.11.011.

Gaia, X. (n.d.). Gaia-X architecture, Document 21.12. Release Gaia-X European Association for Data and Cloud AISBL ©2021 Gaia-X European Association for Data and Cloud AISBL.

Gazebo. (n.d.). Available at: https://gazebosim.org/.

Goap, A., Sharma, D., Shukla, A. K. and Rama Krishna, C. (2018). An IoT based smart irrigation management system using machine learning and open source technologies. Computers and Electronics in Agriculture 155, 41–49. https://doi.org/10.1016/j.compag.2018.09.040.

Google. (n.d.). Protocol buffers. *Google*. Available at: https://developers.google.com/protocol-buffers/ (accessed 9 March 2022).

Google Cloud AI. (n.d.). And machine learning products. *Google*. Available at: https://cloud.google.com/products/ai.

GSMA. (n.d.). Cellular vehicle-to-everything (C-V2X) - Enabling intelligent transport. Available at: https://www.gsma.com/iot/wp-content/uploads/2017/12/C-2VX-Enabling-Intelligent-Transport_2.pdf (accessed 6 January 2023).

Guo, J., Li, X., Lv, Z., Yang, Y. and Li, L. (2020). Design of real-time video transmission system for drone reliability. *IOP Conference Series: Materials Science and Engineering* 790(1), 012004. https://doi.org/10.1088/1757-899X/790/1/012004.

Haddadin, S. (2015). Physical safety in robotics. In: *Formal Modeling and Verification of Cyber-Physical Systems*. Springer Vieweg, Wiesbaden, pp. 249–271.

Hands Free Farm. (n.d.). Harper Adams University. Available at: https://www.handsfree.farm/ (accessed 7 January 2023).

Harsh, S., Singh, D. and Pathak, S. (2021). Efficient and cost-effective drone–NDVI system for precision farming. *International Journal of New Practices in Management and Engineering* 10(4), 14–19.

Hentati, A. I., Krichen, L., Fourati, M. and Fourati, L. C. (2018). Simulation tools, environments and frameworks for UAV systems performance analysis. In: *14th International Wireless Communications & Mobile Computing Conference (IWCMC)* (Vol. 2018). IEEE Publications, pp. 1495–1500.

HKT. (2020). *Report on C-V2X Technology Trial for Autonomous Vehicle at Hong Kong International Airpport Version 1.1.* Hong Kong Telecommunications Limited.

Available at: https://www.ofca.gov.hk/filemanager/ofca/en/content_669/tr202011_03.pdf (accessed 7 January 2023).

Huang, L., Dou, Y., Liu, Y., Wang, J., Chen, G., Zhang, X. and Wang, R. (2021). Toward a research framework to conceptualize data as a factor of production: The data marketplace perspective. *Fundamental Research* 1(5), 586–594. https://doi.org/10.1016/J.FMRE.2021.08.006.

HuggingFace. (n.d.). Models. Available at: https://huggingface.co/models.

Huhns, M. N. (2012). *Distributed Artificial Intelligence* (vol. I) (Vol. 1). Elsevier.

Hummel, P., Braun, M., Tretter, M. and Dabrock, P. (2021). Data sovereignty: A review. *Big Data and Society* 8(1), 2053951720982012. https://doi.org/10.1177/2053951720982012.

IEEE 802.11. *IEEE Standard for Information Technology - Telecommunications and Information Exchange Between Systems – Local and Metropolitan Area Networks – Specific Requirements – Part 11: Wireless LAN Medium Access Control (MAC) and Physical Layer (PHY) Specifications.*

IEEE 802.11p. (2010). *IEEE Standard for Information technology – Local and metropolitan area networks – Specific requirements – Part 11: Wireless LAN Medium Access Control (MAC) and Physical Layer (PHY) Specifications Amendment 6: Wireless Access in Vehicular Environments*.

IETF Trust. (2007). Hypertext transfer protocol – HTTP/1.1. Available at: https://www.w3.org/Protocols/HTTP/1.1/rfc2616bis/draft-lafon-rfc2616bis-03.html (accessed 11 November 2022).

Iost Filho, F. H., Heldens, W. B., Kong, Z. and de Lange, E. S. (2020). Drones: Innovative technology for use in precision pest management. *Journal of Economic Entomology* 113(1), 1–25.

ISO/IEC. (2016). *Information Technology – Message Queuing Telemetry Transport (MQTT) v3.1.1* (Vol. 20922). Available at: https://www.iso.org/standard/69466.html.

ISO 11783-10. (2015). *ISO 11783-10:2015. Tractors and Machinery for Agriculture and Forestry -- Serial Control and Communications Data Network – Part 10: Task Controller and Management Information System Data Interchange*.

ISO/DIS. (2022). *Extended Farm Management Information Systems Data Interface (EFDI)*. Draft International Standard, Geneva, p. 5231.

iSHARE. (n.d.). iSHARE trust framework. Available at: https://ishareworks.atlassian.net/wiki/spaces/IS/overview?homepageId=70025239 (accessed 17 February 2022).

John, A. J., Clark, C. E. F., Freeman, M. J., Kerrisk, K. L., Garcia, S. C. and Halachmi, I. (2016). Review: Milking robot utilization, a successful precision livestock farming evolution. *Animal* 10(9), 1484–1492. https://doi.org/10.1017/S1751731116000495.

Ju, C., Kim, J., Seol, J. and Il Son, H. I. (2022). A review on multirobot systems in agriculture. *Computers and Electronics in Agriculture* 202, 107336. https://doi.org/10.1016/j.compag.2022.107336.

Kaivosoja, J. (2022). Future possibilities and challenges for UAV-based imaging development in smart farming. In: Lipping, T., Linna, P. and Narra, N. (Eds.), *New Developments and Environmental Applications of Drones*. Springer, Cham. https://doi.org/10.1007/978-3-030-77860-6_6.

Kaivosoja, J., Ronkainen, A., Hautsalo, J., Backman, J., Linkolehto, R., Emeterio, M. S. and Soininen, J. P. (2022). Automatized rapeseed pest detection and management with drones. In: *Robot: Fifth Iberian Robotics Conference*, Zaragoza, Spain, November 2022. https://doi.org/10.1007/978-3-031-21062-4_35.

Kaknjo, A., Rao, M., Omerdic, E., Robinson, L., Toal, D. and Newe, T. (2018). Real-time video latency measurement between a robot and its remote control station: Causes and mitigation. *Hindawi, Wireless Communications and Mobile Computing*, 19 pages, Article ID 8638019. https://doi.org/10.1155/2018/8638019.

Kaloxylos, A., Eigenmann, R., Teye, F., Politopoulou, Z., Wolfert, S., Shrank, C., Dillinger, M., Lampropoulou, I., Antoniou, E., Pesonen, L., Nicole, H., Thomas, F., Alonistioti, N. and Kormentzas, G. (2012). Farm management systems and the future internet era. *Computers and Electronics in Agriculture* 89, 130–144. https://doi.org/10.1016/j.compag.2012.09.002.

Kangunde, V., Jamisola, R. S. and Theophilus, E. K. (2021). A review on drones controlled in real-time. *International Journal of Dynamics and Control* 9(4), 1832–1846.

Khalid, S. (2021). Internet of robotic things: a review. *Journal of Applied Science and Technology Trends* 2(3), 78–90. https://doi.org/10.38094/jastt203104.

Kiljander, J., D'Elia, A., Morandi, F., Hyttinen, P., Takalo-Mattila, J., Ylisaukko-Oja, A., Soininen, J. and Cinotti, T. S. (2014). Semantic interoperability architecture for pervasive computing and internet of things. *IEEE Access* 2, 856–873. https://doi.org/10.1109/ACCESS.2014.2347992.

Köksal, Ö. and Tekinerdogan, B. (2019). Architecture design approach for IoT-based farm management information systems. *Precision Agriculture* 20(5), 926–958. https://doi.org/10.1007/s11119-018-09624-8.

Korf, R. E. (1999). *Artificial Intelligence Search Algorithms*.

Krishnan, A., Swarna, S. and S, B. H. (2020, October). Robotics, IoT, and AI in the automation of agricultural industry: A review. In: *IEEE Bangalore Humanitarian Technology Conference (B-HTC)* (Vol. 2020). IEEE Publications, pp. 1-6.

Kubeflow. (n.d.). Available at: https://www.kubeflow.org/ (accessed 9 March 2022).

Lenz, E. (2004). Networked transport of RTCM via internet protocol (NTRIP) – Application and benefit in modern surveying systems. FIG Working Week 2004. Available at: https://www.fig.net/resources/proceedings/fig_proceedings/athens/papers/ts03/ts03_2_lenz.pdf (accessed 7 January 2023).

Lezoche, M., Hernandez, J. E., Díaz, M. D. M. E. A., Panetto, H. and Kacprzyk, J. (2020). Agri-food 4.0: A survey of the supply chains and technologies for the future agriculture. *Computers in Industry* 117, 103187. https://doi.org/10.1016/j.compind.2020.103187.

Lifschitz, V. (1987). On the semantics of STRIPS. In: *Reasoning About Actions and Plans. Proceedings of the 1986 Workshop*, pp. 1–9.

LinkedData. (n.d.). Available at: https://www.w3.org/wiki/LinkedData (accessed 21 December 2022).

Liu, Y., Peng, Y., Wang, B., Yao, S. and Liu, Z. (2017). Review on cyber-physical systems. *IEEE/CAA Journal of Automatica Sinica* 4(1), 27–40. https://doi.org/10.1109/JAS.2017.7510349.

Lytridis, C., Kaburlasos, V. G., Pachidis, T., Manios, M., Vrochidou, E., Kalampokas, T. and Chatzistamatis, S. (2021). An overview of cooperative robotics in agriculture. *Agronomy* 11(9), 1818. https://doi.org/10.3390/agronomy11091818.

Marvin, H. J. P., Bouzembrak, Y., Van der Fels-Klerx, H. J., Kempenaar, C., Veerkamp, R., Chauhan, A., Stroosnijder, S., Top, J., Simsek-Senel, G., Vrolijk, H., Knibbe, W. J., Zhang, L., Boom, R. and Tekinerdogan, B. (2022). Digitalisation and artificial intelligence for sustainable food systems. *Trends in Food Science and Technology* 120, 344–348.

MAVLink. (n.d.). MAVLink developer guide. Available at: https://mavlink.io/en/ (accessed 9 March 2022).

Merzouki, R., Samantaray, A. K., Pathak, P. M. and Bouamama, B. O. (2013). *Intelligent Mechatronic Systems: Modeling, Control and Diagnosis*. Springer, London, pp. 978–971.

Miranda, J., Ponce, P., Molina, A. and Wright, P. (2019). Sensing, smart and sustainable technologies for Agri-Food 4.0. *Computers in Industry* 108, 21–36. https://doi.org/10.1016/j.compind.2019.02.002.

Mirjalili, S. and Dong, J. S. (2020). *Multi-Objective Optimization Using Artificial Intelligence Techniques*. Springer, Cham.

Mittal, K., Jain, A., Vaisla, K. S., Castillo, O. and Kacprzyk, J. (2020). A comprehensive review on type 2 fuzzy logic applications: Past, present and future. *Engineering Applications of Artificial Intelligence* 95, 103916.

MLFlow. (n.d.). Available at: https://mlflow.org/ (accessed 9 March 2022).

Nagel, L. and Lycklama, D. (2021). Design principles for data spaces. https://doi.org/10.5281/ZENODO.5244997.

Nikolenko, S. I (2021). Synthetic data for deep learning. https://doi.org/10.1007/978-3-030-75178-4.

Nguyen, K. D. and Nguyen, T. T. (2019). Vision-based software-in-the-loop-simulation for Unmanned Aerial Vehicles using gazebo and PX4 open source. In: *International Conference on System Science and Engineering (ICSSE)* (Vol. 2019). IEEE Publications, pp. 429–432.

Ocean Protocol. (n.d.). Tools for the web3 data economy, technical whitepaper version 2020-DEC-09, ocean protocol foundation. Available at: https://oceanprotocol.com/technology/roadmap#papers (accessed 15 February 2022).

Ogaja, C. (2022). *Introduction to GNSS Geodesy: Foundations of Precise Positioning Using Global Navigation Satellite Systems*. Springer, Cham. https://doi.org/10.1007/978-3-030-91821-7.

Ojha, T., Misra, S. and Raghuwanshi, N. S. (2021). Internet of things for agricultural applications: The state of the art. *IEEE Internet of Things Journal* 8(14), 10973–10997. https://doi.org/10.1109/JIOT.2021.3051418.

Oliveira, L. F. P., Moreira, A. P. and Silva, M. F. (2021). Advances in agriculture robotics: A state-of-the-art review and challenges ahead. *Robotics* 10(2), 52. https://doi.org/10.3390/robotics10020052.

Oroojlooy, A. and Hajinezhad, D. (2022). A review of cooperative multi-agent deep reinforcement learning. *Applied Intelligence*, 1–46.

Otto, B., Steinbuss, S., Teuscher, A. and Lohmann, S. (n.d.). IDS reference architecture. Available at: https://internationaldataspaces.org/wp-content/uploads/IDS-Reference-Architecture-Model-3.0-2019.pdf (accessed 9 March 2022).

Pesonen, L., Koskinen, H. and Rydberg, A. (Eds.). (2008). InfoXT - User-centric mobile information management in automated plant production. Nordic innovation. Available at: https://norden.diva-portal.org/smash/record.jsf?pid=diva2%3A707134&dswid=-2998 (accessed 10 November 2022).

Pesonen, L. A., Teye, F. K.-W., Ronkainen, A. K., Koistinen, M. O., Kaivosoja, J. J., Suomi, P. F. and Linkolehto, R. O. (2014). Cropinfra – An Internet-based service infrastructure to support crop production in future farms. *Biosystems Engineering* 120, 92–101.

Peteiro-Barral, D. and Guijarro-Berdiñas, B. (2013). A survey of methods for distributed machine learning. *Progress in Artificial Intelligence* 2(1), 1–11.

Ploennigs, J., Cohn, J. and Stanford-Clark, A. (2018). The future of IoT. *IEEE Internet of Things Magazine* 1(1), 28–33. https://doi.org/10.1109/IOTM.2018.1700021.

Plura, M., Müller, H., Stelzer, M. and Florian, S. (2011). *Mission Control Concepts for Robotic Operations: Existing Approaches and New Solutions*. 11th Symposium on Advanced Space Technologies in Robotics and Automation, The Netherlands.

Puri, V., Nayyar, A. and Raja, L. (2017). Agriculture drones: A modern breakthrough in precision agriculture. *Journal of Statistics and Management Systems* 20(4), 507–518.

PX4 Autopilot. (n.d.a). Basic concepts. Available at: https://docs.px4.io/master/en/getting_started/px4_basic_concepts.html#flight_modes (accessed 9 March 2022).

PX4 Autopilot. (n.d.b). Gazebo simulation. Available at: https://docs.px4.io/master/en/simulation/gazebo.html (accessed 9 March 2022).

Pylianidis, C., Osinga, S. and Athanasiadis, I. N. (2021). Introducing digital twins to agriculture. *Computers and Electronics in Agriculture* 184, 105942. https://doi.org/10.1016/j.compag.2020.105942.

QGroundControl. (n.d.). Available at: http://qgroundcontrol.com/.

Quinonero-Candela, J., Sugiyama, M., Schwaighofer, A. and Lawrence, N. D. (Eds.). (2008). *Dataset Shift in Machine Learning*. MIT Press, Cambridge.

Ray, P. P. (2016). Internet of robotic things: Concept, technologies, and challenges. *IEEE Access* 4, 9489–9500. https://doi.org/10.1109/ACCESS.2017.2647747.

Ray, P. P. (2021). A review on TinyML: State-of-the-art and prospects. *Journal of King Saud University-Computer and Information Sciences* 34, 1595–1623.

Riguzzi, F. (2022). *Foundations of Probabilistic Logic Programming: Languages, Semantics, Inference and Learning*. CRC Press, Boca Raton.

Rizk, Y., Awad, M. and Tunstel, E. W. (2019). Cooperative heterogeneous multi-robot systems: A survey. *ACM Computing Surveys* 52, 2, Article 29. https://doi.org/10.1145/3303848.

Roadmap – Ocean Protocol [WWW Document]. (n.d.). Available at: https://oceanprotocol.com/technology/roadmap#papers (accessed 15 February 2022).

Rodriguez, M. A., Cuenca, L. and Ortiz, A. (2018). FIWARE open source standard platform in smart farming - A review. In: *IFIP Advances in Information and Communication Technology*. Springer, New York LLC, pp. 581–589. https://doi.org/10.1007/978-3-319-99127-6_50.

Rondelli, V., Franceschetti, B. and Mengoli, D. (2022). A review of current and historical research contributions to the development of ground autonomous vehicles for agriculture. *Sustainability* 14(15), 9221. https://doi.org/10.3390/su14159221.

ROS. (n.d.). ROS - Robot operating system. Available at: https://www.ros.org/ (accessed 4 January 2023).

ROS2. (n.d.). ROS (vol. 2) documentations. Available at: https://docs.ros.org/en/foxy/index.html (accessed 4 January 2023).

Ryo, M. (2022). Explainable artificial intelligence and interpretable machine learning for agricultural data analysis. *Artificial Intelligence in Agriculture* 6, 257–265. https://doi.org/10.1016/j.aiia.2022.11.003.

Sejnowski, T. J. (2018). *The Deep Learning Revolution*. MIT Press, Cambridge.

Semantic Web. (n.d.). Available at: https://www.w3.org/standards/semanticweb/ (accessed 23 December 2022).

Sorensen, C. G., Fountas, S., Nash, E., Pesonen, L., Bochtis, D., Pedersen, S. M., Basso, B. and Blackmore, S. B. (2010). Conceptual model of a future farm management information system. *Computers and Electronics in Agriculture* 72(1), 37–47.

Steen, K. A., Villa-Henriksen, A., Therkildsen, O. R. and Green, O. (2012). Automatic detection of animals in mowing operations using thermal cameras. *Sensors* 12(6), 7587–7597. https://doi.org/10.3390/s120607587.

Stülpner, A., Adeili, S., Haidn, B., Dörfler, R. and Bernhardt, H. (2014). Reactions of dairy cows during the operation of a robotic slat cleaner. Livestock and machinery. *Landtechnik* 69(5), 225–231. Available at: https://www.researchgate.net/publication/301361537_Reactions_of_dairy_cows_during_the_operation_of_a_robotic_slat_cleaner.

Tang, Y., Dananjayan, S., Hou, C., Guo, Q., Luo, S. and He, Y. (2021). A survey on the 5G network and its impact on agriculture: Challenges and opportunities. *Computers and Electronics in Agriculture* 180, 105895. https://doi.org/10.1016/j.compag.2020.105895.

Tardieu, H. (2021). Role of gaia-X in the European data space ecosystem. In: Otto, B., ten Hompel, M. and Wrobel, S. (Eds.), *Designing Data Spaces, the Ecosystem Approach to Competitive Advantage*. Springer. https://doi.org/10.1007/978-3-030-93975-5.

The World Bank. (n.d.). Arable land (hectares). Available at: https://data.worldbank.org/indicator/AG.LND.ARBL.HA?most_recent_value_desc=true&view=map (accessed 6 January 2023).

Toosi, A., Bottino, A. G., Saboury, B., Siegel, E. and Rahmim, A. (2021). A brief history of AI: How to prevent another winter (a critical review). *PET Clinics* 16(4), 449–469.

TRUSTS. (n.d.). TRUSTS – Trusted secure data sharing space. Available at: https://www.trusts-data.eu/ (accessed 9 November 2022).

Van der Hoek, W. and Wooldridge, M. (2008). Multi-agent systems. *Foundations of Artificial Intelligence* 3, 887–928.

Vasconez, J. P., Kantor, G. A. and Auat Cheein, F. A. (2019). Human–robot interaction in agriculture: A survey and current challenges. *Biosystems Engineering* 179, 35–48.

Verbraeken, J., Wolting, M., Katzy, J., Kloppenburg, J., Verbelen, T. and Rellermeyer, J. S. (2020). A survey on distributed machine learning. *ACM Computing Surveys (CSUR)* 53(2), 1–33.

Verdouw, C., Tekinerdogan, B., Beulens, A. and Wolfert, S. (2021). Digital twins in smart farming. *Agricultural Systems* 189, 103046. https://doi.org/10.1016/j.agsy.2020.103046.

Verdouw, C., Wolfert, S. and Tekinerdogan, B. (2016). Internet of things in agriculture. *CABI Reviews* 2016, 1–12. https://doi.org/10.1079/PAVSNNR201611035.

Vermesan, O. and Friess, P. (2013). Internet of things: Converging technologies for smart environments and integrated ecosystems. *Internet of Things*, 3–27.

Villa, D., Song, X., Heim, M. and Li, L. (2021). Internet of Robotic Things: Current Technologies, Applications, Challenges and Future Directions Davide. *Department of Electrical and Computer Engineering*. Northeastern University, Boston. https://doi.org/10.48550/arXiv.2101.06256.

Wheare, J. (2018). Mission planning for field robots using symbolic planning and topology. Thesis, Flinders University, Australia. Available at: https://flex.flinders.edu.au/file/1ecfc36e-ec5a-4dab-8358-79a8e4ea41f7/1/ThesisWheare2018OA.pdf (accessed 11 November 2022).

Wiering, M. A. and Van Otterlo, M. (2012). Reinforcement learning. *Adaptation, Learning, and Optimization* 12(3), 729.

Wood, M. F. and DeLoach, S. A. (2000). An overview of the multiagent systems engineering methodology. In: *International Workshop on Agent-Oriented Software Engineering*. Springer, Berlin, Heidelberg, pp. 207–221.

Xu, M., David, J. M. and Kim, S. H. (2018). The fourth industrial revolution: Opportunities and challenges. *International Journal of Financial Research* 9(2), 90–95.

Yang, C., Zhu, Y. and Chen, Y. (2021). A review of human–machine cooperation in the robotics domain. *IEEE Transactions on Human-Machine Systems* 52(1), 12–25.

Yang, Q., Liu, Y., Cheng, Y., Kang, Y., Chen, T. and Yu, H. (2019). Federated learning. *Synthesis Lectures on Artificial Intelligence and Machine Learning* 13(3), 1–207.

Ye, Y., He, L., Wang, Z., Jones, D., Hollinger, G. A., Taylor, M. E. and Zhang, Q. (2018). Orchard manoeuvring strategy for a robotic bin-handling machine. Biosystems Engineering 169, 85–103. https://doi.org/10.1016/j.biosystemseng.2017.12.005.

Zangenehnejad, F. and Gao, Y. (2021). GNSS smartphones positioning: Advances, challenges, opportunities, and future perspectives. *Satellite Navigation* 2(1), 24. https://doi.org/10.1186/s43020-021-00054-y.

Zhou, X., Girdhar, R., Joulin, A., Krähenbühl, P, and Misra, I. (2022). Detecting twenty-thousand classes using image-level supervision. *arXiv preprint arXiv:2201.02605.*

Zhou, Z. H. (2021). *Machine Learning.* Springer Nature, Singapore.

Zidan, J., Adegoke, E. I., Kampert, E., Birrell, S. A., Ford, C. R. and Higgins, M. D. (2020). GNSS vulnerabilities and existing solutions: A review of the literature. *IEEE Access* 9, 153960–153976. https://doi.org/10.1109/ACCESS.2020.2973759.

Chapter 9

Improving fault detection and isolation in agricultural robotics

Nicolas Tricot, Mahmoud Almasri and Roland Lenain, Université Clermont Auvergne, France

1 Introduction
2 Principles of fault diagnosis
3 Methods for fault diagnosis
4 Designing a robust fault detection and isolation system
5 Hybrid diagnostic method
6 Human/robot interaction module
7 Implementing the proposed fault detection and isolation system: key components
8 Modelling robot movement and potential faults
9 Testing fault diagnosis using Kalman filters
10 Testing fault diagnosis using deep learning
11 Building the hybrid diagnostic method and human/robot interaction module
12 Conclusion and future trends
13 Abbreviations
14 Acknowledgements
15 Where to look for further information
16 References

1 Introduction

According to the International Federation of Robotics [1], 31 million service robots were sold worldwide between 2014 and 2017. Wheeled autonomous mobile robots have many applications, such as service, space exploration [2], object transport [3] and agricultural production and management [4, 5]. Currently, agricultural robots assist in harvesting, picking, spraying and control of chemicals [6]. According to Tractica [7], orders for agricultural robots are predicted to increase from 32 000 units in 2016 to 594 000 units per year in 2024, when the global market is expected to reach an annual value of $74.1 billion.

http://dx.doi.org/10.19103/AS.2023.0124.16

However, all robots and therefore more particularly mobile robots will encounter faults during their life [8]. A fault is defined as an unauthorized deviation of at least one characteristic property of a variable from acceptable behavior [9]. A fault can quickly degrade system performance or even damage hardware. It can also require frequent user intervention, making the system less autonomous. Fault detection and isolation (FDI) is therefore important to safety, efficiency of operation, protection of costly equipment from potential damage and reducing the time/cost required in human intervention.

The heart of this chapter details the proposal of a hybrid fault diagnosis method for mobile robots, based on the principles of FDI. It begins with the definition of the principles of fault diagnosis. A comparative analysis of traditional fault diagnosis methods is then proposed. Finally, the implementation of the method and the results obtained are detailed on an agricultural use case.

2 Principles of fault diagnosis

Typical faults that can occur on a robot are varied (Table 1) and can include [10] [11]:

- Mechanical faults, e.g. blockage, breakage or degradation due to friction between components (wheels and chassis for example);
- Electrical faults, e.g. damage to wires or battery failure due to a short circuit on a component (motor or electronic card);
- Navigation faults, e.g. due to an inaccurate digital map of the environment, sensor failure or a bug in the navigation algorithm;
- Collision avoidance faults, e.g. due to a fault in the obstacle avoidance system such as a faulty obstacle detection sensor or a navigation failure;
- Faults due to environmental factors, e.g. slipping in wet and muddy conditions, poor visibility (e.g. due to fog), hidden or hard-to-see obstacles (e.g. holes);
- Faults related to excessive time needed for decision-making (overload of the processor(s)) making the robot too slow to execute a required manoeuvre or action; and

Table 1 Faults considered

Source type	Source of fault
Sensor	Absolute Positioning System (GPS) Inertial Unit (IMU) Wheel odometer Obstacle detection system (Laser) Communication via UWB
External	Slippage
Mechanical	Wheel lock

- Faults relating to overheating due to ambient temperatures and long operating times.

Faults can be classified as additive or multiplicative. For example, equations (1) illustrate a state representation of any system. f_i and f_o are additive fault parameters affecting input and output respectively. Multiplicative faults modify the functions *f* and *h*:

$$\dot{X}(t) = f(t, x(t), u(t)) + f_i \quad (1)$$

$$Y(t) = h(t, x(t), u(t)) + f_o\#$$

A distinction can also be made between a sudden fault, an incipient fault (drift) and an intermittent fault [12]. Figure 1 illustrates these types.

FDI has been the subject of research since the 1970s [13]. Key steps in FDI are:

- Detection: determination of the presence of a fault and the moment of its occurrence, without necessarily knowing its location;
- Isolation: determining the location of the fault and its type;
- Identification: determining the severity and time span of a fault; and
- Accommodation: reconfiguring the system to avoid losing control.

Since FDI is typically implemented after robot design, it is necessary to design diagnostic methods based on the information and data available from existing sensors used by the robot. Diagnostic methods for mobile robots must also meet a range of other criteria:

- A diagnostic method that can be used online: The majority of mobile robots used in agricultural and industrial contexts are autonomous. The diagnosis must therefore be feasible while the robot performs a task.
- A method compatible with non-linear representations of systems, since robot sensors, actuators and dynamic models can be non-linear.

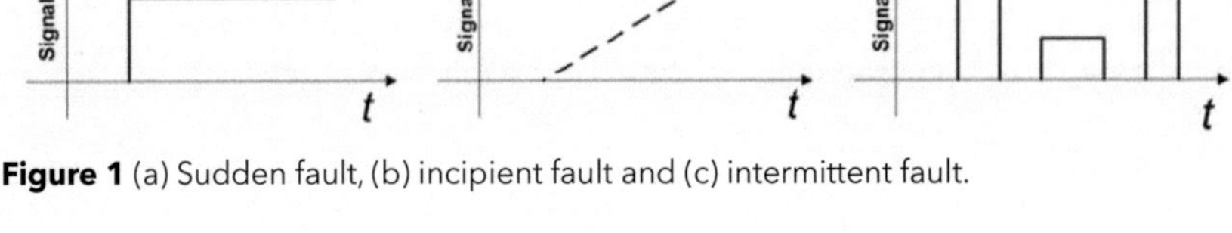

Figure 1 (a) Sudden fault, (b) incipient fault and (c) intermittent fault.

- Rapid diagnosis: To prevent the fault from causing significant damage, it is important to detect and then isolate a fault as quickly as possible, or even anticipate it.
- Accurate diagnosis: Frequently interrupting operations due to false alarms can drastically reduce robot productivity. It is essential to minimize this whilst ensuring real faults are detected.
- Inexpensive diagnosis: Diagnosis must not take up too much calculation time and storage volume (the computing power available on the robot is generally limited). It must also use the minimum number of additional sensors possible.
- Adaptable diagnostics: The diagnostic system must be easily adaptable. Application to robots or in different contexts should be possible with a minimum number of modifications.
- Clear diagnosis: Where a robot is supervised by a remote operator, the diagnostic system must warn the operator in a clear manner to allow effective intervention.

In meeting such criteria, any one diagnostic method has limitations [14]. Research shows that a single method is never 100% reliable [15, 9, 16, 17]. A global diagnostic system is therefore needed combining several diagnostic methods. This should allow for human intervention in offsetting limitations in FDI algorithms [18]. A number of studies have explored types of human-robot interaction in FDI systems [19, 20, 21].

3 Methods for fault diagnosis

Diagnostic methods can be divided into four main categories [22]:

- Model-based methods: Based on a mathematical model of the system, these methods often generate values called residuals. Diagnosis is generally then made through analysis of these residuals (or residues).
- Knowledge-based methods: These methods predict the state of the system from the values of its inputs/outputs and knowledge related to the behavior of the system in normal operating conditions.
- Data-based methods: These methods 'see' the system as a black box: they diagnose faults by analyzing the inputs and outputs of the model.
- Hardware redundancy: This involves duplicating sensors or adding other sensors to measure the same variable.

These types are discussed in more detail in the following sections.

3.1 Model-based diagnostic methods

Model-based methods have been widely studied and used in mobile robotics since the 1970s [23, 24]. The general structure of these methods is shown in Fig. 2. As an example of a model-based approach, the relationships between redundant measurements of different sensors and the motion model of an autonomous, wheeled robot allows generation of residuals. Thresholding of these residues then makes it possible to isolate a fault [25, 26].

Where a system model is simple, model-based techniques can be very precise and efficient (requiring little computing power). The method is ideal for detecting disturbance type faults. However, more complex and non-linear models can make these techniques less precise and more difficult to design.

Depending on how residuals are generated, there are three categories of methods:

- Methods based on observers [27]: Observers are well-known tools in the field of control theory. Residuals are obtained by comparing the estimated value and the corresponding value measured by sensors. They iteratively estimate the complete or partial state of a system and are used in many robotic applications [28].
- Methods based on analysis of parity relations [26]: Diagnosis involves checking the consistency between different variables of a system. Variables are identified and compared from at least two sources (models, sensors, inputs and outputs) to produce residuals. Whilst little computing power is needed, post-processing is needed to obtain satisfactory results while simple thresholding is sensitive to noise. For example, comparing obstacle detection from the LIDAR and the RGB camera.
- Methods using parameter estimation [29]: Parameters can be used to configure diagnostic methods such as recursive estimation of fault parameters [29] or multi-model estimation [30]. However, this requires a preliminary knowledge of faults and their effect on the model.

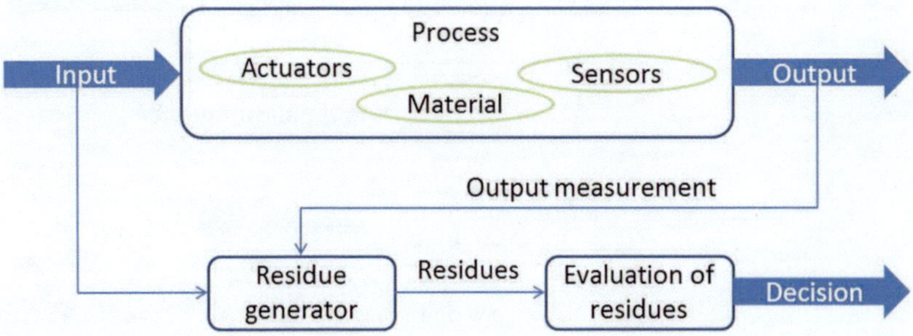

Figure 2 General structure of model-based methods.

There are also graphical model-based methods [31]. Since these can operate offline, they have been proposed as an alternative to fully analytical and numerical models.

Whilst the standard Kalman Filter (KF) is not effective for model-based diagnosis [32], the extended Kalman Filter (EKF), unscented (UKF) and informational filter (IF) have been used successfully, including on non-linear models [3, 33, 34, 35, 36]. These algorithms can be integrated into a Multiple-Model Adaptative Estimation (MMAE) [37, 38]. MMAE can isolate many faults but requires a model in each case. Recursive parameter estimation (RPE) has also been used to ensure fault-tolerant control [39].

3.2 Knowledge-based methods

Knowledge-based methods are based on heuristic system analysis using information extracted from system outputs [40, 41, 2, 42] (Fig. 3). This approach can be advantageous in the case of systems that are difficult to model (e.g. non-linear) but is generally expensive in time and computing power [12]. A good knowledge of the system behavior in each state (failed state, normal state) is important. These methods use methods such as the particle filter (PF) [43] and variants [44] such as the variable resolution PF [45], the Rao-Blackwell version integrating a UKF [41] and the Gaussian PF [46], e.g. in the case of detecting faults on the Mars Rover. Knowledge-based approaches can be combined with model-based approaches [47].

3.3 Data-based methods

Data-based methods can diagnose the state of a system by generally treating it as a black box. The main techniques are illustrated in Fig. 4 [48, 49, 50, 51]. These methods can be combined with knowledge-based or model-based

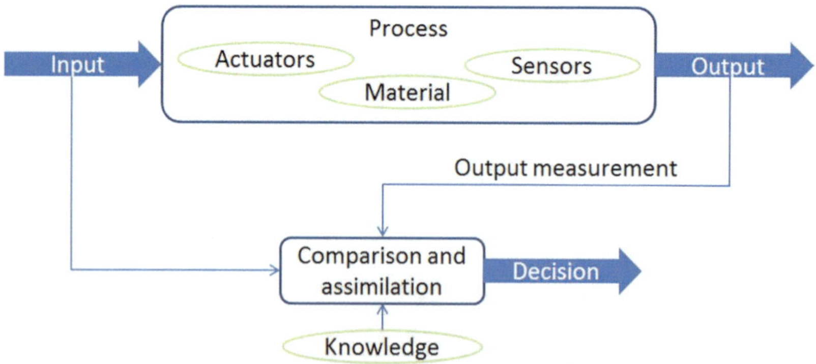

Figure 3 General structure of knowledge-based methods.

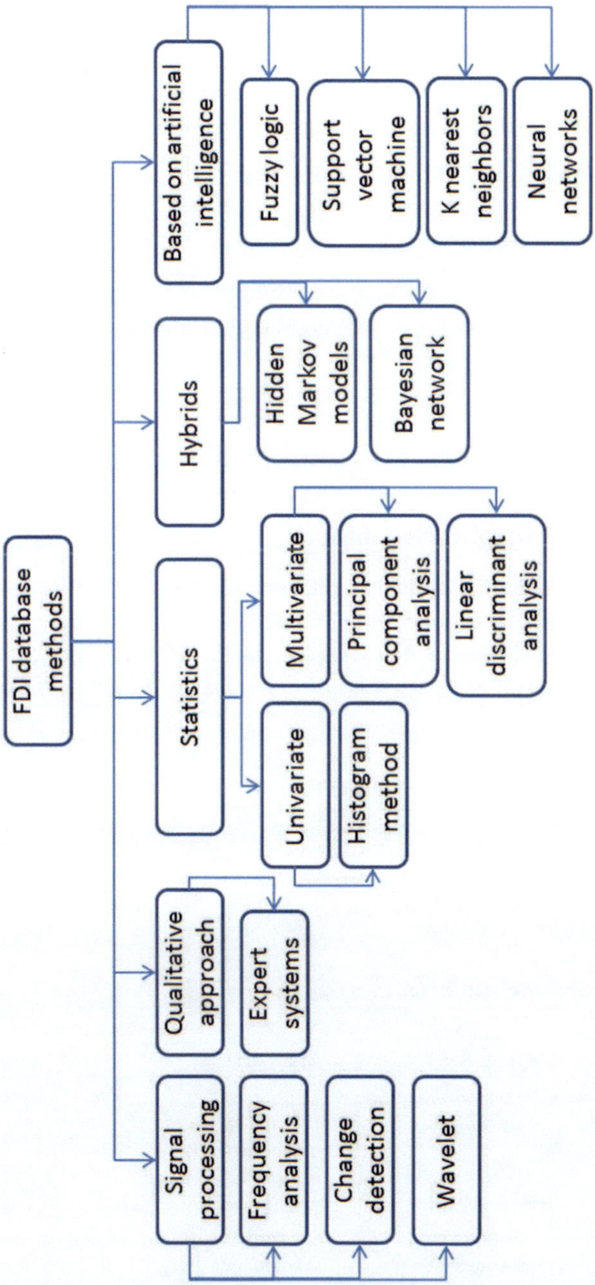

Figure 4 FDI data-based methods.

approaches. They can even be applied to the outputs of other methods since they do not depend on input data. The main drawback is the amount of data and the computation time required for training. Some of the most effective approaches use artificial intelligence techniques such as deep belief networks (DBN)[52] or neural networks [53, 44, 54, 55] which can achieve a fault prediction rate of over 95%. Data-based methods can be used directly for measurement and control or to deal with residuals from model-based approaches.

3.4 Hardware redundancy

Hardware redundancy involves duplicating a sensor (e.g.position sensor and torque sensor) or adding other sensors (e.g. stereo vision and LIDAR for obstacle detection) to measure the same variable [26] (Fig. 5). The system must therefore be able to support redundant sensors. In practice, this method is mainly used to detect and isolate sensor faults. Despite its simplicity, hardware redundancy is expensive and can quickly overload the system.

3.5 Comparing diagnostic methods

Tables 2, 3 and 4 summarize the advantages and disadvantages of each of these methods. It is clear that each method has limitations which prevent it from being applied for all faults. However, it is possible to assign different diagnostic methods to different types of faults. For example, for noise faults on location sensors with a

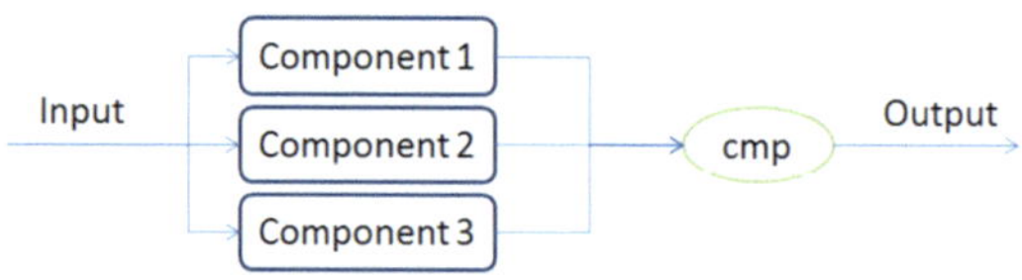

Figure 5 Hardware redundancy.

Table 2 Characteristics of data-based methods

Method	Condition of application	Drawbacks	Advantages
Conventional neural network	Need to have enough data for learning	Dependent on training data and neural network design. Very difficult debugging	Very generic approach
DL	Larger training database than conventional neural network	Database size and training time. Very difficult debugging	Generic approach that has proven its performance in several areas

Table 3 Characteristics of model-based methods

Method	Terms of application	Disadvantage	Advantage
KF	Linear system. Detects disturbance-type faults well	Degradation in performance if the noise is non-Gaussian	Optimal estimator for linear systems with Gaussian noises
EKF	Nonlinear system up to order 2. Detects disturbance type faults well	Computation of the Jacobean matrix expensive in computing time for certain systems	High performance for a first-order nonlinearity system with Gaussian distributions
UKF	Non-linear system up to order 3. Detects disturbance type faults well	Calculation time more important than for an EKF for systems with low order of non-linearity	Converges faster than an EKF for non-linear systems
Parity relationships	Possibility to calculate the same variable with several methods	Not robust to noise	Low computation time
RPE	Known fault architecture	Not robust to noise	Suitable for multiplicative and additive faults
MMAE	Architecture known to the fault	Performance dependent on residue post-treatment	Suitable for multiplicative and additive faults

Table 4 Characteristics of knowledge-based methods

Method	Terms of application	Disadvantage	Advantage
Standard PF	A priori knowledge of the characteristics of the faults	A large number of particles (and computation time) is necessary to detect faults with low probability of occurrence	Theoretically, no approximation performed on the system and it works on any model
Risk-sensitive FP	Limit number of faults that can be monitored	Performance degradation for many similar fault states	Detects unlikely faults better than PF
Variable resolution PF	Presence of similar states	Non-generic method, because relevant only in the presence of similar states	Low computation time and high performance in case of similar transition and observation patterns
Rao-Blackwell PF	Continuous variables: KF conditions	The same drawbacks of KF for estimating continuous variables	Superior performance to PF with the same number of particles
Gaussian PF	Continuous variables: UKF conditions	The same disadvantages of UKF for estimating continuous variables	Better performance than PF with the same number of particles

known pattern, EKF or UKF would be a good choice. Model-based methods are suitable for disturbance type faults (sensors). Knowledge-based methods can diagnose faults such as wheel locking where a robot demonstrates predictable behavior (modelable effect) when experiencing the fault. Localization sensor disturbance faults can be diagnosed by Kalman filters, redundancy relationships and deep learning. Wheel lock faults can be diagnosed by knowledge-based methods, deep learning and redundancy relationships. Laser and UWB faults can be diagnosed by EKF and confidence vector-based methods, respectively. Methods based on artificial intelligence are best suited to deal with a wide range of faults, provided there is enough data for training.

4 Designing a robust fault detection and isolation system

This chapter outlines a methodology to design a global system that can diagnose a range of faults affecting mobile robotics in real time (Fig. 6). It is made up of two main components:

- A Robot Fault Diagnosis Supervision System (S2D2R): It contains two components:
    1. A hybrid diagnostic method (MHD) made up of several diagnostic sub-methods; and
    2. A human/robot interaction module (MIHR) responsible for the link between the MHD and human supervision.
- A group of tools for modelling, parameterization and adaptation of fault diagnostic methods (OMOPAMED2).

OMOPAMED2 allows the construction of the initial MHD (MHD0). The list of diagnostic methods in the MHD may evolve thanks to continuous evaluation due to interaction with the operator via the MIHR.

This approach means different diagnostic methods can be evaluated to identify the best one(s) for particular faults. Their parameters are adjusted automatically in real time to optimize their performance through continuous evaluation by the MHD. Following fault diagnosis, the MIHR communicates key information on the fault, its type and level of diagnosis (detection or isolation) to the human operator to allow intervention as needed.

Key steps in the system are (Fig. 6):

- Identifying fault characteristics: Each fault is studied by determining characteristics such as source, time span, pattern and effects on robot behavior. Identifying fault characteristics allows choice of methods best suited for diagnosis.

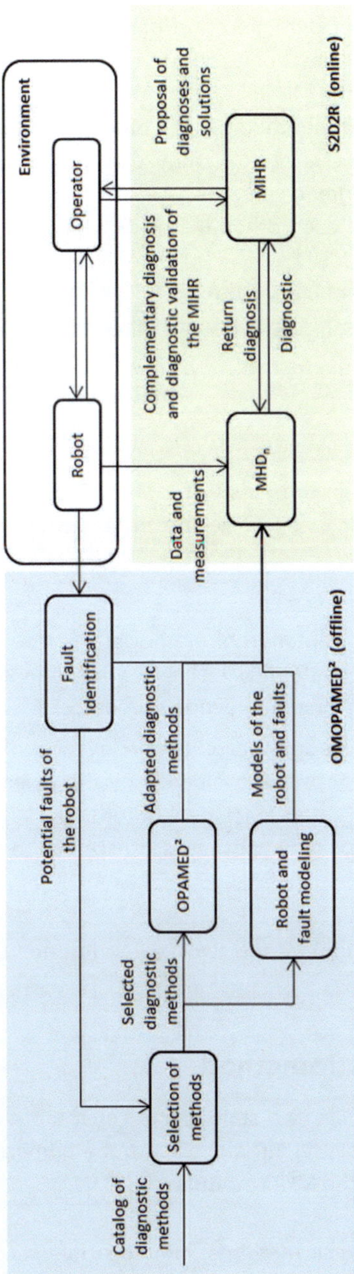

Figure 6 Overall outline of the methodology.

- Selecting diagnostic methods: Following determination of the characteristics of a fault, a search is carried out to identify diagnostic methods capable of detecting and/or isolating that fault. One or more diagnostic methods are then assigned to the fault.
- Initial operation of the MHD:
- Adjustment and optimization of methods: Diagnostic methods are adapted and parameterized to optimize performance.
- Evaluation and integration into the MHD: Following optimization, methods are compared with each other. Methods with the best performance are integrated into the MHD.
- Definition of mode of interaction:
    - Definition of the initial knowledge base: This base defines the rules for the behavior of the method in each situation (detection or isolation of a fault).
- Communication:
    - Communication in the event of a fault: A communication protocol is defined in order to optimize the transmission of information to the operator.
- Evaluation and updating:
    - Evaluation and adaptation of methods: The operator will evaluate the decisions taken by the MHD. This may result in adaptation of method parameters or the initial selection of methods.
- Updating the knowledge base: The relevance of the decision taken following each diagnosis is evaluated by the operator. The knowledge base is modified to better react in the future.
- Validation: The proposed hypotheses are applied to a real case to validate them.

The following sections look at the two key components (the MHD; and the MIHR) in more detail.

5 Hybrid diagnostic method

Figure 7 shows the MHD in its *n* state (i.e. the shape of the MHD after the nth modification triggered by the interaction with the human operator). The three main steps constituting the MHD module are:

- Evaluation of diagnostic methods: After the identification of the applicable methods for each fault, the latter undergo evaluation.

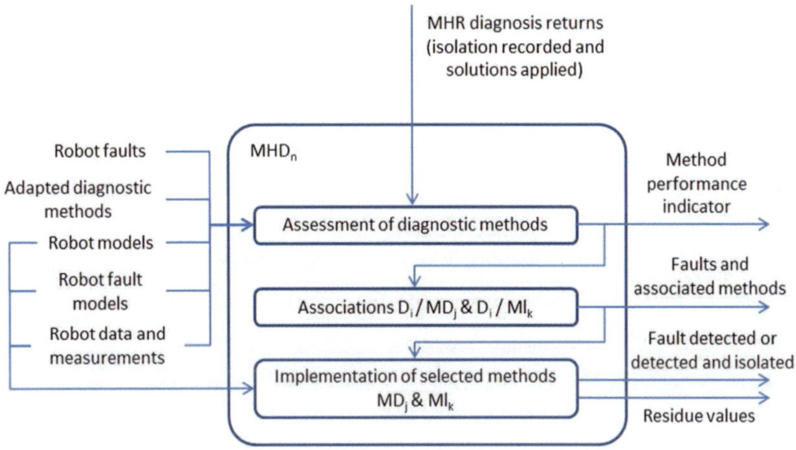

Figure 7 The elements of the MHD in its *n* state.

- Association of methods to faults: The best methods that can isolate or detect each fault are identified. Thanks to this evaluation, a first form of the MHD is established.
- Implementation of the selected methods: The selected methods detect and isolate faults by processing robot data and measurements as well as by using previously established models and knowledge.

The following discussion reviews ways of evaluating diagnostic methods.

Evaluating FDI methods requires a generic performance indicator. This indicator is made up of four sub-indicators:

- Detection performance;
- Isolation performance;
- Cost performance; and
- Forecasting performance.

As an example, a fault may be best detected using method *i* and isolated using method *j*. Method *i* is more accurate in terms of detection: it triggers an alarm faster and it has fewer false alarms than method *j*. However, method *j* is more accurate in terms of isolation: once the fault is detected, it identifies its source more quickly than method *i*.

Table 5 summarizes the structure of this generic performance indicator. The elements of Table 5 are classified into two categories: 'promoting rapid detection' underlined in this table, and 'promoting robustness against false alarms' in bold. These elements are chosen to capture the most important characteristics of different methods.

Table 5 Sub indicators of the performance indicator

Sub indicator	Element
Detection	**False Alarm Rate (FAR)** [56] Missed Detection Rate (MDR) [56] Detection Delay (DD) **Decision Stability (SD)**
Isolation	False Isolation Rate (FIR) Isolation Delay (ID) **Nominal Mode Return Delay (NMRD)**
Cost	calculation Time Per Cycle (TPC) Required Data Storage (RDS) Number of faults processed
Forecast	Diagnosis of unforeseen faults (NonPrev) Error Estimate (EE) Adaptability (Adap)

Implementing this indicator requires following these steps:

- Define the importance of each sub-element based on the user's preferences, depending on the nature of the application;
- Calculate the elements of the sub-indicators;
- Normalize and calculate each sub-indicator by performing a weighted sum.

These are discussed as follows.

Defining the importance of items: Depending on the application, a user may be interested in some elements of the detection and isolation sub-indicators more than others. For example, the proper functioning of the GPS is very important in a GPS tracking application. In this case, we need to quickly isolate any GPS faults.

Calculation of elements: Equations (2) are used to calculate False Alarm Rate (FAR), Missed Detection Rate (MDR) and False Isolation Rate (FIR):

$$FAR = \frac{False\,positives}{Flawless\,data}$$

$$MDR = \frac{False\,negatives}{Data\,with\,fault}$$

$$FIR = \frac{False\,isolation}{Data\,with\,fault} \quad (2)$$

The ability of the method to detect that the normal mode is restored after the appearance of a fault is an important characteristic. The element of the insulation sub-indicator 'time to restore normal mode' corresponds to the average time required by the FDI method to recognize normal mode after a fault. If n is the number of times normal mode is restored, Nominal Mode Return Delay (NMRD) is calculated as follows:

$$NMRD = \sum_{0}^{n} \frac{Recovery\,time}{n} \tag{3}$$

Decision Stability (SD) is detection sub-indicator which measures the oscillations made by the output of the method. The following equation describes the calculation method for this indicator.

$$SD = \frac{\sum_{0}^{n} distance\left(D_t - D_{t-1}\right)}{n} \tag{4}$$

With, D_t and D_{t-1} respectively the decision at time t and t - 1. The sub-elements DD and ID correspond to the average time per iteration taken to detect and isolate the fault.

Cost and forecast sub-indicators are easily determined. Time Per Cycle (TPC) is the average computation time consumed by the method at each iteration. Required Data Storage (RDS) is the required volume of data (in megabytes – MB). The 'number of faults treated' is the number of faults that the method can isolate at the same time. Error Estimate (EE) and 'Diagnosis of unforeseen faults' are Boolean values representing respectively the ability of the method to produce an estimate of the reliability of its decision and to detect faults which were not initially foreseen. Finally, the element 'adaptability' is an empirical estimate of the ease of adaptation of a method to other robots or other faults.

Normalization and calculation: The value of a sub-indicator corresponds to the weighted sum of its normalized elements. Normalization of the elements calculating durations (DD, ID and NMRD) is done using the function defined in the following equation. This function guarantees a normalized value between 0 and 1:

$$Normalized\ value = e^{-0.5\ delay} \tag{5}$$

For the elements representing rates (FAR, FID and MDR), the normalized value is simply 1 – x. This gives a high value for low rates and vice versa. The 'decision stability' element is not directly included in the calculation of the sub-indicator.

The final value of the sub-indicator is obtained by the weighted sum of the elements as shown in equation 6:

$$Sub\,indicator = \frac{\sum(element * weight)}{number\ of\ elements} \tag{6}$$

The weight corresponding to each element is determined when defining the importance of the elements. In the end, each sub-indicator will have a value between 0 and 1. The following equations summarize the process leading to the calculation of the indicator:

$$\varnothing_{method\ evaluation} = \frac{a * \varnothing_{Detection} + b * \varnothing_{Isolation} + c * \varnothing_{Cost} + d * \varnothing_{Forecast}}{a+b+c+d} \tag{7}$$

With $\varnothing_x$ the value of sub-indicator *x*; *a*, *b*, *c* and *d* are weights to be defined according to the task.

As an example, if in an application the detection of a fault is crucial, the isolation is a little less important than the detection, the cost is of medium importance and the prediction is not important, this results in (*a*, *b*, *c*, *d*) = (2, 1.5, 1, 0.5). The $\varnothing_x$ values are calculated as follows:

$$\varnothing_{Detection} = \frac{\alpha_0 * (1-FAR) + \beta_0 * (1-MDR) + \gamma_0 * e^{-0.5*DD} + \sigma * (1-SD)}{4}$$

$$\varnothing_{Isolation} = \frac{\alpha_1 * (1-FIR) + \beta_1 * e^{-0.5*ID} + \gamma_1 * e^{-0.5*NMRD}}{3}$$

$$\varnothing_{Cost} = \frac{\alpha_2 * e^{-40*TPC} + \beta_2 * e^{-0.01*RDS} + \gamma_2 * \frac{N_{faults\ considered} - N_{fault\ treated}}{N_{faults\ considered}}}{3}$$

$$\varnothing_{Forecast} = \frac{\alpha_3 * NonPrev + \beta_3 * EE + \gamma_3 * \frac{Adapt}{10}}{3} \tag{8}$$

The $\alpha, \beta, \gamma\ and\ \sigma$ weights are chosen when implementing the indicator according to the importance of the elements.

Elements representing a rate are normalized by the function (1 − *x*). Consequently, if the rates are zero (FAR = 0 for example), the associated normalized value will be maximized (1).

Values representing delays (e.g. DD) are normalized by the function $e^{-0.5*x}$: the value 0.5 is chosen empirically so that the normalized value is maximum (1). if the delay is zero and it decreases, e.g. for a delay of 1s, the associated value is 0.6. TPC (in seconds) and RDS (in MO) are normalized by similar functions but with a parameter close to difference, ensuring a logical normalization of each value. The normalized value of TPC is 0.45 for a TPC of 0.02s, it is 0.9 for a TPC of 0.003. The normalized value of RDS is 0.13 for an RDS of 200MO, it is 0.9 for an RDS of 10MO.

6 Human/robot interaction module

Following the decision taken by the MHD, the MIHR intervenes to ensure optimal interaction with a human operator as well as continuous improvement of the MHD. The MIHR is shown in Fig. 8. Using a case-based reasoning algorithm (CBR) integrating a Knowledge Base (BdC), MIHR offers solutions for isolated faults and helps the operator to isolate a fault if not isolated by the MHD.

A modular interface manages interaction with the human operator to provide her/him with the relevant information and to allow him to propose solutions and isolations. Figure 7 shows how the MIHR works. There are three cases of fault diagnosis:

- Fault isolated;
- Fault detected but not isolated; and
- Fault signaled by the operator but not detected by the system.

These cases each require a different approach. The following section reviews key steps in the process.

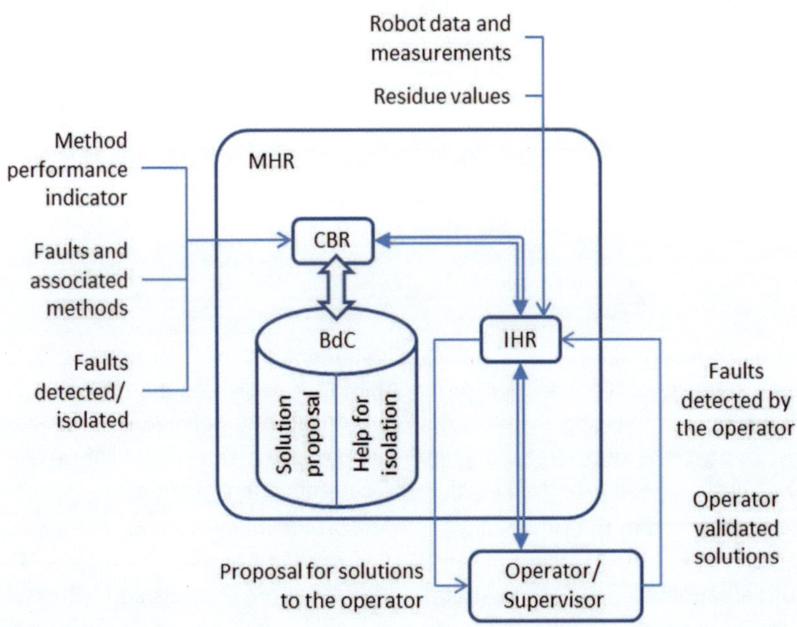

Figure 8 The MIHR.

6.1 Fault isolation and criticality assessment

A fault can be isolated directly by the MHD or with the help of the expertise of the operator. In both cases, following its isolation, a criticality study is carried out. The criticality of a fault depends on four elements [21]:

- Importance/criticality of component: Depending on the robot's mission, it may rely on some components (sensors) more than others. For example, in a GPS trajectory tracking application, the GPS is more important than the wheel encoders. Consequently, a fault on the GPS will be considered more critical than a fault on the encoders.
- Magnitude of the fault: The influence of the fault on the behavior of the robot is considered during the criticality calculation. For example, 20 cm deviation error on position measurement is more serious than 5 cm error.
- Frequency of the alarm: When the fault repeats, it will be considered more serious than a one-time fault.
- Level of confidence in the decision: Some diagnostic methods incorporate a level of confidence in a decision. If so, this element is taken into account during the criticality study. A lower level of confidence implies a lower level of criticality in the event of a fault.

The Dempster-Shafer (DS) theory [21] can be used to assess criticality. This theory allows information to be merged to produce a conclusion about the situation. The basic formula is defined as known as Dempster's combination rule which is a generalization of Bayes' theorem. This rule ignores conflicts through a normalization factor and emphasizes agreement between multiple sources:

$$m(A) = \frac{1}{1-k} \sum_{B_i \wedge C_j \wedge D_l \ldots = A} m_1(B_i).m_2(C_j).m_3(D_l)\ldots$$

$$k = \sum_{B \wedge C \wedge D \ldots = \varnothing} m_1(B_i).m_2(C_j).m_3(D_l)\ldots \quad (9)$$

In this equation, $m(A)$ is the mass of an element A of all the parts (e.g. A can refer to the state 'Very serious' or 'negligible'). It expresses the proportion of all the proofs asserting that *A* is the current state. By definition, the mass of the empty set is zero $m(\varnothing)=0$ and the sum of the masses of all subsets is 1, $\sum_{\Omega} m(A)=1$ i.e. the probability that the current state is none of the possible states is zero, and the probability that the current state is one of the possible cases is 1. *k* is the measure of the level of conflict between the masses. The normalization factor *1 – k* makes it possible to ignore conflicts and assign any mass involved in a conflict to 0. As a result, this gives counter-intuitive results when faced with significant conflicts, which is not the case here.

We define four levels of criticality:

- Low: The fault is negligible; it does not influence the robot's mission.
- Medium: The fault can disrupt the robot without necessarily causing the mission to fail.
- High: The fault may cause the robot's mission to fail.
- Fatal: The fault is dangerous: It can cause damage or make the robot uncontrollable.

For each type of fault, we choose the values of the masses corresponding to each level of criticality as detailed in an example in Table 6.

A blocking mechanical fault damages the equipment, indicating high criticality. If it exists, the main sensor is one on which the robot relies to complete a task. An important sensor may record important information but it is not essential to the mission. A secondary sensor has less effect on the task than an important sensor. Some sensors may not be used during a specific application. In the case of identification of an unforeseen fault, the masses of all the criticality states are equal.

Figure 9 shows the variation of the functions of belonging to the states of criticality according to the value of the corresponding indicator. When the value of the performance indicator increases from 0.8 to 1, the mass corresponding to the fatal state increases from 0 to 1 and that corresponding to the high state decreases from 1 to 0. The associated mathematical functions are stated in the equations:

$$m_{fatal} = \begin{cases} 5x - 4 \, if \; x > 0.8 \\ 0 \, otherwise \end{cases}$$

$$m_{high} = \begin{cases} -5x + 5 \, if \; x \notin [0.8,1] \\ 5x - 5 \, if \; x \notin [0.6,0.8] \\ 0 \, otherwise \end{cases}$$

Table 6 Example of masses associated with criticality states according to the type of fault

Type of fault	Masses associated with states of criticality (fatal high medium weak)
Blocking mechanical fault	(0.8 0.1 0.05 0.05)
Main sensor	(0.8 0.1 0.05 0.05)
Important sensor	(0.1 0.8 0.07 0.03)
Secondary sensor	(0.03 0.07 0.8 0.1)
Sensor not used	(0.01 0.01 0.18 0.8)
Unknown	(0.25 0.25 0.25 0.25)

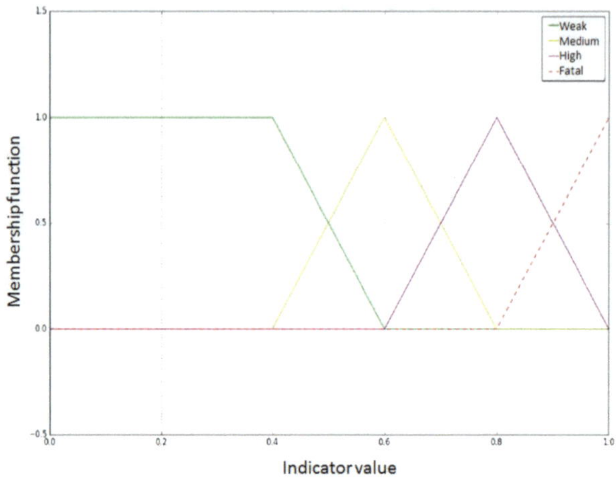

Figure 9 Membership functions of criticality states.

$$m_{medium} = \begin{cases} -5x+4 if\ x \notin [0.6,0.8] \\ 5x-2 if\ x \notin [0.4,0.6] \\ 0 otherwise \end{cases}$$

$$m_{weak} = \begin{cases} -5x+3 if\ x \notin [0.4,0.6] \\ 5x-2 if\ x \notin 0.4 \\ 0 otherwise \end{cases} \quad (10)$$

where x represents in these equations the value of the associated indicator, $x \notin [0,1]$.

6.2 Proposal for a solution

Following the isolation of a fault and assessment of its criticality, a solution is proposed. Depending on the level of criticality, this solution may be applied directly or after validation by the operator. In all cases, the operator assesses the relevance of the solution applied. If the solution is not adequate, he can propose another one or declare a non-isolated fault detected. The knowledge base is then updated to improve future decisions.

The Case-Based Reasoning (CBR) algorithm is effective for such problems [20]. This algorithm copies human behavior by remembering similar previous situations (cases) encountered and comparing them to the current situation with the aim of building a new solution which, after validation, will add to the database. There are four main steps:

- The search for similar cases: For this, we need rules to define the similarity between cases.
- Reuse: After identifying the most similar case, using this case as the solution for the current case (with possible adaptation).
- Revision: Check if the candidate solution corresponds to the current case. If the solution is validated, we proceed to its application; otherwise, either another case is chosen or the human operator proposes a solution.
- Updating the database: Also called the learning stage, this stage consists of adding experience feedback to the database. The case (the data), the proposed solution and the review evaluation are added to the database to enrich it.

CBR requires four types (containers) of knowledge [56]:

- Vocabulary: this includes the variables necessary to describe cases. These variables must satisfy two conditions: (1) be useful for retrieving similar cases, and (2) be sufficiently discriminating to avoid retrieving cases which are too different and which could lead to false solutions.
- Similarity measure: An adequate method for comparison must be chosen according to the number of variables constituting a case and the types of data recorded.
- Adaptation: This includes the knowledge necessary for the continuous updating of the CBR. This step can be entrusted to the operator.
- Cases: Representing the knowledge that the system acquires during use as well as the initial knowledge recorded in the database.

Key issues relating to these are discussed in more detail below.

Key variables used in case vocabulary are: The values of the performance indicators of all the methods contributing to this decision, i.e. the diagnostic sub-methods that can diagnose the current status; Residuals of sub-methods that can diagnose the current state, i.e. the residuals which detect the presence of the fault diagnosed; The values of the performance sub-indicators corresponding to these sub-methods. Additional information on the fault: its location (mechanical, sensor or external), its nature (loss, step, disturbance or drift) and its severity. The proposed solution is also saved with its evaluation which can be +1 if this solution has been deemed adequate or −1 otherwise.

The similarity measure must be both fast and accurate. Considering the dimension of data and their different types, it is better to avoid classical classification methods of type *K* nearest neighbors. An appropriate method involves giving each solution in the database a score based on: (1) the resemblance between the cases (in the database) containing this solution and the current case and (2) on the evaluation of these cases. The final score for a solution is calculated in the equation:

$$score(s) = \frac{\sum_{Database} Sim(X, Yi) * e_i}{N_s} \tag{11}$$

With e_i the evaluation of the case Yi present in the database and N_s the number of data (Yi) in the database having contributed to the calculation of the $score(s)$; $Sim(X,Y)$ is the measure of similarity between the two cases X (current) and Yi (present in the database), its formula is

$$Sim(X, Y) = \frac{\sum_i w_i sim(x_i, yi)}{\sum_i w_i} \tag{12}$$

With w_i the weight of an element of the data vector. To calculate it, we apply the formula $w_i = 1 - \frac{rank_i - 1}{\max(Rank)}$ where $rank_i$ is the rank assigned to an element i after ranking all the elements in descending order. Table 7 shows the ranks and weight associated with each element.

As for the local similarity measure sim(x; y), this is adapted to the type of elements being compared:

- The values of the performance indicator and the measure of criticality (Dempster) are numerical values. Their similarity is measured by the Euclidean distance between the two compared values.
- The distance between two decisions taken by the MHD or two decisions taken by a sub-method is worth 0 if the two decisions are identical, 0.5 if the two decisions belong to the same group of statuses (e.g. each method detects the fault but in the wrong place), 1 otherwise.
- For alphabetical data, the measurement of the distance is binary (0 if two identical data, 1 otherwise) except for the similarity between the nature of

Table 7 Ranks and weights of information in the solution prediction CBR

Element	Rank	Weight
Fault location	1	1
Nature of the fault	1	1
Criticality of the fault	1	1
MHD decision	1	1
Decision of sub-methods	2	2/3
Detection rate of sub-methods involved	2	2/3
Isolation rate of sub-methods involved	3	1/3

Table 8 Distance between the natures of the faults adopted for the similarity calculation

Nature	Drift	Step	Loss	Disturbances
Drift	0	0.5	1	0.3
Step	0.5	0	1	0.3
Loss	1	1	0	1
Disturbances	0.3	0.3	1	0

the faults which is drawn from Table 8. In this table, the distance between two faults of very different types is equal to 1. It is zero if the two types compared are identical. The drift and the disturbance are more similar to each other than the bias and the drift, for this, the distance attributed between the first two is 0.3, for the last two it is 0.5.

The more the current case resembles a positively evaluated case in the database, the more the solution adopted in this case will be favoured. The solution with the highest score will be offered.

The Adaptation step saving the current case as well as the decision taken by the operator concerning the proposed solution, i.e. acceptance or rejection of the proposed solution. If the operator refuses the proposed solution, the case will be saved with an evaluation of −1. The operator can then propose a solution which, if validated, will be recorded in the database in a new case evaluated positively. We describe, later in this chapter, how the operator can propose a solution.

6.3 Dealing with a fault detected but not isolated

If a fault is detected but not isolated, the decision support mode is activated. Relevant information is extracted from the MHD. Information is sent to the operator to help him understand the situation and isolate the fault. Another CBR is used to help in isolation. Its algorithm is identical to the CBR described earlier, though with different vocabulary differs. The vocabulary of this CBR is composed of the decision of the MHD, the decision of the different diagnostic methods and the values of the performance indicator of all the sub-methods.

The definition of the starting cases is more complicated in this CBR. It is possible to define them thanks to a pre-simulation under optimal conditions. This step aims to propose an isolation in the case where the sub-methods of the MHD do not manage to find the source of the fault or when the confidence rates of these are not acceptable. If the isolation of the fault is successful, the procedure is initiated to propose a solution. Otherwise, the operator decides whether to stop the robot or not.

6.3.1 Adjustment of sub-methods

After evaluation by the operator, the values of the associated performance indicators are adjusted. The evaluation history is saved on a sliding window of size n (in our case $n = 100$). Once every n iterations, the performance indicator values of the methods are adjusted based on this history.

Consider a method m dealing with a default d. Let $G = (g_1, g_2, \ldots, g_n)$ be the history of good decisions such that $g_k = 1$ gk = 1 if, at the iteration $t_{current}$ k, the decision taken was evaluated as a good decision (fault d well isolated), $g_k = 0$ otherwise; similarly, $F_a = (f_{a1}, f_{a2}, \ldots, f_{an})$ is the history of false alarms, $F_i = (i_{a1}, f_{i2}, \ldots, f_{in})$ the history of false isolations and $M_d = (m_{d1}, m_{d2}, \ldots, m_{dn})$ the history of failed detections. The effect of each phenomenon (good decision, false detection, false isolation and missed detection) is calculated as in the equation:

$$e(G) = +\sum_i e^{-i\lambda} g_i$$

$$e(F_a) = -\sum_i e^{-i\lambda} f_{a_i}$$

$$e(F_i) = -\sum_i e^{-i\lambda} g_i$$

$$e(M_d) = -\sum_i e^{-i\lambda} md_i \quad (13)$$

- λ is the forgetting factor. The value of the indicator τ of the method m associated with the default d is adjusted in this way:

$$\tau_k - \tau_{k-1} + \frac{e(G)}{2} + e(f_a) + \frac{e(f_i)}{2} + e(m_d) \quad (14)$$

If the performance of a method has become poor, an automatic adjustment is possible. For example, if a method based on thresholding (such as KF) suffers from a high rate of false alarms, an increase in the threshold could reduce them.

6.4 Fault identified by the operator

If the operator notices a fault which the MHD has not noticed, the same procedure as that described for a non-isolated detected fault is followed, i.e. the decision support mode is activated. The CBR first predicts a fault which is validated or rejected by the operator. The solution proposal help mode is activated based on the decision made in the isolation step. Finally, feedback is recorded in the databases.

6.5 Interaction with the human operator

The proposed solution interacts with the human operator in real time. Humans have three essential missions:

- Supervision: Supervise the robot, and this in all cases.
- Contribution to fault isolation: Isolate the fault in the event of a non-isolated detected fault and assess the insulation in the event of an isolated fault.
- Contribution to proposing a solution: evaluate the solution proposed by the MHD and propose a solution if necessary.

Figure 10 shows the transitions between the different modes from the operator's point of view.

The monitoring mode is initially active. In this mode, the operator has access to basic information, i.e. the robot's position, its speed and the command. However, the operator has the right to request more information: residuals capturing faults, measurement of sensors and value of performance indicators. At any time, the operator can report a fault. In this case, the isolation aid mode is activated. It either isolates the specific fault or predicts the type and nature of the fault. A solution is then proposed by the MHD automatically and the solution proposal help mode is activated. Depending on the level of criticality of the fault, the application of the solution requires operator validation.

7 Implementing the proposed fault detection and isolation system: key components

This section reviews how the proposed diagnostic approach might be implemented and tested in practice. It primarily uses simulation conditions to test the approach.

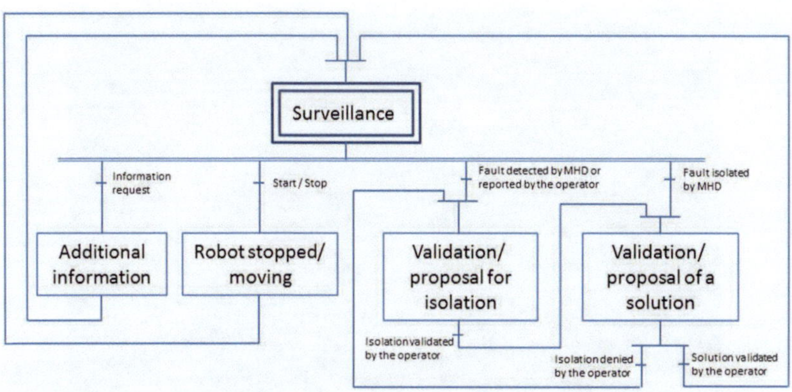

Figure 10 Modes and transitions from the operator's point of view.

Tests were carried out on two skid-steering type test robots: the 'Jaguar' (Fig. 15) (manufactured by Dr Robot) and the Effibot 3 (manufactured by Effidence). Cornering is achieved by giving different speeds to the wheels on each side of the robot. They have four 80 W motors: one for each wheel with the two motors on each side operated by the same controller. Robots can reach a maximum speed of 3 m/s. They can communicate with a remote-control station via an integrated Wireless module. Robots are equipped with a GPS, an IMU (inertial unit which provides angular and linear acceleration, the angular speed and orientation of the robot) and 4 odometers (which give the speed of the wheels). They can be equipped with an audio/video camera, a laser scanner that detects obstacles in its field of vision (Lidar type), a 3D laser scanner (Velodyne) and a 3D camera (camera type Kinect).

These sensors are today used for realizing farming operations such as path tracking, edge following [57] or furrow pursuing [58] for example.

The simulation uses the kinetic version of ROS (Robot Operating System) [59]. This is a set of open-source libraries, software and tools for the development of robotics applications. It includes a 3D simulator called 'Gazebo' which is widely used to simulate different robots in a virtual environment [60]. The simulator can test identification of faults such as wheel wear or locking, failure of a motor, or even faults linked to the environment such as the presence of obstacles or slippery zones. Figure 11 shows an example of the Gazebo environment. It shows a unicycle-type robot (skid-steering) on flat ground, surrounded by several objects. The area in red is a slip zone. The area in green contains two obstacles (holes).

ROS incorporates a specific communication architecture. It provides communication between different processes (nodes) through communication channels (topics). Nodes wishing to provide data notify the ROS 'master' to post

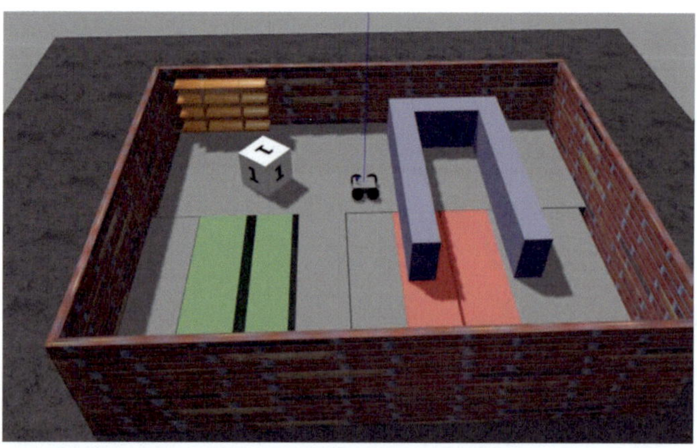

Figure 11 Example of an environment generated by Gazebo.

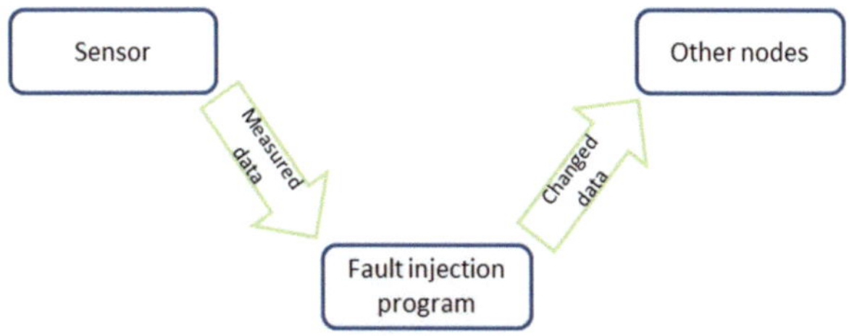

Figure 12 Injection of faults on sensor measurements.

to a topic. Nodes which wish to receive this data notify the master that they want to listen to this topic. This configuration makes it possible to intercept sensor data and modify it with third-party programs to inject sensor faults into the simulation. These programs listen for topics coming from the sensors, modify them to inject faults into them, and resend them again, as shown in Fig. 12.

We can inject the data with faults of different types:

- Disturbances (via random signals);
- Bias (by adding a constant value);
- Drifts (by adding an increasing value); and
- Blockages (by setting a value).

In the case of the GPS signal for example, a disturbance would be akin to noise in the signal, a bias to a shift between the real position of the robot and its estimated position, a drift to a divergence between its two positions and a blockage at an estimated position frozen while the robot moves. In all these cases, the robot's successful completion of the agricultural task is impossible.

7.1 First scenario

A first test scenario is defined to test the responses of the algorithms to various faults. The control signal is plotted in Fig. 13. Blue indicates linear speed control and green indicates angular speed control. The black rectangles separate the command combinations. In part 1, $w = 0$ and v toggle between successive positive and negative values, i.e. the robot moves forward and backward in a straight line. In part 2, $v = 0$ and w switches between successive positive and negative values, i.e. the robot turns right and left. In part 3, v and w switch between different values: here the robot performs circular trajectories. Finally, in part 4, the robot performs a sinusoidal trajectory.

This same succession of commands was replayed for each mode (normal, faults) in order to study the behavior of the diagnostic methods under different conditions. A fault is injected every 15 s and lasts 15 s before normal mode is restored. The total duration of the scenario is approximately 10 min.

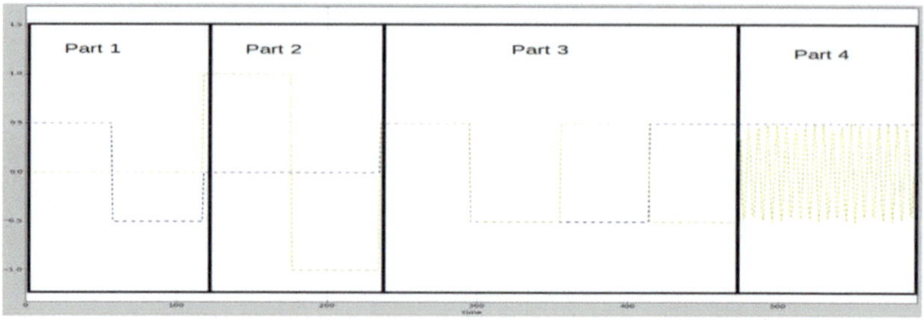

Figure 13 The control signal of scenario 1.

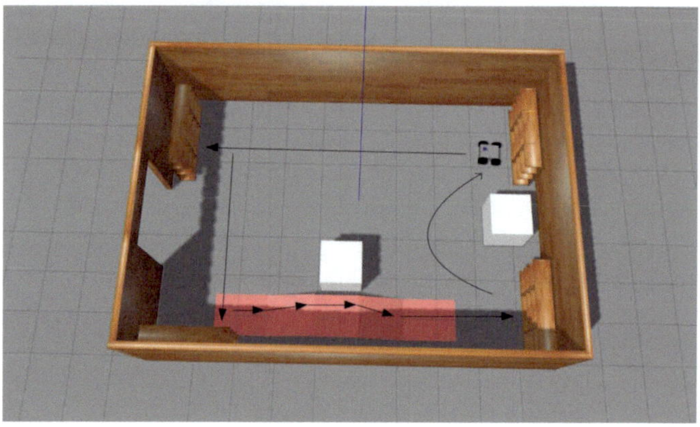

Figure 14 Scenario 2.

Figure 15 Jaguar robot.

7.2 Second scenario

Another scenario is shown in Fig. 14. In this scenario, the robot moves between four shelves. During the mission, the robot must climb a ramp (in red) and avoid an obstacle (white cube) by turning and moving forward. In an agricultural context, this scenario would be similar to moving a robot in a greenhouse for example. It completes this task at low speed and then at high speed. In each scenario, a fault is injected or removed every 30 s. The results are saved and then evaluated at the end of the scenario.

8 Modelling robot movement and potential faults

The Newtonian (unicycle) kinematic model is expressed in equation below (with *x*, *y* and θ the coordinates of the center of the robot in the 2D plane - position and orientation):

$$\dot{x} = \upsilon\cos(\theta)$$

$$\dot{y} = \upsilon\sin(\theta)$$

$$\dot{\theta} = \omega \quad (15)$$

Where υ and ω are the linear and angular velocity commands.

Considering that υ_L and υ_R are respectively the speed of the left wheels and that of the right wheels and that (υ , ω) is the signal of the control, the equations express the relation between the control and the speed of the wheels:

$$\upsilon = \frac{\upsilon_L + \upsilon_R}{2}$$

$$\omega = \frac{\upsilon_R - \upsilon_L}{L} \quad (16)$$

With *L* the length of the axle (or the distance between the right and left wheels). Starting from these equations and considering the idea that each parameter can be calculated or obtained via different methods, equations below are derived using the notations in Fig. 16.

$$\omega = \frac{\upsilon_r - \upsilon_l}{2} \qquad \dot{\theta} = \frac{\upsilon_r - \upsilon_l}{L} \qquad \upsilon = \frac{V_r + \upsilon_l}{2} \qquad \omega = \dot{\theta}$$

$$\upsilon_g = \frac{V_r + \upsilon_l}{2} \qquad \upsilon_{r1} = \upsilon + \frac{L\omega}{2} \qquad \upsilon_{r2} = \upsilon + \frac{L\omega}{2} \qquad \upsilon_{r1} = \upsilon_{r2}$$

$$\upsilon_{l1} = \upsilon - \frac{L\omega}{2} \qquad \upsilon_{l2} = \upsilon - \frac{L\omega}{2} \qquad \upsilon_{r1} = \upsilon_g + \frac{L\dot{\theta}}{2} \qquad \upsilon_{l1} = \upsilon_{l2}$$

$$\upsilon_{r2} = \upsilon_g + \frac{L\dot{\theta}}{2} \qquad \upsilon_{l1} = \upsilon_g - \frac{L\dot{\theta}}{2} \qquad \upsilon_{l2} = \upsilon_g - \frac{L\dot{\theta}}{2} \qquad \upsilon = \upsilon_g$$

(17)

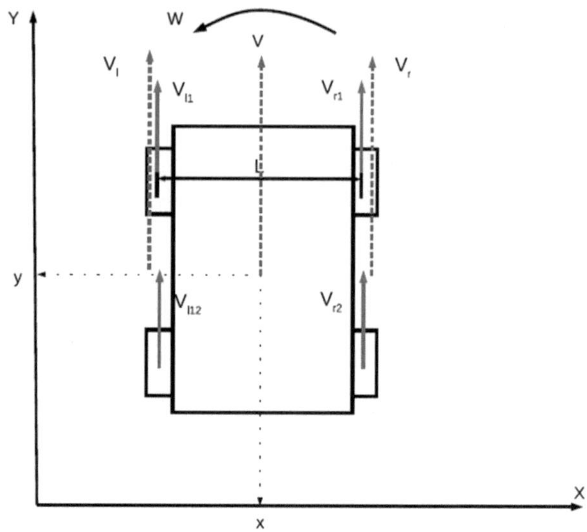

Figure 16 Parameter representations on the robot.

Where ω and θ are angular velocities in rad/s respectively given by the control signal and measured by the IMU. The numbers 1 and 2 in υ_{x1} and υ_{x2} refer to the front and rear wheels on each side of the robot. υ and υ_g are linear velocities in m/s respectively given by the control signal and measured via GPS thanks to equation:

$$\upsilon_g = \frac{\sqrt{(\delta x_g)^2 + (\delta y_g)^2}}{\delta t} \tag{18}$$

Where, δx_g and δy_g represent the distance traveled by the robot in the *XY* absolute 2D frame. The residuals of the redundancy relations are obtained by subtracting the two sides of the equalities of equations 18.

The thresholding of these residuals allows the detection and isolation of faults of the elements (sensor and control) contributing to the development of these equations. Table 9 shows a truth table for a FDI using parity relationship thresholding. The residuals are numbered from 'a' to 'p' as in equations (x signifies an overflow).

A fault on each of the elements of the first line induces the exceeding of the threshold of a combination of residuals. Identifying the combination of residues exceeding their thresholds will result in the default FDI. An additional phase of filtering by a low pass filter (Butterworth) makes it possible to reduce the undesirable oscillations in the residuals.

Table 9 FDI using analytic redundancy relationships

Residues	IMU	GPS	Command	Left front wheel	Left rear wheel	Right front wheel	Right rear wheel
A			X	X	X	X	X
B	X			X	X	X	X
C				X	X	X	X
D		X		X	X	X	X
E			X	X			
F			X		X		
G			X			X	
H			X			X	
I	X	X		X			
J	X	X			X		
K	X	X				X	
L	X	X					X
M	X		X				
N				X	X		
O						X	X
P		X	X				

The fault injected into the measurements of the location sensors (GPS, IMU, odometers) takes the form of an additive Gaussian noise with mean

- $\mu = 0$ and standard deviation $\sigma = 0.5$. Its probability density function is written in equation 19.

$$[H]P(z)=\frac{1}{\sigma\sqrt{2\pi}}e^{-\frac{(z-\mu)^2}{2\sigma^2}} \tag{19}$$

Regarding wheel lock, the kinematic model of the robot taking into account the effect of this type of fault is presented. An error on each wheel will result in different values of $\beta_v, \beta_\omega, \beta_{v2}$ and $\beta_{\omega 2}$.

Considering the dynamics of the robot in Fig. 17, $\overrightarrow{F_i}$ and $\overrightarrow{R_i}$ are respectively active and reactive produced forces (resulting from friction). By applying Newton's law, when one of the wheels is not locked:

$$\Sigma\left(\overrightarrow{F_i}+\overrightarrow{R_i}\right)=\vec{F}+\vec{R}=m.\overrightarrow{a_n} \tag{20}$$

Where $\overrightarrow{a_n}$ corresponds to the normal acceleration of the robot (without fault).

If wheel i is blocked, F_i will be set to zero and R_i will increase by r_0, to an unknown value. A modified equation projected onto $\vec{X}$ will lead to:

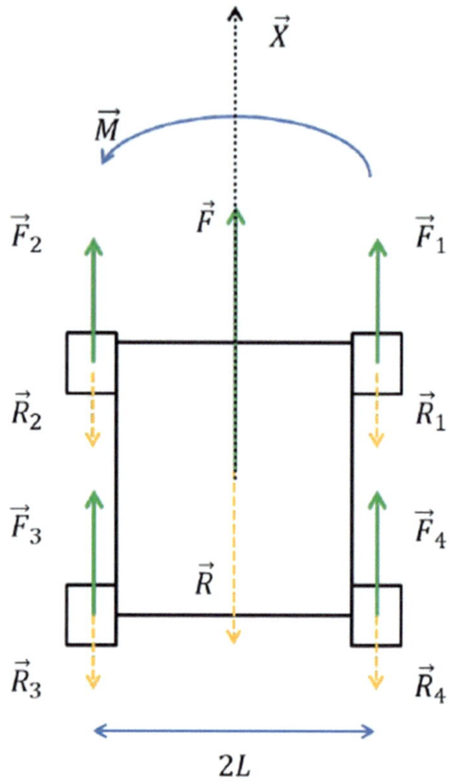

Figure 17 Forces and moments.

$$\alpha_1 F + \alpha_2 R = m.a_b \tag{21}$$

Where α_1 and α_2 are unknown parameters and a_b is the robot acceleration value for the same inputs, under wheel lock. By subtracting these equations, we get:

$$F(\alpha_1 - 1) + R(\alpha_1 - 2) = m.(a_b - a_n) \tag{22}$$

We obtain after integration:

$$\upsilon_b = \frac{\int \left(F(\alpha_1 - 1) + R(\alpha_1 - 2)\right) dt}{m} + \upsilon_n \tag{23}$$

Where υ_b and υ_n are the linear speed of the robot in normal mode and the corresponding value in locked wheel mode. Without loss of generality, assuming a constant control signal, the active and reactive forces become constants. *F* and *R* depend on the control signals υ and ω. Thus, we can write:

$$\upsilon_b = \beta_\upsilon \upsilon + \beta_\omega \omega + \upsilon_n \tag{24}$$

Similarly, from the equation,

$$2L\left(\overrightarrow{F_r} - \overrightarrow{F_l} + \overrightarrow{R_l} - \overrightarrow{R_r}\right) = \vec{M} = I\ddot{\vec{\theta}} \tag{25}$$

With:

$$\overrightarrow{F_r} = \overrightarrow{F_1} + \overrightarrow{F_4}; \overrightarrow{F_l} = \overrightarrow{F_2} + \overrightarrow{F_3}; \overrightarrow{R_r} = \overrightarrow{R_1} + \overrightarrow{R_4}; \overrightarrow{R_l} = \overrightarrow{R_2} + \overrightarrow{R_3} \tag{26}$$

where $\vec{M}$ the resulting torque, *I* the inertial matrix of the robot and $\ddot{\vec{\theta}}$ the angular acceleration. Following the same reasoning, we get:

$$\omega_b = \beta_{\upsilon 2} \upsilon + \beta_{\omega 2} \omega + \omega_n \tag{27}$$

Thus, by implementing the fault parameter equations in the kinematic model of the robot and by considering $\upsilon = \upsilon_n$ and $\omega = \omega_n$, the robot model corresponding to the wheel lock fault is given by:

$$x_{t+1} = x_t + \cos\theta_t(\upsilon + \beta_\upsilon \upsilon + \beta_\omega \omega + \omega_n)dt$$

$$y_{t+1} = y_t + \sin\theta_t(\upsilon + \beta_\upsilon \upsilon + \beta_\omega \omega + \omega_n)dt$$

$$\theta_{t+1} = \theta_t + (\omega(1 + \beta_{\omega 2}) + \beta_{\upsilon 2}\upsilon)dt \tag{28}$$

This model is adopted for wheel lock faults. The default parameters to be estimated are β_υ, $\beta_{\upsilon 2}$, β_ω and $\beta_{\omega 2}$.

This way of injecting a fault is valid in simulation and on the real robot. Simulation is favored to avoid overheating of the motors due to a locked wheel.

9 Testing fault diagnosis using Kalman filters

Both extended (EKF) and unscented (UKF) versions of the Kalman filter are applied. These filters implement the Newtonian model of the robot.

Figure 18 represents the Mahalanobis distance (residual, in green) generated by the EKF on the pose-provider PP2 (fusion of IMU, odometer and GPS data). The broken horizontal line (red) is the detection threshold (here at 0.2). The abscissa axis represents time. In this simulation, we introduce a fault (noise with a standard deviation of 0.1) over a period of 0.5 s every 1 s.

The vertical dotted lines define the beginning and the end of the fault injection (here four injections are represented). The detection threshold is fixed at 0.2. It is clearly noticed that this EKF detects the repetitive and intermittent disturbances introduced. The oscillation of the residue during fault detection is a consequence of the random nature of the fault.

Figure 19 illustrates GPS fault isolation using this method. The dots in green (connected by the green curve) represent the response of the method. The red dashed line represents the real, faulty or normal state. The two broken lines in green and blue represent the linear and angular command signals. This figure shows that the method accurately and reliably isolates GPS faults.

The same method was then tested on other faults (two filters, one on each PP, remaining on high level data). Slippage, locked wheel and coasting (passive motor) are detectable only by giving the robot a relatively fast turn command ($\upsilon > 1.5$ m/s; $\omega > 1$ rd/s). The slip detection delay can be more than 3 s depending on the order. Table 10 summarizes the results obtained by testing this method on these three faults.

Figures 20 show the FDI of GPS and IMU faults using analytical redundancy. In Fig. 20a, the method isolates GPS faults but takes too long to recover after the fault disappears. In the case of IMU disturbances, the algorithm fails to isolate faults as shown in Fig. 20b. We notice detections for which the nature of the fault is unknown (Unknown in the figure), which corresponds to an unknown combination of residual overflows. Theoretically, monitoring analytical redundancy (or parity) relationships allows FDI of IMU and GPS disturbance faults. But when applying this method, the results were not comparable to those of the other methods given in this section.

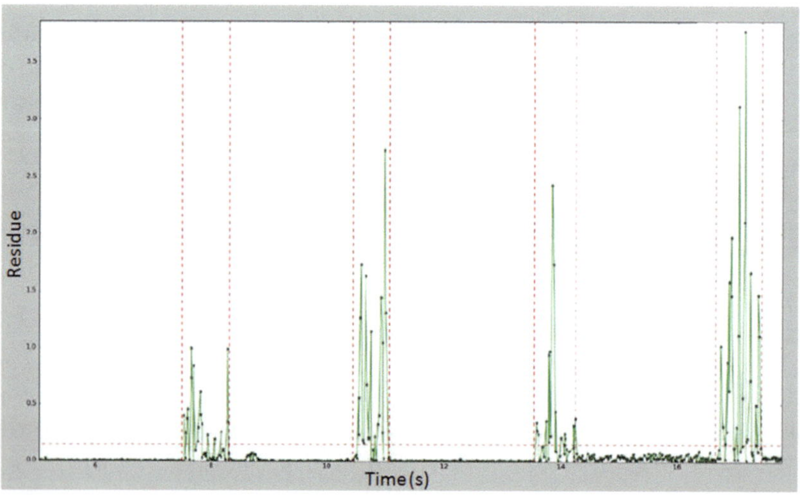

Figure 18 Detection of IMU disturbances by EKF on PP.

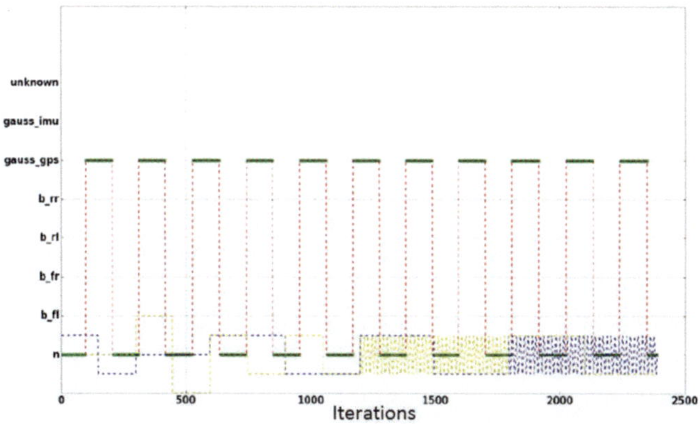

Unknown: unknown fault
Gauss IMU: IMU Disturbance (Gaussian noise)
Gauss_gps: GPS Disturbance (Gaussian noise)
B_rr : right rear wheel lock
B_rl : left rear wheel lock
B_fr: right front wheel lock
B_fl: left front wheel lock
N: Normal mode

Figure 19 Detection of GPS faults by the method of 2 EKF.

Table 10 Detection of slip, locked wheel and freewheel faults from EKF and UKF filters

	UKF on PP1	UKF on PP2	EKF on PP1	EKF on PP2
Slippage	No detection	Detection when turning, detection time >3 s	No detection	Detection when turning, detection time >3 s
Locked wheel	Only at high speed	Only at high speed	Only at high speed	Only at high speed
Freewheel	Only at high speed	Only at high speed	Only at high speed	Only at high speed

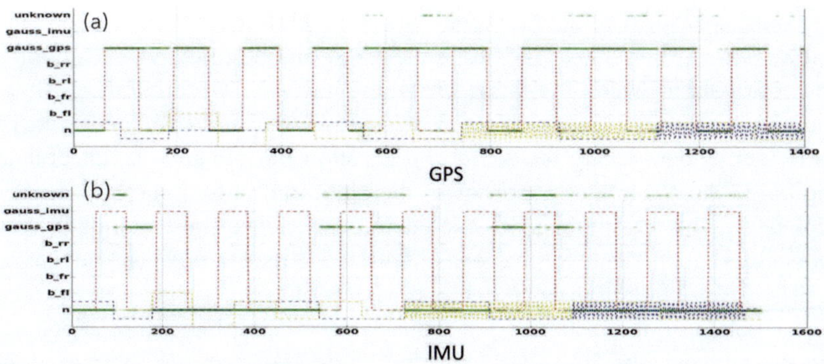

Figure 20 FDI using analytical redundancy equation thresholding.

10 Testing fault diagnosis using deep learning

Two models are applied to the residuals. These two models C and E-C reacted well to these two faults as shown in Fig. 21. The Butterworth filter used to filter the training data is more severe than the one used in real time for classification. The first has a cutoff frequency of 2 Hz and it is third degree while the second has a cutoff frequency of 3 Hz and it is second degree. This further promotes the rapid detection of faults by increasing the sensitivity of the algorithms to disturbances.

The processing by thresholding of the residuals of the analytical redundancy does not make it possible to have an acceptable performance in FDI. In fact, the simple thresholding of these residuals does not make it possible to derive all the useful information from them, contrary to deep learning models. A fault would not necessarily appear on the residues as a pure increase in values, it may appear as a succession of changes on one or more residues.

10.1 Detecting wheel locking faults

The wheel lock fault is processed by four algorithms. The results of these algorithms are shown in the following paragraphs. The values corresponding to each mode are shown in Table 11.

Due to the mechanical configuration of the robot, it is clear from Table 11 that locking the wheels on the same side (right or left) have a similar effect on the parameters $\beta_{\upsilon 2}$ and $\beta_{\omega 2}$, while locking the rear wheel is more powerful. This means that locking a rear wheel has a greater effect on robot dynamics than locking a front wheel.

Contrary to the methods based on the estimation of parameters, the application of this algorithm does not require knowledge of the parameters of the faults. As shown in Table 9, monitoring residuals from analytical redundancy allows FDI of wheel faults. Indeed, a blockage on one of the wheels modifies the residuals to which the value of its angular velocity contributes. After a filtering used to reduce the effect of noise, the thresholding made it possible to detect wheel locking faults, but the isolation performance is not optimal. Figure 22 illustrates the responses of the algorithm to four successive wheel lock faults. This algorithm isolates faults but experiences false alarms for front wheel lock faults. While for the rear wheels, the faults are detected but are not well isolated. In addition, we note the presence of false alarms.

In conclusion, thresholding of redundancy relationships is not effective in isolating these faults. These residues must be treated in a more complex way in order to extract the useful information.

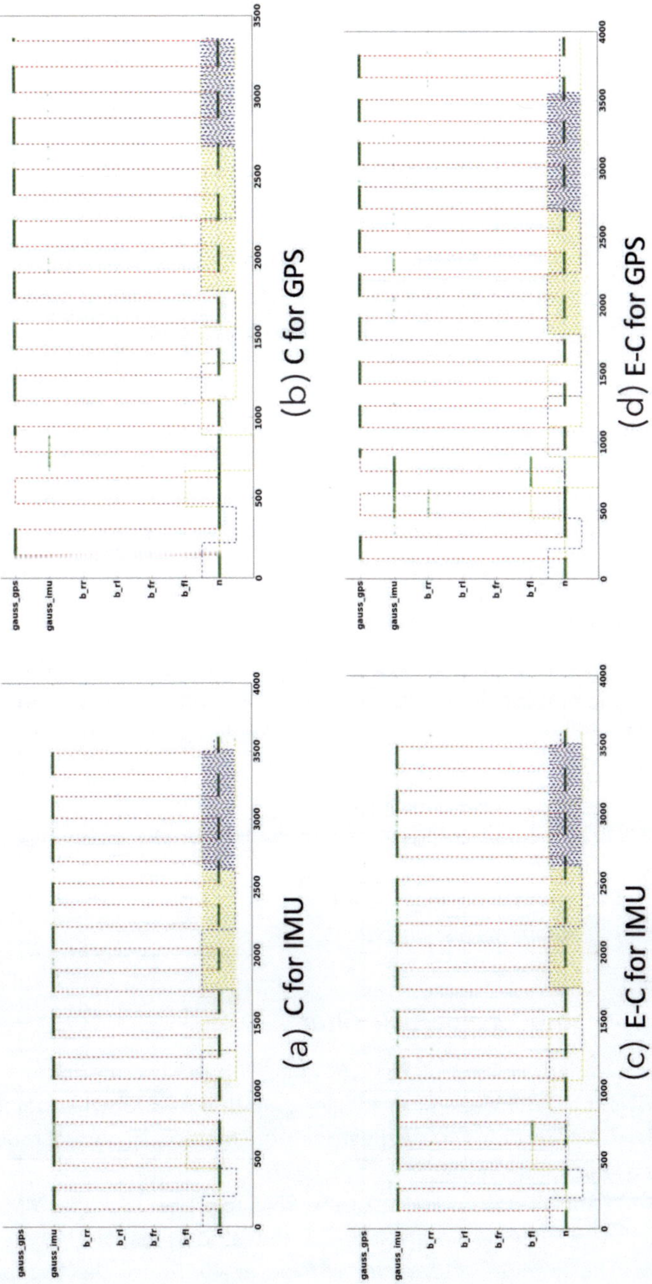

Figure 21 Results of the application of deep learning models on GPS and IMU faults.

Table 11 Identified parameters

Mode	β_υ	β_ω	$\beta_{\upsilon 2}$	$\beta_{\omega 2}$
N	0.03	−0.08	−0.013	−0.3
Right front wheel lock	0.05	−0.3	−0.55	−0.2
Left front wheel lock	0	−0.3	0.5	−0.2
Left rear wheel lock	0	−0.2	0.7	−0.6
Right rear wheel lock	−0.2	−0.3	−0.85	−0.7

Following the offline estimation of the parameters β , the same database is reused with an offline RLS to create another database containing the recursively estimated parameters for each mode. The latter is used to train the LDA algorithm. The results of the application of this method for the FDI of wheel lock faults are shown in Fig. 23. According to these figures, the detection and isolation performance of the rear wheels is better than that of the front wheels, i.e. that there appear to be fewer failed detections for faults on the rear wheels. Both cases do not present false alarms. In conclusion, this algorithm is robust to false alarms, but it takes time before detecting or isolating the fault.

10.2 Isolating wheel lock faults

A MMAE diagram is applied for wheel lock fault isolation. The configuration of the UKF is done empirically for using a genetic algorithm for *Q* (rather than the classic trial-and-error method in order to make its construction more reliable and automated) and according to the characteristics of the sensors for *R*. The size of the residual blocking window before filtering and processing them corresponds to 30 iterations. Figure 24 shows the response of this algorithm to wheel locking faults.

The rear wheel locking faults are better isolated than those of the front wheels. Some false alarms also appear.

10.2.1 Application of a particle filter

The result of the application of PFM [61] merging the two algorithms PF and MMAE shows that fault detection is faster than that of RPE while being more robust to false alarms than MMAE. Restoring the method to normal mode after the end of a fault seems fast.

Figure 25 shows the response of the PFM to wheel lockups. MMAE and RPE have the same parameterization as the algorithms of the preceding paragraphs. The number of particles is fixed at 40. As for the two previous

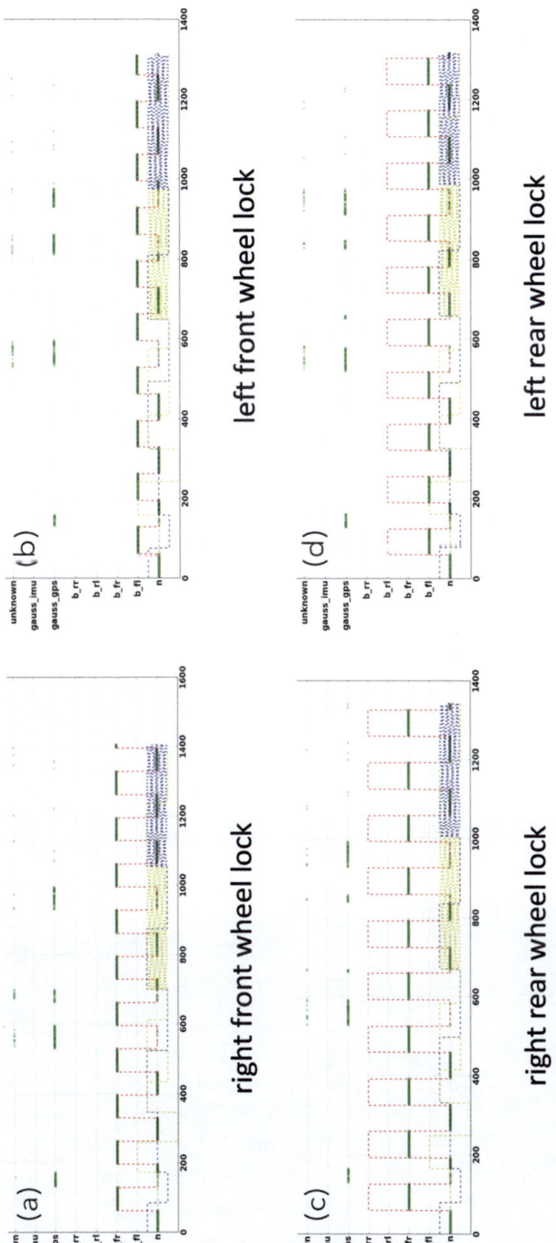

Figure 22 Results of applying analytical redundancy to wheel lock.

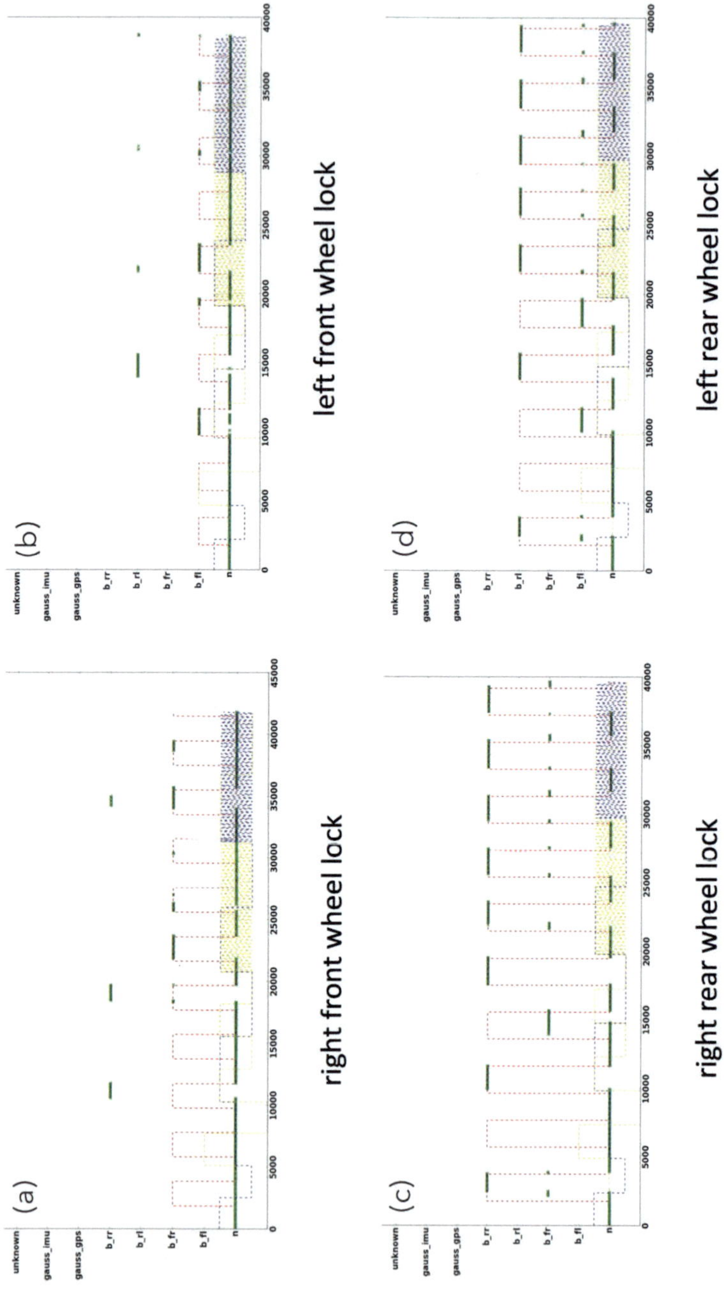

Figure 23 Results of applying RPE to wheel lock.

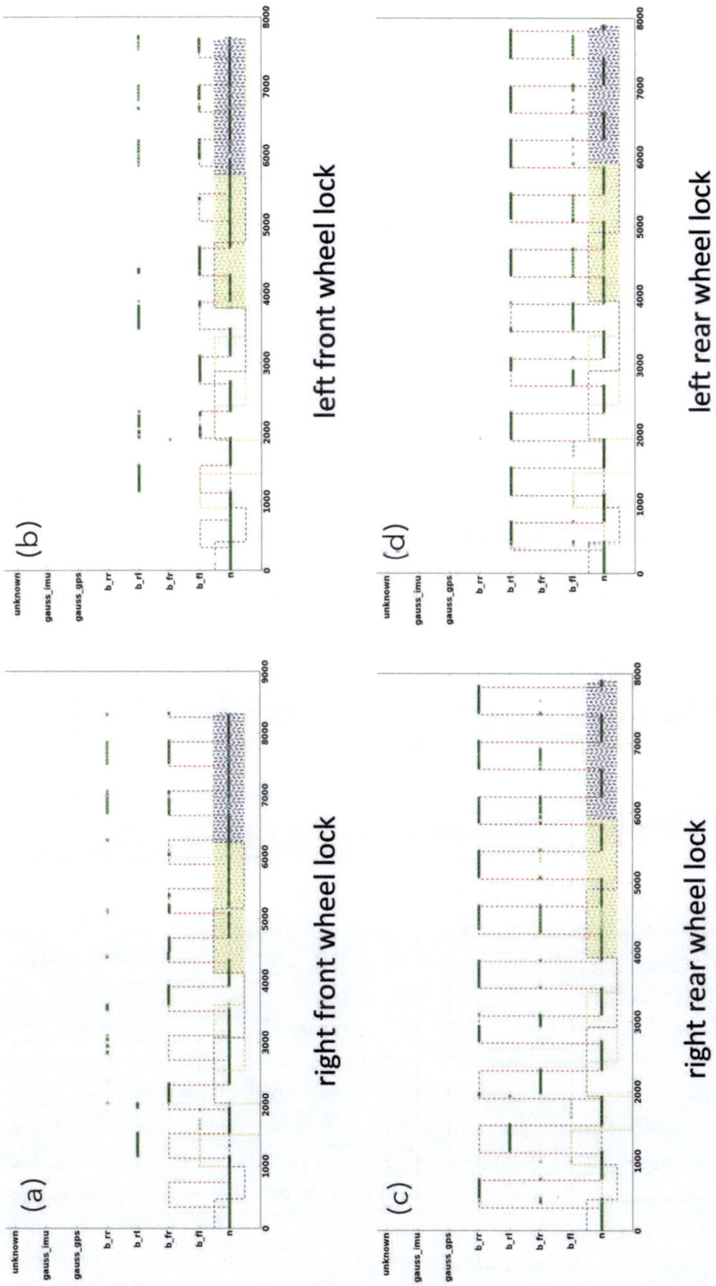

Figure 24 Results of the application of MMAE on wheel lock.

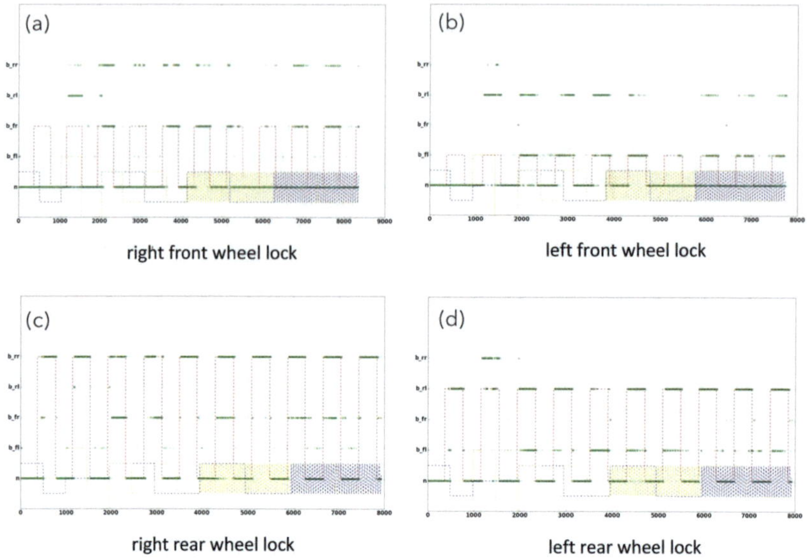

Figure 25 Results of the application of PFM on wheel lock.

algorithms, the isolation of faults on the rear wheels is more efficient than that on the front wheels.

10.2.2 The performance of deep learning in detecting wheel lock faults

After training, the two deep learning models are applied to wheel lock faults. Figure 26 shows the response of the two proposed models to wheel locking faults. The first four figures present the results of the Classifier (C). The others represent those of the encoder followed by a classifier (EC). These results show that the two algorithms were able to diagnose these faults. The classifier exhibits fewer false alarms compared to EC, while EC appears to detect and isolate faults faster.

To evaluate and then objectively compare these results, we are going to apply the performance indicator proposed. This evaluation will allow us to define an initial form of the distribution of methods in the MHD.

The performance indicator proposed is applied to all the diagnostic methods tested. The detection and isolation sub-indicators of a method depend on the fault considered, while the other two sub-indicators, namely those of 'cost' and 'forecast', do not depend on the fault. The evaluation of the cost and forecast sub-indicators for the diagnostic methods considered is described in Table 12.

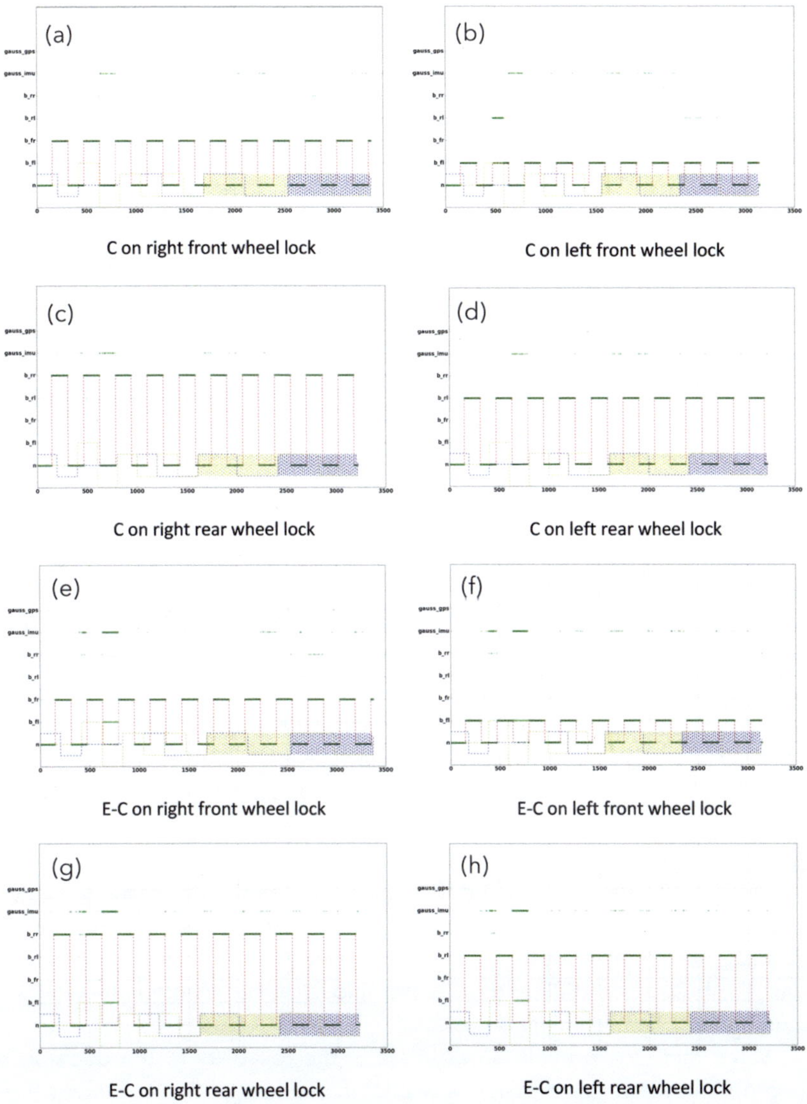

Figure 26 Results of the application of deep learning models on wheel lock.

The calculation time per cycle (TPC) presented in this table corresponds to the average execution time per cycle calculated at the end of the scenario. The calculation time of all the methods approached allows their execution with a frequency of 50 Hz. The PF is the greediest since it merges two other algorithms. Redundancy relationships are the least greedy, just ahead of Kalman filters.

Table 12 Values of sub-indicators independent of faults

	Cost			Forecast		
Method	TPC (s)	RDS (pts/mode)	Number of faults processed	NonPrev	EE	Adap/10
EKF	10^{-3}	0	3	1	1	9
Relations	10^{-4}	0	7	1	0	7
C	10^{-3}	20 000	7	0	1	1
EC	10^{-3}	20 000	7	0	1	1
MMAE	5.10^{-3}	100	4	1	1	3
RPE	10^{-4}	300	4	0	0	3
PF	10^{-2}	300	4	1	1	2

Artificial intelligence models have a computation time comparable to that of Kalman filters.

Regarding the amount of data to be stored, the method of two KFs and the relations of analytical redundancy do not require storing data. On the contrary, the other proposed methods require storing data for training. The methods that require the most are the two models of deep learning where 20 000 input lines per mode considered are necessary for the learning phase. Methods based on parameter estimation need about 100 points per considered mode in order to estimate the default parameters offline. Finally, RPE and PF need an additional 200 points per mode in order to train the LDA.

The number of faults (considered) treated for each method is indicated in this table (Table 12). Both AI models (C and EC) and redundancy relationships benefit from the largest number of isolable faults.

Methods based on Kalman filtering can detect disturbances caused by unforeseen faults without knowing their source. This is also the case for analytical and PFM redundancy which includes Kalman filters. On the other hand, the behavior of RPE and deep learning is unpredictable in the event of an unforeseen fault from the outset.

Some methods provide an estimate of the reliability of the decision or the probability that this decision is wrong. In the case of the methods whose decision is taken thanks to the thresholding of a signal, the distance between this signal and the threshold is an indication of the reliability of the decision.

The greater the distance is, the safer the decision is. Deep intelligence algorithms also provide an estimate on the probability of each mode. As for the PF, the number of particles belonging to each mode constructs an estimate of the probability of error.

11 Building the hybrid diagnostic method and human/robot interaction module

In the light of the results obtained, we define in this paragraph an initial form of the diagnostic methods constructing the MHD. The results of the indicator favouring detection are used.

First, for detecting and isolating GPS faults, the two Kalman filter method has the highest performance values.

In the case of the IMU, the method of the two filters is the best in detection but it is not so in isolation. These are the two models of deep learning that give the best results. KF is adopted for detection but once the fault is detected, deep learning will be used to isolate an IMU fault.

For wheel lock, deep learning model C has the best performance in detection and isolation. This method will be chosen to diagnose wheel lock faults. Figure 27 depicts the initial distribution of diagnostic methods to be implemented in MHD.

Diagnostic sub-methods are coded under ROS on separate nodes sending their decisions and residuals to the central MHD node. Since the two case-based reasoning (CBR) algorithms used for the isolation aid and the solution proposal have data in common, a single database (of SQL type) shared between these two algorithms is constructed. The tables forming this database are optimized to reduce the volume of stored data as shown in Fig. 28.

Thus, three tables are defined. The first (indic) contains the values of the different performance sub-indicators for all the sub-methods. It is linked, via an identifier, called id, to a second array (dec_res) containing the decisions

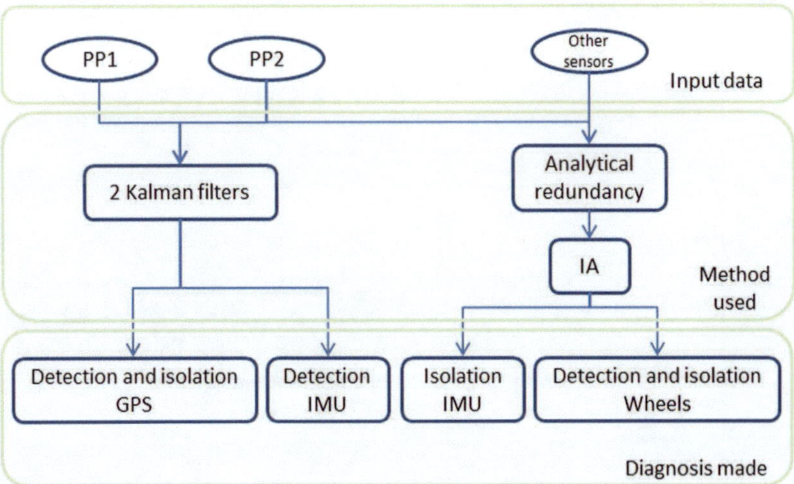

Figure 27 Initial distribution of diagnostic methods for MHD (MHD0).

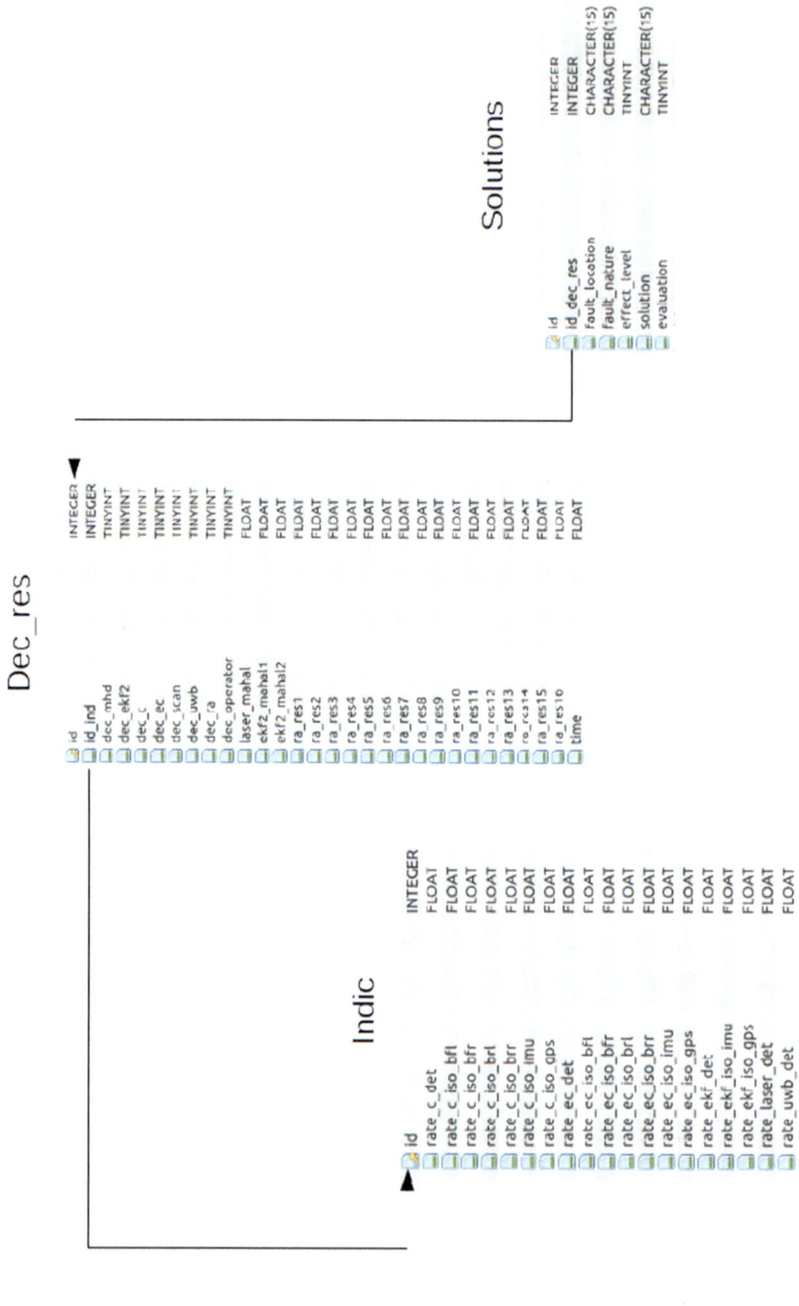

Figure 28 Database SQL tables.

and residuals generated by these methods. This table is linked to a third table (solutions) containing information relating to the proposed solutions.

Indicator values change less often than other variables. This architecture will allow us to save memory space by simply linking these values to the other variables. Therefore, we avoid inserting redundant data. The time is also saved in the dec_res array, it is used to remove the last n rows of data if needed.

In Fig. 27, 'rate_m_det' corresponds to the detection rate of sub-method m; 'rate_m_iso_d' is the isolation rate of the sub-method m by isolating the fault d; 'dec_m' is the decision of method m (or operator); 'f_mahal' is the Mahalanobis distance (residual) of the filter f; 'ra_resx' is the residue number *x* of the analytical redundancy equations. The last 'solutions' table contains information used to propose solutions. In addition to the two identifiers, this table is composed of: the location of the fault (sensor, mechanical or external), the nature of the fault (drift, bias, loss or disturbance), the effect or level of criticality of the fault, the corresponding solution and the evaluation of this solution.

To communicate with the operator, a graphical interface respecting the guidelines is developed.

Information related to diagnosis and supervision is first classified into three priority groups:

1 High priority: Information belonging to this group is essential information and must be transmitted as quickly as possible to the operator in order to prevent the situation from deteriorating. We have chosen to show this type of information permanently and to make it readable more quickly than other information by displaying intuitive pictograms. For example, information on the presence of a fault and information on its criticality are high priority information.
2 Medium priority: Information belonging to this priority group is useful information but is not critical. We have chosen to show this type of information in permanently without adding attractive effects (e.g. colours, blinking). For example, the position and the speed of the robot are medium priority information.
3 Low priority: Not learning information belonging to this priority group right away is not a problem. We have chosen to display this information only at the request of the operator. For example, residual values of diagnostic methods should not be displayed on the monitor screen all the time because they are not useful when the robot is running normally. However, the operator can ask to see some of these residues if he suspects the presence of a fault.

Thus, the highest priority information is displayed in such a way that it is perceived and understood more quickly than the others. The membership

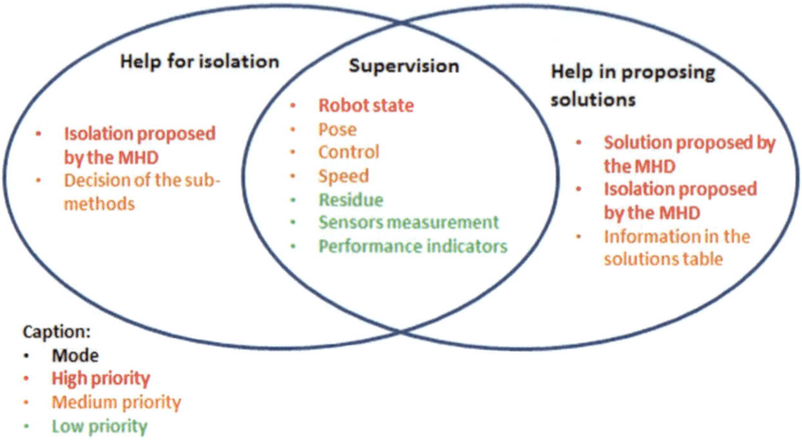

Figure 29 Membership of information to priority groups according to the current mode.

of information in priority groups varies depending on the situation. Figure 29 shows the priority of information according to the current mode of operation.

Three modes are considered:

- Monitoring: No fault is detected. The operator monitors the robot.
- Insulation aid: A fault is detected without being isolated or the operator has signaled a fault. The operator contributes to fault isolation.
- Help in proposing a solution: A fault is isolated. The operator contributes to the proposal of a solution.

Monitoring mode is continuously activated. The state of the robot, which corresponds to the decision taken by the MHD, is a high priority information there; the position of the robot in the map, the command and the speed of the robot are medium priority information. Residues of diagnostic methods, sensor measurements and performance indicators are low priority information.

As for the isolation aid mode, the decisions of the diagnostic sub-methods therein are medium priority information, and the isolation offered by the MHD is high priority information.

Finally, in the solution proposal help mode, the solution proposed by the MHD is high priority information. The information in the 'solutions' table in Fig. 28 is medium priority information.

11.1 Assessing the performance of the hybrid diagnostic method

In order to evaluate the proposed algorithm, the first form of the supervision system is implemented under the same conditions under which the

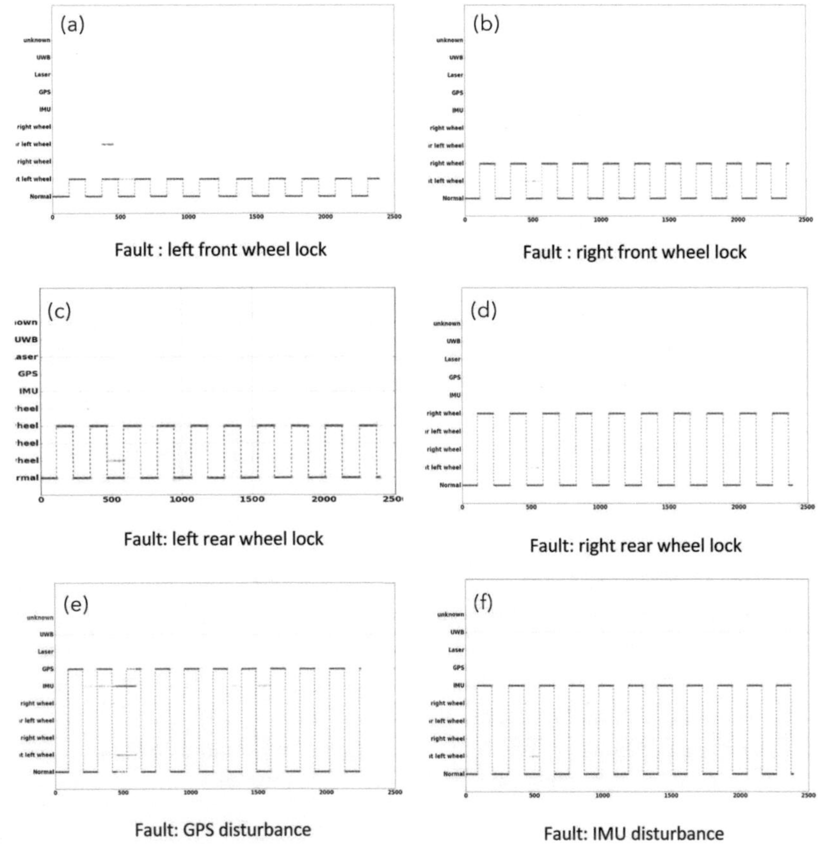

Figure 30 MHD response to faults.

sub-methods were tested. The results of the application of MHD in the cases of each of the mechanical and sensor faults (diagnosed by several methods) are shown in Fig. 30. This figure shows that, for each fault, the performance of the MHD is comparable to that of the best methods dealing with the fault. The results obtained for the faults of UWB and of the laser are not illustrated because they are exactly identical to the sub-methods which treat them.

Table 13 summarizes the performance of the calculated MHD for each fault. The laser and UWB faults being treated by a single diagnostic method, the performance of the MHD in these two cases is equal to that of the methods treating these faults (flexible EKF and confidence vector).

These results correspond to the initial form of the MHD described in Fig. 26. They do not take into account the improvements resulting from interaction with humans (adjustment of performance indicators and refusal of poor isolation by the operator).

To test CBR and operator interaction, the following steps are followed:

1. The database is initiated empirically with a row for each fault/solution.
2. The stop solution is chosen for wheel locking faults.
3. The solution to intervene is chosen for the faults of the location sensors (IMU and GPS).
4. Ignore is used as a solution for the normal case.
5. The left front wheel lock fault scenario is tested first. The supervision system automatically suggests the right stop solution.
6. The scenarios for the locking faults of the other wheels are then tested. For each fault, the right solution is proposed by the CBR from the first isolation. Ten lines of data are recorded for each fault.
7. The IMU fault scenario is tested. Following the first isolation, the CBR proposes the stop solution which should not normally correspond to this fault. The error is due to the instability of the CBR due to insufficient data. Indeed, the database contains too much data corresponding to wheel locks which, although they are not exactly similar to the GPS fault situation, are quite close to it. After five iterations, we refused the stop solution and proposed the intervene solution instead. The system then automatically proposes the right solution to intervene following the isolation of the GPS fault. The solution to intervene is proposed following isolation of the IMU fault. The solution proposal window is launched automatically as well as that of additional information by showing the residual which corresponds to the IMU-command comparison. This residue actually undergoes visible disturbances from the triggering of the fault.
8. The GPS fault scenario is played next. Since the GPS fault is closer to the IMU fault than to a locking of the wheels, the CBR correctly proposes the solution to intervene for this fault. Ten data are then saved in the database.
9. As shown in Fig. 30e, false detections and false isolations occur at each appearance of a false diagnosis (detection or isolation) we refused the diagnosis and proposed the correct one. After five corrections for each state (normal and GPS), the CBR directly proposes the good state following the refusal of the bad decision. The CBR proposes good isolation of the GPS fault contrary to the isolation proposed by the MHD and refused by the operator.

In short, in order to stabilize, the CBR needs $n_{min} = 10$ relevant data for each of the modes. If the number of data constituting the database is modest (<100 points per mode), a balance between the number of data per mode is required, i.e. the difference between the number of data representing a mode m_i and that representing m_j is less than the number of m_{min} data (mode having the minimum number of data in the database). The need for this balance becomes less important by increasing the size of the data belonging to each mode.

Generally, the more relevant data the database contains, the more robust the algorithm will be in the face of data imbalance and human error.

The performance of the isolation aid and solution proposal (CBR) algorithms is difficult to assess due to the evolution of the knowledge base. But, from 100 data per mode, the system will be robust and the proposals relevant.

As for the isolation support CBR, if the decision (mode) drawn by this algorithm is different from that of the MHD, the isolation support mode is triggered. The operator will then have the necessary information to decide on the mode.

To conclude, the supervision system and the integrated diagnostic methods satisfy a priori the specifications defined at the beginning of this chapter. They meet the following criteria:

- Online: The supervision system operates in parallel with the robot.
- In real time: The fault is isolated quickly after it is triggered.
- Compatibility with nonlinearity: The methods used are compatible with the nonlinearity of the robot model.
- Quick and accurate diagnosis: Table 13 shows the detection and isolation indicators corresponding to MHD. These values show that the MHD diagnoses the various faults in a powerful way, i.e. the faults are detected quickly and accurately.
- Inexpensive diagnostics: as shown in Table 12, the various diagnostic methods contributing to the construction of the MHD have a computation time per cycle (TPC) of the order of 10 3. Only deep learning models require a large amount of data for training.
- Adaptable diagnosis: Methods based on artificial intelligence are more difficult to adapt than other approaches because they require a new database for training. The adaptation of the other approaches is simpler.

Table 13 Evaluation of the MHD against faults

	Detection		Isolation	
Fault	Det fav	Rob fav	Det fav	Rob fav
Left front wheel	0.946	0.963	0.914	0.898
Right front wheel	0.948	0.967	0.927	0.904
Left rear wheel	0.947	0.965	0.927	0.905
Right rear wheel	0.948	0.969	0.926	0.904
GPS	0.905	0.904	0.861	0.839
IMU	0.958	0.974	0.889	0.854

Operator interaction (CBR, GUI) is easily adaptable (1) to the operator (experience level) and (2) to the task (other diagnostic applications).

- Reactive diagnosis: The diagnosis system always keeps the operator informed of the situation. The operator is warned following the isolation of a fault and, feedback informs him in the case of an action (application of a solution, insertion in the database).

12 Conclusion and future trends

Given the autonomy required of mobile robots used in agricultural and other applications, faults during their lifespan are inevitable. The complexity of a mobile robot and the variability of the environment mean many potential sources for faults and a high potential risk of faults occurring. FDI systems are therefore key to efficient robot operation in challenging conditions such as agricultural operations.

Diagnostic methods must operate in real time and must know how to manage the non-linearity of the system. Diagnosis must be rapid and precise. Knowing that the storage volume and the computing capacity implemented in robot controllers are generally limited, these methods should not be expensive in terms of computation time. Finally, the diagnostic system must be adaptable to different operating conditions and must interact effectively with human operators monitoring robot activity.

The FDI solution suggested in this chapter seeks to meet all these requirements. As noted, it is made up of two main components:

- A Robot Fault Diagnosis Supervision System (S2D2R):
  1. An MHD made up of several diagnostic sub-methods.
  2. An MIHR responsible for the link between the MHD and human supervision.
- A group of tools for modelling, parameterization and adaptation of fault diagnostic methods (OMOPAMED2).

The solution includes a fault diagnosis supervision system incorporating different individual diagnostic methods. Diagnostic methods fall into three categories: model-based (EKF, UKF, and analytical redundancy), knowledge-based (PFM, MMAE, and RPE), and data-based (deep learning). It also includes a set of tools for modelling, parameter setting and adaptation to optimize the ability of individual diagnostic methods to identify potential faults. Finally, the solution incorporates a human/robot interaction module. This integrates the operator into the diagnostic loop by proposing/validating solutions but also by correcting, if necessary, the decisions made by the MHD. The history of decisions taken by the operator is stored in a knowledge base which is integrated into a case-based reasoning algorithm in order to continuously improve performance.

This solution has been implemented and tested in a simulation environment focused on skid-steering faults (such as wheel locking). This includes applying individual diagnostic methods, evaluating them using performance indicators, building the initial MHD and creating a knowledge base. This process of implementation shows that the proposed system meets key requirements in selecting the right diagnostic method to rapidly and accurately identify a fault. This suggests the potential for the solution both to be applied to other types of fault as well as incorporate other diagnostic methods such as unsupervised learning algorithms and other types of robot design. Other areas of development include improving the interface with the human operator (e.g. to account for differing levels of operator expertise and experience), the development of the knowledge base and how effectively it improves system performance over time.

13 Abbreviations

Adap Adaptability
BdC Knowledge -base
C Classification
CBR Case-based reasoning
CNN Convolutional neural network
DBN Deep belief network
DD Detection delay
DL Deep learning
DS Dempster–Shafer
E-C (Auto)Encoder and classification
EE Error estimate
EKF Extended Kalman filter
FAR False alarm rate
FDI Fault detection and isolation
FIR False isolation rate
GPS Global positioning system
ID Isolation delay
IF Informational filter
IMU inertial measurement unit
KF Kalman filter
LDA Linear discriminant analysis
MD Detection method
MDR Missed detection rate
MHD Hybrid diagnostic method

MHD0 initial shape of the MHD
MI Isolation method
MIHR Human/robot interaction module
MMAE Multiple-model adaptative estimation
NMRD Nominal mode recovery delay
NonPrev Diagnosis of unforeseen faults
OMOPAMED2 Tools for modelling, parameterization and adaptation of diagnostic methods to faults
PF Particle filter
PP Pose provider
RDS Required data storage
ROS Robot operating system
RLS Recursive least square
RPE Recursive parameter estimation
S2D2R Robot fault diagnosis supervision system
SD Decision stability
UKF Unscented Kalman filter
UWB Ultra-wideband

14 Acknowledgements

This research work was funded by the 16-IDEX-0001 initiative of the French government IDE-ISITE (CAP 20–25) with the support of the "Auvergne-Rhône-Alpes" regional council and support of the European Union via the FEDER program.

15 Where to look for further information

For further information, you can visit the website https://www6.clermont.inrae.fr/tscf/.

16 References

1. Stavrou, D., Eliades, D. G., Panayiotou, C. G. and Polycarpou, M. M. Fault detection for service mobile robots using model-based method. *Autonomous Robots* 40(2): 383-394; 2016.
2. Dearden, R., Willeke, T., Hunter, F., Simmons, R., Verma, V. and Thrun, S. Real-time fault detection and situational awareness for rovers : report on the mars technology program task. In 2004 IEEE Aerospace Conference Proceedings (IEEE Cat. No.04TH8720). Volume 2. IEEE, USA. pp. 826–840; 2004.
3. Pratama, P. S., Setiawan, Y. D., Kim, D. H., Jung, Y. S., Kim, H. K., Kim, S. B., Jeong, S. K. and Il Jeong, J. I. Fault detection algorithm for automatic guided vehicle based on multiple positioning modules. In *Advances in Computing, Communications and Informatics (ICACC)*. IEEE Publications, India. pp. 751–757; 2014.

4. Fue, K. G., Porter, W. M., Barnes, E. M. and Rains, G. C. An extensive review of mobile agricultural robotics for field operations: focus on cotton harvesting. *AgriEngineering* 2(1): 150–174; 2020.
5. Luckcuck, M., Farrell, M., Dennis, L. A., Dixon, C. and Fisher, M. Formal specification and verification of autonomous robotic systems: A survey. *ACM Computing Surveys* 52(5): 1–41; 2020. Available at: arXiv :1807.00048.
6. Kushwaha, H., Sinha, J. P., Khura, T., Kushwaha, D., Ekka, U., Purushottam, M. and Singh, N. Status and scope of robotics in agriculture; 2016.
7. Tractica report for agricultural robots. 2019. Available at: www.tractica.com/newsroom/pressreleases/ agricultural-robot-revenue-to-reach-74-1-billion-worldwide-by-2024/.
8. Carlson, J. and Murphy, R. R. Reliability analysis of mobile robots. In *Proceedings - IEEE International Conference on Robotics and Automation*. Volume 1. IEEE, Taiwan. pp. 274–281; 2003.
9. Isermann, R. Model-based fault-detection and diagnosis-status and applications. *Annual Reviews in Control* 29(1): 71–85; 2005.
10. Agricultural machinery and tractors | Safety of highly automated agricultural machines - principles for design. ISO 18497; 2018.
11. Earth-moving machinery and mining | Autonomous and semi-autonomous machine system safety. ISO 17757:2017.
12. Hutter, F. and Dearden, R. The gaussian particle filter for diagnosis of non-linear systems. *IFAC Proceedings Volumes* 36(5): 909–914; 2003.
13. Fragkoulis, D. Détection et Localisation des Défauts Provenant des Capteurs et des Actionneurs: Application sur un Système Non Linéaire, page 177; 2008.
14. Almasri, M., Tricot, N. and Lenain, R. *Toward an Optimal Assignment of Diagnosis Method to Mobile Robots Faults*. Volume 5. IEEE, Thessaloniki, Greece; 2018.
15. Helbing, G. and Ritter, M. Deep Learning for fault detection in wind turbines. *Renewable and Sustainable Energy Reviews* 98: 189–198; 2018.
16. Liu, W., Wang, Z., Liu, X., Zeng, N., Liu, Y. and Alsaadi, F. E. A survey of deep neural network architectures and their applications. *Neurocomputing* 234: 11–26; 2017.
17. Liu, Y. and Bazzi, A. M. A review and comparison of fault detection and diagnosis methods for squirrel-cage induction motors: state of the art. *ISA Transactions* 70:400–409; 2017.
18. Berg, J., Lottermoser, A., Richter, C. and Reinhart, G. Human-robot- interaction for mobile industrial robot teams. *Procedia CIRP* 79: 614–619; 2019.
19. Figueroa Rivera, L. J., Ramos Ruiz, A. Y. and Chandrasekaran, B. A robotic control system for fault tolerance and safety using human robot interaction. In *2019 IEEE 9th Annual Computing and Communication Workshop and Conference (CCWC)*. IEEE, Las Vegas, NV, pp. 0435–0439; 2019.
20. Boral, S., Chaturvedi, S. K. and Narayana Achutha Naikan, V. N. A. A case-based reasoning system for fault detection and isolation: a case study on complex gearboxes. *Journal of Quality in Maintenance Engineering* 25(2): 213–235; 2019.
21. Sun, Z. H., Liu, R. and Ming, X. A fault diagnosis and maintenance decision system for production line based on human-machine multi-information fusion. In *Proceedings of the 2018 Artificial Intelligence and Cloud Computing Conference on ZZZ - AICCC '18*. ACM Press, Tokyo, Japan, pp. 151–156; 2018.
22. Zhang, Y. and Jiang, J. Bibliographical review on reconfigurable fault-tolerant control systems. *Annual Reviews in Control* 32(2): 229–252; 2008.

23. Strano, S. and Terzo, M. Review on model-based methods for on-board condition monitoring in railway vehicle dynamics. *Advances in Mechanical Engineering* 11(2): 168781401982679; 2019.
24. Zhong, M., Xue, T. and Ding, S. X. A survey on model-based fault diagnosis for linear discrete time-varying systems. *Neurocomputing* 306: 51–60; 2018.
25. Khalastchi, E. and Kalech, M. A sensor-based approach for fault detection and diagnosis for robotic systems. *Autonomous Robots* 42(6): 1231–1248; 2018.
26. Lu, Y., Collins, E. G. Jr. and Selekwa, M. F. Parity relation based fault detection, isolation and reconfiguration for autonomous ground vehicle localization sensors. Technical Report, DTIC Document; 2004.
27. Avram, R. C., Zhang, X. and Muse, J. Quadrotor sensor fault diagnosis with experimental results. *Journal of Intelligent and Robotic Systems* 86(1): 115–137; 2017.
28. Lenain, R., Deremetz, M., Braconnier, J.-B., Thuilot, B. and Rousseau, V. Robust sideslip angles observer for accurate off-road path tracking control. *Advanced Robotics* 31(9): 453–467; 2017.
29. Nabi, A. and IJETT. Application of least squares parameter estimation techniques in fault detection. *International Journal of Engineering Trends and Technology - IJETT*; 2013. https://ijettjournal.org/archive/ijett-v4i3p241.
30. Almasri, M., Tricot, N. and Lenain, R. *Multiple Model Adaptive Estimation for Blocked Wheel Fault Detection on Mobile Robots*. Volume 29. IFAC, Warsaw, Poland; 2018.
31. Ould Bouamama, B., Biswas, G., Loureiro, R. and Merzouki, R. Graphical methods for diagnosis of dynamic systems: review. *Annual Reviews in Control* 38(2):199–219; 2014.
32. Kalman, R. E. A new approach to linear filtering and prediction problems. *Journal of Basic Engineering* 82(1): 35–45; 1960.
33. Al Hage, J., Mafrica, S., El Badaoui El Najjar, M. and Ruffier, F. Informational framework for minimalistic visual odometry on outdoor robot. *IEEE Transactions on Instrumentation and Measurement* 68(8): 2988–2995; 2018.
34. Daass, B., Pomorski, D. and Haddadi, K. Using an adaptive entropy-based threshold for change detection methods - Application to fault-tolerant fusion in collaborative mobile robotics. In *2019 6th International Conference on Control, Decision and Information Technologies (CoDIT)*. IEEE, France. pp. 1173–1178; 2019.
35. Alkaya, A. Unscented Kalman filter performance for closed-loop nonlinear state estimation: a simulation case study. *Electrical Engineering* 96(4): 299–308; 2014.
36. D'Alfonso, L., Lucia, W., Muraca, P. and Pugliese, P. Mobile robot localization via EKF and UKF: a comparison based on real data. *Robotics and Autonomous Systems* 74: 122–127; 2015.
37. Mellah, S., Graton, G., El Adel, E. M., Ouladsine, M. and Planchais, A. On fault detection and isolation applied on unicycle mobile robot sensors and actuators. In 2018 7th International Conference on Systems and Control (ICSC). IEEE Publications, Valencia, pp. 148–153; 2018.
38. Fengyun, Z. and Xuesong, X. X. Fault diagnosis of mobile robot based on variable structure multiple model algorithm. In *The 27th Chinese Control and Decision Conference (2015 CCDC)*. Association for Computing Machinery, USA. pp. 2986-2992; 2015.

39. Kanev, S. K., Verhaegen, M. H. G. and Nijsse, G. A method for the design of fault tolerant systems in case of sensor and actuator faults, pp. 837–842; 2001. https://ieeexplore.ieee.org/document/7076015.
40. Zhao, Q. and Xu, Z. Design of a novel knowledge-based fault detection and isolation scheme. *IEEE Transactions on Systems, Man, and Cybernetics, Part B (Cybernetics)* 34(2): 1089–1095; 2004.
41. De Freitas, N., Dearden, R., Hutter, F., Morales-Menendez, R., Mutch, J. and Poole, D. Diagnosis by a waiter and a mars explorer. *Proceedings of the IEEE* 92(3): 455–468; 2004. DOI: 10.1109/JPROC.2003.823157.
42. Plagemann, C., Stachniss, C. and Burgard, W. Efficient failure detection for mobile robots using mixed-abstraction particle filters. In *European Robotics Symposium 2006*. Springer, Berlin. 22, pp. 93–107; 2006.
43. Mien Van, S. S. G. and Ceglarek, D. Fault estimation and accommodation for virtual sensor bias fault in image-based visual servoing using particle filter. *IEEE Transactions on Industrial Informatics* 14(4): 1312–1322; 2017.
44. Duan, Z., Cai, Z. and Min, H. Robust dead reckoning system for mobile robots based on particle filter and raw range scan. *Sensors* 14(9): 16532–16562; 2014.
45. Verma, V., Thrun, S. and Simmons, R. Variable resolution particle filter. In *Proceedings of the International Joint Conference on Artificial intelligence (IJCAI)*, pp. 976-984; 2003.
46. Verma, V., Gordon, G., Simmons, R. and Thrun, S. Real-time fault diagnosis [robot fault diagnosis]. *IEEE Robotics and Automation Magazine* 11(2): 56-66; 2004.
47. Sun, Y.-S., Ran, X. R., Li, Y.-M., Zhang, G. C. and Zhang, Y.-H. Thruster fault diagnosis method based on Gaussian particle filter for autonomous underwater vehicles. *International Journal of Naval Architecture and Ocean Engineering* 8(3): 243-251; 2016.
48. Nor, N. Md., Hassan, C. R. C. and Azlan Hussain, M. A review of data-driven fault detection and diagnosis methods: applications in chemical process systems. *Reviews in Chemical Engineering* 36(4); 2019.
49. Habibi, H., Howard, I. M. and Simani, S. Reliability improvement of wind turbine power generation using model-based fault detection and fault tolerant control: a review. *Renewable Energy* 135: 877–896; 2019.
50. Newhart, K. B., Holloway, R. W., Hering, A. S. and Cath, T. Y. Datadriven performance analyses of wastewater treatment plants: a review. *Water Research* 157: 498–513; 2019.
51. Alauddin, Md., Khan, F., Imtiaz, S. and Ahmed, S. A bibliometric review and analysis of data-driven fault detection and diagnosis methods for process systems. *Industrial and Engineering Chemistry Research* 57(32): 10719–10735; 2018.
52. Gao, Z., Ma, L. and Wang, J. Fault tolerant control method for displacement sensor fault of wheel-legged robot based on deep learning. In *2018 WRC Symposium on Advanced Robotics and Automation (WRC SARA)*. IEEE, China. pp. 147–152; 2018.
53. Diryag, A., Mitic, M. and Miljkovic, Z. Neural networks for prediction of robot failures. *Proceedings of the Institution of Mechanical Engineers Part C-Journal of Mechanical Engineering Science*, 228(8): 1444–1458; 2014. WOS:000336919200014.
54. Jin, X., Fan, J. and Chow, T. W. S. Fault detection for rolling-element bearings using multivariate statistical process control methods. *IEEE Transactions on Instrumentation and Measurement* 68(9): 3128–3136; 2018.

55. Akai, N., Morales, L. Y., Murase, H. and Murase, H. Simultaneous pose and reliability estimation using convolutional neural network and Rao-Blackwellized particle filter. *Advanced Robotics* 32(17): 930–944; 2018.
56. Richter, M. M. On the notion of similarity in case-based reasoning. In *Proceedings of the ISSEK94 Workshop on Mathematical and Statistical Methods in Artificial Intelligence*. Springer, 363, pp. 171–183; 1995.
57. Iberraken, D., Gaurier, F., Roux, J.-C., Chaballier, C. and Lenain, R. Autonomous vineyard tracking using a four-wheel-steering mobile robot and a 2D LiDAR. *AgriEngineering* 4(4): 826–846; 2022. https://doi.org/10.3390/agriengineering 4040053.
58. Pierre, C., Lenain, R., Laneurit, J. and Rousseau, V. A multi-control strategy to achieve autonomous field operation. *AgriEngineering* 4(3): 770–788; 2022. https://doi.org /10.3390/agriengineering4030050.
59. Filliat, D. and Mobile, R. Engineering school. Robotique Mobile, ENSTA ParisTech, page 9; 2011.
60. Farinelli, A., Boscolo, N., Zanotto, E. and Pagello, E. Advanced approaches for multirobot coordination in logistic scenarios. *Robotics and Autonomous Systems* 90: 34–44; 2017.
61. Almasri, M., Tricot, N. and Lenain, R. Parameter estimation based-FDI method enhancement with mixed particle filter. *Neurocomputing* 403: 441–451; 2020.

Part 3

Technologies: actuation

Chapter 10

Advances in mobility platforms for agricultural robots

Renato Vidoni and Giovanni Carabin, Free University of Bozen-Bolzano, Italy; Giuseppe Quaglia and Andrea Botta, Politecnico di Torino, Italy; and Giulio Reina and Rocco Galati, Politecnico di Bari, Italy

1 Introduction
2 Agricultural unmanned ground vehicles: mobility classification
3 Wheeled mobile robots
4 Tracked mobile robots
5 Legged mobile robots
6 Hybrid and other unconventional unmanned ground vehicles mobility platforms for agriculture applications
7 Conclusion and future trends
8 Where to look for further information
9 References

1 Introduction

The use of robots in agriculture is important for precision agriculture (PA) (Cariou et al., 2009; Blackmore, 2016), since it allows repetitive tasks to be accomplished with high and consistent precision throughout the working day. Agricultural robots are of increasing interest in robotic research and, to date, a variety of commercial robots or robotized platforms are available (Bergerman et al., 2016). Looking at the agricultural robotics literature, two main key developments can be identified:

- Unmanned aerial vehicles (UAVs), i.e. flying drones, with sensors used at a distance from the plants for remote sensing and certain basic crop management activities; and
- Ground rovers with sensors used in close proximity to the plants for all-around local monitoring and multi-task specialization.

Because unmanned ground vehicles (UGVs) are deployed locally and operate directly where monitoring and action are needed, they can fill the monitoring

http://dx.doi.org/10.19103/AS.2023.0124.08

resolution gap typical of remote sensing by UAVs. Furthermore, UGVs are not restricted to remote sensing. They can undertake proximal sensing and even perform tasks such as crop fertilization, pesticide application, weeding and harvesting. Since conventional tractors and other agricultural machinery (autonomous or not) are typically large and heavy, they tend to compact and degrade soil, as well as making fields more difficult to traverse (Tazzari et al., 2020). Soil compaction is a critical threat to soil quality, especially in the layers beneath the usual primary tillage depth, i.e. in the subsoil.

As a result, agricultural unmanned ground vehicles (AUGVs) are being designed with decreased bulk, weight and energy consumption to avoid unwanted soil compaction while still being able to achieve a range of PA tasks. Unlike UAVs, where the focus is mostly on sensor applications, the design of AUGVs is focussed more on the wider range of tasks the platform is required to undertake including sensing, monitoring, harvesting, sampling, planting, spraying, etc. (Cuesta et al., 2004; Low and Wang, 2008).

If AUGVs are classified by their size (see Fig. 1), some trends in functionality, architecture, and features become evident. A substantial percentage of AUGVs for precision agriculture are small, electrically powered robots that only perform crop or soil monitoring tasks. Due to their limited size and power, they rarely perform tasks beyond sensing. They require less financial investment, are more agile, have no specific energy supply issues, and can be adapted to a wide range of crops.

In contrast, there are robots of a size and power comparable to conventional tractors and other agricultural machinery. Here the main design trend, in particular, among major agricultural machinery manufacturers, is to add autonomous features to conventional agricultural vehicle architectures (e.g. tractors and combine harvesters) without redesigning the machinery itself. Automation of agricultural tractors and other vehicles has been done mainly by exploiting modern sensor-related technologies, i.e. machine vision, real-time kinematic GPS, and geometric direction sensors among others. These solutions generally consist in the installation of systems like a controller (computer), a steering control device, a localization and a safety system on modern commercial tractors where ISOBUS control technology is available (ISO, 2017). Newer vehicle designs are targeting electrically powered solutions, but many current designs involve hybridization, for instance combining diesel and electrical power, to provide adequate power to the system. Although farmers are more familiar with retaining conventional vehicle designs, which allow the use of a wide range of currently available pulled or carried implements, these systems still cause compaction, tend to have high costs, and tend to be effective only in very large open fields.

Between these two extremes, medium-sized robots potentially represent the best of the two worlds and include the most innovative designs. This

Figure 1 Examples of different AUGVs grouped by size. From top to bottom: small, medium, and large size systems.

category of robots can generally both monitor field conditions through remote and proximal sensing and carry out operations on the field. Some robot designs, e.g. in vineyards and orchards, are highly specialized and limited to very specific tasks. Their design is entirely determined by the requirements of the specific task. In contrast, other robots are multi-purpose platforms designed to perform several operations and require more versatile and modular designs.

This chapter discusses the topic of the mobility of agricultural unmanned ground vehicles. It starts with an introduction to the main concepts needed for a proper classification of mobile robots. It then presents the main characteristics and features of the different locomotion systems and includes a discussion of case studies on AUGVs with advanced mobility capabilities. Finally, future trends are discussed.

2 Agricultural unmanned ground vehicles: mobility classification

Mobility is one of the main aspects of the mobile robot design, as the platform manoeuvrability, controllability, efficiency, and stability in different terrain conditions depend directly on the chosen locomotion system (Rubio et al., 2019). Since potential operating conditions are very diverse, a comprehensive comparison of ground mobile robot mobility systems with an emphasis on agricultural applications is challenging. A common way to classify the mobility of AUGVs or any terrestrial mobile robotic platform for that matter is based on their locomotion principle and (kinematic) architecture.

The locomotion principle refers to the way the platform interacts with the terrain both to discharge the traction power and to control the motion direction. From the motion point of view, holonomic or non-holonomic motions can be achieved. In the first case, the number of controllable degrees of freedom (DoFs) is equal to the total workspace DoFs (i.e. in ground vehicles this DoFs number corresponds to 3). An example of holonomic locomotion is represented by legged robots that can move and rotate freely in any direction. Whereas, in non-holonomic motions, the controllable DoFs are less (e.g. a tracked robot can move forward/backward and rotate, but cannot move sideways). A commonly adopted categorization for locomotion of ground mobile platforms in the literature has four main categories:

- Wheeled;
- Tracked;
- Legged; and
- Hybrid.

This last category accounts for, e.g. legged-wheeled or tracked-legged systems (Bruzzone and Quaglia, 2012). Figure 2 shows different AUGVs from the Field

Figure 2 Mobile robotic platforms available at the Field Robotics Lab of the Free University of Bolzano, Italy (UniBZ, 2022). From left to right: ROVO tracked robot from Mattro (HAWE Mattro, 2022); Husky wheeled robot from Clearpath (Clearpath Robotics, 2022); customized UGV prototype realized from Losi Desert Buggy XL-E; A1 legged robot from Unitree (Unitree, 2022).

Robotics Laboratory of the Free University of Bolzano, Italy. Looking again at the AUGVs shown in Fig. 1 and those discussed in recent reviews (Oliveira et al., 2021), agricultural UGVs fall mainly in the first two categories of systems with wheels or tracks, yet some recent research programs target legged or hybrid systems.

The mechanical architecture of AUGVs concerns both the way the locomotion elements are arranged and the chassis configuration required to ensure the stability and functionality of the platform, i.e. the type of chassis and suspension. The interaction of the chassis with the terrain is mediated by the set of mechanical and non-mechanical elements that constitute the suspension system. A classification in this case is much more complex than with the method of locomotion since many designs have been developed, depending on the purpose for which the vehicle has been designed.

To present and discuss different mobility solutions, we will use the concepts shown in Fig. 3 where solutions for locomotion and architecture are summarized together with the main options for powering AUGVs. Each design option is then discussed in more detail.

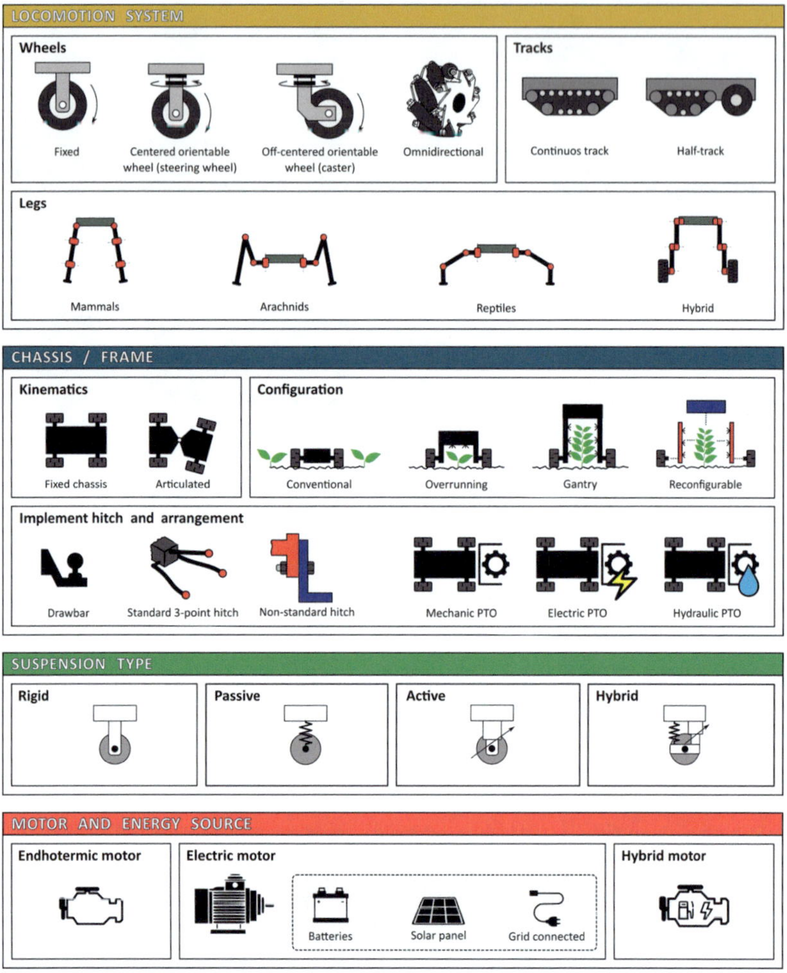

Figure 3 Solutions adopted in AUGVs regarding the locomotion system, chassis architecture - in terms of kinematics, configuration, and possibility to integrate external implements - suspension type, traction motor technology and an energy source.

2.1 Locomotion

As noted previously, the four main locomotion categories are wheeled, tracked, legged and hybrid UGVs. Different locomotion-related solutions can be found for each of the first three and, therefore, also for the latter category. Starting with wheeled robots, if we look beyond the number of wheels, the main methods of installation (see Fig. 3) are as follows:

- Fixed-steering axis wheels installed to ensure a wheel plane perpendicular to the platform plane and having a fixed orientation;
- Centred-orientable and off-centred orientable (castor/caster) wheels, i.e. with an orientation angle that can change; and
- Omnidirectional (e.g. Swedish, mecanum, and omni) wheels that work on a fixed plane and enable a holonomic drive system. They are designed to allow movements in any direction on a sufficiently smooth, rigid, clean, and flat surface and, thus, are not practical for AUGVs.

Looking at tracked robots, the main solutions are full and half-track designs.

If we focus on legged systems, solutions can be evaluated by looking at the number of legs, i.e. 2, 3, 4, 6, 8, as well as at the kinematic design. We need to refer to the number and kind of joints and links per leg and their configuration, e.g. bioinspired from mammals, arachnids, reptiles, and knee inwards or outwards (Tedeschi and Carbone, 2014).

2.2 Suspension

The suspension system of AUGVs and mobile robots in general includes a set of mechanical and non-mechanical elements that allow the interaction of the frame with the terrain. The type of suspension can be divided into four main types (see Fig. 3):

- Rigid, i.e. wheels are rigidly connected to the chassis. Robots with this type of suspension are more suitable for locomotion on regular terrain, e.g. the Husky unmanned ground vehicle from Clearpath Robotics (Clearpath Robotics, 2022) (see Fig. 2);
- Passive, i.e. robots employ passive elements such as springs, shock absorbers, or simply an articulated system, e.g. the Thorwald platform from Saga Robotics (Saga Robotics, 2022) (Fig. 1l);
- Active, i.e. wheels and chassis are connected by electric, hydraulic, or pneumatic actuators, and use some type of control system (Fernandes and Garcia, 2018); and
- Hybrid, where the suspension system contains both passive and active elements, e.g. Agri.q from Politecnico di Torino in Italy (see Section 3.3 Case study 1).

2.3 Chassis and implement coupling

The chassis for a mobile robotic platform fulfils the dual function of containing the locomotion components and acting as a stable support for any payload or additional pulled or carried implement. The chassis can have different shapes and sizes, depending on the main task or tasks it is designed for. Focusing on agricultural applications, the type of crop seems to be an important design parameter in chassis design. Four main solutions have been adopted (see Fig. 3):

- Conventional: the chassis has a closed shape and integrates sensors and implements with a range of action covering the area around the robot, e.g. the Vitirover mover robot (Vitirover, 2022) (Fig. 1e).
- Overrunning: the chassis has a closed shape in general with a greater width than height. It integrates sensors and implements that operate in an area below the robot, e.g. the vegetable weeding robot Naïo Dino from Naïo technologies (Naïo, 2016b) (Fig. 1j).
- Gantry: the chassis has an open shape in general with a greater height than width so that a row crop can pass through. In this way, integrated sensors and implements operate on both sides of the row crop, e.g. the Bakus electric vineyard robot from Vitibot (Vitibot, 2018) (Fig. 1g).
- Reconfigurable: the chassis is modular and can be adapted to operate in different conditions, assuming a conventional, overrunning, or gantry configuration, e.g. the Thorwald platform from Saga Robotics (Saga Robotics, 2022) (Fig. 1l).

Linked to this, the chassis can have different levels of mobility:

- Fixed: the chassis is a unique rigid body on which the locomotion system is connected (with or without a suspension scheme), e.g. the Naïo Oz, farming assistant from Naïo Technologies (Naïo, 2016a) (Fig. 1a).
- Articulated: the chassis is made up of at least two parts connected together by means of an 'articulation' joint to ensure a certain amount of mobility and steerability. The joint has usually 2 DoFs, one relating to the yaw (rotation around the vertical axis), to steer the vehicle, and the other DoF relates to the roll (rotation around the front-to-back axis), to allow the system to more flexibly respond to uneven terrain (Carabin et al., 2016; Vidoni et al., 2016). Solutions with 1 DoF, i.e. with freedom in the roll motions, are also used for off-road and/or agricultural vehicles, e.g. the Metron teleoperated vehicle from Reform (REFORM, 2022).

Different configurations suit different combinations of factors such as type of terrain, sensor and implement array and type of operation for which the system is designed.

Given factors such as changing tasks across the growing season, it is advantageous for AUGVs to be able to interface with different carried or pulled equipment and implements, including implements designed for conventional tractors. The ability to attach different implements makes it possible to have an adaptable, multipurpose machine of which costs can be amortized over a shorter period. This will be beneficial when technology adoption in agricultural practice is concerned. The capability to carry or pull implements depends on the presence or absence of suitable coupling systems. These coupling systems can be categorized into three main groups:

- Non-standard coupling is where the implements are generally designed to be used with a specific robot and cannot be shared with other tractors or robots. In this case, the coupling can be rigid, or it can have some mobility.
- Draw-bar is used when the equipment consists of a simple towed trailer or a towed implement. In this case, the implement can be shared between different systems, and the only limitations are the maximum power output and the relative dimensions and weight of the robot and the trailer in relation to the traction capabilities.
- Standard 3-point hitch is nowadays provided for all agricultural tractors and implements. It allows to create a carried coupling that admits small oscillation motions in the horizontal and vertical planes to adjust the implement position and orientation with respect to the soil. As in the previous case, the only limitations are the maximum power output and the relative weight and dimensions of the robot and the carried implement.

Beyond the coupling system, implements often require power from outside since in general they do not include an independent motor or energy source. For implements currently used to outfit common tractors, this is achieved by attaching them to the tractor hydraulic circuit and/or to the PTO (power take-off) shaft, which, with its rotation, transfers mechanical energy from the engine. These solutions can also be applied in robotic systems, especially if they are powered by an endothermic engine. In the case of an electrically powered robot, the PTO consists of a dedicated electric motor driving the PTO shaft also referred to as e-PTO, likewise the hydraulic circuit is powered by an electric pump. This transformation between different types of energy, which involves inevitable losses, is necessary to ensure compatibility with the current implement technology. In any case, this is gradually becoming less and less necessary because on the one hand, implements exclusively dedicated to robotic platforms already integrate electric motors and actuators that require only an electrical connection. On the other hand, the sector of conventional machine implements is also beginning to move toward transitional hybrid solutions (e.g. PTO that drives an electric generator to provide power to the various electric actuators) that will in time lead to the preponderant use of electric power.

3 Wheeled mobile robots

Wheels represent the most commonly used locomotion system for robotic mobile platforms in agriculture due to their simple design, efficiency, and reliability. They provide many advantages such as

- ease of design, manufacture, and installation;
- low production costs;
- possibility to achieve high speed and high manoeuvrability with low power consumption;
- simplicity and thus reliability (low number of moving parts);
- reduced wear and tear on the surface where they move compared to other systems; and
- no balance issues.

Their main disadvantages are their poor capability in navigating over obstacles, e.g. sharp surfaces, rocky terrain or areas with low friction and their vulnerability to slippage. Robotic locomotion in outdoor, off-road non-compacted terrain frequently results in deformation of the terrain. Wheeled robots often experience wheel slippage as well as sideslip while traversing rough terrain such as agricultural fields. This means, in the case of slippage, that the drive torque is transmitted inefficiently, making it extremely difficult if not impossible for the robot to move. This phenomenon also causes significant damage to the ground. In the case of sideslip, on the other hand, the robot also moves in its lateral direction, causing the wheels to slide laterally and thus damaging the ground. Furthermore, controlling the robot's trajectory becomes more complex because the vehicle exhibits an oversteer or an understeer behaviour. Steering and driving manoeuvrability therefore become important to compensate for slippage (Nagatani et al., 2016), increasing autonomous driving control complexity. In addition, appropriate control that adapts drive torque to terrain characteristics to minimise slip and maximize traction efficiency can greatly improve vehicle performance.

Wheeled vehicles are also more vulnerable to being stuck in some conditions, as a result of contact of the frame with the ground. Two types of failure cases may occur (Mastinu and Plöchl, 2014):

- Nose-in failure; and
- Hang-up failure.

The former is related to the vehicle's approach and departure angles, i.e. the maximum angles of the obstacle that a vehicle can overcome, for both approach and departure, without striking the obstacle. For a vehicle with good off-road abilities, both values should be considered in the design phase. They should be in the range of 30°-40° or more. Hang-up failure is when the undercarriage comes in contact with the ground and is related to mobile platform wheelbase and clearance. Generally speaking, the number of wheels, orientation type, and arrangement, as well as the driving mechanism used, are the main elements to be considered in that case.

3.1 Classification based on the number of wheels

The number of wheels can be used to classify and evaluate wheeled robots. Excluding single-wheeled (unicycle) robots which in the authors' opinion are not suitable for agricultural tasks, we can group robots as follows:

- Two-wheeled robots: using an inverted pendulum layout, equilibrium is achieved in dynamic conditions using zero-speed auxiliary legs or stabilization rotors. Rarely adopted for autonomous robots and for agricultural purposes.
- Three-wheeled robots: usually two independently driven, fixed-steering axis wheels and an idle caster wheel. They require simple control with the possibility to spin around a vertical axis. This kind of design is not always suitable for unstructured environments (like most agricultural terrain) because of poor stability in the presence of sloping and uneven surfaces. However, there are designs of this type like the small-size Fendt Xaver sowing robot (Fendt, 2022), Fig. 1b.
- Four-wheels or more: these solutions allow improved stability and stability margins. We can assess a UGV to be statically stable by considering the vehicle's centre of gravity (CoG) (Nagatani et al., 2016). In a four-wheel design, the CoG of the robot can fall within the rectangle created by the four wheels instead of the triangle represented by a three-wheel design (Vidoni et al., 2015). A greater number of wheels improves traction, obstacle-overcoming capabilities and a better distribution of vertical load weight but involves greater complexity in the vehicle chassis. However, having a hyperstatic layout which maintains all wheels in contact with the ground requires a more complex chassis and suspension system. Many solutions are possible, as can be seen in Figs 1a, c, e, f, g, h, l, j, n, p, q, r, and t.

It should be noted that, to reduce shocks and reduce the need for suspension systems, most wheeled autonomous agricultural machines are fitted with deformable tyres where usually larger sizes are inflatable and smaller sizes are not. A suspension mechanism can be useful when driving over uneven surfaces to allow all the wheels to maintain contact with the ground which allows the generation of traction, braking and cornering forces. In addition, if equipped with dissipative elements, suspension systems mitigate the effects of motion on the robot body (Chung and Iagnemma, 2016).

3.2 Mobility configurations

If the mobility configuration is the focus, the main solutions (see Fig. 4) (Khan et al., 2021) are as follows:

- Skid steering (Fig. 4a): there is no explicit steering of the wheel axles. Steering is achieved by producing a differential thrust between the right and left sides of the mobile robot, causing the wheels to slip or skid, resulting in a change in direction. The two wheels on each side could be powered by a single motor or independently. Maximum speed cannot be achieved during steering. Pros and cons are as follows:
- Pros: high mobility (i.e. zero-turn radius steering), simple mechanical structure, large traction for manoeuvring on rough surfaces, larger payload capabilities, and few actuators; and
- Cons: complex system modelling due to significant wheels slip and sideslipe that heavily affect the robot handling, rapid wearing out of tires due to skidding, erosion of terrain, and large power requirements in particular for accomplishing zero-turn radius.

This is a possible simple and robust drive mechanism for agricultural applications, even if not very precise when traversing non-compact terrain.

- Differential steering (Fig. 4b): closely related to skid steering, differential steering is achieved with two independently driven wheels rotating about the same axis, as well as one or more caster wheels (or low-friction sliders) to keep the robot horizontal. Steering is also obtained by producing a differential thrust between the two wheels. Pros and cons are as follows:
- Pros: low number of actuators, simple structure, and high mobility (turning on the spot); and

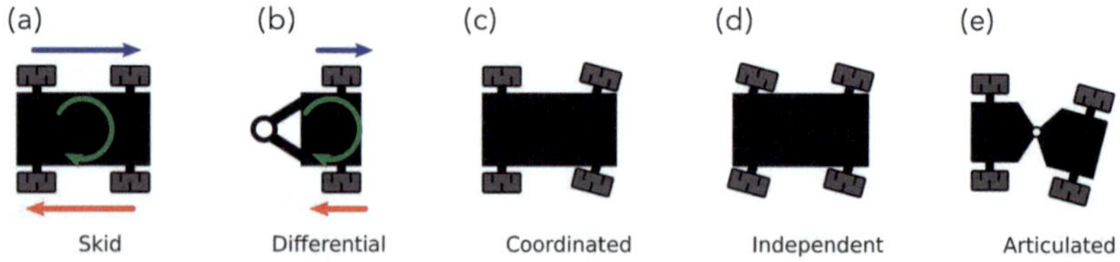

Figure 4 Driving mechanisms.

- Cons: difficulties while facing obstacles, bumps or irregularities, caster wheels affect handling precision, complex system modelling due to castor wheels behaviour.

Examples of AUGVs customized for skid or differential steering are the Kilter AX-1 (see Fig. 1m), and the AgBot II (Bawden et al., 2017) using differential steering through 2 front fixed wheels and two rear castor wheels.

- Coordinated steering (Fig. 4c): the most common arrangement of this type consists of a mechanical coupling that is used to synchronize the turning of two or more wheels subject to a desired kinematic geometry (e.g. Ackermann steering and dual Ackermann steering) and two fixed-steering axis-active wheels usually at the rear of the platform. If mechanical coupling is not used and the related wheels are independently driven, coordinated steering is achieved through a mechatronic system, creating a coordinated steering solution that could also be classified as a type of independent steering. Pros and cons are as follows:
  - Pros: simplicity and low number of actuators as well as medium control and mechanical complexity; and
  - Cons: turning radius is large; steering is difficult on non-compact and loose ground.

In agricultural applications, this driving mechanism is used when adapting commercial farm machines and tractors, such as the CNH autonomous tractor (Fig. 1p) and John Deere GridCon (Fig. 1 (r)). A more unique example is the Sitia Trektor UAGV (Fig. 1f).

- Independent steering (Fig. 4d): each wheel assembly is explicitly controlled in its orientation angle and rotation speed (e.g. active caster wheels). This increases complexity in terms of wheel coordination and position accuracy. However, manoeuvrability is improved thanks to the fact that any angle movements can be achieved by aligning all the wheels at a common angle, i.e. crab steering. Pros and cons are as follows:
  - Pros: good manoeuvrability especially in rough terrains, low energy consumption, capacity for holonomic omnidirectional movement (i.e. it can move in any direction, and it is not limited to follow curved trajectories); and
  - Cons: actuation number (eight needed for a four-wheeled vehicle), complexity and accuracy of coordination control.

Many agricultural prototypes adopt this driving mechanism thanks to its versatility. Examples include AUGVs like Ladybird (Underwood et al., 2015), with an omnidirectional, independent steering scheme as well as the robot by Bakker (Bakker et al., 2010) Vitibot Bakus (Fig. 1g), and Naïo Dino (Fig. 1j).

- Articulated steering (Fig. 4e): this mobility configuration requires a non-rigid chassis, i.e. an articulated design where a part of the chassis and the wheels attached to it steer about a pivot point. The heading of the robot alters by 'folding' the hinged elements. The articulation joint can be either actuated or passive; this means that the steering action is achieved by locking the wheels on one half-chassis unit and driving the wheels on the other. An additional articulation is usually provided around the roll or pitch axes. This makes it possible to mitigate the dynamic effects of steering by improving the terrain adaptability of the robot. Pros and cons are as follows:
    - Pros: manoeuvrability and efficiency increase with the number of articulated joints. Small turning radius; and
    - Cons: increased control complexity.

This kind of steering solution is sometimes adopted on small agricultural vehicles and tractors as well as off-road vehicles. Examples include the ARTI prototype from Stanley Innovation, Inc. (Stanley Innovation, 2022).

Each kind of mobility system thus has its own advantages and disadvantages and there is no one solution that achieves optimal controllability, manoeuvrability, and stability.

3.3 Case study 1: Agri.Q – a sustainable rover for precision agriculture

Agri.Q is an articulated, wheeled mobile robot developed at the Politecnico of Torino to be used in orchards, olive groves, and esp. vineyards. The choice of vineyards is for two main reasons:

- Piemonte, a north-western region of Italy where Politecnico di Torino is located, has a long and distinguished tradition of winemaking.
- Vineyards are typically profitable and follow very strict production quality standards, making use of robots in PA applications potentially attractive in both operational and commercial terms.

As well as Piemonte, other neighbouring Northern Italian regions, namely, Valle d'Aosta and Liguria, are also classic wine-growing locations that produce many well-known wines, frequently in challenging terrain (from a robotics point of view). Vineyard sites can be found at elevations higher than 500 m, with grapes grown on slopes greater than 30% on steep hills, terraces, or embankments (Corinto and Pioletti, 2019). Conventional mechanized agricultural machinery can only operate where slopes do not exceed 30% and vine rows are wide enough (Corinto and Pioletti, 2019).

The Agri.Q rover is especially developed to fill this gap in precision farming applications in vineyards with a focus on low weight, compactness, manoeuvrability in rough terrain, and sustainability. Its mobility allows it to carry various instruments and sensors for activities such as field mapping and crop monitoring. Its design allows collaborative operations with flying drones which are widely used in agriculture.

Figure 5a depicts Agri.Q and its main components. Agri.Q is an articulated mobile robot composed of two modules:

- the front module, which houses the robot control unit; and
- the back module.

(a)

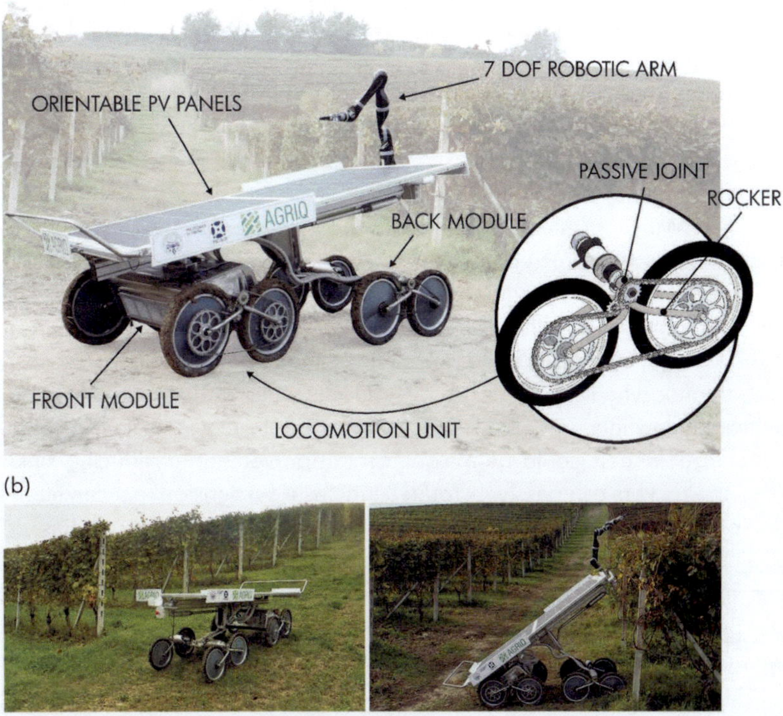

Figure 5 Agri.q AUGV: (a) main parts. (b) robot orienting its photovoltaic panels providing a flat landing surface or extending the robotic arm's workspace.

Each module has two drive units consisting of two wheels, driven by a single DC motor through a chain drive, assembled on a rocker that is free to rotate about a passive revolute joint. As a result, each module is an independent skid-steered component. The back module is connected to the front module through a series of passive joints that enable relative yaw and roll motions between the modules. These passive DoFs, together with the locomotion units' passive rocker, allow the articulated robot to navigate rough terrain because the wheels' contact with the ground is always guaranteed.

Agri.Q also has a linkage mechanism to control the pitching motion of the photovoltaic (PV) panels mounted on top. The PV panels can also be rotated about their roll axis through an additional actuated joint. As a result, the PV panels can self-orient to optimize the gathering of sun rays during the auto-charging phase as well as providing an always level drone landing platform (Fig. 5b). Finally, a 7DoF robotic arm is fixed on the platform to allow different tasks to be performed (Fig. 5b). Details of the Agri.Q design can be found in Quaglia et al. (2019), Cavallone et al. (2021), and Visconte et al. (2021).

The Agri.Q locomotion unit design uses an eight-wheel architecture with each pair supported by a rocker. As with a track system, this allows the normal forces acting on the ground to be distributed over a larger contact surface. This means the vehicle is less likely to become stuck in soft ground and soil compaction is also reduced. However, overall traction efficiency is better than with a four-wheeled rover. Also, this design also reduces wheel slippage during turning compared to a tracked system of similar size, which can increase soil erosion and landslips on sloping soils exposed to wind and rain (Braunack, 1986a,b; Braunack and Williams, 1993; Ayers, 1994; Prosser et al., 2000; Li et al., 2007; Molari et al., 2012). Certainly, the characteristics of the terrain have an effect, but given its extremely low weight, the high number of contact points, a distributed traction and various passive DoFs, Agri.Q is able to move easily even over complex terrain to traverse. Although thicker tyres could distribute the robot's load on the ground even better at the price of a bulkier and heavier robot, thinner wheels performed very well because the weight per wheel is significantly low in vineyard tests. However, if the terrain is particularly soft or loose, larger wheels could be a significant improvement in reducing sinking. At last, the rockers can passively rotateacting as a suspension that can filter out vibrations and oscillations caused by ground obstacles and irregularities.

Since the two locomotion units can be controlled independently from each other, appropriate traction allocation strategies can be employed to improve robot performance while optimising power consumption. For example, driving the rear unit is usually not required on flat ground so it can be turned off in those instances. However, on impervious inclines or when negotiating an obstacle, the rear unit provides a significant contribution to robot mobility.

The actual robot measures about 1 × 2 m in size and weighs 112 kg. These characteristics, together with its articulated design, allow the Agri.Q to navigate a vineyard with ease, without damaging the soil. It can run for about 4.5 h without using its photovoltaic panels that allow for doubling operation time (Quaglia et al., 2020; Cavallone et al., 2021; Botta and Cavallone, 2022). The robot uses autonomous navigation based on a ROS framework (Open Robotics, 2022), a robotic arm to provide an effective mobile manipulator, and a set of sensors to perform monitoring activities. The robotic arm is mainly used to perform sample collection, or more generally pick-and-place operations, to assess the growth progress of the crop. In this respect, an interchangeable tool for cutting and collecting grape samples is currently being developed. As far as sensors are concerned, Agri.Q can become a mobile platform for the main types of sensors used in agriculture. This translates into not only the possibility of implementing cameras operating in different frequency spectrums to estimate the state of plants but also the possibility of integrating proximity sensors for assessing the condition of plants. Alternative traction control strategies are also currently being developed in order to make full use of the traction of both modules to enable even more effective navigation in difficult terrain (e.g. slip control to maximize the force exchanged on the ground, or a kind of torque vectoring to improve robot handling) but at the same time to reduce energy consumption.

4 Tracked mobile robots

Given their large ground contact surface, tracked mobile robots (TMRs) are suited to overcome obstacles and uneven and soft terrain (Bruzzone and Quaglia, 2012). However, usually they move more slowly and consume more energy than wheeled robots. Changing direction requires a skidding turn, requiring a large portion of the track to slide against the ground surface. In general, a tracked system requires more power from the battery pack to rotate the track than a wheeled robot of the same weight and dimensions, resulting in higher energy consumption. Control of TMRs is usually more challenging than other steering systems because the effect of the skid steering largely depends on the characteristics of the topsoil. TMRs perform noticeably poor on surfaces with low friction. In the case of varying friction forces between soil and tracks, steering may become quite variable and requires proper sensing for feedback control when navigating a terrain. As a result, odometry-based position estimation systems may become completely unreliable.

TMRs typically have a limited top speed, depending on the width and length of the rubber track material used to prevent damage to the tracks. Due to the amount of movement from the idle wheels, and the mechanisms required to move the vehicle on tracks, heat can be generated across the rubber used for the track belt. The downward force and weight of each wheel also force the rubber to deform and change its shape. The more rubber compresses and decompresses, the more kinetic energy is generated and leads to a rise in the rubber temperature; this is particularly true for high-thickness and soft rubber belts.

TMRs are subject to vibration, given the lateral track profile of a polygon with moving vertices. They are rarely fitted with damping systems given the more complex design. This limits maximum speed and reduces mechanical efficiency, especially when driving on non-loose surfaces. The larger footprint offered by tracked robots allows the downward forces to be distributed over a larger area, thus reducing rut formation and compaction.

TMRs can be classified according to the number and layout of the tracks. The two main types are full and half-track. A simpler configuration involves two parallel tracks with differential steering. To improve the capacity to climb obstacles and handle uneven terrains, a design with more than two tracks with relative passive mobility can be adopted.

Different configurations in the track design can be considered, such as simple tracks, tracks with improved mobility (i.e. tank or inverted), active and passive flippers as shown in Fig. 6. These configurations can be used to overcome higher obstacles (i.e. tank and flipper configuration) or to allow installing the motor at a higher point to increase the platform clearance (i.e. inverted).

In Fig. 6, configurations (c), (d), and (e) allow the use of the upper wheels as drive sprockets in order to lift the driving motor from the ground, which is particularly useful when the tracked robot has to operate in dusty and rough environments or in presence of wet soil. Tracked systems can be equipped with a suspension system. The passive suspension system of the vehicle performs multiple tasks such as maintaining the contact between the rubber tracks and the terrain surface, providing vehicle stability, and protecting the vehicle frame from shocks generated by the unevenness of the terrain (Galati and Reina, 2019). The support polygon is a key element in static stability. In the case of a tracked vehicle, it is equal to the horizontal projection of the convex shape of the contact points between tracks and terrain.

The larger footprint area of tracked robots allows less soil compaction compared to geometrically equivalent wheeled vehicles. However, when measuring the contact pressure of a generic track system, spikes in pressure can be observed under each idle ground wheel. This phenomenon must be taken into account when the robot needs to move on wet or soft soils because

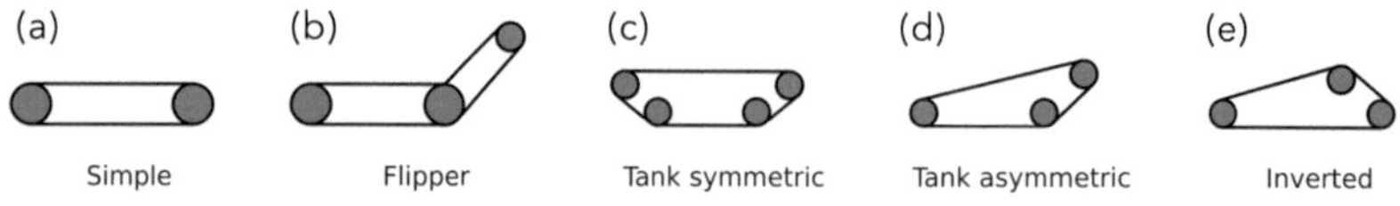

Figure 6 Main track configurations.

the terrain will be damaged by the higher contact pressure right under the ground wheels. Uneven terrains with rocks, debris, and other obstacles can affect the overall stability and balance of the robot, degrading the output quality of onboard sensors such as LiDARS, gyroscopes, and cameras.

Examples of tracked mobile platforms developed or customized for agricultural applications include Byelab (Bietresato et al., 2016; Vidoni et al., 2017), Vibro Crop Robotti from Kongskilde (Kongskilde - Agriculture, 2022), Slopehelper from PeK Automotive (PeK Automotive, 2022) (see Fig. 1o), and AgBot from AgXeed (AgXeed, 2022) (see Fig. 1s).

4.1 Case study 2: Polibot tracked robot

The farmer robot Polibot, completely custom-built at the Politecnico of Bari's Robotic Mobility Lab (Grazioso et al., 2022), is shown in Fig. 7. Polibot employs a skid-steering system that is widely used on tracked vehicles such as tanks and bulldozers as well as on some four- and six-wheeled vehicles like Husky (Clearpath Robotics, 2022). The track on either side can be driven at various speeds in forward and reverse while all drive and support wheels on a side are driven at the same rate, and there is no explicit steering mechanism. As the name suggests, steering is accomplished by actuating the track on either side at a different rate or in a different direction, causing the tracks to slip, or skid, on the ground. Each track sprocket is driven by a 24V DC brushed motor with a gearbox providing a maximum output torque of 50 Nm and a maximum angular velocity of 80 rpm for a total output power of about 750 W.

Skidding has some disadvantages including track wear. Skid-steering vehicles are usually designed for off-road and heavy-duty applications which means the reduced friction of a non-paved surface helps to reduce the wear of the rubber belt. Polibot features an innovative undercarriage configuration with belts made of natural rubber vulcanized together with Kevlar fibre bars to reduce wear and tear resistance with a minimal weight. Because it cannot be precisely controlled, another major drawback of skidding is the negative effect it has on tracking vehicle positioning using odometry (Reina and Galati, 2016). For this reason, Polibot is equipped with a minimum sensor set to correct the position estimation. This includes two quadrature rotary optical encoders mounted on each gearbox shaft, two electrical current sensors, an RTK GPS, and a three-axis inertial measurement unit with a magnetometer for the heading following the NED (North, East, Down) reference frame.

Each undercarriage includes a large rear wheel (3 - in Fig. 7) with a diameter of 200 mm and a couple of two smaller idle front wheels (1) with a diameter of 120 mm pointing forward. This configuration allows the robot to keep its balance when additional tools, i.e. lawn mower unit, weed cutter, rotary hoes, and sensor frame, are mounted on top (see Fig. 7b right). It can also overcome obstacles greater than two-thirds of the diameter of the rear wheel since the upper front wheel leans on it, providing extra grip (Ugenti et al., 2022). As an example of Polibot mobility performance, Fig. 7b shows the robot during the negotiation of a vertical wall.

To improve the ground pressure and the non-uniform weight distribution caused by pressure spikes from wheels, a passive suspension system is used to minimize the vibrations over the main frame caused by track-terrain interactions by maintaining the contact between the rubber tracks and the terrain surface. The center idle wheels (2) ensure a more even pressure distribution on the ground. This passive suspension system enables the whole track assembly to move vertically relative to the vehicle chassis in order to provide good performance on uneven terrain. The track assembly is able to

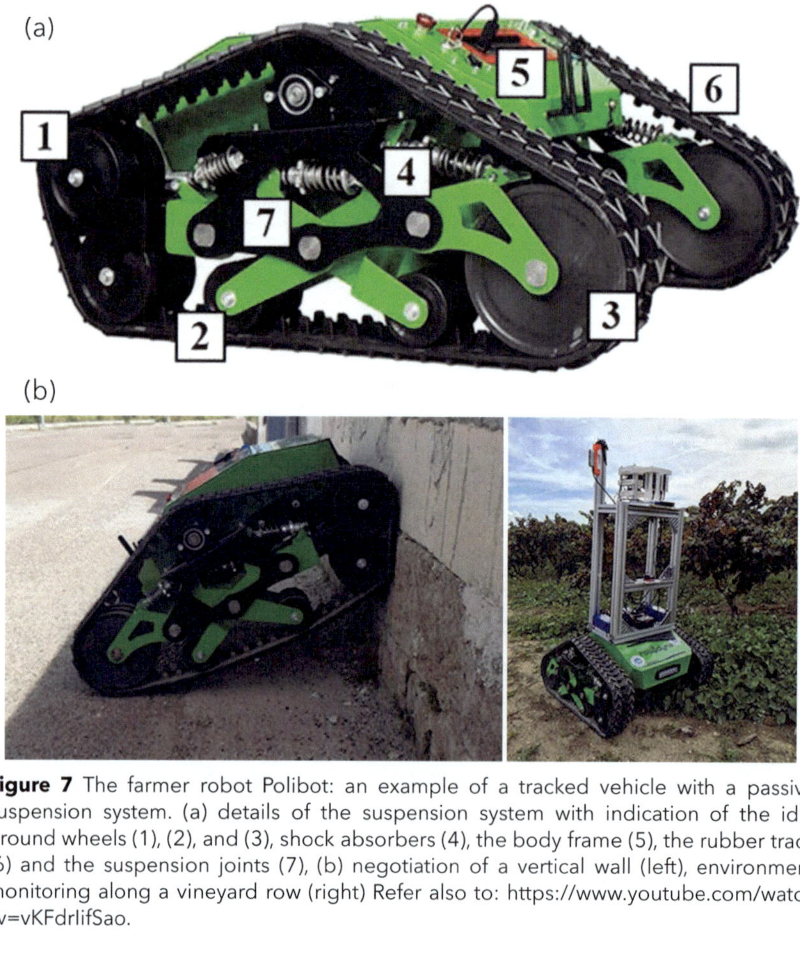

Figure 7 The farmer robot Polibot: an example of a tracked vehicle with a passive suspension system. (a) details of the suspension system with indication of the idle ground wheels (1), (2), and (3), shock absorbers (4), the body frame (5), the rubber track (6) and the suspension joints (7), (b) negotiation of a vertical wall (left), environment monitoring along a vineyard row (right) Refer also to: https://www.youtube.com/watch?v=vKFdrlifSao.

compensate for depressions or bumps in agricultural terrains, preventing the frame of the vehicle from tilting with each minor change in the terrain. When an idle wheel contacts a bump, the suspension mechanism is able to allow sufficient vertical motion to stop the wheel from continuing to move upward, which would otherwise cause large vertical acceleration of the chassis along the z-axis.

The system uses a set of three shock absorbers with preload adjustment, as well as the option to change the elastic stiffness, to better adapt to terrain features. The suspension bracket used for the centre idle wheels is based on fork-type construction. The center shock absorber is compressed when the ground pressure increases by changing the distance between both wheels, while the rear and front wheels move upward, lowering the center of gravity of the robot.

Since Polibot uses a passive suspension system, its vibration response on the terrain is quite different from the response generated by a standard wheeled vehicle or one with a rigid tracked undercarriage. In order to model Polibot performance, it is very important to study its vibration response. In linear system theory, a direct relationship between input and output signals exists. It is possible to define a transfer function for the vehicle that considers the input representing terrain irregularities and generates an output related to the vibration of the vehicle. Assuming a simplified single-degree-of-freedom model for the vehicle, with both the input and output values expressed in terms of displacements and the vibration of the sprung mass as output, the frequency response function can be defined as the ratio of the output to input under steady-state conditions. It is possible to show how the output power spectral density is associated with the input power spectral density through the square of the modulus of the transfer function (Galati and Reina, 2019). The power spectral density defines how the power of a signal is distributed over the frequency range and it is strictly correlated to the interaction between the terrain profile and the track belt and the belt and the track sprocket. In contrast to wheeled vehicles which usually present only a single peak in their frequency response, Polibot shows four distinct and separate peaks and four odd harmonics in total as shown in Fig. 8, where the vibration response has been analysed for different kinds of terrains: sand, gravel, mud and asphalt. The knowledge about the type of the traversed terrain can be beneficial for a vehicle since it can deal with its environment more efficiently in order to better support precision farming tasks. Moreover, the locomotion performance can be optimized in terms of traction and power consumption (e.g. battery autonomy for electric vehicles) by adapting control and planning strategies to site-specific environments.

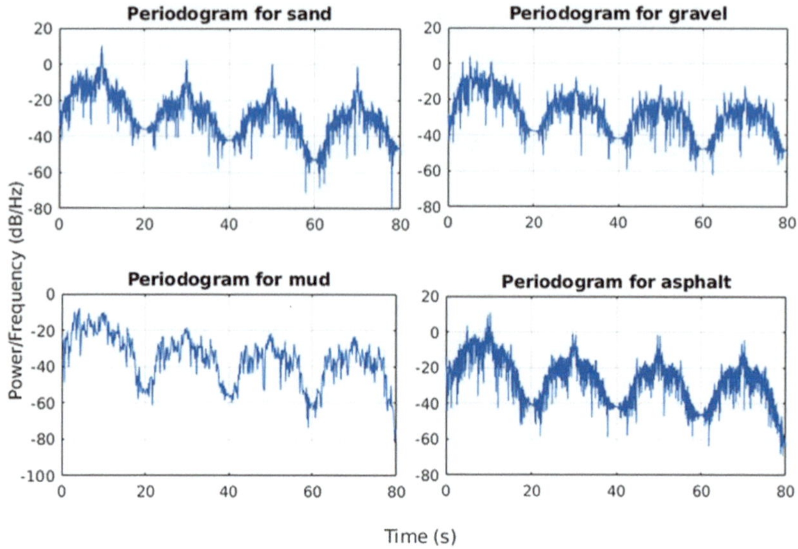

Figure 8 Power spectral density values for sand, gravel, mud, and asphalt.

5 Legged mobile robots

Though wheels generally provide faster and more convenient navigation in the field, robots with legs may have advantages, e.g. on loose or steep ground, areas with rubble, or where obstacles such as branches and stems are an issue (Fue et al., 2020). Though wheeled and tracked robots can work on flat terrain, most cannot work in complex or hazardous environments. In contrast, legged robots can be superior to wheeled and tracked robots because they can explore terrain in a similar way to humans (Biswal and Mohanty, 2021). Legged robots can

- navigate surfaces that are inaccessible to robots with wheels or tracks;
- step over some obstacles whereas wheeled systems must travel over or avoid them; and
- step over locations to be avoided like crops.

Legged locomotion depends on factors such as posture, number of legs, and leg functionality. LMRs have more potential to cope with almost any earth's surface. Quadruped robots are the most suitable choice among legged robots in terms of mobility and stability. Four robot legs are more easily designed, controlled, and maintained compared to two or six legs. Where payload and safety are an issue, as in agricultural applications, quadrupeds maintain better static balance and allow a better choice of foot placement.

However, leg-based robots are more complex from the mechanical and computational point of view in maintaining equilibrium, orientation, and optimising (energy) efficiency. Gait planning is a particular challenge. If the kinematic architecture of the legs is the focus, the configuration and the joint/ actuator arrangement can be used to classify LMRs (Tedeschi and Carbone, 2014; Gonzales-De Santos et al., 2020).

Four- or six-legged robots can be bioinspired with a configuration arrangement designed to mimic an animal's gait such as reptiles, mammals, or arachnids (see Fig. 2). In the case of a mammalian configuration, with the body above the legs, power consumption is reduced as is the overall footprint, affecting overall (static) stability. Arachnid and reptile configurations have legs placed on the sides of the body, putting more focus on the way knees are designed and operated. Joint/actuator arrangement is key for the leg working area, with a focus on simplifying the structure, e.g. minimising the number of actuators. The common way to design a leg of a 2-n robot is to adopt a 4, 3, or 2 DoF solution which mimics hip, knee, and ankle joints in fully actuated or underactuated configurations (see Fig. 9).

Even if they are more adaptable, legged robots for agricultural applications are quite rare and represented mostly by small-size AUGVs (advanced) prototype solutions, sometimes customized from an existing system.

An example of a six-legged robot used for precision harvesting purposes is the six-legged dandelion-picking robot (Garimella and Revzen, 2021). This is based on the BigANT robot, a multilegged system with low-degree-of-freedom legs, adapted for picking dandelions. The design provides effective mobility, stability, and low complexity. The chassis is rigid and a flower-picking appendage is attached to it. Since leg movement is versatile, the robot can be programmed to perform a 'swooping' motion allowing the end-effector to pluck the flowers. Another example of an AUGV with six legs is 'Prospero', an autonomous micro-planter system with crab bio-inspired kinematics, a rigid

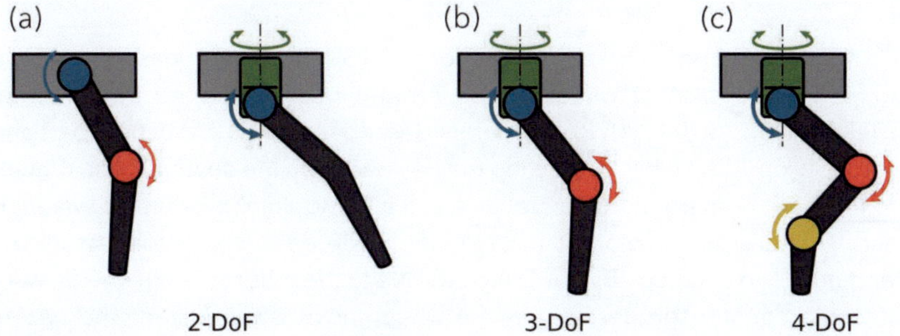

Figure 9 Examples of frequently adopted joint arrangements in legged robots.

chassis, and 3 DoF per leg based on the idea of deploying multiple small machines rather than one large one (Bayer, 2021).

There are also commercial systems that could be customized to perform agricultural tasks. In most of the cases, they are quadruped systems with a mammal configuration and with solutions inspired by existing designs such as Big-Dog (Raibert et al., 2008) from Boston Dynamics, a robot that can run at >6 km/h, climb slopes up to 35° and carry approx. 150 kg. Other examples include Spot, also from Boston Dynamics (Boston Dynamics, 2019), ANYmal C from ANYbotics (ANYbotics, 2019), and A1 from UniTree (Unitree, 2022) (see Fig. 2). An example of an agricultural application is the use of the four-legged robot Spot to herd sheep (Rocos, 2020).

6 Hybrid and other unconventional unmanned ground vehicles mobility platforms for agriculture applications

According to Gonzales-De-Santos et al. (2020), the majority of existing solutions based on wheeled, tracked, or legged UGVs lack flexibility. They often focus only on a specific application or crop, e.g. orchard or vineyard, and a particular type of terrain. Hybrid UGVs are a possible solution to merge the benefits of different classes of UGVs. These include systems where legs are coupled with either wheels or tracks. They can be categorized following the classification shown in Fig. 3.

These mobile platforms have a structure that depends on the number of legs, their configuration, and joint arrangement as well as the kind of wheel or track used. An example of a hybrid robot developed for agricultural purposes is BoniRob (Ruckelshausen et al., 2009), a four-wheeled-legged platform with independent and steerable legs. The legs are made of a link connected, on one side, to the chassis through a 1 DoF hip joint and, on the other side, to the 2 DoF steering wheel. Another example is the Harry robot concept and related developments proposed by the Small Robot Company for a four-wheeled-legged AUGV with an arachnid configuration (Fig. 10a) (Small Robot Company, 2017).

An interesting and unconventional UGV platform based on two independently rotating wheels is the Amphibious Unmanned Vehicle from SeTel, Fig. 10b, (SeTeL, 2022). The patent describes wheels that act as floats and provide propulsion while protecting the payload from the environment. Wheels are free to rotate around a central axis and a 'pendulum' propulsion system is used. This basically exploits a heavy mass inside each wheel connected to a pendulum-like structure. By applying a torque to the mass through one or more motors, a counter torque that lets the wheel move is generated. The system has a low CoG and is intrinsically stable. The payload can be contained in the

Figure 10 (a) Harry from Small Robot Company and (b) WineGRover from Setel (note: permission requested but not yet granted).

wheels and is thus protected. This solution has been tested in a vineyard after being customized as a rover for precision agriculture named WineGRover.

7 Conclusion and future trends

Thanks to developments in production and digital technologies, agriculture is facing a revolution in precision agriculture. Intelligent and advanced platforms need to be developed with special characteristics able to cope with agricultural fields and different crops. Unmanned ground vehicles or robots for agricultural purposes are one of the options and many different solutions are now available, particularly in terms of mobility. If size is taken as a criterion, there are small, medium, and large AUGVs. If the type of mobility is used, wheeled, tracked, legged, and hybrid are the main solutions. Small-size platforms often use tracked or wheeled skid-steering architectures to achieve simple, reliable, and cost-effective robots. Very few designs implement unusual locomotion systems such as legged or rolling locomotion. Medium-size platforms usually employ all-steering-wheel architectures and are sometimes articulated to improve their manoeuvrability in tight spaces. Large-size AUGVs are usually wheeled with a car-like architecture, or they employ tracks if they are particularly heavy. If the frame is considered as the key design criterion, overrunning or gantry solutions, in particular for crop-specific applications like vineyard precision agriculture, have been the preferred configuration.

Service robots for agriculture represent an increasing market and much still needs to be done to develop effective, reliable, and affordable mobility solutions for, in particular, small- to medium-sized farms. Thanks to electrification or hybridization of platforms, as well as the availability of new technologies and sensors, mobility platforms such as over-actuated systems, including legged, hybrid independently driven, wheel-legged systems, and articulated chassis, could represent a solution given their versatility and adaptability to different situations, crops, and terrains.

Future research needs to focus on platform flexibility, i.e. the ability to adapt to different scenarios and activities. This includes being reconfigurable, e.g. with adaptable frame dimensions to cope with different row distances as well as being able to accommodate different agricultural implements. It also includes ease of manoeuvrability and steering, e.g. the ability to turn on small radii while coping with slopes and different terrains. To reach such a goal, the starting point for creating successful advanced mobility platforms for the agriculture of the future can be provided by independent steering configurations possibly linked to legs to allow, e.g. zero turning radii, reconfigurable or articulated chassis to adapt with different working scenarios, as well as configuration to connect or disconnect sensors and implements.

8 Where to look for further information

The following references provide a good overview of different subjects touched upon in this chapter:

- Bergerman, M., Billingsley, J., Reid, J. and van Henten, E. (2016). 'Robotics in Agriculture and Forestry', in Siciliano, B. and Khatib, O. (eds) Springer Handbook of Robotics. Cham: Springer International Publishing (Springer Handbooks), pp. 1463–1492. doi:10.1007/978-3-319-32552-1_56.
- Rubio, F., Valero, F. and Llopis-Albert, C. (2019). 'A review of mobile robots: Concepts, methods, theoretical framework, and applications', International Journal of Advanced Robotic Systems, 16(2), p. 1729881419839596. doi:10.1177/1729881419839596.
- Mastinu, M. and Plöchl, M. (2014). Road and off-road vehicle system dynamics handbook. Boca Raton, Florida: Crc Press, Taylor & Francis Group.
- Bruzzone, L. and Quaglia, G. (2012). 'Locomotion systems for ground mobile robots in unstructured environments', Mechanical Sciences, 3(2), 49–62.
- Oliveira, L. F. P., Moreira, A. P. and Silva, M. F. (2021). 'Advances in Agriculture Robotics: A State-of-the-Art Review and Challenges Ahead', Robotics, 10(2), 52. doi:10.3390/robotics10020052.
- Müller, C., Graf, B. and Pfeiffer, K. (Eds.): World Robotics 2021 – Service Robots, IFR Statistical Department, VDMA Services GmbH, Frankfurt am Main, Germany, 2021. ISBN 978-3-8163-0747-1.

Key organisations and conferences are as follows:

- GOFAR: Global Organization for Agricultural Robotics (https://www.agricultural-robotics.com/).
- FIRA International Forum of Agricultural Robotics (https://www.fira-agtech.com/c/fira).

- AgriExpo, Farm robots https://www.agriexpo.online.
- Future Farming https://www.futurefarming.com/.
- ATLAS webpage https://www.atlas-h2020.eu/.

Centres of expertise include:

- UK Agri-Tech Centres https://www.agritechcentres.com/.
- Lincoln Agri-Robotics (LAR) https://lar.lincoln.ac.uk/.
- Australian Centre of Field Robotics (ACFR) https://www.sydney.edu.au/.

Laboratories in which the authors are involved:

- FiRST lab – Field Robotics laboratory at the Free University of Bolzano (I) https://firstlab.projects.unibz.it/.
- Pic4Ser – Politecnico Interdipatimental Center for Service Robots at the Politecnico di Torino (I) https://pic4ser.polito.it/.
- Robotic Mobility Lab at the Politecnico of Bari (I) https://www.vagostudio.com/giulio/.

9 References

AgXeed (2022). AgXeed [online] Available at: https://agxeed.com/ (Accessed: 29 April 2022).

ANYbotics (2019). ANYbotics | Autonomous Legged Robots for Inspection and More. [online] Available at: https://www.anybotics.com/ (Accessed: 29 April 2022).

Ayers, P. D. (1994). Environmental damage from tracked vehicle operation, *Journal of Terramechanics* 31(3), 173–183. doi:10.1016/0022-4898(94)90014-0.

Bakker, T., Asselt, K., Bontsema, J., Müller, J. and Straten, G. (2010). Systematic design of an autonomous platform for robotic weeding, *Journal of Terramechanics* 47(2), 63–73. doi:10.1016/j.jterra.2009.06.002.

Bawden, O., Kulk, J., Russel, R., McCool, C., English, A., Dayoub, F., Lehnert, C. and Perez, T. (2017). Robot for weed species plant-specific management. *Journal of Field Robotics* 34, 1179–1199. https://doi.org/10.1002/rob.21727.

Bayer (2021). Automated agricultural helpers - ripe for robots. [online] Available at: https://www.bayer.com/en/agriculture/article/ripe-robots (Accessed: 29 April 2022).

Bergerman, M., Billingsley, J., Reid, J. and van Henten, E. (2016). Robotics in agriculture and forestry. In: Siciliano, B., Khatib, O. (Eds). *Springer Handbook of Robotics*. Cham: Springer International Publishing, pp. 1463–1492. doi:10.1007/978-3-319-32552-1_56.

Bietresato, M., Carabin, G., D'Auria, D., Gallo, R., Ristorto, G., Mazzetto, F., Vidoni, R., Gasparetto, A. and Scalera, L. (2016). A tracked mobile robotic lab for monitoring the plants volume and health. In Conference Proceedings of the 12th IEEE/ASME International Conference on Mechatronic and Embedded Systems and Applications (MESA 2016), Article number 7587134. doi:10.1109/MESA.2016.7587134.

Biswal, P. and Mohanty, P. K. (2021). Development of quadruped walking robots: a review, *Ain Shams Engineering Journal* 12(2), 2017–2031. doi:10.1016/j.asej.2020.11.005.

Blackmore, S. (2016). Towards robotic agriculture. *Proceedings of the SPIE*, 8-15. doi:10.1117/12.2234051.

Boston Dynamics (2019). *Boston Dynamics Is Changing Your Idea of What Robots Can Do. | Boston Dynamics*. [online] Available at: https://www.bostondynamics.com/ (Accessed: 29 April 2022).

Botta, A. and Cavallone, P. (2022). Robotics applied to precision agriculture: the sustainable Agri.q Rover Case Study'. In: Quaglia, G., Gasparetto, A., Petuya, V. and Carbone, G. (Eds). Proceedings of I4SDG Workshop 2021. Cham: Springer International Publishing (Mechanisms and Machine Science), pp. 41-50. doi:10.1007/978-3-030-87383-7_5.

Braunack, M. V. (1986a). Changes in physical properties of two dry soils during tracked vehicle passage, *Journal of Terramechanics* 23(3-4), 141-151. doi:10.1016/0022-4898(86)90003-0.

Braunack, M. V. (1986b). The residual effects of tracked vehicles on soil surface properties, *Journal of Terramechanics* 23(1), 37-50. doi:10.1016/0022-4898(86)90030-3.

Braunack, M. V. and Williams, B. G. (1993). The effect of initial soil water content and vegetative cover on surface soil disturbance by tracked vehicles, *Journal of Terramechanics* 30(4), 299-311. doi:10.1016/0022-4898(93)90017-R.

Bruzzone, L. and Quaglia, G. (2012). Review article: Locomotion systems for ground mobile robots in unstructured environments, *Mechanical Sciences* 3(2), 49-62.

Carabin, G., Gasparetto, A., Mazzetto, F. and Vidoni, R. (2016). Design, implementation and validation of a stability model for articulated autonomous robotic systems, *Robotics and Autonomous Systems* 83, 158-168. doi:10.1016/j.robot.2016.05.008.

Cariou, C., Lenain, R., Thuilot, B. and Berducat, M. (2009). Automatic guidance of a four-wheel-steering mobile robot for accurate field operations, *Journal of Field Robotics* 26(6-7), 504-518. doi:10.1002/rob.20282.

Cavallone, P., Visconte, C., Carbonari, L., Botta, A. and Quaglia, G. (2021). Design of the mobile robot Agri.q. In: Venture, G. (Eds). *ROMANSY 23 - Robot Design. Dynamics and Control*. Cham: Springer International Publishing (CISM International Centre for Mechanical Sciences), pp. 288-296. doi:10.1007/978-3-030-58380-4_35.

Cavallone, P., Botta, A., Carbonari, L., Visconte, C. and Quaglia, G. (2021). The Agri.q Mobile Robot: Preliminary Experimental Tests'. In: *Mechanisms and Machine Science* Niola, V. and Gasparetto, A. (Eds) *Advances in Italian Mechanism Science*. Cham: Springer International Publishing, 524-532. doi:10.1007/978-3-030-55807-9_59.

Chung, W. and Iagnemma, K. (2016). Wheeled robots. In: Siciliano, B. and Khatib, O. (Eds). *Springer Handbook of Robotics*. Cham: Springer International Publishing, pp. 575-594. doi:10.1007/978-3-319-32552-1_24.

Clearpath Robotics (2022). *Clearpath Robotics: Mobile Robots for Research & Development*, Clearpath Robotics Available at: https://clearpathrobotics.com/ (Accessed: 29 April 2022).

Corinto, G. L. and Pioletti, A. M. (2019). Viticulture and landscape in the Italian northwestern alpine region, *Geography Notebooks* 2(2), 2. doi:10.7358/gn-2019-002-copi.

Cuesta, F., Gomez-Bravo, F. and Ollero, A. (2004). Parking maneuvers of industrial-like electrical vehicles with and without trailer, *IEEE Transactions on Industrial Electronics* 51(2), 257-269. doi:10.1109/TIE.2004.824855.

Fendt (2022). Fendt Xaver Available at: https://www.fendt.com/int/fendt-xaver (Accessed: 29 April 2022).

Fernandes, H. R. and Garcia, A. P. (2018). Design and control of an active suspension system for unmanned agricultural vehicles for field operations, *Biosystems Engineering* 174, 107–114. doi:10.1016/j.biosystemseng.2018.06.016.

Free University of Bolzano. (2022). First lab: field robotics South-Tyrol laboratory. [online] Available at: https://firstlab.projects.unibz.it/it/home/ (Accessed: 29 April 2022).

Fue, K. G., Porter, W., Barnes, E. and Rains, G. (2020). An extensive review of mobile agricultural robotics for field operations: focus on cotton harvesting, *AgriEngineering* 2(1), 150–174. doi:10.3390/agriengineering2010010.

Galati, R. and Reina, G. (2019). Terrain awareness using a tracked skid-steering vehicle with passive independent suspensions, *Frontiers in Robotics and AI* 6, 46. doi:10.3389/frobt.2019.00046.

Garimella, S. S. and Revzen, S. (2021). Dandelion-picking legged robot, Available at: arXiv:2112.05383 [cs, eess] [Preprint] Available at: http://arxiv.org/abs/2112.05383 (Accessed: 29 April 2022).

Gonzalez-De-Santos, P., Fernández, R., Sepúlveda, D., Navas, E. and Armada, M. (2020). Unmanned ground vehicles for smart farms. In: Amanullah, A. (Ed.). *Agronomy - Climate Change & Food Security*. London: IntechOpen. doi:10.5772/intechopen.90683.

Grazioso, A., et al. (2022). Multibody simulation of a novel tracked robot with innovative passive suspension. Submitted to the 4th IFToMM Italy Conference (IFIT 2022), Naples.

ISO (2017). *International Organization for Standardization. Tractors and Machinery for Agriculture and Forestry – Serial Control and Communications Data Network*. (ISO 11783-1:2017).

Khan, R., Malik, F. M., Raza, A. and Mazhar, N. (2021). Comprehensive study of skid-steer wheeled mobile robots: development and challenges, *Industrial Robot: The International Journal of Robotics Research and Application* 48(1), 142–156. doi:10.1108/IR-04-2020-0082.

Kongskilde – Agriculture (2022). *Kongskilde - Agriculture*. [online] Available at: https://www.kongskilde.com/in/en/en/Agriculture (Accessed: 29 April 2022).

Li, Q., Ayers, P. D. and Anderson, A. B. (2007). Modeling of terrain impact caused by tracked vehicles, *Journal of Terramechanics* 44(6), 395–410. doi:10.1016/j.jterra.2007.09.001.

Low, C. B. and Wang, D. (2008). GPS-based path following control for a car-like wheeled mobile robot with skidding and slipping, *IEEE Transactions on Control Systems Technology* 16(2), 340–347. doi:10.1109/TCST.2007.903100.

Mastinu, G. and Plöchl, M. (2014). *Road and Off-Road Vehicle System Dynamics Handbook*. Boca Raton: CRC Press, Taylor and Francis Group.

Mattro, H. (2022). HAWE Mattro GmbH - HAWE Hydraulik. [online] Available at: https://www.hawe.com/daughter-company/hawe-mattro-gmbh/ (Accessed: 29 April 2022).

Molari, G., Bellentani, L., Guarnieri, A., Walker, M. and Sedoni, E. (2012). Performance of an agricultural tractor fitted with rubber tracks, *Biosystems Engineering* 111(1), 57–63. doi:10.1016/j.biosystemseng.2011.10.008.

Nagatani, K., Ishigami, G. and Okada, Y. (2016). Modeling and control of robots on rough terrain. In: Siciliano, B., Khatib, O. (Eds). *Springer Handbook of Robotics*. Cham: Springer International Publishing, pp. 1267–1284. doi:10.1007/978-3-319-32552-1_50.

Naïo (2016a). Autonomous Oz weeding robot, *Naïo Technologies*. [online] Available at: https://www.naio-technologies.com/en/oz/ (Accessed: 15 November 2021).

Naïo (2016b). DINO vegetable weeding robot for large-scale vegetable farms, *Naïo Technologies*. [online] Available at: https://www.naio-technologies.com/en/dino/ (Accessed: 15 November 2021).

Oliveira, L. F. P., Moreira, A. P. and Silva, M. F. (2021). Advances in agriculture robotics: A state-of-the-art review and challenges ahead, *Robotics* 10(2), 52. doi:10.3390/robotics10020052.

Open Robotics (2022). ROS [online] Available at: https://www.ros.org/ (Accessed: 29 April 2022).

PeK Automotive (2022). Slopehelper, How Agriculture Robot Works Available at: http://slopehelper.com/ (Accessed: 29 April 2022).

Prosser, C. W., Sedivec, K. K. and Barker, W. T. (2000). Tracked vehicle effects on vegetation and soil characteristics, *Journal of Range Management* 53(6), 666–670. doi:10.2307/4003164.

Quaglia, G., Visconte, C., Sabatino Scimmi, L., Melchiorre, M., Cavallone, P., Pastorelli, S. (2019). Design of the positioning mechanism of an unmanned ground vehicle for precision agriculture. In: Uhl, T. (Ed.) *Advances in Mechanism and Machine Science*. Cham: Springer International Publishing, pp. 3531–3540. doi:10.1007/978-3-030-20131-9_348.

Quaglia, G., Visconte, C., Carbonari, L., Botta, A. and Cavallone, P. (2020). Agri.q: a sustainable rover for precision agriculture. In: Visa, I., Duta, A. (Eds). *Solar Energy Conversion in Communities*. Cham: Springer International Publishing, pp. 81–91. doi:10.1007/978-3-030-55757-7_6.

Raibert, M., Blankespoor, K., Nelson, G. and Playter, R. (2008). BigDog, the rough-terrain quadruped robot, *IFAC Proceedings Volumes* 41(2), 10822–10825. doi:10.3182/20080706-5-KR-1001.01833.

REFORM (2022). REFORM - metron P48 RC Available at: https://www.reform.at/en/products/metron/metron-p48 (Accessed: 29 April 2022).

Reina, G. and Galati, R. (2016). Slip-based terrain estimation with a skid-steer vehicle. *Vehicle System Dynamics* 54(10), 1384–1404.

Rocos (2020). Autonomous farm work - enter the robots. [online] Available at: https://www.youtube.com/watch?v=RBLnAhzPpTQ (Accessed: 29 April 2022).

Rubio, F., Valero, F. and Llopis-Albert, C. (2019). A review of mobile robots: concepts, methods, theoretical framework, and applications, *International Journal of Advanced Robotic Systems* 16(2). doi:10.1177/1729881419839596.

Ruckelshausen, A., Biber, P., Dorna, M., Gremmes, H., Klose, R., Linz, A., Rahe, R., Resch, R., Thiel, M., Trautz, D. and Weiss, U. (2009). BoniRob: an autonomous field robot platform for individual plant phenotyping, *Precision Agriculture* 9, 841–847.

Saga Robotics (2022). Thorvald Available at: https://sagarobotics.com/thorvald-platform/ (Accessed: 29 April 2022).

SeTeL (2022). Amphibious rover. [online] Available at: http://setelgroup.it/current-projects/amphibious-rover/ (Accessed: 29 Apr. 2022).

Small Robot Company (2017). *Small Robot Company*. [online] Available at: https://www.smallrobotcompany.com/ (Accessed: 29 April 2022).

Stanley Innovation (2022). Stanley innovation | Custom mobile robots, Stanley Innovation, Inc. [online] Available at: https://stanleyinnovation.com/ (Accessed: 29 April 2022).

Tazzari, R., Mengoli, D. and Marconi, L. (2020). Design concept and modelling of a tracked UGV for orchard precision agriculture. In: IEEE International Workshop on Metrology for Agriculture and Forestry (MetroAgriFor) (vol. 2020), pp. 207–212. doi:10.1109/MetroAgriFor50201.2020.9277577.

Tedeschi, F. and Carbone, G. (2014). Design issues for hexapod walking robots. *Robotics* 3(2), 181–206. doi:10.3390/robotics3020181.

Ugenti, A., Galati, R., Mantriota, G. and Reina, G. (2022). Kinematic modelling of a high mobility tracked robot. In: Niola, V., Gasparetto, A., Quaglia, G. and Carbone, G. (eds). *Advances in Italian Mechanism Science. IFToMM Italy 2022.* Mechanisms and Machine Science 122, Cham: Springer. https://doi.org/10.1007/978-3-031-10776-4_5.

Underwood, J. P., Calleija. M., Zachary, T., Hung, C., Nieto, J. I., Fitch, R. C. and Sukkarieh, S. (2015). Real-time target detection and steerable spray for vegetable crops. Proceedings of the International Conference on Robotics and Automation: Robotics in Agriculture Workshop, Seattle, WA, USA 2015.

Unitree (2022). Unitree Available at: https://www.unitree.com/products/a1 (Accessed: 29 April 2022) (vol. A1). [online].

Vidoni, R., Bietresato, M., Gasparetto, A. and Mazzetto, F. (2015). Evaluation and stability comparison of different vehicle configurations for robotic agricultural operations on side-slopes, *Biosystems Engineering* 129, 197–211. doi:10.1016/j.biosystemseng.2014.10.003.

Vidoni, R., Carabin, G., Gasparetto, A. and Mazzetto, F. (2016). Stability analysis of an articulated agri-robot under different central joint conditions, *Advances in Intelligent Systems and Computing* 417, 335–346.

Vidoni, R., Gallo, R., Ristorto, G., Carabin, G., Mazzetto, F., Scalera, L. and Gasparetto, A. (2017). Byelab :an agricultural mobile robot prototype for proximal sensing and precision farming. *Proceedings of the ASME International Mechanical Engineering Congress and Exposition (IMECE 2017)* 2017, 4A, doi:10.1115/IMECE2017-71216.

Visconte, C., Cavallone, P., Carbonari, L., Botta, A. and Quaglia, G. (2021). Design of a mechanism with embedded suspension to reconfigure the Agri_q locomotion layout, *Robotics* 10(1), 15. doi:10.3390/robotics10010015.

Vitibot (2018). Bakus S, Vitibot Available at: https://vitibot.fr/vineyards-robots-bakus/vineyard-robot-bakus-s/?lang=en (Accessed: 15 November 2021).

Vitirover (2022). Home | Grassing management | Automated mower robot. [online] Available at: https://www.vitirover.fr/en-home (Accessed: 29 Apr. 2022).

Chapter 11

Advances in grasping techniques in agricultural robots

George Kantor and Francisco Yandun, Carnegie Mellon University, USA

1 Introduction
2 Robotic manipulators
3 Mechanics of grasping
4 Gripper design
5 Systems integration
6 Real-world applications
7 Conclusion
8 Where to look for further information
9 References

1 Introduction

According to estimates of the Food and Agriculture Organization (FAO), by 2050 global agriculture production has to increase by about 60–70% in order to feed the estimated world population. In this scenario, modern agriculture not only faces a series of challenges to adopt new technologies that allow it to reach such a productivity level but also improves sustainability, resilience and profitability. This is particularly critical for high-value crops that require highly specialized manual labor, whose availability is being reduced due to the aging of the workers or their shift to other jobs with better wages and working conditions.

In this scenario, the research and development of automated and semi-automated robots to perform tasks like mapping, monitoring or phenotyping have produced effective solutions for management and crop assessment, yield estimation, among other tasks (Narvaez et al., 2017; Oliveira et al., 2021). While these perception capabilities provided important breakthroughs for modern agriculture, there are specific applications that require the robots to actuate on the canopy or to manipulate the objects around. Such a physical interaction requires robotic manipulators and grippers to be designed as adaptive and compliant systems not to only pick and place objects but also to ensure a proper handling that maintains the quality of the product and the health of the vegetation.

http://dx.doi.org/10.19103/AS.2023.0124.09

In contrast to the manufacturing industry where robots are widely used to do repetitive tasks in assembly lines, agricultural applications pose extra complications in terms of sensing and actuation capabilities. For example, a robot operating outdoors needs to deal with wind, occlusions or variable lightning conditions that make perception and planning difficult. Furthermore, the objects to be grasped can have different mechanical characteristics, such as shapes and colors, and can be wet (slippery), adhesive or very delicate. For these reasons, the development of robotic manipulators for tasks like harvesting, pruning, thinning, conveying, among others has been very limited. However, there is a considerable body of work in the research and development of robots to perform activities using specialized gripper and manipulator designs, sensory integration and grasping strategies. This chapter reviews the most recent developments in this important research topic and its application in the industry.

At this point, it is important to make some definitions of terms that we will use throughout the chapter in the context of robotics and agricultural applications:

- Manipulator: A robotic arm composed of joints and links that has multiple degrees of freedom (DOF).
- Gripper: End of the arm or end-effector tool mounted in a manipulator that can handle, carry or place objects.
- Manipulation: The act of changing the position and orientation of objects that are grabbed by the robot's gripper.
- Grasping: Using an end-effector tool (i.e. gripper) to obtain complete control of the movement of the object.

2 Robotic manipulators

A robotic manipulator is a mechanism that comprises multiple rigid segments linked with joints driven by motors. They usually contain end effectors that are tools attached to the last link, which is in charge of gripping and grasping tasks. Robotic manipulators have been used in a wide variety of industrial and research applications like manufacturing, assembly lines, aerospace, among others for improving the efficiency and productivity of the processes. For these industries, there is a variety of commercial devices in the market that provide robust and reliable products like Kuka, Universal Robots, ABB, Franka Emika and UFactory. Agriculture has not been the exception in the use of these robotic devices in tasks like pruning, harvesting, spraying and weed control (Abdeetedal, 2017; Tinoco et al., 2021; Vision Robotics Corp., 2020).

Given the unstructured nature of the agricultural environment and the delicacy of the plants, the literature also reports on custom designs of manipulators. Concepts like compliance, workspace, actuation type and dynamics have to be addressed while designing. For example, the kinematic design and simulation of a manipulator for selective fruit harvesting and precision spraying of pesticides were presented by Baur (2015). It provides a comprehensive overview of the design process that includes analysis of the workspace, the actuation principles and the hardware and control architectures. Furthermore, some authors report the integration of commercial manipulators in systems that add a degree of freedom to expand the workspace, as reported by Silwal et al. (2021). In the same way, integrating mobile bases extends the mobility of the overall system, allowing for a free movement in an extended space. Figure 1 shows some examples of custom designs of manipulators and end effectors for agricultural tasks.

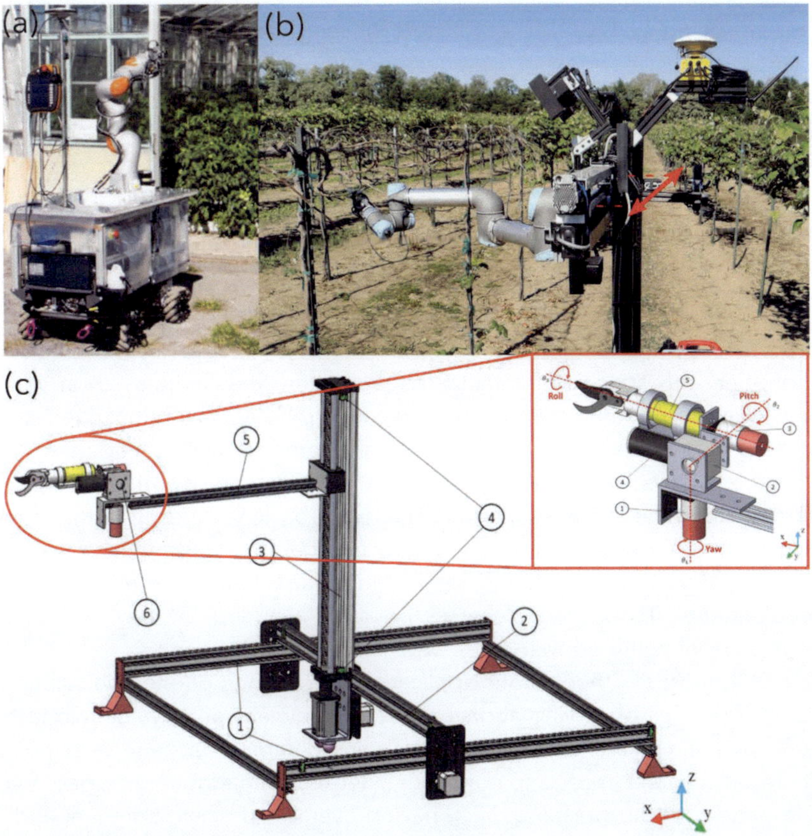

Figure 1 The design of robotic manipulators for agricultural tasks consists of giving extra degrees of freedom to commercial devices or designing custom robots, according to the task to be performed. (a) Commercial manipulator mounted on a mobile base to detect pests in a greenhouse (Martin et al., 2021). (b) A commercial manipulator mounted on a linear slide to obtain a 7 DOF system for automated pruning (Silwal et al., 2021). (c) A cartesian manipulation system and its end effector (Zahid et al., 2020). Figure (c) reproduced with permission from Elsevier.

3 Mechanics of grasping

According to Fantoni et al. (2014), the grasping process consists of six phases: approaching, coming into contact, increasing the force, securing, moving and releasing. A robot emulating skilled workers needs to execute these phases with the same dexterity and care as a human. For example, depending on the fruit or vegetable to be handled, humans can pick it with two fingers or the complete hand. In contrast, robots need actuators or grippers that suit the specific task, executing these phases with customized hardware and software designs. They include perception systems and actuators that account for several factors, namely the relationship between the mechanical properties of the objects to be manipulated, their geometry, maturity level and the environment in which they are being cultivated. These factors influence the kinematic and dynamic analyses not only during the gripper's design process but also in its overall performance. Furthermore, they may be used to transform the crops in a way that its architecture becomes more amenable to introduce and deploy robotic solutions.

3.1 Problem formulation

The highly heterogeneous characteristics of agricultural processes and products make the design and development of robotic grippers a customized process. The same task executed for different fruits or vegetables would need a gripper with specific characteristics. The same case applies even when working with the same product, but executing different tasks (e.g. harvesting or sorting). In either case, the overall design process starts recognizing key operational, mechanical and perception requirements by (i) observing a human doing the same task; (ii) extracting the basic features in terms of the visual appearance of the product, interaction geometry and mechanics and (iii) analyzing the dynamics of the grasping interaction and the compliance requirements. These variables can be later included in the mechanical development of the gripper.

As in most of other robotic applications, the design of effective solutions comes from first observing and studying how humans perform certain tasks. For example, Fig. 2 shows a human grabbing different fruits and vegetables. There are different aspects to note here in terms of the ergonomics, geometry and pattern of the grasping. First, having over 30 muscles, 5 fingers and 3 phalanges per finger allows a human to hold any medium-sized object firmly and securely. While such an amount (or more) of joints and links would be desirable in a gripper, several tradeoffs in terms of mechanical design and complexity of control are considered when designing fingers, as will be detailed in Section 4. Another aspect that comes into play in the problem formulation is the fruit shape. As the second column of Fig. 2 shows, grape clusters are highly

Figure 2 Samples of a human hand grasping different fruits and vegetables. Depending on the activity and the object shape, the way it is grabbed can be significantly different in terms of morphology and pressure applied.

irregular, which makes the grasping approach significantly different from the other objects.

Finally, it is worth noting that the fragility of the object also imposes several constraints on the grabbing pattern and the compliance requirements. Grasping the tomato with excessive force will cause a textural failure and the decrease in the product quality. Likewise, despite having roughly the same shape (and consequently a similar grabbing pattern), grasping the egg and the avocado require different firmness and compliance levels. Naturally, the maturity stage of the fruit also plays an important role when assessing these variables.

The type of analysis described earlier has been used by several works that also accounted for ergonomics and geometry variables to delineate guidelines for the design of grippers handling agricultural products (J. Li et al.,

2016; Hou et al., 2021). Particularity, X. Chen et al. (2019) studied the effect of fruit characteristics, hand dimensions and slip parameters in the effective grasping of fruits for sorting purposes. These variables were later used to fit a logistic regression model to predict the probability of stable grasping with two fingers for apples, mangoes and melons. In addition to these human taxonomy variables, the growing environment of the product is also another important factor to consider at this stage. Particularly, the posture of the plants and their arrangement in the canopy has also been accounted for during the synthesis of fruit-harvesting robots (Li et al., 2019).

While the common approach is to design the robot (or the gripper more specifically for this chapter subject), as a function of the crop or the targeted task, there is a rising trend in using a sort of closed loop where not only the environment influences the gripper design but also the environment itself can be modified for an optimal robot performance (Bloch et al., 2018). For example, van Herck et al. (2020) presented a methodology to modify the architecture of crops in a way that provides a favorable environment for robotic harvesting. A case study in a sweet pepper cultivar, which was conducted over 3 years, evaluated the effects of changing illumination, climate control and cultivation techniques in the performance of harvesting robot. Architectures that yielded sweet peppers to be visible and reachable in the canopy were found to significantly influence the robot's operation.

3.2 Kinematics

In general, the mechanisms and characteristics of human hand grabbing are factors that robotic grippers - ideally - resemble. Among these, grasp ergonomics and object geometry are key to delineate the configuration of the gripper, its design and control strategy. These two are deeply related since how an object is handled depends on its geometry. In this section, we revisit the gripper kinematics assuming the perception system provides an accurate model of the object and its pose with respect to the object (perception details are revisited in Section 5.1). It is worth noting that various perception systems produce point clouds that can be later converted to finer (and more continuous) surface representations (i.e. meshes) (Guo et al., 2020). For this reason and without loss of generality, we will also assume that this model is a point cloud, when pertinent.

Most fruits and vegetables can be modeled in a basic form using primitive shapes like cylinders and spheres. Spherical shapes can be described with two variables: their center location and radii. Cylinders, on the other hand, can be parametrized as function of their radii and the direction of their longitudinal axis. Using these simplified representations has proven to be useful for finding the best grasping postures in known databases. For example, Guo et al. (2020)

represent oranges, cucumbers, strawberries and other fruit and vegetables as cylinders, cones or spheres. The point cloud of the object was then matched with templates of these geometrical primitives. Subsequently, certain descriptors or keypoints representing the best points of contact in these surfaces were calculated. Finally, the appropriate grasping configuration is determined using GaspIt! (Miller and Allen, 2004). In a similar way, several works report the use of software-aided modeling and synthesis of grippers to obtain optimal kinematic configurations for grasping certain vegetables (Ronzhin et al., 2022; Russo et al., 2017). For example, Fig. 3 shows three kinematic schemes whose parameters can be optimized using the software AgroGripModeling (Ronzhin et al., 2022).

As mentioned before, the kinematic design of grippers for picking fruits can be driven by the analysis of human behavior. Factors like the use of opposable thumbs, orientation of the object or the desired action affect and direct how the hand grabs it (Lee and Jung, 2014). Following this idea, researchers can find a scientific guidance on the synthesis of the geometry (among other elements) design of grippers for agricultural applications. For example, Li et al. (2019) used a statistical analysis to find the relationship between the growing environment, fruit size, human body characteristics and choice of grasp type. Experiments in tomatoes showed that simultaneously wrapping the fingers, palm and thumb around the fruit is the most frequent grasp type (see Fig. 2, second row first column) for an average size tomato. Using this information, the researchers can design grippers with multiple fingers that approximate this grasping pattern. Other approaches can use larger datasets that contain hand movements and grasp kinematics for the human hand (Saudabayev et al., 2018; Jarque-Bou et al., 2020). While most of the work using such databases

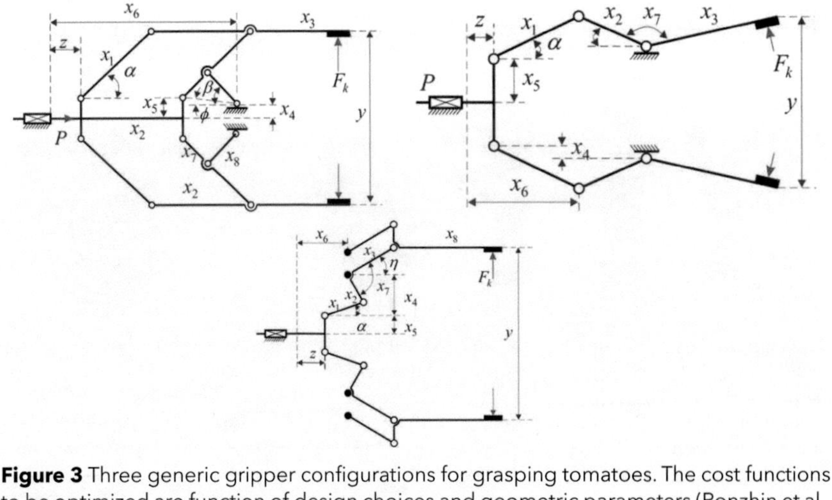

Figure 3 Three generic gripper configurations for grasping tomatoes. The cost functions to be optimized are function of design choices and geometric parameters (Ronzhin et al., 2022). Figure reproduced with permission from Springer.

is related to rehabilitation, it has potential to be used in agricultural research since plenty of the primitives of a human grasping objects in daily living can be used for grabbing fruits and vegetables.

3.3 Dynamics

Fruits, vegetables and food, in general, have different mechanical characteristics that have to be accounted for, in order to handle them in a proper way. Thus, the gripper needs to apply a force that not only ensures a secure grasp but also prevents any bruising or damage in the object. Researchers have tried to evaluate parameters like applied force, bending of the fingers or location of contact points using instrumented gloves worn by people when grabbing spherical objects like apples, oranges and tomatoes (see Fig. 4). After analyzing two different ways of grasping the objects on the basis of the parameters like force, bending, pressure, finger cooperation and acceleration, Zheng et al. (2021) found that grasping the objects from the top requires a more complex grip mechanism compared with grasping from the side. Another type of work models fruits and vegetables as mechanical systems composed of elastic and viscous elements (i.e. spring-damper systems) that can suffer elastic or plastic deformations (Zhang et al., 2018; An et al., 2020; Ji et al., 2021). These models can be later used to improve the design or the grasping control strategy by predicting the force required to pick fruits and vegetables without producing plastic deformations that will diminish the product quality.

In tasks like harvesting that involve not only picking the fruit but also detaching from the tree, the analysis of the dynamics plays a crucial role. Understanding the tension and force required when twisting, horizontally

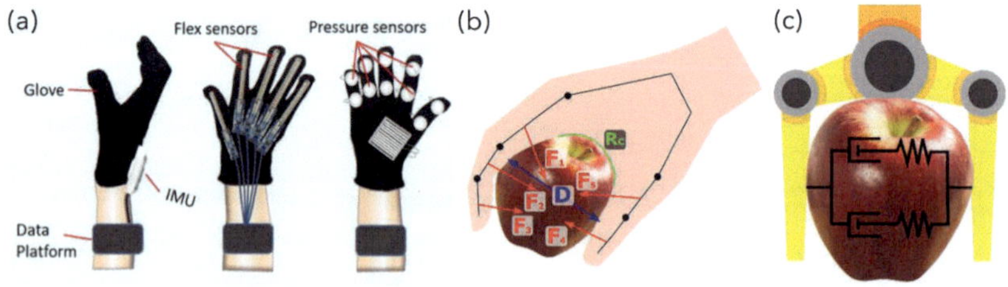

Figure 4 The dynamic analysis of the physical interaction between the gripper and the object provides design guidelines for actuators and the morphology of the gripping. (a) Gloves with force, bending and pressure sensors to create grasping maps and aid in the design of grippers for delicate fruits (Zheng et al., 2021). (b) Optimal grasping designs can be obtained from analyzing the geometric and mechanical variables of the fingers when holding an object. (c) Modeling the behavior of the objects as spring-damper systems allows to design grippers to avoid bruising or damaging them (Ji et al., 2021). Figure c reproduced with permission from Elsevier.

bending or vertically pulling at collection time can be helpful to design grippers that effectively harvest apples (Bu et al., 2020). Another important variable is the location of the contact points between the fingers (or claws) of the gripper and the object. In this context, stability tests that accounted for the trade-off between the number of contact points and the grasping stability were presented by Li et al. (2013). Regarding secure grasping, slip is another variable that plays an important role to determine the optimal gripping force required to grab and grasp different fruits (Tian et al., 2018).

Simulations and finite element analysis have also been employed to estimate the intensities of forces and stresses endured by the fruits and the gripper during various handling processes (Zulkifli et al., 2020; Cai and Shi, 2020). This type of simulation also aided in the design of rigid and soft grippers to securely grab the fruits under different loads and configurations (Zhou et al., 2021; Li et al., 2021; Song et al., 2021).

4 Gripper design

Companies like Festo, Schunk, Robotiq, SMC, among others, produce and sell industrial-grade grippers for various applications like manufacturing and assembly. However, when it comes to agricultural applications, the research work mostly has focused on designing custom devices that can work for a specific set of products or workspace characteristics. For example, a pneumatic gripper can provide a good grasping for apples but would not be very effective for grapes. In this section, we will revisit the design choices commonly used to address problems in agricultural applications like harvesting, handling and transportation.

4.1 Actuation mechanisms

Given the complexity in the design and control of fully actuated grippers, researchers have focused on the use of servo or stepper motors that move tendons that achieve a higher cost-complexity-benefit balance (at the expense of losing precision or dexterity) (Choi et al., 2017). These underactuated grippers have been widely used in applications like food handling, picking or harvesting (Su et al., 2023; X. Song et al., 2016). Particularly, Silwal et al. (2017) deployed and validated in the field an apple-picking gripper with a pulley system that pulled the tendons - made of fishing lines - using two servo motors. Results showed an 85% of successful detachments of the fruits in canopies with planar architecture. Another work presented a low-cost and portable gripper that used a gearbox to directly transmit the movement of the servomotor to the fingers (Denarda et al., 2021). The field tests of this design showed that it is suitable to grasp lightweight thin structures like the stems of flowers. Figure 5 depicts the application of these works.

Figure 5 Grippers actuated using servo motors used in harvesting tasks for apples (Silwal et al., 2017) and saffron (Denarda et al., 2021).

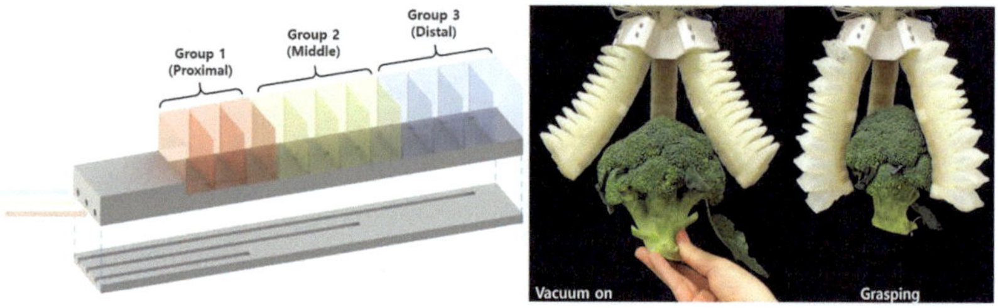

Figure 6 Soft gripper actuated with air. Different air streams in the phalanges of the fingers provide variable stiffness to increase compliance (Song et al., 2021).

While electric actuation is the most utilized in general, some works also report the use of pneumatic grippers with suction cup mechanisms to grab the objects. For example, Vu and Ronzhin (2020) and Chiu et al. (2013) present the design of a four-finger gripper that includes a vacuum suction nozzle to improve the grip. A single linear motion transmitted by a rack and a pinion controls the movement of the fingers and the nozzle. Despite both works presenting similar designs, only Chiu et al. (2013) tested the gripper with real tomatoes (Fig. 6). Aside from research work, companies like Soft Robotics (SoftRobotics, 2022) are manufacturing and commercializing pneumatic grippers that can adapt to different types of fruit, vegetables or even animal products.

4.2 Finger design

Fingers provide a versatile way to decide – at the design level – the number of contact surfaces that could satisfy the requirements of precise and effective grasping without damaging the products. As mentioned before, the fingers are usually fully actuated with individual motors or underactuated employing gear systems and tendon-like joints (Yan et al., 2019; Russo et al., 2017). Additionally, depending on the kinematic and ergonomic analysis at the design stage, the number of fingers can be decided based on the requirements like power consumption, firmness of the grasping or desired points of contact. For instance, fully actuated anthropomorphic grippers can provide precise and high manipulability, with the potential of operating in food and fruit processing lines (Yang et al., 2021). However, few works in the state of the art report their use in practical agricultural applications (when compared with underactuated fingers). One of the reasons could be related to the passive compliance characteristic of most of the underactuated grippers. For example, Davidson and Mo (2015) presented the design of a gripper with three fingers placed in a Y-shaped distribution. Each finger has a fixed base and two joints to extend the adaptability of the gripper for the fruit shape. While the presented design looks promising for picking apples, more tests are necessary to find the 'right' amount of force required to ensure a proper grasp of the fruit. The passive compliance incorporated in this design allowed a deflection in the distal flexure in the fingers when an unexpected collision took place (Silwal et al., 2017). Another work reported the use of a three-finger gripper to handle strawberries. The design included a polyurethane foam in each finger to improve the grip and to distribute the forces during handling, thus preventing the damage of the fruit (Dimeas et al., 2015). While not exactly a gripper design, researchers have also evaluated the use of claw-like end effectors to pick fruits (Cai and Shi, 2020). However, the main drawback of this design lies in the requirement of a highly precise localization of the fruit, otherwise the claw can destroy the fruit when closing. Figure 7 shows some of the grippers that use fingers in different configurations to grasp fruits and vegetables.

As previously discussed, the design of grippers uses a human expert to evaluate all the variables and account for the design criteria. However, recently, other approaches, not directly applied to agriculture, that aim to automate the overall design process itself have been proposed (Honarpardaz et al., 2017). In particular, Hazard et al. (2018) used a series of optimization steps to find the optimal structure of the gripper (i.e. the number of joints per finger) for grasping objects with different shapes. Despite being intended for industrial applications, these approaches show a high potential for designing custom grippers based on high-level user requirements (e.g. for grasping spherical fruits, with specific moving patterns).

4.3 Soft grippers

Using soft and deformable materials for gripper manufacturing has shown great convenience when handling delicate objects in different industries. Variables like stiffness or compliance level (which are constrained by the actuation method and fabrication materials) greatly define the suitability of a soft gripper for agricultural applications. For example, the gripper presented by Firouzeh and Paik (2017) provides multiple grasp models and adjustable

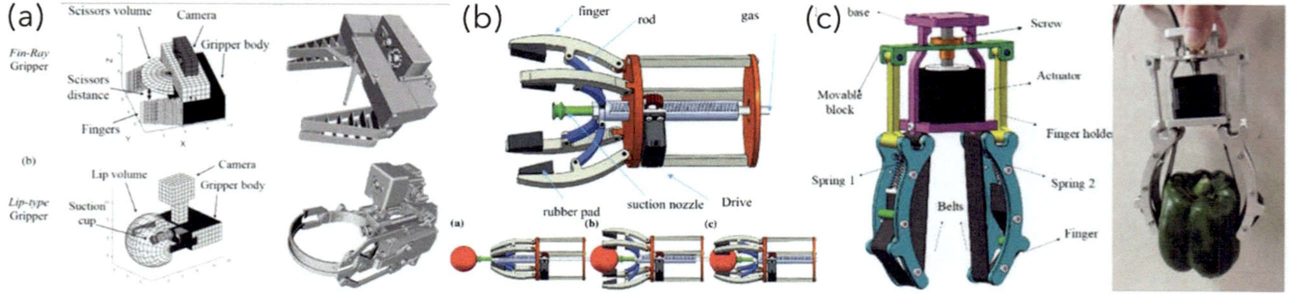

Figure 7 Different types of finger designs. (a) Grippers with local sensing arrays and custom designs have been tested in simulation prior to their practical assembly (Eizicovits et al., 2016). (b) Gripper designs combining electrical and pneumatic actuators provide a more secure grasping of spherical fruits and vegetables (Vu and Ronzhin, 2020). (c) Designs also aim to provide a firm but adaptive way to grasp the fruits and vegetables, preventing slip and damage (Xie et al., 2020). Figure (a) reproduced with permission from Elsevier. Figure (b) reproduced with permission from Springer.

stiffness; however, these properties change as a function of the temperature of the fingers. This makes the control slow and difficult, limiting their practical application in the field. In contrast, Bauer et al. (2022) presented the design of a five-finger robotic soft hand built with low-cost flexible filament that can be printed in three-dimension (3D). The position of the fingers is controlled with tendons that are actuated with servo motors. Although the gripper was tested with a variety of objects, the authors discuss the potential use in agriculture due to the quality of the grasping and the durability and robustness of the materials.

When dealing specifically with fruit and vegetable grasping, several works report the design and evaluation of soft grippers (Navas et al., 2021). Low-cost solutions of 3D-printed prototypes for fruit picking have been presented as proof of concepts that show adaptability, stability, and compliance when handling fruits, without damaging them (Li et al., 2021; Liu et al., 2018). Other more elaborated designs used fingers designed using a customized particle jamming approach. In order to grasp objects with irregular shapes, the fingers can modify their curvature by having different airflow streams in their proximal, middle and distal sections. Gripping stability tests carried out in different fruits and vegetables showed the effectiveness and adaptability of the gripper in agriculture (Song et al., 2021).

Aside from research work, companies like Soft Robotics (SoftRobotics, 2022) are manufacturing and commercializing pneumatic grippers that can adapt to different types of fruits, vegetables or even animal products.

5 Systems integration

According to Fantoni et al. (2014), their grasping process consists of six phases: approaching, coming into contact, increasing the force, securing, moving and releasing. In general, all these phases require the use of a series of perception and control algorithms for processing the sensory data and giving the proper control commands to the actuators to successfully complete the task. In particular, agricultural settings present an extra layer of complexity as the robots are usually operating in outdoors with cluttered scenes and delicate objects. This section will review the most used sensors, planning and control strategies that researchers have used to grasp fruits and vegetables.

5.1 Perception

In order to approach the fruit in a way that ensures a proper contact, the robot needs to have an understanding of the geometry of the surrounding scene and the pose of the objective. The capability of spectral and visual cameras to capture texture and geometric characteristics of the environment has made them the most popular sensors in manipulation and grasping (C. Wouter Bac et al., 2014; Kadeghe G Fue et al., 2020). If they use a stereo configuration or active lightning principles, the robot can also retrieve 3D awareness of the scene. In this context, Hemming et al. (2014) and Eizicovits et al. (2016) used time of flight (ToF) and color cameras mounted on the gripper to create graspability maps (i.e. spatial representations of the grasping quality for different approaching poses) for two gripper configurations. In another work, a couple of stereo cameras with active lightning were used by Silwal et al. (2021) to create a 3DI structure of dormant vine trees for pruning applications. More recently, machine learning strategies have also been proposed to use RGB-D images for estimating a proper grasping pose, as reported by X. Wang et al. (2022). This study proposes a geometry-aware architecture to segregate fruits and obstacles in the environment and estimate the most effective grasping pose to harvest apples. Each network builds on top of known semantic segmentation and point cloud detectors (i.e. YOLACT and PointNet), fine-tuned for working with apples in cluttered canopies. Tests in real working conditions report success detection rates between 56% and 81.3%. However, the perception system suffers when the lightning differs from the training images.

In the process of automatically grasping delicate objects like fruits and vegetables, the grasping action needs to be exerted in a way that ensures a firm grip without compromising the quality of the product. For this reason, at the contact stage, a critical piece of feedback comes from force, pressure and bending sensors mounted in the gripper. A common approach to understand how these variables interact and distribute during the grasping process is to use sensorized gloves. They produce tactile maps that are very useful to understand the distribution of the forces in different parts of the hand, allowing to optimally place the sensors during the design stage (Sundaram et al., 2019; J. Wang et al., 2021). For example, Zheng et al. (2021) designed an e-glove to measure and map the tactile information during the picking and grasping of round fruits like apples. The glove was equipped with bending and force sensors. An additional inertial sensor was used complementary to estimate the posture of the hand. Results showed correlation between variables like pose of the hand and fingers, hand bending and force applied, providing a theoretical guidance for the design and control of robotic grippers. In fact, when it comes to working prototypes, tactile sensors that produce force feedback have been used in harvesting applications, as reported by Zhou et al. (2021). A total of 24 piezo-resistive sensors, mounted in the fingers of a soft robotic gripper, were used to classify four types of contacts: null grasp, good grasp, finger interference grasp and obstacle-obstructed grasp. This perception pipeline allowed us to successfully grasp apples in laboratory experiments. Furthermore, Kultongkham et al. (2021) used a customized force feedback sensing scheme for a pneumatic soft gripper. The prototype was tested for grasping tomatoes, showing a firm grip, without affecting the quality of the vegetables.

5.2 Motion planning and control

The robotic reaching, grasping and manipulating of crops involve planning and control problems that are highly complex given the conditions of the environment and the constraints it imposes on the robot. Entanglement of the branches, occlusion of the objects of interest in the canopy and motion of the objects due to wind are some of the factors that add complexity for the development of planning and control strategies. Without loss of generality, we will refer to harvesting, but all the control methods discussed here are also valid for other manipulation and grasping tasks. Harvesting fruits can be interpreted as a sequence comprising six steps: (i) scout the canopy in order to locate individual fruits, (ii) select one, (iii) place the tool in a way that it is feasible to reach the fruit (without damaging the tree), (iv) move toward the cutting point, (v) make the cut and (vi) grab the fruit to place it on the container. Partitioning the global task in this way allows to explore different approaches to address each one.

In order to address each of these intermediate steps, classic and learning-based approaches have been employed. Classic methods employ control theory and path, or trajectory planning methods based on the mathematical models of the system (that comprises the robot and its surroundings) that account for its kinematics, dynamics and search algorithms. This approach implies a parametric description of the robot and its interaction with the environment (Tao et al., 2017). For example, Killpack et al. (2016) used a model predictive controller (MPC) in a mobile manipulator to reach desired positions in an environment that contained cluttered foliage. Furthermore, a heuristic-based planner was used by Tanigaki et al. (2008) to avoid collision with the tree trunk in cherry harvesting tasks. Similarly, a visual servoing system for reaching fruits in occluded simulation environments was presented by Zapotezny-Anderson and Lehner (2019) and Dewi et al. (2018). Furthermore, a balanced bi-rapid exploring random tree planning algorithm to reach the desired reaching configuration of a nine DOF manipulator was used by Cornelis Wouter Bac (2015) for reaching sweet pepper harvesting. For the grasping stage, methods based on fuzzy logic or admittance and impedance control have been proposed (Dimeas et al., 2015; Wen et al., 2020; Ji et al., 2021).

The learning-based approaches mainly consist of the use of neural networks that 'learn' to represent the robot and its interaction with the environment using a black box model that takes as input the sensory data and outputs the control actions (Bezak et al., 2014; Watson et al., 2017). In this framework, recent works have reported the use of reinforcement learning (RL) strategies (Sutton and Barto, 2018) to solve complex manipulation and grasping problems with prominent results in several applications (Shahid et al., 2020; Nguyen and La, 2019; P. Chen and Lu, 2021). As reinforcement learning approaches are generally trained in simulation, their deployment in real conditions is recognized as a limiting factor, especially for agriculture (Rizzardo et al., 2020). However, a recent work reported the use of an RL policy to control a seven DOF manipulator to reach the pruning location in vines (Yandun et al., 2021). Tests in laboratory settings reported promising results for simple tree structures.

6 Real-world applications

As mentioned before, the complexity of the agricultural environments (that consist in occluded, dynamic scenes) makes the transition from laboratory tests and prototypes to real-world applications really difficult. Despite these challenges, several works report applications and developments working in real conditions, mainly in controlled environments like greenhouses and nursery farms (Zhang et al., 2020).

6.1 In-field tasks

Harvesting and pruning are the tasks that have been widely addressed for certain crops, where the shape and hardness of the objects allow for a robot to manipulate and grasp them without affecting their quality. For example, in the late 1990s, a system to bulk harvesting apples was presented by Peterson et al. (1999). It used a robotized gantry system to move in the *x*, *y* and *z* axes and a hydraulic end effector to reach the fruits in an Empire/M9 tree orchard. While this platform had a human operator to select the place to make the cut of the apple cluster, he used a camera that counted on the end effector to identify the cutting position. To limit the effect of the lightning, a neoprene drapery was placed over the row. The same approach to mitigate the change in illumination was used by Baeten et al. (2008) in their vision system for harvesting individual apples. The actuation system consisted of a six DOF industrial-grade robot and a soft gripper with a suction mechanism. The field experiments showed that about 80% of the apples in the row were detected and harvested. In a more recent work, a six DOF manipulator mounted on a mobile base was used to harvest apples in the field, as described by X. Wang et al. (2022). A deep learning approach was used for perception and a soft end effector, reaching up to 85% of the successful harvesting rate in various natural scenes outdoor. However, this work emphasizes that illumination changes are still a drawback in the visual sensing pipeline. Even though these works span 20 years of difference, they still struggle with the same perception challenges. Another example is given by Vision Robotics Corp. (2020), which was an attempt to develop and commercialize a robotic pruning system for vines. They also addressed the background and illumination challenges by employing a canvas that completely covered the vines during imaging. To address these issues, an illumination invariant camera system and novel approaches for perception and manipulation were used by Silwal et al. (2021) for pruning vines. A seven DOF manipulator equipped with a custom cutting system was mounted on a mobile robot to prune 20 vines. A total of 87% of the canes were pruned successfully, with an average cycle time of 213 s/vine from two sides and 137 s/vine from one side.

Aside from fruits, other works also report the use of manipulators and custom grippers to harvest and pick specialty crops or other plants, in general. For example, a robotized module for harvesting sweet peppers in a greenhouse was employed and evaluated by C Wouter Bac et al. (2017). The design of the manipulator allowed for seven DOF and the end effector consisted of a customized pneumatic gripper. The testing conditions consisted of crops with occluded object clusters and leafy canopies, which resulted in up to 6% of success in the harvesting rate. For harvesting tomatoes growing in a similar greenhouse environment, a Selective Compliance Assembly Robot Arm (SCARA) was equipped with a massive gripper that reached 50% of success (Kondo et al., 2010). However, the size and complexity of the gripping mechanism seem unpractical to operate without damaging the trees. In another harvesting application, a mobile platform with a dedicated manipulator and end effector was also proposed by Kadeghe Goodluck Fue (2020). The overall system comprises a mobile platform and several sensors for autonomous navigation and perception, in general. The manipulation system consisted of a three DOF cartesian configuration with hydraulic actuation. The gripper was designed as a rotating brush roll that grabbed the cotton balls and a vacuum system transported them to the container. Results of the field tests showed the reachability limitations of the three DOF and the difficulty in having a precise position control. However, the system showed a picking success ratio up to 78.58%, which is promising. Pruning and harvesting are not the only activities that have been addressed using robotic manipulation systems. The design and field evaluation of a customized gripper for sampling sorghum stalks was proposed by Mueller-Sim et al. (2017). The manipulator used a SCARA-style configuration, with a vertical prismatic joint and two horizontal revolute joints. The end effector contained a penetrometer that allowed us to measure the strength of the stalks. Field tests showed a grasping success rate of 100% when the plants were static. However, the system performance decreased when the canopy was agitated by the wind.

Out-of-the-box thinking approaches also have been presented in the literature. The concept of an unmanned aerial vehicle (UAV) equipped with a custom gripper to prune tree branches near power lines was presented by Molina and Hirai (2017). The overall system comprises two parts, an UAV and the pruning mechanism (PM). The UAV will be in charge of flying near the trees and allowing the gripper to attach to the branches. A crawling system and a circular saw mounted on the gripper will allow to move the PM in the branches to perform several cuts. This work is only a concept proposal but poses interesting ideas and challenges for pruning tasks. In contrast, a similar mechanism for crawling and cutting coconuts was proposed and implemented by Megalingam et al. (2020). The climbing structure consisted of a ring-shaped body with a spring-wheels system that allowed the robot to grab and climb the tree trunk. A circular saw teleoperated from the ground allowed the robot to make the cutting of the fruit when the desired height was reached. The system was tested in real trees, obtaining success rates up to 100%, but was found to be slower than a human worker. However, the advantages in terms of safety make this robot an interesting development for coconut harvesting.

6.2 Food handling and transportation

In contrast with other transportation and manufacturing industries, handling food encompasses a series of challenges in additional to the technical ones described before. Sanitation, quality control or safety are some of the additional standards that robots working in this industry have to comply. Furthermore, the majority of the food products are non-rigid, and their geometry may change as a function of the handling process itself. According to Chua et al. (2003), the product manipulation behavior is affected mainly by four factors: (i) environmental conditions, (ii) compliance, (iii) fragility and (iv) contamination susceptibility. While certain robotic solutions can be transferred for other sectors, these special requirements lead to the development of unique devices that can improve the efficiency of the overall process in different industries (Lien, 2013).

Meat, fish and poultry processing manipulators have been proposed since the late 90s. For example, a prototype of a washable manipulation system for sheep and lamb carcass processing was proposed by Taylor and Templer (1997). The robot body was manufactured with aluminum and stainless steel for corrosion resistance. The end effector used a pneumatic cutting tool to perform pre-pelting operations on a sheep carcase prior to other manual actions. In a more recent work, a two-robot approach was used to manipulate elastic objects by Navarro-Alarcon et al. (2016). Laboratory experiments in meat showed the method has the potential to handle deformable objects. With respect to the fishing industry, an interesting work describing the design of an underwater robot system for sea cucumber fishing was presented by Shi et al. (2022). The robot included a multi-DOF manipulator with a mechanical claw to catch the animals. While only simulation tests were provided, the overall idea has potential to be applied in the field of aquatic products. The poultry industry has also had research works in the design of systems, mainly for evisceration and deboning (Liyanage et al., 2010; Ahlin, 2022).

In terms of transportation, a key advantage of robots over human-based workflows is the ability to perform repetitive tasks without the uncertainty that labor shortages produce. This is also true in food transportation systems, where robotic prototypes have been proposed for assembly lines of several products like egg packaging, bakery, pasta production, among others (Gray and Davis, 2013; Jiang et al., 2011). Recently, the use of machine learning strategies and RL, in particular, have gained attraction to develop manipulation and gripping strategies to handle and cut food (Kroemer, 2021; Mitsioni et al., 2019; Padalkar et al., 2020). Most of the previously mentioned work consists of prototypes or research work with limited commercial scope. However, companies like Soft Robotics (Whitesides, 2018) provide automated picking solutions for a variety of assembly lines.

Figure 8 summarizes some of the works that have been designed, evaluated and deployed in real-field conditions. Even though they are still prototypes, they provide interesting insights into the state of the art in field robotics and the challenges for going beyond laboratory conditions.

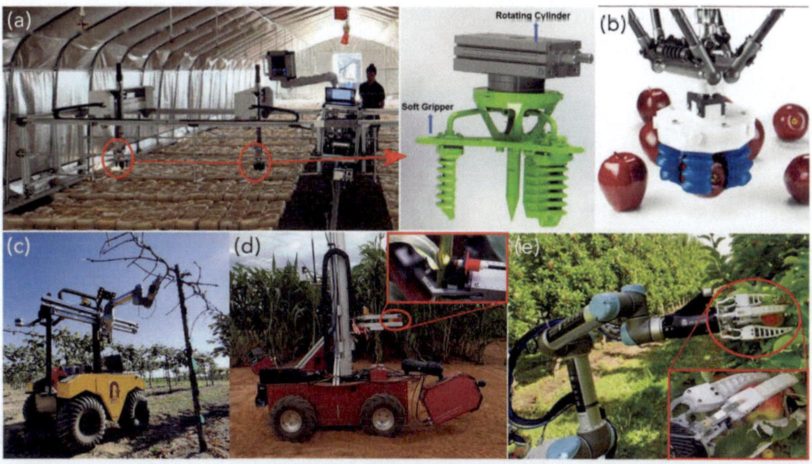

Figure 8 In complement to laboratory tests, robots operating in field and industrial lines have also been developed. Harvesting, pruning, sampling and transportation are examples of tasks to be automated with custom manipulators and grippers. (a) An autonomous robot for harvesting oyster mushrooms in a greenhouse, developed by Rong et al. (2021). (b) Commercial grippers are available in the market to be employed in assembly or transportation lines (Whitesides, 2018). (c) An autonomous mobile manipulator to prune vines in field conditions was developed and evaluated by Silwal et al. (2021). (d) Mueller-Sim et al. (2017) designed and deployed in sorghum fields The Robotanist, a mobile robot with an end effector capable to measure the stiffness of the stalks. (e) A commercial manipulator equipped with a soft gripper to harvest apples (X. Wang et al., 2022). Figure (b) reproduced with permission from John Wiley and Sons. Figure (e) reproduced with permission from Elsevier.

6.3 Case of study: design of a gripper for sorghum breeding

This section describes the design and evaluation of a grasping mechanism to deploy a rind penetrometer in sorghum stalks. The measurement from this sensor is useful to plant breeders and geneticists because it provides an estimate of stalk strength – an important breeding trait. The gripper is mounted as the end effector of a custom manipulator, which in turn is mounted on a mobile base in order to explore the environment and take several samples of different plants. The perception system uses a stereo camera in order to detect the stalks and guide the deployment of the penetrometer. The description of each of these subsystems is out of the scope of this chapter, but the interested reader can refer to Parhar et al. (2018), Jenkins and Kantor (2017), and Mueller-Sim et al. (2017).

The sole purpose of the end effector is to deploy a rind penetrometer. The end effector must accommodate different stalk diameters and strengths as well as be robust to manipulator positioning inaccuracies. The exact requirements are outlined in Table 1.

The process of autonomously applying a rind penetrometer requires two operations: (i) positioning a support finger behind the stalk and (ii) pushing the needle into the stalk. Several mechanisms were investigated for these two operations.

A system that plunges a needle into a stalk requires linear motion, and traditional methods of converting motor torque into linear motion include the following: Acme screw, ball screw and rack and pinion. Unfortunately, there currently do not exist off-the-shelf linear actuators that meet the combined requirements of travel length, torque output and speed, and a rack and pinion would not easily fit within the desired arm link geometry. Maxon Motor Corporation fabricates ball screws and Acme screws that are integrated with motor gearboxes, and ball screws offer very high efficiency in a small package. The final design for the plunger mechanism consists of a Maxon EC16 BLDC motor and a Maxon GP16S stainless steel ball screw.

Table 1 Interpreting the adequacy of dairy farm management from cow milk yields

Mass	Less than 2 kg
Sealing	Mechanisms concealed from the environment
Penetrometer accuracy	Measure forces up to 200 N with 0.25 N resolution
Travel length	80 mm minimum travel
Travel accuracy	Measure stalk diameters with 1 mm accuracy
Speed	Full travel in less than 1 s
Torque	Apply force greater than 200 N
Design	Mechanically simple and easy to disassemble

Range in average herd milk yields on tropical Asian dairy farms.

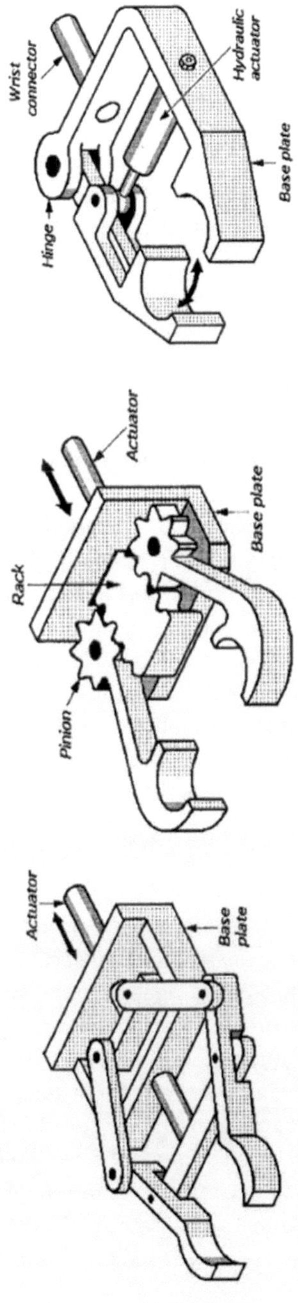

Figure 9 Grasping mechanisms evaluated for stalk grasping.

The secondary mechanical operation requires a finger or multiple fingers to be positioned behind the stalk to support it as the needle penetrates. Several designs are depicted in Fig. 9.

A coupled sliding mechanism offers the advantages of extreme simplicity, compact geometry and full concealment. Delrin was selected as the optimal material for a sliding mechanism due to its wear resistance and very low coefficient of friction on steel. Figure 10 shows the estimated slip point between Delrin and a steel pin in the coupled sliding mechanism, which was calculated at 78.5°. Empirical testing was performed with hand-actuated prototypes to confirm this slip point, and the true slip point was determined to be approximately 70°.

A v-block was designed to center the stalk relative to the end-effector before the needle touches the wall of the stalk. Four pins allow the v-block to slide toward the main Delrin block as the gripper closes around a stalk. This both conceals the needle until contact with the stalk and provides a method of determining stalk diameter. Successful closure of the gripper around a stalk pushes the v-block back toward the arm link, which completes a circuit. Diameter estimation is accomplished by reading the encoder position of the actuation motor when the v-block circuit closes.

The constrained packaging of the gripper necessitated a very small load cell. The load cell that best fit within the design constraints was a button-style load cell manufactured by Futek, Inc. that has a diameter of 9.5 mm. The LLB215 model registers a maximum force of 222 N (50 lbf) and has a non-repeatability of 0.1% of rated output. An instrumentation amplifier manufactured by LabJack Inc. was mounted inside the sliding Delrin block to minimize transmission length and noise on the differential signal. The final gripper design is shown in Fig. 11. The process of actuating the Delrin plunger block forward and backward causes the gripper fingers to open and close. The gripper fingers have a fixed pivot point but slide on the Delrin block.

Testing and validation of the automated penetrometer were performed in Clemson, SC. An exceptionally dry summer caused many of the sorghum stalks to be too thin for the penetrometer measurement, so testing was performed on 46 hand-selected stalks greater than 20 mm in diameter. Each stalk was punctured once with the automated penetrometer and six times with the manual penetrometer. The mean of the manual penetrometer measurements for each stalk were treated as ground truth. Figure 12 shows a histogram of the number of measurements that fell above or below the mean.

A two one-sided equivalency test shows the degree to which the two sets of measurements belong to the same dataset. As the name implies, this test consists of two separate equivalency tests, each with a null hypothesis. If the

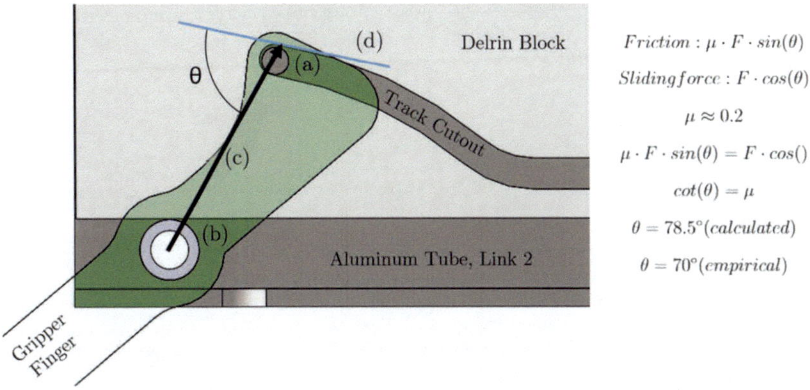

Figure 10 Visualization of the slip point between (a) a steel pin mated to the gripper finger and a slot in the Delrin plunger. The steel pin is attached to the rotational joint (b) which applies a force (c) to the Delrin block (d). The image shows a section view of the front portion of the final gripper design.

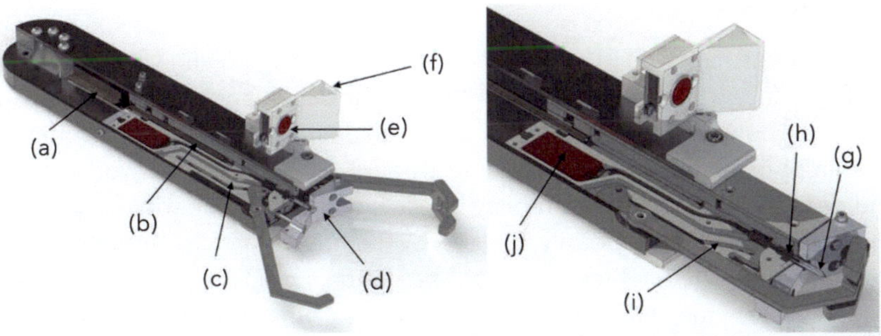

Figure 11 Cutaway of the custom gripper: (a) Maxon Motor EC16 brushless DC motor; (b) Maxon Motor GP16S ball screw; (c) acetal plunger block; (d) v-block for centering plants; (e) Code Laboratories DUO3d stereo camera; (f) sun shield; (g) needle for penetrating stalk walls; (h) Futek llb215 load cell; (i) track for actuating fingers; (j) instrumentation amplifier and A2D converter.

null hypothesis can be rejected from both tests, then one can conclude within a confidence level that the two datasets come from the same distribution. The data from the penetrometer test allows us to reject with 99% confidence the null hypothesis that population means differ more than ±15% from the average force reading.

To get further details of this case of study, please refer to Jenkins and Kantor (2017).

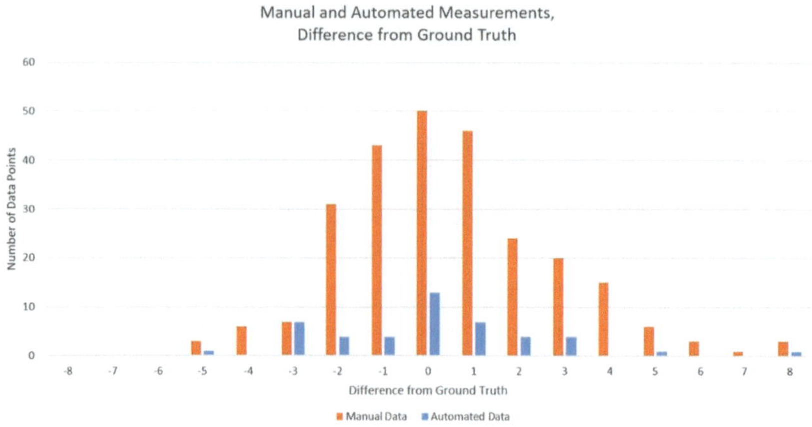

Figure 12 Histogram of penetrometer measurements and their difference from ground truth.

7 Conclusion

This chapter reviewed the advances in robotic manipulation and grasping in agriculture with an emphasis on the recent developments but also giving historical context. We reviewed the design techniques for the mechanical design of manipulators, which are mostly aimed to increase the workspace of the systems and its maneuverability in occluded scenes. In terms of grippers, the design choices are heavily inspired by the human patterns when grasping different plants, vegetables and fruits. The design considerations also were also dependent on the shape, texture and stiffness of the product to be grasped. Electrical drives are preferred to achieve accurate positioning, while pneumatic hydraulic devices provide a stronger grip, compromising the accuracy. It is interesting to note that aside from the force applied, the geometry and the pose of the gripper when making contact with the fruits or vegetables are also important variables to account.

In terms of perception and control, the field conditions and the work environment create important challenges that have been addressed by building customized sensors or algorithms based on visual or force feedback. Visual sensors provide rich geometrical and texture information but suffer when illumination changes in the scene. Force sensors, on the other hand, should be correctly placed on the grippers to provide meaningful information before, during and after grasping. The information from these sensors has been used in control algorithms including classical approaches (e.g. model predictive control) and machine learning based approaches (e.g. reinforcement learning).

Finally, we revisited some of the practical applications of the robotic technologies for practical manipulation, grasping, handling and transportation of food. Although the research work in the area is tested and evaluated in laboratory conditions, there are some works that report the use of manipulators and grippers at an industrial grade, mainly for production and packaging lines. It is interesting to note that aside from research, there are relatively new companies that are commercializing soft grippers and robotic solutions to enable the picking and packaging of individual and bulk food products.

8 Where to look for further information

The company SoftGripping has interesting technical notes about the design and use of their products in the food industry, some examples can be found here:

- https://soft-gripping.com/discover/.

For the early practitioners, the website "instructables" has several tutorials on the design and construction of rigid and soft grippers. They can be further optimized for agricultural applications but it is a good starting point:

- https://www.instructables.com/3D-Printed-Gripper/.
- https://www.instructables.com/Air-Powered-Soft-Robotic-Gripper/.

CAD communities offer a more advanced resource for designing custom grippers. They usually provide assembly files of grippers at different level of complexity. Some examples can be found at:

- https://grabcad.com/library/design-of-robot-gripper-1.
- https://cults3d.com/en/3d-model/gadget/inex-sm-gripper-xt-for-robo-creator-xt.

In terms of regulation on the use of robots and automated machines in the food industry, ABB, Siemens, Bosch and other manufacturers have released white papers that summarize their operational details and regulations when handling food:

- https://library.e.abb.com/public/c8d0f0cc8054d7fdc125799f0059d330/Sanitary%20Robotic%20Food%20Packaging%20White%20Paper.pdf.
- https://search.abb.com/library/Download.aspx?DocumentID=9AKK106713A7973&LanguageCode=en&DocumentPartId=&Action=Launch.
- https://www.siemens.com/global/en/products/automation/process-instrumentation/solutions-food-beverage.html.
- https://sps.honeywell.com/us/en/industries/manufacturing/food-beverage.

9 References

Abdeetedal, M. 2017. *Robotic Manipulation of Environmentally Constrained Objects Using Underactuated Hands*. Doctoral dissertation, The University of Western Ontario (Canada).

Ahlin, K. 2022. The robotic workbench and poultry processing 2.0. *Animal Frontiers: the Review Magazine of Animal Agriculture* 12(2): 49–55.

An, X., Li, Z., Zude-Sasse, M., Tchuenbou-Magaia, F. and Yang, Y. 2020. Characterization of textural failure mechanics of strawberry fruit. *Journal of Food Engineering* 282: 110016.

Bac, C. W. 2015. *Improving Obstacle Awareness for Robotic Harvesting of Sweet-Pepper*. Wageningen University and Research.

Bac, C. W., Van Henten, E. J., Hemming, J. and Edan, Y. 2014. Harvesting robots for high-value crops: state-of-the-art review and challenges ahead. *Journal of Field Robotics* 31(6): 888–911.

Bac, C. W., Hemming, J., Van Tuijl, B. A. J., Barth, R., Wais, E. and van Henten, E. J. 2017. Performance evaluation of a harvesting robot for sweet pepper. *Journal of Field Robotics* 34(6): 1123–39.

Baeten, J., Donné, K., Boedrij, S., Beckers, W. and Claesen, E. 2008. Autonomous fruit picking machine: A robotic apple harvester. In: *Field and Service Robotics: Results of the 6th International Conference*, 531–39. Springer Berlin Heidelberg.

Bauer, D., Bauer, C., Lakshmipathy, A. and Pollard, N. 2022. Fully printable low-cost dexterous soft robotic manipulators for agriculture. In: *AI for Agriculture and Food Systems Workshop at AAAI*. https://aiafs-aaai2022.github.io/papers/.

Baur, J. 2015. *Agricultural Manipulators*. Technische Universität München.

Bezak, P., Bozek, P. and Nikitin, Y. 2014. Advanced robotic grasping system using deep learning. *Procedia Engineering* 96: 10–20.

Bloch, V., Degani, A. and Bechar, A. 2018. A methodology of orchard architecture design for an optimal harvesting robot. *Biosystems Engineering* 166: 126–37.

Bu, L., Hu, G., Chen, C., Sugirbay, A. and Chen, J. 2020. Experimental and simulation analysis of optimum picking patterns for robotic apple harvesting. *Scientia Horticulturae* 261: 1–9.

Cai, W. and Shi, Y. 2020. Simple and reliable: mechanics analysis for robotic arm of strawberry picking robot. In: 2020 7th International Forum on Electrical Engineering and Automation (IFEEA), 806–10.

Chen, P. and Lu, W. 2021. Deep reinforcement learning based moving object grasping. *Information Sciences* 565: 62–76.

Chen, X., Li, Z., Wang, Y. and Liu, J. 2019. Effect of fruit and hand characteristics on thumb--index finger power-grasp stability during manual fruit sorting. *Computers and Electronics in Agriculture* 157: 479–87.

Chiu, Y.-C., Yang, P.-Y. and Chen, S. 2013. Development of the end-effector of a picking robot for greenhouse-grown tomatoes. *Applied Engineering in Agriculture* 29(6): 1001–9.

Choi, M., Myoung-Su, D.H. L., Park, H., Kim, Y.-J., Jang, G.-R., Shin, Y.-D., Park, J.-H., Baeg, M.-H. and Bae, J.-H. 2017. Development of multi-purpose universal gripper. In: 2017 56th Annual Conference of the Society of Instrument and Control Engineers of Japan (SICE), 1421–4.

Chua, P. Y., Ilschner, T. and Caldwell, D. G. 2003. Robotic manipulation of food products: a review. *Industrial Robot* 30(4): 345–54.

Davidson, J. R. and Mo, C. 2015. Mechanical design and initial performance testing of an apple-picking end-effector. In: ASME International Mechanical Engineering Congress and Exposition, 57397:V04AT04A011.

Denarda, A. R., Manuello Bertetto, A. M. and Carbone, G. 2021. Designing a low-cost mechatronic device for semi-automatic saffron harvesting. *Machines* 9(5): 94.

Dewi, T., Risma, P., Oktarina, Y. and Muslimin, S. 2018. Visual servoing design and control for agriculture robot; a review. In: 2018 International Conference on Electrical Engineering and Computer Science (ICECOS), 57-62.

Dimeas, F., Sako, D. V., Moulianitis, V. C. and Aspragathos, N. A. 2015. Design and fuzzy control of a robotic gripper for efficient strawberry harvesting. *Robotica* 33(5): 1085–98.

Eizicovits, D., van Tuijl, B., Berman, S. and Edan, Y. 2016. Integration of perception capabilities in gripper design using graspability maps. *Biosystems Engineering* 146: 98–113.

Fantoni, G., Santochi, M., Dini, G., Tracht, K., Scholz-Reiter, B., Fleischer, J., Kristoffer Lien, T., Seliger, G., Reinhart, G., Franke, J., Nørgaard Hansen, H. and Verl, A. 2014. Grasping devices and methods in automated production processes. *CIRP Annals* 63(2): 679–701.

Firouzeh, A. and Paik, J. 2017. An under-actuated origami gripper with adjustable stiffness joints for multiple grasp modes. *Smart Materials and Structures* 26(5): 1–10.

Fue, K. G. 2020. *Development of the Autonomous Cotton Harvesting Robot*. Doctoral Dissertation, University of Georgia.

Fue, K. G., Porter, W. M., Barnes, E. M. and Rains, G. C. 2020. An extensive review of mobile agricultural robotics for field operations: focus on cotton harvesting. *AgriEngineering* 2(1): 150–74.

Gray, J. O. and Davis, S. T. 2013. Robotics in the food industry: an introduction. In: *Robotics and Automation in the Food Industry*. Elsevier, 21–35.

Guo, N., Zhang, B., Zhou, J., Zhan, K. and Lai, S. 2020. Pose estimation and adaptable grasp configuration with point cloud registration and geometry understanding for fruit grasp planning. *Computers and Electronics in Agriculture* 179: 105818.

Hazard, C., Pollard, N. and Coros, S. 2018. Automated design of manipulators for in-hand tasks. In: 2018 IEEE-RAS 18th International Conference on Humanoid Robots (Humanoids), 1-8.

Hemming, J., Bontsema, J., Bac, C. W., Edan, Y., van Tuijl, B. A. J., Barth, R. and Pekkeriet, E. J. 2014. *CROPS: Intelligent Sensing and Manipulation for Sustainable Production and Harvesting of High Value Crops, Clever Robots for Crops: Final Report Sweet-Pepper Harvesting Robot*.

Herck, L. van, Kurtser, P., Wittemans, L. and Edan, Y. 2020. Crop design for improved robotic harvesting: A case study of sweet pepper harvesting. *Biosystems Engineering* 192: 294–308.

Honarpardaz, M., Tarkian, M., Ölvander, J. and Feng, X. 2017. Finger design automation for industrial robot grippers: a review. *Robotics and Autonomous Systems* 87: 104–19.

Hou, Z., Li, Z., Fadiji, T. and Fu, J. 2021. Soft grasping mechanism of human fingers for tomato-picking bionic robots. *Computers and Electronics in Agriculture* 182: 106010.

Jarque-Bou, N. J., Atzori, M. and Müller, H. 2020. A large calibrated database of hand movements and grasps kinematics. *Scientific Data* 7(1): 12.

Jenkins, M. and Kantor, G. 2017. Online detection of occluded plant stalks for manipulation. In: 2017 IEEE/RSJ International Conference on Intelligent Robots and Systems (IROS), 5162–67.

Ji, W., Zhang, J., Xu, B., Tang, C. and Zhao, D. 2021. Grasping mode analysis and adaptive impedance control for apple harvesting robotic grippers. *Computers and Electronics in Agriculture* 186: 106210.

Jiang, H., Tang, N., Huang, D. and Zhang, C. 2011. The design and analysis of automatic egg packing manipulator. In: 2011 International Conference on Control, Automation and Systems Engineering (CASE), 1–3.

Killpack, M. D., Kapusta, A. and Kemp, C. C. 2016. Model predictive control for fast reaching in clutter. *Autonomous Robots* 40(3): 537–60.

Kondo, N., Yata, K., Iida, M., Shiigi, T., Monta, M., Kurita, M. and Omori, H. 2010. Development of an end-effector for a tomato cluster harvesting robot. *Engineering in Agriculture, Environment and Food* 3(1): 20–4.

Kroemer, O. 2021. Playing with food: learning food item representations through interactive exploration. In: Experimental Robotics: The 17th International Symposium, 19, 309–22.

Kultongkham, A., Kumnon, S., Thintawornkul, T. and Chanthasopeephan, T. 2021. The design of a force feedback soft gripper for tomato harvesting. *Journal of Agricultural Engineering* 52(1), 1–7.

Lee, K. S. and Jung, M.-C. 2014. Common patterns of voluntary grasp types according to object shape, size, and direction. *International Journal of Industrial Ergonomics* 44(5): 761–8.

Li, J., Karkee, M., Zhang, Q., Xiao, K. and Feng, T. 2016. Characterizing apple picking patterns for robotic harvesting. *Computers and Electronics in Agriculture* 127: 633–40.

Li, L., Tian, W., OuYang, Z., Yu, S. and Sun, W. 2021. A rigid-flexible coupling three-finger soft gripper for fruit picking. In: 2021 40th Chinese Control Conference, 4068–72.

Li, Z., Li, P., Yang, H. and Wang, Y. 2013. Stability tests of two-finger tomato grasping for harvesting robots. *Biosystems Engineering* 116(2): 163–70.

Li, Z., Miao, F., Yang, Z., Chai, P. and Yang, S. 2019. Factors affecting human hand grasp type in tomato fruit-picking: A statistical investigation for ergonomic development of harvesting robot. *Computers and Electronics in Agriculture* 157: 90–7.

Lien, T. K. 2013. Gripper technologies for food industry robots. In: *Robotics and Automation in the Food Industry*. Elsevier, 143–70.

Liu, C.-H., Chiu, C.-H., Chen, T.-L., Pai, T.-Y., Chen, Y. and Hsu, M.-C. 2018. A soft robotic gripper module with 3D printed compliant fingers for grasping fruits. In: 2018 IEEE/ASME International Conference on Advanced Intelligent Mechatronics (AIM), 736–41.

Liyanage, M. H., Krouglicof, N. and Gosine, R. 2010. High speed electro-hydraulic actuator for a Scara type robotic arm. In: 2010 IEEE/RSJ International Conference on Intelligent Robots and Systems, 470–6.

Martin, J., Ansuategi, A., Maurtua, I., Gutierrez, A., Obregón, D., Casquero, O. and Marcos, M. 2021. A generic ROS-based control architecture for pest inspection and treatment in greenhouses using a mobile manipulator. *IEEE Access* 9: 94981–95.

Megalingam, R. K., Manoharan, S. K., Mohandas, S. M., Vadivel, S. R. R., Gangireddy, R., Ghanta, S., Kumar, K. S., Teja, P. S. and Sivanantham, V. 2020. Amaran: an unmanned robotic coconut tree climber and harvester. *IEEE/ASME Transactions on Mechatronics* 26(1): 288–99.

Miller, A. T. and Allen, P. K. 2004. Graspit! A versatile simulator for robotic grasping. *IEEE Robotics & Automation Magazine* 11(4): 110–22.

Mitsioni, I., Karayiannidis, Y., Stork, J. A. and Kragic, D. 2019. Data-driven model predictive control for the contact-rich task of food cutting. In: 2019 IEEE-RAS 19th International Conference on Humanoid Robots (Humanoids), 244–50.

Molina, J. and Hirai, S. 2017. A grasping-climbing mechanism for pruning tree-branches using a multirotor helicopter. In: Proceedings of the JSME Annual Conference on Robotics and Mechatronics (Robomec) 2017.

Mueller-Sim, T., Jenkins, M., Abel, J. and Kantor, G. 2017. The Robotanist: A ground-based agricultural robot for high-throughput crop phenotyping. In: 2017 IEEE International Conference on Robotics and Automation (ICRA), 3634–9.

Narvaez, F. Y., Reina, G., Torres-Torriti, M., Kantor, G. and Cheein, F. A. 2017. A survey of ranging and imaging techniques for precision agriculture phenotyping. *IEEE/ASME Transactions on Mechatronics* 22(6): 2428–39. https://doi.org/10.1109/TMECH.2017.2760866.

Navarro-Alarcon, D., Yip, H. M., Wang, Z., Liu, Y.-H., Zhong, F., Zhang, T. and Li, P. 2016. Automatic 3-d Manipulation of Soft Objects by Robotic Arms with an Adaptive Deformation Model. *IEEE Transactions on Robotics* 32(2): 429–41.

Navas, E., Fernández, R., Sepúlveda, D., Armada, M. and Gonzalez-De-Santos, P. 2021. *"Soft Grippers for Automatic Crop Harvesting: A Review." Sensors*. Multidisciplinary Digital Publishing Institute. https://doi.org/10.3390/s21082689.

Nguyen, H. and La, H. 2019. Review of deep reinforcement learning for robot manipulation. In: 2019 Third IEEE International Conference on Robotic Computing (IRC), 590–5.

Oliveira, L. F. P., Moreira, A. P. and Silva, M. F. 2021. Advances in agriculture robotics: A state-of-the-art review and challenges ahead. *Robotics* 10(2): 52.

Padalkar, A., Nieuwenhuisen, M., Schneider, S. and Schulz, D. 2020. Learning to close the gap: combining task frame formalism and reinforcement learning for compliant vegetable cutting. In: ICINCO, 221–31.

Parhar, T., Baweja, H., Jenkins, M. and Kantor, G. 2018. A deep learning-based stalk grasping pipeline. In: 2018 IEEE International Conference on Robotics and Automation (ICRA), 1–5.

Peterson, D. L., Bennedsen, B. S., Anger, W. C. and Wolford, S. D. 1999. A systems approach to robotic bulk harvesting of apples. *Transactions of the ASAE* 42(4): 871–6.

Rizzardo, C., Katyara, S., Fernandes, M. and Chen, F. 2020. The importance and the limitations of sim2real for robotic manipulation in precision agriculture. ArXiv Preprint ArXiv:2008.03983. In 2nd Workshop on Closing the Reality Gap in Sim2Real Transfer for Robotics. RSS.

Rong, J., Wang, P., Yang, Q. and Huang, F. 2021. A field-tested harvesting robot for oyster mushroom in greenhouse. *Agronomy* 11(6): 1210.

Ronzhin, A., Ngo, T., Vu, Q. and Vinh Nguyen, A. 2022. *Results of Modeling and Optimization of the Robotic Gripper Configuration Ground and Air Robotic Manipulation Systems in Agriculture*, 153–75.

Russo, M., Ceccarelli, M., Corves, B., Hüsing, M., Lorenz, M., Cafolla, D. and Carbone, G. 2017. Design and test of a gripper prototype for horticulture products. *Robotics and Computer-Integrated Manufacturing* 44: 266–75.

Saudabayev, A., Rysbek, Z., Khassenova, R. and Varol, H. A. 2018. Human grasping database for Activities of Daily Living with depth, color and kinematic data streams. *Scientific Data* 5(1): 180101.

Shahid, A. A., Roveda, L., Piga, D. and Braghin, F. 2020. Learning continuous control actions for robotic grasping with reinforcement learning. In 2020 IEEE International Conference on Systems Man, and Cybernetics (SMC), 4066–72.

Shi, X., Zhu, C. and Lu, P. 2022. Design and control of underwater robot system for sea cucumber fishing. *International Journal of Advanced Robotic Systems* 19(1): 17298806221077624.

Silwal, A., Davidson, J. R., Karkee, M., Mo, C., Zhang, Q. and Lewis, K. 2017. Design, integration, and field evaluation of a robotic apple harvester. *Journal of Field Robotics* 34(6): 1140–59. https://doi.org/10.1002/rob.21715.

Silwal, A., Yandun, F., Nellithimaru, A., Bates, T. and Kantor, G. 2021. Bumblebee: A path towards fully autonomous robotic vine pruning. *ArXiv Preprint ArXiv:2112.00291*. Field Robotics. 2(1): 1667–1696.

Softrobotics 2022. Food automation solutions for single and bulk picking. April 2022. Available at: https://www.softroboticsinc.com/.

Song, E. J., Lee, J. S., Moon, H., Choi, H. R. and Koo, J. C. 2021. A multi-curvature, variable stiffness soft gripper for enhanced grasping operations. *Actuators* 10(12): 316.

Song, X., Shan, H., Liu, H. and Guo, J. 2016. An underactuated end-effector design for fruit picking. In: 2016 23rd International Conference on Mechatronics and Machine Vision in Practice (M2VIP), 1–5.

Su, C., Wang, R., Lu, T. and Wang, S. 2023. SAU-RFC hand: A novel self-adaptive underactuated robot hand with rigid-flexible coupling fingers. *Robotica* 41(2): 511–29.

Sundaram, S., Kellnhofer, P., Li, Y., Zhu, J. Y., Torralba, A. and Matusik, W. 2019. Learning the signatures of the human grasp using a scalable tactile glove. *Nature* 569(7758): 698–702.

Sutton, R. S. and Barto, A. G. 2018. *Reinforcement Learning: an Introduction*. MIT Press.

Tanigaki, K., Fujiura, T., Akase, A. and Imagawa, J. 2008. Cherry-harvesting robot. *Computers and Electronics in Agriculture* 63(1): 65–72.

Tao, Y., Zhou, J., Wang, M., Zhang, N. and Meng, Y. 2017. An optimum strategy for robotic tomato grasping based on real-time viscoelastic parameters estimation. *International Journal of Advanced Robotic Systems* 14(4): 1729881417724190.

Taylor, M. G. and Templer, R. G. 1997. A washable robot suitable for meat processing. *Computers and Electronics in Agriculture* 16(2): 113–23.

Tian, G., Zhou, J. and Gu, B. 2018. Slipping detection and control in gripping fruits and vegetables for agricultural robot. *International Journal of Agricultural and Biological Engineering* 11(4): 29–34.

Tinoco, V., Silva, M. F., Santos, F. N., Rocha, L. F., Magalhães, S. and Santos, L. C. 2021. A review of pruning and harvesting manipulators. In 2021 IEEE International Conference on Autonomous Robot Systems and Competitions (ICARSC), 155–60.

Vision Robotics Corp. 2020. *Intelligent Autonomous Grapevine Pruner*.

Vu, Q. and Ronzhin, A. 2020. A model of four-finger gripper with a built-in vacuum suction nozzle for harvesting tomatoes. In: Proceedings of the 14th International Conference on Electromechanics and Robotics "Zavalishin's Readings", 149–60.

Wang, J., Li, B., Li, Z., Zubrycki, I. and Granosik, G. 2021. Grasping behavior of the human hand during tomato picking. *Computers and Electronics in Agriculture* 180: 105901.

Wang, X., Kang, H., Zhou, H., Au, W. and Chen, C. 2022. Geometry-aware fruit grasping estimation for robotic harvesting in apple orchards. *Computers and Electronics in Agriculture* 193: 106716.

Watson, J., Hughes, J. and Iida, F. 2017. Real-World, real-time robotic grasping with convolutional neural networks. In: Annual Conference Towards Autonomous Robotic Systems, 617–26.

Wen, R., Yuan, K., Wang, Q., Heng, S. and Li, Z. 2020. Force-guided high-precision grasping control of fragile and deformable objects using Semg-based force prediction. *IEEE Robotics and Automation Letters* 5(2): 2762–9.

Whitesides, G. M. 2018. Soft robotics. *Angewandte Chemie* 57(16): 4258–73.

Xie, Y., Zhang, B., Zhou, J., Bai, Y. and Zhang, M. 2020. An integrated multi-sensor network for adaptive grasping of fragile fruits: design and feasibility tests. *Sensors* 20(17): 1–23. https://doi.org/10.3390/s20174973.

Yan, W., Deng, Z., Chen, J., Nie, H. and Zhang, J. 2019. Precision grasp planning for multi-finger hand to grasp unknown objects. *Robotica* 37(8): 1415–37.

Yandun, F., Parhar, T., Silwal, A., Clifford, D., Yuan, Z., Levine, G., Yaroshenko, S. and Kantor, G. 2021. Reaching pruning locations in a vine using a deep reinforcement learning policy. In: 2021 IEEE International Conference on Robotics and Automation (ICRA), 2400–6.

Yang, H., Wei, G., Ren, L., Qian, Z., Wang, K., Xiu, H. and Liang, W. 2021. A low-cost linkage-spring-tendon-integrated compliant anthropomorphic robotic hand: MCR-hand III. *Mechanism and Machine Theory* 158: 104210.

Zahid, A., Mahmud, Md. S., He, L., Choi, D., Heinemann, P. and Schupp, J. 2020. Development of an integrated 3R end-effector with a Cartesian manipulator for pruning apple trees. *Computers and Electronics in Agriculture* 179: 105837.

Zapotezny-Anderson, P. and Lehner, C. 2019. Towards active robotic vision in agriculture: A deep learning approach to visual servoing in occluded and unstructured protected cropping environments. In: 6th IFAC Conference on Sensing, Control and Automation Technologies for Agriculture AGRICONTROL, 120–5.

Zhang, B., Xie, Y., Zhou, J., Wang, K. and Zhang, Z. 2020. State-of-the-art robotic grippers, grasping and control strategies, as well as their applications in agricultural robots: a review. *Computers and Electronics in Agriculture* 177: 105694.

Zhang, B., Zhou, J., Meng, Y., Zhang, N., Gu, B., Yan, Z. and Idris, S. I. 2018. Comparative study of mechanical damage caused by a two-finger tomato gripper with different robotic grasping patterns for harvesting robots. *Biosystems Engineering* 171: 245–57.

Zheng, W., Xie, Y., Zhang, B., Zhou, J. and Zhang, J. 2021. Dexterous robotic grasping of delicate fruits aided with a multi-sensory e-glove and manual grasping analysis for damage-free manipulation. *Computers and Electronics in Agriculture* 190: 106472.

Zhou, H., Wang, X., Kang, H. and Chen, C. 2021. A tactile-enabled grasping method for robotic fruit harvesting. ArXiv Preprint ArXiv:2110.09051.

Zulkifli, N., Hashim, N., Harith, H. H. and Shukery, M. F. M. 2020. Finite element modelling for fruit stress analysis - A review. *Trends in Food Science & Technology* 97: 29–37.

Chapter 12

Advances in soft grasping in agriculture

Ali Leylavi Shoushtari, Wageningen University & Research and Wageningen Robotics, The Netherlands

1 Introduction
2 Challenges of action in agricultural robotics
3 How soft robotics can address challenges of action
4 Developing an assessment matrix for soft grippers
5 Performance analysis of current soft agri-food grippers
6 Performance analysis: dealing with object variations
7 Designing the ideal versatile gripper
8 Case study: the use of soft robotics in intercropping
9 Conclusion
10 Where to look for further information
11 References

1 Introduction

Despite remarkable progress in agri-food robotics, no commercially available autonomous harvesters have been developed to date due to gaps in scientific and technological knowledge (Adamides and Edan, 2023; Bac et al., 2014; Droukas et al., 2023; Kumar and Mohan, 2022; Martin et al., 2022; Wang et al., 2022). In the process of automating agricultural tasks, robotic technologies face a wide variety of challenges. From a robotics point of view, these challenges can be classified as follows:

- Challenges of perception; and
- Challenges of action (CoA).

The core of the perception challenge is to comprehend, reason, and make sense of the unknown and cluttered environment surrounding the robot, e.g. on arable farms and in greenhouses. The challenges of action (CoA) concern agile maneuvering in and physically interacting with an unstructured environment characterized by high diversity and uncertainty. From a robotics point of view, the former class of challenges is formulated as computer vision

http://dx.doi.org/10.19103/AS.2023.0124.10

and artificial intelligence (AI) problems, while the latter is framed as a set of manipulation, grasping, and navigation problems. Six decades of research have enabled computer vision and AI (Szeliski, 2011) to be applied to agricultural tasks, while robotic grasping, manipulation, and navigation require further research.

Conventional robots are usually made up of rigid materials that limit their ability to adapt to objects and obstacles. The physical interaction of robots with their environment is inevitable, whether in locomotion or manipulation, or even in the form of collision, potentially making rigid robots unsafe. Adaptability and safety, in particular, are limiting factors for rigid robots, especially when implemented in a real-world scenario and in an unstructured environment with high diversity and uncertainty, as is the case in agriculture. Furthermore, maintaining the quality of often delicate products that robotic systems handle in agricultural tasks is a particular challenge. For this reason harvesting using rigid robots (in this case with rigid grippers) has not seemed a feasible option.

Soft robots can be defined as robots and machines that are 'primarily composed of easily deformable matter such as fluids, gels, and elastomers that match the elastic and rheological properties of biological tissue and organs' (Majidi, 2014). There are machines and robots made of materials that have much less elasticity than elastomers but which can still provide large deformations due to their particular geometry. Compliant mechanisms are an example of this group of robots or mechanisms (Howell et al., 2013). The change in material used from one that is rigid to a deformable and stretchable material provides robots and machines with an adaptable shape and a locomotion (Ansari et al., 2015) or manipulation strategy for a broad range of tasks and environments (Majidi, 2014).

Soft robotics is a relatively new yet promising field of research aimed at enhancing robot safety and adaptability while maintaining simplicity and affordability. In agricultural robotics, crops are generally vulnerable and delicate, have high variation (i.e. different shapes, sizes, surface textures and roughness, and softness), and are often affected by environmental conditions such as moisture and dust, which makes them challenging to handle using conventional rigid robots. Soft robotics can significantly contribute to robot versatility, making it an ideal solution for agricultural robotic challenges. This chapter aims to understand how and to what extent soft robotics can enhance the versatility of robots, especially in grasping and manipulation.

Section 1 focuses on the types of challenges that any robotic system would face in agricultural tasks. The chapter then reviews how soft robotics can help deal with these challenges. The chapter includes a case study which, in a nutshell, shows how soft agricultural robotic research is carried out. The chapter concludes with a summary and a short list of sources of further information about agricultural robotics, including review papers and open-source websites.

2 Challenges of action in agricultural robotics

The challenges of action (CoA) in agricultural robots are as follows:

- High diversity in agricultural products: Products all have different pose (position and orientation), shape, size, mechanical properties (structural stiffness), and surface properties, i.e. roughness and adhesion of crops.
- Diversity between different crop variants: Even crops that genetically belong to the same family can have different variants with quite diverse shapes and sizes. For example, cucurbit crops can have several variants, including cucumber, melon, watermelon, and squash, which have completely different shapes and sizes (Fig. 1a–d).
- Diversity within a crop variant (inherent diversity): Although crops within a variant look similar, they are not identical. This diversity is also time-dependent and dynamic. Figure 2a shows the diversity in shape and

Figure 1 The high diversity of shape and size of cucurbit crops. Panels a to d depict the diversity within different cucurbit variants, i.e. cucumber, melon, watermelon, and squash. Source: This figure is taken directly from Pan et al. (2020).

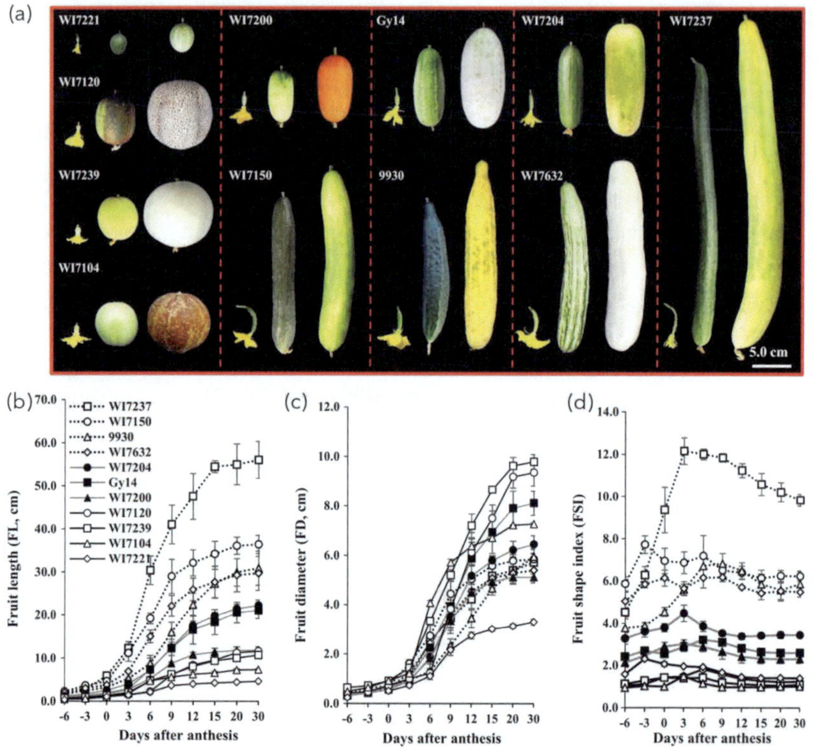

Figure 2 Growing dynamics of cucumbers with different genetic backgrounds. (a) Images of ovary and fruit of 11 cucumbers depicting the dynamics of growth at 0, 12, and 30 days after pollination. (b-d) The quantifications of size and shape dynamics, respectively. Source: This figure is taken directly from Pan et al. (2020).

size of 11 different variants of cucurbits. The quantified dynamics of size and shape variation can be seen in Fig. 2b, 2c and 2d, respectively.

- Safety/quality requirements in the handling of crops: Crops are commercial products and their quality should be maintained by the machines that interact with them. This means that robots are required to satisfy a high level of safety and quality during crop management and harvesting.
- Task complexity: Agricultural tasks are multifaceted. Harvesting requires not only grasping objects but also detaching and holding them robustly while dealing with dynamic perturbances during manipulation. This challenge requires robots (i.e. grippers) to be able to deal with a high variation in force and torque.
- Dense vegetation environments: Agricultural environments typically challenge the maneuverability of robotic systems. A bulky robot will not be agile and maneuverable enough to carry out its task within an

economically feasible amount of time while avoiding collisions with plants and other objects.

The following sections discuss ways of accounting for these challenges both in assessing the current range of soft gripper designs and in developing improved and more versatile future designs.

3 How soft robotics can address challenges of action

Soft robotics is a relatively new field of research aimed at making robots safer and more adaptive. High adaptability is useful, particularly for robotic grasping where robots are required to conform to objects with high variations (challenge 1). Moreover, when the objects to be handled are delicate/fragile, then safety (in handling) is also an important requirement (challenge 2). This means that safety and adaptability are the main requirements of agri-food robotics that increase demand for use of soft grippers.

During the last three decades, a large number of soft robotic grippers have been designed and developed for crop harvesting and handling (Fig. 3), mainly focused on the two challenges already listed. Meanwhile, two further challenges (challenges 3 and 4) need to be overcome in order for a soft gripper to be suitable for agricultural tasks such as greenhouse harvesting. Despite their high adaptability, most soft grippers are not strong enough to handle objects of various weights and are unable to deal dynamically with perturbations in load during harvesting (challenge 3). Although some soft grippers have sufficient strength to deal with variable loads, this strength is usually high along the axis of the gripper, while it is low along other axes, which make the gripper flexible along those axes.

Looking at Fig. 4b as a schematic representation of the strength of soft grippers, generally the gripper has high strength along the y axis and is adaptive under lateral forces applied along the x and z axis. A versatile robotic gripper in agriculture is required to be able to exert sufficient torque and force on crops in order to hold, detach, and manipulate them. Figure 4a shows such requirements for a manual harvesting process. Rigid grippers in slippage-free cases are more successful at applying such force and torque since their rigid links or bodies can endure the forces received during physical interaction with objects. Most soft grippers tend to stiffen in just one direction or plane (i.e. the plane in which the bending happens) and lacks the supporting stiffness in other directions or planes. As a result, their strength is maximized only in the bending direction or plane and minimized in the other planes. Soft grippers, therefore, bend once they undergo lateral loads and stress induced by object weight or during pulling, flicking, twisting, bending or accelerating, and deaccelerating. Figure 4b provides a schematic representation of the heterogeneous strength

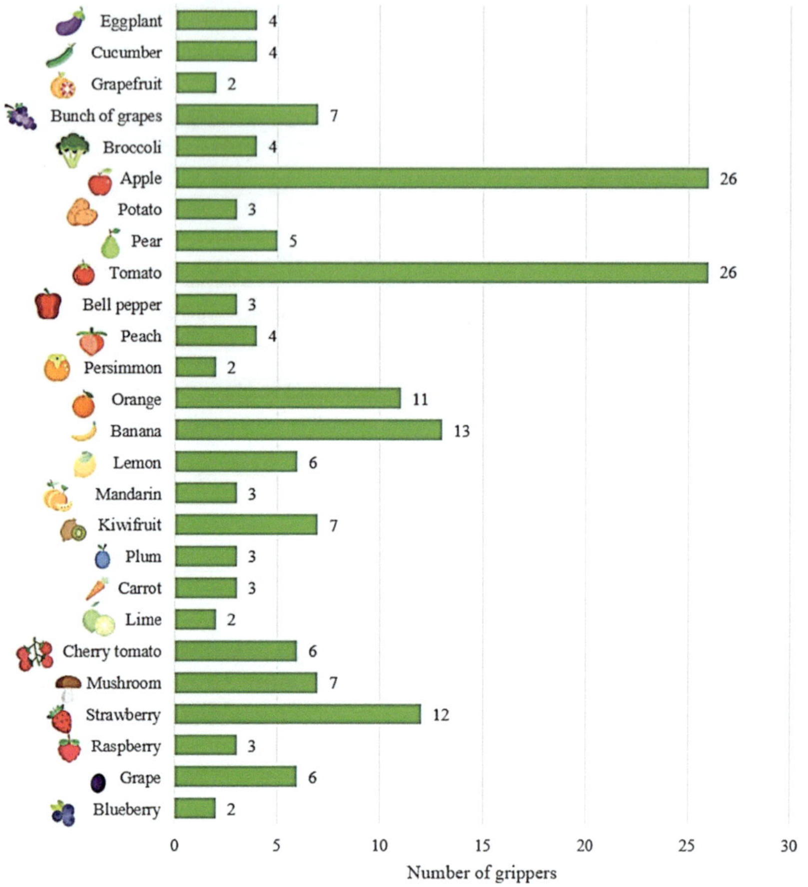

Figure 3 Number of soft grippers used to harvest various crops. Source: Adapted from: Elfferich et al. (2022).

of soft grippers in which the strength is maximized only along the grasping direction, i.e. Y in this example.

The abovementioned issue with soft grippers has been addressed in a recent study (Fig. 4c) in which researchers added an adaptive rigid bar mechanism to the soft fingers to limit the lateral bending of the finger and as a result obtained higher strength in the bending direction (Zhu et al., 2023). However, although these types of solutions have significantly increased the strength of soft grippers, they have also made them larger and heavier due to the size and weight of components (i.e. mechanical mechanisms and their actuators) added to the soft gripper. In contrast, challenge 4 demands thin and lightweight robotic grippers. Although ultra-thin, high-strength, and even

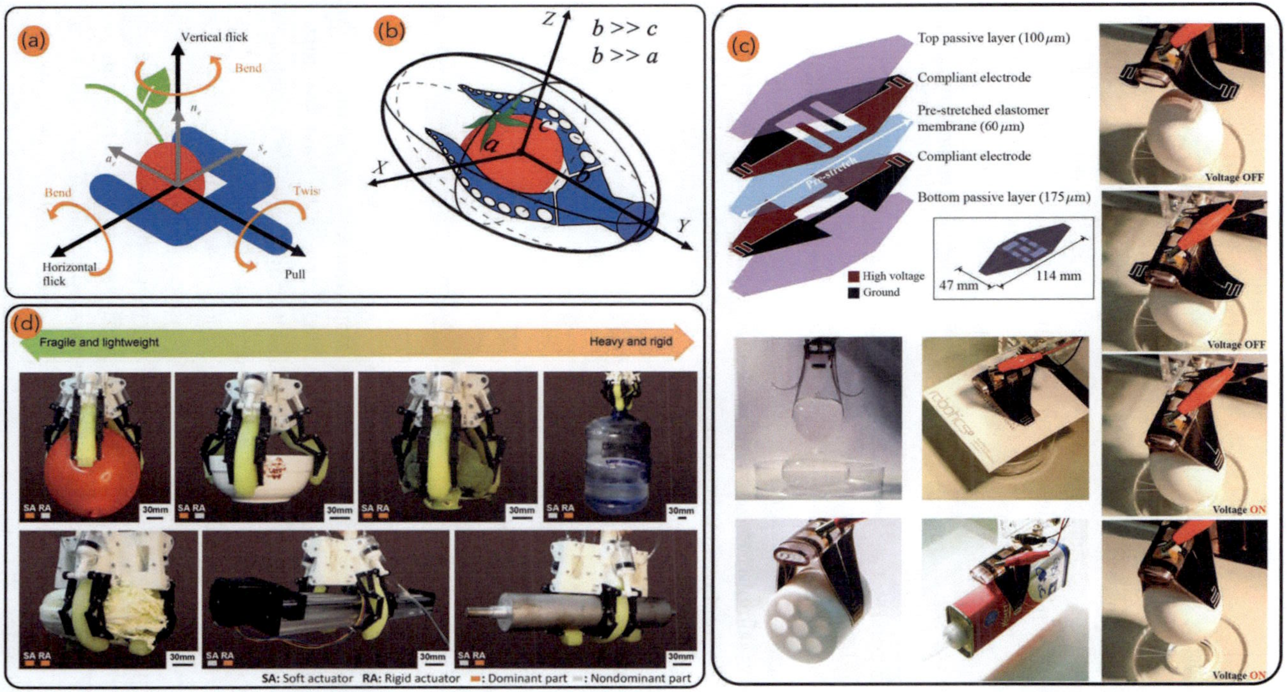

Figure 4 (a) Different ways of detaching crops from peduncle. [adapted from Elfferich et al. (2022)]. (b) Hypothetical ellipsoid of the strength of a soft gripper where a, b, and c represent the strength of the gripper in three different directions. The maximum strength is always in the direction of the bending (Y). (d) An example of a soft gripper that can apply grasping force in one direction when it would bend under lateral loads. (c) A soft gripper that is mechanically supported with the rigid bar mechanisms in order to deal with the lateral forces. Source: Figures (b) and (c) are adapted from Shintake et al. (2016b) and (Zhu et al., 2023), respectively.

ultra-fast soft grippers exist (Shintake et al., 2016a), they are challenged by the issue of strength heterogeneity (Fig. 4d).

4 Developing an assessment matrix for soft grippers

The abovementioned soft robotics solutions illustrate the multidimensional challenges in gripping options for agri-food applications. While one solution may enhance the performance of the gripper in one dimension, it tends to limit performance in other dimensions. This means that multifaceted criteria are required able to encompass the full range of challenges. The authors have developed a performance matrix to account for different factors affecting gripper performance (Fig. 5). This is based on grasping performance and variations in object handling.

Gripper factors to be considered are strength and bulkiness, related to challenges 3 and 4, respectively. In particular, an ideal soft gripper for agricultural tasks should be strong enough to handle loads and dynamic perturbations and be thin and agile enough to maneuver in dense vegetation. Strength can be defined as the maximum force and torque that a gripper can handle in different directions (XYZ directions in Fig. 4b). The strength factor of the gripper can be defined based on the payload to gripper weight ratio pwr in Eqn (1). The bulkiness factor can be represented by the ratio of gripper thickness t_g to the object's width/thickness t_o as expressed in Eqn (2).

$$pwr = \frac{Payload}{W_g} \qquad (1)$$

$$tr = \frac{t_o}{t_g} \qquad (2)$$

Soft grasping should also be carried out both quickly and safely so that automated harvesting is economically feasible. Safety is a qualitative factor; hence, it is weighted based on an expert's opinion range from 0 to 1 and follows the type of grasping mechanisms they are using, i.e. Bernoulli mechanism (Petterson et al., 2010), mechanical interlocking (Leylavi Shoushtari et al., 2019), friction-based, controlled adhesion, and controlled stiffness (Zhang et al., 2020; Shintake et al., 2018). The Bernoulli mechanism and the adhesion-based mechanism are ranked as the safest for grippers, followed by controlled stiffness, mechanical interlocking, and friction-based grasping at the other end of the spectrum of safe grasping mechanisms. Speed is also defined based on the time required for grasping.

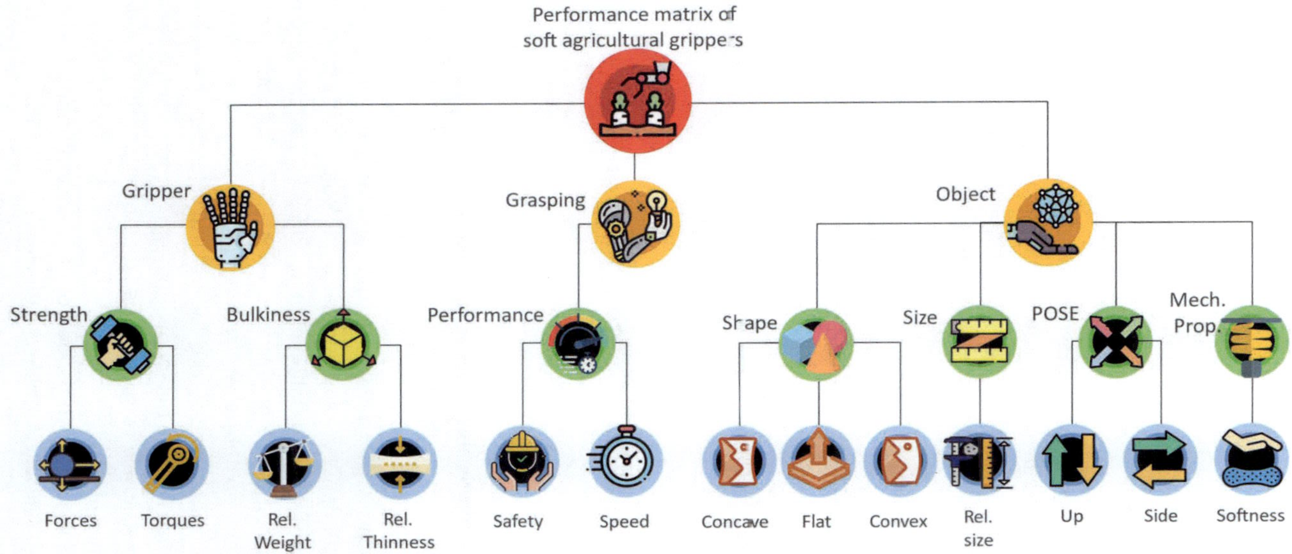

Figure 5 The performance matrix for a soft robotic gripper designed for agricultural applications. It embeds three folded aspects: gripper parameters, i.e. strength and bulkiness, grasping performance, and object variations. The gripper strength can be measured based on the maximum force/torque before the slippage of the object and the bulkiness could be measured by relative weight and thickness of the object to the gripper. Safety and speed are chosen as the most important indices representing the performance of grasping the crops. The shape, size, pose, and the mechanical properties are the indices representing the variability of the crops/objects to be handled. (Icons are downloaded from FlatICON (n.d.).)

Finally, the ideal soft gripper should be able to handle variations in crops, i.e. shape, size, pose, and their mechanical properties. Shape variations are defined as flat, convex, and concave. Size can be measured based on the aspect ratio of the maximum length of the object to the length of the gripper. Pose can be defined as whether the major part of the grasped object is on the side of the gripper (i.e. the weight of the object acts as a lateral load) or the grasped object is facing up or down (i.e. the gripper is either approaching from the top or from the bottom). Finally, the mechanical properties of the object can be simplified as the range of the stiffness of objects that the gripper can carry. The range of stiffness is defined based on Young's modulus, from the softest to the most rigid object as grasped by the gripper. Practically speaking, grasping a rigid object is feasible for soft grippers; the real challenge is when the object is extremely soft (softer than the gripper). This means that the range of stiffness of the objects that the gripper can grasp depends on Young's modulus of the softest object grasped by the gripper. A softness index for a gripper could be developed as the inverse of Young's modulus of the softest object that the gripper can handle as expressed in Eqn (3):

$$S_{index} = \frac{1}{E_{softest}} \qquad (3)$$

5 Performance analysis of current soft agri-food grippers

Although numerous soft grippers have not been developed specifically for – or integrated in – agricultural tasks, in order to demonstrate the effectiveness of soft grippers in agriculture, this chapter focuses only on those soft grippers designed for handling agricultural products. The authors have analyzed a database of soft grippers used in crop harvesting and crop handling using their performance matrix. This database is the same as that used in the study by Elfferich et al. (2022). Of 79 studies that used 'soft robotic grippers' and similar keywords [see Table 1 of (Elfferich et al., 2022)], 11 addressing aspects mentioned in the performance matrix were chosen. The grippers' scores for each parameter were then normalized with respect to the maximum scores of the given parameter for all evaluated grippers. The values of shape and pose were not normalized as they are binary parameters. The following three figures show the scores given to all 11 gripper designs for strength and bulkiness, grasping performance, and object variation.

Figure 6a shows the safety and speed of the selected grippers. The HASL-based gripper (Hydraulically Amplified Self-Healing electrostatic actuators) shown in Fig. 6b scores as the fastest gripper (Acome et al., 2018) together with the ElectroAdhesion-based gripper shown in Fig. 6c, since both of these grippers are based on the principles of electrostatic actuation, which acts rapidly (Shintake et al., 2016b; Cacucciolo et al., 2019). Such an actuation mechanism enables them to grasp in about 200 ms.

While the HASL-based gripper scores lower on safety (due to friction and mechanical interlocking-based grasping), the ElectroAdhesive gripper is the safest since it uses surface adhesion to grasp objects. Figure 6c illustrates how such a mechanism allows the gripper flaps to attach strongly to the surface of the crop, without applying any normal force which is generally the main cause of bruising and skin damage. Alongside this, a Bernoulli principle-based gripper

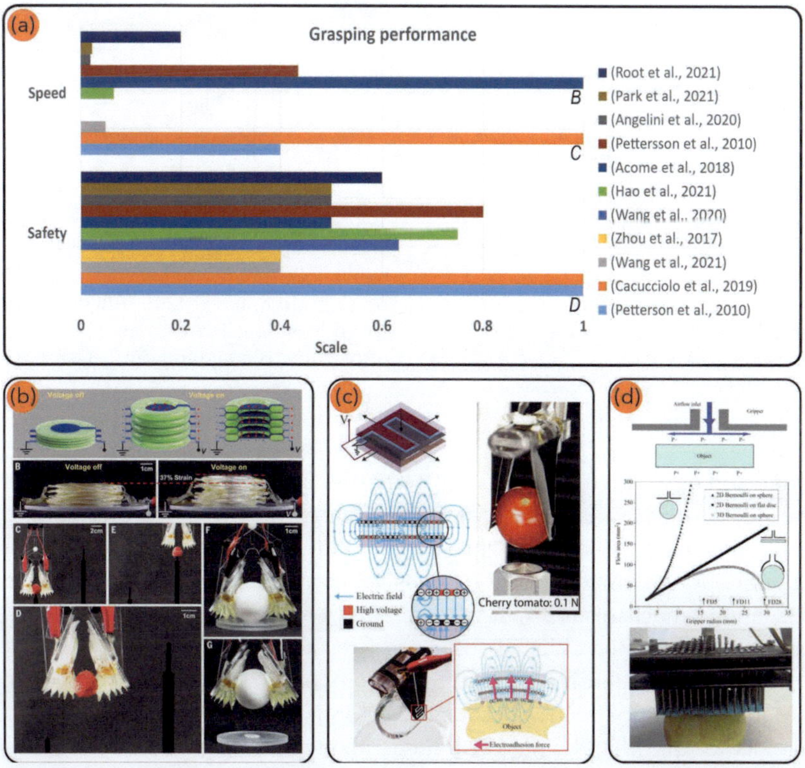

Figure 6 Grasping performance of 11 grippers. (a) Speed and safety scores. (b–d) Grippers with the top scores. (b) From top to bottom: HASL gripper mechanisms consisting of five ring-shaped HASL actuators, the strain gained when the voltage applied, and picking and placing a strawberry and an egg [adapted from Acome et al. (2018)]. (c) Electro-adhesive-based gripper, its mechanism, and grasping time [adapted from Cacucciolo et al. (2019) and Shintake et al. (2016b)]. (d) A Bernoulli principle-based gripper, its mechanism, performance, and deformation in interaction with the objects [adapted from Petterson et al. (2010)].

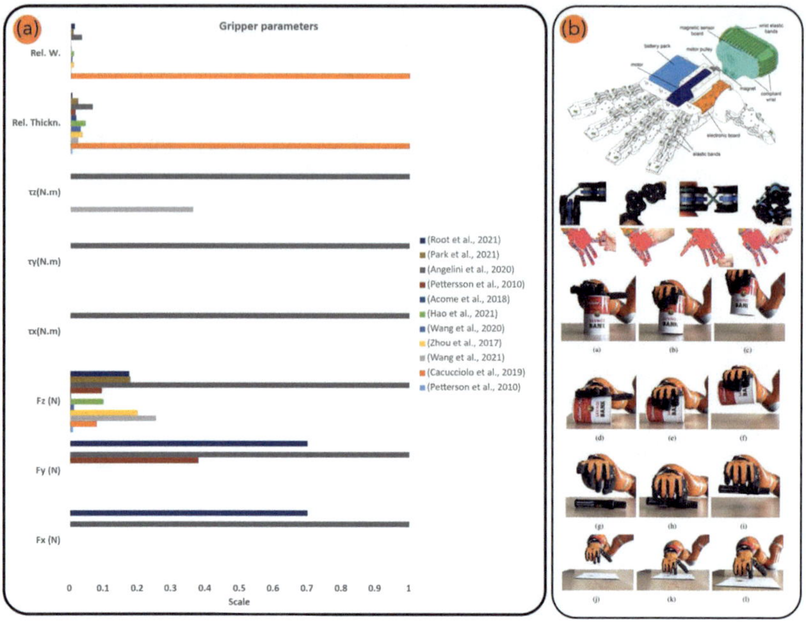

Figure 7 (a) Gripper parameters, i.e. relative weight and thickness, and maximum forces and torques in three dimensions. (b) Top to bottom: the Pisa/IIT SoftGripper; the stretchable ligaments make the gripper adaptive in interaction with its environment; grasping different objects using a synergy control. Photo is adapted from Catalano et al. (2014).

(Petterson et al., 2010) is also identified as one of the safest grippers since, like the ElectroAdhesive gripper, it uses an attachment mechanism to grasp objects. This attachment mechanism works based on the pressure drop between the gripper and the object caused by positive air pressure. The advantage of this attachment mechanism is that there is always a gap between the object and gripper, meaning the object is suspended, resulting in contactless grasping.

The ElectroAdhesive gripper (Shintake et al., 2016b) ranks as the lightest and thinnest gripper (Fig. 7a), capable of handling objects 1000 times heavier than its weight while being 172.5 times thinner than the object. Such a thin design allows it to be integrated in picking crops from a pile in a situation in which the gripper needs to be thin enough to get around the object. The Pisa/IIT SoftGripper is a tendon-driven anthropomorphic hand, which takes advantage of an adaptive ligament constraining system and is ranked as the strongest gripper (Fig. 7b). It is capable of handling loads in a vertical direction (z) up to 200 N, lateral loads of up to 50 N, and enduring torque up to 1, 0.5, and 2 Nm around the *x*, *y*, and *z* axis.

6 Performance analysis: dealing with object variations

Figure 8 shows the extent to which each soft gripper is capable of handling object variations. The figure shows that all 11 grippers are capable of grasping objects facing up (*up pose*) as well as objects with convex shapes. Indeed, for the other grippers, we can clearly see that testing (i.e. grasping objects that are round and facing up) can be used to show their performance. It is clear that these two variations (i.e. up pose and convexity) are not challenging for soft grippers. However, soft grippers do struggle with other object variations such as side pose, concave shapes, object softness, relative size, and flat shape.

6.1 Side pose

All grippers with a purely soft body, i.e. the ElectroAdhesive gripper (Cacucciolo et al., 2019), the adaptive soft hand (Zhou et al., 2017), the HASL-based gripper (Acome et al., 2018), and the dual-mode soft gripper for food packaging (Wang et al., 2020) are incapable of grasping objects from the side. The reason for this is that pure soft bodied grippers do not have sufficient support stiffness. On the other hand, hybrid rigid-soft bodied grippers, such as the soft grasping and conveying mechanism (Fig. 8h; Root et al., 2021), the stiffness-controlled gripper (Fig. 8b; Park et al., 2021), the Pisa/IIT SoftGripper (Fig. 8c; Angelini et al., 2020), the magnetorheological gripper (Fig. 8g; Pettersson et al., 2010), the multimodal enveloping gripper (Fig. 8d; Hao et al., 2021), and the circular shell gripper (Wang et al., 2021) are capable of side grasping.

Side grasping is also a challenge for the adaptive Bernoulli gripper (Fig. 6d) (Petterson et al., 2010) for different reasons. Despite the fact that Bernoulli grippers have been used in food industries before (Davis et al., 2008; Erzincanli and Sharp, 1997), they have always been used in a horizontal orientation. Indeed, in this position, the gripper performs best as the sheer force is minimized. Using a Bernoulli gripper for side grasping entails.

6.2 Object softness

The only two grippers that can handle objects as soft as Tofu have a very low stiffness profile in order to handle such a product (Park et al., 2021; Zhou et al., 2017). The soft segmented gripper (Zhou et al., 2017) uses an ultra-soft pillared interface to safely grasp and accommodate Tofu in a mechanical interlocking regime (Fig. 8f, on the right). Such a passive approach works until the shore hardness of the interface is lower than the object to be handled. Shore hardness is a measure of the hardness of a material, specifically its resistance to indentation or penetration by a hard object. However, if the object is relatively softer than the gripper's interface, then grasping could be challenging.

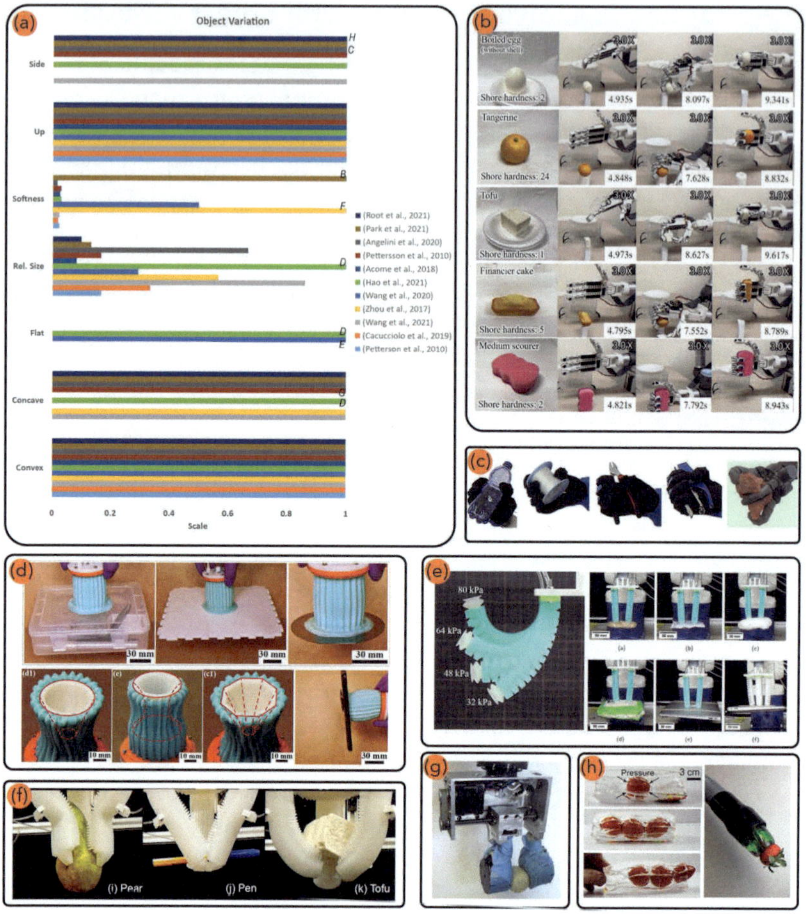

Figure 8 (a) The scores associated with the object variation. For each variation challenge factor (those for which just part of the gripper scored (1) two selected grippers with score 1 are shown with the capital letters on the bar. The capital letters refer to the visual representation of the gripper shown in B–H sections. (b) Variable stiffness anthropomorphic gripper with stiffness control, capable of grasping extra soft (Tofu), concave objects and grasping them from the side. Adapted from: Park et al. (2021). (c) Pisa/IIT SoftGripper capable of grasping concave and unstructured objects and side grasping. Adapted from: Santina et al. (2018). (d–e) A multimodal enveloping soft gripper (Hao et al., 2021) and a dual-mode soft gripper (Wang et al., 2020) both capable of grasping concave and flat objects and grasping from the side. Images of sections D and E are adapted from Hao et al. (2021) and Wang et al. (2020), respectively. (f) A soft gripper with enhanced object adaptation capable of grasping concave and ultra-soft objects. The image of section F is adapted from Zhou et al. (2017). (g and h) An adaptive magnetorheological gripper designed for food handling (Pettersson et al., 2010) and a soft bio-inspired mechanism for grasping, catching, and conveying (Root et al., 2021), both of which are capable of side grasping and lifting objects with concave shapes. Images of sections G and H are adapted from. Images of B-H are adapted from Pettersson et al. (2010) and Root et al. (2021), respectively.

In contrast, the anthropomorphic gripper with controlled stiffness (Park et al., 2021) can handle Tofu despite not having a soft interface by employing an adaptive stiffness approach. It enables the robotic gripper to automatically adjust the stiffness of its fingers to the stiffness and shore hardness of an unknown object once the gripper interacts with it. In particular, the peak force at fingers f_p is correlated with the shore hardness scale h_s (of the objects) and the damping ratio ζ_d of the object as expressed by Eqs (4) and (5). Using the resulting correlation between peak force and shore hardness h_s, as expressed by Eq. (5), allows the robot to estimate the hardness of an unknown object using the force feedback received at its fingertip. The estimated shore hardness is used in Eqn (4) to calculate the damping ratio. The resultant damping ratio is used by an impedance controller to adjust the stiffness of the gripper. To achieve this purpose, 21 objects with quite different shore hardness have been used for these experiments (Fig. 9a). Their standard shore hardness has been measured (Fig. 9b) and used for the correlation (Fig. 9c).

$$\zeta_d = \alpha_1 h_s^2 - \alpha_2 h_s + \alpha_3 \qquad (4)$$

$$h_s = \beta_1 e^{\beta_2 f_p} \qquad (5)$$

6.3 Flat objects

When it comes to handling flat objects, only those grippers that take advantage of suction-based adhesion succeed (Fig. 8a). Other types of adhesion such as electroadhesion or gecko-inspired dried adhesion cannot handle flat and non-lightweight objects due to the peeling-off effect (Ruotolo et al., 2021; Shintake et al., 2016a). The peeling-off effect could be solved for gecko-inspired adhesive pads by having a transmission mechanism to convert the normal load (which is an effect of object weight and causes the peeling phenomenon) to shear loads, which reinforces the attachments (Hawkes et al., 2016; Jiang et al., 2017).

By contrast, suction cup-based adhesion works best on flat (and non-rough) surfaces (due to guaranteed sealing) with normal loads and might fail due to the shear forces. Indeed, the two grippers successful at grasping flat surfaces were tested in this class (Fig. 8d, top row and Fig. 8e). Using a suction-based gripper for side grasping would be quite challenging since the weight of the object applies to the gripper as a shear load. Such a situation requires a stronger suction cup, equivalent to having a bigger suction cup or higher suction force, which is the case for the multimodal enveloping gripper (Hao et al., 2021; shown in the bottom right photograph of Fig. 8d). It is worth mentioning that the dual-mode gripper is unable to hold flat objects as shown in Fig. 8e in the side grasping scenario since it does not have rigid support. In

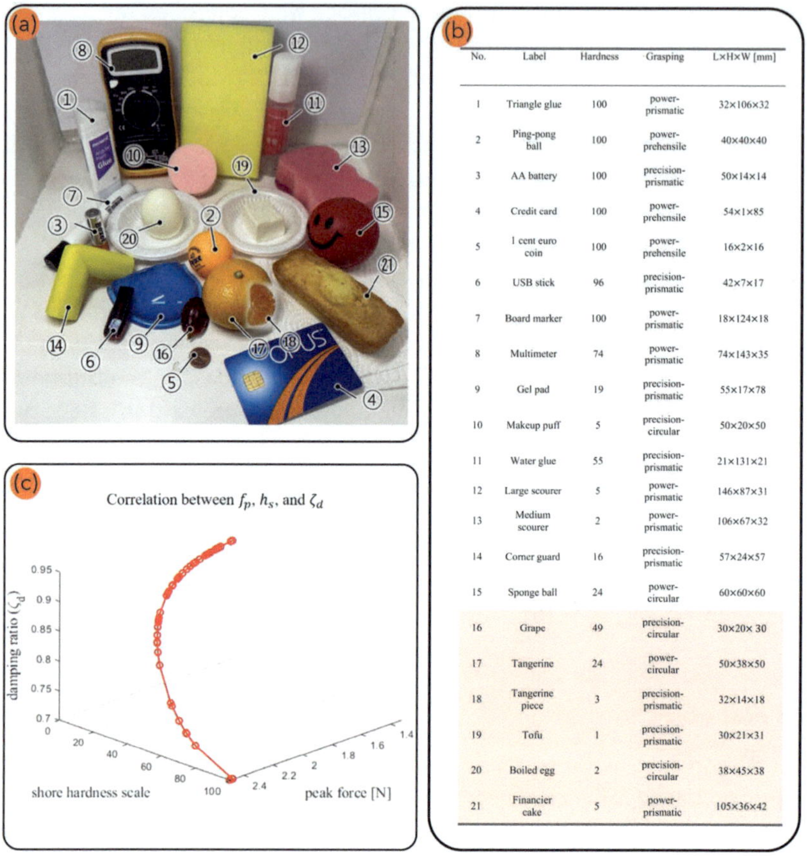

No.	Label	Hardness	Grasping	L×H×W [mm]
1	Triangle glue	100	power-prismatic	32×106×32
2	Ping-pong ball	100	power-prehensile	40×40×40
3	AA battery	100	precision-prismatic	50×14×14
4	Credit card	100	power-prehensile	54×1×85
5	1 cent euro coin	100	power-prehensile	16×2×16
6	USB stick	96	precision-prismatic	42×7×17
7	Board marker	100	power-prismatic	18×124×18
8	Multimeter	74	power-prismatic	74×143×35
9	Gel pad	19	precision-prismatic	55×17×78
10	Makeup puff	5	precision-circular	50×20×50
11	Water glue	55	precision-prismatic	21×131×21
12	Large scourer	5	power-prismatic	146×87×31
13	Medium scourer	2	power-prismatic	106×67×32
14	Corner guard	16	precision-prismatic	57×24×57
15	Sponge ball	24	power-circular	60×60×60
16	Grape	49	precision-circular	30×20× 30
17	Tangerine	24	power-circular	50×38×50
18	Tangerine piece	3	precision-prismatic	32×14×18
19	Tofu	1	precision-prismatic	30×21×31
20	Boiled egg	2	precision-circular	38×45×38
21	Financier cake	5	power-prismatic	105×36×42

Figure 9 (a) Different objects used in the shore hardness-peak force and damping ratio experiment. (b) Their standardized shore hardness. (c) Correlation between the three parameters. Adapted from: Park et al. (2021).

summary, without adhesion, it would be impossible to grasp flat or relatively big objects. This point was previously made as a design choice for versatile grippers (Langowski et al., 2020).

Bernoulli grippers are known as contactless grippers and are used to handle flat objects (Brun and Melkote, 2006; Journee et al., 2011). Although they are not used widely in handling crops, they have great potential for agricultural cases. One reason to why this integration has not taken place is that this category of grippers works best when the surface is perfectly flat and this is not generally the case for agricultural products. In fact, with slight modifications to deal with this issue, Bernoulli-based grippers have proven successful in handling non-flat objects (Petterson et al., 2010).

6.4 Concave objects

In this section, we analyze groups of grippers that have demonstrated both successful and unsuccessful grasping of concave objects. This analysis is based on the grasping mechanism (i.e. Bernoulli principle, controlled adhesion, i.e. surface adhesion or suction-base adhesion, controlled stiffness, mechanical interlocking, and controlled friction) and the materials of the gripper body (i.e. pure soft, hybrid soft-rigid) since these parameters are relevant in this context. The following will explain how these two factors can affect grasping of concave objects.

Four grippers (the adaptive Bernoulli, ElectroAdhesive, HASL-based, and dual-mode soft grippers) were unable to grasp concave (and non-lightweight) objects. The adaptive Bernoulli gripper (Petterson et al., 2010) is designed to approach objects from the top and cannot reach the side of the object where the concavity is a feature. Although the gripper is capable of deforming and conforming to the shape of the object, the deformation is insufficient to reach the side of the object. The ElectroAdhesive grasper (Cacucciolo et al., 2019) relies on a large surface contact area, which cannot be achieved when the object has a concave shape. By contrast, this gripper performs very well if the object is flat on the side or is convex. The dual-mode gripper (Wang et al., 2020) and the HASL-based gripper (Acome et al., 2018) both use friction-based and/or mechanical interlocking grasping mechanisms to pick up objects while having pure soft bodies.

Seven grippers demonstrated successful grasping of concave objects (Angelini et al., 2020; Hao et al., 2021; Park et al., 2021; Pettersson et al., 2010; Root et al., 2021; Wang et al., 2021; Zhou et al., 2017). Likewise, the anthropomorphic gripper (Park et al., 2021) and Pisa/IIT Soft gripper (Angelini et al., 2020) also use friction-based and mechanical interlocking grasping mechanisms while successfully handling concave objects. They both have a hybrid soft-rigid structure that enables them to grasp concave-shaped objects. Moreover, having stiffness control capability is of added value in handling such objects.

The magnetorheological gripper and the circular shelled gripper also have hybrid soft-rigid materials while using the controlled stiffness approach for grasping (Fig. 8). In both cases, having a rigid substrate allows their soft compartment to stiffen while they conform to the object. In particular, their rigid support allows soft pouches to be pushed towards the object (during actuation) that conform to the concavity of the object, which creates a large surface contact area. During the stiffening phase, the grippers can apply a well-distributed force to the concave objects, thanks to these large surface contact areas.

The same physics underlie the grasping of toroidal hydrostat and multimodal soft grippers, i.e. large surface contact area followed by stiffening.

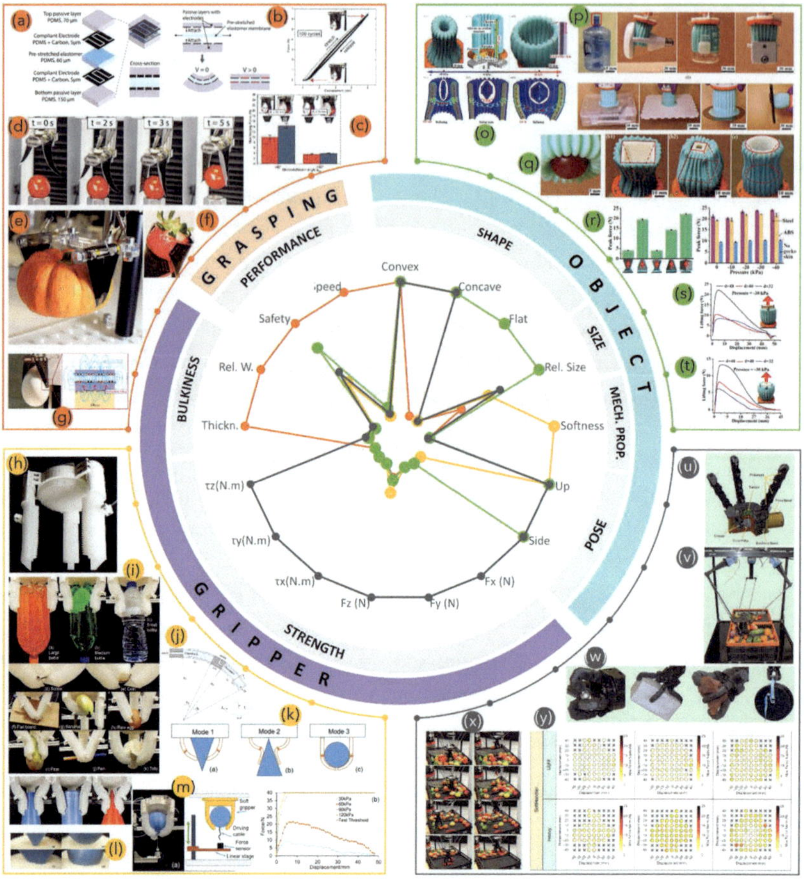

Figure 10 In the center: a graph illustrates the performance matrix measuring grippers' factors, i.e. strength and bulkiness, grasping performance, and the variation in objects that can be handled. The orange, yellow, green, and gray lines show four grippers with the highest scores (Cacucciolo et al., 2019; Zhou et al., 2017; Hao et al., 2021; Angelini et al., 2020, respectively). These grippers, their grasping performance, and the different objects that they handled are shown in four quarters, i.e. top left, bottom left, top right, and bottom right, respectively. The top left quarter depicts the electro-adhesion gripper: (a) mechanism underlying the dielectric-based actuation, (b) on-off cycle, (c) holding force objects with different size and electroadhesion angle, (d) rapid grasping of a cherry tomato, (e) grasping of a relatively large size pumpkin, (f) holding a strawberry safely, and (g) holding an egg while showing the mechanism of electroadhesion [images in sections (a)-(d) are adapted from Cacucciolo et al. (2019)]. The bottom left quarter shows (h) a three-finger gripper with a soft palm, (i) holding an object of various sizes, shapes, and weights and ultra-soft (tofu) using the segmented finger, (j) via different grasping modes (k). (l)-(m) the payload and the soft palm interacting with different objects, respectively. The top right quarter illustrates the multimodal enveloping soft gripper and its working principle, (o) holding a large object (with respect to the size of the gripper), a flat object, and grasping object from the side (p). (q) This gripper enveloping a cherry tomato and also grasping standard objects with convex and concave shapes. (r) Performance of

The only difference is that the grippers are using pure soft bodies to conform to the object and stiffen it. The toroidal hydrostat gripper uses an inversion mechanism known as a sloughing mechanism (Sadeghi et al., 2013) to drag a soft object in a ring pouch via rolling, while the multimodal gripper uses a soft shell structure to collapse in a programmed way to the object. The last gripper in this group (Zhou et al., 2017) uses a different mechanism to adapt to concave shapes. It takes advantage of a segmented finger capable of multiple bending curvatures to conform to objects with both concavity and convexity, such as pears (Fig. 8f). As with other grippers, it has a very large surface contact area with objects, which helps the gripper to apply a well-distributed grasping force during the actuation.

7 Designing the ideal versatile gripper

Analyzing the results to date, we can see that none of the grippers are able to satisfy all the requirements associated with agricultural tasks. Each design satisfies just one or some of the requirements and fails on the other aspects. The reason is simple: their designs are not meant to satisfy all agricultural task requirements. When it comes to design choice, soft robotics have a great deal to offer, despite current examples not yet meeting all potential requirements. One of the challenges with respect to grasping objects from the side is a lack of support stiffness, which could be resolved by jamming mechanisms (Brancadoro et al., 2020; Ibrahimi et al., 2021) or any other soft robotic solution that offers selective stiffness (Visentin et al., 2021). Some examples of such a combined design aim to compensate for the flaws of grippers designed for a certain purpose, such as increasing strength (Yang et al., 2019; Zhou et al., 2020). However, regardless of the combination of technology or gripper mechanisms, the final design will not satisfy all the requirements if they are not taken into account right from the beginning.

Studying natural grippers could help with the design of the ideal versatile gripper, either via inspiration or following biomechanical principles. Indeed, organisms in their natural habitats develop functionalities in order to adapt to their environments as a generic survival approach. For example, the

Figure 10 (*Continued*)

different grasping modalities, (s)-(t) grasping performance of object with different size and shape. The bottom right illustrates the SoftHandler grasping system, (u) and (v) capable of handling quite heavy and unstructured objects (w). (x) Picking and placing agricultural products and (y) force distribution at gripper's palm while handling objects of various sizes (i.e. small, medium, and large) and weights (i.e. light and heavy).

development of opposable thumbs helped chimps climb thin branches to get to fruits that grew at the tip. This could be inspiration for a more optimal versatile gripper. Moreover, studying the neuromusculoskeletal structure of chimpanzee hands could tell us much about their grasping mechanism. Researchers have shown that natural grippers use a hybrid grasping mechanism in order to deal with unpredictability and variations in their environment (Langowski et al., 2020), which is another excellent approach to enhance versatility.

To summarize what we have seen to date in reviewing previously mentioned grippers, not one single grasping mechanism is able to handle the set of requirements of a natural task, i.e. picking up a crop. Figure 10 uses a radar diagram to show how we can use all the soft robotic technologies we have reviewed to date to cover all requirements. Its illustration identifies a minimum number of complementary grasping technologies among the set of 11 grippers. Complementary grasping technologies mean that when a grasping technology fails to meet some requirements, another can compensate.

These technologies are as follows:

- Electroadhesion soft gripper (showed in orange in Fig. 10): it has low bulkiness, i.e. relative weight and thickness, and high grasping performance, i.e. safety and speed;
- The three-finger gripper with soft palm: it scores high when grasping soft objects;
- The multimodal enveloping soft gripper, which scored high in dealing with object variation, i.e. shape size and pose; and
- The SoftHandler: it is a tendon-based gripper, scored high in strength.

Hypothetically, the ideal gripper would be a hybrid design of such grasping technologies. Whether the hybridization of these specific technologies is feasible is a design research question and is addressed in the case study in Section 3.

8 Case study: the use of soft robotics in intercropping

A very good use-case of soft robotics being applied in agriculture is harvesting in intercropping system, as it has a wide range of requirements. The case study is taken from the 'Universal Soft Robotic Harvester for autonomous intercropping system' (USOROH) project (Leylavi Shoushtari and van Henten, 2020) and is a collaboration between Wageningen University and China Agricultural University as part of the Agricultural Green Development Plan. The purpose of this project is to enhance the versatility of soft grippers (through a hybrid design) to deal with the high variation of crops in the intercropping system. In particular, versatility is defined based on adaptability to the shape, size, weight, and mechanical properties of the objects to be handled. Each soft grasping technology has a list of pros and cons regarding adaptability. Through a hybrid design, their drawbacks can be traded-off or ideally canceled out.

Intercropping is a farming practice in which a set of crops with mutual benefits are planted next to each other, resulting in using fewer or no pesticides and fertilizers and enhancing soil health (Vandermeer, 1989). Having different crops in the same field is always a challenge when it comes to harvesting since it has both categories of diversity:

- Diversity within a crop variant; and
- Diversity between different types of crops.

Agricultural technologies are mainly designed for monocultures which are now seen as less sustainable, while traditional farming practices such as intercropping or mixed cropping have been proven to be more sustainable, particularly in improving soil health. Such farming practices have been abandoned in part due to intensive labor requirements for planning, crop maintenance, and harvesting.

Following a systematic approach (Fig. 11), a set of functional requirements for a gripper (i.e. grasping, stiffness and force adaptability) can be derived from the challenges of harvesting on an intercropping farm. Current soft grasping technologies can meet these requirements for individual crops but not simultaneously for two or more crops. Existing soft grasping technologies are assessed based on different functional requirements and a set of candidate technologies are chosen for hybridization. Three different grasping technologies, i.e. controlled stiffness, controlled adhesion, and grasping by actuation have been assessed (Shintake et al., 2018). Our aim is to combine these different technologies to conceptualize a design for the ideal versatile gripper. Knowing that each of these technologies has its own pros and cons, the candidate technologies are grouped based on the various task requirements set out in the assessment matrix.

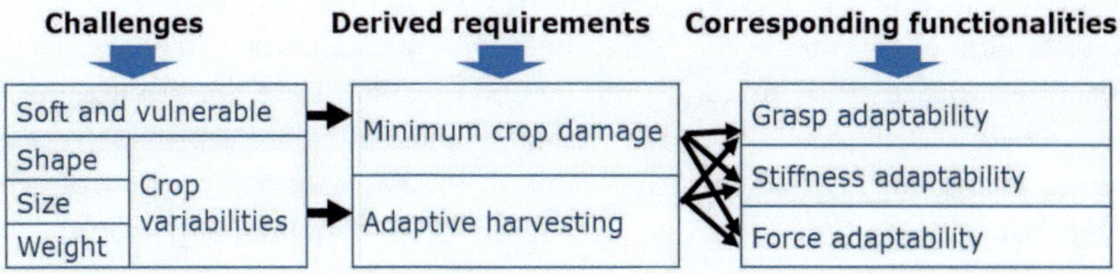

Figure 11 The chain of challenges, requirements, and functionalities for harvesting tasks in an intercropping system.

9 Conclusion

There is high demand for automated harvesting since this accounts for a significant proportion of production costs – around one-third of the entire cost of producing high-value crops (Bac et al., 2014). However, current rigid robots, machines, and automation technologies are not yet ready to meet the high requirements of mixed-crop harvesting.

This chapter has addressed the fact that integrating current soft grasping technologies in agri-food tasks (particularly in the case of harvesting) is a multifaceted research problem. Aspects such as gripper parameters (i.e. strength and bulkiness), grasping performance, and object variations need to be taken into account simultaneously. To determine to what extent soft grasping technologies are capable of being integrated at harvest, these triple aspects of each technology need to be assessed. This chapter has provided an assessment matrix which accounts for these three basic factors. Two parameters to assess soft grippers are strength and bulkiness. The former can be measured based on the maximum force/torque before slippage of the object and the latter by relative weight and thickness of the object to the gripper. Grasping performance is assessed based on safety and speed as the most relevant factors in crop handling. The shape, size, pose, and mechanical properties are all indices representing variability of the crops/objects to be handled. There were two main challenges in extracting data from the literature for use in performance assessment:

- There are insufficient data on the assessment matrix.
- Some measures (safety, shape, and roughness) are qualitative, which makes it difficult for benchmarking.

Comparing the different grasping technologies showed the grippers alongside each other. These results revealed that certain features of the objects (i.e. flat and concave shapes, softness, and side pose) are quite challenging for grippers, as very few scored highly. We also discussed why these grippers outperform their counterparts in such challenging cases.

The chapter then outlined a hypothetical design of an ideal gripper for a harvesting task, comprising a hybrid design of soft grasping technologies that complement each other in relation to the assessment matrix. The chapter then described a case study featuring the USOROH to illustrate the application of such a hybrid design. In a broader context, soft robotics could provide a safer, more flexible, and hence inclusive solution for harvesting mixed crops. The safety aspect discussed in this chapter refers only to maintaining crop quality during robotic handling, but soft robots also provide fairly safe interactions with humans too. I-support is a good (non-agricultural) example that demonstrates soft robots' safety in human–robot interactions (Arleo et al., 2020; Zlatintsi et al.,

2020). Soft robotics also unlocks access to a range of affordable and available materials that are incompatible with rigid robots (Rus and Tolley, 2015). Moreover, having access to certain manufacturing technologies, such as 3D printing and casting, makes it even easier to design, customize, or reproduce a soft robot. Platforms such as the Soft Robotics toolkit (Soft Robotics Toolkit, n.d.) and FlowIO (Shtarbanov, 2021) are very good examples of open access soft robotics, helping to democratize soft robotics as a solution for all users, particularly farmers.

Soft robotics could potentially offer safe, versatile, affordable, and accessible solutions for harvesting in mixed crop farms. This chapter has addressed the technological gap between the actual performance of soft robotics and its potential for agriculture, particularly sustainable farming practices such as intercropping. In order to turn this potential into action, further research is required, such as the USOROH project, to take maximum advantage of current soft robotic technologies and sciences to develop tailored solutions for sustainable farming practices.

10 Where to look for further information

The following articles provide a good overview of the subject:

- Elfferich, J. F., Dodou, D. and Della Santina, C. (2022). Soft robotic grippers for crop handling or harvesting: a review. *IEEE Access* 10(June), 75428–75443. https://doi.org/10.1109/ACCESS.2022.3190863.
- Shintake, J., Cacucciolo, V., Floreano, D. and Shea, H. (2018). Soft robotic grippers. *Advanced Materials* 30(29). https://doi.org/10.1002/adma.201707035.
- Langowski, J. K. A., Sharma, P. and Leylavi Shoushtari, A. (2020). In the soft grip of nature. *Science Robotics* 5(49), 3–6. https://doi.org/10.1126/scirobotics.abd9120.
- Bac, C. W., van Henten, E. J., Hemming, J. and Edan, Y. (2014). Harvesting robots for high-value crops: state-of-the-art review and challenges ahead. *Journal of Field Robotics* 31: 888–911. https://doi.org/10.1002/rob.21525.
- Koerhuis, R. (n.d.). Wageningen University on a quest to develop the best robotic grippers. *Future Farming*. https://www.futurefarming.com/crop-solutions/wageningen-university-on-a-quest-to-develop-the-best-robotic-grippers/.
- Zhang, B., Xie, Y., Zhou, J., Wang, K. and Zhang, Z. (2020). State-of-the-art robotic grippers, grasping and control strategies, as well as their applications in agricultural robots: a review. *Computers and Electronics in Agriculture* 177(April), 105694. https://doi.org/10.1016/j.compag.2020.105694.

Magazine of Future Farming:

- https://www.futurefarming.com/crop-solutions/wageningen-university-on-a-quest-to-develop-the-best-robotic-grippers/.

Website of Dutch Soft Robotics Community:

- https://dutchsoftrobotics.nl/.

Open source soft robotics technologies:

- https://softroboticstoolkit.com/.
- https://www.softrobotics.io/.
- http://opensoftmachines.com/.

11 References

Acome, E., Mitchell, S. K., Morrissey, T. G., Emmett, M. B., Benjamin, C., King, M., Radakovitz, M. and Keplinger, C. (2018). Hydraulically amplified self-healing electrostatic actuators with muscle-like performance. *Science* 359(6371), 61–65. https://doi.org/10.1126/science.aao6139.

Adamides, G. and Edan, Y. (2023). Human–robot collaboration systems in agricultural tasks: a review and roadmap. *Computers and Electronics in Agriculture* 204, 107541. https://doi.org/10.1016/j.compag.2022.107541.

Angelini, F., Petrocelli, C., Catalano, M. G., Garabini, M., Grioli, G. and Bicchi, A. (2020). SoftHandler: an integrated soft robotic system for handling heterogeneous objects. *IEEE Robotics and Automation Magazine* 27(3), 55–72. https://doi.org/10.1109/MRA.2019.2955952.

Ansari, Y., Shoushtari, A. L., Cacucciolo, V., Cianchetti, M. and Laschi, C. (2015). Dynamic walking with a soft limb robot. *Lecture Notes in Computer Science (Including Subseries Lecture Notes in Artificial Intelligence and Lecture Notes in Bioinformatics)* 9222, 13–25. https://doi.org/10.1007/978-3-319-22979-9_2.

Arleo, L., Stano, G., Percoco, G. and Cianchetti, M. (2020). I-support soft arm for assistance tasks: a new manufacturing approach based on 3D printing and characterization. *Progress in Additive Manufacturing* 89, 01234567. https://doi.org/10.1007/s40964-020-00158-y.

Bac, C. W., van Henten, E. J., Hemming, J. and Edan, Y. (2014). Harvesting robots for high-value crops: state-of-the-art review and challenges ahead. *Journal of Field Robotics* 31(6), 1–17. https://doi.org/10.1002/rob.

Brancadoro, M., Manti, M., Tognarelli, S. and Cianchetti, M. (2020). Fiber jamming transition as a stiffening mechanism for soft robotics. *Soft Robotics* 7(6), 663–674. https://doi.org/10.1089/soro.2019.0034.

Brun, X. F. and Melkote, S. N. (2006). Evaluation of Handling Stresses Applied to EFG Silicon Wafer Using a Bernoulli Gripper. In: *2006 IEEE 4th World Conference on Photovoltaic Energy Conference* 1346–1349.

Cacucciolo, V., Shintake, J. and Shea, H. (2019). Delicate yet strong: characterizing the electro-adhesion lifting force with a soft gripper. RoboSoft IEEE International

Conference on Soft Robotics. 108–113. https://doi.org/10.1109/ROBOSOFT.2019.8722706.

Catalano, M. G., Grioli, G., Farnioli, E., Serio, A., Piazza, C. and Bicchi, A. (2014). Adaptive synergies for the design and control of the Pisa/IIT SoftHand. *International Journal of Robotics Research* 33(5), 768–782. https://doi.org/10.1177/0278364913518998.

Davis, S., Gray, J. O. and Caldwell, D. G. (2008). An end effector based on the Bernoulli principle for handling sliced fruit and vegetables. *Robotics and Computer-Integrated Manufacturing* 24(2), 249–257. https://doi.org/10.1016/j.rcim.2006.11.002.

Droukas, L., Doulgeri, Z., Tsakiridis, N. L., Triantafyllou, D., Kleitsiotis, I., Mariolis, I., Giakoumis, D., Tzovaras, D., Kateris, D. and Bochtis, D. (2023). A survey of robotic harvesting systems and enabling technologies. *Journal of Intelligent and Robotic Systems* 107(2), 21. https://doi.org/10.1007/s10846-022-01793-z.

Elfferich, J. F., Dodou, D. and Santina, C. D. (2022). Soft robotic grippers for crop handling or harvesting: a review. *IEEE Access* 10(June), 75428–75443. https://doi.org/10.1109/ACCESS.2022.3190863.

Erzincanli, F. and Sharp, J. M. (1997). Meeting the need for robotic handling of food products. *Food Control* 8(4), 185–190. https://doi.org/10.1016/S0956-7135(97)00047-9.

Flat, I. C. O. N. (n.d.). *No Title*. Available at: https://www.flaticon.com/.

Hao, Y., Biswas, S., Hawkes, E. W., Wang, T., Zhu, M., Wen, L. and Visell, Y. (2021). A multimodal, enveloping soft gripper: shape conformation, bioinspired adhesion, and expansion-driven suction. *IEEE Transactions on Robotics* 37(2), 350–362. https://doi.org/10.1109/TRO.2020.3021427.

Hawkes, E. W., Jiang, H. and Cutkosky, M. R. (2016). Three-dimensional dynamic surface grasping with dry adhesion. *International Journal of Robotics Research* 35(8), 943–958. https://doi.org/10.1177/0278364915584645.

Howell, L. L., Magleby, S. P. and Olsen, B. M. (2013). *Handbook of Compliant Mechanisms*. John Wiley & Sons.

Ibrahimi, M., Paternò, L., Ricotti, L. and Menciassi, A. (2021). A layer jamming actuator for tunable stiffness and shape-changing devices. *Soft Robotics* 8(1), 85–96.

Jiang, H., Hawkes, E. W., Fuller, C., Estrada, M. A., Suresh, S. A., Abcouwer, N., Han, A. K., Wang, S., Ploch, C. J., Parness, A. and Cutkosky, M. R. (2017). A robotic device using gecko-inspired adhesives can grasp and manipulate large objects in microgravity. *Science Robotics* 2(7), 1–12. https://doi.org/10.1126/scirobotics.aan4545.

Journee, M., Chen, X., Member, S., Robertso, J., Jermy, M. and Sellier, M. (2011). An investigation into improved non-contact adhesion mechanism suitable for wall climbing robotic applications. In: *2011 IEEE International Conference on Robotics and Automation* 4915–4920.

Kumar, M. S. and Mohan, S. (2022). Selective fruit harvesting: research, trends and developments towards fruit detection and localization: a review. 095440622211284. https://doi.org/10.1177/09544062221128443.

Langowski, J. K. A., Sharma, P. and Leylavi Shoushtari, A. L. (2020). In the soft grip of nature. *Science Robotics* 5(49), 3–6. https://doi.org/10.1126/scirobotics.abd9120.

Leylavi Shoushtari, A. and van Henten, E. (2020). *Universal Soft Robotic Harvester for Autonomous Intercropping System*. Website: https://dutchsoftrobotics.nl/projects/universal-soft-robotic-harvester-autonomous-intercropping-system.

Leylavi Shoushtari, A. L., Naselli, G. A., Sadeghi, A. and Mazzolai, B. (2019). INFORA: a novel inflatable origami-based actuator. In: Proceedings of the - IEEE International Conference on Robotics and Automation. 7415–7420. https://doi.org/10.1109/ICRA.2019.8794422.

Majidi, C. (2014). Soft robotics: A perspective: current trends and prospects for the future. *Soft Robotics* 1(1), 5–11. https://doi.org/10.1089/soro.2013.0001.

Martin, T., Gasselin, P., Hostiou, N., Feron, G., Laurens, L., Purseigle, F. and Ollivier, G. (2022). Robots and transformations of work in farm: a systematic review of the literature and a research agenda. *Agronomy for Sustainable Development* 42(4). https://doi.org/10.1007/s13593-022-00796-2.

Pan, Y., Wang, Y., McGregor, C., Liu, S., Luan, F., Gao, M. and Weng, Y. (2020). Genetic architecture of fruit size and shape variation in cucurbits: a comparative perspective. In: *TAG. Theoretical and Applied Genetics. Theoretische und Angewandte Genetik*. Springer 133(1), 1–21. https://doi.org/10.1007/s00122-019-03481-3.

Park, H., Kim, M., Lee, B. and Kim, D. (2021). Design and experiment of an anthropomorphic robot hand for variable grasping stiffness. *IEEE Access* 9, 99467–99479. https://doi.org/10.1109/ACCESS.2021.3094060.

Petterson, A., Ohlsson, T., Caldwell, D. G., Davis, S., Gray, J. O. and Dodd, T. J. (2010a). A Bernoulli principle gripper for handling of planar and 3D (food) products. *Industrial Robot* 37(6), 518–526. https://doi.org/10.1108/01439911011081669.

Pettersson, A., Davis, S., Gray, J. O., Dodd, T. J. and Ohlsson, T. (2010b). Design of a magnetorheological robot gripper for handling of delicate food products with varying shapes. *Journal of Food Engineering* 98(3), 332–338. https://doi.org/10.1016/j.jfoodeng.2009.11.020.

Root, S. E., Preston, D. J., Feifke, G. O., Wallace, H., Alcoran, R. M., Nemitz, M. P., Tracz, J. A. and Whitesides, G. M. (2021). Bio-inspired design of soft mechanisms using a toroidal hydrostat. *Cell Reports Physical Science* 2(9), 100572. https://doi.org/10.1016/j.xcrp.2021.100572.

Ruotolo, W., Brouwer, D. and Cutkosky, M. R. (2021). From grasping to manipulation with gecko-inspired adhesives on a multifinger gripper. *Science Robotics* 6(61), eabi9773. https://doi.org/10.1126/scirobotics.abi9773.

Rus, D. and Tolley, M. T. (2015). Design, fabrication and control of soft robots. *Nature* 521(7553), 467–475. https://doi.org/10.1038/nature14543.

Sadeghi, A., Tonazzini, A., Popova, L. and Mazzolai, B. (2013). IEEE International Conference on Robotics and Automation, 3457–3462.

Santina, C. D. Della, Piazza, C., Grioli, G., Catalano, M. G. and Bicchi, A. (2018). Toward dexterous manipulation with augmented adaptive synergies: the Pisa/IIT SoftHand 2. *IEEE Transactions on Robotics* 34(5), 1141–1156. https://doi.org/10.1109/TRO.2018.2830407.

Shintake, J., Cacucciolo, V., Floreano, D. and Shea, H. (2018). Soft robotic grippers. *Advanced Materials* 30(29), e1707035. https://doi.org/10.1002/adma.201707035.

Shintake, J., Rosset, S., Schubert, B., Floreano, D. and Shea, H. (2016a). Versatile soft grippers with intrinsic electroadhesion based on multifunctional polymer actuators. *Advanced Materials* 28(2), 231–238. https://doi.org/10.1002/adma.201504264.

Shintake, J., Rosset, S., Schubert, B., Floreano, D. and Shea, H. (2016b). Versatile soft grippers with intrinsic electroadhesion based on multifunctional polymer actuators. *Advanced Materials* 28(2), 231–238. https://doi.org/10.1002/adma.201504264.

Shtarbanov, A. (2021). FlowIO development platform–The pneumatic "raspberry pi" for soft robotics. In *Extended abstracts of the 2021 CHI conference on human factors in computing systems* 1–6.

Soft Robotics Toolkit (n.d.).

Szeliski, R. (2011). *Computer Vision*. https://doi.org/10.1007/978-1-84882-935-0.

Vandermeer, J. H. (1989). *The Ecology of Intercropping*. Cambridge University Press. https://doi.org/10.1017/CBO9780511623523.

Visentin, F., Murali Babu, S. P., Meder, F. and Mazzolai, B. (2021). Selective stiffening in soft actuators by triggered phase transition of hydrogel-filled elastomers. *Advanced Functional Materials* 31(32), 2101121. https://doi.org/10.1002/adfm.202101121.

Wang, Z., Kanegae, R. and Hirai, S. (2021). Circular shell gripper for handling food products. *Soft Robotics* 8(5), 542–554. https://doi.org/10.1089/soro.2019.0140.

Wang, Z., Or, K. and Hirai, S. (2020). A dual-mode soft gripper for food packaging. *Robotics and Autonomous Systems* 125, 103427. https://doi.org/10.1016/j.robot.2020.103427.

Wang, Z., Xun, Y., Wang, Y. and Yang, Q. (2022). Review of smart robots for fruit and vegetable picking in agriculture. *International Journal of Agricultural and Biological Engineering* 15(1), 33–54. https://doi.org/10.25165/j.ijabe.20221501.7232.

Yang, Y., Zhang, Y., Kan, Z. and Zeng, J. (2019). Hybrid jamming for bioinspired soft robotic fingers. 1–17. https://doi.org/10.1089/soro.2019.0093.

Zhang, B., Xie, Y., Zhou, J., Wang, K. and Zhang, Z. (2020). State-of-the-art robotic grippers, grasping and control strategies, as well as their applications in agricultural robots: a review. *Computers and Electronics in Agriculture* 177(April), 105694. https://doi.org/10.1016/j.compag.2020.105694.

Zhou, J., Chen, S. and Wang, Z. (2017). A soft-robotic gripper with enhanced object adaptation and grasping reliability. *IEEE Robotics and Automation Letters* 2(4), 2287–2293. https://doi.org/10.1109/LRA.2017.2716445.

Zhou, J., Chen, Y., Hu, Y., Wang, Z., Li, Y., Gu, G. and Liu, Y. (2020). Adaptive variable stiffness particle phalange for robust and durable robotic grasping 1–15. https://doi.org/10.1089/soro.2019.0089.

Zhu, J., Chai, Z., Yong, H., Xu, Y., Guo, C., Ding, H. and Wu, Z. (2023). Bioinspired multimodal multipose hybrid fingers for wide-range force, compliant, and stable grasping. *Soft Robotics* 10(1), January, 30–39. https://doi.org/10.1089/soro.2021.0126.

Zlatintsi, A., Dometios, A. C., Kardaris, N., Rodomagoulakis, I., Koutras, P., Papageorgiou, X., Maragos, P., Tzafestas, C. S., Vartholomeos, P., Hauer, K., Werner, C., Annicchiarico, R., Lombardi, M. G., Adriano, F., Asfour, T., Sabatini, A. M., Laschi, C., Cianchetti, M., Güler, A., Kokkinos, I., Klein, B. and López, R. (2020). I-SUPPORT : a robotic platform of an assistive bathing robot for the elderly population. *Robotics and Autonomous Systems* 126(April), 103451.

Chapter 13

Advances in agricultural unmanned aerial vehicles: focus on sensing applications

Tarin Paz-Kagan, Ben-Gurion University of the Negev, Israel*

1 Introduction
2 Unmanned aerial vehicle remote sensing sensors
3 Platforms for precision agriculture
4 Flight planning and pre-processing
5 Application in precision agriculture
6 Conclusion and future trends
7 Where to look for further information
8 References

1 Introduction

The expected growth in the world population will increase the demand for food, requiring more food production to expand by 70% from today to 2050 (Hunter et al., 2017). Given the challenges imposed by climate change, meeting this demand will necessitate informed and precise approaches as agricultural land and resources become increasingly limited. Precision agriculture (PA) integrates technologies and data to support smart decisions, which include automation, robotics, big data analysis, information, and communication technologies, and sensor development, with artificial intelligence (AI) for more sustainable agricultural management (Muchiri and Kimathi, 2022). PA practices use parameters summarized as the 4 Rs: apply different agricultural inputs in the Right place, with the Right sources, at the Right intensity, and, finally, at the Right time (Mulla, 2013). PA aims to improve agricultural production, use efficiency, and reduce environmental pollution while minimizing costs by addressing the spatial and temporal variability in the field (Freidenreich et al., 2019; Tang et al., 2021; Naiqian Zhang et al., 2002), instead of treating agricultural areas homogeneously. The agricultural field is usually divided into subunits or management zones (MZs) receiving adjusted inputs (Pederi and Cheporniuk, 2015). MZs are sub-units characterized by homogeneous characteristics that

* Tarin Paz-Kagan, The Jacob Blaustein Institutes for Desert Research, Ben-Gurion University of the Negev, Israel.

http://dx.doi.org/10.19103/AS.2023.0124.11

should be managed differently from neighboring units or zones (Freidenreich et al., 2019; Termin et al., 2023; Ye et al., 2020; Naiqian Zhang et al., 2002). Thus, PA studies the within-field spatial and temporal variation of crop and soil characteristics to develop spatial decision support systems (SDSS), which assist in increasing productivity, allocating resources, and avoiding food waste while adapting to climate change.

Remote sensing (RS) technology allows for monitoring crop and soil conditions using different platforms, from satellites to unmanned aerial vehicles (UAVs) and ground sensors, with each having its advantages and limitations for environmental monitoring (Olson and Anderson, 2021). In recent years, satellite data acquisition has dramatically improved by developing third-generation satellite systems with higher spatial, temporal, and spectral resolutions. Next-generation satellite systems use narrower spectral bands and a daily revisit time to acquire information on the earth's surface. These advancements enable the retrieval of various biophysical and biochemical soil properties and plant trait assessments (Berger et al., 2020; Wang et al., 2017; Xuefeng et al., 2016). These earth observation systems include satellites such as QuickBird, GeoEye, WorldView, VENµS, Planet-Scope and Sentinel-2, which are multispectral sensors, and satellites such as EnMAP, PRISMA, which are hyperspectral sensors. However, optical satellite data still have spatial and temporal resolution limitations and are strongly affected by atmospheric conditions that cannot be ignored.

The UAV platform, integrated with various sensors, enables more frequent and faster data acquisition, making the UAV platform more valuable for monitoring farming systems than most traditional agronomic methods or RS platforms. The procedure involves sensing, recording, processing, and analyzing data from various sensors and UAV platforms (Prado Osco et al., 2019). Considering the weaknesses of satellite platforms, more small commercial UAVs with multiple sensors are being used to provide measurements that complement satellite data. UAV platforms can provide high-resolution data that answers the need for real-time and quick monitoring of agricultural areas (Ochoa and Guo, 2019; Zhang et al., 2021). Thus, UAVs are an intermediate application between satellites and ground field measurements, capturing high-resolution fine-scale data (Zhou et al., 2021). Furthermore, UAVs have many additional advantages, such as less dependence on weather conditions and georeferenced images than satellite systems (Zhang et al., 2021; Handique et al., 2017; Saldana Ochoa and Guo, 2019). However, their limitations include battery life, operational speed, monitoring scale, data acquisition efficacy, and operational time. Another crucial aspect is the regulations and policies regarding UAV operations.

The use of UAVs in PA has increased steadily since the beginning of the twenty-first century (del Cerro et al., 2021). The estimated industry market of UAVs is about US$127 billion, and the agricultural UAV sector is estimated to be 32.4 US$ billion, which is about 25% of the global UAV market (Mazur et al., 2016). Notably, UAVs allow for better monitoring of agricultural systems, filling the gap of satellite earth observation systems that only offer a general overview at the field scale with limited continuous tracking (Milics, 2019). UAVs provide agronomists, companies, and growers with high-resolution technology, high mobility, and accurate, detailed imaging (Kim et al., 2019). Although UAVs have yet to be fully established in PA practices, they are starting to play an increasingly important role, helping farmers with real-time monitoring and management strategies for more efficient and sustainable agricultural practices (Delavarpour et al., 2021). In this chapter, we aim to present an overview of the advantages and limitations of UAV-RS platforms and their applications in PA. To properly review these applications, we formulated the following questions: What are the available UAV sensors for PA applications? How should one select the best UAV platform for different PA tasks? What needs to be considered in flight planning and in performing the necessary pre-processing to obtain high data quality? What are the main applications of UAV-RS in PA? Finally, we will discuss the future applications and challenges that lie ahead.

2 Unmanned aerial vehicle remote sensing sensors

In recent decades, sensor development and improvement have made an efficient, rapid, reproducible, and relatively inexpensive method for soil and plant monitoring at a regional scale (Gu et al., 2021). Advances in sensing technologies and the increased availability of UAVs have led to the recognition of UAV-based RS as a complementary or alternative method to laboratory analysis of plant leaf tissue and soil samples. UAV-based RS uses two main types of sensors: 'passive' and 'active'. Passive optical sensors use spectral measurements of the interaction between plant and soil electromagnetic radiation to obtain information on the state of the agricultural field. The radiation absorbed by the plant is related to various plant traits, such as plant pigment, structure, nutritional content, and water content, all of which are expressed by a spectral wavelength associated with the plant's physical and chemical composition (Fig. 1). The passive remote sensing sensors collect radiation reflected or emitted from an object that covers the spectral range from the visible spectral to the shortwave infrared (VIS-NIR SWIR) (ranging from 0.4 μm to 2.5 μm) and the thermal spectral infrared (TIR) (ranging from 8 μ m to 14 μm) (Berger et al., 2022). The visible spectral range (0.4-0.7μm) mainly reflects plant pigment and chlorophyll content, while the near-infrared range (0.7-1 μm) mainly reflects plant cell structure. The shortwave infrared spectral range (1-2.5 μm) detects plant water content and chemical composition. The TIR spectral region detects plant and soil surface temperature by measuring the radiation emitted from the surface. Objects with a temperature above 0 K (−273°C or −459°F) emit radiation, and the amount of radiation is

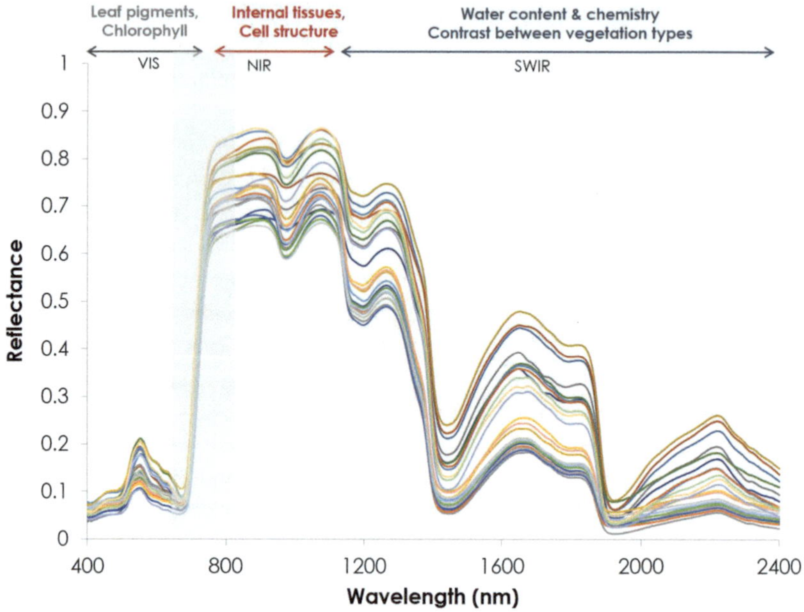

Figure 1 The spectral signal of different crop species in the VIS-NIR-SWIR spectral region (0.4–2.5 μm or 400-2500 nm); the gray area represents the red edge spectral region (0.6–0.8 μm or 600-800 nm).

a function of the surface's emissivity translated to surface temperature (Vargas et al., 2020). The higher the object's temperature, the greater the radiation intensity emitted by it. Another important spectral region for plant monitoring is the red edge, which is used by solar-induced fluorescence (SIF) and is characterized by a wavelength ranging between 0.6 μm and 0.8 μm, related to plant photosynthetic activity (Gerhards et al., 2019). SIF sensors, which measure plant status and proxy for plant photosynthetic activity, are available with very dense and narrow spectral wavebands (Bendig et al., 2021). SIF is a crucial technique for understanding photosynthetic mechanisms and plant responses to stress conditions and tolerance (Murchie and Lawson, 2013), drought survival (Jansen et al., 2009), and disease detection and resistance (Rousseau et al., 2013).

Soil sensing enables the detection of spectral signatures related to soil absorption and reflection features (known as chromophores; Fig. 2). These spectral absorption features are caused by the vibration responses of soil properties and molecular structures to electromagnetic signals (Davies and Calvin, 2017; Levi et al., 2022; Nocita et al., 2015; Paz-Kagan et al., 2015; Viscarra Rossel et al., 2016). Studies on soil sensing have increased in recent years, demonstrating the efficiency of detecting different soil properties (Bogrekci and Lee, 2005; Freschet et al., 2011; Gholizadeh et al., 2020; Romsonthi et al.,

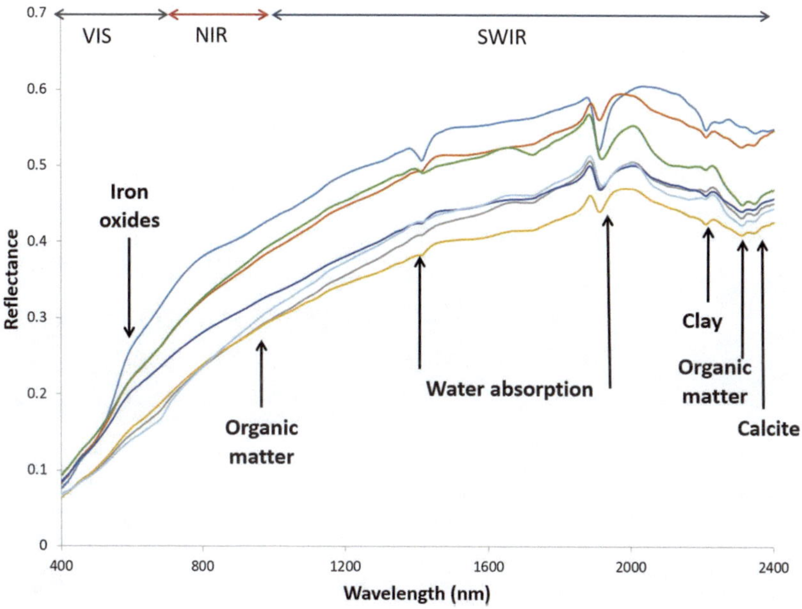

Figure 2 The spectral signals of different soil types in the VIS-NIR-SWIR spectral region (0.4–2.5 μm or 400-2500 nm).

2018). Soil spectral characteristics include the biological, chemical, and physical properties reflected or emitted by a specific soil type (Gholizadeh et al., 2018; Levi et al., 2021). These spectral characteristics have been extensively used to monitor soil organic matter, salinity, moisture, nutritional content, and texture and were used to assess overall soil health (Donager et al., 2018; Gholizadeh et al., 2013; Hu et al., 2019; Paz-Kagan et al., 2014). Several studies have shown that optical remote sensing, mainly hyperspectral in the SWIR region, provides most of the information required for soil property assessment (Chabrillat et al., 2019; Levi et al., 2022; Rossel et al., 2016). Thus, higher spectral band number and wider electromagnetic spectral regions will usually improve the sensors' detection capability of plant traits or soil properties (Gholizadeh et al., 2018).

Light detection and ranging (LiDAR) are active remote sensing systems that can collect detailed plant morphology and soil structure data (Madec et al., 2017). LiDAR technology uses laser pulses to measure distances and intensity accurately by calculating the distance between objects and the sensor (Lin, 2015). There are various LiDAR sensors: (1) UAV laser scanning (ULS), (2) airborne laser scanning (ALS) mounted on aircraft, (3) terrestrial laser scanning (TLS) and mobile laser scanning (MLS) from a static ground platform or a moving ground platform, and (4) satellites LiDAR systems (SLS). The sensor selection depends on the spatial resolution required and on the size of the laser footprint (point density and scanning geometry) and coverage. LiDAR sensors provide 3D point

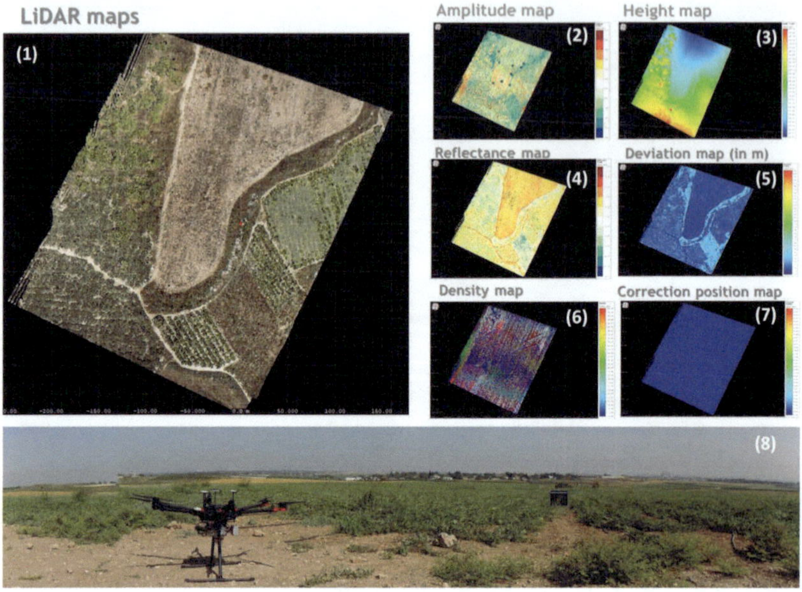

Figure 3 LiDAR measurements contain several dimensions of information: (1) overview of the agricultural field with point cloud data and all the additional output products of amplitude (2), height (3), reflectance (4), deviation (5), point density (6), and correction position maps (7). A photograph of the agricultural field and the UAV with the Rigel-Mini UAV LiDAR system captures these data (8).

clouds with accurate X, Y, and Z coordinate systems (i.e. latitude, longitude, and height). LiDAR measurements contain several additional dimensions of information regarding location, the object height, amplitude, reflectance, point density, and the recorded intensity of each point, which is associated with the target's surface reflections (Fig. 3). Thus, it is used to classify and distinguish between objects. Multiple returns provide a large dataset on the objects within the laser footprint, representing different structural ground cover and elements (Camarretta et al., 2020).

ULS systems produce very dense point cloud data with hundreds of points per square meter (Fig. 4), which can be used to extract digital elevation models (DEM), digital surface models (DSM), and top canopy height models (TCH). LiDAR data can evaluate structural parameters, such as tree height, plant growth, volume, yield, foliage, fruit detection, and leaf area index (LAI) (Holmgren et al., 2022; Itakura and Hosoi, 2018; Rivera et al., 2023). It can also be used to evaluate soil erosion, land degradation, soil salinity, and moisture content with relatively high accuracy (Jiang et al., 2022; Neugirg et al., 2015). ULS-LiDAR technology detects variations and irregularities in plants or soil at a millimeter scale, making it possible to identify surface degradation and plant

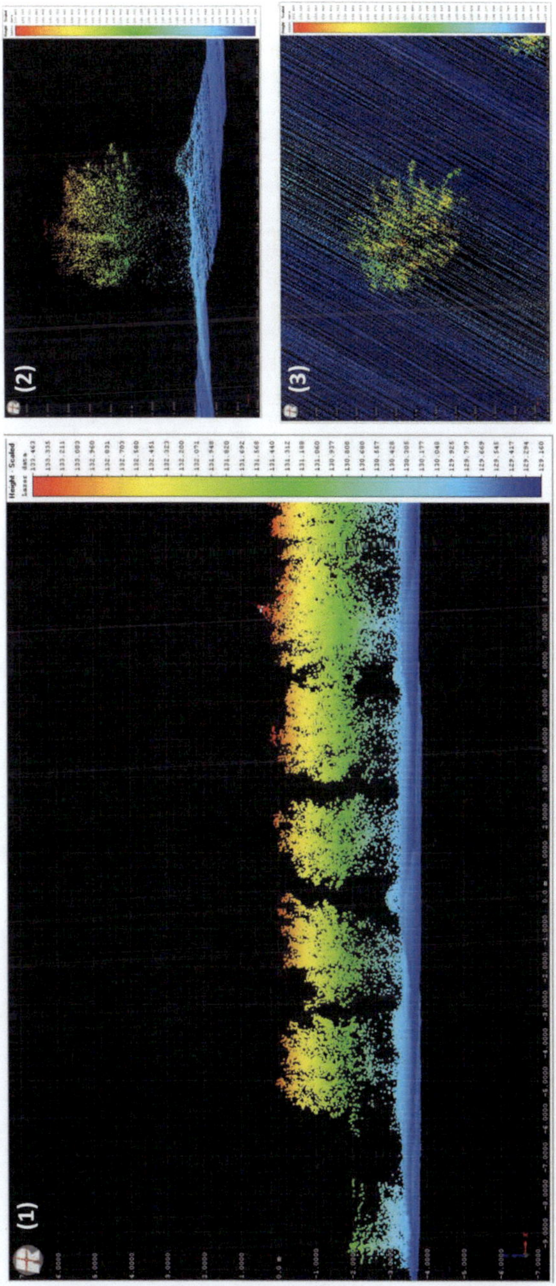

Figure 4 An example of data derived from Rigel-Mini UAV LiDAR in a young almond orchard in Israel. The color gradient represents the height of the object in the three-dimensional image. (3) Side view of the orchard, (2) zoom side view of a specific tree within the orchard, and (1) top view of a tree within the orchard.

Table 1 The different sensors and their common applications in UAV-precision agriculture (PA)

Sensor/camera	Spectral bands	Spectral range	Type of sensors	PA application
RGB visible spectrum	3	R, G, B	Optical	SfM-based 3D photogrammetry, change detection, and high-resolution imaging
Multispectral (broadband)	4–10	400–900 nm	Multiple wavelengths	Plant mapping based on vegetation indices, water quality, classification studies, and weed detection
Hyperspectral (narrow band)	10–more	400–1000 nm 1000–2500 nm	Analyzing the shape of the study surface spectrum	Plant physiology and phenology mapping, plant phenotyping, soil quality, mineral and nutritional mapping, pest and pathogen detection, plant stress, yield estimation, water states, and water quality
Thermal	1	7–14 mm	Surface temperature	Plant stress, soil moisture, object temperature, and animal detection
Fluorescence - SIF	10–more	0.4–0.8 nm	Plant photosynthetic activity	Photosynthetic activity and thermography
LiDAR		100–more points per m	Surface structure	3D digital terrain and surface mapping, canopy structure, yield assessment, growth models, plant structure, and erosion studies

structure (Rivera et al., 2023). Additionally, ULS-LiDAR is used for the precision automation of agricultural machinery for navigating farming systems (Jiang et al., 2022).

The sensors utilized for PA studies are chosen based on their suitability for the specific application, considering factors such as spectral or spatial requirements and cost (Tmušić et al., 2020). Four passive sensors cover most UAV-PA research: RGB (red, green, and blue), multispectral, thermal, and hyperspectral cameras (Table 1). In addition, there is an active three-dimensional (3D) LiDAR sensor. RGB cameras typically have a high spatial resolution with a small pixel size but a relatively low spectral resolution and are commonly used for orthophotos or for structure formation (SfM) photogrammetry to generate high-resolution point cloud data (Awais et al., 2021; Gómez-Candón et al., 2014). Multispectral sensors have a higher spectral resolution, with 4–10 bands, and are usually used to calculate soil, water, or vegetation spectral indices in the a spectral range of 0.4-1 μm. Hyperspectral cameras cover the full spectrum with hundreds of narrow spectral bands in the VIS-NIR region (0.4-1 μm). Line scanning sensors are the hyperspectral cameras most often used with UAVs, primarily for more chemical-based detection. Some hyperspectral UAV systems also integrate with the SWIR (1–2.5 μm) spectral region, although they remain costly and complex to handle (Adão et al., 2017). Furthermore, utilizing hyperspectral cameras on UAVs is still not straightforward and requires specialized image processing and pre- and post-processing (Maes and Steppe, 2018, 2019). UAV-SIF sensors capture a spectral range of 0.4-0.8 μm, providing high-resolution data, at the canopy level, of plant stress, with an accurate estimation of carbon assimilation rates and earlier detection of plant stress than other reflectance spectral regions (Bandopadhyay et al., 2020). Thermal cameras typically capture only one band measured in the long-wave infrared (8–14 μm) region, which is useful for extracting canopy and soil temperature (Fig. 5). Table 1 provides specific information on the common applications and methods used for each type of sensor.

The significant advantages of UAV applications include the increasing availability of lightweight sensors and the simultaneous integration of several sensors (Palubinskas and Reinartz, 2011). Thus, the combination of data obtained from two or more sensors could assist in monitoring different agricultural aspects (Fig. 6). Passive and active multi-sensor synergies have been identified for studying soil and plant status in PA and breeding projects (Gerhards et al., 2019; Herrmann et al., 2020; Yang et al., 2017). Integrating LiDAR data with optical sensors has huge advantages for PA, facilitating data collection on plant traits and structural aspects. In many research areas, thermal information has been integrated with multispectral or hyperspectral information or integrated with LiDAR. The use of information collected from these sensors in the decision-making process is promising but needs to be more developed

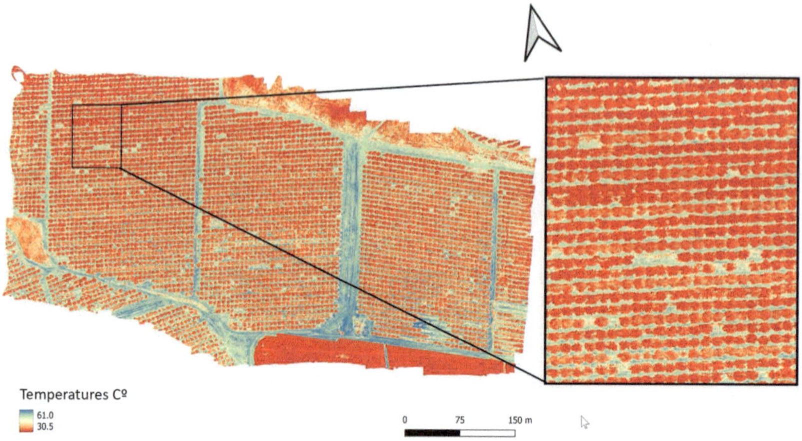

Figure 5 An example of data derived from thermal-UAV sensors in an almond orchard in Israel. The color gradient represents the plant and soil surface temperatures (green color - soil temperature and red color - tree temperature).

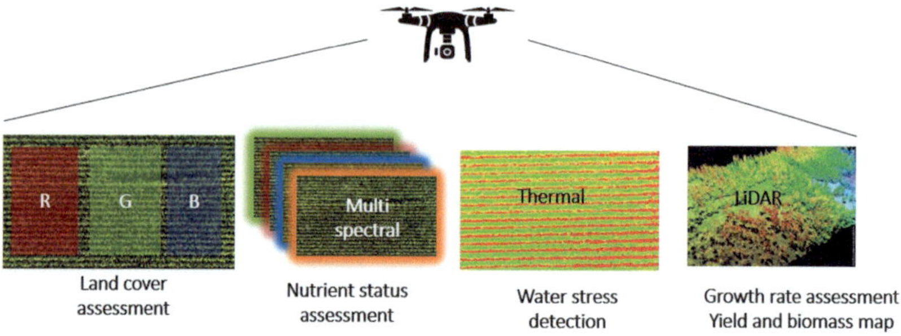

Figure 6 An overview of the uses of UAVs with sensor fusion for different agricultural applications. R-G-B: Red-Green-Blue.

and deserves further attention (Maes and Steppe, 2019). Furthermore, the main gap in data fusion is the development of fully automated systems to process all data, including flight preparation, data processing, and interpretation, for faster real-time management decisions (Maes and Steppe, 2019). Overcoming these challenges required a growing demand for high-quality and automated data, integration with new technologies and sensor fusion to improve safety, and enhanced data processing efficiency and reliability.

3 Platforms for precision agriculture

UAV classification has multiple criteria, including wing type, maximum payload (i.e. takeoff weight), operational systems, and autonomy levels (Korchenko and Illyash, 2013; Torun, 2000). UAVs have two main wing types,

Table 2 The different platforms and their advantages (marked +) and limitations (marked −), including flight time, area coverage, and payload

Platform	Type	Advantages and disadvantages	Flight time, payload, and coverage
Fixed wing	Similar to airplanes	Cover larger areas in a single flight (+), reduced flight time, and flight speed (+) Takeoff and landing require an experienced pilot (−) Regulations of flight altitude (−) Limitation in controlling flight speed (−) Problems in detecting small objects (−)	Long flight time (hours) Coverage of large areas depending on flight altitude Can travel greater distances with high payload (22–23 kg)
Multi-rotary wing	Quadcopter (four rotors), hexacopter (six rotors), octocopter (eight rotors)	Flexibility and ease of use (+) Stability of flight is high (+) Possibility for low flight altitude and low speed (+) Hovering flights (+), lower area coverage in a long time (−), wind affects the UAV stability (−), a trade-off between flight time and weight, and low battery lifetime (−)	Short flight time (20–40 min) Coverage of small areas depending on flight altitude Quadcopter has a smaller payload (1.25–2.6 kg)
Helicopter		Cover larger areas (+) reduced flight time of (+) higher flight speed (+) Takeoff and landing require an experienced pilot (−) Regulations of flight altitude (−) Problems in detecting small objects (−)	Long flight time (hours) Coverage of large areas depending on flight altitude Can travel greater distances with high payload (22–23 kg)

fixed and rotary, detailed in Table 2 and Fig. 7 (Li and Yang, 2012). Fixed-wing UAVs resemble airplanes with aerodynamic lifting force and high speed, are typically larger than rotary wings, and are mainly used for imaging and spraying herbicides and fertilizers over a large area. Unlike rotary-wing UAVs, they require a higher altitude to ensure safe flights due to their inability to reduce speed (Pounds et al., 2012), making them unsuitable for static flights (Pederi and Cheporniuk, 2015). Although fixed-wing UAVs usually have a higher payload capacity and longer range than rotary-wing UAVs, flying regulations for UAVs often limit their advantages (Kim et al., 2019). Rotary-wing UAVs, or drones, can be divided into two main types: helicopter and multi-rotor. The first rotary-wing type is similar to helicopters but has large propellers. Multi-rotor models are named based on the number of rotors they have, such as quadcopters (four rotors), hexacopters (six rotors), and octocopters (eight rotors) (Gómez-Candón et al., 2014; Kim et al., 2019).

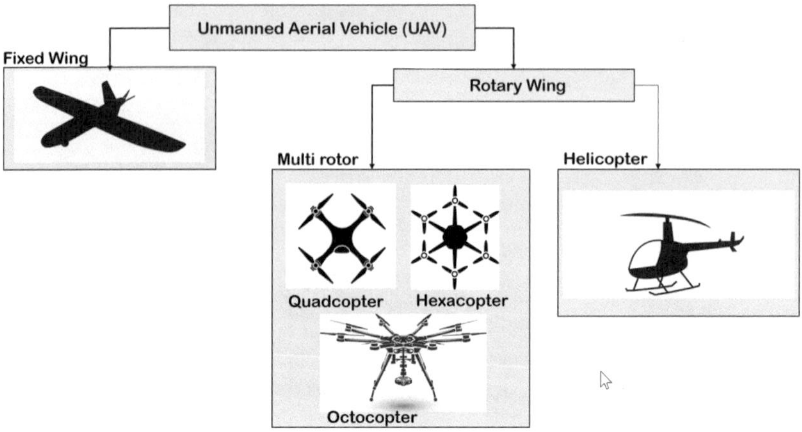

Figure 7 Schematic flowchart of the available type of UAV according to their wing type, with an example of the different UAV models.

Rotary-wing UAVs have high maneuverability, including rapid rotation performance with hovering flight, which is a significant advantage over fixed-wing UAVs (Muchiri and Kimathi, 2022). They also have the added benefits of high flight performance at low altitudes, relatively low-risk flights, vertical lift-off, full automation, and stability in various weather conditions. Multi-rotor manufacturers have increased the number of rotors from 4 to 8, allowing for higher overall payload capability (Pounds et al., 2012). Thus, when selecting a suitable UAV platform for agricultural applications, there is always a trade-off between operating parameters, including maximum flight time and max payload capacity for sensor carrying (del Cerro et al., 2021). Thus, increasing flight time through more powerful batteries or fuel tanks with a higher payload decreases the sensor payload capacity due to weight restrictions. Consequently, quadcopters and hexacopters, which are smaller, are mainly used for imaging, carrying a payload of 1.25–2.6 kg for precise applications. In comparison, fixed and rotary-wing UAVs and helicopters can carry larger payloads (22–23 kg) and are used for longer distances (Kim et al., 2019).

4 Flight planning and pre-processing

Multiple interacting factors influence the quality of UAV-based output images. Section 4 will provide an overview of the parameters that must be considered during flight preparation and the pre-processing aspect, including geometric rectification and radiometric calibration. Flight planning involves considerations such as image overlap, flight altitude, speed, flight timing and direction, solar elevation, radiometric quality, and sensor specifications, which collectively determine the spatial resolution and quality of the data acquisition (Barreto et al., 2019; Whitehead and Hugenholtz, 2014). For example, spatial resolution can be determined by the focal length and distance from the target (Herban et al., 2023); however, mountainous regions need to consider topographical aspects such as terrain and slope. Reducing flight height will affect the spatial resolution (i.e. pixel size), leading to smaller coverage areas and therefore requiring more flights to cover the studied area (Tmušić et al., 2020). The drawbacks of a prolonged mission time length include a possible increase in cloudiness and sun angle variability, affecting the radiometric correction and increasing the spectral variability. Moreover, the flight height affects the target size selection that must be detected in the image (Herban et al., 2023; Tmušić et al., 2020). Consequently, there is a need to consider the cross-interactions between study weather and conditions and flight operations that affect the data processing. Thus, flight planning can be an essential part of helping to improve data quality and image resolution (Tu et al., 2020).

During the flight, accurate measurements and several standardization targets are highly recommended. Proper flight planning and geo-rectifying the UAV position help to provide high-quality data to produce the most suitable imagery (Tu et al., 2020). For high radiometric accuracy, a global navigation satellite system (GNSS) is used as a receiver mounted on the UAV, which also includes the use of a ground base station or the service of operating reference stations providing correction data to the GNSS receiver (Stöcker et al., 2017). In addition, it is recommended to have ground control points (GCPs) with maximum visibility, which could include GCP targets using existing stationery or natural objects (Fig. 8). To generate georeferenced and spatial distributions (geospatial products), a minimum of three GCPs is recommended (Tmušić et al., 2020). The selection of the optimal GCPs number and its spatial distribution is an important issue in georeferencing processes and the accuracy of the RTK-GPS receiver (James et al., 2017).

Moreover, radiometric calibration targets are used during the flight operation to normalize sensor-generated spectral data. Several spectral targets are usually used, including reflectance targets, characteristic spectral targets, coded targets, etc. (Tmušić et al., 2020). The selection of spectral targets, their configuration, location, and number depend on the study area, the sensors used, and the study's purpose, similar to selecting GCP targets (Hakala et al., 2018). The most common, basic radiometric correction includes white and black panels or grayscale radiometric targets. For example, multiple light-to-dark grayscale targets are used to quantify the calibration curve for each pass to reduce changes in illumination during flight time (Tmušić et al., 2020). In addition to calibration targets, radiometric correction can be applied by combining irradiance measurements by upwelling sensors or spectrometers for object radiance measurements integrated with the UAV system during flight or on the ground (Herban et al., 2023). The calibration target or the upwelling sensors can be used separately or in combination (Tmušić et al., 2020). Alternatively, radiometric calibration targets with different brightness levels are used, and their reflectance values can be measured with a field spectrometer at different angles to evaluate the reflection levels (Johansen et al., 2018).

4.1 Image pre-processing

Pre-processing of UAV data usually includes two steps: geometric rectification and radiometric calibration. First, the data captured from the different sensors must go through geometric correction, a process called 'image stitching', which integrates the single images captured into one larger image, which is termed orthomosaic (Olson and Anderson, 2021). The images' alignments are done using the GCPs and the GNSS systems as 'key points' of interest (Fig. 8). As the images are stitched into an orthomosaic, they are aligned to their appropriate geographic location utilizing the image's accurate geographic metadata. The data products can include several outputs, depending on the selected sensors, such as color mosaics, thermal mosaics, SfM-photogrammetric models, and point clouds (Barreto et al., 2019). The captured images can be processed on-site or later with personal computers with the appropriate software (Kontogiannis and Ekaterinaris, 2013).

The second step involves radiometric correction, which accounts for the spectral signal of the captured data. The spectral data of an object is influenced by various factors, such as absorption and scattering caused by atmospheric conditions during flight and the sensing system itself, including

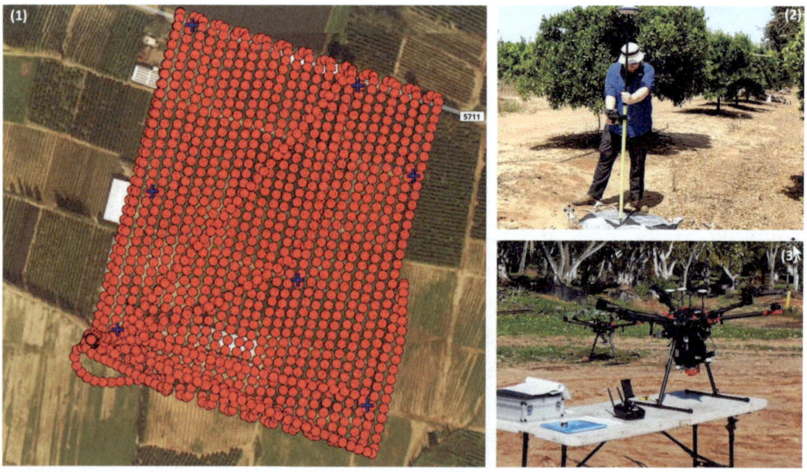

Figure 8 An example of a flight plan with the location of GCPs marked in blue (1). Marking GCPs with GPS-RTK for geo-rectifying the UAV position (2). Example of DJI-600 systems with GNSS receiver and upwelling sensors on top of the UAV (3).

its geometric measurement (Tu et al., 2018). To obtain reliable radiometric images, it is necessary to quantify and normalize these radiometric effects (Tmušić et al., 2020). Aasen et al. (2018) outlined several essential steps to achieve radiometrically and spectrally consistent data. These steps include (1) characterizing and calibrating the sensor periodically in the laboratory, both spectrally and radiometrically; (2) generating reflectance factors through radiometric normalization based on scene targets, achieved by using radiometric targets or irradiance measurements from a secondary sensor for validation, and (3) generating metadata for radiometrically corrected images (Tmušić et al., 2020).

In addition to the orthomosaic, to obtain a 3D digital model, SfM-photogrammetry has been developed to represent the geometry of objects in the image (Westoby et al., 2012). SfM-photogrammetry is used for creating 3D models from a sequence of overlapping 2D images. It has become a popular and low-cost application in recent years due to its ability to deal with disordered and heterogeneous images, requiring no prior knowledge (Schläpfer et al., 2015; Westoby et al., 2012). SfM algorithms are obtained with high point density for UAVs; it is usually possible to get a detailed point cloud model and extract the digital terrain, surface models (DTM and DSM), canopy height models, and orthoimages from it (Singh and Frazier, 2018). The SfM model's quality depends on the overlapping levels between the images, which should be planned before the flight. Several studies have presented comparisons between the results of UAV-SfM and LiDAR, indicating the pros and cons of the two techniques (Hemati et al., 2021; James et al., 2017; Li et al., 2020; Westoby et al., 2012). Thus, flight planning and pre-processing activities are important steps for spectral image quality assessment, resulting in a more reliable model for PA application.

5 Application in precision agriculture

5.1 Abiotic and biotic stress detection

Advances in optical and active RS systems have provided opportunities and tools to enhance our understanding of general plant stress detection and related physiological and phenological processes (Berger et al., 2022). Plant stress detection relies on analyzing the spectral signals emitted by plants to identify how crops respond to various stress, for example, the changes in spatial and temporal patterns of crop water availability. Abiotic stress factors typically include water stress, extreme temperatures, radiation, nutrient shortage, soil salinity, and soil pollution from heavy metals (Larcher, 2003). Biotic factors influence plant responses to other co-occurring organisms, such as weeds, insect pests, and pathogens. Biotic stress involves competition and diseases that affect plants' physiological and phenological responses (Berger et al., 2022; Madani et al., 2019). The interactions between biotic and abiotic factors can determine the quality and quantity of yield and can affect crop survival or success (Berger et al., 2022). However, the combination of stressors often leads to diverse physiological responses in plants, making it difficult to identify the causes of stress (Kranner et al., 2010). In the following subsections, we will discuss the capabilities of UAV-based RS technology for PA applications in estimating plant stress responses and present commonly used methodologies.

5.2 Water: precision irrigation

Water often limits crop production, and any water stress can negatively impact plant growth and development. Thus, to address this challenge, precision irrigation techniques have been developed to enhance water use efficiency and reduce water resource consumption in agriculture (Abioye et al., 2020). Precision irrigation addresses water scarcity and prevents soil salinity and nutrient losses. Canopy water content (CWC) is a commonly used metric to monitor canopy-level water conditions. It is calculated as the amount of water per unit area of the ground surface. CWC depends on the plant's water status, growth, and developmental stages. Various RS methods have been proposed to assess water use, including thermal, multi, and hyperspectral sensors.

Plant temperature as an indicator for stress detection has long been recognized as an indicator of plant water availability (Cohen et al., 2005; Gates, 1964). Transpiration cools the leaf by removing latent heat from the plant to the atmosphere through water vapor (Wasson et al., 2017). In turn, leaf transpiration is regulated by stomatal conductance (Jarvis and Mcnaughton, 1986). Plants tend to close their stomata in response to water stress, reducing stomatal conductance and preventing excessive water loss in their tissues (Hsiao, 1973). Even a small amount of transpiration can cause a significant drop in plant temperature, which, under high temperatures, can threaten the plant's survival due to water stress. Thermal infrared (TIR) remote sensing applications (LWIR; 8–14 μm) can potentially acquire spatial information on canopy temperature. A thermal camera based on UAVs was extensively studied, allowing a single operator to measure remotely sensed temperature over large areas (Berni et al., 2009). The implicit advantage of thermal imaging is its capacity to measure canopy temperature as an indicator of water conditions and status. The acquisition time of canopy temperature is greatly important due to the daily fluctuation of canopy temperature and the effect of direct exposure to sunlight. Thermal sensing is based on the relationship between transpiration, an energy-demanding process that linearly decreases the leaf surface temperature (Maes and Steppe, 2019). Multiple studies have demonstrated the association between stomatal conductance and surface temperature data (Adeyemi et al.,

2018; Berni et al., 2009; Sepaskhah and Kashefipour, 1994). Plant response to environmental conditions varies among plant species by stomatal regulation; these must also be considered when plant temperature is suggested as a proxy for plant water status (Ballester et al., 2013). The surface temperature is usually normalized by calculating the crop water stress index (CWSI), a well-known thermal index (Jones, 2013).

In addition to thermal sensing, multi and hyperspectral sensing techniques have been employed to develop water vegetation indices (VIs). These VIs are based on the concept and combination of surface reflectance values at two or more wavelengths, designed to highlight specific spectral water properties in the plant. Typically, water stress indices are measured in the shortwave infrared (SWIR) reflectance data, which contain specific absorption features related to plant water content at wavelengths such as 970, 1180, 1245, 1760, 1950, 2050, 2145, and 2173 nm (Fensholt et al., 2010). Another important spectral region is the red edge region (680–760 nm), which shows an increase in response to low leaf water content (Carter, 1991, 1993; Liu et al., 2004). This increase is often referred to as a 'blue shift', which occurs when the reflectance is plotted against the wavelength, and there is a shift toward the shorter wavelengths of the red-infrared transition curve (also known as the red-edge inflection point or REIP) in stressed plants (Curran et al., 1989; Rock et al., 1988).

Multiple studies have demonstrated the ability of RS applications to develop precision irrigation using UAVs (Bwambale et al., 2022; Nhamo et al., 2020). These studies include using UAVs to detect water stress in orchards and crop fields (Abioye et al., 2020). For instance, regulated deficit irrigation (RDI) was applied in an Orri mandarin crop, which was assessed for water status using UAVs equipped with thermal infrared (TIR) and multispectral cameras to reduce the incidence of fruit splitting (Rozzen, 2021). Canopy water content (CWC) was predicted based on four UAV data acquisition surveys conducted on different dates throughout the year and was compared with field water potential measurements. Figure 9 displays the experimental results of the integrated random forest (RF) model for CWC prediction, which includes the CWSI and other water VIs. The RF model results showed a correlation of R^2 = 0.9 and an root mean square error (RMSE) of 0.67. The OSAVI index contributed the most to the model performance, whereas the CWSI displayed the lowest response to the RDI application on the four sampling dates. This may be related to the isohydric behavior of citrus trees, but further investigation is needed. The researchers concluded that the temporal variability and large sample size from the predicted CWC values enabled the identification of differences in tree water status between irrigation treatments. Several studies have recommended integrating multiple sensor data throughout the growing season to capture the temporal variability representing dynamic crop water demand fluctuations, thus improving precision irrigation. Future research could involve calculating

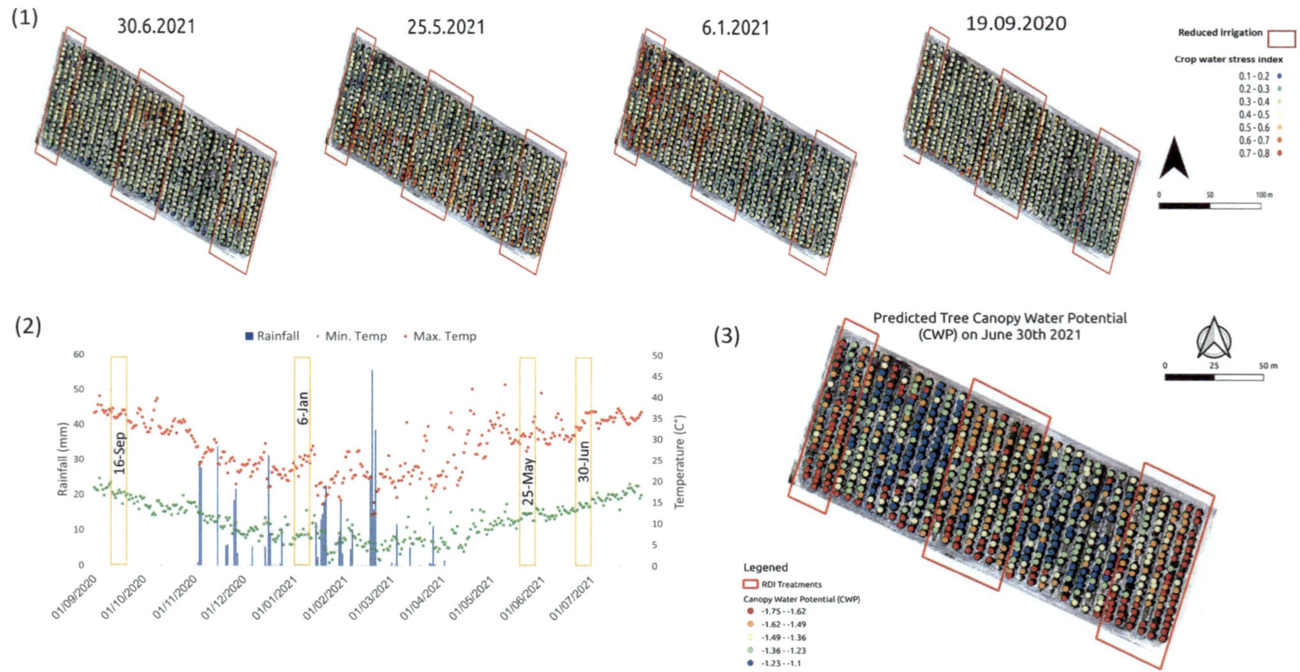

Figure 9 (1) CWSI model of the trees in the Orri mandarin orchard for the four sampling dates, (2) rainfall (mm), minimal and maximal temperatures (C°) throughout the data acquisition surveys and subsequent leaf water potential and the remote sensing measurements, (3) prediction map of average canopy water potential (CWP) values across all trees in the orchard for 30 June 2021 (Rozzen, 2021).

thermal and spectral related to water stress and assigning economic limitations for different crops and their growth stages to optimize irrigation timings (Olson and Anderson, 2021). Finally, hyperspectral imaging in the SWIR spectral region could be extensively used for determining plant water conditions and may be an important aspect of ongoing research due to the high spectral wavelength range it captures (Olson and Anderson, 2021).

5.3 Nutrients: precision fertilization

Crops require essential nutrients, including the three main nutrients of nitrogen (N), phosphorous (P), and potassium (K) to be added to the soil in order to maintain their commercial viability. Therefore, future farming practices must rely on the precise application of sustainable resources to optimize productivity while preventing environmental pollution. The global efficiency of fertilizer use is relatively low, with N at 47% (Lassaletta et al., 2014), P at 20% (Syers et al., 2008), and K at 19% (Dhillon et al., 2019), resulting in high loss to the surrounding environment (Olson and Anderson, 2021). N is often considered the most growth-limiting factor and critical nutrient for plant growth in soil (Freidenreich et al., 2019) and supplemented by N-fertilization to minimize yield production and financial loss (Christensen and Peacock, 2000; Sete et al., 2019). However, excessive N fertilizer application may cause nutrient imbalances and contribute to groundwater contamination by nitrate (NO_3^-) leaching and N_2O and CH_4 emissions into the atmosphere (Cameira and Mota, 2017; Colaço and Molin, 2017; Cui et al., 2020; Termin et al., 2023; Wang et al., 2013). Overuse of fertilizer application can stimulate plant growth, decrease yield quantity and quality, and increase production costs (Zulfiqar et al., 2019). As a result, growers are encouraged to implement various cultivation strategies to reduce N leaching into the environment.

UAV nutrient stress detection studies have mostly focused on N content through RS techniques (Liu et al., 2018), with only a few addressing P and K (Severtson et al., 2016). Several wavelengths related to N content, such as 850 nm and 1510 nm, are linked to plant N–H bonds, which are associated with the amount of N present in plant protein in leaves (Herrmann et al., 2010). However, the standard method for monitoring N content is developed based on spectral nitrogen indices (NIs) rather than using these single wavelengths related to N content. The NIs are based on enhancing the N signal and enabling monitoring of N content's spatial and temporal estimation (Thilakarathna and Raizada, 2018). A common way to develop NIs is by differentiating the reflectance values of two spectral bands related to a phenomenon (Herrmann et al., 2010). The first approach directly links plant reflectance to the N status, usually through a specific spectral band. The second approach relates surface reflectance to leaf chlorophyll content and only indirectly to N content.

For example, experiments examined the effect of different fertilization levels on the physiological activity of almond trees in the Gilat Research Center at the Agricultural Research Organization (ARO) in Israel. The research conducted in 2020 took place in tanks (lysimeters) and followed the development of 21 almond trees with different growth parameters (Sperling et al., 2019; Paz-Kagan et al., 2020). The experiment included various nitrate fertilization levels (ranging from 0 to 150 mg/l), isolating plant responses to N over or under application. Responses were based on variables measured including mineral uptake, non-structural carbohydrates, photosynthesis, growth rate based on LAI, and yield (Omondi et al., 2019; Paz-Kagan et al., 2020). The study examined the relationship between the spectral average of canopy N and chlorophyll content and a laboratory analysis based on the partial least squares regression (PLS-R) model. Four monthly UAV flights monitored tree development, carrying multispectral sensors with five spectral bands ranging from 0.4-0.9 μm wavelengths. The nitrogen measurements revealed that high N levels in irrigation water or excessive seasonal harm the development and fertility of almond trees, even though this was not reflected in the nitrogen measurements in the leaves. As the growing season progressed, it became apparent that trees with excess nitrogen experienced energy deficiencies and reduced growth (Fig. 10). The PLS-R model predicted the chlorophyll content with an accuracy of $R^2 = 0.81$ and nitrogen with an $R^2 = 0.8$ for the four image dates (Fig. 11). The nutritional deficit or overapplication could be detected with ML models based on spectral indices with relatively good accuracy, using only five spectral bands (RGB, NIR, and red edge) and NIs. Most SSNM studies use empirical regression models to relate the RS-observed signal to chlorophyll or N content. However, these could be improved by integrating the phenological and structural aspects of crop development. Future work should study the relationship between yield and canopy nutritional content, which could help in guiding fertilization applications and reducing environmental pollution (Maes and Steppe, 2019). Furthermore, future research could assess the compositions of other nutrients in plants and soil using hyperspectral UAV-RS technology.

5.4 Weeds: precision weed management

Biotic factors, including weeds, pests, and diseases, greatly reduce crop yield, with an estimated global production loss of about 40% annually (Oerke, 2006). The standard approach for managing weeds is using highly effective chemical herbicides, due to their low human labor and time requirements, thus ensuring optimal yield (Moretti et al., 2021). However, the negative health repercussions of herbicides have become common knowledge, leading to demands for strict regulation and enforcement of laws relevant to their use (Westwood et al., 2018). Some herbicides have already been subject to stringent regulatory

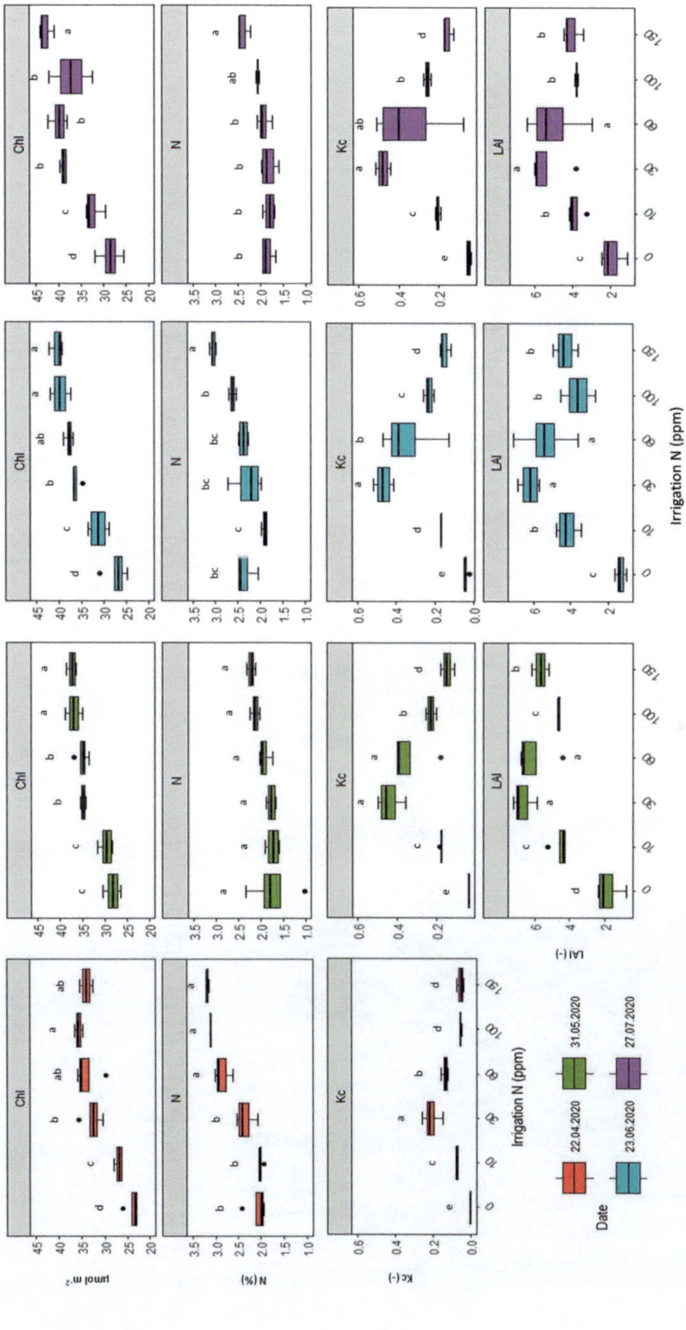

Figure 10 The physiological measurements on four sampling dates in a fertilization experiment in almonds in the Gilat Research Center (April–July 2020, represented in different colors). General measurement identification of tree responses to N application levels based on chlorophyll values, leaf nitrogen concentrations (N% dry wight), crop condition (Kc), and LAI values. The lysimeter experiment illustrates the changes in temporal dynamics of eco-physiological parameters at different fertilization levels.

Figure 11 (1) An example of the EVI2 vegetation index taken on 31 May 2020 to evaluate the spectral variation in the canopy of the experimental trees. PLS-R prediction models for chlorophyll (2) and nitrogen content in the canopy (3). The model was based on physiological laboratory measurements and the average sample tree canopies' spectral signatures at the different nitrogen fertilization levels. PLS-DA model for separating the spectral signatures between the four sampling dates. The control group without N application (0 ppm) is well distinguished from the other fertilization levels.

measures, while others will be in the future, and such restrictions will have local and immediate knock-on effects on yield. Therefore, the economic profitability of crops is expected to affect food security in the long term (Jabran and Chauhan, 2018). As a result, farmers are increasingly motivated to introduce non-chemical weed control tools and to develop integrated weed management (IWM) approaches that can help maintain crop health and productivity. These approaches aim to restore the competitiveness between the crop and infesting weeds, which depends on weed density, emerging times, and the competition duration (Rask and Kristoffersen, 2007). Exploiting these alternative weed control methods can help mitigate the over-reliance on herbicides (Lati et al., 2021). Precision weed management (PWM) aims to detect and monitor the within-field variability of weed infestation to develop site-specific weed management applications (Bajwa et al., 2015).

UAVs are an ideal platform for weed detection tasks. They can accurately map weed infestations within fields, reduce the need for ground-based weed detection, and enable targeted herbicide spraying instead of field-scale applications (Huang et al., 2018). Integrating UAV technology into PWM can greatly reduce herbicide applications and improve efficiency while minimizing negative effects on the environment, beneficial insects, and wildlife (Hunter III et al., 2020). PWM can target early-season weed control, a critical time for weed management, as crops have not yet matured (Singh et al., 2020). Weed detection with UAVs has been successfully applied in maize, sugarcane, beans, and other crops. Most studies investigating PWM have used optical sensors with multi and hyperspectral sensing (Olson and Anderson, 2021). However, a few studies have used high-resolution RGB cameras for weed detection (Bah et al., 2018). Recent work has combined RGB images with SfM data to detect crop height and classify weeds accurately, highlighting the advantages of multi-sensors of integrative several-parameter approaches to achieve high-accuracy classifications (Olson and Anderson, 2021). Machine learning (ML) and AI have also played an important role in weed detection, with different techniques of pixel and object-based classifications. The timing of weed detection is crucial and directly affects the accuracy of image analysis, which must be tested for various weeds (Olson and Anderson, 2021). Another important aspect of weed detection is the spatial resolution of the image. High-resolution images captured at low altitudes provide higher detection accuracy. However, this requires longer flight times to cover large areas as well as higher image processing computational requirements. Future research in this field could explore the use of spectral, thermal, and textural parameters for weed detection, higher sensor resolutions, and the performance of ML models.

5.5 Insects and pathogens

Insects are the second-largest cause of crop yield reduction among the biotic stresses, with potential losses of up to 18%, while pathogens are the third most significant cause of yield loss, up to 16% worldwide (Olson and Anderson, 2021). Insects harm crops through both direct feeding and indirect damage. This can result in reduced LAI, pigment discoloration, lower photosynthetic rates, and other negative effects. Precision control applications for insects and pathogens could be implemented by monitoring and detecting their damage (Olson and Anderson, 2021). This approach can reduce the need for large-scale pesticide applications and lessen environmental pollution. Insects and pathogens can damage crops at different developmental stages and may even cause sudden crop death (Jain et al., 2019). Therefore, maps of insect and pathogen infestations can provide valuable information for estimating pesticide requirements and costs and informing decisions about crop rotations and replanting.

Most UAV applications that focus on detecting insect stress use optical sensors, such as color or multispectral sensors (Marston et al., 2020; Stanton et al., 2017), with relatively little use of thermal cameras. In contrast to insect detection, pathogen detection primarily utilizes thermal sensors due to their ability to detect abnormalities in plant water transport systems (Zarco-Tejada et al., 2012). Recent insect detection studies have started using direct detection methods conducted by low-altitude flights with super high spatial resolution, enabling the detection of insects or the preliminary damage caused by them (Bouroubi et al., 2018; Zhang et al., 2018). Hyperspectral sensors can detect insect and pathogen stress better than multispectral cameras (Ning Zhang et al., 2018). This technology enables the identification of critical pre-outbreak pathogen phases, even when symptoms may not be reliably detectable by color or multispectral cameras. Thus, early detection of disease lifecycles may be more feasible using hyperspectral imaging than lower spectral resolution images (Olson and Anderson, 2021). However, the higher cost of this technology may be less practical for agricultural applications at present.

5.6 Phenotyping

Agricultural systems must adapt to the impacts of climate change and address resistance in pest and disease control to enhance agricultural productivity. One possible solution to meet the increasing demand for food production is to improve crop genetics through plant breeding, which can enhance yield, resistance, and biotic and abiotic stress resilience. Precision technologies such as UAV-RS can rapidly gather the phenotypic information needed to optimize agricultural production and make management and breeding decisions. While the development of crop gene sequencing has advanced quickly in the last two decades, there is still a lack of knowledge regarding translating associated phenotypes into desirable traits at the field scale (Furbank and Tester, 2011). Thus, it is crucial to understand better the connection between plant observable characteristics (phenotype) and genetics (genotype) (Chawade et al., 2019; Yang et al., 2017). This will enable the development of improved-yield cultivars with higher agricultural production that are stress-tolerant plants (Yang et al., 2017). High-throughput phenotyping platforms (HTPPs) offer reliable and automatic multifunctional platforms that enhance insight into genomic information (Li et al., 2014). While HTPPs have made significant progress, mostly in fully automated greenhouses or growth chambers with tight environmental controls (Yang et al., 2017), the results do not represent field conditions, making it difficult to interpret the knowledge into field scale. Field-based phenotyping (FBP) focuses on practical approaches to precise crop cultivation at the field scale, providing a more realistic approach to PA (Yang et al., 2017). The future detection of pests, weeds, and pathogen resistance is promising for revolutionizing PA-related phenotyping. The phenotyping method may vary based on the crop, trait, phenological growth stages, and available resources. FBP techniques can enable the screening of more samples with higher accuracy, which is expected to reduce costs and improve genomic selection intensity and precision. Identifying crop traits (i.e. biomass, yield potential, and tolerance to different abiotic or biotic stresses) requires considering the interaction between genetic and environmental factors (Yang et al., 2017).

UAVs are important and affordable FBP since they enable low-altitude and flexible flight capturing high-resolution images. They provide a low-cost approach to meeting the critical requirements of high spatial resolutions for FBP (Deng et al., 2018). In a study conducted by Condorelli et al. (2018), it was shown that UAV spectral indices were used to associate with drought-adaptive traits in durum wheat (*Triticum turgidum L.*) by identifying quantitative trait loci (QTL). This approach accounted for 89.6% of the phenotypic variance in a wheat panel (Olson and Anderson, 2021). FBP was also tested for detecting heat stress in different grapevine variants based on thermal sensing. Heat events can extend the growing season, leading to an increased risk of yield loss (sunburn, shriveling, etc.), and may, lowers the grape quality (e.g. fruit sweetness), by altering the aroma and sugar accumulation in the fruits and the ethanol level in the produced wine. Heat-induced stress can require increased inputs such as irrigation and pesticide application, leading to negative economic and environmental consequences. The research aimed to develop tools for monitoring and predicting crop heat-induced stress responses of six grapevine cultivars. The study found that all varieties had increased transpiration at noon, which decreased in the late afternoon. The stomatal conductance and photosynthetic assimilation rate decreased from morning to late afternoon for all varieties except Sangiovese (Gashu et al., 2022). The researchers detected highly significant differences in evapotranspiration between cultivars using thermal imagery (Reta et al., 2023; Reta et al., under revision). The study showed that cultivars such as Tempranillo and Pinot Noir could not buildup Brix to meet the harvesting requirements and demonstrated between the two experimental vineyards, characterized by an average temperature difference of 2°C, which was accurately detected by UAV thermal imagery (Fig. 12). Therefore, thermal imagery can be a useful monitoring method for viticulturists. UAVs are increasingly important for crop phenotyping in large plots and field trials, enabling near real-time monitoring. Sensor integration within the UAV-based fusion system can assist in monitoring different characteristics in parallel. Therefore, developing a UAV-based fusion platform could contribute to FBP, but it requires the development of appropriate platforms and protocols for assessing FBP.

5.7 Growth vigor and yield assessment

Accurate and early yield prediction is crucial for the agricultural sector and vital to crop production and health. Traditionally, yield monitoring has relied on previous yield history and seasonal conditions to guide management decisions for the next season (Olson and Anderson, 2021). However, crop yield is highly heterogeneous and influenced by multiple factors, such as environmental, management, and genetic aspects. To address this, crop yield assessment has been extensively studied using UAV optical remote sensing (Maes and Steppe, 2019). The combination of UAV-based imagery with ML offers the possibility of accurate estimates and can reduce or eliminate the need for a ground-based survey (Maimaitijiang et al., 2020). As the crop develops through its lifecycle, the specific phenological timing of yield assessment becomes increasingly important (Xiang Zhou et al., 2017). Therefore, selecting appropriate parameters for predicting yields, such as particular VIs, phenological stage, and structural parameters, is essential. Using UAVs for yield prediction has shown different levels of accuracy by applying RGB-SfM-derived data that capture top canopy height and canopy cover, VIs, or multispectral imagery (Maimaitijiang et al., 2020). As seen with satellite data, multiple images taken throughout the growing season perform better than a single measurement. Most yield prediction studies with UAVs have mainly focused on developing empirical regression models for testing the relationship between yield harvest and spectral or structural data (Gong et al., 2018). However, several alternative models have been suggested for estimating yield based on crop growth models and structural characteristics (Maes and Steppe, 2019). The crop's growth stage can be a good indicator of plant yield, development, and biomass. 3D-LiDAR data, or SfM, characterizes crop height, volume, and light intercept to improve yield

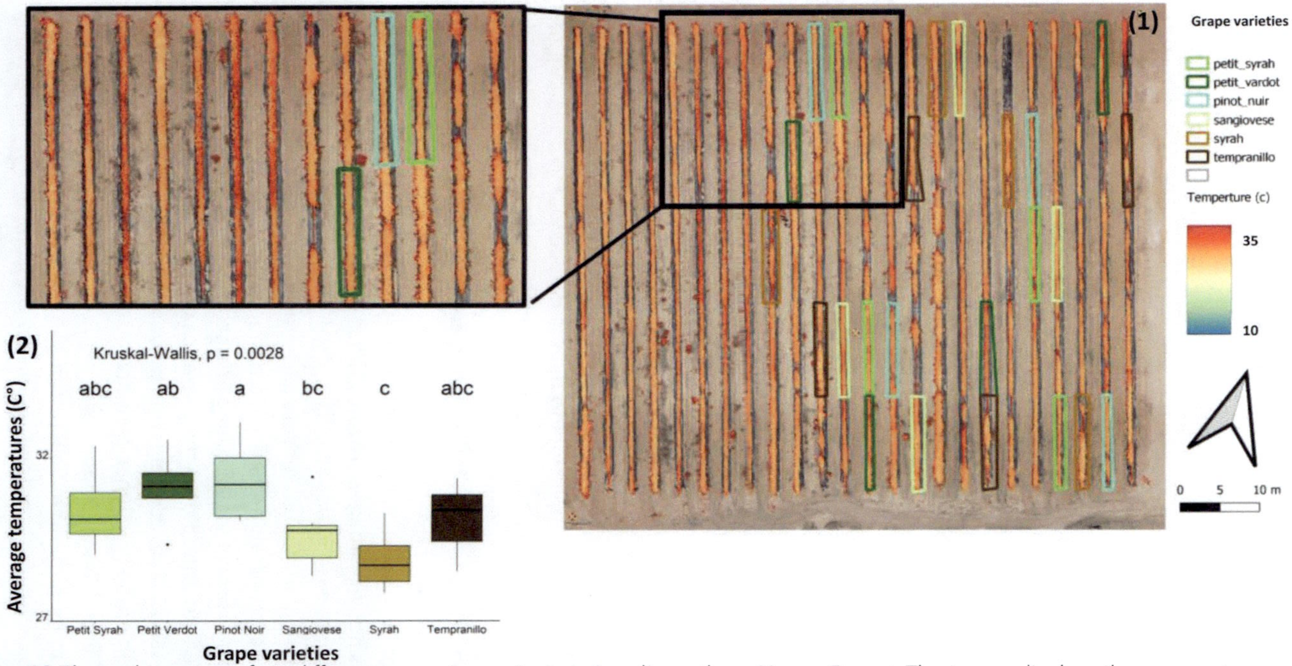

Figure 12 Thermal imaging of six different grapevine variants in Israel's northern Negev Desert. The image displays the canopy temperature response of the variants one day after experiencing heat stress (1). The boxplot illust-ates the differences in average canopy temperatures among the six grape variants (2).

prediction (Apolo-Apolo et al., 2020). Monitoring plant growth based on structural aspect as height, cover, and LAI derived from SfM photogrammetric imagery provides a good estimate of plant development in croplands and orchards. Several studies have compared the advantages and limitations of SfM and LiDAR data regarding growth and yield assessment, indicating that SfM is a good alternative to the more expensive UAV-based LiDAR (Murray et al., 2020). However, additional research is needed to define the optimal sensor configurations, the timing of flights, and the timing of the phenological stage and to adjust crop growth models for optimal use of UAV information to predict yields accurately (Maes and Steppe, 2019).

5.8 Precision livestock farming

Precision livestock farming (PLF) refers to the continuous real-time monitoring of grazing animals using sensor systems and computer algorithms to detect early signs of problems and develop management strategies for rangeland management (Tzanidakis et al., 2023). These technologies include automated weighing systems, body temperature monitoring, GPS tracking of the animal's location, and RS and geographic information system (GIS) applications for evaluating and optimizing pasture. One of the most promising research aspects in PLF is using UAVs, which are utilized for monitoring, detecting, and tracking grazing animals. UAVs can alert farmers about abnormal activity or potential threats to the animals. Several tasks performed by UAVs significantly contribute to livestock management, including animal detection and counting, real-time tracking of grazing animals, welfare risk management, and monitoring the health and behavior of animals in rangelands. Another aspect is monitoring rangeland plants' condition by assessing pastures' quality and quantity using multi and hyperspectral sensors (Alanezi et al., 2022; Adar et al., 2022, 2023).

Numerous studies have been conducted on animal detection and counting using UAV images. These studies initially involved targeted manual analysis of animal detection (Goolsby et al., 2016), which was later improved by incorporating additional sensors and processing approaches. Furthermore, image segmentation algorithms have dramatically enhanced animal detection and counting processes. Exciting research has focused on monitoring behavioral changes in cattle in the presence of UAVs. For example, Abdulai et al. (2021) assessed livestock's heart and movement rates under varying time and flight conditions. Research has recently included animal welfare management when operating UAVs over rangeland fields (Schroeder et al., 2020). The focus has shifted to monitoring animal health and welfare, often involving collecting information on animal temperature, blood pressure, and other physiological factors to detect animal stress (Alanezi et al., 2022). UAVs integrated with other automation technologies and livestock monitoring sensors, particularly in large, dispersed pastures, could have a significant advantage in monitoring animal welfare. This automated monitoring system continuously captures environmental factors, animal welfare, and physiology to assist in rangeland management. UAVs enable efficient tracking, increased production efficiency, reduced environmental impact, and improved animal welfare (Lovarelli et al., 2020). However, concerns have been raised regarding drones' potential to cause abnormal livestock behavior. PLF using UAVs faces several challenges, including technical challenges related to data analysis and interpolation, sensor selection, and large-scale image processing. These challenges require appropriate tools and strategies for research within a reasonable timeframe. Additionally, there are challenges concerning government policies, economic constraints, and environmental issues in pasture areas (Alanezi et al., 2022).

6 Conclusion and future trends

This book chapter provides a comprehensive overview of the successful implementation of UAV-RS technology in various PA applications. These applications encompass precision fertilization, irrigation management, weed, pest, and pathogen control, as well as the development of crop growth, health, and damage detection maps. The substantial interest and existing literature on UAV-RS technology in PA demonstrate its rapidly expanding importance. With reduced costs, improved technology, usability, and ML capabilities, the utilization of UAVs in the agricultural industry will continue to grow, accompanied by enhanced capabilities and increased implementation. Historically, the primary limitation of UAV applications in PA was data collection, which acted as the main bottleneck. However, this challenge has shifted to pre- and post-processing and analysis of large volumes of data, including multi-sensor images obtained from UAV platforms (Olson and Anderson, 2021). The time required between data collection and processing has become a significant factor. While ML technologies have begun to address this limitation, they also possess inherent constraints. They can sometimes function as black boxes during processing, analysis, and real-time data quality assessment (Olson and Anderson, 2021). Future technological advances will facilitate daily data collection, including flight operations, enabling independent real-time data analysis. This advancement will allow the development of fast and accurate SDSS and enable tasks, such as spraying or fertilizing, to be carried out by UAVs without human intervention. Furthermore, smaller robots and machines that can easily navigate through crops can be directed by UAVs to collect information. This would relieve the user from investing time in these tasks, allowing them to access the data effortlessly. However, several challenges must be addressed for UAVs to become fully established in PA practices.

In terms of sensor data, the fusion of passive and active sensors could be further developed to enhance reliability and improve the accuracy of data collection. Integrating structural crop parameters, such as crop volume and height, LAI, with plant spectral traits could benefit PA applications (Olson and Anderson, 2021). To enhance accuracy, these parameters can be combined with higher-performing spectral bands or indices during specific stages of the crop's growth cycle or by considering its phenology. Plant phenological stages are crucial for generating comprehensive and robust data products, enabling multi-temporal and yearly monitoring within modeling algorithms. These models can then establish economic thresholds for assessing agricultural performance. While hyperspectral cameras have the potential to capture higher spectral resolution compared to multispectral cameras, their current high cost restricts their widespread use. Therefore, reducing the cost of hyperspectral cameras would facilitate broader adoption and yield benefits in terms of crop trait assessment.

Several factors require consideration when addressing the challenges of implementing operational UAVs in agriculture. These include operational issues and economic factors, such as the cost of large data storage, software design, and the qualifications of personnel, needed to operate and process the substantial amount of collected data. Furthermore, manual analysis of such big data may be impractical, necessitating developing high-performance real-time data analysis algorithms. Another crucial aspect is the imposition of government regulations driven by security concerns by law enforcement agencies. These regulations pose significant operational barriers and problems for many stakeholders. For example, current laws heavily restrict UAV operations in controlled and uncontrolled airspace. The primary challenge lies in finding an optimal balance between the demands of various actors (Stöcker et al., 2017). On the other hand, end users have their specific needs and market interests based on their priorities. In conclusion, a substantial amount of literature written in recent years indicates that UAVs will likely continue to play a critical role in PA by providing valuable field data. As UAV technologies continue to advance and barriers related to sensor development, cost, computing power, algorithms, and software diminish, we can expect a more widespread adoption as researchers and producers push the boundaries of PA.

7 Where to look for further information

Recent years have seen a significant increase in publications, including review articles, studies in the field of PA with UAVs, and books, all focused on the use of UAVs for agriculture or environmental monitoring. We will suggest several recently published books on using UAVs for agricultural and environmental monitoring. These books cover various aspects of UAV technology, applications, and data analysis in agriculture and environmental sciences. For further reading on the topics related:

1 *Unmanned Aerial Systems for Monitoring Soil, Vegetation, and Riverine Environments* (Manfreda and Eyal, 2023). This book is the outcome of 5 years of dedicated work conducted as part of the COST Action HARMONIOUS. This initiative brought together over 200 researchers and technicians from 36 countries within Europe and beyond. The primary goal was to promote advanced monitoring strategies, establish standardized practices, and disseminate the latest UAS (Unmanned Aerial Systems) methodologies across a global network. Within the book, readers will find valuable guidelines, technical insights, and practical experiences to assist researchers, educators, and practitioners in enhancing monitoring efficiency through UAS.
2 *Unmanned Aerial Vehicle Applications In Agriculture* (Avtar and Watanabe, 2020). The book delves into both the conceptual and operational aspects of UAVs and their significant roles in agriculture and the environment. This book is not intended for engineers seeking to design UAVs for specific applications. Instead, its primary focus lies in exploring the practical use of UAVs across various applications, including PA, forest management, crop monitoring, fertilization management, plant health assessment, land management, road patrolling, 3D modeling, topographic surveys, and more.
3 *Precision Agriculture Technology for Crop Farming* (Zhang, 2016). This book provides an overview of PA technologies, including the use of UAVs, in crop farming. It covers data collection, analysis, and decision-making processes.
4 *UAV sensors for environmental monitoring* (Toro and Tsourdos, 2018). This book covers a range of topics related to UAV remote sensing, including sensors, data acquisition, and analysis methods for environmental monitoring.

8 References

Aasen, H., Honkavaara, E., Lucieer, A. and Zarco-Tejada, P. J. (2018). Quantitative remote sensing at ultra-high resolution with UAV spectroscopy: a review of sensor technology, measurement procedures, and data correction workflows. *Remote Sensing* 10(7), 1091.

Abdulai, G., Sama, M. and Jackson, J. (2021). A preliminary study of the physiological and behavioral response of beef cattle to unmanned aerial vehicles (UAVs). *Applied Animal Behaviour Science* 241, 105355.

Abioye, E. A., Abidin, M. S. Z., Mahmud, M. S. A., Buyamin, S., Ishak, M. H. I., Abd Rahman, M. K. I. A., Otuoze, A. O., Onotu, P. and Ramli, M. S. A. (2020). A review on monitoring and advanced control strategies for precision irrigation. *Computers and Electronics in Agriculture* 173, 105441.

Adão, T., Hruška, J., Pádua, L., Bessa, J., Peres, E., Morais, R. and Sousa, J. (2017). Hyperspectral imaging: a review on UAV-based sensors, data processing and applications for agriculture and forestry. *Remote Sensing* 9(11), 1110.

Adar, S., Sternberg, M., Argaman, E., Henkin, Z., Dovrat, G., Zaady, E. and Paz-Kagan, T. (2023). Testing a novel pasture quality index using remote sensing tools in semiarid and Mediterranean grasslands. *Agriculture, Ecosystems and Environment* 357, 108674.

Adar, S., Sternberg, M., Paz-Kagan, T., Henkin, Z., Dovrat, G., Zaady, E. and Argaman, E. (2022). Estimation of aboveground biomass production using an unmanned aerial vehicle (UAV) and VENμS satellite imagery in Mediterranean and semiarid rangelands. *Remote Sensing Applications: Society and Environment* 26, 100753.

Adeyemi, O., Grove, I., Peets, S., Domun, Y. and Norton, T. (2018). Dynamic modelling of the baseline temperatures for computation of the crop water stress index (CWSI) of a greenhouse cultivated lettuce crop. *Computers and Electronics in Agriculture* 153, 102–114.

Alanezi, M. A., Shahriar, M. S., Hasan, M. B., Ahmed, S., Sha'aban, Y. A. and Bouchekara, H. R. E. H. (2022). Livestock management with unmanned aerial vehicles: a review. *IEEE Access* 10, 45001–45028.

Apolo-Apolo, O. E., Martínez-Guanter, J., Egea, G., Raja, P. and Pérez-Ruiz, M. (2020). Deep learning techniques for estimation of the yield and size of citrus fruits using a UAV. *European Journal of Agronomy* 115, 126030.

Avtar, R. and Watanabe, T. (2020). *Unmanned Aerial Vehicle: Applications in Agriculture and Environment*. Switzerland: Springer.

Awais, M., Li, W., Cheema, M. J. M., Hussain, S., Shaheen, A., Aslam, B., Liu, C. and Ali, A. (2021). Assessment of optimal flying height and timing using high-resolution unmanned aerial vehicle images in precision agriculture. *International Journal of Environmental Science and Technology* 1–18.

Bah, M. D., Hafiane, A. and Canals, R. (2018). Deep learning with unsupervised data labeling for weed detection in line crops in UAV images. *Remote Sensing* 10(11), 1690.

Bajwa, A. A., Mahajan, G. and Chauhan, B. S. (2015). Nonconventional weed management strategies for modern agriculture. *Weed Science* 63(4), 723–747. https://doi.org/10.1614/WS-D-15-00064.1.

Ballester, C., Jiménez-Bello, M. A., Castel, J. R. and Intrigliolo, D. S. (2013). Usefulness of thermography for plant water stress detection in citrus and persimmon trees. *Agricultural and Forest Meteorology* 168(1), 120–129. https://doi.org/10.1016/j.agrformet.2012.08.005.

Bandopadhyay, S., Rastogi, A. and Juszczak, R. (2020). Review of top-of-canopy sun-induced fluorescence (SIF) studies from ground, UAV, airborne to spaceborne observations. *Sensors* 20(4), 1144.

Barreto, M. A. P., Johansen, K., Angel, Y. and McCabe, M. F. (2019). Radiometric assessment of a UAV-based push-broom hyperspectral camera. *Sensors* 19(21), 4699.

Bendig, J., Chang, C. Y., Wang, N., Atherton, J., Malenovský, Z. and Rascher, U. (2021). Measuring solar-induced fluorescence from unmanned aircraft systems for operational use in plant phenotyping and precision farming. In: *IEEE International Geoscience and Remote Sensing Symposium IGARSS* (vol. 2021, pp. 1921–1924). IEEE Publications.

Berger, K., Verrelst, J., Féret, J. B., Wang, Z., Wocher, M., Strathmann, M., Danner, M., Mauser, W. and Hank, T. (2020). Crop nitrogen monitoring: recent progress and principal developments in the context of imaging spectroscopy missions. *Remote Sensing of Environment* 242, 111758.

Berger, K., Machwitz, M., Kycko, M., Kefauver, S. C., Van Wittenberghe, S., Gerhards, M., Verrelst, J., Atzberger, C., van der Tol, C., Damm, A., Rascher, U., Herrmann, I., Paz, V. S., Fahrner, S., Pieruschka, R., Prikaziuk, E., Buchaillot, M. L., Halabuk, A., Celesti, M., Koren, G., Gormus, E. T., Rossini, M., Foerster, M., Siegmann, B., Abdelbaki, A., Tagliabue, G., Hank, T., Darvishzadeh, R., Aasen, H., Garcia, M., Pôças, I., Bandopadhyay, S., Sulis, M., Tomelleri, E., Rozenstein, O., Filchev, L., Stancile, G. and Schlerf, M. (2022). Multi-sensor spectral synergies for crop stress detection and monitoring in the optical domain: a review. *Remote Sensing of Environment* 280, 113198.

Berni, J. A. J., Zarco-Tejada, P. J., Sepulcre-Cantó, G., Fereres, E. and Villalobos, F. (2009). Mapping canopy conductance and CWSI in olive orchards using high resolution thermal remote sensing imagery. *Remote Sensing of Environment* 113(11), 2380–2388.

Bogrekci, I. and Lee, W. S. (2005). Spectral phosphorus mapping using diffuse reflectance of soils and grass. *Biosystems Engineering* 91(3), 305–312.

Bouroubi, Y., Bugnet, P., Nguyen-Xuan, T., Gosselin, C., Bélec, C., Longchamps, L. and Vigneault, P. (2018). Pest detection on UAV imagery using a deep convolutional neural network. In: Proceedings of the International Society of Precision Agriculture.

Bwambale, E., Naangmenyele, Z., Iradukunda, P., Agboka, K. M., Houessou-Dossou, E. A. Y., Akansake, D. A., Bisa, M. E., Hamadou, A. H., Hakizayezu, J., Onofua, O. E. and Chikabvumbwa, S. R. (2022). Towards precision irrigation management: a review of GIS, remote sensing and emerging technologies. *Cogent Engineering* 9(1), 2100573.

Camarretta, N., Harrison, P. A., Bailey, T., Potts, B., Lucieer, A., Davidson, N. and Hunt, M. (2020). Monitoring forest structure to guide adaptive management of forest restoration: a review of remote sensing approaches. *New Forests* 51(4), 573–596.

Cameira, M. D. R. and Mota, M. (2017). Nitrogen related diffuse pollution from horticulture production–mitigation practices and assessment strategies. *Horticulturae* 3(1), 25.

Carter, G. A. (1991). Primary and secondary effects of water content on the spectral reflectance of leaves. *American Journal of Botany* 78(7), 916–924.

Carter, G. A. (1993). Responses of leaf spectral reflectance to plant stress. *American Journal of Botany* 80(3), 239–243.

Curran, P. J. (1989). Remote sensing of foliar chemistry. *Remote Sensing of Environment* 30(3), 271–278.

Chabrillat, S., Gholizadeh, A., Neumann, C., Berger, D., Milewski, R., Ogen, Y. and Ben-Dor, E. (2019). Preparing a soil spectral library using the Internal Soil Standard (ISS) method: influence of extreme different humidity laboratory conditions. *Geoderma* 355, 113855. https://doi.org/10.1016/j.geoderma.2019.07.013.

Chawade, A., van Ham, J., Blomquist, H., Bagge, O., Alexandersson, E. and Ortiz, R. (2019). High-throughput field-phenotyping tools for plant breeding and precision agriculture. *Agronomy* 9(5), 258.

Christensen, L. P. and Peacock, W. L. (2000). Mineral nutrition and fertilization. *Raisin Production Manual*, 102–114.

Cohen, Y., Alchanatis, V., Meron, M., Saranga, Y. and Tsipris, J. (2005). Estimation of leaf water potential by thermal imagery and spatial analysis. *Journal of Experimental Botany* 56(417), 1843–1852.

Colaço, A. F. and Molin, J. P. (2017). Variable rate fertilization in citrus: A long term study. *Precision Agriculture* 18(2), 169–191.

Condorelli, G. E., Maccaferri, M., Newcomb, M., Andrade-Sanchez, P., White, J. W., French, A. N., Sciara, G., Ward, R. and Tuberosa, R. (2018). Comparative aerial and ground based high throughput phenotyping for the genetic dissection of NDVI as a proxy for drought adaptive traits in durum wheat. *Frontiers in Plant Science* 9, 893.

Cui, M., Zeng, L., Qin, W. and Feng, J. (2020). Measures for reducing nitrate leaching in orchards: a review. *Environmental Pollution* 263(B), 114553.

Davies, G. E. and Calvin, W. M. (2017). Mapping acidic mine waste with seasonal airborne hyperspectral imagery at varying spatial scales. *Environmental Earth Sciences* 76(12), 1–14. https://doi.org/10.1007/s12665-017-6763-x.

del Cerro, J., Cruz Ulloa, C., Barrientos, A. and de León Rivas, J. (2021). Unmanned aerial vehicles in agriculture: A survey. *Agronomy* 11(2), 203.

Delavarpour, N., Koparan, C., Nowatzki, J., Bajwa, S. and Sun, X. (2021). A technical study on UAV characteristics for precision agriculture applications and associated practical challenges. *Remote Sensing* 13(6), 1204.

Deng, L., Mao, Z., Li, X., Hu, Z., Duan, F. and Yan, Y. (2018). UAV-based multispectral remote sensing for precision agriculture: A comparison between different cameras. *ISPRS Journal of Photogrammetry and Remote Sensing* 146, 124–136.

Dhillon, J. S., Eickhoff, E. M., Mullen, R. W. and Raun, W. R. (2019). World potassium use efficiency in cereal crops. *Agronomy Journal* 111(2), 889–896.

Donager, J. J., Sankey, T. T., Sankey, J. B., Sanchez Meador, A. J., Springer, A. E. and Bailey, J. D. (2018). Examining forest structure with terrestrial lidar: suggestions and novel techniques based on comparisons between scanners and forest treatments. *Earth and Space Science* 5(11), 753–776.

Fensholt, R., Huber, S., Proud, S. R. and Mbow, C. (2010). Detecting canopy water status using shortwave infrared reflectance data from polar orbiting and geostationary platforms. *IEEE Journal of Selected Topics in Applied Earth Observations and Remote Sensing* 3(3), 271–285.

Freidenreich, A., Barraza, G., Jayachandran, K. and Khoddamzadeh, A. A. (2019). Precision Agriculture Application for Sustainable Nitrogen Management of Justicia brandegeana Using Optical Sensor Technology. *Agriculture* 9(5), 98.

Freschet, G. T., Barthès, B. G., Brunet, D., Hien, E. and Masse, D. (2011). Use of near infrared reflectance spectroscopy (NIRS) for predicting soil fertility and historical management. *Communications in Soil Science and Plant Analysis* 42(14), 1692–1705. https://doi.org/10.1080/00103624.2011.584597.

Furbank, R. T. and Tester, M. (2011). Phenomics–technologies to relieve the phenotyping bottleneck. *Trends in Plant Science* 16(12), 635–644.

Gashu, K., Song, C., Dubey, A. K., Acuña, T., Sagi, M., Agam, N., Bustan, A. and Fait, A. (2022). The effect of topo-climate variation on the secondary metabolism of berries in white grapevine varieties (Vitis vinifera). *Frontiers in Plant Science* 13, 847268.

Gates, D. M. (1964). Leaf temperature and transpiration 1. *Agronomy Journal* 56(3), 273–277. https://doi.org/10.2134/agronj1964.00021962005600030007x.

Gerhards, M., Schlerf, M., Mallick, K. and Udelhoven, T. (2019). Challenges and future perspectives of multi-/Hyperspectral thermal infrared remote sensing for crop water-stress detection: a review. *Remote Sensing* 11(10), 1240.

Gholizadeh, A., Borůvka, L., Saberioon, M. and Vašát, R. (2013) Visible, near-infrared, and mid-infrared spectroscopy applications for soil assessment with emphasis on soil organic matter content and quality: state-of-the-art and key issues. *Applied Spectroscopy* 67(12), 1349–1362. https://doi.org/10.1366/13-07288.

Gholizadeh, A., Saberioon, M., Ben-Dor, E. and Borůvka, L. (2018). Monitoring of selected soil contaminants using proximal and remote sensing techniques: background, state-of-the-art and future perspectives. *Critical Reviews in Environmental Science and Technology* 48(3), 243–278.

Gholizadeh, A., Saberioon, M., Rossel, R. A. V., Boruvka, L. and Klement, A. (2020). Spectroscopic measurements and imaging of soil colour for field scale estimation of soil organic carbon. *Geoderma* 357, 113972.

Gómez-Candón, D., De Castro, A. I. and López-Granados, F. (2014). Assessing the accuracy of mosaics from unmanned aerial vehicle (UAV) imagery for precision agriculture purposes in wheat. *Precision Agriculture* 15(1), 44–56.

Gong, Y., Duan, B., Fang, S., Zhu, R., Wu, X., Ma, Y. and Peng, Y. (2018). Remote estimation of rapeseed yield with unmanned aerial vehicle (UAV) imaging and spectral mixture analysis. *Plant Methods* 14(1), 70.

Goolsby, J., Jung, J., Landivar, J., McCutchoon, W., Lacewell, R., Duhaime, R. and Schwartz, A. (2016). Evaluation of unmanned aerial vehicles (UAVs) for detection of cattle in the cattle fever tick permanent quarantine zone. *Subtropical Agriculture and Environments* 67, 24–27.

Gu, H., Lin, Z., Guo, W. and Deb, S. (2021). Retrieving surface soil water content using a soil texture adjusted vegetation index and unmanned aerial system images. *Remote Sensing* 13(1), 145.

Hakala, T., Markelin, L., Honkavaara, E., Scott, B., Theocharous, T., Nevalainen, O., Näsi, R., Suomalainen, J., Viljanen, N., Greenwell, C. and Fox, N. (2018). Direct reflectance measurements from drones: sensor absolute radiometric calibration and system tests for forest reflectance characterization. *Sensors* 18(5), 1417.

Hemati, M., Hasanlou, M., Mahdianpari, M. and Mohammadimanesh, F. (2021). A systematic review of landsat data for change detection applications: 50 years of monitoring the earth. *Remote Sensing* 13(15), 2869. https://doi.org/10.3390/rs13152869.

Herban, S., Manfreda, S., Tmušić, G., Maltese, A. and Brook, A. (2023). Protocols for UAS-based observation. In: *Unmanned Aerial Systems for Monitoring Soil, Vegetation, and Riverine Environments* (pp. 37–69). Elsevier.

Herrmann, I., Karnieli, A., Bonfil, D. J., Cohen, Y. and Alchanatis, V. (2010). SWIR-based spectral indices for assessing nitrogen content in potato fields. *International Journal of Remote Sensing* 31(19), 5127–5143.

Herrmann, I., Bdolach, E., Montekyo, Y., Rachmilevitch, S., Townsend, P. A. and Karnieli, A. (2020). Assessment of maize yield and phenology by drone-mounted superspectral camera. *Precision Agriculture* 21(1), 51–76.

Holmgren, J., Lindberg, E., Olofsson, K. and Persson, H. J. (2022). Tree crown segmentation in three dimensions using density models derived from airborne laser scanning. *International Journal of Remote Sensing* 43(1), 299–329.

Hsiao, T. C. (1973). Plant responses to water stress. *Annual Review of Plant Physiology* 24(1), 519–570.

Hu, J., Peng, J., Zhou, Y., Xu, D., Zhao, R., Jiang, Q., Fu, T., Wang, F. and Shi, Z. (2019). Quantitative estimation of soil salinity using UAV-borne hyperspectral and satellite multispectral images. *Remote Sensing* 11(7), 736.

Huang, Y., Reddy, K. N., Fletcher, R. S. and Pennington, D. (2018). UAV low-altitude remote sensing for precision weed management. *Weed Technology* 32(1), 2–6.

Hunter, J. E., Gannon, T. W., Richardson, R. J., Yelverton, F. H. and Leon, R. G. (2020). Integration of remote-weed mapping and an autonomous spraying unmanned aerial vehicle for site-specific weed management. *Pest Management Science* 76(4), 1386–1392.

Hunter, M. C., Smith, R. G., Schipanski, M. E., Atwood, L. W. and Mortensen, D. A. (2017). Agriculture in 2050: recalibrating targets for sustainable intensification. *BioScience* 67(4), 386–391.

Itakura, K. and Hosoi, F. (2018). Automatic individual tree detection and canopy segmentation from three-dimensional point cloud images obtained from ground-based lidar. *Journal of Agricultural Meteorology* 74(3), 109–113.

Jabran, K. and Chauhan, B. S. (2018). Overview and significance of non-chemical weed control. In: Jabran, K., B. S. B. T.-N.-C. and Chauhan, W. C. (Eds). *Non-chemical Weed Control* (pp. 1–8). Academic Press. https://doi.org/10.1016/B978-0-12-809881-3.00001-2.

Jain, A., Sarsaiya, S., Wu, Q., Lu, Y. and Shi, J. (2019). A review of plant leaf fungal diseases and its environment speciation. *Bioengineered* 10(1), 409–424.

James, M. R., Robson, S., d'Oleire-Oltmanns, S. and Niethammer, U. (2017). Optimising UAV topographic surveys processed with structure-from-motion: ground control quality, quantity and bundle adjustment. *Geomorphology* 280, 51–66.

Jansen, M., Gilmer, F., Biskup, B., Nagel, K. A., Rascher, U., Fischbach, A., Briem, S., Dreissen, G., Tittmann, S., Braun, S., De Jaeger, I., Metzlaff, M., Schurr, U., Scharr, H. and Walter, A. (2009). Simultaneous phenotyping of leaf growth and chlorophyll fluorescence via GROWSCREEN fluoro allows detection of stress tolerance in Arabidopsis thaliana and other rosette plants. *Functional Plant Biology* 36(11), 902–914.

Jarvis, P. G. and Mcnaughton, K. G. (1986). Stomatal control of transpiration: scaling up from leaf to region. *Advances in Ecological Research* 15(3), 1–49. https://doi.org/10.1016/S0065-2504(08)60119-1.

Jiang, S., Wang, S., Yi, Z., Zhang, M. and Lv, X. (2022). Autonomous navigation system of greenhouse mobile robot based on 3D lidar and 2D lidar SLAM. *Frontiers in Plant Science* 13, 815218.

Johansen, K., Raharjo, T. and McCabe, M. F. (2018). Using multi-spectral UAV imagery to extract tree crop structural properties and assess pruning effects. *Remote Sensing* 10(6), 854.

Jones, H. G. (2013). *Plants and Microclimate: A Quantitative Approach to Environmental Plant Physiology* (3rd edn.). Cambridge University Press. https://doi.org/10.1017/CBO9780511845727.

Kim, J., Kim, S., Ju, C. and Son, H. I. II (2019). Unmanned aerial vehicles in agriculture: a review of perspective of platform, control, and applications. *IEEE Access* 7, 105100–105115.

Kontogiannis, S. G. and Ekaterinaris, J. A. (2013). Design, performance evaluation and optimization of a UAV. *Aerospace Science and Technology* 29(1), 339–350.

Korchenko, A. G. and Illyash, O. S. (2013). The generalized classification of unmanned air vehicles. In: 2013 IEEE 2nd International Conference Actual Problems of Unmanned Air Vehicles Developments Proceedings (APUAVD) (vol. 2013, pp. 28–34). IEEE.

Kranner, I., Minibayeva, F. V., Beckett, R. P. and Seal, C. E. (2010). What is stress? Concepts, definitions and applications in seed science. *New Phytologist* 188(3), 655–673.

Larcher, W. (2003). *Physiological Plant Ecology: Ecophysiology and Stress Physiology of Functional Groups*. Springer Science & Business Media.

Lassaletta, L., Billen, G., Grizzetti, B., Anglade, J. and Garnier, J. (2014). 50 year trends in nitrogen use efficiency of world cropping systems: the relationship between yield and nitrogen input to cropland. *Environmental Research Letters* 9(10), 105011.

Lati, R. N., Rosenfeld, L., David, I. B. Ben and Bechar, A. (2021). Power on! Low-energy electrophysical treatment is an effective new weed control approach. *Pest Management Science* 77(9), 4138–4147. https://doi.org/10.1002/ps.6451.

Levi, N., Hillel, N., Zaady, E., Rotem, G., Ziv, Y., Karnieli, A. and Paz-Kagan, T. (2021). Soil quality index for assessing phosphate mining restoration in a hyper-arid environment. *Ecological Indicators* 125, 107571. https://doi.org/10.1016/j.ecolind.2021.107571.

Levi, N., Karnieli, A. and Paz-Kagan, T. (2022). Airborne imaging spectroscopy for assessing land-use effect on soil quality in drylands. *ISPRS Journal of Photogrammetry and Remote Sensing* 186, 34–54.

Li, T., Zhang, B., Xiao, W., Cheng, X., Li, Z. and Zhao, J. (2020). UAV-based photogrammetry and LiDAR for the characterization of ice morphology evolution. *IEEE Journal of Selected Topics in Applied Earth Observations and Remote Sensing* 13, 4188–4199.

Li, X. and Yang, L. (2012). Design and implementation of UAV intelligent aerial photography system. In: 4th International Conference on Intelligent Human-Machine Systems and Cybernetics (vol. 2). IEEE Publications.

Lin, Y. (2015). LiDAR: an important tool for next-generation phenotyping technology of high potential for plant phenomics? *Computers and Electronics in Agriculture* 119, 61–73.

Liu, S., Li, L., Gao, W., Zhang, Y., Liu, Y., Wang, S. and Lu, J. (2018). Diagnosis of nitrogen status in winter oilseed rape (Brassica napus L.) using in-situ hyperspectral data and unmanned aerial vehicle (UAV) multispectral images. *Computers and Electronics in Agriculture* 151, 185-195.

Lovarelli, D., Bacenetti, J. and Guarino, M. (2020). A review on dairy cattle farming: is precision livestock farming the compromise for an environmental, economic and social sustainable production? *Journal of Cleaner Production* 262, 121409.

Madani, B., Mirshekari, A. and Imahori, Y. (2019). Physiological responses to stress. In: *Postharvest Physiology and Biochemistry of Fruits and Vegetables* (pp. 405-423). Elsevier.

Madec, S., Baret, F., De Solan, B., Thomas, S., Dutartre, D., Jezequel, S., Hemmerlé, M., Colombeau, G. and Comar, A. (2017). High-throughput phenotyping of plant height: comparing unmanned aerial vehicles and ground LiDAR estimates. *Frontiers in Plant Science* 8, 2002.

Maes, W. H. and Steppe, K. (2019). Perspectives for remote sensing with unmanned aerial vehicles in precision agriculture. *Trends in Plant Science* 24(2), 152–164. https://doi.org/10.1016/J.TPLANTS.2018.11.007.

Maimaitijiang, M., Sagan, V., Sidike, P., Hartling, S., Esposito, F. and Fritschi, F. B. (2020). Soybean yield prediction from UAV using multimodal data fusion and deep learning. *Remote Sensing of Environment* 237, 111599.

Manfreda, S. and Eyal, B. D. (2023). Unmanned aerial systems for monitoring soil, vegetation, and riverine environments. Netherlands: ElsevierRadarweg.

Marston, Z. P. D., Cira, T. M., Hodgson, E. W., Knight, J. F., MacRae, I. V. and Koch, R. L. (2020). Detection of stress induced by soybean aphid (Hemiptera: Aphididae) using multispectral imagery from unmanned aerial vehicles. *Journal of Economic Entomology* 113(2), 779–786.

Mazur, M., Wisniewski, A. and McMillan, J. (2016). *Clarity from above: PwC Global Report on the Commercial Applications of Drone Technology*. Warsaw: Drone Powered Solutions, PriceWater house Coopers.

Milics, G. (2019). Application of UAVs in precision agriculture. In: Palocz-Andresen, M., Szalay, D., Gosztom, A., Sípos, L. and Taligás, T. (Eds). *International Climate Protection*. Cham: Springer. 93–97. https://doi.org/10.1007/978-3-030-03816-8_13.

Moretti, M. L., Bobadilla, L. K. and Hanson, B. D. (2021). Cross-resistance to diquat in glyphosate/paraquat-resistant hairy fleabane (Conyza bonariensis) and horseweed (Conyza canadensis) and confirmation of 2,4-D resistance in Conyza bonariensis. *Weed Technology* 35(4), 554–559. https://doi.org/10.1017/wet.2021.11.

Muchiri, G. N. and Kimathi, S. (2022). A review of applications and potential applications of UAV. In: Proceedings of the Sustainable Research and Innovation Conference (pp. 280–283).

Mulla, D. J. (2013). Twenty five years of remote sensing in precision agriculture: key advances and remaining knowledge gaps. *Biosystems Engineering* 114(4), 358–371.

Murchie, E. H. and Lawson, T. (2013). Chlorophyll fluorescence analysis: a guide to good practice and understanding some new applications. *Journal of Experimental Botany* 64(13), 3983–3998.

Murray, J., Fennell, J. T., Blackburn, G. A., Whyatt, J. D. and Li, B. (2020). The novel use of proximal photogrammetry and terrestrial LiDAR to quantify the structural complexity of orchard trees. *Precision Agriculture* 21(3), 473–483.

Neugirg, F., Kaiser, A., Schmidt, J., Becht, M. and Haas, F. (2015). Quantification, analysis and modelling of soil erosion on steep slopes using LiDAR and UAV photographs. *Proceedings of the International Association of Hydrological Sciences* 367, 51–58.

Nhamo, L., Magidi, J., Nyamugama, A., Clulow, A. D., Sibanda, M., Chimonyo, V. G. P. and Mabhaudhi, T. (2020). Prospects of improving agricultural and water productivity through unmanned aerial vehicles. *Agriculture* 10(7), 256.

Nocita, M., Stevens, A., van Wesemael, B., Aitkenhead, M., Bachmann, M., Barthès, B., Ben Dor, E., Brown, D. J., Clairotte, M., Csorba, A., Dardenne, P., Demattê, J. A. M., Genot, V., Guerrero, C., Knadel, M., Montanarella, L., Noon, C., Ramirez-Lopez, L., Robertson, J., Sakai, H., Soriano-Disla, J. M., Shepherd, K. D., Stenberg, B., Towett, E. K., Vargas, R. and Wetterlind, J. (2015). Chapter Four. Soil spectroscopy: an alternative to wet chemistry for soil monitoring. In: Donald, L. S., et al. (Ed.). *Advances in Agronomy* (Vol. 132, pp. 139–159). Academic Press. Available at: https://doi.org/10.1016/bs.agron.2015.02.002.

Ochoa, K. S. and Guo, Z. (2019). A framework for the management of agricultural resources with automated aerial imagery detection. *Computers and Electronics in Agriculture* 162, 53–69.

Oerke, E.-C. (2006). Crop losses to pests. *The Journal of Agricultural Science* 144(1), 31–43.

Olson, D. and Anderson, J. (2021). Review on unmanned aerial vehicles, remote sensors, imagery processing, and their applications in agriculture. *Agronomy Journal* 113(2), 971–992.

Omondi, J. O., Lazarovitch, N., Rachmilevitch, S., Yermiyahu, U. and Sperling, O. (2019). High nitrogen availability limits photosynthesis and compromises carbohydrate allocation to storage in roots of Manihot esculenta Crantz. *Frontiers in Plant Science* 10, 1041.

Palubinskas, G. and Reinartz, P. (2011). Multi-resolution, multi-sensor image fusion: general fusion framework. In: *Joint Urban Remote Sensing Event* (vol. 2011, pp. 313–316). IEEE Publications.

Paz-Kagan, T., Shachak, M., Zaady, E. and Karnieli, A. (2014). A spectral soil quality index (SSQI) for characterizing soil function in areas of changed land use. *Geoderma* 230–231, 171–184. https://doi.org/10.1016/j.geoderma.2014.04.003.

Paz-Kagan, T., Schmilovitch, Z. E., Yermiyahu, U., Rapaport, T. and Sperling, O. (2020). Assessing the nitrogen status of almond trees by visible-to-shortwave infrared reflectance spectroscopy of carbohydrates. *Computers and Electronics in Agriculture* 178, 105755.

Paz-Kagan, T., Zaady, E., Salbach, C., Schmidt, A., Lausch, A., Zacharias, S., Notesco, G., Ben-Dor, E. and Karnieli, A. (2015). Mapping the spectral soil quality index (SSQI) using airborne imaging spectroscopy. *Remote Sensing* 7(11), 15748–15781.

Pederi, Y. A. and Cheporniuk, H. S. (2015). Unmanned aerial vehicles and new technological methods of monitoring and crop protection in precision agriculture. In: IEEE International Conference Actual Problems of Unmanned Aerial Vehicles Developments (APUAVD) (vol. 2015, pp. 298–301). IEEE Publications.

Pounds, P. E. I., Bersak, D. R. and Dollar, A. M. (2012). Stability of small-scale UAV helicopters and quadrotors with added payload mass under PID control. *Autonomous Robots* 33(1–2), 129–142.

Prado Osco, L., Marques Ramos, A. P., Roberto Pereira, D., Akemi Saito Moriya, É., Nobuhiro Imai, N., Takashi Matsubara, E., Estrabis, N., de Souza, M., Marcato Junior, J., Gonçalves, W. N., Li, J., Liesenberg, V. and Eduardo Creste, J. (2019). Predicting canopy nitrogen content in citrus-trees using random forest algorithm associated to spectral vegetation indices from UAV-imagery. *Remote Sensing* 11(24), 2925.

Rask, A. M. and Kristoffersen, P. (2007). A review of non-chemical weed control on hard surfaces. *Weed Research* 47(5), 370–380. https://doi.org/10.1111/j.1365-3180.2007.00579.x.

Reta, K., Yishai, N., Naftali, L. and Aaron, F. (under review). Practices of canopy management in vineyards to improve grape yield and quality under climate change influence.

Reta, K., Caras, T., Acuña, T., Paz-Kagan, T., Lazarovitch, N. and Fait, A. (2023). Physiological responses of red grapevine varieties to heatwave: insights from monitoring physiological activities and drone-based thermal imagery analysis. 44th World Congress of Vine and Wine 2023 in Cadiz-Jerez, Spain.

Rivera, G., Porras, R., Florencia, R. and Sánchez-Solís, J. P. (2023). LiDAR applications in precision agriculture for cultivating crops: a review of recent advances. *Computers and Electronics in Agriculture* 207, 107737.

Rock, B. N., Hoshizaki, T. and Miller, J. R. (1988). Comparison of in situ and airborne spectral measurements of the blue shift associated with forest decline. *Remote Sensing of Environment* 24(1), 109–127.

Romsonthi, C., Tawornpruek, S. and Watana, S. (2018). In situ near-infrared spectroscopy for soil organic matter prediction in paddy soil, pasak watershed, Thailand. *Plant, Soil and Environment* 64(2), 70–75. https://doi.org/10.17221/716/2017-PSE.

Rossel, R. A. V., Behrens, T., Ben-Dor, E., Brown, D. J., Demattê, J. A. M., Shepherd, K. D., Shi, Z., Stenberg, B., Stevens, A., Adamchuk, V., Aïchi, H., Barthès, B. G., Bartholomeus, H. M., Bayer, A. D., Bernoux, M., Böttcher, K., Brodský, L., Du, C. W., Chappell, A., Fouad, Y., Genot, V., Gomez, C., Grunwald, S., Gubler, A., Guerrero, C., Hedley, C. B., Knadel, M., Morrás, H. J. M., Nocita, M., Ramirez-Lopez, L., Roudier, P., Campos, E. M. R., Sanborn, P., Sellitto, V. M., Sudduth, K. A., Rawlins, B. G., Walter, C., Winowiecki, L. A., Hong, S. Y. and Ji, W. (2016). A global spectral library to characterize the world's soil. *Earth-Science Reviews* 155, 198–230.

Rousseau, C., Belin, E., Bove, E., Rousseau, D., Fabre, F., Berruyer, R., Guillaumès, J., Manceau, C., Jacques, M. A. and Boureau, T. (2013). High throughput quantitative phenotyping of plant resistance using chlorophyll fluorescence image analysis. *Plant Methods* 9(1), 17.

Rozzen, S. (2021). Assessing the effects of different irrigation regimes on fruit splitting incidence and tree water status in mandarins using remote sensing. Master's thesis, Robert, H. Smith Faculty of Agriculture, Food & Environment, The Hebrew University of Jerusalem.

Schläpfer, D., Richter, R. and Feingersh, T. (2015). Operational BRDF effects correction for wide-field-of-view optical scanners (BREFCOR). *IEEE Transactions on Geoscience and Remote Sensing* 53(4), 1855–1864. https://doi.org/10.1109/TGRS.2014.2349946.

Schroeder, N. M., Panebianco, A., Gonzalez Musso, R. and Carmanchahi, P. (2020). An experimental approach to evaluate the potential of drones in terrestrial mammal research: a gregarious ungulate as a study model. *Royal Society Open Science* 7(1), 191482.

Sepaskhah, A. R. and Kashefipour, S. M. (1994). Relationships between leaf water potential, CWSI, yield and fruit quality of sweet lime under drip irrigation. *Agricultural Water Management* 25(1), 13–21.

Sete, P. B., Comin, J. J., Nara Ciotta, M., Almeida Salume, J., Thewes, F., Brackmann, A., Toselli, M., Nava, G., Rozane, D. E., Loss, A., Lourenzi, C. R., da Rosa Couto, R. and Brunetto, G. (2019). Nitrogen fertilization affects yield and fruit quality in pear. *Scientia Horticulturae* 258, 108782.

Severtson, D., Callow, N., Flower, K., Neuhaus, A., Olejnik, M. and Nansen, C. (2016). Unmanned aerial vehicle canopy reflectance data detects potassium deficiency and green peach aphid susceptibility in canola. *Precision Agriculture* 17(6), 659–677.

Singh, K. K. and Frazier, A. E. (2018). A meta-analysis and review of unmanned aircraft system (UAS) imagery for terrestrial applications. *International Journal of Remote Sensing* 39(15–16), 5078–5098.

Singh, V., Rana, A., Bishop, M., Filippi, A. M., Cope, D., Rajan, N. and Bagavathiannan, M. (2020). Unmanned aircraft systems for precision weed detection and management: prospects and challenges. *Advances in Agronomy* 159, 93–134.

Sperling, O., Karunakaran, R., Erel, R., Yasuor, H., Klipcan, L. and Yermiyahu, U. (2019). Excessive nitrogen impairs hydraulics, limits photosynthesis, and alters the metabolic composition of almond trees. *Plant Physiology and Biochemistry* 143, 265–274.

Stanton, C., Starek, M. J., Elliott, N., Brewer, M., Maeda, M. M. and Chu, T. (2017). Unmanned aircraft system-derived crop height and normalized difference vegetation index metrics for sorghum yield and aphid stress assessment. *Journal of Applied Remote Sensing* 11(2), 26035.

Stöcker, C., Nex, F., Koeva, M. and Gerke, M. (2017). Quality assessment of combined IMU/GNSS data for direct georeferencing in the context of UAV-based mapping. *The International Archives of the Photogrammetry, Remote Sensing and Spatial Information Sciences* XLII-2/W6, 355–361.

Syers, J. K., Johnston, A. E. and Curtin, D. (2008). Efficiency of soil and fertilizer phosphorus use. *FAO Fertilizer and Plant Nutrition Bulletin, 18* 108, 5–50.

Tang, Y., Dananjayan, S., Hou, C., Guo, Q., Luo, S. and He, Y. (2021). A survey on the 5G network and its impact on agriculture: challenges and opportunities. *Computers and Electronics in Agriculture* 180, 105895.

Termin, D., Linker, R., Baram, S., Raveh, E., Ohana-Levi, N. and Paz-Kagan, T. (2023). Dynamic delineation of management zones for site-specific nitrogen fertilization in a citrus orchard. *Precision Agriculture* 24(4), 1570–1592.

Thilakarathna, M. S. and Raizada, M. N. (2018). Challenges in using precision agriculture to optimize symbiotic nitrogen fixation in legumes: progress, limitations, and future improvements needed in diagnostic testing. *Agronomy* 8(5), 78.

Tmušić, G., Manfreda, S., Aasen, H., James, M. R., Gonçalves, G., Ben-Dor, E., Brook, A., Polinova, M., Arranz, J. J., Mészáros, J., Zhuang, R., Johansen, K., Malbeteau, Y., de Lima, I. P., Davids, C., Herban, S. and McCabe, M. F. (2020). Current practices in UAS-based environmental monitoring. *Remote Sensing* 12(6), 1001.

Toro, F. G. and Tsourdos, A. (2018). *UAV Sensors for Environmental Monitoring*. Switzerland: MDPI.

Torun, E. (2000). *UAV Requirements and Design Consideration*. Turkish Land Forces Command, Ankara (Turkey).

Tu, Y.-H., Phinn, S., Johansen, K. and Robson, A. (2018). Assessing radiometric correction approaches for multi-spectral UAS imagery for horticultural applications. *Remote Sensing* 10(11), 1684.

Tzanidakis, C., Tzamaloukas, O., Simitzis, P. and Panagakis, P. (2023). Precision livestock farming applications (PLF) for grazing animals. *Agriculture* 13(2), 288.

Vargas, J. Q., Bendig, J., Mac Arthur, A., Burkart, A., Julitta, T., Maseyk, K., Thomas, R., Siegmann, B., Rossini, M., Celesti, M., Schüttemeyer, D., Kraska, T., Muller, O. and Rascher, U. (2020). Unmanned aerial systems (UAS)-based methods for solar induced chlorophyll fluorescence (SIF) retrieval with non-imaging spectrometers: state of the art. *Remote Sensing* 12(10), 1624.

Viscarra Rossel, R. A., Behrens, T., Ben-Dor, E., Brown, D. J., Demattê, J. A. M., Shepherd, K. D., Shi, Z., Stenberg, B., Stevens, A., Adamchuk, V., Aïchi, H., Barthès, B. G., Bartholomeus, H. M., Bayer, A. D., Bernoux, M., Böttcher, K., Brodský, L., Du, C. W., Chappell, A., Fouad, Y., Genot, V., Gomez, C., Grunwald, S., Gubler, A., Guerrero, C., Hedley, C. B., Knadel, M., Morrás, H. J. M., Nocita, M., Ramirez-Lopez, L., Roudier, P., Campos, E. M. R., Sanborn, P., Sellitto, V. M., Sudduth, K. A., Rawlins, B. G., Walter, C., Winowiecki, L. A., Hong, S. Y. and Ji, W. (2016). A global spectral library to characterize the world's soil. Elsevier B.V. *Earth-Science Reviews* 155, 198–230. https://doi.org/10.1016/j.earscirev.2016.01.012.

Wang, J., Shen, C., Liu, N., Jin, X., Fan, X., Dong, C. and Xu, Y. (2017). Non-destructive evaluation of the leaf nitrogen concentration by in-field visible/near-infrared spectroscopy in pear orchards. *Sensors* 17(3), 538.

Wang, L., Butcher, A. S., Stuart, M. E., Gooddy, D. C. and Bloomfield, J. P. (2013). The nitrate time bomb: a numerical way to investigate nitrate storage and lag time in the unsaturated zone. *Environmental Geochemistry and Health* 35(5), 667–681.

Wasson, T., Choudhury, T., Sharma, S. and Kumar, P. (2017). Integration of RFID and sensor in agriculture using IOT. In: International Conference On Smart Technologies For Smart Nation (SmartTechCon) (vol. 2017, pp. 217–222). IEEE Publications.

Westoby, M. J., Brasington, J., Glasser, N. F., Hambrey, M. J. and Reynolds, J. M. (2012). 'Structure-from-Motion' photogrammetry: A low-cost, effective tool for geoscience applications. *Geomorphology* 179, 300–314. https://doi.org/10.1016/J.GEOMORPH.2012.08.021.

Westwood, J. H., Charudattan, R., Duke, S. O., Fennimore, S. A., Marrone, P., Slaughter, D. C., Swanton, C. and Zollinger, R. (2018). Weed management in 2050: perspectives on the future of weed science. *Weed Science* 66(3), 275–285. https://doi.org/10.1017/wsc.2017.78.

Whitehead, K. and Hugenholtz, C. H. (2014). Remote sensing of the environment with small unmanned aircraft systems (UASs), part 1: A review of progress and challenges. *Journal of Unmanned Vehicle Systems* 2(3), 69–85.

Xuefeng, L., Qiang, L., Shaolan, H., Shilai, Y., Deyu, H., Zhitao, W., Rangjin, X., Yongqiang, Z. and Lie, D. (2016). Estimation of carbon and nitrogen contents in citrus canopy by low-altitude remote sensing. *International Journal of Agricultural and Biological Engineering* 9(5), 149–157.

Yang, G., Liu, J., Zhao, C., Li, Z., Huang, Y., Yu, H., Xu, B., Yang, X., Zhu, D., Zhang, X., Zhang, R., Feng, H., Zhao, X., Li, Z., Li, H. and Yang, H. (2017). Unmanned aerial vehicle remote sensing for field-based crop phenotyping: current status and perspectives. *Frontiers in Plant Science* 8, 1111.

Ye, X., Abe, S. and Zhang, S. (2020). Estimation and mapping of nitrogen content in apple trees at leaf and canopy levels using hyperspectral imaging. *Precision Agriculture* 21(1), 198–225.

Zarco-Tejada, P. J., González-Dugo, V. and Berni, J. A. J. (2012). Fluorescence, temperature and narrow-band indices acquired from a UAV platform for water stress detection using a micro-hyperspectral imager and a thermal camera. *Remote Sensing of Environment* 117, 322–337.

Zhang, Q. (Ed.). (2016). *Precision Agriculture Technology for Crop Farming*. (1st ed.). Boca Raton, FL: CRC Press. https://doi.org/10.1201/b19336.

Zhang, C., Valente, J., Kooistra, L., Guo, L. and Wang, W. (2021). Orchard management with small unmanned aerial vehicles: a survey of sensing and analysis approaches. *Precision Agriculture* 22(6), 2007–2052.

Zhang, D., Zhou, X., Zhang, J., Lan, Y., Xu, C. and Liang, D. (2018). Detection of rice sheath blight using an unmanned aerial system with high-resolution color and multispectral imaging. *PLoS ONE* 13(5), e0187470.

Zhang, N., Wang, M. and Wang, N. (2002). Precision agriculture–a worldwide overview. *Computers and Electronics in Agriculture* 36(2–3), 113–132.

Zhang, N., Zhang, X., Yang, G., Zhu, C., Huo, L. and Feng, H. (2018). Assessment of defoliation during the Dendrolimus tabulaeformis Tsai et Liu disaster outbreak using UAV-based hyperspectral images. *Remote Sensing of Environment* 217, 323–339.

Zhou, X., Zheng, H. B., Xu, X. Q., He, J. Y., Ge, X. K., Yao, X., Cheng, T., Zhu, Y., Cao, W. X. and Tian, Y. C. (2017). Predicting grain yield in rice using multi-temporal vegetation indices from UAV-based multispectral and digital imagery. *ISPRS Journal of Photogrammetry and Remote Sensing* 130, 246–255.

Zhou, X., Yang, L., Wang, W. and Chen, B. (2021). UAV data as an alternative to field sampling to monitor vineyards using machine learning based on UAV/Sentinel-2 data fusion. *Remote Sensing* 13(3), 457.

Zulfiqar, F., Navarro, M., Ashraf, M., Akram, N. A. and Munné-Bosch, S. (2019). Nanofertilizer use for sustainable agriculture: advantages and limitations. *Plant Science* 289, 110270.

Part 4

Social, ethical and economic aspects

Chapter 14

Regulatory frameworks and standards for agricultural robotics in the European Union

Andrea Bertolini[1] and Rocco Limongelli, Scuola Superiore di Studi Universitari e di Perfezionamento Sant'Anna - Istituto Dirpolis, Italy

1 Introduction
2 Product safety regulation and standardization
3 The European Union directive on liability for defective products
4 Data privacy and data ownership
5 The European Union's Artificial Intelligence Act
6 Conclusion
7 References

1 Introduction

Arguably, the most significant impact that robotics and automation had on the agri-food industry are (1) precision agriculture, used, for example, to remove weeds, (2) self-driving field machinery, and (3) autonomous fruit and vegetable harvesting (Bergerman et al., 2016). Innovation has become a safe harbor for business to shield from increasing costs and regain a competitive edge. In the United States, the automatization of tomato harvesting over 35 years resulted in a drop of labor by 92% while tomato harvesting soared by more than double (Christiaensen et al., 2021). Nonetheless, adoption is somewhat slowing down. In 2016, circa 5300 milking robots were sold - while about 5860 units were sold in 2015, while the sale of field robots decreased in 2016 to 6000 units.[2] Robotics applications to dairy farming are now well established, while other agricultural applications such as fruit picking are not creating traction due to the complexity of technology and the low cost of human labor.[3]

[1] Andrea Bertolini, LL.M. (Yale), Ph.D. (SSSA) is an Associate Professor of Private Law at Scuola Superiore Sant'Anna, Rocco Limongelli is a Ph.D. candidate at Scuola Superiore Sant'Anna. Rocco Limongelli has received funding from the European Union's Horizon 2020 research and innovation programme under the Marie Skłodowska-Curie grant agreement No 955778. AB is primarily responsible for §§ 1, 3, 6; RL for §§2,4,5.
[2] International Federation of Robotics. (2017) Executive summary world robotics 2017 service robots.
[3] International Federation of Robotics. (2020) World Robotics 2020 Report International Federation of Robotics 24 September 2020.

http://dx.doi.org/10.19103/AS.2023.0124.18

It is the expert's consensus that legal frameworks strongly influence the acceptance and adoption of robotics within the agri-food industry (AFI), as standards are helpful in regulating robotics in the AFI because they give specific and technical guidance of the kind engineers are more accustomed to handle, as opposed to complex legal concepts that often times need to be interpreted in light of the peculiarities of the given case (Von Veltheim and Heise, 2020). In fact, concepts, such as product liability and safety, require complex analyses, tailored to specific applications, such as AFI, and may leave some degree of uncertainty with respect to potential scenarios and outcomes in an *ex ante* perspective (Garske et al., 2021).

It has been argued that fully autonomous farms are a faraway reality as they lack a legal framework, which would allow a transition from human to robotic labor (Cheein et al., 2015). Although interest in new technologies is high, farmers generally seem reluctant to adopt new technologies until more certainty about the legal context is achieved (Kester et al., 2013). Efforts have been made to develop the discipline of Robot Law as an organized and systematic field of law; however, we are not quite there yet (Palmerini et al., 2014). One of the principle problems in providing a comprehensive legal framework for robotics lies in the fact that robotics is a broad domain encompassing a wide array of applications that display no real commonality on a technological level (Bertolini, 2013). This entails that even their regulation becomes ever more complex, if at all possible, when approached top-down and with a one-size-fits-all solution. There has been some interest worldwide in the development of a legal framework for agribots. Most notably, we must mention Japan's adoption of the *Test codes for the safety of robot tractors & robot rice transplanters*, which provides safety guidelines for the safe usage of robot tractors.[4] While, in Australia, there is no specific regulation aimed at agribots, the main document to look at is the industry developed *Code of Practice for Agricultural Mobile Field Machinery with Autonomous Functions in Australia*.[5] Lastly, in the United States of America, there is no trace of plans to regulate agribots, although the State of California is stating that *beginning rulemaking at this juncture would be premature* and has granted leave to experiments with agribots.[6]

This chapter aims at giving an overview of the current European Union (EU) legal framework for agribots, focusing on some of the most current issues. This will entail discussing product regulation and standardization (2. PRODUCT SAFETY REGULATION AND STANDARDIZATION), civil liability (3. THE EUROPEAN UNION DIRECTIVE ON LIABILITY FOR DEFECTIVE PRODUCTS), and data (4. DATA PRIVACY AND DATA OWNERSHIP). Lastly, the chapter will

[4] See de Bourgogne (2021). Smart Farming Technology in Japan and Opportunities for EU Companies. *ECOS*.
[5] See Rose et al. (2021). Responsible development of autonomous robotics in agriculture. *Nature Food*, *2*(5), 306-309.
[6] See Petition File No. 596 Proposed Decision (ca.gov).

look at the novel draft AI Act[7] and address whether or not it applies to agribots and if so, to which extent.

2 Product safety regulation and standardization

With the introduction by the EU of the *New Approach*[8] – followed later by the complementary *New Legislative Framework Approach,*[9] – a new framework was developed for ensuring harmonization in relation to the safety of products sold within the EU, consisting of general norms working together with product-specific legislation. Thus, the key pieces of legislation regulating product safety in Europe vary according to the type of product and are represented by a broad set of directives and regulations, which set out the health and safety requirements for products to be sold within the EU, albeit the General Product Safety Directive (GPSD)[10] provides for general requirements that can be outlined in three points: (1) ensure that products placed on the market are safe; (2) inform consumers of any risks associated with the products supplied; and (3) take corrective action when products prove to be unsafe (Argiros, 1994).

Nonetheless, as mentioned above, product-specific legislation must also be complied with since compliance with the General Product Safety Directive on its own does not suffice. Specifically in relation to the AFI, the following legislations will most commonly be applicable[11]: the Machinery Directive,[12] the Low Voltage Directive,[13] and the Agricultural and Forestry Machinery Regulation.[14]

It should be noted that regulations[15] and directives[16] are different instruments despite both being primary legislation; in fact, a regulation is immediately applicable (as is), while a directive will require to be transposed

[7] Proposal for a Regulation of The European Parliament and of The Council Laying Down Harmonised Rules On Artificial Intelligence (Artificial Intelligence Act) and Amending Certain Union Legislative Acts EUR-Lex - 52021PC0206 - EN - EUR-Lex (europa.eu).
[8] Council Resolution of 7 May 1985 on a new approach to technical harmonization and standards.OJ C 136, 4 June 1985, p. 1–9.
[9] See New legislative framework (europa.eu).
[10] Directive 2001/95/EC of the European Parliament and of the Council of 3 December 2001 on general productSafety.
[11] See COVR toolkit, accessible at https://www.safearoundrobots.com/toolkit/documentfinder.
[12] Directive 2006/42/EC of the European Parliament and of the Council of 17 May 2006 on machinery, and amending Directive 95/16/EC (recast).
[13] Directive 2014/35/EU of the European Parliament and of the Council of 26 February 2014 on the harmonization of the laws of the Member States relating to the making available on the market of electrical equipment designed for use within certain voltage limits.
[14] Regulation (EU) No 167/2013 of the European Parliament and of the Council of 5 February 2013 on the approval and market surveillance of agricultural and forestry vehicles.
[15] Article 288 of the Treaty on the Functioning of the European Union provides that regulations 'shall have general application. [they] shall be binding in its entirety and directly applicable in all Member States', simply meaning that all regulations published on the Official Journal of the European Union, will become binding in all Member States from the date specified in the regulation or on the twentieth day following its publication.
[16] See Article 288 of the Treaty on the Functioning of the European Union 'A directive shall be binding, as to the result to be achieved, upon each Member State to which it is addressed, but shall leave to the national authorities the choice of form and methods'; although the failure of a MS in implementing a certain directive by the given deadline will result in vertical direct effect, meaning that an individual will still have a right to enforce a non-implemented directive against the state or an emanation of the state.

into national legislation. This, in turn, allows for minimal discrepancy among Member States (MS).

Following the introduction of the New Approach, the EU has put harmonization at the forefront of its objectives. Product safety regulations define mandatory requirements that all manufacturers need to comply with in order to distribute their products within the EU market. However, it should not be mistaken for technical standards, often referred to as *norms*. Those, in fact, constitute no legal norm and impose no obligation upon manufacturers. Compliance is thus merely voluntary and serves the purpose of defining the criteria the industry conforms to as well as eases the identification of the state of the art useful to certify their safety, as required by relevant product safety regulations. Nonetheless, compliance with applicable directives and regulations may always be proven by resorting to alternative tests and criteria than those provided by standards; hence, there is no legal obligation to respect existing standards (Delimatsis (Ed.), 2015).

Indeed, a standard is defined as a 'document, established by consensus and approved by a recognized body, that provides, for common and repeated use, rules, guidelines or characteristics for activities or their results, aimed at the achievement of the optimum degree of order in a given context', furthermore 'Standards should be based on the consolidated results of science, technology and experience, and aimed at the promotion of optimum community benefits'.[17]

Standards are useful instruments that aim at codifying good practices with the aid of industry experts and academics - standards being among the most important outputs of research. Well-structured and recognized, they contribute to the development of national and international regulations and currently play a key role within the EU and its MS policymaking (Xenias, 2012).[18]

Standards also tend to increase competition and benefit economies as a whole by reducing production costs and fostering the adoption of new technologies, steering further business investment toward R&D (Bekkers et al., 2006).

Robots within the AFI rely on a range of advanced technologies, including GPS, lidar, and computer vision - these technologies are constantly evolving. Therefore, this makes it difficult to establish standards that will remain relevant and effective over time. In order to address this issue, standards are being developed with the aim of being flexible and adaptable, allowing for changes in technology over time and diminishing the risk of obsolescence of such standards. There are numerous standards relating to robotics applications in the AFI - see the Table 1 - and although not mandatory, the use of standards

[17] ISO/ IEC Guide 2:2004, definition 3.2

[18] See also COM(2004) 674 final at paragraph 2.1 https://eur-lex.europa.eu/legal-content/EN/TXT/PDF/?uri=CELEX:52004DC0674&from=EN, - Standardization is in fact 'an integral part of the Council's and the Commission's policies to carry out "better regulation", to increase competitiveness of enterprises and to remove barriers to trade at international level'.

Table 1 ISO Standards applicable to agribots

STANDARDS[a]	
ISO/TS 15066	*Robots and robotic devices - Collaborative robots*
EN ISO 10218-1	*Robots and robotic devices - Safety requirements for industrial robots - Part 1: Robots*
EN ISO 10218-2	*Robots and robotic devices - Safety requirements for industrial robots - Part 2: Robot systems and integration*
ISO/TR 20218-1	*Robotics - Safety design for industrial robot systems - Part 1: End-effectors*
EN ISO 13855	*Safety of machinery - Positioning of safeguards with respect to the approach speeds of parts of the human body*
ISO 25119-1	*Tractors and machinery for agriculture and forestry - Safety-related parts of control system - Part 1: General principles for design and development*
ISO 25119-2	*Tractors and machinery for agriculture and forestry - Safety-related parts of control system - Part 2: Concept phase*
ISO 25119-3	*Tractors and machinery for agriculture and forestry - Safety-related parts of control system - Part 3: Series development, hardware, and software*
ISO 25119-4	*Tractors and machinery for agriculture and forestry - Safety-related parts of control system - Part 4: Production, operation, modification, and supporting process*
EN ISO 13849-1	*Safety of machinery - Safety-related parts of control systems - Part 1: General principles for design*
EN ISO 13849-2	*Safety of machinery - Safety-related parts of control systems - Part 2: Validation*
EN ISO 12100:2011	*Safety of machinery - General principles for design - Risk assessment and risk reduction*
EN ISO 13850	*Safety of machinery - Emergency stop function - Principles for design*
EN IEC 60204-1	*Safety of machinery - Electrical equipment of machines - Part 1: General requirements*
EN IEC 62061	*Safety of machinery - Functional safety of safety-related electrical, electronic, and programmable electronic control systems*
EN ISO 13854	*Safety of machinery - Minimum gaps to avoid crushing of parts of the human body*
EN ISO 13857	*Safety of machinery - Safety distances to prevent hazard zones being reached by upper and lower limbs*
EN ISO 14118	*Safety of machinery - Prevention of unexpected start-up*
EN IEC 62046	*Safety of machinery - Application of protective equipment to detect the presence of persons*
EN ISO 18497	*Agricultural machinery and tractors - Safety of highly automated agricultural machines - Principles for design*
EN ISO 4254	*Agricultural machinery - Safety - Part 1: General requirements*
EN ISO 13851	*Safety of machinery - Two-hand control devices - Principles for design and selection*

[a] See COVR toolkit, accessible at https://www.safearoundrobots.com/toolkit/documentfinder.

may be beneficial, for example, for reputational and marketing reasons, to grant access to protected markets or to take part in public procurement, or simply to enhance efficiency, coordination, and cooperation (Fosch Villaronga and Golia, 2019), therefore providing a competitive edge. In fact, while it is accepted that the use of standards can be a first step in the development of an interdisciplinary and comprehensive legal framework to regulate robotics, access to ISO standards is limited by the payment of a fee, and the complexity of such standards would usually require the employment of expert consultants, effectively cutting off smaller actors which cannot afford compliance with standards.

A partial exception is harmonized standards, introduced with the New Approach as a tool for further harmonization. Harmonized standards are published in the Official Journal of the EU and are developed following an official request by the European Commission[19] to a recognized European Standards Organization, being CEN, CENELEC, and ETSI.[20] However, while adherence to the so-called harmonized standards will result in a presumption of conformity with relevant EU law (Hughes, 2009) – harmonized standards will state under Annex Z to which EU law they relate and provide a presumption of conformity – compliance with such standards still remains voluntary.

The ISO 12100 *Safety of Machinery – General Principles for Design – Risk Assessment and Risk Reduction* can be considered as one of the most important standards for modern industrial applications. This standard has been mandated by the EC and approved by CEN in October 2010. Compliance with this standard presumes conformity with the Machinery Directive,[21] which aims at ensuring machinery's safety within the EU.

The ISO 12100 provides a classification of related safety standards as type A, type B, and type C. These types of standards refer to different levels of detail and specificity. Type A standards are the least specific and are often introductory in nature, in that they provide a general overview of the principles, concepts, and practices related to a particular field. In contrast, type B standards are more specific as they provide detailed guidance and recommendations on specific aspects of a particular field. In fact, type B standards usually include technical specifications, performance criteria, and design guidelines and are used to support the implementation of type A standards. Lastly, type C standards, being the most specific, provide detailed technical specifications and requirements for a particular product, process, or service and often include test methods, performance criteria, and design guidelines. Type C standards are used

[19] See the database of mandates to view standardization requests issued by the European Commission.
[20] See Harmonised Standards (europa.eu).
[21] DIRECTIVE 2006/42/EC OF THE EUROPEAN PARLIAMENT AND OF THE COUNCIL of 17 May 2006 on machinery, and amending Directive 95/16/EC (recast).

to ensure the consistent quality and performance of a particular product or service.

The purpose of ISO 12100 - being a type A standard - is that of providing designers with a framework and guidance for the development of safe machinery, in the sense that the machine shall have the ability to safely perform its intended functions during its life cycle.

For instance, the ISO18497 Agricultural Machinery and Tractors - Safety of Highly Automated Agricultural Machines - Principles for Design, will be a relevant type B standard, when dealing with the design of highly automated machines and vehicles in agricultural field operations. In fact, we find that this applies to 'mobile vehicle or machine with or without on-board operator allowing highly automated operation'.[22] ISO 18497 is an international standard that specifies the requirements for the performance and safety of autonomous mobile robots (AMRs) used in industrial applications. This standard provides guidelines for the design, development, manufacture, and use of AMRs and sets out the requirements for their safe operation and interaction with other systems and equipment. It covers a wide range of safety-related issues, including the design and construction of the robot itself, the provision of emergency stop functions, the protection of workers from hazardous movements, and the use of sensors and other components to detect and respond to changes in the environment. The standard also includes requirements for the performance and reliability of AMRs, including guidelines for their accuracy, repeatability, and reliability in various industrial applications. It provides a basis for the evaluation of AMRs and helps to ensure that they are designed and manufactured to meet the needs of industry and society. ISO 18497 is designed to be used by manufacturers, system integrators, and end-users of AMRs and provides a common basis for the design, evaluation, and certification of these robots. It is intended to promote the safe and responsible use of AMRs and to help ensure that these robots are used in a way that is consistent with industry best practices and the needs of society.

Besides conformity to standards and legislation, the essential requirement to sell a product in the EU is the CE marking, which must be obtained before a product can be placed on the market. The CE mark was introduced following the implementation of the New Approach, as per the EU's effort on harmonization, which aims at achieving the free movement of goods across the EU; in fact, the CE mark has also been dubbed as a *product's passport* (Tricker, 2000; Wende, 2015). To affix the CE mark on a product, the relevant regulation, directive, or harmonized standard - applicable to the product in question - must be identified. The manufacturer can then complete a self-declaration, in which it will be stated that the product is in conformity with all relevant and

[22] ISO 18497 at paragraph 3.7.

applicable EU laws.[23] Although it is necessary to note that for products that present increased risks of danger to people or things, only a notified body will be able to issue a declaration of conformity.[24]

However, the CE mark is not a final signature of safety. The GPSD established the Rapid Exchange of Information System (RAPEX),[25] which allows the exchange of information on unsafe products within the EU, resulting in rapid recall of defective and non-compliant products. In fact, even where all the conformity steps have been taken, and the product has reached the stage of having the CE mark affixed, this will not exclude liability for product defects as provided by the Product Liability Directive,[26] which will still protect customers when injured by a defective product regardless of whether such defect was due to the manufacturer negligence. In fact, such liability can neither be limited nor excluded.

3 The European Union directive on liability for defective products

Robots' applications are extremely diverse; therefore, the creation of a set of liability rules for torts involving robots is surely an undesirable solution, as robotics applications should be addressed individually by identifying a specific distinctive trait, which will trigger – as necessary – a change in the legal analysis (Bertolini, 2013).

While product safety regulation aims at ensuring the products distributed on European markets reach a minimum required level of safety ex-ante, product liability operates ex-post to hold the manufacturer accountable for damage caused by defective products. This entails that a fully certified product, lawfully distributed onto European markets after having satisfied all requirements posed by applicable product safety legislation, might nonetheless be considered defective forcing the manufacturer to compensate for the damage caused.

The fundamental rationale being that, since the producer benefits from the distribution of his products for profit, he shall also internalize associated costs (externalities), in the form of damages suffered by those who used the product itself or where exposed to its use. Pursuant to art. 4 of the EU Product Liability Directive[27] (henceforth referred to as PLD), the claimant – who may well not be in any prior legal (e.g. contractual) relationship with the producer – needs to demonstrate he suffered a damage that was the consequence of (causal

[23] See Manufacturers (europa.eu) for more information.
[24] See Notified bodies (europa.eu) for more information.
[25] See Safety Gate for dangerous non-food products (europa.eu) for more information.
[26] Council Directive 85/374/EEC of 25 July 1985 on the approximation of the laws, regulations and administrative provisions of the Member States concerning liability for defective products OJ L 210, 7 August 1985, p. 29–33.
[27] Council Directive 85/374/EEC of 25 July 1985 on the approximation of the laws, regulations, and administrative provisions of the Member States concerning liability for defective products OJ L 210, 7 August 1985, p. 29–33.

nexus[28]) the defective nature of the product. Defectiveness is to be understood as the 'lack of safety which a person is entitled to expect'(art.6). Traditionally, scholars and courts have identified three different forms of defectiveness: (1) a manufacturing defect – when the single specimen deviates from the prototype (a coca-cola bottle that explodes due to excessive pressure); (2) an information defect – whenever the user is not sufficiently warned against the intrinsic risks of the product –; and (3) design defect – if the way the product is conceived, and its functioning is deemed unreasonably dangerous, as opposed to other possible alternatives. Against the latter, the producer possesses the so-called development risk defense (art. 7(e)), whereby he may not be held liable and called in to compensate damages, if it may be argued that the existing technical and scientific knowledge would not allow him to conceive the product otherwise. It shall then be noted that, with respect to damages, only those caused to other things but not to the defective good itself may be compensated (art. 9(b)).

This EU directive is often deemed ineffective in providing adequate protection to the victims, as demonstrated by the very limited number of cases litigated on its grounds across MS,[29] and the limited number of cases in which claimants successfully obtained compensation. Similarly, its qualification as a strict (i.e. non-fault-based) liability rule[30] is also disputed. Indeed, courts such as the Italian Corte di Cassazione[31] question such a categorization and define the PLD as a presumed-fault liability rule. Indeed, thanks to the development risk defense, the producer may escape liability by demonstrating that they designed their products in light of the most advanced technological and scientific knowledge available at that time. Such a judgment, however, closely resembles that required by a rule of fault, whereby the court needs to determine what a model agent ought to have done in similar circumstances and that the specific agent negligently failed to comply with such a standard of conduct. In such a perspective, the rule laid out by the PLD displays evident commonalities with a fault-based liability rule, much more than a strict liability one. Finally, it is disputed that non-embedded software could be covered by the directive – since it might not fall in the definition of product as provided by art. 2 of the PLD – yet efforts are taking place to revise the directive and extend that notion so as

[28] In law, the causal nexus refers to the causal connection or relationship between an event (cause) and a result (effect) that is legally recognized as the basis for liability. In other words, it refers to the link between an act or omission and the harm or loss suffered by someone else. Beebee et al., p. (2009). *The Oxford handbook of causation*. Oxford University Press.

[29] See Ernst&Young, Technopolis, and VVA. (2018). *Evaluation of Council Directive 85/374/EEC on the approximation of laws, regulations and administrative provisions of the Member States concerning liability for defective products*, p.22, and Bertolini (2020). *Artificial Intelligence and Civil Liability*. E. Parliament.

[30] A strict liability rule, as opposed to a fault-based one (or rule of negligence), does not require the claimant to demonstrate that the agent is at fault – i.e. he was supposed to behave otherwise – in order to hold the latter liable and force him to compensate damages. Giliker (2017). *Research Handbook on EU Tort Law*. Edward Elgar Publishing. Strict liability rules are deemed – for this very reason – to be more favourable to the victim, since they do simplify the burden of proof they need to meet in order to obtain compensation.

[31] Corte di Cassazione, September 1st 2013, n. 13458, in Foro Italiano, 2013, I, 2118.

to encompass AI-based applications that do not possess a physical body. This piece of legislation clearly pursues technological neutrality (i.e. the broadest application across the most diverse range of products) yet fails to achieve it, as is demonstrated by the clear clustering of cases in a few domains, primarily raw materials and pharmaceutical products.[32]

In light of this EU framework, the manufacturer of any robotics device could be called in to compensate for damages arising from the functioning of the device. Increasing autonomy and the preeminent role of the device in handling a given task – alongside a human operator – theoretically expands the application of said norms and the subsequent exposure of the producer of the device to possible claims for damages suffered by third parties. Autonomous tractors are the typical example where said norms would be applied, often – even if not always – replacing the liability of the human while sharing the task with the machine.

However, other liability rules – typically identifiable at the MS level – will overlap with the PLD, in as much as there will also be legal grounds to trace the liability to a given human or legal entity (e.g. the employer in a labor contract) due to the application of different sets of norms (e.g. employers responsibility, drivers', and owners' responsibility in traffic accidents, medical responsibility, etc.). General liability rules will also typically trace the responsibility back to someone who negligently caused harm while operating the machine. However, contractual relationships could also lead to the identification of parties that hold a position of responsibility, irrespective of their direct involvement in the operation of the system.[33]

Most relevantly, however, the European Commission, in light of the recommendations received by the Expert Group on Civil Liability New Technologies formation[34], is considering adopting a new piece of legislation, specifically to regulate the civil liability for damages emerging from the use of AI-based systems. A proposal from the European Parliament dating 20 October 2020[35] identifies the *operator* – defined as 'both the frontend and the backend operator as long as the latter's liability is not already covered by Directive 85/374/EEC' (art. 3) – as the primary responsible party for damages arising from the use of high-risk applications – those will be specifically enumerated

[32] See Ernst&Young, Technopolis, & VVA. (2018). *Evaluation of Council Directive 85/374/EEC on the approximation of laws, regulations and administrative provisions of the Member States concerning liability for defective products*, p.19-20, and Bertolini (2020). *Artificial Intelligence and Civil Liability*. E. Parliament.

[33] Such is typically the case of the owner of a vehicle circulating on public roads (e.g. pursuant to art. 2054 of the Italian Civil Code, or art. 2 of Loi n. 85-677 of 5 July 1985, 'Tendant à l'amélioration de la situation des victimes d'accidents de la circulation et à l'accélération des procédures d'indemnisation', also known as 'Loi Badinter'). Similarly, the responsibility of the employer for damages caused by his employee to third parties, in light of his role in organizing the entire entrepreneurial activity (e.g. pursuant to art. 2049 of the Italian civil code), as well as toward the employee himself (e.g. art. 2087 of the Italian civil code).

[34] Expert Group on Liability and New Technologies. (2019). *Report on Liability for Artificial Intelligence and other emerging digital technologies* (E. Commission, Ed.). European Commission. https://doi.org/10.2838/573689.

[35] European Parliament. (2020). *Civil liability regime for artificial intelligence. European Parliament resolution of 20 October 2020 with recommendations to the Commission on a civil liability regime for artificial intelligence (2020/2014(INL))*. European Parliament.

in an annex to be updated regularly - on a strict liability basis (art. 4), as well as from the use of low-risk applications - which is a residual category, not more specifically defined - on a presumed fault basis (art. 8).[36]

The proposal, when enacted as presented, would certainly reinforce the responsibility both of the manufacturer as well as of the different parties involved in the deployment and use of the AI-based system. Agricultural robotics applications would not be an exception in this sense, while it is disputable whether they would be classified as high risk (as, for instance, driverless vehicles will most likely be) or rather low risk, since they do not operate in densely populated areas with numerous non-sophisticated parties. Those liability rules are also associated with duties to acquire insurance that burden the operators (art. 4(4)), so that, in case they were liable, the insurance companies would be called in to compensate damages, thence ensuring a full recovery of the losses suffered by the victim. Even in such a scenario, however, the most delicate part would entail the exact assessment of a causal nexus between a given risk - under the control of a given operator - and the material harm caused. The issue of multiple causality is, in fact, intrinsic to advanced technologies where multiple parties collaborate and contribute toward the functioning of a device that often is considered to be opaque and not transparent with respect to its internal processes leading to a given outcome (also referred to as a black box).[37]

The tendency is, however, clearly to shift from the fault-based liability of a human agent to that of the manufacturer or business user and deployer of the technology on the basis of the possibility for those parties to control a specific risk associated with the operation of the system, rather than considerations of negligence.[38]

4 Data privacy and data ownership

Farming fields deploying high-tech applications can yield a vast quantity of valuable data including but not limited to weather conditions, crop production, soil conditions, and GPS data. These data can be *mined* - i.e. exploited to interpret or extract information (Witten et al., 2017) - in order to drive financial decisions, making them commercially valuable assets. In fact, data collected by sensors on the fields may be used to reliably predict volumes of crop production (Colombo-Mendoza et al., 2022) or predict wine quality based on meteorological events and other agricultural attributes (Last et al.;

[36] If fault is presumed, the claimant needs not demonstrate it in order to obtain compensation from the defendant. However, the defendant may escape his duty to compensate by demonstrating that the conduct he maintained was diligent (i.e. not negligent, not at fault) in as much as he behaved to the standard of an ideal model agent in identical circumstances, or his failure to comply with such a superior standard of conduct could not be blamed upon him. Van Dam (2013). *European tort law*. OUP Oxford. A strict standard of liability, instead, does not allow the defendant to escape liability by demonstrating he is not at fault.

[37] AI (2019). High-level expert group on artificial intelligence. In (pp. 6): European Commission.

[38] For a more detailed discussion, allow reference to *Artificial Intelligence does not exist! Defying the technology-neutrality narrative, in the regulation of civil liability for advanced technologies* (Bertolini, 2022).

Shanmuganathan et al. (2010). Thus, two legal issues can be identified as being that of data privacy – the protection of personal data under EU law – and data ownership.

In the EU, data protection is regulated by the EU General Data Protection Regulation (GDPR) 2016, which was enacted in substitution to the Directive 96/46/EC. The choice to implement a regulation rather than a directive lies in the fact that while a regulation is directly applicable law of each and every EU Member State (MS), a directive needs to be implemented in national law by each MS, possibly causing fragmentation due to the different implementation strategies of each MS – a clear pitfall of Directive 96/46/EC that lacked proper harmonization, which, in turn, created legal uncertainty preventing the free flow of data within the EU (De Hert and Papakonstantinou, 2012).

The GDPR aims at solving the above-mentioned problems, creating legal certainty on the protection of *personal data* – defined under Article 4 of the GDPR as 'any information relating to an identified or identifiable natural person ('data subject'); an identifiable natural person is one who can be identified, directly or indirectly, in particular by reference to an identifier such as a name, an identification number, location data, an online identifier or to one or more factors specific to the physical, physiological, genetic, mental, economic, cultural or social identity of that natural person'.

Furthermore, the GDPR sets an obligation for the processing of data to be lawful. Lawful processing can be achieved by the application of one of the lawful bases for processing under Article 6(1), being (1) consent, (2) contract, (3) legal obligation, (4) vital interests, (5) public task, and (6) legitimate interest. Nonetheless, (1) consent still plays a central role within the private sector – where (2) contracts cannot be utilized on a lawful basis. Thus, it was deemed necessary to expand on the concept of valid consent through Articles 7 and 8, in contrast with other legal grounds only mentioned in Article 6 (Kuner et al., 2020). In fact, where the ground for processing is that of consent, under Article 7 of the GDPR, it must be demonstrated that the data subject[39] has freely consented to the processing of his or her personal data and that the right to withdraw at any time such consent is provided.

In a scenario where an agribot is deployed in a field, the GDPR may come into play – an example can be where the agribot requires an operator to be in charge for safety reasons; therefore, if GPS data are collected, it could be linked directly to an individual, making it personal data. However, it is argued that personal data collected through agribot applications are minimal, have no particular value, and are rather simply incidental to the other data – i.e. soil, weather, crop production, and the like. In these instances, in order to avoid any redress for breach of the GDPR, some simple precautions may be put in place

[39] Under Article 4 (1) of the GDPR a 'data subject' is any identified or identifiable natural person.

to make sure that no data collected can be directly traced back to a specific individual.

A clear example of data ownership is, for example, the instance where a company deploys an agribot in a field and gathers information from multiple runs – for instance by employing machine learning techniques – should the same company be allowed to deploy an agribot in a competing farmer's field, where that agribot will deliver better results thanks to the data gathered on the previous runs. This clearly is a matter of data ownership.

When dealing with *digital assets* such as data, the question is whether they can be classified as legal objects capable of ownership. In a strict view of legal theory, data cannot be owned, as ownership: 'is contrary to well-established legal precedent in the United States, United Kingdom, and many other jurisdictions, which for good reasons do not recognize property interests in mere facts or information' (Contreras et al., 2018). In fact, data ownership is not and cannot be regulated by traditional concepts of property ownership, which have been developed throughout the years – even centuries – with physical assets being the legal object on which such doctrine has been built (Van Erp, 2020a). Nonetheless, the current trend is to depart from established doctrinal concepts and mold them into novel legal objects such as data. In this sense, it is accepted that data cannot be owned in the strict legal meaning of the word, rather it can be managed – management, as the expression of access, control, portability, and erasure (van Erp, 2017, 2020b,c).

The EU has moved toward the concept of data management as opposed to ownership, as shown by the European Commission's proposal for a Regulation on harmonized rules on fair access to and use of data – also known as the Data Act[40] – adopted on 23 February 2022. The Data Act aims at regulating the accessibility of data to users – both businesses and consumers. One of the objectives of the Data Act is to ensure access and portability of 'data generated by the use of a product or related service, irrespective of its nature as personal data, of the distinction between actively provided or passively observed data, and irrespective of the legal basis of processing' as provided in Recital 31, effectively filling the gaps left by the GDPR (Graef and Husovec, 2022). In fact, the Data Act will not apply only to personal data as the GDPR does but will apply to *data* meaning any digital representation of acts, facts, or information and any compilation of such acts, facts, or information, including in the form of sound, visual, or audiovisual recording – a very broad definition given under Article 2(1) of the Data Act.

[40] Proposal for a REGULATION OF THE EUROPEAN PARLIAMENT AND OF THE COUNCIL on harmonised rules on fair access to and use of data (Data Act) Data Act: Proposal for a Regulation on harmonised rules on fair access to and use of data | Shaping Europe's digital future (europa.eu).

The Data Act will further set an unfair term restriction insofar as when data need to be made available upon request, this is done in a transparent, fair, and reasonable manner.

Furthermore, the Data Act will also introduce a non-compete requirement under Articles 6(2)(e) and 4(4), respectively, barring manufacturers and users from using or causing the data originated to be used to develop a product that competes with the product that originated said data.

The Data Act addresses some of the problems posed earlier, by creating a legal framework for the management of data, once again reaffirming the established position that data cannot be owned. However, the Data Act will still need to be debated by the EU legislators before it comes into force.

5 The European Union's Artificial Intelligence Act

The EC on 25 April 2018 published a communication titled *Artificial Intelligence for Europe,*[41] setting out the EU's AI strategy. Later, on 21 April 2021, the EC released one of its most ambitious proposals: the AI Act, a forward-thinking piece of legislation made up of an Explanatory Memorandum, 89 Recitals, 12 Titles, 85 Articles, and 9 Annexes. In giving the reasons for the proposal, the EC refers to the EU's commitment to becoming a market leader in relation to AI by making sure that AI developed and marketed within the EU is trustworthy. The AI Act is still currently a draft proposal and will need to be approved by the EU Parliament before becoming a binding regulation.

The AI Act is the first piece of legislation in Europe, to give a comprehensive and binding definition of AI as: 'software that is developed with one or more of the techniques and approaches listed in Annex I and can, for a given set of human-defined objectives, generate outputs such as content, predictions, recommendations, or decisions influencing the environments they interact with'.[42]

Furthermore, under Annex I of the AI Act, for the purpose of defining AI, the following techniques and approaches apply: '(a) Machine learning approaches, including supervised, unsupervised and reinforcement learning, using a wide variety of methods including deep learning; (b) Logic- and knowledge-based approaches, including knowledge representation, inductive (logic) programming, knowledge bases, inference and deductive engines, (symbolic) reasoning and expert systems;(c) Statistical approaches, Bayesian estimation, search and optimization methods'.[43]

[41] Communication from the Commission to the European Parliament, the European Council, the Council, the European Economic and Social Committee, and the Committee of the Regions Artificial Intelligence for Europe, 25 April 2018. Available at: https://eur-lex.europa.eu/legal-content/EN/TXT/?uri=COM%3A2018%3A237%3AFIN.

[42] Article 3(1) of the AI Act.

[43] Annex I of the AI Act. There are several proposals from the European Parliament to amend the AI Act. Therefore, the final text might differ in some aspects from the one reported here.

As not all AI applications are comparable, the EC has classified AI – within and for the purposes of the AI Act – in three categories: low risk, high risk, and banned.

Low risk. Under Article 52 of the AI Act, the low-risk classification will include any AI that does not fall within the other categories and that (i) is intended to interact with natural persons; (2) uses emotion recognition or biometric categorization technologies; or (3) generates the so-called *deep fake*.

In all the above cases, the use of AI must be flagged to make sure that a natural person is, respectively, informed that (1) he is interacting with an AI system; or (2) he is exposed to emotion recognition or biometric categorization technologies; or (3) the content has been generated by AI.

High risk. Under Article 6(1) of the AI Act, an AI system is to be considered high-risk, if both of the following conditions are met: '(a) the AI system is intended to be used as a safety component of a product, or is itself a product, covered by the Union harmonisation legislation listed in Annex II; and (b) the product whose safety component is the AI system, or the AI system itself as a product, is required to undergo a third-party conformity assessment with a view to the placing on the market or putting into service of that product pursuant to the Union harmonisation legislation listed in Annex II'.[44]

Furthermore, pursuant to Article 6(2) of the AI Act, any of the AI systems as listed under Annex III of the AI Proposal are automatically high-risk. A wide array of AI systems are included in Annex III, which have been listed based on their application: biometric identification and categorization of natural persons, management and operation of critical infrastructure, education and vocational training, employment, workers management and access to self-employment, access to and enjoyment of essential private services and public services and benefits, law enforcement, migration, asylum, and border control management, and lastly administration of justice and democratic processes.[45] It is also noted that the category with the most listed high-risk AI system is that of Law Enforcement and Border Control – underlying the human-centric approach putting fundamental rights at the forefront of this piece of legislation. In order to allow for the AI Act to dynamically adapt to the fast-evolving technologies, the EC has the power to amend Annex III in order to add new AI systems that should be considered high-risk or remove obsolete ones that no longer pose substantial risks.[46]

Banned. Most notably, the AI Act will ban the placing on the market of AI systems that use subliminal techniques or exploit vulnerable persons in order

[44] Article 6(1) of the AI Act.
[45] Annex III of the AI Act.
[46] Article 7 of the AI Act.

to materially distort a person's behavior in a manner that causes or is likely to cause that person or another person's physical or psychological harm.

Although there might be some very peculiar and odd cases where agribots might fall within one of the AI Act's protected categories,[47] it is apparent that unless the text is amended before coming into force or once in force, Annex III is amended to include specific AI systems used in agriculture, and then, it would be difficult to argue that the AI Act will be feeding into the regulatory framework for agri-food robotics.

Lastly, Articles 53 and 54 of the AI Act provide for the setup of AI regulatory sandboxes - again a factor that will likely contribute to innovation. In fact, regulatory sandboxes have shown great success in Japan. In 2003, the Japanese cabinet approved the first 'Special Zone for Robot Development and Practical Testing (Tokku)', leading to the creation of many more Special Zones such as Fukuoka, Osaka, Gifu, Kanagawa, and Tsukuba.[48] Weng and others suggest that regulatory sandboxes such as Tokku zones have been shown to contribute to increased safety and easing of ethical disputes - regulatory sandboxes can help to better regulate AI around the world (Weng et al., 2015).

This provides an opportunity for governments to incentivize the uptake of high-tech field equipment by farmers, in order to allow for the development of a stronger framework and incentivize the adoption of new technologies.

6 Conclusion

Technological innovation is very much shaped by regulation, this is also relevant in the agricultural sector. Applicable European regulation already today influences the way technology is developed and used. However, at the current time, prominent proposals are under consideration, which address all fields, ranging from product certification to liability and data use. This will potentially profoundly alter current balances, requiring attentive assessments to determine what kind of technological solutions ought to be favored, as well as which economic models will be preferable to favor their diffusion and uptake.

Above all, how liability will be apportioned among those that are involved in the development and use of agribots, as well as how access to data will be denied or granted, and under which conditions, to those that could benefit

[47] For example, where an agribot is equipped with a chatbot functionality for tech support, it will fall within the Low-Risk category of the AI Act. Another instance where agribots may fall within the remits of the AI Act is if they are equipped with systems to track employee performance, see AI Act Annex III paragraph 4 (b) *AI intended to be used for making decisions on promotion and termination of work-related contractual relationships, for task allocation and for monitoring and evaluating performance and behavior of persons in such relationships.*

[48] Japan's Robot Policy and the Special Zone for Regulating Next Generation Robots, 2015. Tech and Law Center. URL http://techandlaw.net/japans-robot-policy-and-the-special-zone-for-regulating-next-generation-robots/ (accessed 9 July 2022).

and be capable of using it will determine the economic incentives each party is facing in developing or adopting said technologies.

7 References

AI, H. (2019). *High-level Expert Group on Artificial Intelligence*. European Commission (p. 6).

Argiros, G. (1994). EEC directive on general product safety, the. *Legal Issues of European Integration* 21, 125–154.

Beebee, H., Hitchcock, C. and Menzies, P. (2009). *The Oxford Handbook of Causation*. Oxford University Press.

Bekkers, R. N. A., Blind, K., Coenen, H., Iverson, E., Jacobs, K. and Hossain, K. (2006). *Case Studies on the Interface Between Research and Standardisation, and Case Studies on Patent Pools as a Coordination Mechanism*. INTEREST Consortium.

Bergerman, M., Billingsley, J., Reid, J. and van Henten, E. (2016). Robotics in agriculture and forestry. In: Siciliano, B., Khatib, O. (Eds). *Springer Handbook of Robotics* (pp. 1463–1492). Springer International Publishing. https://doi.org/10.1007/978-3-319-32552-1_56.

Bertolini, A. (2013). Robots as products: the case for a realistic analysis of robotic applications and liability rules. *Law, Innovation and Technology* 5(2), 214–247. Available at: https://doi.org/10.5235/17579961.5.2.214.

Bertolini, A. (2020). *Artificial Intelligence and Civil Liability*. E. Parliament.

Bertolini, A. (2022). Artificial Intelligence does not exist! Defying the technology-neutrality narrative, in the regulation of civil liability for advanced technologies. *Europa e Diritto Privato*.

Cheein, F. A., Herrera, D., Gimenez, J., Carelli, R., Torres-Torriti, M., Rosell-Polo, J. R., Escola, A. and Arno, J. (2015). *Human-Robot Interaction in Precision Agriculture: Sharing the Workspace with Service Units*.

Christiaensen, L., Rutledge, Z. and Taylor, J. E. (2021). Viewpoint: the future of work in agri-food. *Food Policy* 99, 101963. https://doi.org/10.1016/j.foodpol.2020.101963.

Colombo-Mendoza, L. O., Paredes-Valverde, M. A., Salas-Zárate, M. D. P. and Valencia-García, R. (2022). Internet of things-driven data mining for smart crop production prediction in the peasant farming domain. *Applied Sciences* 12(4), 1940. https://doi.org/10.3390/app12041940.

Contreras, J. L., Rumbold, J. and Pierscionek, B. (2018). Patient data ownership. *JAMA* 319(9), 935. https://doi.org/10.1001/jama.2017.21672.

Delimatsis, P. (Ed.). (2015) *The Law, Economics and Politics of International Standardisation*Cambridge University Press. Available at: https://doi.org/10.1017/CBO9781316423240.

de Bourgogne, R. M. (2021). *Smart Farming Technology in Japan and Opportunities for EU Companies*. Ecos.

De Hert, P. and Papakonstantinou, V. (2012). The proposed data protection Regulation replacing Directive 95/46/EC: a sound system for the protection of individuals Directive 95/46/EC. *Computer Law and Security Review* 28(2), 130–142. Available at: https://doi.org/10.1016/j.clsr.2012.01.011.

Ernst&Young, Technopolis, & VVA (2018). *Evaluation of Council Directive 85/374/EEC on the Approximation of Laws, Regulations and Administrative Provisions of the Member States Concerning Liability for Defective Products*.

European Parliament (2020). *Civil Liability Regime for Artificial Intelligence. European Parliament Resolution of 20 October 2020 with Recommendations to the Commission on a Civil Liability Regime for Artificial Intelligence*. European Parliament (p. 2020/2014(INL)).

Expert Group on Liability and New Technologies (2019). *Report on Liability for Artificial Intelligence and Other Emerging Digital Technologies* (E. Commission, Ed.). European Commission. https://doi.org/10.2838/573689.

Fosch Villaronga, E. and Golia, A. J., Jr. (2019). Robots, standards and the law: rivalries between private standards and public policymaking for robot governance. *Computer Law and Security Review* 35(2), 129–144. https://doi.org/10.1016/j.clsr.2018.12.009.

Garske, B., Bau, A. and Ekardt, F. (2021). Digitalization and AI in European agriculture: A strategy for achieving climate and biodiversity targets? *Sustainability* 13(9), 4652. https://doi.org/10.3390/su13094652.

Giliker, P. (2017). *Research Handbook on EU Tort Law*. Edward Elgar Publishing.

Graef, I. and Husovec, M. (2022). Seven things to improve in the data act. [SSRN Scholarly Paper]. *SSRN Electronic Journal*, (ID 4051793). Available at: https://papers.ssrn.com/abstract=4051793.

Hughes, S. E. (2009). Chapter 8. Codes, standards and documentation. In: Hughes, S. E. (Ed.) *A Quick Guide to Welding and Weld Inspection* (pp. 108–124). Woodhead Publishing. Available at: https://doi.org/10.1016/B978-1-84569-641-2.50008-8.

Kester, C., Griepentrog, H. W., Hörner, R. and Tuncer, Z. (2013). A survey of future farm automation: a descriptive analysis of survey responses. 9th Conference on Precision Agriculture (ECPA): Lerida, Spain.

Kuner, C., Bygrave, L. A. and Docksey, C. (2020). *The EU General Data Protection Regulation (GDPR) the EU General Data Protection Regulation (GDPR): A Commentary* Kuner, C., Bygrave, L. A., Docksey, C. and Drechsler, L. (Eds). Oxford University Press. https://doi.org/10.1093/oso/9780198826491.001.0001.

Last, M., Elnekave, S., Naor, A. and Schoenfeld, V. Predicting wine quality from agricultural data with single-objective and multi-objective data mining algorithms. In: *Recent Advances in Data Mining of Enterprise Data: Algorithms and Applications* (323–365). https://doi.org/10.1142/9789812779861_0007.

Palmerini, E., Azzarri, F., Battaglia, F., Bertolini, A., Carnevale, A., Carpaneto, J., Cavallo, F., Di Carlo, A., Cempini, M., Controzzi, M., Koops, B.-J., Lucivero, F., Mukerji, N., Nocco, L., Pirni, A., Shah, H., Salvini, P., Schellekens, M. and Warwick, K. (2014). *Guidelines on Regulating Robotics* [Deliverable of the FP7 research project Robolaw].

Rose, D. C., Lyon, J., de Boon, A., Hanheide, M. and Pearson, S. (2021). Responsible development of autonomous robotics in agriculture. *Nature Food* 2(5), 306–309.

Shanmuganathan, S., Sallis, P. and Narayanan, A. (2010). Data mining techniques for modelling the influence of daily extreme weather conditions on grapevine, wine quality and perennial crop yield. Second International Conference on Computational Intelligence, Communication Systems and Networks, CICSyN 2010, Liverpool, UK, 28-30 July, 2010.

Tricker, R. (2000). *CE Conformity Marking: and New Approach Directives*. Elsevier.

Van Dam, C. (2013). *European Tort Law*. Oxford University Press.

van Erp, S. (2017). Ownership of data and the numerus clausus of legal objects. In: Uncitral (Ed.). *Modernizing International Trade Law to Support Innovation and Sustainable Development* (vol. 4). United Nations.

Van Erp, S. (2020a). Data regulation: A race to...? *European Journal of Comparative Law and Governance* 7(4), 335–338. https://doi.org/10.1163/22134514-00704004.

van Erp, S. (2020b). Management as ownership of data. In: Lohsse, S., Schulze, R. and Staudenmayer, D. (Eds). *Data as Counter-performance - Contract Law 2.0?: Münster Colloquia on EU Law and the Digital Economy V* (1 ed., pp. 77–94). Nomos Verlagsgesellschaft mbH & Co. KG. https://doi.org/10.5771/9783748908531-77.

van Erp, S. (2020c). Who "Owns" the data in a coronavirus Tracing (and/or Tracking) App? In: Nicolussi, A., Wendehorst, C., Hondius, E., Zoll, F., Silva, M. S. and Coderch, P. S. (Eds). *Coronavirus and the Law in Europe* (pp. 131–156). Intersentia: https://doi.org/10.1017/9781839701801.008.

Von Veltheim, F. R. and Heise, H. (2020). The AgTech startup perspective to farmers ex ante acceptance process of autonomous field robots. *Sustainability* 12(24), 10570. https://doi.org/10.3390/su122410570.

Wende, S. (2015). The CE marking as a passport to the European market [IP corner]. *IEEE Consumer Electronics Magazine* 4(4), 125–137. https://doi.org/10.1109/MCE.2015.2463412.

Weng, Y. H., Sugahara, Y., Hashimoto, K. and Takanishi, A. (2015). Intersection of "tokku" special zone, robots, and the law: A case study on legal impacts to humanoid robots. *International Journal of Social Robotics* 7(5), 841–857. https://doi.org/10.1007/s12369-015-0287-x.

Witten, I. H., Frank, E., Hall, M. A. and Pal, C. J. (2017). Chapter 1. What's it all about? In: Witten, I. H., Frank, E., Hall, M. A. and Pal, C. J. (Eds), *Data Mining* (4th edn.) (pp. 3–41). Morgan Kaufmann. Available at: https://doi.org/10.1016/B978-0-12-804291-5.00001-5.

Xenias, A. (2012). The new global rulers: The privatization of regulation in the world economy [book review]. *Political Science Quarterly* 127(3), 498–499.

Chapter 15

Economics of agricultural robotics

James Lowenberg-DeBoer, Harper Adams University, UK*

1 Introduction
2 History of adoption of agricultural robotics
3 Expected benefits of agricultural robotics
4 Potential agricultural robotics in low- and middle-income countries
5 Broader implications of agricultural robotics for the farm sector
6 Social impact
7 Policy, regulatory and institutional issues
8 Conclusion
9 Where to look for further information
10 Acknowledgements
11 References

1 Introduction

The world is in the early stages of a wave of robotics in agriculture. As with previous waves of agricultural technology, farmers and agribusinesses are in the process of identifying which robotic technologies are worthwhile. The general objective of this chapter is to describe the economic potential for the widespread adoption of agricultural robotics worldwide including low- and middle-income countries. The study is of interest to farmers, agribusiness people, agricultural researchers, farm machine manufacturers, agricultural policy makers and members of the general public who have an interest in food security, the environment and rural economies.

For this chapter, the word 'robot' refers to a machine capable of autonomous operation without direct human intervention. The word robot tends to be used in the media and by the general public for any device capable of autonomous operation. Robots are often anthropomorphized as mobile and speaking but might take a wide variety of forms (e.g. stationary and mute). More technical discussions tend to use the terms like 'autonomous machine' or 'autonomous

* Corresponding author contact information – jlowenberg-deboer@harper-adams.ac.uk.

http://dx.doi.org/10.19103/AS.2023.0124.19

equipment', which are defined as mechanical and electrical devices that can perform certain functions without direct interaction with a human operator.

Innovations in agricultural technology have the potential to improve food security, food quality, and quantity of food produced, reduce the environmental footprint of agriculture, and help societies achieve food sovereignty goals, but in market economies, those technologies are only used if they have substantial benefits for farmers. In many cases, those on-farm benefits are mainly monetary but can include reduced workload, more flexible schedule, risk mitigation, quality and nutritional improvements, and enhanced farmer and farm family wellbeing. With every new wave of agricultural technology farmers and agribusinesses must sort out those technologies that help them solve their problems from those that solve the problems of others. Technologies may be introduced for a wide range of reasons. Researchers and technology developers often innovate to solve their understanding of the farmer's problems or to achieve their notions of public goods. Governments and non-governmental civil society organizations may advocate, subsidize and promote new technologies to achieve public goods that may or may not have farm-level advantages. Manufacturers and retailers usually introduce new technology to increase their profits. Farmers and agribusinesses use many sources of information to identify those technologies for the definitive test that is performance in on-farm use. Those sources include research results, the farm press, social media, participation in field days and farm shows, discussion with friends, family and neighbours, and government and non-governmental extension programs. It is the professional responsibility of agricultural economists, rural sociologists and other social scientists to provide information to help farmers, agribusiness and those who advise them to sort through the flood of new technology.

Agricultural work is often perceived as physically challenging drudgery. Consequently, 'automation' has been a goal from the earliest days of agriculture. Tools and machines were developed to make that work easier and more effective. The first steps of that process were manual (e.g. hoes, shovels and rakes). Subsequent steps included machines (e.g. ploughs, seeders and harrows) pulled by traction animals (e.g. horses, cattle, donkeys and camels) and more complex mechanical equipment powered by internal combustion engines (e.g. tractors and combined harvesters). In the future, mechanical equipment might be powered by electricity generated from wind and solar installations, hydrogen, methane or other renewable power sources.

Technology choice is a long-term interest of many of those who have studied the economics of agriculture at least as far back as the Physiocrats in Eighteenth Century France who compared the benefits of oxen and horses for tillage (Neill, 1948). There is a rich research literature analysing agricultural technology choices (e.g. Cochrane, 1958; Feder and Umali, 1993; Feder et al., 1985; Lee, 2005; Doss, 2006). Microeconomic theory has argued that the choice of production technology in any sector should be utility-maximizing (e.g. Henderson and Quandt, 1958), which farm management experts have agreed (e.g. Boehlje and Eidman, 1984; Kay et al., 2020). Choosing technology to maximize profit is usually the easiest analysis to implement, but utility theory indicates that there are many other factors (e.g. the value of leisure time, risk, capital and other resource constraints and transaction costs). While most new technology must at least cover costs to be widely adopted, in some cases those other sources of utility are more important than profit maximization in the choice among technologies that cover costs.

An important aspect of the study of agricultural technology has been documenting the patterns of technology adoption. One of the first studies of agricultural technology adoption was the work on hybrid maize in the USA by Grilliches (1957). Subsequently, there were adoption studies of tractors (e.g. Clarke, 1991), conservation tillage (e.g. Nelson, 1997) and many other technologies. The adoption of precision agriculture technology is relatively well documented. Lowenberg-DeBoer and Erickson (2019) review data on the adoption of precision agriculture (PA) technologies worldwide and show that some, such as Global Navigation Satellite Systems (GNSS), have been among the most rapidly adopted agricultural technologies in history, while others, such as variable rate technology for fertilizer, have lagged. Lowenberg-DeBoer (2019a) hypothesizes that economic benefits have been a good predictor of long-run adoption of PA technology, but the short-run adoption patterns are more difficult to predict because they depend on a multitude of factors including the education level of farmers, credit availability, marketing of the technology and social pressure in rural communities. Tey and Brindal (2022) did a meta-analysis of PA adoption studies and showed that economic benefits are the most reliable predictor of PA technology adoption.

The information gap addressed by this chapter is the lack of a good overview of the economics of agricultural robotics agriculture. Research and commercialization of agricultural robots has advanced rapidly since Lowenberg-DeBoer et al. (2019a) did their review of the economics of crop robots. Milking robots have been commercially available since 1992 and have been widely adopted in Northern Europe and other parts of the industrialized world, followed by robots for animal feeding, manure handling and intensive livestock tasks. Information on the economics of livestock robotics is scattered in numerous studies. The general objective is to describe the economic potential for the widespread adoption of agricultural robotics in agriculture worldwide including low- and middle-income countries. While robots can be used to collect information for advancing precision agriculture, the focus of this chapter is on the use of robots to accomplish physical farming tasks (e.g. seeding, pesticide application, fertilizer spreading and harvest). The specific objectives are as follows:

- Review the adoption history of robotic technology in agriculture;
- Summarize the benefits expected from agricultural robotics;
- Discuss the implications of robotics for the agricultural sector, especially in terms of farm structure, farmer skills and agricultural institutions;
- Identify the likely impact of robots in agriculture on the distribution of income and rural standards of living; and
- Summarize policy, regulation and institutional issues related to the adoption of robots in agriculture.

2 History of adoption of agricultural robotics

Multiyear use of a new technology is the best indicator that at least some farmers and agribusinesses have found the new technology to be beneficial. Even though the history of agricultural robotics is short, it can provide some insights into the economics and potential adoption patterns for the future. This section will provide a brief overview of the adoption of agricultural robotics by farmers and agribusinesses.

The context of the adoption of agricultural robotics is a long history of innovation by farmers, blacksmiths, engineers and scientists with the goal of producing more food with less human effort (Diamond, 1998; Smith and Marx, 1998; Tudge, 1999). In most cases, this innovation historically was driven by two forces: (1) the earth's resources are fixed, but the growing human population required more food and (2) developments in other human activities provided ideas and innovations that could be adapted for agricultural use. For example, the large workhorses that were a primary source of agricultural power before motorized mechanization were originally bred in the late Middle Ages for military use. With the development of crossbows and guns, knights needed heavier armour and consequently stronger horses were needed to carry that heavier armour. Only later did farmers realize that those large horses also enabled them to do more work in a day than the oxen, smaller horses and ponies that were previously used. The growth of manufacturing in the nineteenth century increased the demand for labour, led to the growth in wage rates and, as workers migrated to the cities, it made it hard to find agricultural workers. The steam and internal combustion engines developed for industrial purposes were adapted to make the remaining few better-paid farm workers more productive. Agricultural robotics is largely built on technologies developed for space and military use. For example, the ideas for GNSS grew out of the American space program and were developed by the US military before being released for civilian use. Similarly, UAVs and satellite remote sensing were first developed for military use. The motivation for adapting these technologies for agricultural use is familiar from earlier waves of agricultural innovation. Farmers needed to produce more food with fewer workers and fewer resources. But with agricultural robotics, the environmental motivation has become more urgent. Using fewer resources in production is not just cost savings but also a reduced environmental burden on the planet.

Table 1 lists selected milestones in agricultural robotics. The dates, countries and technologies listed are intended to be indicative of the general adoption patterns but will be discussed for years by technology historians. No technology springs fully developed from the laboratory or design studio to the farm. It is an iterative process with basic research opening new opportunities for technology development, applied research to show the potential for application of this new science, technology development that converts scientific ideas into usable commercial products and entrepreneurship that takes those potentially commercial technologies from the factory to the farm. Sometimes each step takes years, and there are many false starts along the way. In many cases, there are parallel developments in different countries and by several companies or research organizations. The list in Table 1 has attempted to list the first mover for each technology, but dating technology introduction is not always simple. It is not always clear when a technology moves from being a scientific discovery, to a prototype, to the beta test stage and from there to being a standardized commercial product.

2.1 Robotics in livestock production

The most common robotic technology in agriculture is milking robots, which allow cows to be milked without direct human involvement. Traditionally,

Table 1 Selected milestones in agricultural robotics

Year	Technology or activity	Company or Organization	Country	Reference
1983	Executive order that allowed civilian use of the Global Position System (GPS)	US government	USA	Brustein, 2014 Rip and Hasik, 2002
1983	UAV fertilizer and pesticide application	Yamaha	Japan	Sheetz, 2018
1992	Milking robot	Lely	Netherlands	Lely, 2022 Sharipov et al., 2021
2011	Weeding robot	Ecorobotix Naïo Technologies	Switzerland France	Ecorobotix, 2022 Naïo, 2022
2017	First fully autonomous field crop production	Harper Adams University	UK	Hands Free Hectare, 2018
2018	Autonomous chaser bin	Smart Ag	USA	Smart Ag, 2018
2022	Autonomous large scale tractors	John Deere	USA	John Deere, 2022

milking was done by hand. The conventional machine milking uses a vacuum technology that mimics a calf sucking but still requires a human operator to place the milk machine on the cow (or other dairy animal) and remove it when milking is completed. Because the udder shape differs slightly from animal to animal, attaching the milk machine and removing it could not be a simple mechanical process. Electronic tags (EID) allow a milking robot to access a database of udder coordinates for specific cows and place the milking equipment correctly (Knight, 2020). The adoption of milking robots is around 30% of dairy farms in Iceland and Sweden and more than 20% in countries, such as Belgium and the Netherlands. Adoption has been low in other major dairy countries, such as Canada and the United Kingdom (7%), the United States (3%) and Australia and New Zealand (less than 1%) (Eastwood and Renwick, 2020). Milking robots are on their way to becoming a widespread practice in industrialized countries for medium and larger dairy herds (i.e. >50 cows), but the transition will take time because of the infrastructure replacement cycle for milking facilities. For example, in France, only 10% of dairy farms used robot milking in 2018, but 70% expected to install robots when they replaced milking facilities (Lachia, 2018). Many of the milking systems are linked to automated feeding of varying amounts of concentrates to cows based on milk production (Ordolff, 2001). Other digital automated technologies in livestock agriculture include poultry feeding systems based on bird weight, egg counting and computerized control of ventilation based on temperature and humidity (Banhazi et al., 2012).

2.2 Autonomous crop machines

For decades, universities and research institutes have had prototype autonomous crop machines that were demonstrated on parking lots and football pitches. A few were even evaluated in the field for specific crop operations (Lowenberg-DeBoer et al., 2019a). Hands Free Hectare in 2017 marked a turning point because it was the first public demonstration of using autonomous crop machines throughout the growing season to produce and harvest commercial crops (Hands Free Hectare, 2018). In the last five years, major manufacturers of farm equipment have announced their autonomous machines (Table 1), and there are over 40 start-up companies around the world focused on developing commercial autonomous crop machines.

Because autonomous crop machines started to be commercialized very recently, data on their use is very limited. Weeding robots are being trialled all over Europe, but only in France has the approximate number of robots been made public. Lachia et al. (2019) estimated that there were 150 weeding robots used in 2018 in French agriculture, mainly for weeding organic vegetables and sugar beets. Similarly, in North America, various autonomous crop machines are starting to be commercialized, but quantitative estimates are rare. Erickson and Lowenberg-DeBoer et al. (2021) estimate that 4% of US agricultural input dealerships use robots for crop scouting services and 2% use them for providing weeding services. However, those dealers expect substantial growth by 2024 with 18% expecting to offer robotic crop scouting and 13% expecting to offer robotic weeding by that time. Crop scouting robots are used to gather very detailed information on plant conditions (e.g. weed infestation, insect populations, disease symptoms and nutrient deficiencies). They can be used in combination with remote sensing. The satellite or UAV images provide a general perspective. Robots are sometimes programmed to collect detailed data on anomalies (e.g. areas where crop growth is lagging) identified via remote sensing.

2.3 Uncrewed aerial vehicles

Uncrewed aerial vehicles (UAVs) are also called 'drones'. They are 'robots in the sky'. Like ground-based autonomous machines, UAVs have been a popular topic for agricultural researchers and in the farm media for the last few years. Most UAVs are used for information gathering and that information can increase input use efficiency, but they can also be used to accomplish physical tasks, in particular, to automate input application. In most cases, UAV input application is like map-based variable rate technology. Information gathering is a separate activity. The application map is created by a human operator. The UAV only delivers the input to the site. UAVs are especially useful for spot spraying pesticides or localized fertilizer application. Many industrialized countries regulate UAVs tightly because of concerns about spray drift and possible negative interactions with civil or military aviation. Consequently, UAV input application is often banned or highly regulated. For example, in the UK, UAV spraying herbicides is currently allowed only for applying herbicide to inaccessible locations under limited conditions. Switzerland has led Europe in allowing some more flexible testing of UAVs for input application (Lowenberg-DeBoer et al., 2021). The 2021 CropLife survey shows that 14% of US ag retailers provided UAV input application services that year. By 2024, 29% of those ag input dealers expect to offer UAV input application services (Erickson and Lowenberg-DeBoer, 2021). Anecdotal accounts indicate that UAV input application is quite common in some low- and middle-income countries such as China and Brazil. Kendall et al. (2022) provide survey data from the Hebei and Shandong regions in the North China Plain, which indicates that the only precision agriculture technology used by a substantial number of farmers in that area is UAV spraying. Many technical challenges remain with UAV spraying, especially pesticide drift (Carvalho et al., 2020; Wang et al., 2021).

While data is sparse, the adoption of agricultural robotics in middle-income countries with substantial mechanized agriculture sectors seems to follow the same pattern as adoption in higher-income countries. A few milking robots are being used in middle-income countries. The use of autonomous crop machines is just starting. One exception is the use of UAVs for input application. Because regulation of UAVs is less rigid in some middle-income countries (e.g. China and Brazil), there are indications that in those countries, there is a growing business in UAV spot spraying and site-specific seeding.

The adoption of agricultural robotics in non-mechanized agriculture anywhere in the world is negligible. This non-adoption is largely due to the fact that robotic technology from mechanized agriculture does not transfer easily to non-mechanized farms, and research to adapt the technology for smallholder farms is almost non-existent. No agricultural robotics have been developed and commercialized with the non-mechanized smallholder farmer in mind.

3 Expected benefits of agricultural robotics

Consideration of the expected benefits of agricultural robotics almost always starts with labour costs and labour availability but often quickly moves on to the benefits of greater precision in application, individualized management of animals and plants, more data which can be analysed to fine tune decision, selective harvest and other benefits not related to labour. In this section, an economic analysis of benefits will be presented for two technologies which have attracted the attention of economic researchers: robotic milking and autonomous crop machines.

3.1 Milking robots

Evidence of the monetary benefits of milking robots is mixed. Economic benefits can result from labour savings, up to around 18–30% in some studies, but around 10% on average (see Hansen, 2015), and increased milk production, perhaps of 10–15% per cow (Steeneveld et al., 2012; Hansen, 2015; Drach et al., 2017). Steeneveld et al. (2012), for example, quantified the capital cost of automated milking at €12.71 per 100 kg of milk instead of €10.10 per 100 kg of milk for conventional milking machine systems. However, Steeneveld et al. (2012) also found little difference between the economic performance of robotic milking and conventional systems.

In the long run, the data collected my milking robot systems may have a bigger impact on dairy farms than labour saving. That data can help better match feed and other cow management with individual animal requirements. It may also aid in the early detection of health issues. While labour required to operate robotic milking systems is minimal, human time and effort is needed to interpret the vast amounts of data collected in robotic milking systems. Farmers, as well as workers, can find themselves doing different work rather than less work (Bear and Holloway, 2019; Rose and Chilvers, 2018), and the stress of dealing with the vast quantity of data could negatively impact mental health (Hansen, 2015). The animal welfare implications caused by the changing relationship between stockman and cow (less contact between humans and animals) have also been explored (Butler and Holloway, 2015; Driessen and Heutinck, 2016); Holloway and Bear, 2019), although it is noted that data from robotic milking systems can be used to identify health and welfare issues with stock. The introduction of robotic milking has also been associated with the restructuring of national dairy systems with the total number of farms reduced and the remaining farms getting larger (Tse et al., 2017; Vik et al., 2019). Regardless of the relative efficacy of robotic milking versus conventional systems, the experience of changing farm workflows and structures after implementation provides a precedent for identifying some of the social, ethical and legal implications of robotic systems in arable farming.

The conclusion that emerges from the research is that while the profitability of robotic milking systems varies from farm to farm, overall, they are about a breakeven compared to conventional milking machines, which require a human to attach and remove the milker, but farmers adopted the systems for the more flexible work schedule and quality of life benefits (Straete et al., 2017; Bergman and Rabinowicz, 2013; Castro et al., 2015; Hansen, 2015). Dairy farmers particularly appreciated the ability to spend more time with family and in community activities when milking robots were used. It should be noted that until very recently, most milking robots were installed on medium-sized family run dairy farms (e.g. 100 to 300 cows). Often milking robots were installed as part of an intergenerational transfer on the farm. The younger generation was interested in dairy farming but not eager to take on milking cows two or three times per day for the rest of their working life. There are more recent anecdotal accounts of large 1000+ cow dairies installing robotic milking because of concerns about hired labour availability. The decision to use robotic milking may be quite different on those larger dairy farms.

3.2 Autonomous crop machines

The assessment of the economic benefits of autonomous crop machines often starts with labour saving and then extends to improving timeliness, greater accuracy of input application, reduced soil compaction when using smaller swarm robots and other benefits. Because autonomous crop machines are just starting to be commercialized, all the publicly available economic analysis is extrapolation from research results. Lowenberg-DeBoer et al. (2019) did a systematic review of published literature on the economics of autonomous crop machines. They found 18 studies covering arable and horticultural crops, mostly partial budgeting analyses of automation of one field operation (e.g.

seeding, weeding and harvesting). All those studies found autonomous crop machines to be economically feasible in certain circumstances.

Studies implementing whole farm analysis of the economics of autonomous crop machines have started to appear in the last few years. Shockley et al. (2019) developed a farm linear program analysis based on autonomous crop machine prototypes at the University of Kentucky in the USA. They show the economic feasibility of autonomous crop machines for American maize and soybean farms and highlight the potential for profitable use of autonomous crop machines on small- and medium-sized farms

Table 2 Summary of US whole farm[a] net returns from autonomous machines for maize and soybean cropping

Study	Scenario	USD change in net return[b] with autonomous machines
Shockley et al. (2019)	Baseline with planter, sprayer, N applicator autonomous[c]	−7155
	10% cost reduction	23 774
	7% yield increase	75 156
	Both a 10% cost reduction and a 7% yield increase with autonomous	106 085
Shockley et al. (2021)	Baseline of planter, sprayer, and N applicator autonomous with 100% human supervision[c]	−2399
	100% supervision with 10% cost reduction[d]	27 346
	100% supervision with 7% yield increase[d]	83 392
	100% supervision with 10% cost reduction and 7% yield increase with autonomous4	113 137
	Baseline with planter, sprayer, and N applicator autonomous and a speed limit of 3.2 km/hr[d].	−27 896
	Speed limit with 10% cost reduction[e]	1849
	Speed limit with a 7% yield increase[e]	57 501
	Speed limit with 10% cost reduction and 7% yield increase with autonomous[e]	87 246

[a] Both Shockley et al. studies assume an 850-ha farm in western Kentucky.
[b] For Shockley et al. (2019) and Shockley et al. (2022) Net Return is defined as gross revenue from crops minus seed, chemical, fertilizer and other variable costs, and machinery ownership and operation. Change is defined in comparison with conventional equipment with human operators.
[c] Planting is assumed to be no-till, so no-tillage operations. P and K fertilizer spreading, lime application and harvest are assumed to be done by a contractor. For estimates of increased net return, assume the autonomous retrofit cost from Shockley et al. (2022) (i.e. $13 148).
[d] The 100% human supervision scenario is to test the impact of legal requirements for 100% time human in-field supervision similar to California requirements.
[e] The 3.2 km/hr speed limit is to test the impact of speed limits similar to California requirements.

(Table 2). The analysis indicates that prototype autonomous machines for seeding, spraying and nitrogen fertilizer application are profitable if swarm robotics reduces cost by greater accuracy of input application or by reducing soil compaction.

Lowenberg-DeBoer et al. (2021) used the Hand Free Hectare (HFH) experience at Harper Adams University to estimate parameters for a linear programming analysis of autonomous crop equipment for arable farming in the UK West Midlands. They estimated cost curves (Fig. 1) that show autonomous equipment has the potential to reduce wheat production costs by US$18–US$37 per ton depending on farm size and flatten the cost curve by reducing the economies of size so that smaller-scale farms can come closer to the minimum costs of production. Estimated wheat production costs are reduced on all farm sizes. While net returns are positive for all farm sizes in the study, the return to operator labour, management and risk taking on the smallest UK farm considered is reduced by using the autonomous equipment because the reduction in hired labour cost does not cover the retrofit cost (Table 3). Operators of the smallest farm would be better off with autonomy if they had opportunities to use their time in other enterprises or non-farm activities. For the slightly larger farms, ROLMRT estimates are increased by autonomy. The analysis also shows that autonomy can cut equipment investment by more than half on larger farms by using smaller, low-cost equipment more intensively.

Lowenberg-DeBoer (2019b) used the HFH LP model to ask if autonomous grain carts were introduced into the UK, which farmers should be interested?

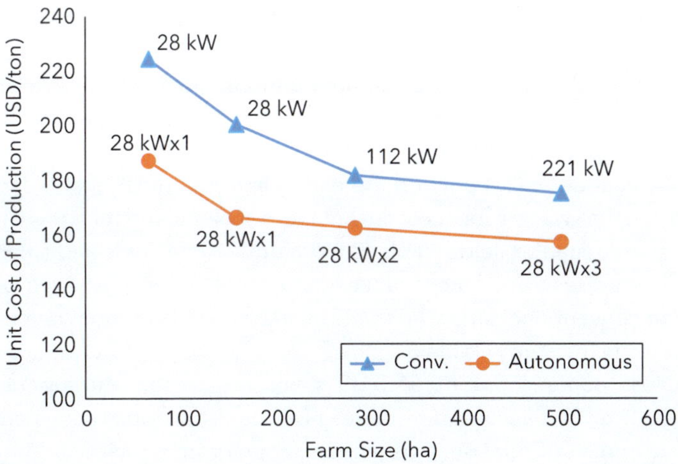

Figure 1 Estimated wheat cost of production in Britain with conventional equipment and retrofitted autonomous machines. The label at each node is the size of the tractor in the equipment set, and the number following the 'x' is the number of robot units used. Adapted from Lowenberg-DeBoer et al., 2021.

Table 3 Summary of UK whole farm economic studies of autonomous machines for wheat, barley and oilseed rape cropping

		USD[a] change in return to operator labour, management and risk taking with autonomous machines by farm size			
Study	Scenario	66 ha	159 ha	284 ha	500 ha
Lowenberg-DeBoer et al. (2021)	All crop machine operations autonomous	−4726	14061	32212	68248
Al Amin et al. (2023)	10 ha triangular fields	−4729	14525	37604	94325
	1 ha triangular fields	−3088	6740	25661	75122
Martian et al. (2023)	All Autonomous, 100% on-site supervision trouble free	−4726	-3034	14998	33883
	All Autonomous, 100% on-site supervision troublesome	−4726	−4395	7976	7212
	All Autonomous, remote supervision trouble free	−4726	13400	31031	64786
	All Autonomous, remote supervision troublesome	−5667	−16296	−13392	−24685

[a] Average exchange rate in 2018 from https://www.exchangerates.org.uk/USD-GBP-spot-exchange-rates-history-2018.html, USD/GBP = 0.7501. Change is defined in comparison with conventional equipment with human operators.

He found that with current wage rates, farmers are usually better off with human grain cart drivers if they can hire them reliably, but when labour availability is in doubt, the solution is to use the autonomous grain cart. In the UK, grain production harvest delays can lead to late seeding of subsequent winter crops and thus disrupt the entire farming system.

Al Amin et al. (2023) built on the earlier analysis for UK arable farms and showed that the swarm robot cost advantage is accentuated on farms with small and irregularly shaped fields (Table 3). With both 10 ha fields and smaller 1 ha fields, all farm sizes using autonomous equipment had lower wheat production costs than conventional farms. ROLMRT increased with autonomy on the three larger farms in the study (Table 3).

Because some countries and US states require on-site supervision of autonomous crop machines, Lowenberg-DeBoer et al. (2021) considered the economic impact of rules such as human supervision regulations. They found that in many cases when 100% time human supervision is required, the farmer is better off using conventional equipment. With current technology, if the human must be in the field, he or she can usually just as well drive the equipment.

Maritan et al. (2022) examined what human supervision of autonomous crop machines would be economically optimal if not required by law or regulation. That study shows that remote supervision (e.g. from the farm office) is optimal only if the autonomous operation is relatively trouble free (Table 3). They emphasize the need for greater robot artificial intelligence (AI) capacity so that the autonomous machine can resolve more issues without human intervention. Shockley et al. (2022) extended the earlier Kentucky analysis to consider the economic impact of autonomous machine speed restrictions. They find that for maize and soybean farms, autonomous crop machine speed restrictions like those in the US state of California (i.e. not over 3.2 kph) can make autonomous crop machines unprofitable (Table 2).

All of the crop robot economic studies cited have assumed that farmers own the robots as they have traditionally owned most conventional farm equipment. That assumption is used in preliminary studies because the economics of ownership are relatively simple; the farmer makes an initial investment and depreciates the robot's cost over several years. Robot rental, contracting and 'farming-as-a-service' approaches have potential advantages, but estimating the cost to farmer requires assumptions about the rate of return required by the company providing the service, market size, capacity utilization, the ownership of intellectual property developed by the robot AI and other business specifics.

Experience with the adoption of precision agriculture and other agricultural technology, and the economic analyses of crop robotics suggest that the adoption path may differ depending on the type of farming and the agricultural landscape. The economic case for autonomy is most compelling for horticulture because of the high-value crops and difficulties in hiring the seasonal manual labour required. Robots will probably be adopted quickly for seeding and weeding annual vegetables, but robotic harvesting of relatively delicate fruits and vegetables entails engineering challenges that are not easily solved. For arable crops, there will probably be two adoption pathways: (1) in those areas with large rectangular fields where motorized mechanization was very successful (e.g. US Midwest and Great Plains, Canadian Prairies, Brazilian Cerrados, Argentine Pampas and Australia), large-scale co-robotic smart equipment with some continuing on-site human operators is likely, (2) those areas with small, irregularly shaped fields (e.g. most of western Europe, eastern USA and much of Asia) where motorized mechanization was less successful are likely to choose small swarm robots, which, in some cases, will work without on-site human supervision. The ability of swarm robots to work efficiently in small, irregularly shaped fields gives them a major economic advantage in that topography, which may help farmers overcome the cost of transitioning to a very different way of organizing field work. The transition to swarm robotics would be facilitated if small, adaptable, low-cost robots could be mass produced.

There are several reasons why large-scale co-robotic smart equipment is likely to be used in areas with large, rectangular fields. A key reason is that swarm robotics reduce but do not eliminate economies of size. The cost of the hardware and software needed for autonomy is almost the same for any size of equipment. With large-scale equipment, some human operators are likely to remain in the field in a co-robotic arrangement because of the safety risks posed if large autonomous equipment malfunctions. Lastly, motorized mechanization in these areas is very successful, and it would probably be easier for most farmers to transition to large-scale co-robotic equipment than to switch to swarm robotics. The autonomous chaser bin (i.e. grain cart pulled by an autonomous tractor) is a good example of this type of co-robotic technology. A human operator in the combine harvester signals to the driverless autonomous chaser bin when unloading grain is required. That human operator is in the field and can stop the autonomous chaser bin if there is a malfunction.

Farm labour is increasingly difficult to hire almost everywhere in the world, including in developing countries. Rural young people in the developing world often see opportunities in the cities leaving farm work to children and their elders. Development of small, low-cost autonomous crop machines for use in small- and medium-scale farms has been proposed as part of the solution (e.g. Tarannum et al., 2015; Reddy et al., 2016; Valle and Kienzle, 2020; Al-Amin and Lowenberg-DeBoer, 2021), but unfortunately, no publicly available economic analysis has yet been done on the use of autonomous crop machines in the developing world. Setting aside for the moment the engineering challenges of developing low-cost autonomous crop machines and the entrepreneurial challenges of supplying them to farmers sustainably, the emerging literature on the economics of autonomous crop machines highlights some aspects of swarm crop robotics that would be of interest to developing country farmers including the following:

- Reducing human labour required for crop production with a modest equipment investment;
- Flattening the cost curve (Fig. 1) and reducing economies of size so that minimum production costs can be achieved at a smaller farm size than with conventional mechanization; and
- Ability to farm small irregularly shaped fields cost-effectively. This avoids the need to reshape rural landscapes and disrupt communities to create large rectangular fields on which conventional mechanization is most efficient.

4 Potential agricultural robotics in low- and middle-income countries

Most of the agricultural robotics adoption and economic analyses summarized so far in this report have been for mechanized agriculture in industrialized or middle-income countries, but to achieve the food security and environmental goals noted in the introduction, the use of agricultural robotics must reach smallholders in low- and middle-income countries, especially those on non-mechanized farms.

4.1 UAVs for smallholder farms

UAV services are being used by non-mechanized farmers in Asia and Africa, but the number of farmers and the area managed with UAVs is not well documented. Low-cost donor subsidized or venture capital-funded drone spraying is sometimes offered in developing countries, but at the current time, commercial services that cover all costs are often out of the range of small farmers (e.g. Njagi, 2019; Chikasha and Chipadza, 2021). In South Africa, there are even some doubts about whether larger mechanized farmers can afford UAV spraying (Daniel, 2021). It is possible that re-engineering to reduce costs, mass production and innovative business models can make these technologies more affordable for non-mechanized farmers.

There have been research interests and some business start-ups focused on supplying UAV spraying services on smallholder farms in Africa (e.g. Ayamga et al., 2021; Yawson and Frimpong-Wiafe, 2018). Unfortunately, robust data are not available on how widespread that practice is in the developing world.

For farms of any size, the advantages of UAV input application include targeting specific areas instead of spraying whole fields, application to fields too wet for equipment, application to inaccessible remote, steep areas and application in standing crops without damage to crops from equipment movement. For small holder farmers who would otherwise make pesticide application with a backpack sprayer, the use of a UAV potentially reduces pesticide exposure. However, there are many challenges to overcome in UAV input application including systems to refill spray tanks, fertilizer bins or seed hoppers, battery life, pesticide regulation label rates for spot application, training of users and drift to non-target areas (Carvalho et al., 2020). The profitability of UAV input application depends on the cost of the spraying service, effectiveness of the application given drift, input savings due to spot application, and improved yields because of reduced damage from the ground-based machines compared to alternatives (e.g. backpack sprayer, tractor-mounted or towed sprayer, fertilizer spreader or seeder, and application with crewed aircraft). Because smallholder farmers are unlikely to own UAVs, the cost of the UAV application service is crucial.

4.2 A vision for low-cost autonomous crop machines

For many researchers, research funders, entrepreneurs, politicians and venture capitalists, autonomous crop machines for smallholder farmers are an oxymoron. One of the main barriers to innovation in robotics for smallholders is cost. Autonomous machines are perceived as expensive. The development and spread of mobile phones provide a counter example. The first mobile phones introduced in the 1970s were heavy, clumsy and cost about US$3000. Many people imagined that they would always be a toy for the ultra rich. Now mobile phones are sold in shops and markets throughout the developing world, often for less than US$20. Technology improvement and high-volume manufacturing made the mobile phone much less expensive, and the prepaid business model fits the budgets and cashflow of many in the developing world. Mobile phones paved the way for the introduction of smart phones, which are increasingly used for PA apps. The combination of technology change, mass production and innovative business models could do the same for autonomous crop machines.

To develop practical agricultural tools that could achieve worldwide adoption, scientists, engineers and technology developers usually need a vision for the technology and design criteria. One vision is of a small wheeled autonomous crop machine with AI that could learn to seed, weed and harvest for the price of a motorbike. Some smallholder farm families in low-income countries have motorbikes, so that is a useful price point (US$500–US$1000) for an autonomous crop machine that could be widely adopted. While a leg robot might be useful in fields (i.e. it could step over obstacles), with current technology, leg robots usually cost several multiples of a wheeled robot of the same size. The ability of the autonomous crop machine to learn using AI would make mass production possible. Producing specialized robots for each crop and agro-ecology would be a high-cost low-volume business. A plausible business model is that a manufacturer delivers a generic autonomous machine which is taught what it needs to do (perhaps by working alongside a human). Appropriate tools for the autonomous machines would be adapted to the task. Some of those tools might be locally manufactured. The autonomous machine would be GNSS enabled to allow it to create maps (e.g. soil colour, soil strength based on the force required for hoeing and the yield from the plant-by-plant harvest). There are several possible energy sources for the autonomous machines (e.g. fuel cells, solar electricity and methane). To make use of the autonomous machines more affordable, especially initially when it is unfamiliar, rental or fee-for-service farm work might be implemented.

With the generic autonomous crop machine, many other types of digital automation become possible. For example, with a crop sensor, the autonomous machine might determine the fertilizer needs of individual plants and incorporate the required fertilizer in the soil at the base of each plant. This would be essentially what human farmers now do when microdosing fertilizer (Aune et al., 2017). To add soil capacity or yield goal information to this AI fertilizer decision process, the autonomous machine might use previously recorded soil, plant and yield maps. With robust and inexpensive sensors, the autonomous machine might also determine the presence of insects or plant diseases and apply insecticides or fungicides as needed. Weeds could be controlled mechanically or with targeted herbicide applications.

This vision of robotics for smallholder farmers represents an enormous engineering and entrepreneurial challenge, but it can be envisioned with current technology and may be facilitated by innovations. The millions of small farmers in the developing world should be an enticing market if someone could show businesses that there is a feasible technology and that there is a market. This would be a classic mass market business strategy as outlined by Prahalad (2004) in the book *Fortune at the Bottom of the Pyramid*. In agriculture, it would be similar to research, technology development and entrepreneurship that spread of hermetic grain storage through Africa and South Asia (Noughoheflin et al., 2017). Before the Purdue Improved Crop Storage (PICS) bag, manufacturers were reluctant to invest in grain storage innovations for small holder farmers because of the perceived lack of buying power. After PICS sold millions of bags in 30+ countries, there are many imitators and competitors.

5 Broader implications of agricultural robotics for the farm sector

Agricultural technologies often have economic and social implications that extend far beyond their farm-level benefits and costs. For example, motorized mechanization of agriculture often resulted in farm size expansion and rural depopulation, with the associated decline in rural political and economic influence. A more positive example is the introduction of hybrid maize in the USA resulted in the geographic expansion of the 'Cornbelt' to areas where the growing season was too short or summer rainfall too low for open-pollinated maize (Hart, 1986; Green et al., 2018). This was possible because hybridization gave breeders greater control over the maturity, drought tolerance and other agronomic characteristics of the crop. That expansion of the geographic area where maize could be grown in turn led to the growth of maize processing and intensive livestock production in those new maize production areas. Similarly, if low-cost agricultural robots like those described in Section 4 were developed and widely commercialized, the currently available research suggests that they could have major economic and social implications including the following:

- Impact of swarm robotics on farm structure – Small swarm robots can be almost constant returns to scale. With them, small farms can achieve the minimum cost of production, and larger farms can add more autonomous units that produce at that same minimum cost level. This would reduce economies of scale in agricultural production and eliminate one of the major motivations for farm size expansion. Whether the economics of size and scope in input purchasing, marketing, finance and other farm management functions continue to drive farm size increases probably depends as much on cultural factors, legal structures, and regulatory constraints, as on the production profitability. By rapidly adopting swarm robots, areas in middle- and low-income countries now dominated by manually operated smallholder farms or slightly larger farms using animal traction may avoid the social disruption of farm size expansion and rural depopulation. By reducing drudgery, increasing profitability and enhancing the image of agriculture as a high-tech industry, swarm robotics has the potential to help rural communities retain their young people and even attract talent from elsewhere.

- Ability to farm small irregularly shaped fields efficiently with swarm robots – In industrialized countries with a legacy of medium and small farms, motorized mechanization frequently led abandonment of small irregularly shaped fields or the transition to less intensive uses, such as rural residences or hobby farming. For example, this occurred in the eastern USA in the early twentieth century. Countries which tried to maintain a small farm structure were saddled with massive farm subsidy costs. This occurred in many European countries. The introduction of swarm robots may allow commercial agriculture to reclaim some of those abandoned small, irregularly shaped fields, which, in some cases, have other economic advantages, such as good soils, reliable rainfall and proximity to markets. Small farm subsidy programs may become less costly as swarm robotics helps reestablish profitability for agriculture in small, irregularly shaped fields. In areas dominated by manual smallholder farms or slightly larger animal traction-powered farms, conventional wisdom encourages motorized mechanization to deal with labour shortages and to improve productivity. Those areas may be able to skip the motorized mechanization step and move directly to robotics, avoiding the need to reshape the rural landscape into larger fields. This may also have environmental benefits in that small, irregularly shaped fields have greater biodiversity than large rectangular fields (Batáry et al., 2017; Fahrig et al., 2015; Flick et al., 2012; Firbank et al., 2008; Lindsay et al., 2013).
- The introduction of swarm robots could radically alter the farm equipment market structure – The core customers of the current group of major farm equipment manufacturers are a relatively small number of large farms. This leads to a 'high touch' marketing and service strategy for sophisticated products via intensive interaction with that relatively small group. In contrast, the swarm robotics vision outlined in Section 4 would require mass marketing of low-cost standardized products to millions of small- and medium-sized farmers. The optimal business model for this swarm robot market is not yet determined, but as with the prepaid business model used by mobile phone companies in low- and middle-income countries, it

may be quite different from the current business model. This change in the customer base and the business model may lead to changes in the farm equipment market structure. It will create opportunities for entrepreneurs who have the technical capacity to develop low cost, reliable autonomous machines and link that technology with innovative business models.

Of course not any of these outcomes are inevitable. They depend on many factors including the exact characteristics of the technology, the legal and regulatory frameworks, business decisions by major corporations and start-up companies, social media reactions and cultural attitudes about robotics in agriculture. Innovations using AI often depend on the reliable availability of high-speed internet access. Governments and civil society can encourage positive outcomes from agricultural robotics through digital infrastructure, appropriate legal and regulatory approaches, public sector research and education.

6 Social impact

The first question asked in most public discussions of agricultural robotics relates to the perception that this technology will eliminate many rural livelihoods, leading to farming communities with a few relatively wealthy farmers and many unemployed former farm workers who would either live in poverty or move to the cities. The loss of jobs to automation may occur most prominently in fruit and vegetable production where manual methods are still widely used, even in industrialized countries. However, because many industrialized countries have depended on migrant labour for fruit and vegetable production, this is not primarily a domestic problem for them, but rather a problem for the sending country. For industrialized countries, reduction in international migrant labour could help resolve political problems created by immigration and medical/biosecurity issues linked to international movement of workers during disease outbreaks. For the sending countries, the loss of migrant farm jobs is a mixed outcome because those migrant jobs were often not good jobs. They often did not pay particularly well. They often forced people to be away from their families for long periods. They often did not have health or social benefits. However, they were jobs and did provide incomes to people who often had few other options. Agricultural robotics will create a challenge in the low- and middle-income countries that depend on remittances from migrant farm workers to create other economic opportunities.

In the large-scale commercial arable farming sector, robotics is likely to mean many workers change their responsibilities, but unlikely to mean the loss of many jobs. In that sector, the major loss of jobs already occurred with motorized mechanization and chemical weed control. With the higher productivity linked to agricultural robotics, farm workers who adapt and retrain could achieve higher incomes and better standards of living. For example, a former tractor driver may supervise a swarm of autonomous crop machines or retrain to do robot training, maintenance or repairs. With the reduction in economies of size, some former farm workers may find entrepreneurial opportunities in small and medium robotic enterprises. From a social perspective, changing jobs, retraining, and adapting to automated working and entrepreneurship can be very stressful even in growing economies with a long-term demand for labour (Charlton et al., 2022).

For small- and medium-scale arable farms, robotics could create opportunities as well as challenges. Those farms could use robotics to lower their costs of production and be more economically competitive, but even with lower costs, their farm scale may not provide an acceptable standard of living. They could use the labour saved to expand farm size, find off farm employment or add farm enterprises.

In contrast to the widespread perception of the loss of jobs due to robotics in agriculture, there is the possibility of entrepreneurial opportunities created when the availability of human labour is no longer the binding constraint for agriculture. For example, one of the main constraints to organic or biodynamic farming in industrialized countries is the cost of labour. If organic growers could rely on autonomous weeding machines to control weeds and AI to identify plant diseases and suggest biological remedies, organic production could expand rapidly. In industrialized countries, many consumers would prefer to buy organic products, but they do not want to pay a premium. With robotics, organic production could undercut the costs of conventional methods and become the standard. Similarly, robotics could revive the production of nutrient-dense heirloom crops that were difficult to mechanize. For example, when maize production was mechanized, hybrids were developed with ears all at about the same height on the stalk to facilitate harvest; in the plant breeding process for harvest efficiency and other mechanization traits, nutritional and culinary diversity was lost. Autonomous machines with AI could be developed to harvest traditional maize varieties with ears at different heights. Similarly, the mechanization of tomato harvest required varieties that would ripen evenly. In the process of developing tomatoes for mechanical harvest and the other requirements of long-distance supply chains, nutritional and flavour were lost. Selective harvest with autonomous machines could allow commercial production of flavourful heirloom varieties. It could also create opportunities for the production of botanicals with valuable aromatic or medicinal properties, even though those crops require very intense management. Some of those entrepreneurial opportunities could be generated in low- and middle-income countries.

The common science fiction plot of robots replacing human workers and destroying livelihoods is not inevitable with the introduction of agricultural robotics. If low-cost, highly effective agricultural robotics becomes as ubiquitous as mobile phones in low- and middle-income countries are now, then with the enabling digital infrastructure, legal, regulatory and cultural environment, there is the potential for sustainable rural economic development based on intensive agriculture. Whether low- and middle-income countries gain or lose depends on how they manage the transition. Countries that build the physical, economic and social infrastructure for agricultural robotics stand to benefit. Countries that ignore the challenge may lose the low-wage manual agricultural employment that they now have, but not see the development of higher-wage agricultural opportunities with robotics. History suggests that the international community can help countries prepare, but it cannot oblige them to recognize the opportunity.

7 Policy, regulatory and institutional issues

Anticipating the issues that will arise with the introduction of new technology is very difficult because the future use of those innovations and the human reactions are not completely known. In general, people develop new uses for technology that are often far different from the intent of researchers or technology developers. For example, tractors were originally invented to replace draft animals for field work, but the availability of mobile mechanical power with rubber tires, hydraulics, electronics and power-take-off led to the development of unanticipated uses (e.g. direct seeding and conservation tillage, harvesting and packaging forage on-the-go and transporting crops to market). Regulations and standards need to be developed for the safe use of agricultural robots similar to the rules and guidelines for manufacturing automation and robotics. Some specific policy, regulatory and institutional issues that have been anticipated for agricultural robotics by researchers, farm equipment company staff, entrepreneurs, farmers and civil society include the following:

- Appropriate guidance on human supervision of autonomous crop machines – The EU and the US state of California currently require on-site human supervision of autonomous crop machines in most cases. Research shows that with current technology, the requirement of 100% time on-site human supervision of autonomous machines substantially cuts into the economic benefit of their use (Lowenberg-DeBoer et al., 2021). In many cases, if the human must be in the field, they may as well drive the equipment. Discussions are on-going about what should determine the level of human supervision. Maritan et al. (2022) show that the economically optimal supervision of autonomous crop machines in the absence of regulation depends largely on the frequency of human intervention required and on the placement of the supervisor (e.g. on-site or remote). Beyond the economic issues in supervision, health and safety concerns are often expressed, especially in relatively densely populated countrysides like those in most of Europe. In response to those concerns, the British Standards Institute (BSI) has organized an effort to create and autonomous agricultural machine code of practice for the UK. Factors that might influence appropriate supervision include the following:

- Size of the autonomous machine – Small swarm robots have less potential for causing harm than some of the large autonomous machines proposed by major farm equipment manufacturers.
- Speed of the autonomous machine – The state of California requires autonomous crop equipment to travel less than 3.2 kph. Under some ISO standards, autonomous machine categories are limited to less than 0.8 kph. Shockley et al. (2022) show that applying such speed limits generally in crop farming would undercut the economic benefit of autonomous machinery.
- Population of the countryside – A malfunctioning autonomous machine is less likely to create a health and safety problem in the sparsely settled Outback of Australia than it would in relatively densely populated rural areas in Great Britain.
- Site preparation – Signage, fencing and other site preparation might be used to prevent injury or death of workers, rural residents, companion animals, livestock and wildlife.
- Community preparation – Should rural communities near farms where autonomous machines are in use be notified? Who should be notified (e.g. everyone, those who sign up for the phone or internet-based alert system)? How should they be notified? Should they be notified only if the autonomous machines are working without on-site human supervision?
- Autonomous machine benefits beyond labour saving – For example, if the use of swarm robots reduces soil compaction, increases soil health and facilitates higher yields, then a higher level of human supervision can be economically justified.

- Training required for human supervisors of robots for both crops and livestock – What should the supervisors be on the alert for? How should they report incidents of human–robot interactions? This topic occupies a major portion of the Australian Autonomous Agricultural Machine Code of Practice (GPA et al., 2021).
- Agricultural robotics often requires internet access especially if AI is used – Internet access allows easy updating of software, reduces computational capacity needs by cloud computing and facilitates access to remote sensing and other public databases. Internet access in rural areas worldwide is often sparse and expensive, and it is particularly spotty in

low- and middle-income countries. Policies to encourage the development of rural digital infrastructure could include low-interest loans for rural internet providers and the formation of communications cooperatives that offer data services and subsidies.

- Data privacy and security – Agricultural robots would collect massive amounts of data on both crop and livestock farms. Some of that data may raise privacy issues for the farmer, farm family and others. Other data may be proprietary information for the farm or company. Rules need to be clear on who owns the data, who controls it, and how it is to be handled.
- Theft prevention – In countries where rural crime is common, the theft of small robots working alone in isolated fields is a frequently mentioned concern. Should robots have RFID locator chips implanted in them like pets and some domestic livestock are required to have in some countries? How should the resale market for used robots be regulated to make selling of stolen robots difficult?
- AI – While most of the agricultural robots currently in the commercialization pipeline have very little decision-making capacity, in the longer run, AI is an essential part of what will make agricultural robots useful. AI will allow robots to deal with many of the unexpected obstacles, thereby reducing human supervision needs. AI will help identify and target pests. Machine learning is an essential part of what will make AI useful, but it is also what makes it potentially dangerous because the manufacturer and human supervisor have little control over what it learns. There is also the question of who owns the knowledge generated by machine learning (e.g. manufacturer, farmer and the contractor supplying robot services).
- Technical training and retraining – Supervising agricultural robots, maintenance and repair of the machines and working with AI to resolve agronomic and livestock production problems are not in the skill set of most people in the agricultural sector today, especially on small farms in low-income countries. What training is needed to supervise agricultural robots? Should programs for robot training, maintenance and repair be started now, so that when the technology becomes common, there is a capacity to maintain and repair it? Should crop and livestock consultants be trained to use the data collected by robots and educated on how to interact with their AI systems?
- Public sector research and education – In the last two centuries, the basic scientific knowledge responsible for many agricultural advances has been developed and collected at universities and other public sector research organizations. Agricultural higher education is based on that scientific knowledge. Most robots in agriculture will probably be privately owned, by companies or individuals, and the data collected will be proprietary. In theory, if that farm data could be collected and analysed, it could lead

to unanticipated breakthroughs in crop and livestock production, with implications for food security, human health and safety, environmental management, biodiversity and other public concerns. Under what circumstances should public sector researchers have access to the agricultural data collected by privately owned agricultural robots?

- Policies to encourage agricultural robotics where it would have public good benefits – Some aspects of agricultural robotics have public good benefits (e.g. farming small and irregularly shaped fields with higher biodiversity, reducing pesticide use, avoiding the disruption of rural landscapes and communities to create large fields for efficient operation of motorized mechanization). Some of those public goods will be generated by private decisions given the right legal and regulatory guidance, but in some cases, it may be useful to encourage agricultural robot technologies. For example, where upfront investment and retraining transition costs are substantial, public subsidies might encourage farmers to re-equipment their farms with robots, instead of acquiring or continuing with motorized mechanization. The movement of agricultural research and educational institutions to the development and use of robots could be encouraged.

8 Conclusion

This report documents that robotics has been used successfully in agriculture for several decades (e.g. robotic milking), and technology is in the pipeline to make robotics ubiquitous on farms worldwide (e.g. mobile autonomous crop equipment). The discussion of benefits of agricultural robotics usually starts with reducing labour costs and coping with labour availability bottlenecks, but quickly moves on to other benefits, including greater food security, improving the quantity and quality of food production, increasing efficiency on small- and medium-sized farms, greater accuracy of input application, reduced soil compaction with small swarm robots, allowing field operations at times and places where they are difficult manually or with mechanical technology (e.g. wet soils, steep hillsides), profitable farming for small and irregularly shaped fields and automating the collection of crop and livestock data. By rethinking and re-engineering the underlying science, many of the benefits of agricultural robotics could be made available to medium and small farms in low- and middle-income countries. For example, the development of low-cost crop robots that could learn to seed, weed and harvest would help resolve labour constraints on manually operated smallholder farms and provide a basis for sensor-based fertilizer and pesticide application. If such low-cost, reliable and effective robots were developed and widely commercialized, it could radically change the farm sector. Numerous options exist to facilitate farmer access to robotic technology. Robots might be owned by farmers, rented or provided as an Uber-like service organized on a mobile phone app. The dominance of large-scale farms using motorized mechanical technology would diminish, and medium and small farms everywhere would have a greater possibility of success. Robotics has the potential to eliminate some manual farm worker livelihoods, and this is a specific problem for countries supplying migrant agricultural workers to more developed regions, but the technology also has the potential to create higher skilled, better paid employment in rural areas (e.g. supervising robots, training, maintenance and repair) and entrepreneurial opportunities. Realization of the potential benefits of agricultural robotics would require better digital infrastructure in rural areas, an appropriate legal and regulatory framework, facilitating digital entrepreneurship, retraining workers, revising technical educational curricula, attention to data security and policies to encourage agricultural robotics where it would have public good benefits.

9 Where to look for further information

Agricultural robotics and automation are changing rapidly as new technology is introduced and new business models are developed. Updating information on the economics of agricultural robotics and automation is an ongoing task. Beyond the scientific journals and other sources cited in the references, some sources which regularly have new information on this topic include:

- Future Farming (https://www.futurefarming.com/newsletter/) – The best single source of updates on commercial introduction of agricultural automation and robotic technology is the Future Farming newsletter. It also often has reports of farmer and agribusiness experience with the technologies.
- International Society of Precision Agriculture (ISPA – https://www.ispag.org/) – Conferences under the ISPA umbrella are often where new research on the economics of agricultural automation and robotics is reported. Those conferences include the International Conference on Precision Agriculture (ICPA) held in odd-numbered years and the European Conference on Precision Agriculture held in even-numbered years. Some papers and abstracts from the ICPA and ECPA can be accessed via the ISPA website. Many papers on the economics of agricultural automation and robotics are presented at ISPA-related conferences before they are published in the journals cited in this chapter.
- Global Institute for Agri-Tech Economics (GIATE – https://www.harper-adams.ac.uk/research/giate/) – GIATE annually organizes a symposium on agri-tech economics which attracts presentations on economics of agricultural automation and robotics work in progress. Papers and abstracts

from that symposium are posted on the GIATE website. In addition, GIATE posts articles, papers and presentations on agri-tech economics prepared by Harper Adams University staff.

- International Forum for Agricultural Robotics (French acronym is FIRA - https://www.agricultural-robotics.com/fira) - FIRA is organized by the Global Organization for Agricultural Robotics (GOFAR). It is a good source of information on the agricultural robotics industry. World FIRA, organized annually in Toulouse, France, attracts robot manufacturers, agricultural businesses, producers, investors, suppliers, entrepreneurs and researchers from all over the world. A North American focused event has been organized in California in recent years. Videos, reports and other information from the FIRA conferences are posted on their website.
- Grain Producers Australia (GPA - https://www.grainproducers.com.au/codeofpractice) - GPA and their collaborators introduced the first code of practice for autonomous agriculture machines. Updates are posted on the GPA site.
- British Standards Institute (BSI - https://knowledge.bsigroup.com/products/use-of-autonomous-mobile-machinery-in-agriculture-and-horticulture-code-of-practice?version=standard) - BSI has led the world in developing standards for autonomous crop machines for agriculture in densely populated rural areas. Updates on the BSI standards are found on the website.

10 Acknowledgements

Funding from the Food and Agriculture Organization (FAO) of the United Nations in doing this study is acknowledged. The document incorporates review comments and suggestions from the 2022 State of Food and Agriculture team including FAO staff and experts.

11 References

Al-Amin, A. and Lowenberg-DeBoer, J. (2021). Agricultural robots may herald a new era to the vision of Digital Bangladesh, *Financial Express*, Dhaka, Bangladesh, 30 Oct. 2021, 4.

Al-Amin, A. K. M. A., Lowenberg-DeBoer, J., Franklin, K. and Behrendt, K. (2023). Economics of field size and shape for autonomous crop machines. In *Precision Agriculture*, Minneapolis, MN. Available at: https://doi.org/10.1007/s11119-023-10016-w.

Aune, J. B., Coulibaly, A. and Giller, K. E. (2017). Precision farming for increased land and labour productivity in semi-arid West Africa: a review, *Agronomy for Sustainable Development* 37(3). https://doi.org/10.1007/s13593-017-0424-z.

Ayamga, M., Tekinerdogan, B. and Kassahun, A. (2021). Exploring the challenges posed by regulations for the use of drones in agriculture in the African context, *Land* 10(2). https://doi.org/10.3390/land10020164.

Banhazi, T. M., Lehr, H., Black, J., Crabtree, H., Schofield, P., Tscharke, M. and Berckmans, D. (2012). Precision livestock farming: an international review of scientific and commercial aspects, *International Journal of Agricultural and Biological Engineering* 5, 1–9.

Batáry, P., Gallé, R., Riesch, F., Fisher, C., Dormann, C. F., Musshoff, O., Császár, P., Fusario, S., Gayer, C., Happe, A. K., Kurucz, K., Molnár, D., Rösch, V., Wietzke, A. and Tscharntke, T. (2017). The former Iron Curtain still drives biodiversity-profit trade-offs in German agriculture, *Nature Ecology and Evolution* 1(9), 1279–1284. https://doi.org/10.1038/s41559-017-0272-x.

Bear, C. and Holloway, L. (2019). Beyond resistance: geographies of divergent more-than-human conduct in robotic milking, *Geoforum* 104, 212–221.

Bergman, K. and Rabinowicz, E. (2013). *'Adoption of the Automatic Milking System by Swedish Milk Producers', Working*, Agri-Food Economics Center, Lund, Sweden, Paper 2013:7.

Boehlje, M. D. and Eidman, V. R. (1984). *Farm Management*, Wiley, New York.

Brustein, J. (2014). GPS as we know it happened because of Ronald Reagan, *Bloomberg News*. Available at: https://www.bloomberg.com/news/articles/2014-12-04/gps-as-we-know-it-happened-because-of-ronald-reagan, accessed 14 April 2022.

Butler, D. and Holloway, L. (2015). 'Technology and Restructuring the Social Field of Dairy Farming: Hybrid Capitals, 'Stockmanship', and automatic milking systems', *Sociologica Ruralis* 56(4), 513–530.

Carvalho, F. K., Chechetto, R. G., Mota, A. A. B. and Antuniassi, U. R. (2020). Challenges of aircraft and drone spray applications, *Outlooks on Pest Management* 31(2), 83–88.

Castro, Á., Pereira, J. M., Amiama, C. and Bueno, J. (2015). Typologies of dairy farms with automatic milking systems in Northwest Spain and Farmer's satisfaction, *Italian Journal of Animal Science* 14(2), 2 https://doi.org/10.4081/ijas.2015.3559.

Charlton, D., Hill, A. and Taylor, J. E. (2022). Automation and social impacts: winners and losers, *Background Paper for the State of Food and Agriculture 2022' FAO Agricultural Development Economics Working Paper 22-09*, Food and Agriculture Organization, Rome. https://doi.org/10.4060/cc2610en.

Chikasha, P. and Chipadza, T. (2021). Drone technology for cost-effective and environment friendlier farming in Zimbabwe. *AfricaGoesDigital*. Available at: http://www.afgoesdigital.com/drone-technology-for-sustainable-farming-in-zimbabwe/, accessed 14 April 2022.

Clarke, S. (1991). New Deal regulation and the revolution in American farm productivity: the case study of the diffusion of the tractor in the Cornbelt, 1920–1940, *Journal of Economic History* 51(1), 101–123.

Cochrane, W. (1958). *Farm Prices: Myth and Reality*, University of Minnesota Press, St. Paul, MN.

Daniel, L. (2021). Crop-spraying drones are slowly taking off in SA - and that's good news for smaller farms, *Business Insider*. Available at: https://www.businessinsider.co.za/drones-as-crop-sprayers-are-becoming-more-common-in-sa-heres-why-thats-good-news-for-small-farms-2021-2#:~:text=The%20price%20of%20these%20specialised,may%20not%20save%20much%2C%20yet, accessed 14 April 2022.

Diamond, J. (1998). *Guns, Germs and Steel: A Short History of Everybody for the Last 13,000 Years*, Vintage Books, London.

Doss, C. R. (2006). Analyzing technology adoption using microstudies: limitations, challenges, and opportunities for improvement, *Agricultural Economics* 34(3), 207–219.

Drach, U., Halachmi, I., Pnini, T., Izhaki, I. and Degani, A. (2017). Automatic herding reduces labour and increases milking frequency in robotic milking, *Biosystems Engineering* 155, 134–141.

Driessen, C. and Heutinck, L. F. M. (2015). Cows desiring to be milked? Milking robots and the co-evolution of ethics and technology on Dutch dairy farms, *Agriculture and Human Values* 32(1), 3–20.

Eastwood, C. and Renwick, A. (2020). Innovation uncertainty impacts the adoption of smarter farming approaches, *Frontiers in Sustainable Food Systems* 4, 1–14.

Ecorobotix (2022). A bit of history. Available at: https://ecorobotix.com/en/a-bit-of-history/, accessed 14 April, 2022.

Erickson, B. and Lowenberg-DeBoer, J. (2021). Survey confirms a data driven market for retailers, *CropLife*, accessed July, 2021. Available at: https://www.croplife.com/precision/2021-precision-agriculture-dealership-survey-confirms-a-data-driven-market-for-retailers/, accessed 14 April 2022 (vol. 2021).

Fahrig, L., Girard, J., Duro, D., Pasher, J., Smith, A., Javorek, S., King, D., Lindsay, K. F., Mitchell, S. and Tischendorf, L. (2015). Farmlands with smaller crop fields have higher within-field biodiversity, *Agriculture, Ecosystems and Environment* 200, 219–234. https://doi.org/10.1016/j.agee.2014.11.018.

Feder, G. and Umali, D. L. (1993). The adoption of agricultural innovations, *Technological Forecasting and Social Change* 43(3–4), 215–239.

Feder, G., Just, R. E. and Zilberman, D. (1985). Adoption of agricultural innovations in developing-countries: A survey, *Economic Development and Cultural Change* 33(2), 255–298.

Firbank, L. G., Petit, S., Smart, S., Blain, A. and Fuller, R. J. (2008). Assessing the impacts of agricultural intensification on biodiversity: A British perspective, *Philosophical Transactions of the Royal Society of London. Series B, Biological Sciences* 363(1492), 777–787. https://doi.org/10.1098/rstb.2007.2183.

Flick, T., Feagan, S. and Fahrig, L. (2012). Effects of landscape structure on butterfly species richness and abundance in agricultural landscapes in eastern Ontario, Canada, *Agriculture, Ecosystems and Environment* 156, 123–133. https://doi.org/10.1016/j.agee.2012.05.006.

Grain Producers Australia (GPA), Society of Precision Agriculture Australia (SPAA) and Tractor and Machinery Association (TMA) (2021). *Code of Practice: Agricultural Mobile Field Machinery with Autonomous Functions in Australia*. ISBN: 978-0-646-83727-7. Available at: https://www.grainproducers.com.au/codeofpractice, accessed 14 April 2022.

Green, T. R., Kipka, H., David, O. and McMaster, G. S. (2018). Where is the USA Corn Belt, and how is it changing?, *Science of the Total Environment* 618, 1613–1618. https://doi.org/10.1016/j.scitotenv.2017.09.325.

Griliches, Z. (1957). Hybrid corn: an exploration in the economics of technological change, *Econometrica* 25(4), 501–522. https://doi.org/10.2307/1905380.

Hands Free Hectare (2018), *Timeline*, Harper Adams University. Available at: http://www.handsfreehectare.com/timeline.html, accessed 14 April 2022.

Hansen, B. G. (2015). Robotic milking-farmer experience and adoption rate in Jaeren, Norway, *Journal of Rural Studies* 41, 109–117.

Hart, J. F. (1986). Change in the Cornbelt, *Geographical Review* 76(1), 51–72.

Henderson, J. and Quandt, R. (1958). *Microeconomic Theory: A Mathematical Approach*, McGraw-Hill Book Company, New York.

John Deere (2022). John Deere reveals fully autonomous tractor at CES 2022. Available at: https://www.deere.com/en/news/all-news/autonomous-tractor-reveal/, accessed 14 April 2022.

Kay, R., Edwards, W. and Duffy, P. (2020). *Farm Management* (9th edn.), McGraw-Hill Education, New York.

Kendall, H., Clark, B., Li, W., Jin, S., Jones, G. D., Chen, J., Taylor, J., Li, Z. and Frewer, L. J. (2022). Precision agriculture technology adoption: a qualitative study of small-scale commercial "family farms" located in the North China Plain, *Precision Agriculture* 23(1), 319–351. https://doi.org/10.1007/s11119-021-09839-2.

Knight, C. H. (2020). Review: sensor techniques in ruminants: more than fitness trackers, *Animal* 14(S1) (Suppl.1), s187–s195. https://doi.org/10.1017/S1751731119003276.

Lachia, N. (2018). Usage des robots en agriculture, *Observatoire des Usages de l'Agriculture Numérique*. Available at: https://agrotic.org/observatoire/2018/10/09/usage-des-robots-en-agriculture/, accessed 12 April 2022.

Lachia, N., Pichon, L. and Tisseyre, B. (2019). A collective framework to assess the adoption of precision agriculture in France: description and preliminary results after two years, in Stafford, J. V. (Ed.) Precision Agriculture '19, Proceedings of the 12th European Conference on Precision Agriculture, Wageningen Academic Publishers, Wageningen, Netherlands, 851–857.

Lee, D. R. (2005). Agricultural sustainability and technology adoption: issues and policies for developing countries, *American Journal of Agricultural Economics* 87(5), 1325–1334.

Lely (2022). Our history. Available at: https://www.lely.com/gb/about-lely/our-company/history/, accessed 14 April 2022.

Lindsay, K. E., Kirk, D. A., Bergin, T. M., Best, L. B., Sifneos, J. C. and Smith, J. (2013). Farmland heterogeneity benefits birds in American Mid-West watersheds, *American Midland Naturalist* 170(1), 121–143. https://doi.org/10.1674/0003-0031-170.1.121.

Lowenberg-DeBoer, J. (2019a). Economics of precision agriculture, in Stafford, J. (Ed.). *Precision Agriculture for Sustainability*, Burleigh Dodds Science Publishing, Cambridge, UK, http://doi.org/10.19103/AS.2017.0032.19.

Lowenberg-DeBoer, J. (2019b). *Making Technology Pay on Your Farm*, Future Farm Technology Expo, Birmingham, UK.

Lowenberg-DeBoer, J., Franklin, K., Behrendt, K. and Godwin, R. (2021). Economics of autonomous equipment for arable farms, *Precision Agriculture* 22(6), 1992–2006. https://doi.org/10.1007/s11119-021-09822-x.

Lowenberg-DeBoer, J., Huang, I. Y., Grigoriadis, V. and Blackmore, S. (2019). Economics of robots and automation in field crop production, *Precision Agriculture*, 1–22. https://doi.org/10.1007/s11119-019-09667-5.

Lowenberg-DeBoer, J., Behrendt, K., Ehlers, M.-H., Dillon, C., Gabriel, A., Huang, I. Y., Kumwenda, I., Mark, T., Meyer-Aurich, A., Milics, G., Olgunju, K. O., Pedersen, S. M., Shockley, J. and Rose, D. (2022). Lessons to be learned in adoption of autonomous equipment for field crops, *Applied Economic Perspectives and Policy* 44(2), 848–864, https://doi.org/10.1002/aepp.13177.

Lowenberg-DeBoer, J. and Erickson, B. (2019). Setting the record straight on precision agriculture adoption, *Agronomy Journal* 111(4), 1552–1569. https://doi.org/10.2134/agronj2018.12.0779.

Maritan, E., Lowenberg-DeBoer, J., Behrendt, K. and Franklin, K. (2023). Economically optimal farmer supervision of crop robots, *Smart Agricultural Technology* 3. https://doi.org/10.1016/j.atech.2022.100110.

Njagi, K. (2019). Ghana's farmers are using drones to boost crop yields - here's how, World Economic Forum. Available at: https://www.weforum.org/agenda/2019/10/ghanas-farmers-drones-boost-crop-yields, accessed 14 April 2022.

Naïo Technologies (n.d.). Naïo technologies agricultural robotics pioneers. Available at: https://www.naio-technologies.com/en/naio-technologies/#:~:text=Founded%20in%202011%2C%20Na%C3%AFo%20Technologies,use%20of%20chemical%20weed%20killers, accessed 14 April 2022.

Neill, T. P. (1948). Quesnay and physiocracy, *Journal of the History of Ideas* 9(2), 153–173. https://doi.org/10.2307/2707226.

Nelson, P. J. (1997). To hold the land: soil erosion, agricultural scientists and the development of conservation tillage techniques, *Agricultural History* 59(Winter,), 71–90.

Noughoheflin, T., Coulibaly, J. Y., D'Alessandro, S., Aitchedji, C. C., Damisa, M., Baributsa, D. and Lowenberg-DeBoer, J. (2017). Management lessons learned in supply chain development: the experience of PICS bags in West and Central Africa, *International Food and Agribusiness Management Review* 20(3), 427–438.

Ordolff, D. (2001). Introduction of electronics into milking technology, *Computers and Electronics in Agriculture* 30(1–3), 125–149. https://doi.org/10.1016/S0168-1699(00)00161-7.

Prahalad, C. K. (2004). *Fortune at the Bottom of the Pyramid: Eradicating Poverty through Profit*, Wharton School Publishing, Upper Saddle River.

Reddy, N. V., Reddy, A. V. V. V., Pranavadithya, S. and Kumar, J. J. (2016). A critical review on agricultural robots, *International Journal of Mechanical Engineering and Technology* 7(4), 183–188. Available at: http://www.iaeme.com/ijmet/issues.asp?JType=IJMET&VType=7&IType=4.

Rip, M. R. and Hasik, J. M. (2002). *The Precision Revolution: GPS and the Future of Aerial Warfare*, Naval Institute Press, Annapolis. ISBN 1-55750-973-5.

Rose, D. C. and Chilvers, J. (2018). Agriculture 4.0: broadening responsible innovation in an era of Smart farming, *Frontiers in Sustainable Food Systems* 2(87). https://doi.org/10.3389/fsufs.2018.00087.

Sharipov, D. R., Yakimov, O. A., Gainullina, M. K., Kashaeva, A. R. and Kamaldinov, I. N. (2021). Development of automatic milking systems and their classification, *IOP Conference Series: Earth and Environmental Science* 659(1), 1–5. https://doi.org/10.1088/1755-1315/659/1/012080.

Sheetz, K. (2018). The Japanese impact on global drone policy and law: why a laggard United States and other nations should look to Japan in the context of drone usage, *Indiana Journal of Global Legal Studies* 25(1), 513–537.

Shockley, J. M., Dillon, C. R. and Shearer, S. A. (2019). An economic feasibility assessment of autonomous field machinery in grain crop production, *Precision Agriculture* 20(5), 1068–1085. https://doi.org/10.1007/s11119-019-09638-w. accessed 7 August, 2020.

Shockley, J., Dillon, C., Lowenberg-DeBoer, J. and Mark, T. (2022). How will regulation influence commercial viability of autonomous equipment in U.S. production agriculture?, *Applied Economic Perspectives and Policy* 44(2), 865–878. https://doi.org/10.1002/aepp.13178.

Smart Ag (2018). Smart Ag unveils AutoCart driverless tractor technology at 2018 farm progress show, *OEM OffHighway*. Available at: https://www.oemoffhighway.com/trends/gps-automation/news/21020794/smart-ag-unveils-autocart-driverless-tractor-technology-at-2018-farm-progress-show, accessed 14 April 2022.

Smith, M. R. and Marx, L. (Eds.) (1998). *Does Technology Drive History? The Dilemma of Technological Determinism*, MIT Press, Cambridge, MA.

Steeneveld, W., Tauer, L. W., Hogeveen, H. and Oude Lansink, A. G. (2012). Comparing technical efficiency of farms with an automatic milking system and a conventional milking system, *Journal of Dairy Science* 95(12), 7391–7398.

Straete, E. P., Vik, J. and Hansen, B. G. (2017). The social robot: A study of the social and political aspects of automatic milking systems, *International Journal on Food System Dynamics*, 220–233. https://doi.org/10.18461/pfsd.2017.1722.

Tarannum, N., Rhaman, M. K., Khan, S. A. and Shakil, S. R. (2015). A brief overview and systematic approach for using agricultural robots in developing countries, *Journal of Modern Science and Technology* 3(1), 88–101.

Tey, Y. S. and Brindal, M. (2022). A meta-analysis of factors drving the adoption of precision agriculture, *Precision Agriculture* 23(2), 353–372. https://doi.org/10.1007/s11119-021-09840-9.

Tse, C., Barkema, H. W., DeVries, T. J., Rushen, J. and Pajor, E. A. (2017). Effect of transitioning to automatic milking systems on producers' perceptions of farm management and cow health in the Canadian dairy industry, *Journal of Dairy Science* 100(3), 2404–2414.

Tudge, C. (1999). *Neanderthals, Bandits and Farmers: How Agriculture Really Began*, Yale University Press, New Haven. ISBN: 0300080247.

Valle, S. S. and Kienzle, J. (2020). *Agriculture 4.0: Agricultural Robotics and Automated Equipment for Sustainable Crop Production*, Food and Agriculture Organization of the United Nations, Rome. Available at: https://www.fao.org/publications/card/en/c/CB2186EN/, accessed 14 April 2022.

Vik, J., Straete, E. P., Hansen, B. G. and Naerland, T. (2019). The political robot - the structural consequences of automated milking systems (AMS) in Norway, *NJAS - Wageningen Journal of Life Sciences* 90–91, 100305.

Wang, C., Herbst, A., Zeng, A., Wongsuk, S., Qiao, B., Qi, P., Bonds, J., Overbeck, V., Yang, Y., Gao, W. and He, X. (2021). Assessment of spray deposition, drift and mass balance from unmanned aerial vehicle sprayer using an artificial vineyard, *The Science of the Total Environment* 777, 146181–146181.

Yawson, G. and Frimpong-Wiafe, B. (2018). The socio-economic benefits and impact study on the application of drones, sensor technology and intelligent systems in commercial scale agricultural establishments in Africa, *International Journal of Agriculture and Economic Development* 6(2), 18–36.

Chapter 16

Social and ethical considerations for agricultural robotics

Kirsten Ayris, University of Reading, UK[1]; and David Christian Rose, Cranfield University, UK[2]*

1 Introduction
2 Responsible robotics: anticipating impacts
3 Inclusion for responsible agricultural robotics
4 Responsible robotics: reflexivity and responsiveness
5 Conclusion
6 Where to look for further information
7 Acknowledgements
8 References

1 Introduction

This chapter explores the social and ethical implications of scaling agricultural robotics. Despite the many promises offered by robots, a large number of social science studies have highlighted their potential to create winners and losers, which complicates the notion that their use will usher in a triumphant fourth agricultural revolution (see reviews by Klerkx et al., 2019; Fielke et al., 2020; also van der Burg et al., 2019). We use a responsible innovation framework to frame discussion of these debates (see e.g. Bronson, 2019; Eastwood et al., 2019; Rose and Chilvers, 2018; Regan, 2019). Responsible innovation contains four components (Stilgoe et al., 2013) - *anticipation, inclusion, reflexivity, responsiveness* - and methods for agriculture specifically have been outlined by Eastwood et al. (2019). We explore the social and ethical implications of agricultural robotics, and more importantly, what we can do to include and listen to a wide range of stakeholders when developing and scaling this technology. When listening to these stakeholders, we should be open to considering alternative visions for the future of agriculture. These could range from the dismissal of robotics as a solution all the way through to an uncritical

* Corresponding author: k.j.ayris@pgr.reading.ac.uk.
[1] School of Agriculture, Policy and Development, University of Reading, Earley, Reading, RB6 6BZ, UK.
[2] School of Water, Energy and the Environment, Cranfield University, College Road, Cranfield, Bedford, MK43 OAL, UK.

http://dx.doi.org/10.19103/AS.2023.0124.20

scaling of existing robots, as well as other states in between, such as changing current approaches to design and scaling.

2 Responsible robotics: anticipating impacts

In this section, we focus on anticipating or foresighting the effects of agricultural robotics on people, production, and the planet; in other words, what impacts can we foresee, both positive and negative, from the scaling of robots. Foresighting exercises more generally have been conducted for digital agriculture (e.g. Ehlers et al., 2022; Fleming et al., 2018), but exercises for robotics have been limited to perspectives, rather than building on empirical data. For all their promises, agricultural robotics and their enabling technologies, such as artificial intelligence (AI), may be a double-edged sword. Daum (2021), for example, argues that robots could lead us to ecological utopia or dystopia depending on how and why they are incorporated into agricultural production systems. Recent scholarship in this area from Sparrow and Howard (2021), Rose et al. (2021), Daum (2021), Ryan et al. (2022), Ryan (2022), amongst other discussions of the ethics of robots and AI in non-agricultural settings (e.g. Winfield, 2019), highlights the need for responsible development.

There are precedents in agriculture to learn from when it comes to the side effects of innovation. At first, we celebrated the rise of modern machinery, such as tractors, as a progressive step away from horse-drawn or human-powered equipment. We praised the value of insecticides, fungicides, and fertilisers as ways of increasing productivity. We promoted the use of genetic modification to boost yields and cut input costs. We glorified the so-called Green Revolution as bringing modern seed varieties and other technologies to boost yields in the developing world. Yet, despite their obvious benefits, all four examples have caused controversy. Modern large, fossil fuel-guzzling machinery has contributed to atmospheric pollution and soil compaction, chemicals from dichlorodiphenyltrichloroethane and neonicotinoids have contributed to the so-called 'Silent Spring' (Carson, 1962) by killing pollinators and polluting rivers, and genetically modified crops were banned in some parts of the world due to a public backlash. The Green Revolution has been criticised for its negative social impacts, as some people benefited more than others (Shiva, 2016; Kumar et al., 2017). None of these precedents should stop us from making transitions in agriculture. Transitions are always normative and disruptive (de Boon et al., 2022a, b), and there are winners and losers from the *status quo*. However, responsible innovation asks us to foresee potential negative consequences and take steps to minimise them.

The social science aspects of digital agricultural technologies have been growing in recent years (Klerkx et al., 2019; Fielke et al., 2020) illustrating their potential to create social and ethical problems. These include issues of power

– for example, the inability of small-scale farmers to invest in new technologies and the fact that digitalisation tends to favour larger-scale farmers (Bronson and Sengers, 2022), the greater ability of existing powerful companies to set trajectories (Birner et al., 2021; Duncan et al., 2021), or the unequal ownership, use, and privacy of data collected from a farm (Wiseman et al., 2019). Concerns also include precision technologies causing further intensification of farming, including larger herds of animals made possible by individual monitoring which could create welfare issues associated with overcrowding (Schillings et al., 2021); the capacity of the existing workforce to adapt to new conditions (Rotz et al., 2019), including advisers (Charatsari et al., 2021); a reduction in farmer autonomy (Brooks, 2021; Duncan et al., 2021); and lack of trust in new technologies (Jakku et al., 2019).

With the precedents in mind, as well as the wider critiques of digitalisation, a diagram by Rose et al. (2021) summarised the potential positive and negative impacts of autonomous robotics in agriculture, building on the work of Sparrow and Howard (2021). This is not replicated here, but Table 1 presents an overview of ten ethical concerns (see Ryan, 2022 for more), with additional insights from subsequent scholarship by Daum (2021), Streed et al. (2021), Tzachor et al. (2022), Ryan (2022), Rose and Bhattacharya (2023), and Tamirat et al. (2023). We also draw insights from a white paper led by the Agri-EPI Centre and 'Hands Free Farm' (2022), which covered issues of health and safety, cybersecurity, connectivity, and data ownership and trust. The risks associated with AI in agriculture are considered important by policy-makers, including in a European Union Joint Research Centre document, which discussed its advantages and disadvantages (Loudjani et al., 2020). The benefits and risks of agricultural automation were also the subject of The Food and Agriculture Organisation of the United Nation's State of Food and Agriculture report in 2022 (FAO, 2022). By mapping the AI in agriculture literature, Ryan (2022) was able to show how 11 overarching ethical principles were discussed, finding that the least mentioned areas were transparency, dignity, and solidarity.

There will be spatial differences across the world for these social and ethical challenges. For example, we know that many parts of the developing world suffer from an intense digital divide (also present in many developed nations to a lesser extent), whereby rural farmers do not enjoy 3G, 4G, or 5G coverage, or if they do, data is unaffordable (Mehrabi et al., 2021). Poorer regions of the world, as well as those places lacking digital skills (including in cybersecurity), will be less able to take advantage of some robotic solutions. Furthermore, many parts of the developing world in particular do not suffer from a rural labour shortage, and so the rise of robotics could be a greater threat to livelihoods here. These ethical challenges are thus place dependent and there are no one-size-fits-all solutions.

Table 1 Ten prominent ethical concerns for agricultural robotics

Theme	Positive impacts	Negative impacts	Further information
Trust and data	Robotics could allow more data to be collected to improve trust in the supply chain	AI lacks emotive state to be trusted, and farmers may not understand how it works or how data is used	Rose and Bhattacharya (2023); Ryan (2022); Tzachor et al. (2022)
Health and safety	Robots could replace dull, dirty, and dangerous jobs and reduce the existing high levels of farm injury	Robots could injure workers or the public accessing farmland (or near farmland)	Agri-EPI Centre/Hands Free Farm (2022); Rose and Bhattacharya (2023); Tamirat et al. (2023)
Cybersecurity	Provides skilled jobs for cybersecurity experts	Robots could be hackable and present a security risk	NCC Group (2019); Sparrow and Howard (2021); Agri-EPI Centre/Hands Free Farm (2022)
Employment	Robots can fill labour shortages and attract skilled workers	Robots could displace labour, and the workforce might not be able to retrain	Rotz et al. (2019)
Energy use	Robots could run on renewable energy	Materials required to build robots could be unsustainable	Streed et al. (2021); Pearson et al. (2022)
Economics and performance	Robots could reduce input costs	Robots may only be investable for richer farmers They may lack reliability	Rose et al. (2021); Rose and Bhattacharya (2023); Tamirat et al. (2023); Sparrow and Howard (2021)
Loss of farmer control	Robots can assist farmers in doing manual tasks	Robots could reduce the autonomy of a farmer	Rose et al. (2021); Ryan (2022)
Connectivity	N/A	Robots may not work in areas of poor connectivity widening the digital divide	Agri-EPI Centre/Hands Free Farm (2022); Rose and Bhattacharya (2023)
Animal welfare	Robotic milking collects data to allow monitoring of individual animals for health and welfare	Robots could lead to larger herd sizes or the further objectification of animals	Schillings et al. (2021)

Theme	Positive impacts	Negative impacts	Further information
Ecological utopia or dystopia?	Robots could allow more targeted and precise use of chemicals (reduced inputs) Robotic ultraviolet-C disease treatment could reduce chemical usage Small robots could facilitate agroecological systems	Robots could further intensify production systems and facilitate existing models of damaging large-scale agriculture	Daum (2021); Sparrow and Howard (2021)

We have little empirical data of farmers encountering these challenges in practice since many autonomous robotic technologies are not implemented at scale. One useful proxy where autonomous robotics, albeit in a fixed form, has been rolled out at scale is robotic milking machines. Research on the impacts of these machines can help us to understand the potential social and ethical challenges that will be encountered in the further scaling of autonomous milking, as well as field robotics. Evidence of outcomes reinforces the concept of robotic technology as a double-edged sword. While farmers have reported improved job satisfaction and more flexible lifestyles, improved animal welfare and profitability (Hansen et al., 2020), and better working conditions including for workers with disabilities (Barrett and Rose, 2022), there have also been changes to the farm industry. In places where there has been widespread scaling, smaller farms have closed and the structure of the industry has moved towards a model of fewer, larger farms (Vik et al., 2019). Concerns have been raised about the effects on the human–animal relationship as farmers spend less time with their stock, potentially raising welfare issues and reducing the enjoyment of the job (Bear and Holloway, 2019). Farmers report being more stressed due to being overwhelmed with the amount of data that the robots are collecting (Barrett and Rose, 2022). With the potential side effects of agricultural robotics in mind, it is important to take steps to include stakeholders in development and implementation, as well as to translate their views into responsive practices.

3 Inclusion for responsible agricultural robotics

Here, we discuss the inclusion strand of a responsible innovation framework. The white paper by the Agri-EPI Centre and 'Hands Free Farm' (2022) on the future of agricultural robotics called for greater inclusion of stakeholders. Several participatory methods are highlighted by Eastwood et al. (2019) in their

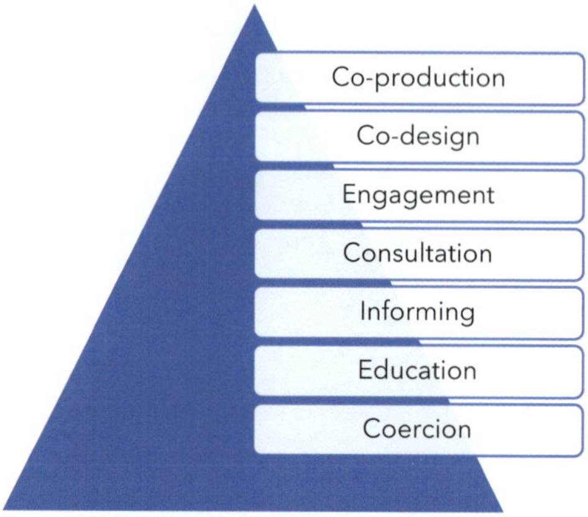

Figure 1 Ladder of participation adapted by the 'Think Local Act Personal Partnership'. Inclusion is more substantive the further up the pyramid.

framework of responsible innovation in agriculture, while Rose and Chilvers (2018) urge us to open up conversations to question underlying assumptions, rather than simply being guided by a user acceptance lens. Activities such as public dialogues and deliberative exercises can open up avenues for greater engagement (Stilgoe et al., 2013; Rose and Chilvers, 2018).

3.1 How do we achieve inclusion with participatory methods?

Inclusion has been identified as a clear point of emphasis for many of the most well-known frameworks for participation (Bell and Reed, 2021). However, definitions of inclusion across frameworks can differ greatly. Arnstein's classic typology (1969) is a seminal work in participation, characterising levels of participation as rungs of a ladder. In this work, success is measured by the empowerment of citizens to be involved in decision-making, with their role ranging from full power to drive decision-making and trajectories (citizen control) through to having no power at all and being 'manipulated' in tokenistic participation exercises. The idea of a spectrum of participation has been adapted by the 'Think Local Act Personal Partnership'[1] (Fig. 1). Here, stakeholders can play a leading role in setting research questions, methodologies, and directions (co-production, co-design) or be tokenistically included in a sham consultation process in which key decisions have already been made (coercion). In order for innovation to be responsible, inclusion must be substantive (Bronson, 2019).

[1] https://www.thinklocalactpersonal.org.uk/_assets/Resources/Coproduction/LadderOfParticipation.pdf

Other models of participation, such as the *Split Ladder* (Hurlbert and Gupta, 2015) and the *Tree of Participation* (Bell and Reed, 2021), can also guide efforts to make inclusion more substantive. Key principles for co-production or other forms of substantive inclusion are early and continued engagement of stakeholders; all affected stakeholders are engaged, stakeholders have decision-making power; conflicts between stakeholders are well managed; trajectories are open to change; and two-way dialogue is established throughout.

More recently, Chilvers and Kearnes (2015) identify inclusion as an aspect of a residual realist approach and advocate for more reflexivity, diversity, and responsibility as a move beyond participatory approaches which may have previously closed down forms of inclusion. But, there are challenges in doing participation. The first such challenge is how to ensure that 'inclusion' within a participatory exercise is truly inclusive. In examples such as the study of Bavarian farmers in Section 3.2 (Spykman et al., 2021), the authors identified a barrier to inclusivity via the self-selecting nature of participants. This highlights a challenge that is particularly prominent for agriculture – how can innovators and developers ensure that inclusion is truly inclusive in a system with so many diverse stakeholders? In agriculture, there are a vast number of possible stakeholders – farmers, farm workers, industry members, governments, rural communities, and consumers. There is heterogeneity within these broad groups, e.g. large/small farmers, rich/poor, digitally skilled/unskilled, varying within regions and across countries. The agricultural sector is one in which everybody has a stake – we all need to eat. Based on this, there will always be fierce debate and discussion over who should be included in any participatory exercise. Answering this question is one of the key problems to be solved in making inclusion truly inclusive.

Participatory approaches have been criticised for failing to engage with wider groups of stakeholders (Cleaver, 1999; Chilvers and Kearnes, 2015), and this issue is particularly striking in the agricultural sector. Approaches should start with a stakeholder mapping exercise (Fig. 2) to determine who is affected by the issue and who has the most influence to effect change, although the latter should not restrict the involvement of less-powerful stakeholders who may be more likely to lose from disruption. Before starting stakeholder engagement, stakeholder groups and individuals should be laid out on the matrix. The most important axis relates to how far stakeholders are affected by the decision (interest) – in farming, this will certainly be farmers, farm advisers, and others in the farming community but also likely consumers and others in the supply chain. None of these stakeholders should be excluded, but we know that some often are, for example farm workers (Burch and Legun, 2021). Also, there is a need to be wary of including too many high-influence, high-interest stakeholders (top right) as compared to the high-interest, low-influence group (bottom right) as

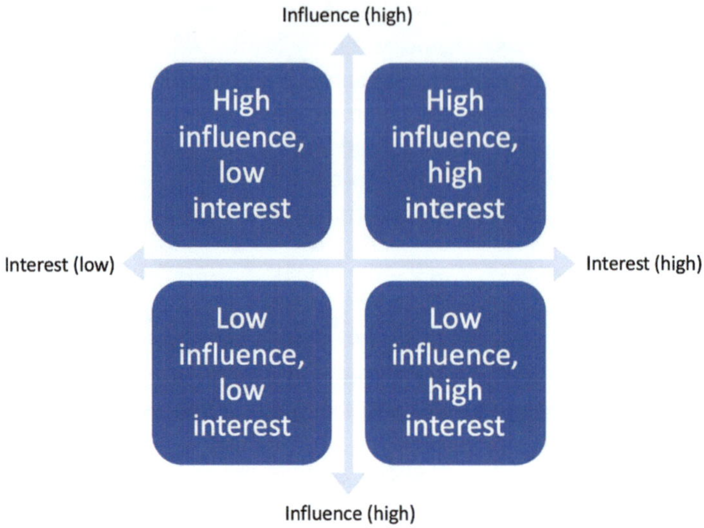

Figure 2 Matrix to guide stakeholder mapping.

this could reinforce power inequalities and make participation less inclusive. Only the bottom left segment could be excluded, although decisions on the level of participation may be determined by the level of time and resources.

Engaging harder-to-reach stakeholders is a challenge (e.g. bottom right), but these people must be included in order for innovation to be inclusive. In agriculture, Hurley et al. (2022) identified why some farmers can be harder to reach, as a result of a digital divide, lack of time, lack of trust in government, and other business factors. They suggested a number of methods for how policy-makers and industry groups could engage these harder-to-reach groups, from working through skilled intermediaries who have developed trust with farmers, to increasing rural connectivity and digital skills, and the establishment of two-way communication so that those who have engaged know how their views were incorporated into policy design.

As many scholars have highlighted, power and its distribution is a key aspect of participation. Where power dynamics are not addressed within a participatory process, that process does not represent typologies of participation as portrayed within many frameworks (Arnstein, 1969; Reed et al., 2018; Bell & Reed, 2021). With the hourglass shape of the food industry, in which there is a small number of multinationals in the middle of the chain and a large number of producers and consumers at either end, addressing power imbalances is crucial to substantive inclusion and participation for the agricultural sector (Patel, 2007, p13; Thompson et al., 2007). However, it is not enough to be aware of power dynamics for the sake of inclusion of stakeholders. It is also crucial to continuously address and manage the issue of power throughout any participatory activity (Reed et al., 2018). For example, it is not enough to include farm workers in a participatory activity without acknowledging the power imbalances that may influence their input or their willingness to take part. Many workers may be on precarious contracts, face language barriers, or feel otherwise unable to express their honest thoughts on a given subject for fear of losing their livelihood (Rye and Scott, 2018; Fiałkowska and Matuszczyk, 2021). Substantive inclusion does not just look like inviting these stakeholders to the table – it also requires protecting their capacity to take part and contribute in an open and robust manner.

A further challenge highlighted by participatory exercises which have thus far taken place for the development of agricultural robotics is framing – in particular, the framing of the 'publics' participating in exercises. The outcomes and perceived success of a participatory exercise are highly interpretable (Lezaun and Soneyrd, 2007) and may depend on the ways in which participant groups have been defined by the goals of the exercise. One pitfall of framing of participation as an exercise in engaging a 'general public' is the trap of believing that members of such a public should have no preconceived notion of the issue they are to discuss: the search for what has been termed both an 'innocent citizen' (Irwin, 2006) and as the classical notion of the privately and internally focussed 'idiot' (Lezaun and Soneyrd, 2007, p280). This framing may be intended to avoid perceptions of bias but excludes those deemed to be high-interest 'activists' with strong opinions formed outside the boundaries of the participatory exercise.

3.2 Inclusion for agricultural robotics to date

To date, participatory approaches used in the development of agricultural robotics have predominantly involved consultation on technical challenges in the field and technical problems in agriculture that require solutions (Rose et al., 2021). These approaches lack the principles of substantive inclusion identified above (Rose and Bhattacharya, 2023). Examples of participation and inclusion in the technical dimensions of agricultural robotic development have thus far employed a range of tools, including the use of augmented reality (Huuskonen & Oksanen, 2019), on-farm trials (Adamides et al., 2017), and simulation events (see Riek, 2012 for methods), farmer surveys (Rose and Bhattacharya, 2023; Spykman et al.; 2021; Tamirat et al., 2023; von Veltheim and Heise, 2021; von Veltheim et al., 2022), public surveys (Pfeiffer et al., 2021; Spykman et al., 2022), and farmer interviews (Redhead et al., 2015). A recent study in four European countries by Tamirat et al. (2023) went further than including only farmers and instead interviewed robotic companies, academics, project managers, public authorities, and environmental conservation societies. Other studies with a focus on technical issues have also integrated some social dimensions into investigations for technical robotic development, for example by gathering data on farm worker attitudes to robotic futures (Baxter et al., 2018).

As robotic technologies for agriculture have shifted closer to being ready for real-world application, some participatory exercises have begun to consider these technologies within the context of wider societal questions and challenges. Ditzler and Driessen (2022) marry the design of agricultural robotics with wider questions around the sustainable future of the agricultural sector. Their approach integrated a series of creative methods to develop robotics as a tool for agroecology, including group discussion, a World Café workshop, a design challenge with undergraduate design students, and on-farm trials. Each of these elements reflexively influenced the next in terms of design and focus, and this reflexivity is highlighted by the authors as one of the key strengths of their study. Other more inclusive deliberation exercises included the process for the Australian code of practice for field robotics (see Section 4) and the Agri-EPI Centre/Hands Free Farm white paper. In addition, lessons will be learned from a robotic co-design project in New Zealand in which multiple academic disciplines are collaborating with stakeholders to develop farm robots (Legun and Burch, 2021). These approaches start to move in the direction of more substantive inclusion needed for responsible innovation. As the development of these robotic technologies progresses, further investigation into the wider implications of their use will be necessary. From the examples above, it is possible to identify some of the key challenges inherent in designing and applying such approaches for agricultural robotics, particularly in the context of achieving substantive inclusion.

3.3 Participation in robotic futures

While the vision of the role of participatory approaches can be debated and disputed, there is clear potential for their use to enhance responsible development and innovation of agricultural robotics. The application of participatory approaches has continued to evolve from the original ladder of participation (Arnstein, 1969), experiencing growing popularity as a replacement for the deficit model during the 1990s following a series of scandals which impacted public trust in science (Gibbons, 1999; Irwin, 2001; Bensaude Vincent, 2014), to the twenty-first century, in which several scholars have proposed revisions to the original frameworks of both practice and evaluation of participation (Rowe & Frewer, 2000; Hurlbert and Gupta, 2015; Chilvers and Kearnes, 2015). With this continued refinement and evolution of participatory processes, there is a need to ensure that 'perfect' is not the enemy of 'good' when it comes to creating more inclusive visions for the development of agricultural robotics. However, it should be noted that participation done badly can actually make approaches less inclusive by reinforcing existing power inequalities.

This section has discussed the inherent challenges in creating substantive inclusion, and it is clear that there is great complexity in achieving 'perfect' inclusion in a system in which the potential pool of stakeholders is so wide. Furthermore, several scholars have highlighted that participation is not always the solution, and while it can be a useful approach for some problems, for others it can make little to no progress (Neef and Neubert, 2011; Hurlbert and Gupta, 2015). Reed et al. (2018) highlight that the normative idea of participation as following a ladder from low to high fails to account for variations in the efficacy of different approaches to participation. A more reflexive approach is here called for in order to achieve better outcomes for decision-making processes (Reed et al., 2018).

A framework for participation in the development of agricultural robotics would benefit from reflecting upon the tensions of attempting participation within a pre-defined goal framework, rather than allowing participants to define their own goals (Bruges and Smith, 2007). Avoiding a predetermined goal allows for the development of a coupled innovation approach (Meynard et al., 2017), which has been shown as a benefit within the development of agroecological tools for sustainability transitions (Salembier et al., 2020; Ditzler and Driessen, 2022). It is important for the ethical development of agricultural robotics that participation is not undertaken for participation's sake. To avoid tokenistic participation, reflecting upon the purpose of the exercise is crucial – facilitators may find that some areas require goals to be defined externally, and in such cases, participatory approaches may not be effective.

4 Responsible robotics: reflexivity and responsiveness

We now turn to reflexivity and responsiveness in a shorter section, as research and action in this area is relatively under-developed given that many robots are only now being scaled on the ground. Reflexivity means listening to the views of stakeholders and acting on them through the development of responsive mechanisms for innovation. Including stakeholders is insufficient if robotic development trajectories are impervious to challenge and change. Here, we report on two initiatives from Australia and the UK to develop codes of practice and standards for agricultural robotics as a means of inspiring further work elsewhere. These are crucial in setting supportive regulations to allow agricultural robotics to be used safely and sustainably, and we encourage them to be developed across the world, targeted at national and regional regulatory environments.

The forebearer of current attempts to develop a code of practice for agricultural robotics comes from Australia. Here the code of practice 'Agricultural Mobile Field Machinery with Autonomous Functions in Australia'[2] was developed through consultation between Grain Producers Australia and the Tractor and Machinery Association – with engagement from machinery manufacturers, the agricultural industry, and in consultation with the Department of Mines, Industry Regulation and Safety, an industry which has utilised autonomous vehicles for some time. It actually goes further than a voluntary code of practice and instead has legal bearing. It is designed to 'operate under the Safe Work Australia Model Work Health and Safety (WHS) law and regulation, Australian and State government agricultural environment legislation and regulations and the Agricultural and Veterinary Chemicals Code Act 1994 and supporting state government legislation and regulations.' It presents the roles and responsibilities of manufacturers, importers, distributors, owners, and operators.

The code of practice covers several areas, including:

1 safety and risk management;
2 information needs, training, and supervision;
3 hazard controls, emergency preparedness, and farm planning;
4 operational management;
5 vehicle transport between fields;
6 maintenance and repair requirements; and
7 emergency management.

In covering these areas, the code of practice sets out expectations on who is responsible for various aspects of health and safety, data privacy, and maintenance, amongst other issues. This gives users the confidence to use autonomous vehicles and sets a clear, enabling regulatory environment. In many places in the world, regulations still restrict the optimal use of agricultural robotics (Lowenberg-DeBoer et al., 2022).

A similar process is now underway in the UK, led by the Agri-EPI Centre and Hands Free Farm team in consultation with industry groups (see Agri-EPI Centre/Hands Free Farm, 2022 white paper). They highlighted key issues such as cybersecurity, health and safety, and data ownership and are taking steps to develop standards and a code of practice to deal with them. This is being done through stakeholder inclusion with groups such as the National Ramblers Association (walkers who use farmland), the Institute of Agriculture Engineers (manufacturers), farming industry groups, the Health and Safety Executive, and

[2] https://www.graincentral.com/wp-content/uploads/2021/08/Code-of-Practice.pdf.

policy-makers. A special working group on smart farming at the ISO is also considering standards for the use of agricultural robotics.

5 Conclusion

Agricultural robotics, like all disruptive technologies, is a double-edged sword. For all the benefits promised to people, production, and the planet, there are potential side effects. To adapt a quote from Daum (2021), robots may offer us agricultural 'utopias' or 'dystopias', depending on how they are used. If robots are used to maintain current unsustainable models of production, or further intensify them, there are great risks. However, if they are used to create a step change in how we do agriculture, including embracing more regenerative or agroecological processes, challenging models of large, fossil fuel-based machinery, and eliminating imprecise use of inputs, they could revolutionise farming for the better. A farming future that achieves more output, with less and more targeted use of inputs, could be facilitated by the use of robotics.

Though the scaling of robots will be disruptive, it does not mean that we should shy away from embracing robots in agriculture; innovation is an inherently normative and disruptive process, and sticking with the *status quo* also presents a set of winners and losers (de Boon et al., 2022a, b). We have shown how a process of responsible innovation could be undertaken to allow stakeholders to co-develop agricultural transition pathways, including interrogating the role of robots within them. This starts with a process of stakeholder mapping to guide substantive inclusion. These stakeholders need to be listened to in a reflexive process and their views taken into account in order to guide development and implementation, using standards, codes of practice, regulation, and other policy instruments to ensure desirable outcomes. We recommend that proponents of agricultural robotics follow such a process from the very start of development and carry it on through and beyond implementation.

6 Where to look for further information

The Food and Agriculture Organization of the United Nations has published the 2022 *The State of Food and Agriculture* report, which addresses agricultural automation for transforming agrifood systems. It is available here: https://www.fao.org/publications/sofa/2022/en/.

For an introduction to the Responsible Innovation framework discussed in this chapter, please see Stilgoe, J., Owen, R. and Macnaghten, P. (2013). Developing a framework for responsible innovation. *Research Policy* 42(9), 1568–1580.

For further discussion of the key ethical considerations in the responsible development of autonomous agricultural robotics, see. Rose, D. C., Lyon, J., de Boon, A., Hanheide, M. and Pearson, S. (2021). Responsible development of autonomous robotics in agriculture. *Nature Food* 2(5), 306–309.

7 Acknowledgements

Kirsten Ayris is funded by UKRI doctoral training grant no: BB/T008776/1 (BBSRC FoodBioSystems DTP). The authors would like to thank the reviewers for their comments, as well as the editors and the publisher for their work on producing this volume.

8 References

Adamides, G., Katsanos, C., Constantinou, I., Christou, G., Xenos, M., Hadzilacos, T. and Edan, Y. (2017). Design and development of a semi-autonomous agricultural vineyard sprayer: Human-robot interaction aspects. *Journal of Field Robotics* 34(8), 1407–1426.

Agri-EPI Centre. (2022). Safe-tech Hackathon: Enhancing the safety and security of autonomous agricultural vehicles. *Hands Free Farm*. Available at: https://agri-epicentre.com/wp-content/uploads/2022/03/Hackathon-whitepaper.pdf.

Arnstein, S. R. (1969). A ladder of citizen participation. *Journal of the American Planning Association* 35(4), 216–224.

Barrett, H. and Rose, D. C. (2022). Perceptions of the fourth agricultural revolution: What's in, what's out, and what consequences are anticipated? *Sociologia Ruralis* 62(2), 162–189. https://doi.org/10.1111/soru.12324.

Baxter, P., Cielniak, G., Hanheide, M. and From, P. (2018). Safe human-robot interaction in agriculture. *Academic Medicine*, 59–60.

Bear, C. and Holloway, L. (2019). Beyond resistance: Geographies of divergent more-than-human conduct in robotic milking. *Geoforum* 104, 212–221.

Bell, K. and Reed, M. (2021). The tree of participation: A new model for inclusive decision-making. *Community Development Journal* 57, 595–614.

Bensaude Vincent, B. (2014). The politics of buzzwords at the interface of technoscience, market and society: The case of "public engagement in science." *Public Understanding of Science* 23(3), 238–253.

Birner, R., Daum, T. and Pray, C. (2021). Who drives the digital revolution in agriculture? A review of supply-side trends, players and challenges. *Applied Economic Perspectives and Policy* 43(4), 1260–1285.

Bronson, K. (2019). Looking through a responsible innovation lens at uneven engagements with digital farming. *NJAS* 90–91(1), 1–6.

Bronson, K. and Sengers, P. (2022). Big tech meets big ag: Diversifying epistemologies of data and power. *Science as Culture* 31(1), 15–28.

Brooks, S. (2021). Configuring the digital farmer: A nudge world in the making? *Economy and Society* 50(3), 374–396.

Bruges, M. and Smith, W. (2007). Participatory approaches for sustainable agriculture: A contradiction in terms? *Agriculture and Human Values* 25(1), 13–23.

Burch, K. A. and Legun, K. (2021). Overcoming barriers to including agricultural workers in the co-design of new AgTech: Lessons from a COVID-19 present world. *Culture, Agriculture, Food and Environment* 43(2), 147–160.

Carson, R. (1962). Silent Spring. Houghton Mifflin Company. The Riverside Press, Boston, Cambridge Massachusetts

Charatsari, C., Lioutas, E. D., Papadaki-Klavdianou, A., Michailidis, A. and Partalidou, M. (2022). Farm advisors amid the transition to Agriculture 4.0: Professional identity, conceptions of the future and future-specific competencies. *Sociologia Ruralis* 62(2), 335–362.

Chilvers, J. and Kearnes, M. (2015). *Remaking Participation: Science, Environment and Emergent Publics*. Abingdon: Routledge.

Cleaver, F. (1999). Paradoxes of participation: Questioning participatory approaches to development. *Journal of International Development* 11(4), 597–612.

Daum, T. (2021). Farm robots: Ecological utopia or dystopia? *Trends in Ecology and Evolution* 36(9), 774–777.

de Boon, A., Sandström, C. and Rose, D. C. (2022a). Governing agricultural innovation: A comprehensive framework to underpin sustainable transitions. *Journal of Rural Studies* 89, 407–422.

de Boon, A., Sandström, C. and Rose, D. C. (2022b). Perceived legitimacy of agricultural transitions and implications for governance: Lessons learned from England's post-Brexit agricultural transition. *Land Use Policy* 116, 106067.

Ditzler, L. and Driessen, C. (2022). Automating agroecology: How to design a farming robot without a monocultural mindset? *Journal of Agricultural and Environmental Ethics* 35(1), 2.

Duncan, E., Glaros, A., Ross, D. Z. and Nost, E. (2021). New but for whom? Discourses of innovation in precision agriculture. *Agriculture and Human Values* 38(4), 1181–1199.

Eastwood, C., Klerkx, L., Ayre, M. and Dela Rue, B. (2019). Managing socio-ethical challenges in the development of smart farming: From a fragmented to a comprehensive approach for responsible research and innovation. *Journal of Agricultural and Environmental Ethics* 32(5–6), 741–768.

Ehlers, M.-H., Finger, R., El Benni, N., Gocht, A., Sørensen, C. A. G., Gusset, M., Pfeifer, C., Poppe, K., Regan, Á., Rose, D. C. and Wolfert, S. (2022). Scenarios for European agricultural policymaking in the era of digitalisation. *Agricultural Systems* 196, 103318.

FAO. (2022). 2022: Leveraging agricultural automation for transforming agrifood systems. *The State of Food and Agriculture*. Available at: https://www.fao.org/publications/sofa/2022/en/.

Fiałkowska, K. and Matuszczyk, K. (2021). Safe and fruitful? Structural vulnerabilities in the experience of seasonal migrant workers in agriculture in Germany and Poland. *Safety Science* 139, 105275.

Fielke, S., Taylor, B. and Jakku, E. (2020). Digitalisation of agricultural knowledge and advice networks: A state-of-the-art review. *Agricultural Systems* 180, 102763.

Fleming, A., Jakku, E., Lim-Camacho, L., Taylor, B. and Thorburn, P. (2018). Is big data for big farming or for everyone? Perceptions in the Australian grains industry. *Agronomy for Sustainable Development* 38(3), 24.

Gibbons, M. (1999). Science's new social contract with society. *Nature* 402(6761), C81–C84.

Hansen, B. G. and Straete, E. P. (2020). Dairy farmers' job satisfaction and the influence of automatic milking systems. *NJAS - Wageningen Journal of Life Sciences* 92, 100328.

Hurlbert, M. and Gupta, J. (2015). The split ladder of participation: A diagnostic, strategic, and evaluation tool to assess when participation is necessary. *Environmental Science and Policy* 50, 100–113.

Hurley, P., Lyon, J., Hall, J., Little, R., Tsouvalis, J., White, V. and Rose, D. C. (2022). Co-designing the environmental land management scheme in England: The why, who, and how of engaging 'harder to reach' stakeholders. *People and Nature* 4(3), 744–757.

Huuskonen, J. and Oksanen, T. (2019). Augmented reality for supervising multirobot system in agricultural field operation. *IFAC-PapersOnLine* 52(30), 367–372.

Irwin, A. (2001). Constructing the scientific citizen: Science and democracy in the biosciences. *Public Understanding of Science* 10(1), 1–18.

Irwin, A. (2006). The politics of talk: Coming to terms with the "new" scientific governance. *Social Studies of Science* 36(2), 299–320.

Jakku, E., Taylor, B., Fleming, A., Mason, C., Fielke, S., Sounness, C. and Thorburn, P. (2019). "If they don't tell us what they do with it, why would we trust them?" Trust, transparency and benefit-sharing in smart farming. *NJAS* 90–91(1), 1–13.

Klerkx, L., Jakku, E. and Labarthe, P. (2019). A review of social science on digital agriculture, smart farming and agriculture 4.0: New contributions and a future research agenda. *NJAS* 90–91(1), 1–16.

Kumar, P., Lorek, T., Olsson, T. C., Sackley, N., Schmalzer, S. and Laveaga, G. S. (2017). Roundtable: New narratives of the green revolution. *Agricultural History* 91(3), 397–422.

Legun, K. and Burch, K. (2021). Robot-ready: How apple producers are assembling in anticipation of new AI robotics. *Journal of Rural Studies* 82, 380–390.

Lezaun, J. and Soneryd, L. (2007). Consulting citizens: Technologies of elicitation and the mobility of publics. *Public Understanding of Science* 16(3), 279–297.

Loudjani, P., Devos, W., Baruth, B. and Lemoine, G. (2020). *AIA: Artificial Intelligence and EU Agriculture*. Joint Research Center, EU Commission.

Lowenberg-DeBoer, J., Behrendt, K., Ehlers, M.-H., Dillon, C., Gabriel, A., Huang, I. Y., Kumwenda, I., Mark, T., Meyer-Aurich, A., Milics, G., Olagunju, K. O., Pedersen, S. M., Shockley, J. and Rose, D. (2022). Lessons to be learned in adoption of autonomous equipment for field crops. *Applied Economic Perspectives and Policy* 44(2), 848–864.

Mehrabi, Z., McDowell, M. J., Ricciardi, V., Levers, C., Martinez, J. D., Mehrabi, N., Wittman, H., Ramankutty, N. and Jarvis, A. (2021). The global divide in data-driven farming. *Nature Sustainability* 4(2), 154–160.

Meynard, J. M., Jeuffroy, M. H., le Bail, M., Lefèvre, A., Magrini, M. B. and Michon, C. (2017). Designing coupled innovations for the sustainability transition of agrifood systems. *Agricultural Systems* 157, 330–339.

NCC Group. (2019). Cyber security in UK agriculture. Available at: https://research.nccgroup.com/wp-content/uploads/2020/07/agriculture-whitepaper-final-online.pdf.

Neef, A. and Neubert, D. (2011). Stakeholder participation in agricultural research projects: A conceptual framework for reflection and decision-making. *Agriculture and Human Values* 28(2), 179–194.

Patel, R. (2007). *Stuffed and Starved: Markets, Power, and the Hidden Battle for the World Food System*. London: Portobello Books.

Pearson, S., Camacho-Villa, T. C., Valluru, R., Gaju, O., Rai, M. C., Gould, I., Brewer, S. and Sklar, E. (2022). Robotics and autonomous systems for net zero agriculture. *Current Robotics Reports* 3(2), 57–64.

Pfeiffer, J., Gabriel, A. and Gandorfer, M. (2021). Understanding the public attitudinal acceptance of digital farming technologies: A nationwide survey in Germany. *Agriculture and Human Values* 38(1), 107–128.

Redhead, F., Snow, S., Vyas, D., Bawden, O., Russell, R., Perez, T. and Brereton, M. (2015). Bringing the farmer perspective to agricultural robots. In: *Proceedings of the Conference on Human Factors in Computing Systems* (Vol. 18), pp. 1067–1072.

Reed, M. S., Vella, S., Challies, E., de Vente, J., Frewer, L., Hohenwallner-Ries, D., Huber, T., Neumann, R. K., Oughton, E. A., Sidoli del Ceno, J. and van Delden, H. (2018). A theory of participation: What makes stakeholder and public engagement in environmental management work? In: *Restoration Ecology*. Blackwell Publishing Inc, p. 26.

Regan, A. (2019). Smart farming. In: *Ireland: A Risk Perception Study With Key Governance Actors: Wageningen Journal of Life Sciences* (Vols. 90–91), pp. 1–10.

Riek, L. D. (2012). Wizard of Oz studies in HRI: A systematic review and new reporting guidelines. *Journal of Human-Robot Interaction* 1(1), 119–136.

Rose, D. C. and Bhattacharya, M. (2023). Adoption of autonomous robots in the soft fruit sector: Grower perspectives in the UK. *Smart Agricultural Technology* 3, 100118.

Rose, D. C. and Chilvers, J. (2018). Agriculture 4.0: Broadening responsible innovation in an era of smart farming. *Frontiers in Sustainable Food Systems* 2, 1–7.

Rose, D. C., Lyon, J., de Boon, A., Hanheide, M. and Pearson, S. (2021). Responsible development of autonomous robotics in agriculture. *Nature Food* 2(5), 306–309.

Rotz, S., Gravely, E., Mosby, I., Duncan, E., Finnis, E., Horgan, M., LeBlanc, J., Martin, R., Neufeld, H. T., Nixon, A., Pant, L., Shalla, V. and Fraser, E. (2019). Automated pastures and the digital divide: How agricultural technologies are shaping labour and rural communities. *Journal of Rural Studies* 68, 112–122.

Rowe, G. and Frewer, L. J. (2000). Public participation methods: A framework for evaluation. *Science, Technology, and Human Values* 25(1), 3–29.

Ryan, M. (2022). The social and ethical impacts of artificial intelligence in agriculture: Mapping the agricultural AI literature. *AI and Society*. https://doi.org/10.1007/s00146-021-01377-9.

Ryan, M., van der Burg, S. and Bogaardt, M. J. (2022). Identifying key ethical debates for autonomous robots in agri-food: A research agenda. *AI and Ethics* 2(3), 493–507.

Rye, J. F. and Scott, S. (2018). International labour migration and food production in rural Europe: A review of the evidence. *Sociologia Ruralis* 58(4), 928–952.

Salembier, C., Segrestin, B., Sinoir, N., Templier, J., Weil, B. and Meynard, J. M. (2020). Design of equipment for agroecology: Coupled innovation processes led by farmer-designers. *Agricultural Systems* 183, 102856.

Schillings, J., Bennett, R. and Rose, D. C. (2021). Animal welfare and other ethical implications of precision livestock farming technologies. *CABI Agriculture and Bioscience* 2(1), 17.

Shiva, V. (2016). *The Violence of the Green Revolution: Third World Agriculture, Ecology, and Politics*. Kentucky: University Press of Kentucky.

Sparrow, R. and Howard, M. (2021). Robots in agriculture: Prospects, impacts, ethics and policy. *Precision Agriculture* 22(3), 818–833.

Spykman, O., Emberger-Klein, A., Gabriel, A. and Gandorfer, M. (2022). Autonomous agriculture in public perception – German consumer segments' view of crop robots. *Computers and Electronics in Agriculture* 202, 107385.

Spykman, O., Gabriel, A., Ptacek, M. and Gandorfer, M. (2021). Farmers' perspectives on field crop robots – Evidence from Bavaria, Germany. *Computers and Electronics in Agriculture* 186, 106176.

Stilgoe, J., Owen, R. and Macnaghten, P. (2013). Developing a framework for responsible innovation. *Research Policy* 42(9), 1568–1580.

Streed, E., Tomlinson, B., Kantar, M. and Raghavan, B. (2021). How sustainable is the smart farm? *LIMITS*. Available at: https://computingwithinlimits.org/2021/papers/limits21-streed.pdf.

Tamirat, T. W., Pedersen, S. M., Ørum, J. E. and Holm, S. H. (2023). Multi-stakeholder perspectives on field crop robots: Lessons from four case areas in Europe. *Smart Agricultural Technology* 4, 100143.

Thompson, J., Millstone, E., Scoones, I., Ely, A., Marshall, F., Shah, E. and Stagl, S. (2007). *Agri-Food System Dynamics: Pathways to Sustainability in an Era of Uncertainty*. STEPS Working Paper 4. Brighton: STEPS Centre.

Tzachor, A., Devare, M., King, B., Avin, S. and Ó hÉigeartaigh, S. (2022). Responsible artificial intelligence in agriculture requires systematic understanding of risks and externalities. *Nature Machine Intelligence* 4(2), 104–109.

van der Burg, S., Bogaardt, M.-J. and Wolfert, S. (2019). Ethics of smart farming: Current questions and directions for responsible innovation towards the future. *NJAS* 90-91(1), 1–10.

Vik, J., Straete, E. P., Hansen, B. G. and Naerland, T. (2019). The political robot – The structural consequences of automated milking systems (AMS) in Norway. *NJAS – Wageningen Journal of Life Sciences* 90–91, 100305.

von Veltheim, F. R. and Heise, H. (2021). German farmers' attitudes on adoption autonomous field robots: An empirical survey. *Agriculture* 11, 216.

von Veltheim, F. R., Theuvsen, L. and Heise, H. (2022). German farmers' intention to use autonomous field robots: A PLS-analysis. *Precision Agriculture* 23(2), 670–697.

Winfield, A. (2019). Ethical standards in robotics and AI. *Nature Electronics* 2(2), 46–48.

Wiseman, L., Sanderson, J., Zhang, A. and Jakku, E. (2019). Farmers and their data: An examination of farmers' reluctance to share their data through the lens of the laws impacting smart farming. *NJAS* 90–91(1), 1–10.

Part 5

Applications

Chapter 17

Advances in the use of robotics in crop phenotyping

M. Wattad, Technion – Israel Institute of Technology, Israel; V. Alchanatis, Volcani Institute – Agricultural Research Organization, Israel; Y. Edan, Ben-Gurion University of the Negev, Israel; S. Shriki and T. Sandovsky, Volcani Institute – Agricultural Research Organization and Ben-Gurion University of the Negev, Israel; and S. Filin, Technion – Israel Institute of Technology, Israel

1 Introduction
2 The role of robotics in phenotyping
3 Two-dimensional plant imaging and analysis for phenotyping
4 Three-dimensional plant imaging and analysis for phenotyping
5 Four-dimensional plant imaging and analysis for phenotyping
6 Case study: single view 3D reconstruction and analysis for phenotyping
7 Case study: abiotic stress
8 Case study: biotic stress
9 Conclusion and future perspectives
10 Where to look for further information
11 References

1 Introduction

The world's population is projected to reach 9.7 billion people by 2050 (United Nations, 2019). This, coupled with a growing demand for a richer diet, is expected to result in a 70% increase in global food demand (Tomlinson, 2012). One of the primary constraints to meeting this demand is abiotic stress such as drought that can substantially reduce yields (Gull et al., 2019). To ensure future food security, it is crucial to develop abiotic stress-resistant plants. Detecting stress is a critical component in breeding stress-resistant cultivars.

A phenotype is the observable characteristics of an organism (e.g. a plant) such as its morphology and development. Plant phenotyping involves documenting the functional body (phenotype) of an organism (e.g. a plant) during its development as a result of the dynamic interaction between its genetic background (genotype) and the physical world in which it develops

http://dx.doi.org/10.19103/AS.2023.0124.21

(environment) (Pieruschka and Schurr, 2022). These interactions determine plant performance and productivity.

Plant phenotyping can be divided into two basic types (Li et al., 2020):

- Phenotyping of plant organs; and
- Phenotyping of the whole plant.

Plant organ phenotypes can be divided into three broad categories:

- Structural;
- Physiological; and
- Temporal.

Structural phenotyping considers morphological organ characteristics such as shape, area and angle of, e.g. stems, leaves, flowers and fruit. A physiological phenotype can be described as a set of characteristics that reflect aspects of plant growth and metabolism such as water content and chlorophyll concentration. Temporal phenotyping involves monitoring how plant organs develop over time and space, including the calculation of elongation or angular trajectory. Since traits change over time, they need to be repeatedly measured at specific intervals so that time-lapse features can be extracted.

Examples of non-complex traits that can be measured include plant height and width, volume measurement and estimation of leaf area (Li et al., 2014; Jiang and Li, 2020). Complex traits include exact leaf area, stem length, internode distance, fruit count or volume. Phenotyping encompasses above-ground plant organs such as leaves (Das Choudhury et al., 2019). It also includes below-ground organs such as roots, though this specialized area of phenotyping is not discussed in this chapter (Ahmed et al., 2016; Naveed et al., 2017). In contrast to measuring specific organs, whole plant phenotyping recognizes that plants are complex organisms with multiple organs that work in conjunction and, as such, should be analyzed holistically. In recent years, image-based two-dimensional (2D) and 3D point cloud-based solutions have emerged as powerful tools for monitoring plant growth and development (Fig. 1). By using technologies such as laser scanning, structure from motion (SfM) and the more recent polarization imaging techniques, high-resolution and detailed 3D representations of plants can be constructed. A variety of plant characteristics, including growth patterns, leaf area, chlorophyll content, structural characteristics and overall architecture, can be measured and analyzed non-destructively. Mapping the growth and spatial arrangement of plant organs such as leaves, stems, petioles and roots over time can be used to identify how plants respond to abiotic and other stresses.

A variety of advanced computational techniques, including machine-learning and computer vision algorithms, can extract valuable insights from

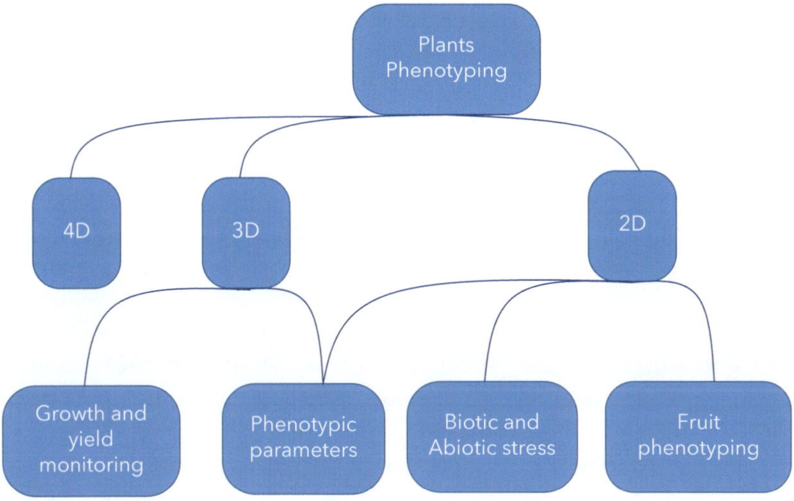

Figure 1 2D, 3D and 4D plant phenotyping.

image-based and 3D point cloud-based data. These algorithms enable automated and accurate measurement of phenotypic traits and stress responses. This makes it easier to identify stress-resistant traits, select promising cultivars and accelerate development of abiotic stress-resistant crops (Das Choudhury et al., 2019; Li et al., 2020).

2 The role of robotics in phenotyping

It is still common to undertake plant phenotyping using manual methods. Physical features such as size, shape-related parameters and color can be measured using tools such as rulers, calipers, clinometers and protractors (Su et al., 2018; Ahn et al., 2020; Zeng et al., 2022; Natukunda et al., 2022). However, such techniques are slow, labor-intensive, difficult to implement at scale, require expert input, are subjective and error-prone and can be destructive (Harberd, 2006; Trachsel et al., 2011; Araus et al., 2012). Recent advances in sensing and robotics technologies and processing algorithms provide opportunities for phenotyping at scale with unparalleled accuracy, speed and degree of detail (Costa et al., 2019). As a result robotics and automation have significantly accelerated plant phenotyping, particularly at the whole-plant level (Dhondt et al., 2013). Using robotic sensing, measurement and analysis techniques, it is possible identify how a plant is performing against predicted growth over its entire growth cycle.

Atefi et al. (2021) describe several indoor and outdoor robotic platforms and technologies for high throughput phenotyping. They divide robotic systems into three sub-systems:

- A sensing system that measures plants and their environment;
- A computational module to interpret the sensed information and make adaptive (or context-specific) decisions; and
- An actuation module to complete certain desired operations (e.g. robotic probing, trait measurement and navigation).

Robotic manipulators and ground-based vehicles are used as platforms to attach different sensors to collect data rapidly and with high repeatability. Robots are able to collect repeated measurements of plant traits within large populations at key points during a growing season, providing the necessary speed and accuracy for these phenotyping tasks (Atefi et al., 2021). Examples include a robotic system for performing in-field hyperspectral leaf imaging able to accurately predict plant nitrogen content (Chen et al., 2021) and a field-based high-throughput phenotyping system for wheat canopy height measurement, using a robot self-propelled platform equipped with LiDAR and spectral sensors (Pérez-Ruiz et al., 2020).

The operation of robots for plant phenotyping can be challenging in complex agricultural environments. The complex, heterogeneous deformable and changing nature of plant properties represents a major challenge for robot vision and sensing systems, as do dynamic and unstructured agricultural environments. Other challenges include the ability to collect and process information in real time, reliable power sources, the ability to operate in potentially challenging environments in terms of terrain and weather conditions as well as the cost of robotic technology (Barth et al., 2014; Park et al., 2020) and the need for accurate phenotype information.

Despite these challenges, agricultural robotics have been developed for different phenotyping tasks (Atefi et al., 2021; Yao et al., 2021) in field applications (e.g. Xiang et al., 2023; Xu and Li, 2022; Virlet et al., 2016), vineyards (Shafiekhani et al., 2017) and greenhouses (e.g. Smitt et al., 2021) for a multitude of plants (e.g. Virlet et al., 2016), roots (Ruppel et al., 2023; Song et al., 2021) and soil (e.g. Iqbal et al., 2020) phenotypes.

Plant phenotyping robotic systems include different types of platforms - ground (e.g. Xu and Li, 2022; Mueller-Sim et al., 2017; Iqbal et al., 2020; Ruckelshausen et al., 2009; Stager et al., 2022; Bao et al., 2021), air (e.g. Chapman et al., 2014 - helicopter; Xia and Yang, 2020 - UAV) and indoor controlled chambers (e.g. Shah et al., 2016; Smitt et al., 2021).

Different sensors have been applied for phenotyping including stereo (e.g. Bao et al., 2019a; Xiang et al., 2023), 3D LIDAR (e.g. Qiu et al., 2019; Nelson et al., 2023), hyperspectral (e.g. Chen et al., 2021; Xie and Yang, 2020) and chemical (e.g. Xiang et al., 2021; Coatsworth et al., 2023; Hu et al., 2023b; Tanner et al., 2022).

Specific attention has been given to fruit detection (for harvesting operations). Hemming et al. (2014) and Arad et al. (2020) both describe robots for harvesting sweet peppers in greenhouses. Features include a six-degree-of-freedom arm, camera systems supported by algorithms to detect ripe fruit and autonomous motion control. Lili et al. (2017) have developed a tomato harvesting robot equipped with a four-wheel steering system, a five-degree-of-freedom harvesting system, a navigation system and a binocular stereo vision system able to demonstrate accurate steering, precise laser navigation, obstacle avoidance and ripe tomato recognition. Xiong et al. (2020) have described an autonomous strawberry harvesting robot equipped with an obstacle-separation algorithm and an enhanced vision system that achieves success rates ranging from 50% to 97.1% when picking clusters of strawberries. In one-arm mode, picking takes approximately 6.1 s, and in two-arm mode, it takes about 4.6 s.

Moreover, plant structure has been evaluated using robotic arms. For example, a four-degree-of-freedom robot (Hejazipoor et al., 2021) was used to calculate plant bulk volume, detect plant stems and perform efficient spraying operations.

3 Two-dimensional plant imaging and analysis for phenotyping

Image-based data collection and processing allow researchers to conduct detailed and accurate non-destructive analyses of plant characteristics such as growth patterns, stress responses and disease symptoms. These techniques allow for fast data collection at high volume in assessing factors influencing growth and development. The principal sensors employed in image-based plant phenotyping are red-green-blue (RGB) cameras, near-infrared (NIR) and hyperspectral (HS) sensors.

RGB cameras can be used for areas such as extraction of phenotypical parameters, leaf and fruit counting as well as identifying stress factors. NIR cameras, operating just beyond human vision in the electromagnetic spectral range, can provide information in such areas as water content, nutrient status and deficiencies as well as early detection of symptoms of abiotic stress or disease. HS cameras take imaging to an even higher level of detail by capturing a wide range of narrow wavelength bands across the electromagnetic spectrum. HS imaging can detect subtle changes in nutrient content, pigment concentrations and, therefore, early stress symptoms. Advanced computer vision algorithms can then be used to extract and analyze the image data, enabling quantitative assessment of various plant traits such as leaf area, chlorophyll content, biomass, photosynthetic activity and stress responses.

3.1 Analysis of biotic and abiotic stress factors

2D sensing is frequently used in the context of identifying and classifying both biotic and abiotic stress factors, encompassing the identification of stressors and diseases (Susič et al., 2018; Abdulridha et al., 2019; Anami et al., 2020). Aiming to identify and categorize biotic stress in plants, advanced algorithms have been developed over the years (Al-Hiary et al., 2011; Mahlein et al., 2012; Lati et al., 2014; Sabrol and Satish et al., 2016; Zhang et al., 2018; McDonald et al., 2022). Early ones have focused on utilizing techniques such as color space transformation, clustering and unsupervised algorithms, which in recent years have been replaced by deep learning (DL)-driven methods. Mohanty et al. (2016) and Sladojevic et al. (2016) were the first to use DL architectures, such as the AlexNet and GoogLeNet for disease classification. Later, Ferentinos (2018) utilized a wider variety of DL networks including the AlexNet, AlexNetOWTBn, GoogLeNet, Overfeat and VGG to identify 26 disease types in 14 types of plants, using training images taken under field conditions to improve performance. Verma et al. (2020) identified diseases in grapes, experimenting with the AlexNet and ResNet18, with the latter exhibiting a greater degree of accuracy. Bi et al. (2020) used a leaf-based system to identify diseases in apple trees by using the mobileNet for real-time analysis, yet achieving a similar accuracy as heavier networks. Nagasubramanian et al. (2019) showed that DL models not only provide high accuracy but also new physiological information that can be incorporated into model prediction. By using convolutional neural networks (CNNs) on hue-saturation-intensity (HIS) values of inoculated and non-inoculated stem images, they detected disease symptoms with an accuracy of 95.7%. Zhang et al. (2020) have employed NIR imaging and 1D HS information to identify bacterial-blight-resistant rice seeds. Discriminant models were built utilizing CNN which achieved the highest classification accuracy. Feng et al. (2022) were able to detect ECD-05 disease in cabbage clubroot by phenotyping above-ground plant parts using NIR HSI combined with CNN, achieving a classification accuracy of over 85%.

Similarly, a great body of work has been devoted to identifying abiotic stressors (Kalaji et al., 2018; Beisel et al., 2018; Al-Tamimi et al., 2022). Behmann et al. (2014) presented an early stress detection in crop plants based on hyperspectral imaging and support vector machine (SVM) machine-learning approach. Zhang et al. (2019) used UAV RGB and thermal images to measure crop water stress in maize by extracting temperature with a proposed red-green ratio index (RGRI). Abdelhakim et al. (2021) employed RGB, infrared (IR) and fluorescence images to determine a wheat genotype's ability to endure drought and heat conditions by combining biomass calculations along with chlorophyll measurements from the fluorescence images. Recent approaches also focused on DL to identify and detect abiotic influences. Based on RGB images, Chandel et al. (2021) used the AlexNet, GoogLeNet and Inception V3 networks to assess water stress. Zhou et al. (2021) studied the ability to estimate flood-induced soybean injury using UAV-captured image features at various flight heights. A five-band multispectral and an IR thermal cameras were used to obtain the data. An off-the-shelve feedforward net was employed to classify soybean breeding plots according to five levels of flooding injury. Azimi et al. (2021) proposed a specialized DL pipeline for analyzing temporal changes in plant visual characteristics due to stress, allowing them to identify three distinct water stress conditions in chickpea plant shoots. Relying on machine-learning strategies, Ahmed et al. (2023) examined the impact of climate change on crop growth of the *Poaceae* family using different machine-learning models long short-term memory (LSTM), SVM and random forest (RF) to classify proteins that are affected by abiotic stresses, primarily in reference to cold, drought, heat and salinity.

3.2 Phenotypic parameters

2D image data also facilitate the extraction of geometric phenotypic parameters and contextual plant-related information. Plant characteristics, such as internode length, height, stem diameter and leaf count, provide valuable insights into plant growth and development for use in phenotyping and other applications. Image processing and graph algorithms were initially used for this purpose (Munns et al., 2010; Alchanatis et al., 2010; Das Choudhury et al., 2017). Choudhury et al. (2016) used a graph-based bi-angular convex hull method to analyze images of whole plants. Bao et al. (2019b) have used graph-based algorithms to process data acquired by a robot equipped with a variety of sensors. DL has now allowed more advanced methods of parameter estimation (Kierdorf et al., 2023). Vit and Bar-Hillel (2019) were able to measure the height of trees and the length, width and aspect ratio of banana leaves from RGB-D images using the faster region-based CNN (R-CNN). Zhou et al. (2021) also used faster R-CNN to identify characteristics such as projection, color, internode length, height, stem diameter and leaf count. Hu et al. (2023a) have used YOLOv5 architecture to determine the length of internodes in crops, achieving an accuracy of 90.5% with an average detection time of 0.02 s. Using spectral information, Sun et al. (2019) have used HS images of rice to identify characteristics such as protein content to support genome-wide association studies (GWAS), whilst Zhou et al. (2020) have been able to accurately classify wheat kernels using NIR images. Koh et al. (2021) used CNN to classify unmanned aerial vehicle (UAV) images of lodged and non-lodged wheat. Chu et al. (2022) used shallow CNNs to analyze HS images in identifying herbicides' type and stress levels in wheat.

Another valuable phenotypical plant growth aspect relates to leaf segmentation and counting, which involves identifying individual and overall numbers of leaves (Scharr et al., 2014; Minervini et al., 2016; Cruz et al., 2016; Gehan et al., 2017). Early segmentation approaches, e.g. Pang et al. (2011) and Scharr et al. (2016), applied unsupervised methods involving seed region growing, distance maps and the watershed transformation. Later ones integrated also optimization-based frameworks into their model (e.g. Wang et al., 2018). The advent of DL-based architectures has significantly increased the ability to identify complex patterns and features from large sets of data (Ubbens and Stavness, 2017; Weyler et al., 2021; Zhang et al., 2021). As an example, Giuffrida et al. (2018) have described a leaf-counting approach using multi-modal 2D images from visible light, fluorescent and NIR imaging, with several modal branches used to extract distinctive leaf features. Dobrescu et al. (2020) were able to extract different leaf traits, leaf numbers, projected leaf area and genotype classification using a multitask DL plant phenotyping model based on a state-of-the-art backbone architecture. They found that training a single model to simultaneously perform multiple tasks can improve the efficiency and accuracy of the phenotyping process by allowing the model to share information and learn common features across tasks. Bhagat et al. (2022) have developed a novel leaf segmentation method that provides insights into leaf area, count, stress and phase of development using the EfficientNet-B4 DL algorithm and Eff-Unet++. Experimental results on three datasets demonstrated that the method outperforms state-of-the-art methods for leaf segmentation.

3.3 Fruit phenotyping

Fruit phenotyping using segmentation and counting to identify features such as size, shape and color plays a crucial role in areas such as crop breeding, yield estimation, quality assessment and disease detection (Okamoto and Lee, 2009; Deepa and Geethalakshmi, 2011; Thendral et al., 2014; Fu et al., 2019). In early work, Patel et al. (2012) proposed detecting fruit using color transformations and edge detection. Diaz-Garcia et al. (2018) have used the persistent homology algorithm to improve the accuracy of shape characterization for quantitative trait locus (QTL) mapping and fruit quality assessment. DL has become the main recent approach (Cen et al., 2016; Gan et al., 2018; Grimm et al., 2019; Gao et al., 2020; Kalantar et al., 2020). As an example, Sa et al. (2016) have used NIR image fusion techniques for fruit detection using faster R-CNN models. Steinbrener et al. (2019) have classified HS images of fruits and vegetables using pre-trained CNNs. Yu et al. (2022) used the U-Net model to segment maize tassels (a particularly complex plant feature to identify accurately) from proximal and UAV-derived RGB images, achieving consistent segmentation accuracy across different tasseling stages, maize varieties and spatial resolutions.

4 Three-dimensional plant imaging and analysis for phenotyping

While 2D images are useful in identifying and measuring a wide range of plant traits, 3D imaging techniques make it possible to measure and track more complex traits such as size, volume, surface area, shape and structure to a high level of precision and accuracy (Lati et al., 2013c; Paulus, 2019). 3D data acquisition techniques include laser scanning, structured light, time-of-flight (ToF) cameras and photogrammetry (Lati et al., 2013a; Roscher et al., 2016a,b; Li et al., 2022a,b; Arad et al., 2020; Rossi et al., 2022). The generated point cloud facilitates the collection of such measurements as length, width, height, inclination and angle and diameter. These acquisition techniques require designated devices or multiple images, which may limit their application. In Section 6, we describe a case study where a single image suffices to generate 3D data of plant leaves with comparable accuracy to the aforementioned strategies.

4.1 Extraction of phenotypical information

To create 3D models of cotton plants Paproki et al. (2012) used multi-view stereo (MVS) techniques from 60-plus separate images. Applying region-growing-based segmentation techniques and modeling fitting on the reconstructed plant, they estimated phenotypic parameters such as stem height, leaf width and leaf length. Chaivivatrakul et al. (2014) utilized ToF cameras and multi-view imaging to reconstruct the 3D shape of corn plants in order to determine their phenotypic parameters. Müller-Linow et al. (2015) also generated sugar beet 3D surface models using MVS to analyze leaf blade angle distribution. Duan et al. (2016) have developed a 3D segmentation method for extracting leaf-related parameters, including number, height, length and angle from MVS-derived point clouds; a region-growing method based on an octree search was used to create 3D models of wheat organs. Elnashef et al. (2019) proposed a tensor-based point cloud classification model to semantically segment the plant into stem and leaf classes, while Zhang et al. (2023) proposed a boundary peeling clustering to isolate overlapping leaves. We make note to a range of other studies making use of MVS techniques (Bailey and Mahaffee, 2017; Elnashef et al., 2019; Atefi et al., 2019; Paulus et al., 2019; Wu et al., 2020), where variables in optimizing accuracy, including number of images and camera view angle, were discussed (Thompson et al., 2019; Qi et al., 2019; Das Choudhury et al., 2017; Li et al., 2022b; Minhui et al., 2021).

3D point cloud-based phenotyping has also benefited in terms of accuracy and efficiency from the integration of DL-based architectures (Turgut et al., 2022; Li et al., 2023). Boogaard et al. (2020), e.g. used the YOLO v3 network to estimate internode length in cucumber plants using images from multiple viewpoints. Li et al. (2022) developed DeepSeg3-Dmaize, a technique for segmenting plant point clouds utilizing high-throughput data acquisition and DL. Multi-view image acquisition was utilized to collect comprehensive data from various perspectives. Six phenotypic properties were successfully extracted using PointNet.

4.2 Growth and yield monitoring

Growth monitoring and yield estimation can be accomplished by analyzing 3D point cloud data, e.g. predicting crop yield based on growth patterns and precise counting of fruits for yield estimation (Lati et al., 2013a,b; Ge et al., 2016; Elnashef et al., 2019; Gan et al., 2018). Kjaer and Ottosen (2015) assessed the use of a commercial 3D NIR laser scanner to monitor the daily plant growth of *Brassica napus* L. (rapeseed). McCormick et al. (2016) used a Kinect depth camera to monitor developmental stages in sorghum, measuring both individual traits such as plant height, leaf angle and length as well as more complex traits like plant compactness. Herrero-Huerta et al. (2018) characterized plant leaf movement using Terrestrial LiDAR (TLiDAR). A 3D grid based on octrees was used to segment individual leaves, and principal component analysis (PCA) was applied to monitor the leaf orientation changes. Using a convex-hull computation, canopy variations of plants were also characterized. Using MVS in breeding programs Zhou et al. (2019) evaluated three machine-learning methods, boosting, K-means and SVM, to segment soybean plants at different growth stages. K-means clustering outperformed other methods in removing background information and separating non-overlapping plants. Using 3D point cloud data obtained by a laser scanner with a built-in camera, Wu et al. (2019) estimated phenotyping parameters for different growth stages for maize plants. Using point cloud clustering, denoizing and adaptive sampling, the algorithm was able to identify morphological features such as leaf length, leaf inclination angle and plant height.

Le Louëdec and Cielniak (2021) used stereo- and ToF cameras mounted on a robotic platform to segment soft fruits in order to estimate their size and shape. They employed a modified PointNet++ model to process information while the Segnet network was used to segment the images. Schunck et al. (2021) described a triangulation laser scanning-based method to identify phenotypic traits and track their development using PointNet and LatticeNet (Qi et al., 2017a,b; Rosu et al., 2019). Gong et al. (2021) proposed an architecture based on PointConv (Wu et al., 2019) with skip connections for segmenting rice ears and stalks. Li et al. (2022a,b) monitored plant growth parameters from a 3D laser-scanned plant point cloud by applying DL semantic segmentation.

5 Four-dimensional plant imaging and analysis for phenotyping

Phenotyping of plant growth dynamics requires tracing their evolution over time. Using 3D point cloud data requires spatially aligning individual plant organs among pointsets. As plants' shape changes, the alignment between different growth stages becomes a problem of non-isometric registration. To accommodate plant form complexity, Chebrolu et al. (2020) organized point cloud correspondence and registration processes using the L1-median algorithm and a hidden Markov model (HMM) for skeleton-based representation (Huang et al., 2013). To avoid unreliable skeletal leaf structures, Magistri et al. (2020) first segmented pointsets into leaf- and stem-related labels using a pre-trained SVM classifier and by applying the DBSCAN algorithm (Ester et al., 1996) separated leaf-labeled points into individual leaf blades. Pan et al. (2021) enhanced L1 skeletonization by utilizing a Laplacian contraction and proposed a hierarchical approach to find correspondence between scans by solving linear assignment problems, initially at the organ level and then at the skeletal point level. The registration was performed using a localized iterative closest point (ICP) algorithm per skeletal node, and the transformation was then extrapolated to the rest of the data.

Recognizing that common skeletal computations, Euclidean proximity measures and non-smooth transformations, do not suit evolving plant structures, Zhang et al. (2023) introduced an embedded deformation graph representation, using plant semantic information and invariant features (Figs 2 and 3). The construction of the EDG was established by first semantically segmenting the stem- and leaf-related data (Fig. 2 left). The authors proposed a native path-seeking solution, partitioning paths into individual and overlapping ones to identify stem and petiole topological structures. Then utilized invariant features to identify the skeletonized leaf form. Entity matching used linear assignment formulation followed by the establishment of key point correspondence. To address form complexity and high degrees of freedom of leaf surface evolution, the authors used the as rigid as possible (ARAP) formulation, which supports modeling the elastic

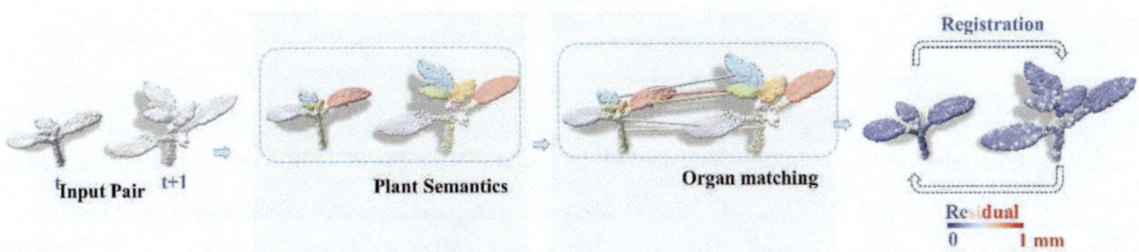

Figure 2 The 4D registration framework, where plants are first segmented into the semantic categories and individual instances, then organ matching and keypoint correspondence is established, followed by regularized ARAP registration.

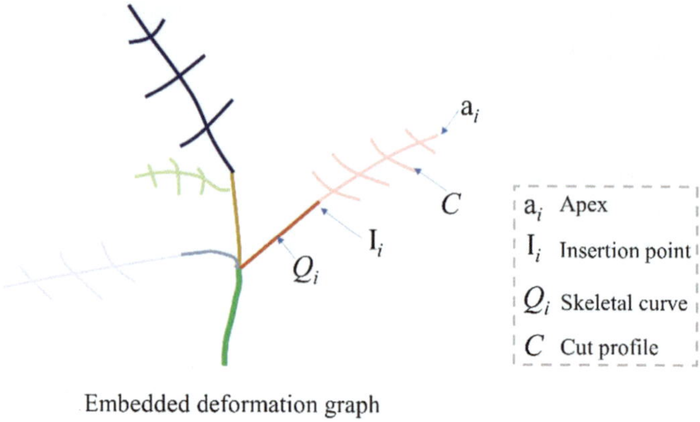

Figure 3 The embedded deformation graph (EDG) representation, formed by the stem and petioles medial axis, geodesic curves along the leaf blades and orthogonal cut profiles along their surface.

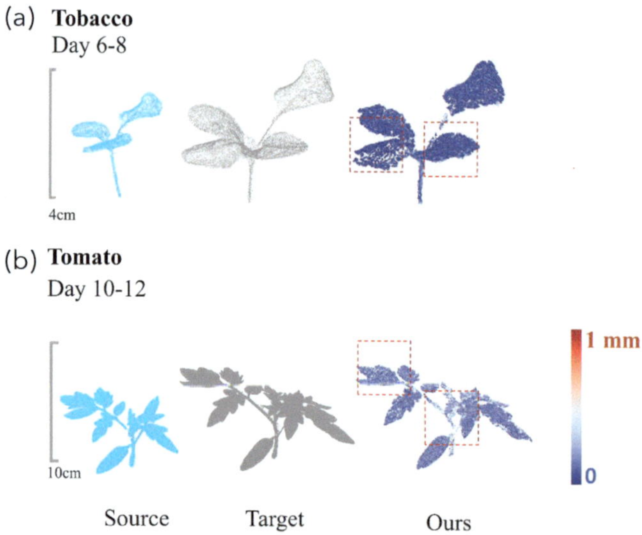

Figure 4 Registration residual on (a) tobacco and tomato (b) samples. Registered scan points are colored by the residual magnitude.

evolution while optimizing deformation according to a local rigidity criterion but introduced a point-to-surface distance regularization constraint to achieve better alignment accuracy. The application of the 3D registration in time is demonstrated in Fig. 4, demonstrating a low-residual and smooth, continuous

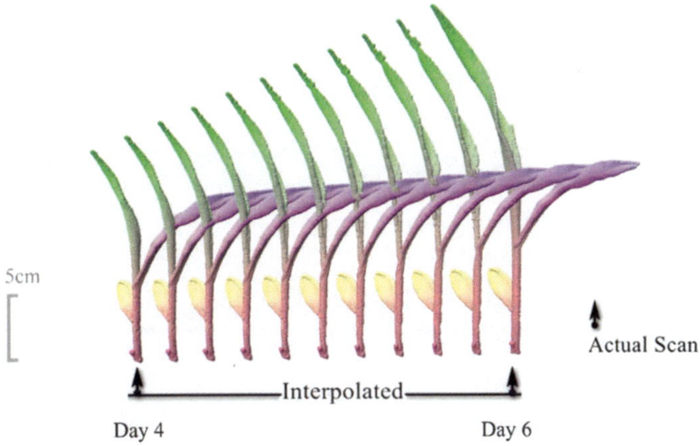

Figure 5 Temporal interpolation of a maize plant observation from day 4 to day 6. Color scheme is by a RGB color cube here representing coordinate in space. It is applied to evaluate the correspondence quality, where continuous mapping suggests correct correspondence.

Figure 6 Plant polarization reflection model and the polarization image acquisition system setup.

registration representation. The smooth transformation model also facilitates dynamic tracking of plant growth (Fig. 5).

6 Case study: single view 3D reconstruction and analysis for phenotyping

To address limitations in the application of traditional 3D sensing techniques due to sensor cost and time needed to acquire and process data, Xue et al. (2023) proposed a novel plant leaf normal estimation model, utilizing the

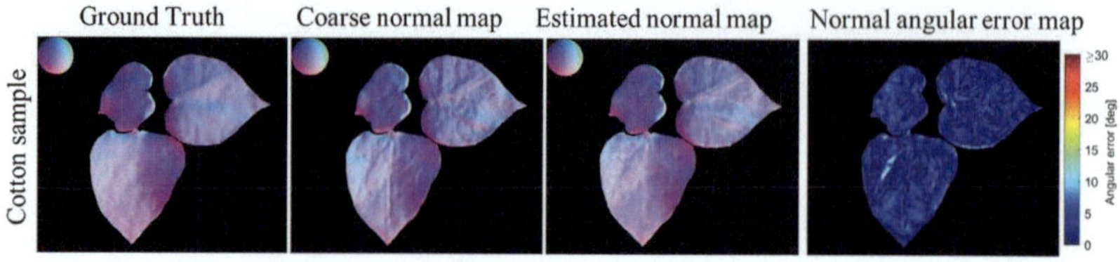

Figure 7 Leaf normal estimation result on a real-world cotton example.

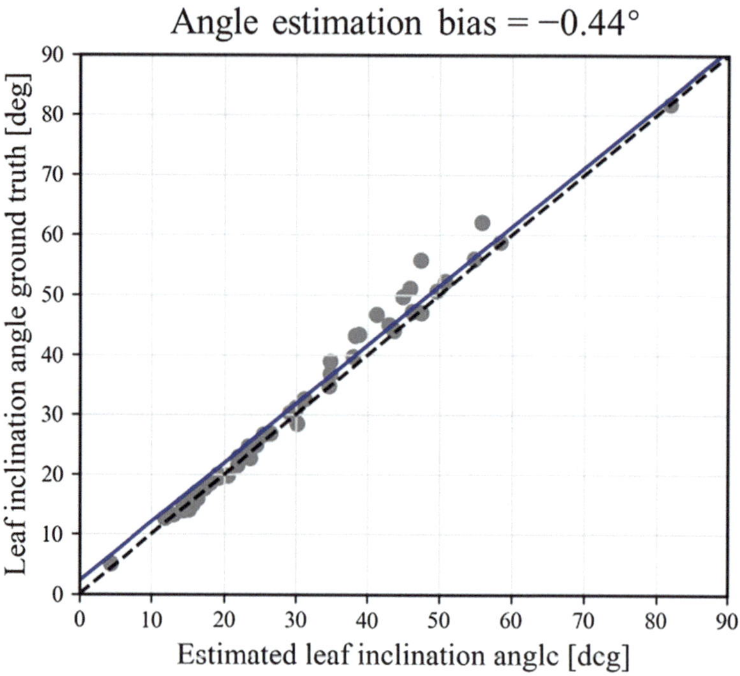

Figure 8 Leaf inclination angle estimation result on the cotton sample.

shape-sensing capability of a single polarization image. They developed a surface polarization reflection model able to process a mixture of diffuse or specular reflections with no additional computation required (Fig. 6). To address the intrinsic normal solution ambiguity associated with the polarimetric reflection, the authors derived a coarse normal map using an off-the-shelf CNN, reducing dependence on external information. As shown in Figs 7 and 8, this method produced accurate and consistent leaf shape normal estimation. Quantitative results showed that the median normal angular error was 5.3°, equivalent or better than one obtained by MVS, and leaf inclination angle bias was −0.44° using only a single image. Their model also exhibited robustness to polarization image noise and the guiding depth map quality.

7 Case study: abiotic stress

This case study reviews models to detect abiotic stress in corn plants using spectral reflectance and HS images of plant leaves (Shriki et al., 2020). Methods to efficiently reduce spatial and wavelength resolutions of HS images are described. First, supervised classification models are described to detect the abiotic stress of corn using spectral reflectance of the leaves. HS images were then used to detect abiotic stress.

7.1 Methods

Abiotic stress was induced by spraying chemicals (herbicides) that inhibit specific plant functions. In a controlled experiment with 140 corn (cultivar VIVANI 700) and sunflower (cultivar Jerusalem Dwarf Yellow Spray) plants, they were sprayed with herbicides in two consecutive years. Leaf spectral reflectance data was collected using a field spectroradiometer and an HS camera. The phenotype of each sample was labeled by an expert agronomist on each sampling day. Ground truth of abiotic stress was evaluated manually on a 1-6 scale by visual inspection by an expert agronomist at each sampling day.

Models were built separately for each crop. Three main approaches were used based on the spectral data employed (Shriki et al., 2020):

- Models based on a single wavelength (equivalent to monochrome data);
- Models based on a small number of wavelengths (equivalent to multispectral data); and
- Models based on the entire spectral signature (high-resolution spectral data).

Four statistical parametric models were constructed, and two machine-learning approaches were explored (Table 1). Principle components analysis (PCA) and partial least squares regression (PLS) were used to address the

Table 1 Statistical parametric models and machine-learning approaches that were explored

Explanatory variables	Classification model
One band	Logistic regression with one wavelength
Small number of bands	Logistic regression with seven wavelengths
Entire spectrum	PCA+ Logistic regression
	PLS-DA
	Random forest
	XGBoost

high dimensionality versus the low sample size and multi-collinearity between variables.

The data acquired in 1 year were used to train the statistical models and the data acquired in the second year were used to test the models' performance (defined as test set). Based on selected PCA components, a logistic regression model was built to identify healthy and stressed plants. In the PLS-DA model (PLS combined with discriminant analysis) the optimal number of latent variables (LVs) was selected. In addition, two decision tree ensemble algorithms were tested (Xgboost, Chen and Guestrin, 2016): random forest with bootstrap aggregation and extreme gradient boosting.

Most of these datasets were extremely imbalanced: around 80% of the plants were labeled under stress and the rest 20% were labeled as healthy plants. To overcome the imbalanced data caused by the lack of healthy plants, a Synthetic Minority Over-Sampling Technique (SMOTE) was used (Juanjuan et al., 2007).

7.2 Results

When HS images were used, a F1 score of 0.98 was achieved in detecting whether a plant was under stress, as compared to an 0.88 F1 score achieved with spectral reflectance signature only (best model using Xgboost).

When seven wavelengths were considered, the wavelengths were selected using the stepwise process. These wavelengths were used for the logistic regression classification models. For corn, they were all within the VIS and the NIR (1st derivative at 562 nm, 794 nm, 683 nm, 656 nm, 535 nm, 703 nm and 542 nm). XGBoost and random forest models reached the highest cross-validation results for the 1st derivative datasets with 0.88 and 0.92 F1 score for corn and sunflower plants, respectively.

An early detection model was evaluated in an attempt to predict the phenotypes' status observed 2 days ahead. The models that exhibited the best results for current time classification also reached the best results for 2 days ahead classification. The best performance for corn plants was achieved by random forest and logistic regression with seven wavelengths (0.73 F1 score).

As described previously, the models which achieved the best performance were based on random forest and logistic regression. When the logistic regression model is considered, the wavelengths that were selected as explanatory variables were also the most important.

Considering that multispectral systems do not have spectral resolution as narrow as 3 nm, but rather in the range of 8–10 nm, feature importance was merged every 8 nm and then the most important wavelengths were selected. The most important spectral ranges were found to be in the green and NIR

Table 2 The most important wavelengths in detection of abiotic stress in corn

	Wavelength (nm)
Logistic regression	558, 790, 678, 654, 534, 702, 542
Random forest	422, 550, 430, 414, 558, 462, 798, 846, 790, 862

Table 3 Classification accuracy of a logistic regression model using the reflectance of ten wavelengths for corn and sunflower raw data sets

	Year 1		Year 2	
	Corn reflectance data	Corn 1st derivative	Corn reflectance data	Corn 1st derivative
F1 score 10 wavelengths model (entire spectrum model)	0.76	0.75 (0.88)	0.91	0.90 (0.79)

ranges (Table 2). Just one model indicated that wavelengths in the SWIR range are important.

The data used to construct the models were pre-processed using smoothing and derivative operators (Stavitzky Golay procedure). For applicative use, this implies that a number of wavelengths around the center wavelength have also to be acquired (11 additional wavelengths in the present work). When just 10 wavelengths are measured, the models are constructed using the reflectance spectrum only. Table 3 shows the classification accuracy of a logistic regression model using the reflectance of ten wavelengths for corn and sunflower raw data sets. The classification performance of the models using the 1st derivative is shown for comparison.

The early prediction models that were developed for intervals of 2 days yielded similar results. The random forest classification model achieved an F1 score of 0.73 and 0.85 for corn and sunflower plants respectively, for a 2 days ahead forecast.

More details can be found at Shriki et al. (2020).

8 Case study: biotic stress

This case study compares leaf point spectra and whole plant HS images in the early detection of Fusarium infection in corn plants (Sandovsky et al., 2021).

8.1 Methods

Commercial corn seeds (7210) were infected and germinated in a greenhouse for a period of 3 weeks. The spectral reflectance of plant leaves was measured 21 days after seeding, when they reached stage V2, according to the developmental stages of growth defined by the BBCH scale (Lancashire et al., 1991).

Then, 22 days after seeding, one day after spectral sampling, plants were taken out of the pots and the disease infection level in the roots was visually assessed by an expert phyto-pathologist, visually inspecting the roots of the plant according to an established protocol (Kokkoris et al., 2019). The disease severity was classified from level 0 to 5 where level 0 represented a healthy and level 5 represented a severely infected plant, according to the state of the roots of each plant (Fig. 9).

All plants were sampled with two systems (Fig. 10):

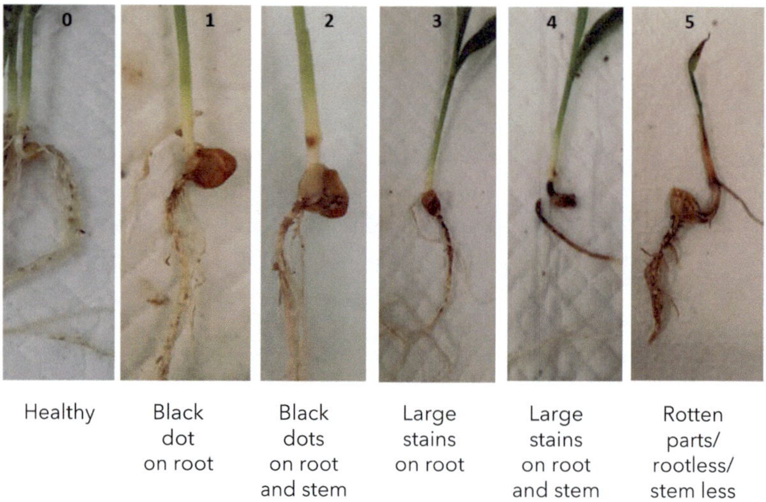

Figure 9 Labeling protocol of plants to different levels of disease according to plant status after 21 days.

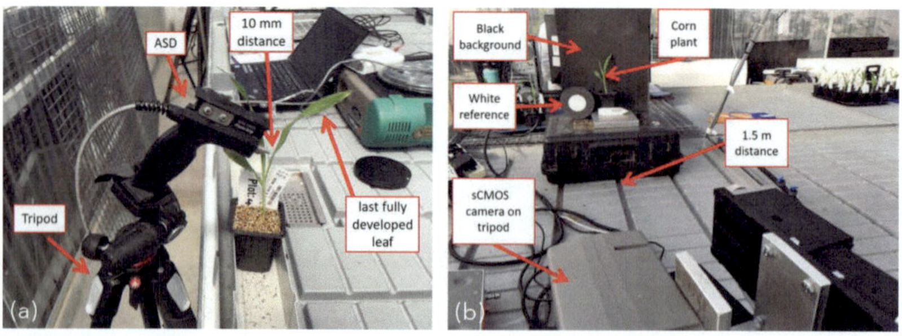

Figure 10 Experimental setting (a) ASD measurement (b) HS camera measurement.

- A spectro-radiometer (FieldSpec 4 hi-res, ASD Inc. Malvern Panalytical, Boulder, Colorado, USA) in the range of 350–2500 nm, with 3 nm spectral resolution in VNIR range) that provided the spectral reflectance of one of the plant's leaves at a single point; and
- An HS camera (V10E pushbroom HS camera (SPECIM, Oulu, Finland) VIS and VNIR ranges, 400–1000 nm) that provided a hyper cube that contained the spectral reflectance of each point in an image of the plant.

The point spectral measurements from the ASD were used to determine the significant wavelengths for Fusarium detection using stepwise regression. Using the derived significant wavelengths, two classification models were developed:

- A logistic regression model that used the point spectral measurements from the ASD; and
- A random forest model that used features from the HS images.

The most significant wavelengths for separating between healthy and infected plants were derived using a stepwise regression model applied with the Akaike Information Criterion (AIC, Yamashita et al., 2007). AIC estimates the relative amount of information lost by a given model: the less information a model loses, the higher the quality of that model. Logistic regression was then used to develop a model to classify healthy and infected plants using the reflectance at the selected wavelengths (model 1).

Raw HS data were obtained in DN (digital number). The first steps in processing the data were to crop the plant from the entire scene and derive the spectral reflectance of each point. Reflectance images of the wavelengths selected by the stepwise regression on the point measurements were calculated and saved as multispectral cubes.

Five different features were calculated for each wavelength of the multispectral image and two additional geometrical features from one of the wavelengths:

1. *Aver* (λ) – average reflectance of all plant pixels in the image at wavelength λ.
2. SD (λ) – standard deviation of all plant pixels in the image at wavelength λ.
3. Coefficient of variation (λ) (CV) – $\frac{\sigma}{\mu}$ the average reflectance of all plant pixels in the image at wavelength λ divided by the standard deviation of all plant pixels in the image at wavelength λ.

4 *Discriminabil* (λ) (*Disc*) - $\frac{\mu_{edges} - \mu_{center}}{\mu_{all}}$ the average reflectance of pixels in the edge of the plant (μ_{edges}) subtraction of the pixels in the center of the plan (μ_{center}) divided by the average reflectance of all plant pixels (μ_{all}) in the image at wavelength λ.
5 E(λ) - $\frac{\mu_{edges}}{\mu_{center}}$ the average reflectance of pixels in the edge of the plant (μ_{edges}) divided by the pixels in the center of the plan (μ_{center}) in the image at wavelength λ.
6 LAI - the number of the plant's pixels.
7 Height/width ratio - the ratio between the height to the width of the plant.

Features 1-3 were selected based on previous recommendations. All features except for *Ave* (λ) use the spatial information contained in the multispectral image. *Discriminab* (λ) and E*R* (λ) were selected considering the Fusarium infection process which starts from the roots and continues to the edges of the leaves. LAI and Height/Width Ratio are related to the shape of the plant (which is the same in all wavelengths), therefore they were calculated once for each plant from the first wavelength (400 nm). The first five features are vectors that contain a value for each wavelength λ. The two remaining features are scalars.

A random forest model (model 2) was used to classify between healthy and infected plants. Each decision tree was built over a random extraction of the observations from the dataset and a random extraction of the features (wavelengths). Since not every tree sees all the features or all the observations, this guarantees that the trees are de-correlated and therefore less prone to over-fitting (Ghojogh and Crowley, 2019).

Due to the small number of samples, the number of disease levels were aggregated from five to two: levels 0 and 1 were aggregated to healthy, and levels 2 to 5 were aggregated to infected.

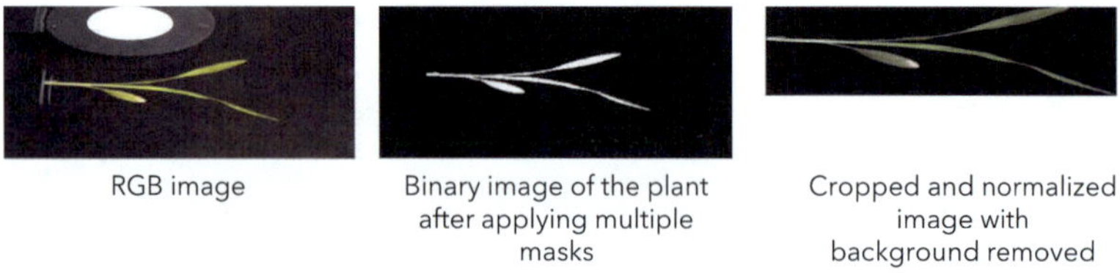

Figure 11 HS images pre-processing.

8.2 Results

Stepwise regression ranked results reveal that the most significant wavelength for detection of Fusarium-infected plants was 400 nm. Additional significant wavelengths were found at 400, 440, 630 and 700 nm, mainly affected by leaf pigment content, and reflectance at 750, 820 and 900 nm, that depends on the leaf biology structure.

The reflectance at the above specific wavelengths was used in a logistic regression model (model 1) to classify between healthy and infected plants. Using the leave-one-out-cross-validation method, the model was able to predict the state of the plants for 76% of the plants. The prediction of the infected plants was consistently better than healthy plants prediction (correctly classified 87% of the infected plants vs. 56% of healthy plants).

Figure 11 shows typical HS images along the pre-processing pipeline, which includes the following:

- Extract the color (RGB) image of the plant;
- Find the plant contour;
- Crop the image to the minimum length and width; and
- Normalize the image and save the multispectral cube with seven wavelengths: 400, 440, 630, 700, 750, 820 and 900 nm.

Table 4 summarizes the classification results of the models with AUC and accuracy depicted for models' comparison. The model which was based on the HS features (model 2) was able to predict the state of the plant for 82% of the plants compared to 76% in model 1. This result is very significant since identification was carried out at an early stage when it was not possible to discriminate between healthy and infected plants without destructively testing the plant (chemical testing or plant extraction and visual assessment of the roots).

Table 4 Summary of classification results of the models

Model	Performance measure			
	AUC	Accuracy		
Model 1	0.729	0.76	Confidence interval	
Logistic regression – point measurements			0.713	0.816
Model 2	0.794	0.82	Confidence interval	
Random forest – features from multispectral imaging			0.782	0.864

The confidence interval of the classification accuracy is shown at a significance level of $\alpha = 0.05$.

Model 2 (based on the features calculated from the multispectral cube) was evaluated using features that were calculated from 50 nm-bandwidth multispectral cube and the 150 nm-bandwidth multispectral cube. The classification results of the model did not change for any of the bandwidths

In summary, early detection of Fusarium infection, without destructive phenotyping, was demonstrated with a classification accuracy of 82%. This was achieved by applying spectral analysis on HS images of corn plants in the V2 stage, when no external visual symptoms of the disease were expressed. The accuracy of the classification models in this work were found to be invariant to spectral bandwidth, thus potentially enabling smaller and lower cost sensors. This will also enable their use with simpler robotic arms to perform the phenotyping task. Dimensionality reduction enables to build a simpler model and can lead to reduced sensor cost. The narrower the bandwidth, the higher the cost of the sensors.

For more robust results, more samples should be collected over the course of the growing stages to provide a better understanding of the spectral discrimination over time. To create a more generic model, it is necessary to collect images in different lighting conditions, from different areas and of different varieties. Increasing the data set will allow us to split the data to train and test set and build a stronger model, instead of the leave one out method that was used in this work.

More details can be found at Sandovsky et al. (2021).

9 Conclusion and future perspectives

Agricultural robotics is a rapidly developing field with a range of developments that can support phenotyping operations. Robotic manipulators and ground-based vehicles are used as platforms to attach different sensors to collect data rapidly and with high repeatability. Robots are able to collect repeated measurements of plant traits within large populations at key points during a growing season, providing the necessary speed and accuracy for these phenotyping tasks.

In this chapter, a wide spectrum of sensing technologies have been presented. Combination of advanced sensors with robotic platforms can provide phenotyping systems with high performance capabilities.

Indoor high-throughput phenotyping systems have become commercially available, with a high degree of automation and robotics, as well as a wide range of sensing possibilities. Nevertheless, field phenotyping systems are much more challenging and less available; most of such systems are custom made and not commercially available. Robotic technologies for autonomous navigation and sensors deployment in such field phenotyping platforms will gradually become available.

Moreover, incorporation of robotic systems in high-throughput field phenotyping systems produces vast amounts of data which require new processing methodologies, e.g. edge computing for mitigating the requirements of data transfer from the field and artificial intelligence methods to extract meaningful information from the acquired data.

10 Where to look for further information

This chapter has focused on the capabilities of sensing systems that can be mounted on robotic platforms for indoor and field phenotyping. Further reading on the robotics components on phenotyping systems can be found at Atefi et al. (2021), Field robots for plant phenotyping, https://edepot.wur.nl/575608. Robotic phenotyping, https://www.lincoln.ac.uk/liat/research/roboticphenotyping/.

11 References

Abdelhakim, L. O. A., Rosenqvist, E., Wollenweber, B., Spyroglou, I., Ottosen, C. O. and Panzarová, K. (2021). Investigating combined drought-and heat stress effects in wheat under controlled conditions by dynamic image-based phenotyping. *Agronomy* 11(2), 364.

Abdulridha, J., Ehsani, R., Abd-Elrahman, A. and Ampatzidis, Y. (2019). A remote sensing technique for detecting laurel wilt disease in avocado in presence of other biotic and abiotic stresses. *Computers and Electronics in Agriculture* 156, 549–557.

Ahmed, B., Haque, M. A., Iquebal, M. A., Jaiswal, S., Angadi, U. B., Kumar, D. and Rai, A. (2023). DeepAProt: Deep learning based abiotic stress protein sequence classification and identification tool in cereals. *Frontiers in Plant Science* 13, 1008756.

Ahn, E., Odvody, G., Prom, L. K. and Magill, C. (2020). Leaf angle distribution in Johnsongrass, leaf thickness in sorghum and Johnsongrass,and association with response to Colletotrichum sublineola. *Scientific Reports* 10 (1), 1-6. https://doi.org/10.1038/s41598-020-79473-x.

Alchanatis, V., Cohen, Y., Cohen, S., Moller, M., Sprinstin, M., Meron, M., Tsipris, J., Saranga, Y. and Sela, E. (2010). Evaluation of different approaches for estimating and mapping crop water status in cotton with thermal imaging. *Precision Agriculture* 11(1), 27–41.

Al-Hiary, H., Bani-Ahmad, S., Reyalat, M., Braik, M. and Alrahamneh, Z. (2011). Fast and accurate detection and classification of plant diseases. *International Journal of Computer Applications* 17(1), 31–38.

Al-Tamimi, N., Langan, P., Bernád, V., Walsh, J., Mangina, E. and Negrão, S. (2022). Capturing crop adaptation to abiotic stress using image-based technologies. *Open Biology* 12(6), 210353.

Anami, B. S., Malvade, N. N. and Palaiah, S. (2020). Classification of yield affecting biotic and abiotic paddy crop stresses using field images. *Information Processing in Agriculture* 7(2), 272-285.

Arad, B., Balendonck, J., Barth, R., Ben-Shahar, O., Edan, Y., Hellström, T., Hemming, J., Kurtser, P., Ringdahl, O., Tielen, T. and van Tuijl, B. (2020). Development of a sweet pepper harvesting robot. *Journal of Field Robotics* 37(6), 1027–1039.

Atefi, A., Ge, Y., Pitla, S. and Schnable, J. (2019). In vivo human-like robotic phenotyping of leaf traits in maize and sorghum in greenhouse. *Computers and Electronics in Agriculture* 163, 104854.

Atefi, A., Ge, Y., Pitla, S. and Schnable, J. (2021). Robotic technologies for high-throughput plant phenotyping: Contemporary reviews and future perspectives. *Frontiers in Plant Science* 12, 611940.

Azimi, S., Wadhawan, R. and Gandhi, T. K. (2021). Intelligent monitoring of stress induced by water deficiency in plants using deep learning. *IEEE Transactions on Instrumentation and Measurement* 70, 1–13.

Bailey B. N. and Mahaffee W. F. (2017). Rapid measurement of the three-dimensional distribution of leaf orientation and the leaf angle probability density function using terrestrial LiDAR scanning. *Remote Sensing of Environment* 194, 63–76. doi: 10.1016/j.rse.2017.03.011.

Bao, Y., Tang, L., Breitzman, M. W., Salas Fernandez, M. G. and Schnable, P. S. (2019a). Field-based robotic phenotyping of sorghum plant architecture using stereo vision. *Journal of Field Robotics* 36(2), 397–415.

Bao, Y., Zarecor, S., Shah, D., Tuel, T., Campbell, D. A., Chapman, A. V. E., Imberti, D., Kiekhaefer, D., Imberti, H., Lübberstedt, T., Yin, Y., Nettleton, D., Lawrence-Dill, C. J., Whitham, S. A., Tang, L. and Howell, S. H. (2019b). Assessing plant performance in the Enviratron. *Plant Methods* 15(1), 117.

Bao, Y., Gai, J., Xiang, L. and Tang, L. (2021). Field robotic systems for high-throughput plant phenotyping: A review and a case study. In: Zhou, J. and Nguyen, H. T. *High-Throughput Crop Phenotyping* (pp. 13–38). Cham: Springer International Publishing.

Behmann, J., Steinrücken, J. and Plümer, L. (2014). Detection of early plant stress responses in hyperspectral images. *ISPRS Journal of Photogrammetry and Remote Sensing* 93, 98–111.

Beisel, N. S., Callaham, J. B., Sng, N. J., Taylor, D. J., Paul, A. L. and Ferl, R. J. (2018). Utilization of single-image normalized difference vegetation index (SI-NDVI) for early plant stress detection. *Applications in Plant Sciences* 6(10), e01186.

Bhagat, S., Kokare, M., Haswani, V., Hambarde, P. and Kamble, R. (2022). Eff-UNet++: A novel architecture for plant leaf segmentation and counting. *Ecological Informatics* 68, 101583.

Boogaard, F. P., Rongen, K. S. A. H. and Kootstra, G. W. (2020). Robust node detection and tracking in fruit-vegetable crops using deep learning and multi-view imaging. *Biosystems Engineering* 192, 117–132.

Cen, H., Lu, R., Zhu, Q. and Mendoza, F. (2016). Nondestructive detection of chilling injury in cucumber fruit using hyperspectral imaging with feature selection and supervised classification. *Postharvest Biology and Technology* 111, 352–361.

Chaivivatrakul, S., Tang, L., Dailey, M. N. and Nakarmi, A. D. (2014). Automatic morphological trait characterization forcorn plants via 3-D holographic reconstruction. *Computers and Electronics in Agriculture* 109, 109–123.

Chandel, N. S., Chakraborty, S. K., Rajwade, Y. A., Dubey, K., Tiwari, M. K. and Jat, D. (2021). Identifying crop water stress using deep learning models. *Neural Computing and Applications* 33(10), 5353–5367.

Chapman, S. C., Merz, T., Chan, A., Jackway, P., Hrabar, S., Dreccer, M. F., Holland, E., Zheng, B., Ling, T. and Jimenez-Berni, J. (2014). Pheno-copter: A low-altitude, autonomous remote-sensing robotic helicopter for high-throughput field-based phenotyping. *Agronomy* 4(2), 279–301.

Chebrolu, N., Labe, T. and Stachniss, C. (2020). Spatio-temporal non-rigid registration of 3-D point clouds of plants. In *IEEE International Conference on Robotics and Automation*, Paris, France. (pp. 3112–3118).

Chen, Z., Wang, J., Wang, T., Song, Z., Li, Y., Huang, Y., Wang, L. and Jin, J. (2021). Automated in-field leaf-level hyperspectral imaging of corn plants using a Cartesian robotic platform. *Computers and Electronics in Agriculture* 183, 105996.

Choudhury, S. D., Stoerger, V., Samal, A., Schnable, J. C., Liang, Z. and Yu, J. G. (2016). Automated vegetative stage phenotyping analysis of maize plants using visible light images. In KDD Workshop on Data Science for Food, Energy and Water, San Francisco, California, USA.

Chu H., Zhang C., Wang M., Gouda M., Wei X., He Y. and Liu Y. (2022). Hyperspectral imaging with shallow convolutional neural networks (SCNN) predicts the early herbicide stress in wheat cultivars. *Journal of Hazardous Materials* 421, 126706. doi: 10.1016/j.jhazmat.2021.126706.

Coatsworth, P., Cotur, Y., Naik, A., Asfour, T., Silva-Pinto Collins, A., Olenik, S., Gonzalez-Macia, L., Bozkurt, T., Chao, D.-Y. and Güder, F. (2023). Time-resolved chemical phenotyping of whole plant roots with printed electrochemical sensors and machine learning. *bioRxiv* 10, eadj6315(2024). doi: 10.1126/sciadv.adj6315.

Costa, C., Schurr, U., Loreto, F., Menesatti, P. and Carpentier, S. (2019). Plant phenotyping research trends, a science mapping approach. *Frontiers in Plant Science* 9, 426195. https://doi.org/10.3389/fpls.2018.01933.

Cruz, J. A., Yin, X., Liu, X., Imran, S. M., Morris, D. D., Kramer, D. M. and Chen, J. (2016). Multi-modality imagery database for plant phenotyping. *Machine Vision and Applications* 27(5), 735–749.

Das Choudhury, S., Goswami, S., Bashyam, S., Samal, A. and Awada, T. (2017). Automated stem angle determination for temporal plant phenotyping analysis. In *Proceedings of the IEEE International Conference on Computer Vision Workshops (ICCVW)*, Venice, Italy (pp. 2022–2029). doi: 10.1109/ICCVW.2017.237.

Das Choudhury, S., Samal, A. and Awada, T. (2019). Leveraging image analysis for high-throughput plant phenotyping. *Frontiers in Plant Science* 10, 508. https://doi.org/10.3389/fpls.2019.00508.

Deepa, P. and Geethalakshmi, S. N. (2011, July). Improved watershed segmentation for apple fruit grading. In *2011 International Conference on Process Automation, Control and Computing* (pp. 1–5). IEEE Publications.

Dhondt, S., Wuyts, N. and Inzé, D. (2013). Cell to whole-plant phenotyping: The best is yet to come. *Trends in Plant Science* 18(8), 428–439.

Diaz-Garcia, L., Covarrubias-Pazaran, G., Schlautman, B., Grygleski, E. and Zalapa, J. (2018). Image-based phenotyping for identification of QTL determining fruit shape and size in American cranberry (Vaccinium macrocarpon L.). *PeerJ* 6, e5461.

Duan, T., Chapman, S. C., Holland, E., Rebetzke, G. J., Guo, Y. and Zheng, B. (2016). Dynamic quantification of canopystructure to characterize early plant vigour in wheat genotypes. *Journal of Experimental Botany* 67(15), 4523–4534.

Elnashef, B., Filin, S. and Lati, R. N. (2019). Tensor-based classification and segmentation of 3-D image-driven data for organ level plant growth analysis. *Computers and Electronics in Agriculture* 156, 51–61. https://doi.org/10.1016/j.compag.2018.10.036. (https://www.sciencedirect.com/science/article/pii/S0168169918306008).

Ester, M., Kriegel, H. P., Sander, J. and Xu, X. (1996). A density-based algorithm for discovering clusters in large spatial databases with noise. In *Proceedings of the*

Second International Conference on Knowledge Discovery and Data Mining (pp. 226–231).

Feng, L., Wu, B., Chen, S., Zhang, C. and He, Y. (2022). Application of visible/near-infrared hyperspectral imaging with convolutional neural networks to phenotype aboveground parts to detect cabbage Plasmodiophora brassicae (clubroot). *Infrared Physics & Technology* 121, 104040. https://doi.org/10.1016/j.infrared.2022.104040

Ferentinos, K. P. (2018). Deep learning models for plant disease detection and diagnosis. *Computers and Electronics in Agriculture* 145, 311–318.

Fu, L., Tola, E., Al-Mallahi, A., Li, R. and Cui, Y. (2019). A novel image processing algorithm to separate linearly clustered kiwifruits. *Biosystems Engineering* 183, 184–195.

Gan, H., Lee, W. S. and Alchanatis, V. (2018). A photogrammetry-based image registration method for multi-camera systems–With applications in images of a tree crop. *Biosystems Engineering* 174, 89–106.

Gan, H., Lee, W. S., Alchanatis, V., Ehsani, R. and Schueller, J. K. (2018). Immature green citrus fruit detection using color and thermal images. *Computers and Electronics in Agriculture* 152, 117–125.

Gao, Z., Shao, Y., Xuan, G., Wang, Y., Liu, Y. and Han, X. (2020). Real-time hyperspectral imaging for the in-field estimation of strawberry ripeness with deep learning. *Artificial Intelligence in Agriculture* 4, 31–38.

Ge, Y., Bai, G., Stoerger, V. and Schnable, J. C. (2016). Temporal dynamics of maize plant growth, water use, and leaf water content using automated high throughput RGB and hyperspectral imaging. *Computers and Electronics in Agriculture* 127, 625–632.

Gehan, M. A., Fahlgren, N., Abbasi, A., Berry, J. C., Callen, S. T., Chavez, L., Doust, A. N., Feldman, M. J., Gilbert, K. B., Hodge, J. G., Hoyer, J. S., Lin, A., Liu, S., Lizárraga, C., Lorence, A., Miller, M., Platon, E., Tessman, M. and Sax, T. (2017). PlantCV v2: Image analysis software for high-throughput plant phenotyping. *PeerJ* 5, e4088.

Ghojogh, B. and Crowley, M. (2019). Instance ranking and numerosity reduction using matrix decomposition and subspace learning. Lecture Notes in Computer Science (including subseries Lecture Notes in Artificial Intelligence and Lecture Notes in Bioinformatics), 11489 LNAI, 160–172. https://doi.org/10.1007/978-3-030-18305-9_13.

Giuffrida, M. V., Doerner, P. and Tsaftaris, S. A. (2018). Pheno-deep counter: A unified and versatile deep learning architecture for leaf counting. *The Plant Journal* 96(4), 880–890.

Gong, L., Du, X., Zhu, K., Lin, K., Lou, Q., Yuan, Z., Huang, G. and Liu, C. (2021). Panicle-3D: Efficient Phenotyping Tool for Precise Semantic Segmentation of Rice Panicle Point Cloud. *Plant Phenomics* doi:10.34133/2021/9838929.

Grimm, J., Herzog, K., Rist, F., Kicherer, A., Toepfer, R. and Steinhage, V. (2019). An adaptable approach to automated visual detection of plant organs with applications in grapevine breeding. *Biosystems Engineering* 183, 170–183.

Gull, M., Sofi, P. A. and Ara, A. (2019). Physiological and biochemical response of cowpea Vigna unguiculata L. landraces of Kashmir valley under water stress. *Electronic Journal of Plant Breeding* 10 (4), 1461–1470.

Hejazipoor, H., Massah, J., Soryani, M., Vakilian, K. A. and Chegini, G. (2021). An intelligent spraying robot based on plant bulk volume. *Computers and Electronics in Agriculture* 180, 105859.

Hemming, J., Bac, C. W., van Tuijl, B. A., Barth, R., Bontsema, J., Pekkeriet, E. J. and Van Henten, E. (2014). A robot for harvesting sweet-pepper in greenhouses. *International Conference of Agricultural Engineering* Zurich. 6–10. https://www.geyseco.es/geystiona/adjs/comunicaciones/304/C01140001.pdflast accessed 7.2.2024.

Herrero-Huerta, M., Lindenbergh, R. and Gard, W. (2018). Leaf movements of indoor plants monitored by terrestrial LiDAR. *Frontiers in Plant Science* 9, 189.

Hu, J., Li, G., Mo, H., Lv, Y., Qian, T., Chen, M. and Lu, S. (2023a). Crop node detection and internode length estimation using an improved YOLOv5 model. *Agriculture* 13(2), 473.

Hu, X. L., Gan, H. Q., Qin, Z. Y., Liu, Q., Li, M., Chen, D., Sessler, J. L., Tian, H. and He, X. P. (2023b). Phenotyping of methicillin-resistant Staphylococcus aureus using a ratiometric sensor array. *Journal of the American Chemical Society* 145(16), 8917–8926.

Huang, H., Wu, S., Cohen-Or, D., Gong, M., Zhang, H., Li, G. and Chen, B. (2013). L1-medial skeleton of point cloud. *ACM Transactions on Graphics* 32, 65–61.

Iqbal, J., Xu, R., Halloran, H. and Li, C. (2020). Development of a multi-purpose autonomous differential drive mobile robot for plant phenotyping and soil sensing. *Electronics* 9(9), 1550.

Jiang, Y. and Li, C. (2020). Convolutional neural networks for image-based high-throughput plant phenotyping: A review. *Plant Phenomics* 2020, 4152816. https://doi.org/10.34133/2020/4152816.

Juanjuan, W., Mantao, X., Hui, W. and Jiwu, Z. (2007). Classification of imbalanced data by using the SMOTE algorithm and locally linear embedding. *International Conference on Signal Processing Proceedings, ICSP* 3, 4129201. https://doi.org/10.1109/ICOSP.2006.345752.

Kalaji, H. M., Rastogi, A., Živčák, M., Brestic, M., Daszkowska-Golec, A., Sitko, K., Alsharafa, K. Y., Lotfi, R., Stypiński, P., Samborska, I. A. and Cetner, M. D. (2018). Prompt chlorophyll fluorescence as a tool for crop phenotyping: An example of barley landraces exposed to various abiotic stress factors. *Photosynthetica* 56(3), 953–961.

Kalantar, A., Edan, Y., Gur, A. and Klapp, I. (2020). A deep learning system for single and overall weight estimation of melons using unmanned aerial vehicle images. *Computers and Electronics in Agriculture* 178, 105748.

Kierdorf, J., Junker-Frohn, L. V., Delaney, M., Olave, M. D., Burkart, A., Jaenicke, H., Muller, O., Rascher, U. and Roscher, R. (2023). GrowliFlower: An image time-series dataset for GROWth analysis of cauLIFLOWER. *Journal of Field Robotics* 40(2), 173–192.

Kjaer, K. H. and Ottosen, C. O. (2015). 3-D laser triangulation for plant phenotyping in challenging environments. *Sensors* 15(6), 13533–13547.

Kokkoris, V., Pogiatzis, A. and Hart, M. M. (2019). Contrasting Common Measures of Arbuscular Mycorrhizal Fungal Root Colonization. *Journal of Microbiological Methods* 167, 105727. https://doi.org/10.1016/j.mimet.2019.105727.

Lancashire, D., Bleiholder, H., Van Den Boom, T., Langeluddeke, P., Stauss, R., Weber, E., Hack, H. and Witzenberger, A. (1991). A uniform decimal code for growth stages of crops and weeds. *Annals of Applied Biology* 119 (3), 561–601. https://doi.org/10.1111/j.1744-7348.1991.tb04895.x.

Lati, R. N., Filin, S. and Eizenberg, H. (2013a). Estimating plant growth parameters using an energy minimization-based stereovision model. *Computers and Electronics in Agriculture* 98, 260–271.

Lati, R. N., Filin, S. and Eizenberg, H. (2013b). Estimation of plants' growth parameters via image-based reconstruction of their three-dimensional shape. *Agronomy Journal* 105(1), 191–198.

Lati, R. N., Filin, S. and Eizenberg, H. (2013c). Plant growth parameter estimation from sparse 3-D reconstruction based on highly-textured feature points. *Precision Agriculture* 14(6), 586–605.

Lati, R. N., Filin, S., Aly, R., Lande, T., Levin, I. and Eizenberg, H. (2014). Using genetically modified tomato crop plants with purple leaves for absolute weed/crop classification. *Pest Management Science* 70(7), 1059–1065.

Le Louëdec, J. and Cielniak, G. (2021). 3-D shape sensing and deep learning-based segmentation of strawberries. *Computers and Electronics in Agriculture* 190, 106374.

Li, D., Cao, Y., Shi, G., Cai, X., Chen, Y., Wang, S. and Yan, S. (2019). An overlapping-free leaf segmentation method for plant point clouds. *IEEE Access* 7, 129054–129070.

Li, D., Shi, G., Li, J., Chen, Y., Zhang, S., Xiang, S. and Jin, S. (2022a). PlantNet: A dual-function point cloud segmentation network for multiple plant species. *ISPRS Journal of Photogrammetry and Remote Sensing* 184, 243–263.

Li, Y., Wen, W., Miao, T., Wu, S., Yu, Z., Wang, X., Guo, X. and Zhao, C. (2022b). Automatic organ-level point cloud segmentation of maize shoots by integrating high-throughput data acquisition and deep learning. *Computers and Electronics in Agriculture* 193, 106702.

Li, Y., Zhan, X., Liu, S., Lu, H., Jiang, R., Guo, W., Chapman, S., Ge, Y., Solan, B., Ding, Y. and Baret, F. (2023). Self-supervised plant phenotyping by combining domain adaptation with 3-D plant model simulations: Application to wheat leaf counting at seedling stage. *Plant Phenomics* 5, 0041. doi: 10.34133/plantphenomics.0041.

Li, Z., Guo, R., Li, M., Chen, Y. and Li, G. (2020). A review of computer vision technologies for plant phenotyping. *Computers and Electronics in Agriculture* 176, 105672.

Lili, W., Bo, Z., Jinwei, F., Xiaoan, H., Shu, W., Yashuo, L., Qiangbing, Z. and Chongfeng, W. (2017). Development of a tomato harvesting robot used in greenhouse. *International Journal of Agricultural and Biological Engineering* 10(4), 140–149.

Magistri, F., Chebrolu, N. and Stachniss, C. (2020). Segmentation-based 4d registration of plants point clouds for phenotyping. In *IEEE International Conference on Intelligent Robots and Systems (IROS)* Las Vegas, NV, USA. (pp. 2433–2439). doi: 10.1109/IROS45743.2020.9340918.

Mahlein, A. K., Oerke, E. C., Steiner, U. and Dehne, H. W. (2012). Recent advances in sensing plant diseases for precision crop protection. *European Journal of Plant Pathology* 133(1), 197–209.

McCormick, R. F., Truong, S. K. and Mullet, J. E. (2016). 3D sorghum reconstructions from depth images identify QTL regulating shoot architecture. *Plant Physiology* 172 (2), 823–834. https://doi.org/10.1104/pp.16.00948.

McDonald, S. C., Buck, J. and Li, Z. (2022). Automated, image-based disease measurement for phenotyping resistance to soybean frogeye leaf spot. *Plant Methods* 18(1), 103.

Minervini, M., Fischbach, A., Scharr, H. and Tsaftaris, S. A. (2016). Finely-grained annotated datasets for image-based plant phenotyping. *Pattern Recognition Letters* 81, 80–89.

Minhui, L., Shamshiri, R. R., Schirrmann, M. and Weltzien, C. (2021). Impact of camera viewing angle for estimating leaf parameters of wheat plants from 3D point clouds. *Agriculture (Switzerland)* 11 (6), 563. https://doi.org/10.3390/agriculture11060563.

Mohanty, S. P., Hughes, D. P. and Salathé, M. (2016). Using deep learning for image-based plant disease detection. *Frontiers in Plant Science* 7, 1419.

Mueller-Sim, T., Jenkins, M., Abel, J. and Kantor, G. (2017, May). The Robotanist: A ground-based agricultural robot for high-throughput crop phenotyping. In *2017 IEEE International Conference on Robotics and Automation (ICRA)* (pp. 3634–3639). IEEE Publications.

Müller-Linow, M., Pinto-Espinosa, F., Scharr, H. and Rascher, U. (2015). The leaf angle distribution of natural plant populations: Assessing the canopy with a novel software tool. *Plant Methods* 11, 11.

Munns, R., James, R. A., Sirault, X. R., Furbank, R. T. and Jones, H. G. (2010). New phenotyping methods for screening wheat and barley for beneficial responses to water deficit. *Journal of Experimental Botany* 61(13), 3499–3507.

Natukunda, M. I., Mantilla-Perez, M. B., Graham, M. A., Liu, P. and Salas-Fernandez, M. G. (2022). Dissection of canopy layer-specific genetic control of leaf angle in sorghum bicolor by RNA sequencing. *BMC Genomics* 23 (1), 95.https://doi.org/10.1186/s12864-021-08251-4.

Nelson, H. J., Smith, C. E., Bacharis, A. and Papanikolopoulos, N. P. (2023). Robust plant localization and phenotyping in dense 3d point clouds for precision agriculture. In *2023 IEEE International Conference on Robotics and Automation (ICRA)* (pp. 9615–9621). IEEE Publications.

Okamoto, H. and Lee, W. S. (2009). Green citrus detection using hyperspectral imaging. *Computers and Electronics in Agriculture* 66(2), 201–208.

Pan, H., Hetroy-Wheeler, F., Charlaix, J. and Colliaux, D. (2021). Multi-scale space-time registration of growing plants. In *Proceedings – 2021 International Conference on 3-D Vision, 3-DV 2021* (pp. 310–319).

Pang, J., Bai, Z. Y., Lai, J. C. and Li, S. K. (2011). Automatic segmentation of crop leaf spot disease images by integrating local threshold and seeded region growing. In *2011 International Conference on Image Analysis and Signal Processing* (pp. 590–594). IEEE Publications.

Paproki, A., Sirault, X., Berry, S., Furbank, R., and Fripp, J. (2012). A novel mesh processing based technique for 3-D plantanalysis. *BMC Plant Biology* 12, 63.

Patel, H. N., Jain, R. K. and Joshi, M. V. (2012). Automatic segmentation and yield measurement of fruit using shape analysis. *International Journal of Computer and Applications* 45(7), 19–24.

Paulus, S. (2019). Measuring crops in 3D: using geometry for plant phenotyping. *Plant Methods* 15, 103. https://doi.org/10.1186/s13007-019-0490-0.

Pérez-Ruiz, M., Prior, A., Martínez-Guanter, J., Apolo-Apolo, O. E., Andrade-Sanchez, P. and Egea, G. (2020). Development and evaluation of a self-propelled electric platform for high-throughput field phenotyping in wheat breeding trials. *Computers and Electronics in Agriculture* 169, 105237.

Pieruschka, R. and Schurr, U. (2022). Origins and drivers of crop phenotyping. In: Walter, A. (Ed.), *Advances in Plant Phenotyping for More Sustainable Crop Production* (pp. 3–28). Cambridge: Burleigh Dodds Science Publishing.

Qiu, Q., Sun, N., Bai, H., Wang, N., Fan, Z., Wang, Y., Meng, Z., Li, B. and Cong, Y. (2019). Field-based high-throughput phenotyping for maize plant using 3D LiDAR point cloud generated with a "Phenomobile". *Frontiers in Plant Science* 10, 554.

Roscher, R., Behmann, J., Mahlein, A. K. and Plümer, L. (2016a). On the benefit of topographic dictionaries for detecting disease symptoms on hyperspectral 3-D plant models. In 2016 8th Workshop on Hyperspectral Image and Signal Processing: Evolution in Remote Sensing (WHISPERS) (pp. 1–5). IEEE Publications.

Roscher, R., Behmann, J., Mahlein, A. K., Dupuis, J., Kuhlmann, H. and Plümer, L. (2016b). Detection of disease symptoms on hyperspectral 3-D plant models. *ISPRS Annals of the Photogrammetry, Remote Sensing and Spatial Information Sciences* 3, 89–96.

Rossi, R., Costafreda-Aumedes, S., Leolini, L., Leolini, C., Bindi, M. and Moriondo, M. (2022). Implementation of an algorithm for automated phenotyping through plant 3-D-modeling: A practical application on the early detection of water stress. *Computers and Electronics in Agriculture* 197, 106937.

Rosu, I.-A., Cazacu, M.-M., Prelipceanu, O. S. and Agop, M. (2019). A turbulence-oriented approach to retrieve various atmospheric parameters using advanced lidar data processing techniques. *Atmosphere* 10 (1), 38. https://doi.org/10.3390/atmos10010038.

Ruckelshausen, A., Biber, P., Dorna, M., Gremmes, H., Klose, R., Linz, A., Rahe, R., Resch, R., Thiel, M., Trautz, D. and Weiss, U. (2009). BoniRob–an autonomous field robot platform for individual plant phenotyping. *Precision Agriculture* 9(841), 1.

Ruppel, M., Nelson, S. K., Sidberry, G., Mitchell, M., Kick, D., Thomas, S. K., Guill, K. E., Oliver, M. J. and Washburn, J. D. (2023). RootBot: High-throughput root stress phenotyping robot. *Applications in Plant Sciences*, 11(6), e11541. doi: 10.1002/aps3.11541.

Sa, I., Ge, Z., Dayoub, F., Upcroft, B., Perez, T. and McCool, C. (2016). Deepfruits: A fruit detection system using deep neural networks. *Sensors* 16(8), 1222.

Sabrol, H. and Satish, K. (2016). Tomato plant disease classification in digital images using classification tree. In *2016 International Conference on Communication and Signal Processing (ICCSP)* (pp. 1242–1246). IEEE Publications.

Sandovsky, T., Edan, Y. and Alchanatis, V. (2021). Early detection of fusarium infection in corn using spectral analysis. Beer-Sheva: Ben-Gurion University of the Negev. https://primo.bgu.ac.il/permalink/972BGU_INST/u9jrac/alma9926664188604361.

Scharr, H., Minervini, M., Fischbach, A. and Tsaftaris, S. A. (2014). Annotated image datasets of rosette plants. In *European Conference on Computer Vision. Zürich, Suisse* (pp. 6–12).

Schunck, D., Magistri, F., Rosu, R. A., Cornelißen, A., Chebrolu, N., Paulus, S., Léon, J., Behnke, S., Stachniss, C., Kuhlmann, H. and Klingbeil, L. (2021). Pheno4D:a spatio-temporal dataset of maize and tomato plant point clouds for phenotyping and advanced plant analysis. *PLoS ONE* 16 (8 August), e0256340. https://doi.org/10.1371/journal.pone.0256340.

Shafiekhani, A., Kadam, S., Fritschi, F. B. and DeSouza, G. N. (2017). Vinobot and vinoculer: Two robotic platforms for high-throughput field phenotyping. *Sensors* 17(1), 214.

Shah, D., Tang, L., Gai, J. and Putta-Venkata, R. (2016). Development of a mobile robotic phenotyping system for growth chamber-based studies of genotype x environment interactions. *IFAC-PapersOnLine* 49(16), 248–253.

Shriki, S., Edan, Y. and Alchanatis, V. (2020). Early abiotic stress detection of corn and sunflower plants using spectral data. Ben-Gurion University of the Negev. https://primo.bgu.ac.il/permalink/972BGU_INST/u9jrac/alma9926564488304361.

Sladojevic, S., Arsenovic, M., Anderla, A., Culibrk, D. and Stefanovic, D. (2016). Deep neural networks based recognition of plant diseases by leaf image classification. *Computational Intelligence and Neuroscience* 2016, 3289801.

Smitt, C., Halstead, M., Zaenker, T., Bennewitz, M. and McCool, C. (2021). Pathobot: A robot for glasshouse crop phenotyping and intervention. In *2021 IEEE International Conference on Robotics and Automation (ICRA)* (pp. 2324–2330). IEEE Publications.

Stager, A., Tanner, H. G. and Sparks, E. (2022). Design and construction of unmanned ground vehicles for sub-canopy plant phenotyping. In: Lorence, A. and Jimenez, M. K. *High-Throughput Plant Phenotyping: Methods and Protocols* (pp. 191–211). New York, NY: Springer US.

Song, Z., Qiu, W. and Jin, J. (2021). MISIRoot: A robotic, Minimally invasive, in situ imaging system for plant root phenotyping. *Transactions of the ASABE* 64(5), 1647–1658.

Steinbrener, J., Posch, K. and Leitner, R. (2019). Hyperspectral fruit and vegetable classification using convolutional neural networks. *Computers and Electronics in Agriculture* 162, 364-372.

Su, W., Zhu, D., Huang, J. and Guo, H. (2018). Estimation of the vertical leaf area profile of corn (Zea mays) plants using terrestrial laser scanning (TLS). *Computers and Electronics in Agriculture* 150, 5–13. https://doi.org/10.1016/j.compag.2018.03.037.

Sun, D., Cen, H., Weng, H., Wan, L., Abdalla, A., El-Manawy, A. I., Zhu, Y., Zhao, N., Fu, H., Tang, J., Li, X., Zheng, H., Shu, Q., Liu, F. and He, Y. (2019). Using hyperspectral analysis as a potential high throughput phenotyping tool in GWAS for protein content of rice quality. *Plant Methods* 15, 54.

Susič, N., Žibrat, U., Širca, S., Strajnar, P., Razinger, J., Knapič, M., Vončina, A., Urek, G. and Stare, B. G. (2018). Discrimination between abiotic and biotic drought stress in tomatoes using hyperspectral imaging. *Sensors and Actuators B: Chemical* 273, 842–852.

Tanner, F., Tonn, S., de Wit, J., Van den Ackerveken, G., Berger, B. and Plett, D. (2022). Sensor-based phenotyping of above-ground plant-pathogen interactions. *Plant Methods* 18(1), 35.

Thendral, R., Suhasini, A. and Senthil, N. (2014). A comparative analysis of edge and color based segmentation for orange fruit recognition. In *2014 International Conference on Communication and Signal Processing* (pp. 463–466). IEEE Publications.

Thompson, A. L., Thorp, K. R., Conley, M. M., Elshikha, D. M., French, A. N., Andrade-Sanchez, P. and Pauli, D. (2019). Comparing nadir and multi-angle view sensor technologies for measuring in-field plant height of upland cotton. *Remote Sensing* 11 (6), 700. https://doi.org/10.3390/RS11060700.

Tomlinson, I. (2012). Doubling food production to feed the 9 billion: a critical perspective on a key discourse of food security in the UK. *Journal of Rural Studies* 29, 81–90. https://doi.org/10.1016/j.jrurstud.2011.09.001.

Turgut, K., Dutagaci, H., Galopin, G. and Rousseau, D. (2022). Segmentation of structural parts of rosebush plants with 3-D point-based deep learning methods. *Plant Methods* 18(1), 20.

Ubbens, J. R. and Stavness, I. (2017). Deep plant phenomics: A deep learning platform for complex plant phenotyping tasks. *Frontiers in Plant Science* 8, 1190.

United Nations, Department of Economic and Social Affairs, Population Division. (2019). World Population Prospects 2019: Highlights (ST/ESA/SER.A/423).

Verma, S., Chug, A. and Singh, A. P. (2020). Impact of hyperparameter tuning on deep learning based estimation of disease severity in grape plant. In *International Conference on Soft Computing and Data Mining* (pp. 161–171). Cham: Springer.

Virlet, N., Sabermanesh, K., Sadeghi-Tehran, P. and Hawkesford, M. J. (2016). Field Scanalyzer: An automated robotic field phenotyping platform for detailed crop monitoring. *Functional Plant Biology* 44(1), 143–153.

Wang, Z., Wang, K., Yang, F., Pan, S. and Han, Y. (2018). Image segmentation of overlapping leaves based on Chan-Vese model and Sobel operator. *Information processing in agriculture* 5(1), 1–10.

Weyler, J., Milioto, A., Falck, T., Behley, J. and Stachniss, C. (2021). Joint plant instance detection and leaf count estimation for in-field plant phenotyping. *IEEE Robotics and Automation Letters* 6(2), 3599–3606.

Wu, S., Wen, W., Xiao, B., Guo, X., Du, J., Wang, C. and Wang, Y. (2019). An accurate skeleton extraction approach from 3-D point clouds of maize plants. *Frontiers in Plant Science* 10, 248.

Wu, S., Wen, W., Wang, Y., Fan, J., Wang, C., Gou, W. and Guo, X. (2020). MVS-Pheno: a portable and low-costphenotyping platform for maize shoots using multiview stereo 3-D reconstruction. *Plant Phenomics* 1848437. doi: 10.34133/2020/1848437.

Xiang, L., Nolan, T. M., Bao, Y., Elmore, M., Tuel, T., Gai, J., Shah, D., Wang, P., Huser, N. M., Hurd, A. M., McLaughlin, S. A., Howell, S. H., Walley, J. W., Yin, Y. and Tang, L. (2021). Robotic Assay for Drought (RoAD): An automated phenotyping system for brassinosteroid and drought responses. *The Plant Journal* 107(6), 1837–1853.

Xiang, L., Gai, J., Bao, Y., Yu, J., Schnable, P. S. and Tang, L. (2023). Field-based robotic leaf angle detection and characterization of maize plants using stereo vision and deep convolutional neural networks. *Journal of Field Robotics* 40(5), 1034–1053.

Xie, C. and Yang, C. (2020). A review on plant high-throughput phenotyping traits using UAV-based sensors. *Computers and Electronics in Agriculture* 178, 105731.

Xiong, Y., Ge, Y., Grimstad, L. and From, P. J. (2020). An autonomous strawberry-harvesting robot: Design, development, integration, and field evaluation. *Journal of Field Robotics* 37(2), 202–224.

Xu, R. and Li, C. (2022). A review of high-throughput field phenotyping systems: Focusing on ground robots. *Plant Phenomics*. doi: 10.34133/2022/9760269.

Xue, F., Elnashef, B., Jin, W. and Filin, S. (2023). Polarimetric monocular leaf normal estimation model for plant phenotyping. *ISPRS Journal of Photogrammetry and Remote Sensing* 202, 142–157.

Yamashita, T., Yamashita, K. and Kamimura, R. (2007). *A Stepwise AIC Method for Variable Selection in Linear Regression*. Taylor & Francis, 0926, 2395–2403. https://doi.org/10.1080/03610920701215639.

Yao, L., Van De Zedde, R. and Kowalchuk, G. (2021). Recent developments and potential of robotics in plant eco-phenotyping. *Emerging Topics in Life Sciences* 5(2), 289–300.

Yu, X., Yin, D., Nie, C., Ming, B., Xu, H., Liu, Y., Bai, Y., Shao, M., Cheng, M., Liu, Y., Liu, S., Wang, Z., Wang, S., Shi, L. and Jin, X. (2022). Maize tassel area dynamic monitoring based on near-ground and UAV RGB images by U-Net model. *Computers and Electronics in Agriculture* 203, 107477.

Zeng, Z.-L., Sun, H., Wang, X.-Q., Zhang, S.-B. and Huang, W. (2022). Regulation of leaf angle protects photosystem i under fluctuating light in tobacco young leaves. *Cells* 11 (2), 252. https://doi.org/10.3390/cells11020252.

Zhang, J., Wang, B., Zhang, X., Liu, P., Dong, Y., Wu, K. and Huang, W. (2018). Impact of spectral interval on wavelet features for detecting wheat yellow rust with hyperspectral data. *International Journal of Agricultural and Biological Engineering* 11(6), 138–144.

Zhang, L., Niu, Y., Zhang, H., Han, W., Li, G., Tang, J. and Peng, X. (2019). Maize canopy temperature extracted from UAV thermal and RGB imagery and its application in water stress monitoring. *Frontiers in Plant Science* 10, 1270.

Zhang, J., Yang, Y., Feng, X., Xu, H., Chen, J. and He, Y. (2020). Identification of bacterial blight resistant rice seeds using terahertz imaging and hyperspectral imaging combined with convolutional neural network. *Frontiers in Plant Science* 11, 821.

Zhang, L., Xia, C., Xiao, D., Weckler, P., Lan, Y. and Lee, J. M. (2021). A coarse-to-fine leaf detection approach based on leaf skeleton identification and joint segmentation. *Biosystems Engineering* 206, 94–108.

Zhang, T., Elnashef, B. and Filin, S. (2023). Spatio-temporal point cloud registration of plants non-rigid structures. *ISPRS Journal of Photogrammetry and Remote Sensing* 205, 263–283.

Zhou, J., Fu, X., Zhou, S., Zhou, J., Ye, H. and Nguyen, H. T. (2019). Automated segmentation of soybean plants from 3-D point cloud using machine learning. *Computers and Electronics in Agriculture* 162, 143–153.

Zhou, L., Zhang, C., Taha, M. F., Wei, X., He, Y., Qiu, Z. and Liu, Y. (2020). Wheat kernel variety identification based on a large near-infrared spectral dataset and a novel deep learning-based feature selection method. *Frontiers in Plant Science* 11, 575810.

Zhou, J., Mou, H., Zhou, J., Ali, M. L., Ye, H., Chen, P. and Nguyen, H. T. (2021). Qualification of soybean responses to flooding stress using UAV-based imagery and deep learning. *Plant Phenomics*. https://doi.org/10.34133/2021/9892570.

Zhou, S., Chai, X., Yang, Z., Wang, H., Yang, C. and Sun, T. (2021). Maize-IAS: A maize image analysis software using deep learning for high-throughput plant phenotyping. *Plant Methods* 17(1), 48.

Chapter 18

Advances in the use of robots in field crop cultivation

Avital Bechar, Agricultural Research Organization, Israel; and Dionysis Bochtis, Institute for Bio-Economy and Agri-Technology (iBO), Greece

1 Introduction
2 Requirements for robots involved in field cultivation
3 Enabling technologies for in-field robots
4 Applications: transplanting
5 Applications: monitoring and control of weeds, plant diseases and pests
6 Applications: harvesting
7 Conclusion and future trends
8 References

1 Introduction

A key economic goal of agriculture is to achieve optimal yields while minimizing labor costs (Alwis et al., 2022). One way of achieving this has been through increasing mechanization and automation of farming operations which has a long history (Fig. 1) (Nof, 2022). This process is now entering the digital age (Sørensen et al., 2011) with a pressing need to enhance agricultural productivity even further to address increasing competition for land, water, energy and labor resources (Bochtis and Moustakidis, 2021). A particular issue in field crop production is, despite advances in mechanization, a continued dependence on manual labor. In addition to the cost of labor, which can account for 25–40% of overall production costs (Bechar and Eben-Chaime, 2014), farms face challenges such as fluctuations in availability of labor, variable levels of training and expertise as well as health and safety issues associated with demanding physical tasks in potentially dangerous environments (Perez-Ruiz et al., 2014, Riemer and Bechar, 2016).

To address these challenges, and enabled by advances in digital technologies, considerable recent research efforts have been directed toward the integration of labor-saving robots and intelligent automation systems into

http://dx.doi.org/10.19103/AS.2023.0124.22

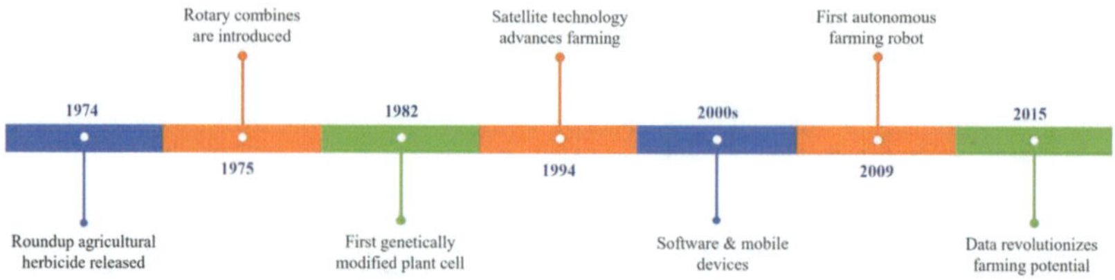

Figure 1 Evolutionary change in agriculture: the past, present and future. Source: Adapted from: Bochtis and Moustakidis (2021).

Table 1 The world domains from a robotic point of view.

		Environment	
		Structured	**Unstructured**
Objects	**Structured**	Industrial/service domains	Military/space/underwater domains
	Unstructured	Medical/social Domains	Agricultural domain

Source: Adapted from: Bechar and Vigneault (2016).

agricultural field operations (Bac et al., 2016). The potential impact of new technology can be seen in the case of precision agriculture where developments in sensors, data analysis and variable rate delivery technologies have made it possible, via more targeted application, to maintain yields with reductions in nitrogen fertilizer application of over 30% (Tremblay et al., 2010). Robotics offers the potential to further improve resource use and labor productivity as well as safety and quality.

The research of autonomous vehicles in agriculture started in field crops and although initial research goes back as early as 1960s, current adoption of robots in agricultural production remains limited (Benos et al., 2020). Unlike highly controlled industrial settings, e.g. robots face highly unstructured and dynamic environments in agricultural settings as well as complex, variable, changing and often delicate target objects such as fruit (Table 1) (Bechar and Vigneault, 2016). This is particularly the case for field crops. These range from annual, arable crops (such as cereals), typically grown on a large scale in dense patterns in open fields, through perennial, horticultural crops which may be grown in widely spaced, relatively uniform rows in fields (as well as in semi-controlled greenhouse conditions). The relatively 'unstructured' open conditions in which field crops are grown pose significant challenges for robots which operate best in controlled environments with standardized objects to process (Moysiadis et al., 2020).

Unlike industrial settings, where objects can be brought to static robots for processing, agricultural robots for field crop operations also typically need to be mobile to carry out operations, adding a further set of technological challenges and costs to overcome (Canning et al., 2004). A final issue is cost. Industrial robots in industrial sectors can be highly productive and cost-effective

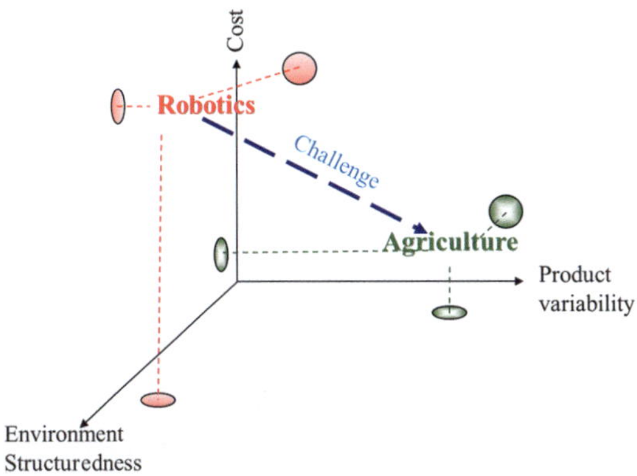

Figure 2 The domain of robotics for field crops with the three dimensions: product variability, environment structuredness and cost. The blue line represents the gap robotics will need to cover for successful application in agriculture. Source: Adapted from: Bechar (2021).

by performing a uniform task more or less continuously throughout the year to produce a high-value product such as a car. In contrast, agricultural robots for field crops may need to perform a range of complex and varying tasks at differing times of the year on relatively low-value agricultural produce (in terms of price per unit), with correspondingly poor cost-benefit ratios (Bechar and Vigneault, 2016). The challenges of unstructured environments which increases in field crop production, product variability and cost make it difficult to translate established robotic technologies (e.g. in manufacturing) to agriculture (Fig. 2) (Vidoni et al., 2015). Successful application of robotics in field crops is feasible only if costs are lower than existing solutions, where robots can deliver a higher standard of performance (e.g. in terms of precision (more efficient resource use), productivity, consistency and product quality), and where conditions make existing solutions difficult to implement safely and effectively (e.g. due to challenging terrain).

This chapter provides an overview of the research and developments in agricultural robotics, focusing on field crops.

2 Requirements for robots involved in field cultivation

Agricultural field robots operating in unstructured environments require two distinct sets of capabilities. The first encompasses robot functionalities relating to the environment in which it operates, including

- sensing capabilities (Mann et al., 2016);
- self-localization, navigation and path planning (Li et al., 2023);
- mobility, steering, and control (Lipinski et al., 2016);

- obstacle avoidance (Gonzalez-de-Soto et al., 2016, Khadatkar et al., 2023); and
- decision support systems for managing unexpected obstacles events, providing a certain level of autonomy (Van Henten et al., 2013).

The second set relates to product handling and the ability to carry out specific agricultural tasks such as planting, harvesting and sorting (Arikapudi and Vougioukas, 2023, Wu et al., 2022), spraying (Asaei et al., 2019), weeding (Wu et al., 2020), disease monitoring (Liang et al., 2020) and pruning (Bechar et al., 2014). These tasks typically have to be tailored to specific products/tasks such as harvesting of melons (Bechar et al., 2009), rice seedling (Li et al., 2023), transplanting (Zhao et al., 2022) and weeding (Bručienė et al., 2022).

These capabilities must be effectively integrated as sets of primary tasks (e.g. disease monitoring) supported by secondary tasks (e.g. path planning) which enable the primary task to be performed (Fig. 3) (Schor et al., 2016). In addition, robotic systems need to be sufficiently robust to deal with difficult terrain and potentially harsh weather conditions. As aforementioned, an important additional factor is cost: robots must not be so expensive to purchase and operate so they can ensure economic return on individual farms (e.g. in saving on labor and input costs).

We define three levels of complexity in robot technology, depending on the degree of interaction between robot and plant. At a lower level, there is no physical contact between robot and plant, e.g. data collection tasks using visual and other sensors such as detection of diseases and pests or signs

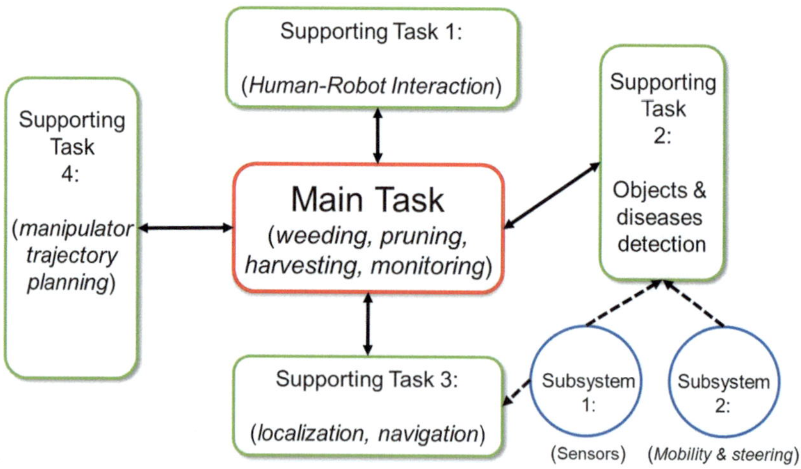

Figure 3 Structure of task sub-systems in an agricultural robot. Solid arrows represent commands, data and information transfer; dashed arrows represent conceptual connections. In parentheses are examples for agricultural robot 'main tasks', 'supporting tasks' and subsystems (Bechar and Vigneault, 2016).

of abiotic stress (Freitas et al., 2020), transportation of produce or materials (Guzman et al., 2016), as well as material deposition tasks like fertilizer or pesticide application (Bechar et al., 2008). The middle level of complexity requires physical contact between robot and plant but does not involve handling of plants. Typical tasks at this level include selective mechanical weeding which physically damages weeds but does not require handling them (McCool et al., 2018, Raja et al., 2020, Tillett et al., 2008), as well as tasks like tillage.

The highest level of robot-plant interaction involves complex handling of produce such as harvesting leaf crops that require sophisticated decision-making processes and careful handling of produce without causing damage or reducing quality. These require specialized manipulators tailored specifically for agricultural tasks (Bručienė et al., 2022, Gao et al., 2016, Khadatkar et al., 2023).

3 Enabling technologies for in-field robots

A key technology in agricultural robotics is sensor systems (Bechar and Cohen, 2022) consisting internal and external sensors. These sensors serve to support various aspects of the robotic system, including decision-making processes, operational activities, task execution and performance evaluation.

Internal sensors are responsible for measuring the state of various components within the system. For instance, encoders provide feedback on joint or wheel angles, accelerometers capture linear accelerations or inertia, and gyroscopes measure rotational accelerations. These internal sensors are commonly utilized for dead reckoning purposes, aiding in the estimation of the system's position and orientation.

Environmental parameter sensors (external sensors) encompass a wide range of technologies, including infrared (IR), near-infrared (NIR), X-ray, fluorescence, acoustic, optical, 2D and 3D vision sensors, laser radar (LIDAR), ultrasonic waves or hyperspectral imaging technologies. There have been important developments in LIDAR (Underwood et al., 2015) and in advanced Global Navigation Satellite Systems (GNSS) such as Real-Time Kinematic (RTK) GPS and Differential GPS (DGPS) which allow highly accurate in-field positioning, though the high cost of such technologies and their ability to cope with changing weather conditions (such as cloud and fog) remain limiting factors (Rovira-Más et al., 2015). A range of sensor technologies can be combined to provide precise mapping of environments and target objects.

A key function of sensors is to allow a robot to navigate a path where it is needed for a particular task. The fundamental challenge in path planning involves determining an optimal path from a source point to a destination point while avoiding collisions with obstacles (Bochtis et al., 2010). Early systems relied on determining a preestablished path based on either local positioning system (LPS) stations or GNSS, though with the disadvantage of potentially not being able to cope with unforeseen obstacles or changing conditions (Lipinski et al., 2016). An alternative approach relies on camera-based sensors and computer vision techniques for navigation (Hiremath et al., 2014) (Bietresato et al., 2016, Choi et al., 2015). While this approach allows the machine to customize and adapt its path and has been successfully implemented, e.g. in navigating row crops, it is more technically challenging to implement (Blok et al., 2019, Li et al., 2023, Utstumo et al., 2015).

More recent techniques to interpret and act on sensor data include genetic algorithms, fuzzy logic, reinforcement learning, machine learning and artificial neural networks (ANNs) ((Auat Cheein et al., 2016). Kim et al. (2020) used a patch-based convolutional neural network (CNN) for image classification, path area segmentation, cropping image patches, generating a path score map using a four-layer CNN for tree and path classification, path area segmentation and target path detection using boundary line determination. Penizzotto et al. (2015) have designed a mobile robot for use in olive groves using laser and odometer sensors combined with an extended Kalman filter for data analysis.

Other developments in robotics technologies include creating simulated environments in which to test potential solutions. Developments include a 'Simulation Environment for Precision Agriculture Tasks using Robot Fleet' (SEARFS) using a virtual 3D crop field to trial different ways of carrying out agricultural tasks by an autonomous fleet of robots (Emmi et al., 2013). Wang et al. (2016) have developed a 3D visualization model using the Unified Robot Description Format (URDF) and Rviz to facilitate motion planning

There are two scenarios where it may be advantageous to use multiple robots to undertake a common task. The first scenario arises when a task (e.g. dealing with emerging weeds or harvesting a crop) is of sufficient scale and needs to be undertaken within such a tight timetable that is beyond the capacity of a single robot (Conesa-Munoz et al., 2015). The second scenario is when differing tasks need to be executed by different machines simultaneously, e.g. in harvesting cereals where a harvester and a wagon for grain collection must work in parallel (Emmi et al., 2014). This may require a master–slave robotic system in which the 'master' vehicle assumes the role of decision-making for both vehicles with the 'slave' vehicle communicating relevant information about its location, orientation and operational status. Approaches to robot collaboration include 'Vehicle Routing Problem with Time Windows' methodology (Thayer et al., 2020). Ajidarma and Nof (2021) and Zion et al. (2014) for example have developed control architectures tailored for robotic systems operating collectively, addressing challenges such as code reusability and data distribution within teams of collaborating robots.

Another key concept is a human-robot integrated systems. Applying the Pareto principle to many agricultural tasks often shows that around 80% of a task can be easily automated, while the remaining 20% poses significant challenges (Stentz et al., 2002). This suggests automating the simpler aspects of a task while leaving the complex tasks to human operators and can significantly increase labor productivity while reducing the cost and complexity of a robotic solution. In field crops, enhancing observational and manual capabilities of the human operator while allowing human intervention to offset robot limitations in areas such as navigation, identifying and manipulating target objects (Nof et al., 2013) in tasks such as detecting melons ready for harvest and in pruning operations compared to both traditional manual and fully automated alternatives (Bechar and Edan, 2003; Bechar et al., 2014) and develop algorithms to allow dynamic switching between different collaboration levels in executing melon harvesting task more efficiently (Tkach et al., 2011).

Specific field crops application of human robot collaboration includes development of tractor-towed robot able to identify melons for collection, determine a collecting sequence (Zion et al., 2014) and grasp a target melon for collection (Fig. 4). Building on such work, Nair et al. (2021) have developed a collaborative robot-human system for melon harvesting. Human operators first identify and detach marked ripe melons using pruning shears. In the second step, a wheeled robotic arm system collects harvested melons by tracking their coordinates within the collection area (with this stage traditionally taking up 80% of workers' time).

Figure 4 Field prototype of a multi-arm melon collector. Source: Volcani Institute, Israel.

4 Applications: transplanting

Transplanting plays a pivotal role in the initial stages of crop cultivation and production, encompassing a crucial step for a wide range of crop types and plot configurations. Automation of transplanting processes is cost effective when large-scale, repeated transplanting in the same plot at regular intervals is required, e.g. as for leafy vegetables or herbs, particularly if uniformity of planting is required to optimize growth and final produce quality (Liu et al., 2023). Technologies include an image recognition system able to assess seed age and moisture content in tracking growth for transplanting (Zhao et al. 2022) and a novel pincette-type pick-up apparatus to transplant seedlings without damage (Mao et al., 2014). Hu et al. (2014) developed a high-speed plug seedling transplanting robot, employing a 2-DOF parallel translation mechanism equipped with a pneumatic manipulator. Nagasaka et al. (2009) developed an autonomous six-row rice transplanter capable of traversing straight lines across the field and executing U-turns at each end. Khadatkar et al. (2023) developed a robotic transplanter (Fig. 5) for plug chili plant seedlings consisting of a robot platform, seedling picking mechanism and vehicle movement system with over 90% success in transplanting seedlings with healthy subsequent growth.

Figure 5 Test setup of robotic transplanting with plug seedlings. Source: Adapted from: Khadatkar et al. (2023).

5 Applications: monitoring and control of weeds, plant diseases and pests

Biotic stresses have significant impacts on crop yields. Approximately 40% of global food production is lost due to diseases, insects and weeds (Oerke and Dehne, 2004). However, effective monitoring and control of diseases and pests is a potentially complex, time-consuming and labor-intensive task that often exposes human operators to hazardous chemicals. Automation of pest/disease detection, pesticide application and weeding operations can address these challenges while improving accuracy of detection and precision of application (Jorgensen et al., 2008). The timely and accurate identification of yield-inhibiting stresses and pertinent variables is imperative to enable the prompt and precise implementation of mitigative strategies. Site-specific spraying based on precision agriculture technologies has demonstrated potential herbicide savings ranging from 5% to nearly 90%, depending on spatial and temporal distribution of weeds (Mink et al., 2018, Riar et al., 2011), while other studies have achieved reductions ranging from 5% to 30% in application of fungicides and insecticides (Oberti and Schmilovitch, 2021). These systems have contributed to the development of fully automated, robotic systems for detection and control.

There have been numerous studies into robotic weed detection and control (Gonzalez-de-Soto et al., 2016, Lazar and Bechar, 2021, Pantazi et al., 2016). Hagras et al. (2002) classified field operations into two categories:

- Field-inspection tasks involving localization, recognition and assessment of weed presence and numbers; and
- Field-manipulation tasks involving selective weed removal methods (with minimal environmental impact).

Information regarding weed distribution and density can be obtained through field maps and direct sensory inputs. In addition, thermal, chemical and electrical methods have been investigated for selective in-row weed control (Oberti and Schmilovitch, 2021).

Blasco et al. (2002) for example devised a system using a manipulator equipped with a 15 kV electrical discharge electrode, controlled by a vision system capable of a weed detection rate of 84% while accurately identifying 99% of lettuce plants to avoid. Lazar and Bechar (2021) developed a site-specific electrical weeding robot for field crops (Fig. 6) using a 'normalized crop sample index' (NCSI) and machine learning methods analyzing data from sensing across four spectral bands to correctly identify weeds and reaching misclassification error level of 11% on average.

Tillett et al. (2008) developed a mechanical within-row weed controller for use in cabbage crops with minimal crop damage levels and weed reductions

Figure 6 Site-specific weeding robot for field crops. Source: Volcani Institute, Israel.

of up to 87%. An autonomous robotic weed control system for sugar beet combined a vision module and a rotary hoe for mechanical weed removal between rows and individual plants, eliminating up to 51% of weeds while leaving 99% of crop plants intact (Astrand and Baerveldt, 2005).

Disease detection techniques include both aerial and ground-based mobile robotic platforms (Moshou et al., 2014). Polder et al. (2014) for example developed a platform equipped with multispectral and RGB cameras to detect tulip breaking virus, exceeding human detection of early symptoms with an accuracy rate of 90% and a false positive rate of 10%. Oberti et al. (2016) developed a robotic sprayer for treatment of powdery mildew on grapevines. Their system utilized a 6-DOF robotic manipulator equipped with a precision-spraying end-effector incorporating an integrated disease-sensing system based on R-G-NIR multispectral imaging, able to detect and spray between 85% and 100% of diseased areas and reducing pesticide usage by 65% to 85% compared to conventional spraying.

Lightweight unmanned aerial vehicles (UAVs) equipped with multispectral cameras have been utilized for monitoring virus-borne diseases of grapevines such as grapevine leaf stripe disease (GLSD) (Di Gennaro et al., 2016) and Flavescence dorée disease (Albetis et al., 2017), as well as fungal diseases in wheat (Su et al., 2019) and cucurbits (Kalischuk et al., 2019). Bohnenkamp et al. (2019) compared detection of yellow rust in wheat using hyperspectral cameras mounted on both ground-based and UAV platforms. The ground platform demonstrated significantly enhanced detection accuracy due to the higher spatial resolution of its images (0.4 mm per pixel compared to 8 mm per pixel), indicating the need for further improvements in aerial stabilization and optics technologies in UAV platforms. Despite limited payload capacity, UAVs have also been used for targeted spraying, e.g. tall, densely planted crops or terraced cultivation where tractor-based or manual backpack spraying is more challenging. UAVs typically combine preplanned flight plans with prescription maps and vision-based target recognition for spraying (Dai et al., 2017, Xue et al., 2016).

The EU RHEA project (funded by Horizon 2020) developed an autonomous herbicide patch spraying system combining prescription mapping from remote sensing with real-time detection of weed targets by on-board cameras able to achieve a treatment accuracy of around 99%. The multi-robot system (Fig. 7) included UAVs for remote monitoring (Gonzalez-de-Santos et al., 2017). UAVs are equipped with NIR and RGB cameras (Sigma DP2 Merrill) which communicate with ground-based units (Boomer-3050 tractors) which include an onboard computer, communication apparatus, RTK-GPS receiver, camera system, laser unit and a spraying mechanism featuring a 200 L water reservoir and a 50 L herbicide container. This herbicide application system incorporates a direct injection mechanism, facilitating precise delivery of chemicals through twelve solenoid nozzles, strategically positioned along a 6 m boom, with selective activation. A central mission manager controller and communication infrastructure ensure effective collaborative within the multi-robot fleet.

Figure 7 The cooperative fleet of robots RHEA. The UAV platforms (in the front) have the task of remote monitoring of fields for weed detection; the processed data are sent to autonomous ground platforms (in the back) for physical weeding or selective spraying of herbicides. Source: Dr. Pablo Gonzalez de Santos; http://www.rhea-project.eu/index.php.

Slaughter et al. (2008) for example have developed a robot weed sprayer using rapid switching of solenoid nozzles to enable spot spraying of weed patches using offline infestation maps and real-time optical sensing. Utstumo et al. (2018) have developed a three-wheeled robotic system called Asterix for micro-application of herbicide in row crops using needle nozzles with the machine vision-based controller activating individual nozzles as necessary while navigating a row, allowing over 90% reduction in herbicide usage.

6 Applications: harvesting

Since harvesting task involves handling produce, harvesting row crops pose particular challenges in robotics (Karkee et al., 2021). Despite some progress, robotic harvesters have yet to match the performance and efficiency of human workers. While current systems are typically able to identify 75–85% of fruits ready for picking, skills such as obstacle navigation and reaching, grasping and handling fruits without damage remain challenging.

Developments in sensors such as RTK-GPS, 3D cameras and LADAR, combined with more advanced computing techniques, have led to significant improvements in detection and positioning. Several studies have actively contributed to this domain and even extended these advancements to encompass diverse crop types and agricultural environments (Arikapudi and Vougioukas, 2023, Barnea et al., 2016, Hohimer et al., 2019, Williams et al., 2019). Irie et al. (2009) for example used 3D sensors to measure cross-sectional profiles of asparagus plants to identify target plants for harvesting.

Cho et al. (2002) developed a robotic system for lettuce harvesting, including a 3-DOF manipulator, end-effector, conveyor, air blower, six photoelectric sensors and a fuzzy logic controller, able to achieve a harvesting success rate of 94% with an average harvesting time of approximately 5 s per plant. Other examples include Rath and Kawollek (2009) who developed a robot to harvest gerbera flowers, incorporating algorithms for pedicel identification from digital images and a 6-DOF industrial robot with a pneumatic cutter gripper. Cui et al. (2013) have designed a Cartesian-type robot for harvesting strawberries, integrating color cameras, a harvesting hand, a fiber-optic sensor and a mobile control unit, with an almost 94% success rate in detection of ripe fruit and an average harvesting time of 16.6 s per fruit.

7 Conclusion and future trends

Agricultural robotics has made significant advances in recent decades. There have been important developments in sensing technologies, complex data analysis to allow robots to use inputs from sensors to navigate their way through a complex environment and correctly locate a target as well as in mechanical techniques to complete a wide range of tasks with a high degree of precision. All of these have brought robots closer to replicating and then exceeding the adaptability, accuracy and speed of skilled manual labor.

However, given the complexity and variability of both environments and plants, field crops present a particular challenge in agricultural robotics. As a result, there remains a significant continuing attainment gap, particularly in tasks that require manipulation of a target (such as harvesting). Challenges such as obstacle detection and avoidance are still largely at a research stage. As well as the need for continuing developments in individual functions (such as navigation, object identification and manipulation), there is also a need for further improvements in ensuring different technologies can communicate and are fully integrated to ensure seamless, rapid and precise execution of a task. It is unfortunate that many current initiatives to design particular agricultural robots for field crops are undertaken in relative isolation from one another in terms of objectives, methodologies and technologies deployed, making comparison and technology transfer difficult with the resulting risk of duplicating effort and potentially diluting overall progress in the field. There is a need for example to establish standardized terminologies and methodological frameworks for evaluating and comparing system performance. In addition, current costs of robotic solutions are often still too high to be economically attractive (Bechar and Vigneault, 2016).

A potential interim solution to some of these challenges is greater human-robot collaboration where for example a single human operator is able to supervise multiple machines to complete tasks that might otherwise require teams of people. However, the longer-term trend will be toward full automation of routine tasks such as planting, pruning/thinning, detection and control of diseases, pests and weeds, harvesting and sorting produce. Further advances in the field may include improving technologies for coordinating teams of robots either to perform the same task (e.g. by working in different parts of a field) or to collaborate in the more efficient implementation of a complex task. There will be further improvements in sensor capability and data analysis for decision-making.

Based on the technological advances we are experiencing in the field of agricultural robotics, the characteristics of field crop tasks and the nature of the operating domain (e.g. Table 1), it is likely to speculate that the area of field crop cultivation will be the first to witness agricultural robots in commercial use.

8 References

Ajidarma, P. and Nof, S. Y. 2021. Collaborative detection and prevention of errors and conflicts in an agricultural robotic system. *Studies in Informatics and Control* 30(1), 19–28.

Albetis, J., Duthoit, S., Guttler, F., Jacquin, A., Goulard, M., Poilvé, H., Féret, J.-B. and Dedieu, G. 2017. Detection of Flavescence dorée Grapevine Disease Using Unmanned Aerial Vehicle (UAV) multispectral imagery. *Remote Sensing* 9(4), 308.

Alwis, S. D., Hou, Z., Zhang, Y., Na, M. H., Ofoghi, B. and Sajjanhar, A. 2022. A survey on smart farming data, applications and techniques. *Computers in Industry* 138Article no. 103624.

Arikapudi, R. and Vougioukas, S. G. 2023. Robotic tree-fruit harvesting with arrays of Cartesian arms: A study of fruit pick cycle times. *Computers and Electronics in Agriculture* 211, 17114–17126.

Asaei, H., Jafari, A. and Loghavi, M. 2019. Site-specific orchard sprayer equipped with machine vision for chemical usage management. *Computers and Electronics in Agriculture* 162, 431–439.

Astrand, B. and Baerveldt, A. J. 2005. A vision based row-following system for agricultural field machinery. *Mechatronics* 15(2), 251–269.

Auat Cheein, F. A., Scaglia, G., Torres-Torriti, M., Guivant, J., Javier Prado, A. J., Arno, J., Escola, A. and Rosell-Polo, J. R. 2016. Algebraic path tracking to aid the manual harvesting of olives using an automated service unit. *Biosystems Engineering* 142, 117–132.

Bac, C. W., Roorda, T., Reshef, R., Berman, S., Hemming, J. and Van Henten, E. J. 2016. Analysis of a motion planning problem for sweet-pepper harvesting in a dense obstacle environment. *Biosystems Engineering* 146, 85–97.

Barnea, E., Mairon, R. and Ben-Shahar, O. 2016. Colour-agnostic shape-based 3D fruit detection for crop harvesting robots. *Biosystems Engineering* 146, 57–70.

Bechar, A. 2021. Agricultural robotics for precision agriculture tasks: concepts and principles. In Bechar, A. (Ed.). *Innovation in Agricultural Robotics for Precision Agriculture: A Roadmap for Integrating Robots in Precision Agriculture*. Cham: Springer International Publishing. 17–30.

Bechar, A., Bloch, V., Finkelshtain, R., Levi, S., Hoffman, A., Egozi, H. and Schmilovitch, Z. 2014. Visual servoing methodology for selective tree pruning by human-robot collaborative system. *AgEng 2014*. Zurich, Switzerland.

Bechar, A. and Cohen, Y. 2022. *Robotics and Remote Sensing in Orchards*. CABI Publishing International.

Bechar, A. and Eben-Chaime, M. 2014. Hand-held computers to increase accuracy and productivity in agricultural work study. *International Journal of Productivity and Performance Management* 63(2), 194–208.

Bechar, A., Gan-Mor, S. and Ronen, B. 2008. A method for increasing the electrostatic deposition of pollen and powder. *Journal of Electrostatics* 66(7–8), 375–380.

Bechar, A., Meyer, J. and Edan, Y. 2009. An objective function to evaluate performance of human-robot collaboration in target recognition tasks. *IEEE Transactions on Systems, Man, and Cybernetics, Part C-Applications and Reviews*, 39(6), 611–620.

Bechar, A. and Vigneault, C. 2016. Agricultural robots for field operations: concepts and components. *Biosystems Engineering* 149, 94–111.

Benos, L., Bechar, A. and Bochtis, D. 2020. Safety and ergonomics in human-robot interactive agricultural operations. *Biosystems Engineering* 200, 55–72.

Bietresato, M., Carabin, G., Vidoni, R., Gasparetto, A. and Mazzetto, F. 2016. Evaluation of a LiDAR-based 3D-stereoscopic vision system for crop-monitoring applications. *Computers and Electronics in Agriculture* 124, 1–13.

Blasco, J., Aleixos, N., Roger, J. M., Rabatel, G. and Molte, E. 2002. AE-automation and emerging technologies: robotic weed control using machine vision. *Biosystems Engineering* 83(2), 149–157.

Blok, P. M., Van Boheemen, K., Van Evert, F. K., Ijsselmuiden, J. and Kim, G. H. 2019. Robot navigation in orchards with localization based on Particle filter and Kalman filter. *Computers and Electronics in Agriculture* 157, 261–269.

Bochtis, D. and Moustakidis, S. 2021. Mobile robots: current advances and future perspectives. In Bechar, A. (Ed.). *Innovation in Agricultural Robotics for Precision Agriculture: A Roadmap for Integrating Robots in Precision Agriculture*. Cham: Springer International Publishing. pp. 1–15.

Bochtis, D. D., Sorensen, C. G. and Vougioukas, S. G. 2010. Path planning for in-field navigation-aiding of service units. *Computers and Electronics in Agriculture* 74(1), 80–90.

Bohnenkamp, D., Behmann, J. and Mahlein, A.-K. 2019. In-field detection of yellow rust in wheat on the ground canopy and UAV scale. *Remote Sensing* 11(21), 2495.

Bručienė, I., Buragienė, S. and Šarauskis, E. 2022. Weeding effectiveness and changes in soil physical properties using inter-row hoeing and a robot. *Agronomy* 12(7), 1514.

Canning, J. R., Edwards, D. B. and Anderson, M. J. 2004. Development of a fuzzy logic controller for autonomous forest path navigation. *Transactions of the ASAE* 47(1), 301–310.

Cho, S. I., Chang, S. J., Kim, Y. Y. and An, K. J. 2002. Development of a three-degrees-of-freedom robot for harvesting lettuce using machine vision and fuzzy logic control. *Biosystems Engineering* 82(2), 143–149.

Choi, K. H., Han, S. K., Han, S. H., Park, K.-H., Kim, K.-S. and Kim, S. 2015. Morphology-based guidance line extraction for an autonomous weeding robot in paddy fields. *Computers and Electronics in Agriculture* 113, 266–274.

Conesa-Munoz, J., Gonzalez-De-Soto, M., Gonzalez-De-Santos, P. and Ribeiro, A. 2015. Distributed multi-level supervision to effectively monitor the operations of a fleet of autonomous vehicles in agricultural tasks. *Sensors* 15(3), 5402–5428.

Cui, Y., Gejima, Y., Kobayashi, T., Hiyoshi, K. and Nagata, M. 2013. Study on Cartesian-type strawberry-harvesting robot. *Sensor Letters* 11(6), 1223–1228.

Dai, B., He, Y., Gu, F., Yang, L., Han, J. and Xu, W. 2017. A vision-based autonomous aerial spray system for precision agriculture. In *2017 IEEE International Conference on Robotics and Biomimetics (ROBIO)*, 5–8 December 2017, pp. 507–513.

Di Gennaro, S., Battiston, E., Di Marco, S., Facini, O., Matese, A., Nocentini, M., Palliotti, A. and Mugnai, L. 2016. Unmanned Aerial Vehicle (UAV)-based remote sensing to monitor grapevine leaf stripe disease within a vineyard affected by esca complex. *Phytopathologia Mediterranea* 55, 262–275.

Emmi, L., Gonzalez-De-Soto, M., Pajares, G. and Gonzalez-De-Santos, P. 2014. Integrating sensory/actuation systems in agricultural vehicles. *Sensors* 14(3), 4014–4049.

Emmi, L., Paredes-Madrid, L., Ribeiro, A., Pajares, G. and Gonzalez-De-Santos, P. 2013. Fleets of robots for precision agriculture: a simulation environment. *Industrial Robot: An International Journal* 40(1), 41–58.

Freitas, H., Faical, B. S., Silva, A. V. and Ueyama, J. 2020. Use of UAVs for an efficient capsule distribution and smart path planning for biological pest control. *Computers and Electronics in Agriculture* 173, 105387.

Gao, G. H., Feng, T. X., Yang, H. and Li, F. 2016. Development and optimization of end-effector for extraction of potted Anthurium seedlings during transplanting. *Applied Engineering in Agriculture* 32(1), 37–46.

Gonzalez-De-Santos, P., Ribeiro, A., Fernandez-Quintanilla, C., Lopez-Granados, F., Brandstoetter, M., Tomic, S., Pedrazzi, S., Peruzzi, A., Pajares, G., Kaplanis, G.,

Perez-Ruiz, M., Valero, C., Del Cerro, J., Vieri, M., Rabatel, G. and Debilde, B. 2017. Fleets of robots for environmentally-safe pest control in agriculture. *Precision Agriculture* 18(4), 574–614.

Gonzalez-De-Soto, M., Emmi, L., Perez-Ruiz, M., Aguera, J. and Gonzalez-De-Santos, P. 2016. Autonomous systems for precise spraying–evaluation of a robotised patch sprayer. *Biosystems Engineering* 146, 165–182.

Guzman, R., Navarro, R., Beneto, M. and Carbonell, D. 2016. Robotnik-professional service robotics applications with ROS. In Koubaa, A. (Ed.). *Robot Operating System*. 625, 253–288.

Hagras, H., Colley, M., Callaghan, V. and Carr-West, M. 2002. Online learning and adaptation of autonomous mobile robots for sustainable agriculture. *Autonomous Robots* 13(1), 37–52.

Hiremath, S. A., Van Der Heijden, G. W. A. M., Van Evert, F. K., Stein, A. and Ter Braak, C. J. F. 2014. Laser range finder model for autonomous navigation of a robot in a maize field using a particle filter. *Computers and Electronics in Agriculture* 100, 41–50.

Hohimer, C. J., Wang, H., Bhusal, S., Miller, J., Mo, C. and Karkee, M. 2019. Design and field evaluation of a robotic apple harvesting system with a 3D-printed soft-robotic end-effector. *Transactions of the ASABE* 62(2), 405–414.

Hu, J., Yan, X., Ma, J., Qi, C., Francis, K. and Mao, H. 2014. Dimensional synthesis and kinematics simulation of a high-speed plug seedling transplanting robot. *Computers and Electronics in Agriculture* 107, 64–72.

Irie, N., Taguchi, N., Horie, T. and Ishimatsu, T. 2009. *Asparagus Harvesting Robot Coordinated with 3-D Vision Sensor*. Ieee International Conference on Industrial Technology. 1-3, 408–413.

Jorgensen, R. N., Norremark, M., Sorensen, C. G. and Andersen, N. A. 2008. Utilising scripting language for unmanned and automated guided vehicles operating within row crops. *Computers and Electronics in Agriculture* 62(2), 190–203.

Kalischuk, M., Paret, M. L., Freeman, J. H., Raj, D., Da Silva, S., Eubanks, S., Wiggins, D. J., Lollar, M., Marois, J. J., Mellinger, H. C. and Das, J. 2019. An improved crop scouting technique incorporating unmanned aerial vehicle–assisted multispectral crop imaging into conventional scouting practice for gummy stem blight in watermelon. *Plant Disease* 103(7), 1642-1650.

Karkee, M., Zhang, Q. and Silwal, A. 2021. Agricultural robots for precision agricultural tasks in tree fruit orchards. In Bechar, A. (Ed.). *Innovation in Agricultural Robotics for Precision Agriculture: A Roadmap for Integrating Robots in Precision Agriculture*. Cham: Springer International Publishing. pp. 63-89.

Khadatkar, A., Pandirwar, A. P. and Paradkar, V. 2023. Design, development and application of a compact robotic transplanter with automatic seedling picking mechanism for plug-type seedlings. *Scientific Reports* 13(1), 1883.

Kim, W. S., Lee, D. H., Kim, Y. J., Kim, T., Hwang, R. Y. and Lee, H. J. 2020. Path detection for autonomous traveling in orchards using patch-based CNN. *Computers and Electronics in Agriculture* 175, 11.

Lazar, A. and Bechar, A. 2021. Four-band weed detection using machine learning algorithms based on hyperspectral images. In Stafford, J. V. (Ed.). *ECPA 21, 2021*. Budapest, Hungary: Wageningen Academic, pp. 233-240.

Li, Y., Wu, T., Xiao, Y., Gong, L. and Liu, C. 2023. Path planning in continuous adjacent farmlands and robust path-tracking control of a rice-seeding robot in paddy field. *Computers and Electronics in Agriculture* 210, 107900.

Liang, H., He, J. and Lei, J. J. 2020. Monitoring of corn canopy blight disease based on UAV hyperspectral method. *Guang Pu Xue Yu Guang Pu Fen Xi. Spectroscopy and Spectral Analysis* 40, 1965–1972.

Lipinski, A. J., Markowski, P., Lipinski, S. and PYRA, P. 2016. Precision of tractor operations with soil cultivation implements using manual and automatic steering modes. *Biosystems Engineering* 145, 22–28.

Liu, W., Tian, S., Wang, Q. and Jiang, H. 2023. Key technologies of plug tray seedling transplanters in protected agriculture: a review. *Agriculture* 13(8), 1488.

Mann, M., Zion, B., Shmulevich, I. and Rubinstein, D. 2016. Determination of robotic melon harvesting efficiency: a probabilistic approach. *International Journal of Production Research* 54(11), 3216–3228.

Mao, H., Han, L., Hu, J. and Kumi, F. 2014. Development of a pincette-type pick-up device for automatic transplanting of greenhouse seedlings. *Applied Engineering in Agriculture* 30, 547–556.

McCool, C. S., Beattie, J., Firn, J., Lehnert, C., Kulk, J., Bawden, O., Russell, R. and Perez, T. 2018. Efficacy of mechanical weeding tools: A study into alternative weed management strategies enabled by robotics. *IEEE Robotics and Automation Letters* 3, 1184–1190.

Mink, R., Dutta, A., Peteinatos, G. G., Sökefeld, M., Engels, J. J., Hahn, M. and Gerhards, R. 2018. Multi-temporal site-specific weed control of *Cirsium arvense* (L.) Scop. and *Rumex crispus* L. in maize and sugar beet using unmanned aerial vehicle based mapping. *Agriculture* 8(5), 65.

Moshou, D., Pantazi, X.-E., Kateris, D. and Gravalos, I. 2014. Water stress detection based on optical multisensor fusion with a least squares support vector machine classifier. *Biosystems Engineering* 117, 15–22.

Moysiadis, V., Tsolakis, N., Katikaridis, D., Sørensen, C. G., Pearson, S. and Bochtis, D. 2020. Mobile robotics in agricultural operations: A narrative review on planning aspects. *Applied Sciences* 10(10), 3453.

Nagasaka, Y., Saito, H., Tamaki, K., Seki, M., Kobayashi, K. and Taniwaki, K. 2009. An autonomous rice transplanter guided by global positioning system and inertial measurement unit. *Journal of Field Robotics* 26(6–7), 537–548.

Nof, S. Y. (Ed.). 2022. *Handbook of Automation*. Springer: Berlin.

Nof, S. Y., Cheng, G. J., Weiner, A. M., Chen, X. W., Bechar, A., Jones, M. G., Reed, C. B., Donmez, A., Weldon, T. D., Bermel, P., Bukkapatnam, S. T. S., Cheng, C., Kumara, S. R. T., Bement, A., Koubek, R., Bidanda, B., Shin, Y. C., Capponi, A., Lee, S., Lehto, M. R., Liu, A. L., Nohadani, O., Dantus, M., Lorraine, P. W., Nolte, D. D., Proctor, R. W., Sardesai, H. P., Shi, L., Wachs, J. P. and Zhang, X. C. 2013. Laser and photonic systems integration: emerging innovations and framework for research and education. *Human Factors and Ergonomics in Manufacturing* 23(6), 483–516.

Oberti, R., Marchi, M., Tirelli, P., Calcante, A., Iriti, M., Tona, E., Hočevar, M., Baur, J., Pfaff, J., Schütz, C. and Ulbrich, H. 2016. Selective spraying of grapevines for disease control using a modular agricultural robot. *Biosystems Engineering* 146, 203–215.

Oberti, R. and Schmilovitch, Z. E. 2021. Robotic spraying for precision crop protection. In Bechar, A. (Ed.). *Innovation in Agricultural Robotics for Precision Agriculture: A Roadmap for Integrating Robots in Precision Agriculture*. Cham: Springer International Publishing. 117-150.

Oerke, E.-C. and Dehne, H.-W. 2004. Safeguarding production—losses in major crops and the role of crop protection. *Crop Protection* 23(4), 275–285.

Pantazi, X.-E., Moshou, D. and Bravo, C. 2016. Active learning system for weed species recognition based on hyperspectral sensing. *Biosystems Engineering* 146, 193–202.

Penizzotto, F., Slawinski, E. and Mut, V. 2015. Laser radar based autonomous mobile robot guidance system for Olive Groves navigation. *IEEE Latin America Transactions* 13(5), 1303–1312.

Perez-Ruiz, M., Slaughter, D. C., Fathallah, F. A., Gliever, C. J. and Miller, B. J. 2014. Co-robotic intra-row weed control system. *Biosystems Engineering* 126, 45–55.

Polder, G., Van Der Heijden, G. W. A. M., Van Doorn, J. and Baltissen, T. A. H. M. C. 2014. Automatic detection of tulip breaking virus (TBV) intulip fields using machine vision. *Biosystems Engineering* 117, 35–42.

Raja, R., Nguyen, T. T., Slaughter, D. C. and Fennimore, S. A. 2020. Real-time robotic weed knife control system for tomato and lettuce based on geometric appearance of plant labels. *Biosystems Engineering* 194, 152–164.

Rath, T. and Kawollek, M. 2009. Robotic harvesting of Gerbera jamesonii based on detection and three-dimensional modeling of cut flower pedicels. *Computers and Electronics in Agriculture* 66(1), 85–92.

Riar, D. S., Ball, D. A., Yenish, J. P. and Burke, I. C. 2011. Light-activated, sensor-controlled sprayer provides effective postemergence control of broadleaf weeds in fallow. *Weed Technology* 25(3), 447–453.

Riemer, R. and Bechar, A. 2016. Investigation of productivity enhancement and biomechanical risks in greenhouse crops. *Biosystems Engineering* 147, 39–50.

Rovira-Más, F., Chatterjee, I. and Sáiz-Rubio, V. 2015. The role of GNSS in the navigation strategies of cost-effective agricultural robots. *Computers and Electronics in Agriculture* 112, 172–183.

Schor, N., Bechar, A., Ignat, T., Dombrovsky, A., Elad, Y. and Berman, S. 2016. Robotic disease detection in greenhouses: combined detection of powdery mildew and tomato spotted wilt virus. *IEEE Robotics and Automation Letters* 1(1), 354-360.

Sørensen, C. G., Pesonen, L., Bochtis, D. D., Vougioukas, S. G. and Suomi, P. 2011. Functional requirements for a future farm management information system. *Computers and Electronics in Agriculture* 76(2), 266–276.

Stentz, A., Dima, C., Wellington, C., Herman, H. and Stager, D. 2002. A system for semi-autonomous tractor operations. *Autonomous Robots* 13(1), 87–104.

Su, J., Liu, C., Hu, X., Xu, X., Guo, L. and Chen, W.-H. 2019. Spatio-temporal monitoring of wheat yellow rust using UAV multispectral imagery. *Computers and Electronics in Agriculture* 167, 105035.

Thayer, T. C., Vougioukas, S., Goldberg, K. and Carpin, S. 2020. Multirobot routing algorithms for robots operating in vineyards. *IEEE Transactions on Automation Science and Engineering*. 17(3), pp.1184–1194.

Tillett, N. D., Hague, T., Grundy, A. C. and Dedousis, A. P. 2008. Mechanical within-row weed control for transplanted crops using computer vision. *Biosystems Engineering* 99(2), 171–178.

Tremblay, N., Bouroubi, M. Y., Panneton, B., Guillaume, S., Vigneault, P. and Belec, C. 2010. Development and validation of fuzzy logic inference to determine optimum rates of N for corn on the basis of field and crop features. *Precision Agriculture* 11(6), 621–635.

Underwood, J. P., Jagbrant, G., Nieto, J. I. and Sukkarieh, S. 2015. Lidar-based tree recognition and platform localization in orchards. *Journal of Field Robotics* 32(8), 1056–1074.

Utstumo, T., Berge, T. W. and Gravdahl, J. T. 2015. Non-linear model predictive control for constrained robot navigation in row crops. In *IEEE International Conference on Industrial Technology (ICIT 2015)*, 17–19 March 2015, pp. 357–362.

Utstumo, T., Urdal, F., Brevik, A., Dørum, J., Netland, J., Overskeid, Ø., Berge, T. W. and Gravdahl, J. T. 2018. Robotic in-row weed control in vegetables. *Computers and Electronics in Agriculture* 154, 36–45.

Van Henten, E. J., Bac, C. W., Hemming, J. and Edan, Y. 2013. Robotics in protected cultivation. In *4th IFAC Conference on Modelling and Control in Agriculture, Horticulture and Post Harvest Industry, 27–30 August, 2013, Espoo, Finland*. IFAC, pp. 170–177.

Vidoni, R., Bietresato, M., Gasparetto, A. and Mazzetto, F. 2015. Evaluation and stability comparison of different vehicle configurations for robotic agricultural operations on side-slopes. *Biosystems Engineering* 129, 197–211.

Wang, Z., Gong, L., Chen, Q., Li, Y., Liu, C. and Huang, Y. 2016. Rapid developing the simulation and control systems for a multifunctional autonomous agricultural robot with ROS. Paper presented at the 9th International Conference on Intelligent Robotics and Applications (ICIRA), Hachioji, JAPAN.

Williams, H. A. M., Jones, M. H., Nejati, M., Seabright, M. J., Bell, J., Penhall, N. D., Barnett, J. J., Duke, M. D., Scarfe, A. J., Ahn, H. S., Lim, J. and Macdonald, B. A. 2019. Robotic kiwifruit harvesting using machine vision, convolutional neural networks, and robotic arms. *Biosystems Engineering* 181, 140–156.

Wu, X., Aravecchia, S., Lottes, P., Stachniss, C. and Pradalier, C. 2020. Robotic weed control using automated weed and crop classification. *Journal of Field Robotics* 37(2), 322–340.

Wu, X., Chi, J., Jin, X. Z. and Deng, C. 2022. Reinforcement learning approach to the control of heavy material handling manipulators for agricultural robots. *Computers and Electrical Engineering* 104.

Xue, X., Lan, Y., Sun, Z., Chang, C. and Hoffmann, W. C. 2016. Develop an unmanned aerial vehicle based automatic aerial spraying system. *Computers and Electronics in Agriculture* 128, 58–66.

Zhao, X., Cheng, D., Dong, W., Ma, X., Xiong, Y. and Tong, J. 2022. Research on the end effector and optimal motion control strategy for a plug seedling transplanting parallel robot. *Agriculture* 12(10).

Zion, B., Mann, M., Levin, D., Shilo, A., Rubinstein, D. and Shmulevich, I. 2014. Harvest-order planning for a multiarm robotic harvester. *Computers and Electronics in Agriculture* 103, 75–81.

Chapter 19

Advances in the use of robotics in orchard operations

Manoj Karkee and Qin Zhang, Washington State University, USA; Uddhav Bhattarai, University of California Davis, USA; and Xin Zhang, Mississippi State University, USA

1 Introduction
2 Ground platforms and in-orchard navigation
3 Situation awareness and operation management
4 Actuation
5 Use cases of robotic orchard operations
6 Challenges and opportunities
7 Conclusion
8 Where to look for further information
9 References

1 Introduction

1.1 Background

As discussed earlier in this book, agriculture has witnessed transformative innovation and technology adoption over the past 100 years including the adoption of tractors, harvesting machines (e.g. combine harvesters for corn and soybean), boom chemical sprayers, and auto-steering and selective weeding machines. These mechanized or automated machines relieved a huge population around the world from hard labor in challenging farming environments while also increasing economic activities and skilled jobs in manufacturing, operating, and maintaining these machines. In addition, these revolutionary changes led to increased agricultural productivity and improved produce quality, and expansion of the farmland. According to the World Bank, employment in agriculture decreased from 44% to 27% from 1991 to 2019, while the economic value added from agriculture, forestry, and fishery increased from US$1.1 trillion to US$3.51 trillion in the same duration (World Bank). Acknowledging this contribution, agricultural technology was recognized as one of the greatest engineering achievements (among the top seven) of the twentieth century (Wulf, 2000).

http://dx.doi.org/10.19103/AS.2023.0124.23

The transformation in farming, however, has not fully penetrated to a large segment of agriculture, what is often called specialty crops that include fruit, vegetables, flowers, and nursery crops. Tree fruit crops, for example, still are farmed using a high volume of human labor irrespective of the location, type, and scale of farming across the globe.

1.2 Orchard operations

Conventionally, fruit crops are grown in free-standing, bushy trees with large trunks and canopies without managing their shapes and sizes. For example, apple trees in the natural growing pattern would include one main trunk growing vertically and numerous secondary branches growing out from the main trunk to gradually form a vase shape (Zhang, 2020). In these orchards, field operations such as fruit picking require complex positioning and movement of tall ladders as a substantial proportion of fruit grows in higher and deeper parts of the canopies behind trunks and branches. To improve the overall canopy-light interception, which improves fruit quality and yield, growers in recent decades are adopting narrower, trellis-trained fruit tree canopies that are trained to vertical or angled planes (Fig. 1). In such a modern orchard (often called SNAP – simple, narrow, accessible, productive), tree trunks and/or branches are trained to trellis wires and the canopy height and depth is kept low (using dwarf varieties/rootstock) that presents most of the fruit to the outer part of the canopy thus creating what is called a fruiting wall orchard. These canopies help improve worker efficiency and productivity as farming operations do not require hand penetration into the canopy and a substantial portion of the canopies can be covered from the ground.

Figure 1 An example of a V-shaped fruiting wall (Simple Narrow Accessible, Productive or SNAP) orchard in early growth stage (Picture Credit: Washington State University #AgRobotics Lab; https://labs.wsu.edu/karkee-ag-robotics/). The fruits are managed to be present in the outer side(s) of the canopies to create an effective penetration/approach path for the robotic systems.

Consequently, these labor-friendly modern orchards also facilitated robotic orchard operations because most of the canopy parts (e.g. branches for pruning and fruit for picking) would be accessible from one side of the canopy or the other (particularly in a vertical canopy). However, to create such a canopy system and achieve the desired quantity of high-quality fruit crops, a few farm operations need to take place meticulously. First, trees are carefully planted in the desired fields using real-time kinematic (RTK)-GPS-guided systems to create precisely spaced rows of trees. The trees in modern fruiting wall orchards are planted as close as 12 inches apart to create dense two-dimensional (2D) canopies. For a few years after planting, growing trunks and branches are carefully trained to vertical trellis structures and horizontal trellis wires. Once the trees grow to the desired height and branches fill the desired space, training is repeated only for the new branches or trees that are grown to replace diseased, dead, or otherwise unproductive branches or trees.

Pruning is another important orchard operation that is repeated every year over the entire life cycle of these fruit trees to optimize branch and fruit distribution and overall crop-load. In this operation, diseased, unproductive, or too closely spaced branches need to be trimmed or pruned out selectively, which helps improve light penetration/interception and air flow, optimize the distribution of flowering buds, and minimize bud density. Every spring, when these buds open to flower clusters, farmers then remove a proportion of flowers (flower thinning) so that only a desired number of flowers could be pollinated. However, because of the uncertainty in pollination (which is often achieved using rented beehives or naturally available pollinators) and natural drop or thinning of fruitlets, growers keep higher flower density in the tree canopies than their target fruit density. This strategy, however, often leads to a higher number of fruits per tree than desired to achieve the target fruit size, color, and other quality parameters. To further optimize the crop load, green-fruit (fruitlet) thinning is applied when fruits are still tiny (e.g. in apples, when they are about 1–5 cm in diameter). Once these canopy and crop-load management operations are completed, several other orchard management operations are continued over the next 4–6 months to provide desired amounts of input (e.g. water, nutrient, and pest control chemicals) at the right time and right location. Mature fruit will then be selectively harvested and collected into containers, which are then transported out of the field to packing houses.

1.3 Robotics for orchard operations

Many of these orchard operations including tree training, flower and fruitlet thinning, and harvesting are highly labor intensive, laborious (requiring a high level of physical strength), and pose critical workers' health and safety issues such as ladder falls. Other operations such as pollination are highly impacted by the weather and availability of natural or rented pollinators (e.g. honey bee hives) leading to huge uncertainty and risk of crop loss. As farmers around the world are facing decreasing availability and increasing cost of farm labor, it is highly desirable to use advanced robotic technologies to perform labor-intensive field operations like harvesting and pruning. Similarly, robotic technologies also have a huge potential to improve worker health and safety issues and achieve precision artificial pollination without relying on declining honeybee populations and other natural pollinators. Recognizing these needs and exploiting innovation and advancement in sensing, robotics and artificial intelligence, computational technologies, as well as horticultural systems (e.g. fruiting wall orchards), public and private organizations (including funding agencies) around the world have been, over the last three decades, focusing on developing robotic systems for a wide range of orchard operations including pest control, nutrient application, irrigation, tree training, pruning, thinning (flower and fruitlet), harvesting, and post-harvest logistics in orchards.

A robotic system for orchard operations generally consists of (i) a ground vehicle/platform; (ii) a perception or machine vision system; (iii) a manipulator system; and (iv) an end-effector. The ground vehicle provides a platform for installing other sub-systems (e.g. manipulators) and mobility to the overall robotic system. The ground vehicle may be human operated in the orchards, but technologies have been developed for partial to full autonomy in navigating and steering the vehicle through orchard rows and to make automated turns at the end of the field. RTK-GPS machine, laser sensors, and vision-based guidance have been used along with various positioning and control strategies (e.g. simultaneous localization and mapping) for automated operation of the ground platform in orchards. A perception or vision system provides the backbone of robotic operations in fruit orchards. This system is used to understand and interpret tree canopies to locate target objects (e.g. branches for pruning) and potential obstacles to getting to the target objects (e.g. other branches and trellis wires). Advancements in machine vision technologies have also enabled precision surround awareness of the machine for avoiding ground obstacles and collisions and ensuring human and infrastructure safety. As mentioned before, this technology would also provide the desired input for guiding a vehicle through orchard rows. A manipulator system would provide hardware, control, and software infrastructure to move an end-effector to target locations to perform specific operations (e.g. robot hand for picking apples) carefully maneuvering around obstacles. Finally, an end-effector (also called a tool or hand) desired specifically to perform a given field operation (e.g. a pollen spray nozzle for robotic pollination) is attached at the end of the manipulator to perform specific field operations. Recent advancements in all these component technologies will be discussed in the following sub-sections of this chapter. The integrated systems designed for some of the important orchard operations (e.g. pruning and pollination) will be discussed in Section 6 as case studies. The chapter will then be closed with a reflection on challenges and opportunities for the future.

2 Ground platforms and in-orchard navigation

2.1 Wheel-type platforms and their steering mechanism

One essential function of orchard robots is their mobility. While there are three basic types of mobility mechanisms, i.e. wheel-type, track-type, and bionic leg-type, being developed over the years for off-road vehicles to provide desired mobility on different terrains, wheel-type platforms are most suitable for orchard applications as the soil-bearing capacity in orchards is higher than most other crop fields, in general.

As an orchard robot must always follow a narrow path constrained by fruit trees while performing designated operations, a wheel-mechanism used on orchard robots must provide a good maneuverability to be able to perform orchard operations effectively and efficiently. While there are many different designs that could meet this requirement, many four-wheel mobile platforms choose matured technologies, such as an Ackermann steering mechanism or omnidirectional driving mechanisms, as these systems have been widely adopted on existing field mobile machinery commercially.

The Ackermann mechanism (Fig. 2) is a geometric design of steering linkage to turn inner and outer wheels at appropriate angles, which minimizes the wheel slip during turning. This mechanism has been extensively studied and widely applied to on-highway and off-road vehicles, using either front-wheel or rear-wheel steering designs. Detailed theoretical explanations of this mechanism could be found in many textbooks, such as *Theory of Ground Vehicles* (Wong, 1993).

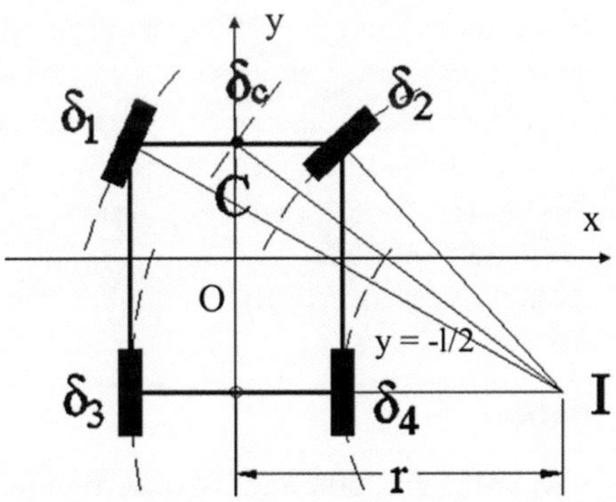

Figure 2 Geometric illustration of Ackermann Steering.

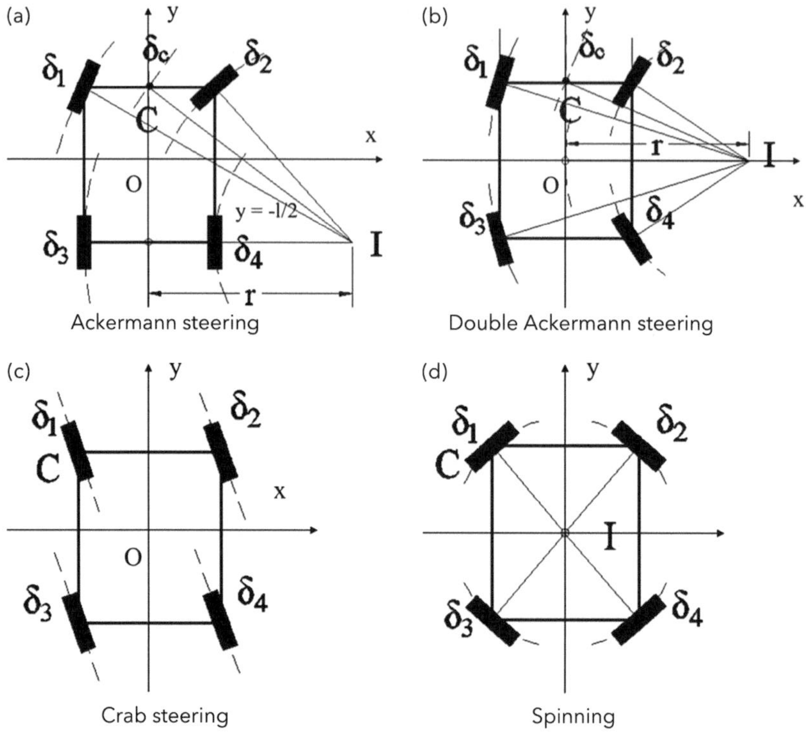

Figure 3 Four ways of steering a ground vehicle using an omnidirectional driving mechanism.

However, the minimal turning radius achieved with an Ackermann mechanism could limit the robot's maneuverability in a constrained orchard environment. Thus, the omnidirectional driving mechanism has found its usage on orchard platforms. A typical omnidirectional wheel driving mechanism often consists of four individually drivable and steerable wheels. While the arrangement of those wheels could be different in dimensions or sizes, the basic structural design of this mechanism is always similar and simple: it often attaches an individually controllable electric or hydraulic motor directly on all wheels and each wheel could be turned independently in terms of both rolling directions and steering angles. Thus, it could implement four ways of steering, namely Ackermann steering, double Ackermann steering, crab steering, and spinning, as illustrated in Fig. 3 (Ye et al., 2017).

2.2 Orchard navigation

Agricultural vehicle navigation and guidance (e.g. commercially available auto-steering systems) often is achieved using RTK-GPS techniques that can provide sub-inch positioning accuracy when GPS signals are unobstructed (Gan-Mor et al., 2007). These technologies have widely been adopted in commercial row crop farms (e.g. corn) to perform various field operations such as planting and harvesting (John Deere, n.d.; CNH Industrial, n.d.). However, in-orchard navigation, i.e. guiding an orchard robot in-between rows of trees and turning around at the end of the tree rows, is quite different from navigating agricultural machinery on open fields as the orchard space is tightly confined by fruit trees. The GPS signals are often less accurate and reliable when driving in a dense orchard environment (Libby and Kantor, 2010; Underwood et al., 2016). Navigation and guidance of a robot in such an environment thus generally rely on the relative positioning of the platform with reference to wall of trees involving other sensing techniques on which the robot needs to perform designated tasks effectively and efficiently.

2.2.1 Vision-based guidance

Different from absolute positioning which determines the position and pose of a robot in reference to the global coordinate system, relative positioning determines the distance and attitude of the robot in relation to either a predetermined reference point (often called the initial position) or simply to the tree walls near the mobile platform (local perception). Dead reckoning techniques (Kao, 1991) could be used in such a situation, which could determine the relative position of a robot platform to its initial position only based on the displacement and attitude of the moving robot measurable using inertial sensors, such as gyroscopes, encoders, and compasses. This method is commonly used in open field (annual crops such as paddy) navigation applications for short-distance guidance, due to its high short-term positioning accuracy, good stability, and high update rate (Kondo et al., 2011). However, as local perception could often provide more accurate relative position information between the moving platform and the surrounding tree walls, it is much more common to use this method to guide an orchard robot along the rows of trees. The sensors used to provide such relative position include stereovision, Laser or Lidar, and ultrasonic sensors (Rovira-Más et al., 2010).

Most of the vision-based guidance systems use either 2D images or three-dimensional (3D) point cloud/RGB-D data to estimate a reference trajectory (e.g. center line representing the center of inter-row spacing) which acts as a feedback for the steering controller to adjust the vehicle heading and consequently the position. The reference trajectory can be computed via the estimation of different features such as vanishing point, tree trunks on both sides of the row, edge detection, and segmentation of ground space in between rows (Sarmento et al., 2021; Sivakumar et al., 2021, Blok et al., 2019; Aghi et al., 2020; Yang et al., 2022). Yang et al. (2022) used semantic segmentation to segment the inter-row path followed by path edge detection to extract the center navigation. Sarmento et al. (2021) used deep-learning-based object detection to detect the grape canopy trunk in 2D images. Information of the detected trunks on both the left and right side of the vehicle was then used to estimate the vanishing point for the canopy row as guidance feedback. Blok et al. (2019) evaluated two localization algorithms: a particle filter and a Kalman filter in Dutch apple orchards for navigational accuracy and robustness based on 2D LIDAR data. Experimental evaluation showed that the particle filter with a laser beam model outperformed a Kalman filter with centerline detection via K-means clustering for navigational accuracy in an orchard environment.

2.2.2 Simultaneous localization and mapping

In some in-field navigation practices, it is also possible to combine both absolute and relative positioning methods to improve the positioning accuracy of a mobile robot in the field. One such approach in mobile robot applications is integrating inertial and vision-based positioning using simultaneous localization and mapping (SLAM) (Aulinas et al., 2008). SLAM is a real-time mapping technique in which the localization of the robotic vehicle and the construction of a map of the previously unknown environment can be solved effectively and simultaneously. Examples of a few successful SLAM-based studies on mobile robot navigation include GMap (Grisetti et al., 2007) and Hector (Kohlbrecher et al., 2014). An integrated navigation method often requires having accurate maps for planning robot paths and accurately guiding the robot through the planned path to conduct field operations (Sujiwo et al., 2016).

In addition to providing a tool for autonomous or semi-autonomous orchard operations such as pruning and harvesting, SLAM technology could also be used in developing autonomous crop monitoring systems (Dos Santos et al., 2015; Karpyshev et al., 2021; EOS Data Analytics (https://eos.com/products/crop-monitoring/)), autonomous chemical application systems (Dai et al., 2017; Autospray Systems (https://autospraysystems.com/); Guss (https://gussag.com/guss/)), and fruit transportation robots (Ye et al., 2017; Augean Robotics (https://burro.ai/)). For example, Washington State University developed a robot called Bin-Dog that can automatically enter the path between rows of trees in orchards, autonomously followed pickers or robots as they picked fruit and presented fruit containers near to them, and brought full bins or containers out of the orchards to a loading station (Ye, et al., 2017; Fig. 4). The Bin-Dog system consisted of a driving and steering system and a bin loading system, and it was operated in both automated and manual modes. The robot navigation system consisted of localization, object detection, task scheduling, and navigation modules. Information from LIDAR, GPS, bin loading sensor, and fork height sensor was processed in a ROS navigation node to

Figure 4 A Bin-Dog robot developed by Washington State University for managing fruit bins in orchards during harvest. Source: Adapted from: Ye et al. (2017).

generate driving-control commands. The robot localization in open space was achieved via GPS, while two laser scanners and inertial measurement unit (IMU) were used as perception sensors for localization in between tree rows (Ye et al., 2017). In another work, Lepej and Rakun (2016) used a SLAM technique for mapping and localization of a mobile robot platform in apple orchards using Laser scanner data. The rotational and translational variation between the successive laser scans were estimated using an image registration technique and were used to create the map of the environment and localize the robot.

3 Situation awareness and operation management

As mentioned earlier, any orchard robots should possess the capabilities of awareness, autonomy, and actuation. Among them, awareness plays an essential role in supporting the autonomy of a robotic machine in completing desirable orchard operations precisely and effectively. Situation (surround) awareness involves understanding the environment in which a robot is operating. A perception system can be used to sense the color, 3D, and other data of the environment and interpret the scene for semantic, texture, geometric, and topological information of the objects in the environment. The output of such a system would provide two basic pieces of information for the robotic system: (i) target locations where the robotic system has to reach (e.g. fruit locations for picking and tree locations that need to visit next) and (ii) potential obstacles between the current position and the target position (e.g. branches and trellis wires in canopies; human or other obstacles between the robot position and position of trees to be visited). Once this information is generated, a path planning and control strategy could be used to generate a safe path to perform the specific task avoiding any injury to humans and minimizing any collisions between the robot and the objects in the environment. The same vision system could be used for assessing the performance of a given operation and to make a decision if another attempt is necessary. In this section, the latest advancements in the sensing and machine vision systems as applied to robotic orchard operation will be discussed, in addition, robot–canopy interaction and strategies that help improve robotic operations in orchards will be discussed.

3.1 Machine vision system

3.1.1 Overview

A machine vision system consists of cameras/sensors (e.g. color/RGB and 3D cameras) and image processing techniques, often utilizing various types of machine learning models such as deep learning, used to interpret and make sense of the environment and detect and localize specific types of objects in the environment such that their properties could be extracted for specific applications (e.g. robotic picking of fruit). In simple terms, a machine vision system functions as the eye and brain of a robotic orchard machine. For specific applications in an orchard environment, the vision systems developed in the past focused heavily on detecting and localizing fruits as well as various canopy parts including trunk, branches, leaves, and trellis wires so that obstacle-free paths could be estimated to approach target objects such as fruit and branches to perform specific field operations such as picking and pruning. With a single RGB camera, objects can be identified and delineated in 2D, which could then be used to estimate object characteristics such as color, shape, size (in pixels), and texture. Using depth or 3D sensing techniques such as a time-of-flight (ToF), a stereo camera, or a laser scanner, the distance from the sensor to the objects of interest can be determined so that 3D properties of the objects such as location and orientation in the space and its physical size can be estimated, which is essential for robotic field operations (Fu et al., 2020).

3.1.2 Machine learning tools for an orchard vision system

Processing of orchard images can be highly challenging because of the complex environment with variable lighting conditions. Therefore, image processing approaches with higher accuracy and robustness are essential to achieve desirable performance in orchard applications. In the past, orchard images have often been analyzed using conventional image processing techniques, such as feature (e.g. color, geometric, texture, or fusion of all) extraction and line/

edge detection. Extracted features could then be used as inputs to supervised or unsupervised machine learning algorithms, such as K-means clustering, Bayesian classifier, and k-Nearest-Neighbor (kNN), so that each object could be categorized into a pre-defined class (Gongal et al., 2015; Silwal et al., 2016). While utilizing these approaches, researchers and developers need to be very meticulous when they define the key features of the target objects because the accuracy is largely dependent on those key features.

The benefit of the conventional image processing techniques is that this process is built using tools based on clear mathematical principles and the results are generally easy to understand and interpret. However, such a processing pipeline often face limitations dealing with images with complex background and uncontrolled and variable lighting conditions (e.g. sunlight with varying directions and intensities), both prevalent in orchard images. Particularly, in orchard conditions, the target objects are often similar to their background (e.g. green apples on green foliage), which can cause severe detection failures. To address these challenges, as discussed before, researchers and engineers have been using end-to-end deep-learning models to process orchard images. Deep-learning, a type of machine learning technique, is a data-driven modeling technique that has recently been used extensively in various industries and businesses with applications ranging from predictive modeling to image processing as it provides more accurate and robust results compared to the same with conventional techniques when sufficient input-output data are available (Alom et al., 2018). With the introduction of relatively more affordable parallel computing hardware (e.g. graphical processing units) in recent years, the development and application of deep-learning-based machine vision systems have accelerated (Kamilaris and Prenafeta-Boldú, 2018).

Particularly, the advent of convolutional neural networks (CNNs) and recurrent neural networks (RNNs) provided a widely applicable deep-learning model for image processing including those of orchards. In such attempts, algorithms such as Faster R-CNN (Ren et al., 2015), Mask R-CNN (He et al., 2017), You Only Look Once or YOLO (Redmon et al., 2016), Single-Shot Detector or SSD (Liu et al., 2016), and different other convolution-based algorithms are used to process RGB images along with depth and near infra-red (NIR) images. A large number of studies reported in recent years used these models to segment out and locate canopy parts such as fruit, branches, trunks, leaves, sky, ground, and other parts/objects in the images acquired in an orchard environment (Zhang et al., 2018; Zhang et al., 2020a,b,c; Zhang et al., 2021b). However, it is still worth mentioning that deep-learning models often require a large and diverse training dataset, require intensive manual annotation, and do not generalize well, especially in a complex agricultural environment.

3.1.2 Obstacle detection in orchards

Various types of obstacles might be present in orchards limiting the maneuvering capability of the robot from one tree to the next as they perform field operations. One of the most important obstacles that may be present in the field is human beings. For safe robot operation, any presence of humans within a critical operation zone (could be pre-defined based on the specific type and speed of robotic operation) must be detected without any false negatives. In addition, other obstacles such as overhanging branches, ladders, or fruit containers could be present in the path a robotic platform might want to take. A surround awareness system developed for a robotic orchard operation must detect these obstacles with the desired speed, accuracy, and robustness.

In the past, the safety concerns of autonomous systems received relatively low attention. Nowadays, it is widely acknowledged that safety measures must be prioritized when such autonomous systems are to be designed for orchard operations. Therefore, obstacle detection and avoidance has been one of the major focuses in terms of risk assessment in agricultural environments (Shutske, 2023; Shutske et al., 2022). Generally, people/humans are treated as moving obstacles in the field conditions, in addition to positive obstacles (e.g. poles and trellis wires), negative obstacles (e.g. ditches), and terrain obstacles (Bai et al., 2023). Freitas et al. (2012) proposed a practical obstacle detection method using the existing, low-cost system in the orchards, including a 3D laser scanner and an IMU. The model took a 3D point cloud as the input and transformed it into a binary assessment of the presence or absence of an obstacle (e.g. people or fruit containers/bins) on the path the vehicle must be moving on. Liu et al. (2022) developed an obstacle avoidance system in orchards using a novel trinocular camera system that can detect the moving row and obstacle (e.g. tree trunks) at the same time. The obstacle avoidance algorithm was developed by coupling the geometric constraint and virtual force field (VFF) method. For obstacle detection, the optical flow method was adopted to segment the trunk region where the motion vectors of pixels were calculated.

3.1.3 Canopy interpretation

Automating various orchard operations requires a robust sensing and machine vision system that can operate in near-real time to implement robotic field operations. Various types of RGB and depth sensors (e.g. consumer RGB-D sensors, stereo cameras, and laser scanners) can be used to collect desired orchard images for canopy interpretation. Past studies on canopy interpretation, often, were conducted with a specific field operation such as harvesting or pruning in mind (Fig. 5). However, they can be organized into four different areas: (i) trunk and branch detection, (ii) fruit detection, (iii) flower detection, (iv)

Figure 5 (a) Fruiting-wall apple tree architecture, (b) semantic segmentation for apple trees for automated training to trellis wires. Source: Adapted from: Majeed et al. (2020), (c) semantic segmentation of sweet cherry trees for pruning. Source: Adapted from: You et al. (2022a), (d) green apple fruit detection and counting for thinning. Source: Adapted from: Lu et al. (2022), (e) weakly supervised deep-learning-based apple flower counting for flower thinning. Source: Adapted from: Bhattarai and Karkee (2022), and (f) instance segmentation of apple flowers for robotic pollination. Source: Adapted from: Mu et al. (2023).

foliage detection, and (v) trellis system detection. In the following paragraphs, the latest advancements in each of these areas will be discussed.

3.1.3.1 Trunk and branch detection

Tree trunk and branch detection provides a foundation for robotic training of trees and branches onto trellis wires and pruning of selected branches/ trunks. In addition, trunk and branch detection is essential for any other robotic operation that needs to avoid collisions with these canopy parts. As pruning and training generally are conducted in the dormant season, many studies in this area have focused on processing dormant season images. Consumer

RGB-Depth cameras (e.g. Zed2, Stereolabs.com) have been the most popular imaging sensors used in these studies.

Researchers from Washington State University (Majeed et al., 2018, 2020) used an encoder-decoder-based deep CNN, SegNet (Badrinarayanan et al., 2017), semantic segmentation of apple canopy images (Fig. 5b). Both daytime and nighttime images were collected with a Kinect V2 sensor (Microsoft Corporation, Redmond, WA, USA) for model training. Four classes of pixels were defined as trunk, branch, trellis wire, and background. Since Kinect V2 can produce both RGB and depth information, the authors compared the segmentation results achieved with simple RGB images and with foreground RGB images in which the image background was removed by applying a depth filter. The results achieved with foreground RGB images were slightly better (Accuracy of 0.97) than the same achieved with the simple RGB images (Accuracy of 0.92).

Ma et al. (2021) also developed a machine vision system using two Azure Kinect ToF cameras (Microsoft Corporation, Redmond, WA, USA) collecting point cloud data from two perspectives. The data were then used to reconstruct the 3D structure of jujube trees. Once 3D reconstruction was completed, all points were passed into SPGNet (Landrieu and Simonovsky, 2018) for semantic segmentation by constructing a super-point graph. Individual pruning branches were then identified using density-based spatial clustering of applications with noise (Landrieu and Simonovsky, 2018). The results showed that they achieved an intersection-over-union (IoU) of 0.85 and 0.76, respectively, in detecting tree trunks and branches.

More recently, researchers from Oregon State University (You et al., 2022a) presented a novel work on sweet cherry tree segmentation using optical flow as a measure of depth. The optical flow was created using motion induced by camera movement. Overall, their segmentation model combined the FlowNet2 (Ilg et al., 2017), which is an existing optical flow network, and pix2pix (Isola et al., 2017) generative adversarial network (GAN) (Goodfellow et al., 2020). They concluded that the fusion of synthetic data from GAN and optical flow data led to a robust tree segmentation result in complex orchard environments with reduced effort in hand labeling of images. In addition to using imaging sensors, Westling et al. (2021) used a light detection and ranging (LiDAR) system to scan fruit trees and to detect and localize branches and trunks. Huge progress has been made in recent years in developing robust branch and trunk detection techniques based on low-cost sensors and deep-learning techniques. However, it is essential to continue advancing machine vision systems for more accurate estimation of branch location, diameter, angle, orientation, density, and age with real-time processing speed so that downstream applications such as robotic training and pruning could be further improved.

3.1.3.2 Flower detection

Detecting and counting flowers and estimating their density is essential for developing robotic flower thinning systems. In addition, flower detection and estimation of their orientation is fundamental to developing robotic pollination of fruit crops. Recent work on flower detection and segmentation also employs various types of deep-learning models. Wang et al. (2018) used Faster R-CNN to detect flowers in mango orchards. Dias et al. (2018a) implemented semantic segmentation of apple flowers using a pipeline involving iterative computation of region proposals, feature extraction using CNN, dimensionality reduction by principal component analysis (PCA), and classification using support vector machine (SVM). Improving previous work in terms of generalizability, Dias et al. (2018b) used CNN for foreground and background segmentation followed by region-growing refinement to perform semantic segmentation for apple flowers. Wu et al. (2020) proposed YOLOv4-based model to identify apple tree flowers, where the redundant and non-useful channels were pruned to reduce the model complexity, which substantially increased inference speed (computational time from 76 ms to 46 ms per image). Li et al. (2022) presented a similar work using both YOLOv3 and YOLOv4 for kiwi fruit flower and bud detection in orchard conditions.

Furthermore, Bhattarai et al. (2020) (Fig. 6) and Tian et al. (2020) used Mask R-CNN and Mask Scoring R-CNN to detect and segment apple flowers. The model proposed by Bhattarai et al. (2020) achieved an average precision

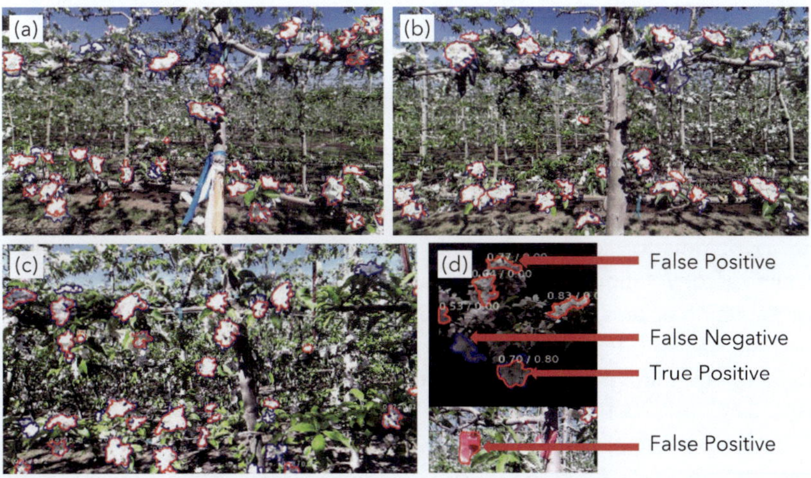

Figure 6 Detection results achieved by Mask R-CNN algorithm compared with ground truth dataset: (a), (b) Scifresh, and (c) Envy (objects inside blue and red polygons indicate ground truth and detection results, respectively); (d) Representative examples of different detection scenarios. Source: Adapted from: Bhattarai et al. (2020).

of 0.86 on the test dataset. To avoid massive hand-annotations for the entire dataset, Bhattarai and Karkee (2022) developed CountNet for flower counting that uses point annotation instead of area annotation (Fig. 5e). The count and density of flowers estimated with this technique can be used for robotic flower thinning.

It is noted that these studies focused on flower cluster detection without regard to identifying the king flowers from the flower clusters. The king flower is one of the flowers in a flower cluster that opens the first in the season and sets the largest fruit (Mu et al., 2023). Therefore, it is important to identify the king flower so that selective thinning (removal or non-king flowers) and precision pollination (pollinating king flowers) could be performed to ensure premium fruit quality. To address this challenge, Mu et al. (2023) recently proposed a mask R-CNN-based algorithm to both identify apple flower buds and detect king flowers within flower clusters. They achieved a precision of 78.2% in detecting apple flower buds whereas the king flower identification accuracy ranged from 66% to 99% during the entire flowering season. The higher values were achieved at the early flowering stages when most of the side flowers in flower buds were not open (e.g. 20–40% flowering stage).

3.1.3.3 Fruit detection

Fruit detection and counting are crucial for developing robotic solutions for green fruit (fruitlet) thinning, precision application of input materials to target fruit, and harvesting the mature fruit. Furthermore, fruit monitoring can benefit fruit yield and quality prediction for the growers. Depending on the purpose and nature of the tasks, object detection and segmentation are the two widely adopted methods in fruit detection (Karkee and Bhattarai, 2022). To incorporate different scenarios in the orchard, Gao et al. (2020) used Faster R-CNN to detect mature apples in four different situations: non-occluded, leaf-occluded, branch/wire-occluded, and fruit-occluded conditions. Wan and Goudos (2020) proposed improved Faster R-CNN including regularization to decay weights for some high dimensional parameters and loss functions to optimize convolution and pooling layers. The improved algorithm was used for detecting apples, mangoes, and oranges and achieved an mAP of >91% in all three datasets. Bargoti and Underwood (2017) used tiled Faster R-CNN by dividing input images into image patches for detecting apples, almonds, and mangoes with an F1 score of >0.9.

Fusion of RGB, NIR, and depth information has been shown to further improve the accuracy of the deep-learning-based fruit detection models. Fu et al. (2020) used Faster R-CNN and depth-filtered RGB-D images to detect apples, which increased the detection accuracy compared to the same with RGB images alone. In the work proposed by Sa et al. (2016), RGB and NIR information were fused in the input level or classification decision level to detect sweet peppers and other crops. Fusion at the classification decision level achieved higher performance (F1 = 0.838) than processing RGB and NIR images separately or fusion applied at the input level. In a similar line of work, Liu et al. (2019) fused RGB and NIR images as image fusion at input. This technique showed improved performance compared to using RGB and NIR images separately.

Mask R-CNN has also been employed to detect and segment fruits and provide masks for each segmented fruit in an orchard environment. Gené-Mola et al. (2020) implemented a Mask R-CNN-based model to segment apples. The detected fruits were then projected onto 3D point clouds based on structure from motion (SfM) to locate the apples in the 3D space. Jia et al. (2020) leveraged Mask R-CNN and a combination of ResNet and DenseNet feature extraction techniques to segment obstructed apples out. Mask R-CNN was also used for orange detection and segmentation by Ganesh et al. (2019).

Faster R-CNN and Mask R-CNN are complex models with multi-stage neural networks requiring powerful computational infrastructure for real-time applications. Relatively less complex models such as YOLO have also been proposed by researchers for fruit detection in fruit tree canopies. Wang and He (2021) demonstrated a channel-pruned YOLO V5s to identify green apple fruitlets when they were small. The authors adopted the channel pruning algorithm to eliminate the number of total channels in the network so that the complexity of the network can be reduced. The proposed method appeared to be highly effective because the F-1 score remained the same between the generic YOLOv5s and channel-pruned YOLOv5s, while the number of model parameters was reduced approximately by 93%. The final channel-pruned model's size was only 9.5% of the generic one, which could potentially be useful for edge deployment and implementation. Additionally, Tian et al. (2019) improved YOLO V3 and DenseNet for feature extraction to detect apples in three growth stages: young apples, expanding apples, and ripe apples. Additionally, the Washington State University research team (Lu et al., 2022) attempted at detecting both immature and mature apples in dense-foliage canopies (Fig. 5d). An attention mechanism module, named convolutional block attention module (CBAM) (Woo et al., 2018), was added to the YOLOv4 network (i.e. canopy-attention-YOLOv4 or CA-YOLOv4) to improve the detection performance during the entire fruit growing season.

It is noted that fruit detection is the most widely studied area in developing robotic solutions for orchards. With deep learning, techniques for mature fruit detection and delineation have been quite mature and reliable even under variable lighting conditions. However, we still face a lot of challenges in detecting and delineating small fruitlet and even larger fruit when they are growing in clusters or are only partially visible. For certain robotic picking approaches (e.g. scissor cutting), detecting fruit stems is essential, which is a challenging problem because stems are thin and often fully or partially occluded by fruit, foliage, or other canopy parts. Future research could focus on these issues so that robotic systems for various orchard operations such as thinning and harvesting could be further advanced.

3.1.3.4 Trellis wire detection

In all robotic orchard operations, robotic end-effectors need to interact with the canopies (discussed later) and penetrate the inner part of the canopies (needed in the modern fruiting wall orchards as well). In this process, the end-effector and manipulator may collide with the trellis wires that are installed in orchards to hold tree trunks and branches. Therefore, it is essential to detect and localize trellis wires so that robot paths could be planned to avoid these wires. This is a challenging problem as trellis wires are thin (a few millimeters) and often occluded by fruit, flowers, branches, leaves, and/or other canopy parts. Only a limited number of studies have been reported in the literature addressing this specific challenge. As mentioned before, Majeed et al. (2018) and Majeed et al. (2020) used an encoder–decoder network to classify dormant canopy images into trellis wire and three other classes at the pixel level. Another study by Bhusal et al. (2022) focused on detecting horizontal trellis wires in apple orchards supporting a robotic harvesting operation. In this study, RGB images were acquired from a commercial apple orchard using RGB images captured with an Intel RealSense 435D camera (Intel Corporation, Santa Clara, CA). The labeled images were then used to train a Mask RCNN-based model for segmenting out visible sections of trellis wires. Exploiting the fact that trellis wires were linearly positioned in tree canopies, the Hough Transform was then used to connect parts of individual trellis wires. The technique resulted in an 81% success rate in detecting the wires, and ~90% of the detected wires were located within 10 mm of the actual position.

3.1.4 Nighttime capability

Synchronized artificial lighting has been used by various researchers to improve light uniformity in the orchard images (Silwal et al., 2021). This strategy, while increasing the complexity a bit, has been reported to improve the outcome of a machine vision system in both detecting and localizing objects of interest (e.g. fruit and branches) and improving the robustness of the system to be applicable to different crop varieties (Silwal et al., 2021). This technique also adds capability for the nighttime operation of the robotic machines, which dramatically increases the economic viability of these expensive machines because the machine could then be used in both day and night shifts thus reducing the number of machines to be used, and consequently reducing initial investment as well as ensuring quick return on investment.

3.2 Human-machine-canopy interaction/collaboration

3.2.1 Human-machine interaction in orchards

Human, machine, and tree canopy are three essential components interacting with each other in orchard mechanization and automation. During orchard management and operations, these three components would have to collaborate with and complement each other for optimal performance. Currently, with human-operated machines, human–machine collaboration is adopted, albeit not optimally, quite commonly in mechanized and automation orchard operations. As more advanced decision-making and robotic systems are developed and adopted, more effective and optimal human–machine collaboration and interaction are expected to take place in orchard operations. A typical example of such a human–machine interaction and collaboration could be described with selective pruning of fruit trees. A robotic pruner (e.g. Section 5.1) could be programed to identify, locate, and prune specific branches of fruit trees. In this process, using the machine vision system input, the robotic pruner could approach the target branch at a certain orientation. A human operator would be monitoring multiple robotic arms operating in this fashion simultaneously using regular or virtual reality displays. In situations when the human operator finds the robotic arm position or orientation slightly off the target due to erroneous position or orientation estimation, the human operator would then provide input to adjust the approach direction to successfully complete the pruning action.

Scientific studies on human–machine or human–robot collaboration in orchard operations are limited. However, there have been some efforts in other agricultural industries. Adamides et al. (2017) studied the human–robot interaction (HRI) experience using a semi-autonomous vineyard sprayer. The interface was evaluated in both laboratory and field conditions. The study found that the use of a PC keyboard appeared to be more efficient than using PS3 gamepad during the collaborative operation. In addition, Huang et al. (2020) carried out an experiment on HRI in a simulated strawberry harvesting field. Results indicated that a human operator often dealt better with false positives than false negatives when the robot was trying to judge if the target strawberry was ready for harvest. Vasconez et al. (2019) presented an excellent work in reviewing and summarizing the challenges for HRI in agriculture, including the worker health and safety issues when HRI frameworks are implemented as accidents might happen due to unintended collisions and human errors or negligence. Therefore, it is crucial to continue comprehensive and trans-disciplinary research and development on intuitive and safe frameworks for human–machine collaboration in orchard operations.

3.2.2 Machine-canopy interaction in orchards

In addition to the interaction and collaboration between the human and machine, machine and tree canopy also interact extensively during the orchard operations. As also discussed elsewhere in the chapter, various mechanized, automated, and robotic machinery have been developed to minimize the involvement of human workers in orchard operations. However, the progress of commercial adoption of those machines is slow, primarily due to the lack of desired performance caused by complex, 5D tree canopy architecture (space, time, and differences between any two tree canopies). Studies have shown that key canopy parameters such as branch length and spacing, branch orientation, and fruit location can highly influence the overall performance of automated machines such as robotic harvesters (Zhang et al., 2020b). Therefore, developing and maintaining machine-friendly crop structures is highly critical to further advance the overall system performance and enhance the viability of commercial adoption of those machines in tree fruit orchards (Zhang, 2020; Zhang et al., 2021b). There are several approaches to maintain a compact tree structure with clear accessibility to canopy parts such as fruit and branches for robotic operations. These approaches include strategic tree training, pruning (dormant and summer), and thinning. More discussion on these aspects of engineering and crop physiology can be found in Robinson et al. (2013) and Lauri and Corelli Grappadelli (2014).

4 Actuation

4.1 Robotic manipulation

Robotic manipulation is the primary actuation element on a robot for implementing any orchard operation. A wide range of manipulators with varying reachability or workspace, obstacle avoidance ability, positioning accuracy, and actuating speed are available for developing robotic orchard operations. These manipulators, in terms of their structural and actuation features, could be classified into serial, parallel, and coordinated manipulators. Soft manipulation technology has also been an area of recent interest, which will be discussed briefly in Section 6.

4.1.1 Serial manipulator

This type of manipulator uses an open-loop kinematic chain mechanism composed of several rods connected in series using driving joints to control the end-effector installed on its actuation end. Those manipulators have attractive features of high flexibility, ample working space, and a relatively simple control, thus being widely applied in many robotic applications including orchard robots. Figure 7 shows one example of a serial manipulator

Figure 7 An example of 6-DOF serial manipulator developed for robotic apple harvesting. Source: Picture credit adapted from: Washington State University #AgRobotics.

of 6 degrees of freedom (6-DOF) being used on an orchard robot (Silwal et al., 2017).

As a widely adopted manipulating mechanism on orchard robots, it often uses a set of inverse kinematic equations to determine the spatial pose of its end-effector as the base for controlling the manipulator. Interested readers could easily find such fundamental equations in many fundamental books on robotics (e.g. Corke, 2017; Karkee and Zhang, 2021). In practice, the 6-DOF mechanism could often be simplified to optimize the functionality and complexity sufficient for specific orchard applications. For example, Hayashi et al. (2002) reduced 1-DOF from a 6-DOF manipulator (Fig. 8a) to improve the efficiency of eggplant harvesting, which resulted in successful picking of 62.5% of fruit. Wang (2010) found that it could further reduce the complexity to 4-DOF (Fig. 8b) for cucumber harvesting and obtained even a better result of 86% picking success.

On the opposite, researchers have also found that a redundant degree of freedom (i.e. total DOF of a manipulator greater than that needed to control its end-effector) would allow the end-effector to reach its target position from multiple directions to enhance the ability of a serial manipulator in dealing with obstacle avoidance. One example of a successful application was the use of a 7-DOF manipulator to obtain an ability to avoid plant stems and leaves to reach target tomato fruits for picking (Kondo et al., 1996).

Serial manipulation has also been widely adopted by commercial manipulator developers. One of the widely used commercially available

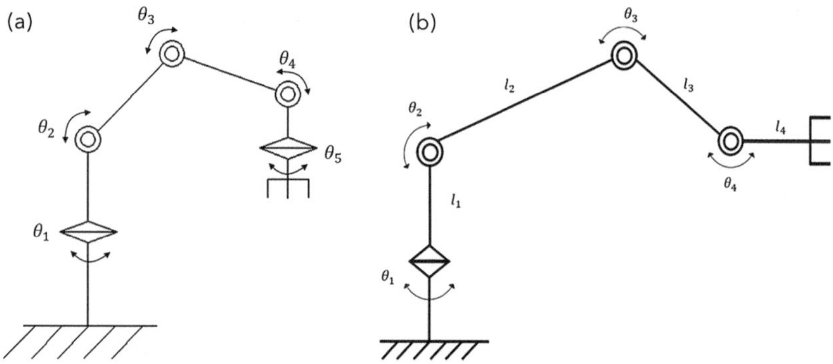

Figure 8 Schematics of two simplified serial manipulators developed for fruit picking; (a) a 5-DOF manipulator for eggplant harvesting. Source: Adapted from: Hayashi et al. (2002); and (b) a 4-DOF manipulator for cucumber harvesting. Source: Adapted from: Wang (2010).

serial manipulators is Universal Robotics UR5e (universal-robots.com/). The manipulator consists of six rotating joints providing 6-DOF with a payload of 5 kg and reachability of 850 mm. Several research programs around the world used this manipulator in developing robotic orchard or vineyard operations. For example, UR5 was used by Botterill et al. (2017) in developing a robotic pruning system for grapevines. Bhattarai et al. (2023a and b) developed a robotic flower thinning and pollination robot using UR5e and evaluated it in an orchard environment. Other serial manipulators available in the market with similar capabilities include xArm6 (ufactory.cc), AUBO i5 (aubo-cobot.com/), IRB 14050 (abb.com), CR7 (https://www.kleineducational.com/), and Rethink Robotics Sawyer collaborative robots. (References to any companies and products here and elsewhere in this chapter only portray the examples of advanced technologies for orchard application. It is not, in any way, an endorsement of the companies mentioned in this chapter or their products by the authors or the authors' institutions.) Despite its widespread application, it is noted that serial manipulators often suffer from the accumulation of errors at individual joints (DOFs) during the operation and tend to create larger errors in end-effector positioning.

4.1.2 Parallel manipulator

Different from the serial manipulator, a parallel manipulator is a mechanism in which the end-effector and the robot frame are connected by two or more parallel kinematic chains or branches. In general, parallel manipulators are lighter with a higher rigidity and load-bearing capacity compared to the serial counterparts. Another important advantage of the parallel manipulator is that it avoids the error accumulation drawback of serial manipulators, which allows it to achieve a higher accuracy in positioning the end-effector installed on it. As a drawback, the control of a parallel manipulator is often more complicated and the manipulator has a smaller maneuverable space than a serial manipulator. One widely applied parallel manipulating mechanism is the 3-DOF Delta mechanism invented by Clavel in 1988 (Zsombor-Murray, 2004) as illustrated in Fig. 9.

To capitalize on high payload, weak kinematic coupling, ease of force control, and ease of installation and operation features, Yang et al. (2022) designed a parallel manipulator based on this Delta mechanism for their apple-picking robot (Fig. 10). Based on their results obtained from a series of validation test, the authors found that a Delta mechanism-based parallel manipulator could carry sufficient weight and achieve a higher apple picking speed than many serial manipulator-based picking robots. Apple picking robots developed by private companies such as Abundant Robotics (now discontinued) also utilized Delta mechanism-based parallel manipulators.

4.1.3 Hybrid manipulator

For obtaining improved maneuverability in a relatively small working space, researchers integrated serial and parallel mechanisms leading to a hybrid manipulator. For example, Li et al. (2020) designed a serial-parallel hybrid articulated manipulator for their orange harvesting robot. As illustrated in Fig. 11, this design used a parallel structure to improve the stiffness of the

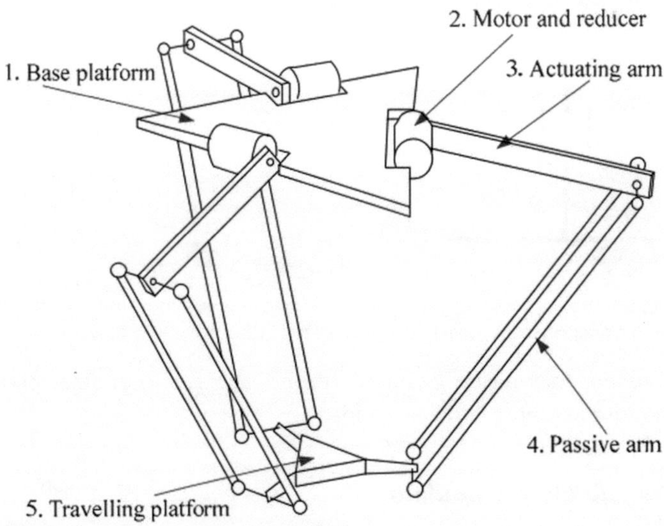

Figure 9 Illustration of the principle of a Delta mechanism for parallel manipulation. Source: Adapted from: Lu and Liu (2016).

Figure 10 Abundant robotics apple picking robot utilizing a parallel manipulator. Source: Adapted from: GoodFruitGrower (2016).

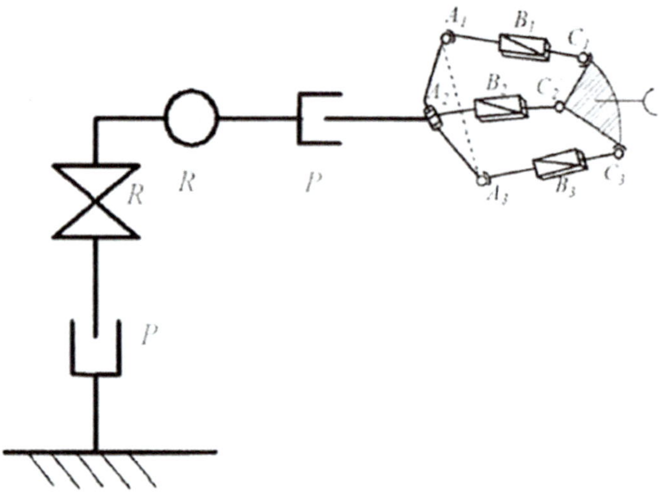

Figure 11 Schematics showing the working principle of a hybrid manipulator designed for orange picking. Source: Adapted from: Li et al. (2020).

manipulator for minimizing positioning error at the end-effector and used a serial structure to gain a larger workspace.

4.1.4 Coordinate manipulator

One manipulating mechanism gaining increased attention in agricultural applications is the coordinate manipulator (e.g. a cartesian manipulator),

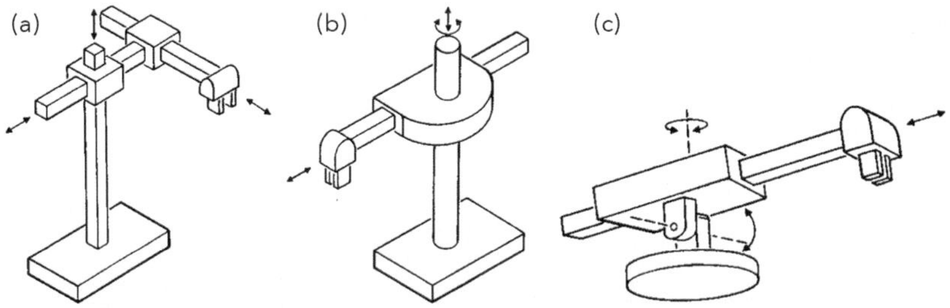

Figure 12 Illustration of the working principles of different coordinate manipulation systems. Source: Adapted from: Lee (1982).

which realizes linear, curvature, or other motion trajectories using a coordinate system. Figure 12 shows a few examples of such mechanisms in cartesian, cylindrical, and polar coordinates (Lee, 1982). As illustrated in the figure, a cartesian coordinate manipulator is mainly composed of three moving pairs on *X*-, *Y*-, and *Z*-directions orthogonal to each other whereas a cylindrical coordinate manipulator replaces the cartesian coordinate *XY*-axis with a rotary joint turning around the base, a linear translational moving pair along? the R-direction, and another translational pair on *Z*-direction. A polar coordinate manipulator is composed of a rotary joint, a pitch joint, and a mobile pair for telescopic operation (Fig. 12).

Among these, cartesian coordinate manipulators, in general, have the highest positioning accuracy and are the simplest systems to control with no coupling limitation. However, this manipulator often requires a larger structure, which limits its flexibility and workspace. The cylindrical coordinate manipulator normally has a smaller structural size and a better obstacle avoidance ability with a position accuracy inferior to that of the cartesian coordinate manipulator. In addition, the reachable space of the manipulator is limited as it cannot reach the space near the column or near the ground. This feature could be attractive to applications with confined working spaces. One important feature of polar coordinate manipulators is their small footprint and compact structure, but they are limited in their capability to avoid obstacles and have relatively lower positioning accuracy.

Both academic researchers and private industries have been utilizing coordinate manipulators (particularly the cartesian system) in their orchard robots. Zahid et al. (2020) used a cartesian robot and a scissor cutter and tested it in pruning apple trees. Fresh Fruit Robotics (Haifa, Israel) has been advancing and testing their robotic apple-picking robot in the USA and Israel. The robot uses three layers of cartesian manipulators (two arms in each layer) on each side of the machine totaling 12 cartesian arms operating simultaneously (Fig. 13). Even though these manipulators have limited capability to maneuver around

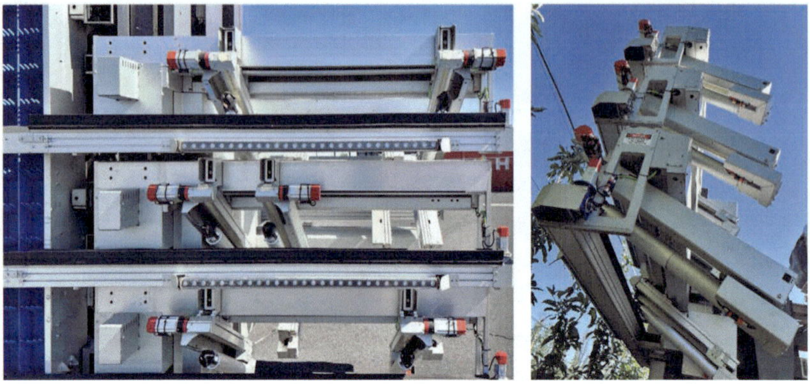

Figure 13 Fresh Fruit Robotics (FFRobotics Inc., Israel) uses 12 cartesian robotic manipulators, six on each side of the machine. Each arm is equipped with an RGB-D camera. Left: front view; Right: side view.

obstacles, their practical adoption for orchard operations such as pruning, thinning, and harvesting could be satisfactory in modern fruiting wall orchards where most of the target objects are linearly accessible, as they offer a great balance between cost, complexity, and performance.

4.2 End-effector technologies

An end-effector (sometimes also called hand or tool), attached at the end of a robotic manipulator, is a critical component of any robotic system designed for orchard operations. This component engages with the object being manipulated (e.g. fruit being picked or branches being pruned) and is designed uniquely to perform a specific orchard operation. In this sub-section, specific design choices and examples of recently developed end-effectors for four specific orchard operations (harvesting, pruning, flower thinning, and pollination) will be discussed briefly.

4.2.1 Harvesting end-effectors

Compared to the end-effector technologies and designs for other robotic operations such as pruning, thinning, and pollination discussed later, there is a much larger variation of the same investigated for fruit picking. These end-effectors can be categorized, at a higher level, into two classes: mechanically operated and air-operated (vacuum). When it comes to mechanical end-effectors, a human-hand-like mechanism with multiple fingers is often used to grasp and pick fruit. Multiple fingers are either operated simultaneously or individually. Fruit picking is accomplished by opening the fingers, getting on to the target fruit, closing the fingers to grab the fruit, and rotating and/or pulling the fruit after it is secured inside the hand. Silwal et al. (2017), Dischinger et al. (2021), and Shi et al. (2018) are some of the example studies recently reported using mechanical fingers for fruit harvesting. To improve the efficiency in holding the fruit with minimal pressure and minimize the number of moving parts, two cup-shaped fingers have been used by Fresh Fruit Robotics (ffrobotics.com) in their full-scale robotic apple harvesting machine. Though less common, scissor cutting of the stem is another technique used in fruit picking. A study from Zhao et al. (2016) is one of the studies using a stem-cutting approach for tomato picking. This technique, however, can be challenging in tree fruit crops where fruits commonly grow in clusters occluding the fruit stems.

Vacuum end-effectors have been designed in two different ways. In the first design, a tube with an opening larger than the diameter of the fruit is used to suck fruit in using high-power vacuum. One or more fruit can quickly be sucked up depending on the size of the vacuum tube. Abundant Robotics (now discontinued) used this technique to design their end-effector for apple harvesting as it offered a fast speed of picking (Fig. 14 – left). In the second design, vacuum tubes with a diameter smaller than the fruit diameter are used to hold and pull the fruit. When in place, the fruit end-effectors can be pulled or rotated to detach the fruit. This end-effector could be called a hybrid as both air and mechanical assists have been used to pick fruit. Recently, Advanced Farm (https://advanced.farm/technology/apple-harvester/) and Zhang et al. (2021a) used such an end-effector design in their studies on harvesting apples.

In addition to the two types of end-effectors described earlier, increasingly, fruit-picking robots are incorporating soft end-effectors, which are designed using special synthetic materials. These materials have the property to facilitate harvesting end-effector design with more compliant contact with the fruit being picked and cause milder impact to fruit compared to the mechanical or vacuum end-effectors. Soft robotic end-effectors can be 3D printed and operated using either air (by creating an air pocket inside end-effector mechanisms such as fingers) or using electric motors like mechanical hands discussed earlier (Hohimer et al., 2019).

4.2.2 Pruning end-effectors

End-effectors for pruning a fruit tree are designed, almost always, to include a pair of scissors or scissor-like cutters or clippers. The scissors could be operated using an electric, hydraulic, and pneumatic actuation system. To help design efficient and effective end-effectors for cutting certain types of branches in certain fruit trees, researchers conducted a cutting force analysis using a wide range of branch sizes (Zahid et al., 2020). Using this information, a specific end-effector could be designed or adopted from existing tools. Some of the latest studies on robotic pruning used commercially available clippers, mostly

Figure 14 Example end-effectors for fruit harvesting and fruit tree pruning; Left - a vacuum end-effector for robotic apple harvesting developed by Abundant Robotics. Source: Picture credit adapted from: goodfruit.com; Right - A cutting shear installed on UR5e robot. Source: Picture credit adapted from: Washington State University.

electrically actuated. For example, Zahid et al. (2020) used an electric cutting shear to prune apple trees. You et al. (2022) also used a similar end-effector in testing their robotic pruner in a cherry orchard (Fig. 14 - right).

4.2.3 Flower thinning end-effectors

Various types of end-effectors have been designed in the past for flower thinning depending on the type of thinning process (mechanical or chemical) and flowering stage (unopened/opened) when the thinning is performed. A commonly used mechanical end-effector for removing fully open flowers includes a center-rotating spindle with a string/brush covering a section of the tree canopy. Hand-held Bloom Buster commercially sold by Automated Ag (https://www.automatedag.com/bloom-bandit) and tractor-operated Darwin Thinner commercially sold by Fruit Tec (https://fruit-tec.com/en/202200-2) are two leading mechanical flower thinning solutions for branch-level and canopy-level flower thinning. Similar ideas have been used by Bhattarai et al. (2023b) to adopt the technology to design a miniature end-effector for precise, selective, and controlled flower thinning with a robot (Fig. 15 - left). Precision application of a flower thinning chemical using a robot is another way to thin fully opened flowers. However, no such study and end-effector design could be found in the literature. Similarly, no study was found on the robotic thinning of unopened flowers. Recently, studies have reported detecting unopened flowers using images acquired from reasonably close distances (Khanal et al., 2023). However, it could be challenging for the imaging sensors to estimate depth because of the small flower size and limited feature information. Future studies could also focus on developing effective end-effectors to facilitate robotic operations in these specific situations. Some of the specific end-effector techniques to be explored include pressurized water, pneumatically

Figure 15 Flower thinning end-effector attached on a tool-tip of UR5e robotic manipulator. Photo credit adapted from: Washington State University #AgRobotics lab. Spray module to spray pollens on kiwi flowers. Source: Adapted from: Williams et al. (2020).

actuated tools, and laser-based flower/stem cutting/removal. To ensure proper thinning, the end-effectors should be precise and accurate as the flowers are tightly located in clusters.

4.2.4 Pollination end-effectors

Different end-effector systems have been reported for robotic pollination of fruit crops such as kiwi, tomato, and bramble flowers using a ground-based robot. These end-effectors can broadly be categorized into contact and non-contact technologies. In a contact approach, the end-effector physically gets in contact with the flowers. Strader et al. (2019) developed a contact end-effector and tested it in a laboratory environment to pollinate Bramble flowers. The end-effector consisted of three linear servos acting as a parallel robot to move a thermoplastic polyurethane-based flexible end-effector plate (TPU95) coated with cotton padding for pollen delivery. In a non-contact approach, pollen solution is created by mixing pollens with water or other solution agents, which is then sprayed onto flower/flower clusters from a specific distance. Williams et al. (2020) developed a pollen sprayer system for kiwi flower pollination (Fig. 15 - right). The end-effector system consisted of 20 air-assisted nozzles linearly spaced 25 mm apart covering 500 mm of kiwi canopies. In another work, Yuan et al. (2016) reported a tomato pollination end-effector system with a nozzle and solenoid valve to control reagent/pollen spray volume and a manual valve to control the water flow. Pollen was sprayed at ~8 cm from the flower cluster with a cycle time of ~15 s.

5 Use cases of robotic orchard operations

As discussed before, researchers and engineers are developing many different kinds of robotic technologies to automate various orchard operations. Some of the major efforts include tree pruning (You et al., 2020, 2022a,b; Zahid et al.,

2020), flower thinning (Bhattarai et al., 2023a), pollination (Bhattarai et al., 2023b; Strader et al., 2019; Williams et al., 2020; Yuan et al., 2016), fruitlet thinning (Hussain et al., 2022), chemical application (Dai et al., 2017), nutrient and water management (Avgoustaki et al., 2022), and fruit harvesting (De-An et al., 2011; Hohimer et al., 2019; Silwal et al., 2017; Zhang et al., 2021a). In this section, recent advancements in automating three specific orchard operations are presented: pruning, flower thinning, and pollination. Advancements in automated or robotic harvesting techniques have been widely discussed elsewhere (Karkee et al., 2018). Similarly, automated chemical application systems have been discussed widely in recent years (Dai et al., 2017; Yallappa et al., 2017; Autospray Systems (https://autospraysystems.com/); Guss (https://gussag.com/guss/)).

5.1 Robotic pruning

For tree fruit crops, such as apples and sweet cherries, selective pruning is one of the most essential annual operations to keep the trees healthy and productive over their long life span. In recent years, it has also been demonstrated that strategic tree pruning can potentially improve the performance of mechanical and robotic fruit harvesting (Zhang et al., 2018). Two types of pruning techniques are generally practiced: dormant (or winter) pruning and summer pruning. In many of the tree fruit crops such as apples, dormant pruning is usually more aggressive to keep only a desired amount of flower buds on individual tree branches such that there can be reduced internal competition between the vegetative and reproductive growth, ensuring reproductive organs to have sufficient nutrition to achieve desired fruit yield and quality. Summer pruning is typically a light operation to remove long and extra slim limbs/twigs to keep the tree size compact and minimize shading on fruit. Therefore, most of the past studies focused on developing robotic solutions for dormant pruning.

Fruit tree pruning is conducted following specific farming strategies to achieve the desired distribution of fruiting sites. These strategies are often dictated by the type of fruit tree architecture and varieties. For example, in Upright Fruiting Off-shoot (UFO) cherries, lateral branches growing out of the vertical stems are removed. Similarly, in a formal tree architecture, where branches are trained horizontally to trellis wires, secondary branches growing out of those primary branches are generally kept short (grower-specified length) and any secondary branches growing vertically upward or downward from the primary branches are removed. In addition, any diseased or unproductive branches (also called blind wood) are removed so that new branches could be grown to fill the canopy space with more productive branches.

In general, these pruning strategies are dictated by geometric and topological parameters of branches such as location, length, spacing, and diameter of branches. For example, if a branch has a diameter bigger than a certain threshold or is longer than a specified length, they are pruned out. Also, if there are two branches closer together than a desired spacing, one of the two branches is pruned (Karkee et al., 2014; Karkee and Adhikari, 2015). These geometric criteria have been applied successfully in developing rules for automating pruning in apple orchards. Studies have suggested that a given branch diameter might be able to support only a certain number of apples to achieve the desired quality (Zhang et al., 2018). This knowledge could also be used to formulate a specific pruning strategy to remove secondary branches so that only a targeted number of fruiting sites are left in the specific canopy regions. These pruning strategies developed and practiced by growers and horticulturists form a knowledge base, which could be built into a decision support system or a rule base for identifying specific branches for robotic pruning of fruit trees. It is noted that research and development in various aspects of tree fruit pruning including rule-base development could benefit from much wider studies reported on pruning in vineyards (e.g. Corbett-Davies et al., 2012; Botterill et al., 2017).

To help implement these pruning strategies or rules automatically, fruit trees have to be represented in 3D space and their geometric and topological information need to be estimated (including trunk and branch size and location) (Karkee and Adhikari, 2015). The vision system discussed in Section 3.1 covered some of the major techniques used in detecting branches along with other canopy parts. Such a system could then be integrated with the decision support system discussed earlier and a manipulator (Section 4.1) and an end-effector (Section 4.3) (often a scissor type) to complete the robotic pruning system. For example, Westling et al. (2021) presented an automated tree pruning procedure using color image processing as well as LiDAR scanning for 3D information. Further details on the recent advancement in robotic pruning with a few examples can be found in Zahid et al. (2021). In the following paragraph, one specific robotic pruning system is discussed.

Researchers at Oregon State University and Washington State University developed and field evaluated an integrated robotic system for pruning cherry trees trained to UFO architecture (You et al., 2022c). The system was built using a commercially available UR5e (Universal-robotis.com) manipulator (6-DOF) and a commercially available electrically actuated pair of scissors (Fig. 16 – left). The manipulator was installed on a linear axis to increase the workspace in the lateral direction (into the tree canopies). An eye-in-hand (end-effector) RGB-D camera was used to acquire and process canopy images. The entire system was mounted on a ground utility vehicle, which was driven manually. The system used a hybrid vision-interaction controller based on training a GAN to segment out the target branches. The segmented branches were then fed into a Mask R-CNN network, i.e. segmentation, so that individual branches could

Figure 16 Integrated orchard robots developed and tested by a collaborative team of researchers from Washington State University, Oregon State University, and Pennsylvania State University; Left - robotic pruning of UFO (upright fruiting offshoot) cherry trees in a research orchard; center - robotic blossom thinning in a commercial apple orchard; and right - robotic pollination in a research apple orchard. Source: Photo credit adapted from: Washington State University #AgRobotics lab.

be identified for pruning. Both simulated and real images were utilized during GAN training allowing for robust results in a wide range of lighting conditions. No 3D information was explicitly used. The field evaluation results showed that 58% of the branches attempted were successfully pruned. The authors claimed that the system could be operated completely autonomously with minimal or no changes in the canopy environment available during the field test.

5.2 Robotic thinning

Generally, fruit trees produce more flowers and fruitlets than desired for optimal crop yield and quality. A higher number of fruits per tree/branch may result in smaller fruit size and lower fruit quality. Flower and fruitlet thinning is a way to mitigate such competition during the fruiting stage by removing a certain number of flowers and/or fruitlets. Currently, chemical flower and fruitlet thinning has been the most common approach (Deckers et al., 2010). However, chemical thinning is not selective, has limited efficacy, and leads to uncertain and variable outcomes. To improve the precision and selectivity, manual flower and fruitlet thinning is practiced by many growers, which is highly labor intensive. To address these challenges, there have been some studies around robotic thinning, but mostly focusing on flower thinning (Bhattarai et al., 2023b; Lyons et al., 2015).

One such study on robotic flower thinning was recently carried out by Washington State University researchers (Bhattarai et al., 2023b). The team developed the robotic flower thinning system using a UR5e manipulator and a custom-developed end-effector (Fig. 16 - center), which was then evaluated in a commercial apple orchard to remove a proportion of flowers from clusters of apple flowers. The system used an Intel RealSence RGB-D camera to collect canopy images and implemented the Mask R-CNN algorithm, i.e. segmentation of flower clusters. Once the flower clusters were segmented, the position and orientation of the flower clusters were estimated in 3D space followed by sequences of clusters for the end-effector to visit. Finally, precision thinning was achieved by navigating a miniature electrically actuated end-effector (spindle string design as discussed in Section 4.2) via UR5e manipulator. Inspired by hand flower thinning, two thinning strategies were developed and implemented in the field: cluster center thinning and cluster boundary thinning. Field evaluation showed that the robotic system was able to successfully thin 67.2% of flowers with the boundary thinning approach and 59.5% of flowers with the center thinning approach. The thinning strategy could be modified and fine-tuned to account for seasonal and varietal differences to achieve the desired thinning results.

5.3 Robotic pollination

For tree fruit crops, pollination quality can greatly affect the overall fruit set and fruit quality at harvest. Natural pollination is often carried out with insects (including honeybees) that are available in orchards. In places where not enough natural pollinators are available, honeybee colonies are hired to pollinate these crops. However, in recent years, honeybee availability has been critically impacted due to colony collapse disorder, a condition causing a rapid decline of beehives (Ellis et al., 2010). Robotic pollination, therefore, is an important area of research, development, and practice to minimize the uncertainty in crop pollination and ensure the long-term sustainability of tree fruit crops.

Studies have shown that mass supplement of pollens into the canopy during the peak bloom could help fruit set (Whiting et al., 2022), whereas more precise robotic pollination could drastically reduce the amount of pollen used while ensuring the desired level of fruit set. Ohi et al. (2018) demonstrated a prototype robotic system for pollinating artificial flowers in a greenhouse environment. A group of researchers from the University of Auckland (Williams et al., 2020) have also been developing a robotic kiwifruit pollinator. Their robotic pollinator could identify about 80% of the target flowers in the orchard environments, however, the final fruit set was found to be below the requirements for commercial production.

Similarly, a robotic pollination system was developed and tested by a team of researchers from Washington State University and Pennsylvania State University recently. Like robotic flower thinning systems discussed earlier, this robot was built using a commercially available UR5e manipulator and a custom-built pollen spray system (Fig. 16 – right). The spray end-effector utilized an electrostatic spray technique to precisely deliver pollens onto target flowers. Pollen spraying was initiated after the spray nozzle was moved to a specified distance and orientation with respect to the target flower clusters. To keep the pollens viable for a longer time and to facilitate spraying, a pollen solution was developed by combining cultivar-specific pollen powder, a proprietary pollen suspension media (Whiting et al., 2022), and water. The pollen suspension media was charged via electrostatic induction using an air-assisted atomization system such that the electrically charged pollens would be attracted by low-impedance earthed flower stigma when sprayed. To complete the robotic system, a machine vision system was developed for flower cluster detection and localization (like the vision system developed for robotic thinning discussed earlier). All these components were then integrated together, and the robotic system was installed on a utility cart for field testing. The first experiment was conducted in a commercial Honeycrisp apple orchard in Naches, WA. Prior to flower opening and after the robotic pollination, the apple tree branches were covered with nets to prevent pollination any natural pollination. Pollen solutions of varying concentrations (1 g or 2 g) were prepared by mixing pollen with 1 L of water and 100 g of suspension media. During full bloom, experiments were conducted by spraying the pollen solutions onto flower clusters for 2 s from ~20 cm using 15 psi spray pressure, which achieved up to 43.1% pollination success. Further experiments are being conducted in different varieties at larger scales.

6 Challenges and opportunities

The current generation of orchard robots faces various technical challenges specific to different component technologies such as vision systems and end-effectors. For example, fruit detection techniques (and for that matter all orchard vision systems) developed in the past are mostly tailored to specific applications under specific field and crop conditions. For more robust and multi-purpose robotic operations, efforts need to be in place for developing more generic orchard vision systems that can handle detecting objects such as fruit of different kinds and in difficult situations such as clustered growth. A robotic fruitlet thinning hand must be precise and accurate so that it can pluck and remove tiny fruitlets from the tree canopies while keeping other closely spaced fruitlets intact. These types of technical challenges and potential opportunities on specific technologies have been further discussed in respective sections (e.g. Section 5). In this section, we focus on presenting some thoughts around higher level challenges and opportunities in orchard robotics.

The sustainability of fruit crop production around the world is challenged by decreasing availability and increasing cost of farm labor to complete various labor-intensive orchard operations. In addition, various orchard operations such as pest control pose critical worker health and safety issues. Other operations such as pollination pose sustainability risks due to bee colony collapse disorder

(Seitz et al., 2015). To address these challenges, researchers and engineers around the world have been developing robotic solutions for various orchard operations for several decades. With recent advancements in artificial intelligence (specifically deep learning), sensing, robotics and computational technologies, and horticultural systems, and the increased support from all stakeholders including farmers, private industries, and investors as well as local and federal agencies, we are making rapid progress in solving this critical challenge the fruit crop industry is facing. Despite all these advancements, we still lack successful adoption of these technologies in commercial orchards. The commercial viability of these technologies has been hindered primarily by (i) lack of speed, accuracy, and robustness of the robotic machines to operate under wide variations in environment and field conditions, crop architectures, and crop varieties; (ii) complexity and cost of the technology; and (iii) socio-economic factors.

First of all, the majority of efforts in developing automated solutions for orchards have been in harvesting. This focused effort by research organizations (De-An et al., 2011; Hohimer et al., 2019; Silwal et al., 2017; Zhang et al., 2021a) and private companies (e.g. ffrobotics.com and advanced.farm) has led to a rapid advancement of robotic fruit picking technologies. As we get closer to the commercial adoption of robotic harvesting solutions, we are addressing a huge challenge faced by the fruit crop industry with their dependence on human labor. However, it is not enough for sustainable fruit crop production as there are many other field operations such as pruning and thinning requiring huge amounts of manual labor in other growing stages (Fig. 17). This figure presents a tentative distribution of labor use in apple production in Washington State

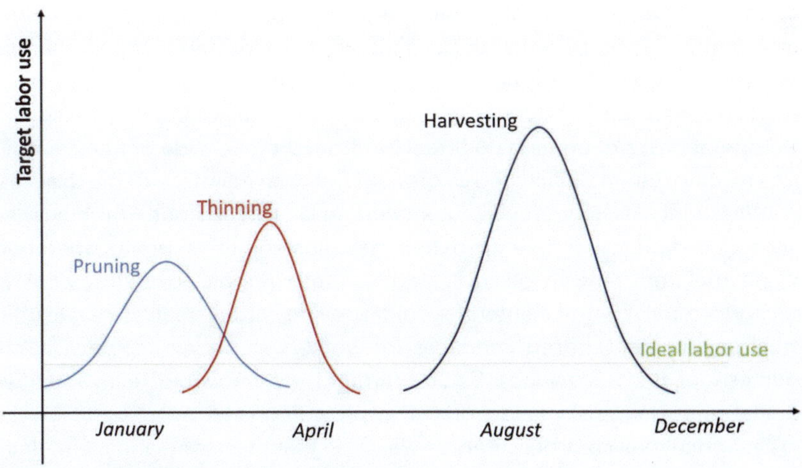

Figure 17 Conceptual (not in scale) presentation of labor demand in producing tree fruit crops such as apples in Washington State, USA.

farms. As can be seen, harvesting requires the highest number of farm workers in apple orchards which is concentrated in a window of about 3 months in Fall. In addition, labor-intensive operations such as pruning and thinning require a seasonal supply of a high number of workers in other short periods while there will be times in the growing season when not many people would be working in the farms. Ideally, growers would like to have a small but constant employment of skilled workers.

To address this challenge, researchers have been developing robotic solutions for many different orchard operations (also discussed in Section 5). In the future, even more emphasis should go into developing robotic solutions for all labor-intensive field operations in various types of fruit crops. Particularly, these efforts have been and will continue to create an opportunity for developing a multi-purpose robotic solution for fruit production. As it could be observed from the discussion in Section 5, it is viable to use the same robotic system for multiple field operations simply by swapping in and out the end-effectors designed for the specific field operation. With this new direction for research and development, it is expected that complex and expensive robotic machines could be used year-round (not just for one operation that runs for a few weeks to a few months), which will improve the return on investment and economic viability of these machines.

One major challenge in commercializing robotic machines is the low overall performance. For example, even with the best modern canopy architecture and the best robot available, we are still harvesting only about 60–70% of apples due to the variability and complexity of the canopy structure and the need for simpler and affordable robot design. Similar limitations exist in other operations such as pruning and fruitlet thinning. When it comes to conventional bushy trees, it may be practically not viable to pick fruit using robots as a huge percentage of the fruit is in the inner parts of the thick canopies. It is expected that close human-machine collaboration will have to be implemented to improve both target object detection and localization accuracy and physically complete the specific field operations, e.g. pruning out branches from the back side of tree canopies that a robotic system with limited degrees of freedom may not be able to access.

When increasingly more automated and autonomous machines are adopted in farming, it is anticipated that most of the farming operations will be run from remote office locations. These remote locations, what we call agricultural control centers, would provide virtual reality (real-like 3D virtual environment)-based interfaces for workers to operate, supervise, and troubleshoot these machines. Such a farming system would relieve workers from laborious and risky field work while providing year-round, better-paying jobs at agricultural control centers as well as in manufacturing and maintaining automated machines. When successful, these farms of the future would help create more vibrant rural communities leading to both sustainable agricultural production systems as well as sustainable human and social systems.

Continued exploration of novel technologies such as soft-robotic manipulator also provides an opportunity for increasing speed and reducing the cost and complexity of orchard robots (Wan et al., 2023). Research teams at Washington State University, and the University of Illinois at Urbana Champaign among others, are focusing on advancing this technology for orchard applications such as fruit picking. As the technology includes soft-fabric based robotic manipulators, it has the potential to reduce the cost of the technology by several folds while also offering a simpler and an easy-to-repair (like a bicycle tire) robotic machine for farming.

Last but not the least, a systems approach with trans-disciplinary expertise including physical sciences (engineering and computer science), biological sciences (e.g. horticulture), and socioeconomic sciences needs to be continued to develop robust and commercially viable robotic solutions for orchard operations that would also maintain not only economic and environmental sustainability but also human and social sustainability by providing safe working environment, continuous employment, and a salary comfortable to support family and social responsibilities.

Finally, it is noted that, in addition to the farming industry, academic researchers and government funding agencies, venture capitalists, startups and established tech and manufacturing companies are also increasingly interested in advancing robotic technologies for farming. Such a wide interest in agricultural technologies has been a new phenomenon; the one that has the potential to help the farming industry move into the new era of farming where decision-making (e.g. which branches to prune out), implementation of the farming decisions (e.g. pruning branches), and assessment of the outcome (e.g. tree growth, and fruit yield and quality assessment) could all be performed by increasingly autonomous/smart machines and systems with minimal human intervention. Such a system could be used to retain the knowledge and experience of the aging farming population around the world and minimize the challenges the next generation of farmers is going to face because of generally limited experience and understanding of the farming systems and processes. When researchers develop and test robotic systems, they often focus on basic performance measures such as speed and efficiency. These measures, however, do not always assess the progress (or the lack of progress) we are making in meeting practical and engineering requirements for robotic systems to be commercially adoptable. Transdisciplinary systems approaches have the potential to benefit research and development teams in better understanding these requirements and assessing the practical viability of orchard robots.

7 Conclusion

In conclusion, the development of robotic systems for orchard operations is challenging due to the intricate decision-making and manipulation processes involving uncertain orchard environments and biological variability. Nonetheless, the continued, focused efforts of researchers and private companies worldwide, combined with the convergence of cutting-edge technologies in machine vision, artificial intelligence, robotics, computational technologies, horticulture, and farming practices, are driving transformative change in orchard operations that traditionally relied on manual labor. While certain challenges remain to be addressed, it is expected that robotic systems will soon become pivotal in various aspects of tree fruit production, with their influence continuously expanding in the future. These robotic solutions have the potential to enhance efficiency, increase productivity, reduce costs, and overall improve orchard operations. Such development signifies a promising future where technology and automation play a vital role in sustainable tree fruit production.

8 Where to look for further information

8.1 Books

- Karkee, M. and Zhang, Q. (2021). *Fundamentals of Agricultural and Field Robotics*. Springer, Cham, Switzerland, ISBN: 978-3-030-70399-8, DOI: 10.1007/978-3-030-70400-1.
- Huang, Y. and Zhang, Q. (2021). *Agricultural Cybernetics*. Springer, Cham, Switzerland, ISBN: 978-3-030-72101-5, DOI: 10.1007/978-3-030-72102-2.
- de Queiroz, D. M., Valente, D. S. M., de Carvalho Pinto, F. D. A., Borém, A. and Schueller, J. K. (2022). *Digital Agriculture*. Springer, Cham, Switzerland, ISBN: 978-3-031-14532-2, DOI: 10.1007/978-3-031-14533-9.
- Karkee, M., Gordón, J., Salato, B. and Whiting, M. D. (2019). Optimizing fruit production efficiencies via mechanization. In: *Achieving Sustainable Cultivation of Temperate Zone Tree Fruits and Berries (vol. 1) - Physiology, genetics and cultivation (Editor: Greg Lang)*. Burleigh Dodds Science Publishing, Cambridge, UK, ISBN: 9780429275517, DOI: 10.1201/9780429275517.
- Zhang, Q., Karkee, M. and Tabb, A. (2019). The use of agricultural robots in orchard management. In: *Robotics and Automation for Improving Agriculture (Editor: John Billingsley)*. Burleigh Dodds Science Publishing, Cambridge, UK (pp. 187–214), ISBN: 9780429266737, DOI: 10.19103/AS.2019.0056.14.

- Karkee, M., Zhang, Q. and Silwal, A. (2021). Agricultural robots for precision agricultural tasks in tree fruit orchards. In: *Innovation in Agricultural Robotics for Precision Agriculture, Progress in Precision Agriculture (Editor: Avital Bechar)*. Springer, Cham, Switzerland (pp. 63–89), ISBN: 978-3-030-77035-8, DOI: 10.1007/978-3-030-77036-5_4.
- Shamshiri, R. R., Hameed, I. A., Karkee, M. and Weltzien, C. (2018). Robotic harvesting of fruiting vegetables: A simulation approach in V-REP, ROS and MATLAB. In *Automation in Agriculture - Securing Food Supplies for Future Generations (Editor: Stephan Hussmann)*. InTech, Croatia, ISBN: 978-953-51-3874-7, DOI: 10.5772/intechopen.73861.
- Karkee, M., Silwal, A. and Davidson, J. R. (2018). Mechanical harvest and in-field handling of tree fruit crops. In: *Automation in Tree Fruit Production, Principles and Practice (Editor: Qin Zhang)*. CABI, Boston, ISBN: 978-1-78064-850-7.
- Zhang, Q., Karkee, M. and Khot, L. R. (2017). Mechanization and automation for apple production. In: *Achieving Sustainable Cultivation of Apples (Editor: Kate Evans)*. Burleigh Dodds Science Publishing, Cambridge (pp. 279–306), ISBN: 9781351114233, DOI: 10.19103/AS.2016.0017.14.
- Zhang, Q. (2022). *Encyclopedia of Smart Agriculture Technologies*. Springer, Cham, Switzerland, DOI: 10.1007/978-3-030-89123-7.

8.2 Journals

- Computers and Electronics in Agriculture (https://www.sciencedirect.com/journal/computers-and-electronics-in-agriculture).
- Journal of Field Robotics (https://onlinelibrary.wiley.com/journal/15564967).
- Journal of the American Society of Agricultural and Biological Engineers (ASABE) (https://www.asabe.org/JA).

8.3 Conferences

- ASABE Annual International Meeting
- CIGR World Congress
- International Conference on Precision Agriculture
- International Federation of Automatic Control (IFAC) AgriControl
- IFAC BioRobotics
- IEEE International Conference on Robotics and Automation
- IEEE International Conference on Intelligent Robots and Systems
- International Society for Horticultural Science (ISHS)
- The European Society of Agricultural Engineers (EurAgEng)

9 References

Adamides, G., Katsanos, C., Constantinou, I., Christou, G., Xenos, M., Hadzilacos, T. and Edan, Y. (2017). Design and development of a semi-autonomous agricultural vineyard sprayer: human-robot interaction aspects. *Journal of Field Robotics* 34(8), 1407–1426.

Aghi, D., Mazzia, V. and Chiaberge, M. (2020). Local motion planner for autonomous navigation in vineyards with a RGB-D camera-based algorithm and deep learning synergy. *Machines* 8(2), 27.

Alom, M. Z., Taha, T. M., Yakopcic, C., Westberg, S., Sidike, P., Nasrin, M. S., van Esesn, B. C., Awwal, A. A. S. and Asari, V. K. (2018). The history began from alexnet: A comprehensive survey on deep learning approaches. *ArXiv Preprint ArXiv:1803.01164*.

Aulinas, J., Petillot, Y., Salvi, J. and Lladó, X. (2008). The SLAM problem: a survey. *Artificial Intelligence Research and Development*, 363–371.

Avgoustaki, D. D., Avgoustakis, I., Miralles, C. C., Sohn, J. and Xydis, G. (2022). Autonomous mobile robot with attached multispectral camera to monitor the development of crops and detect nutrient and water deficiencies in vertical farms. *Agronomy* 12(11), 2691.

Badrinarayanan, V., Kendall, A. and Cipolla, R. (2017). Segnet: A deep convolutional encoder-decoder architecture for image segmentation. *IEEE Transactions on Pattern Analysis and Machine Intelligence* 39(12), 2481–2495.

Bai, Y., Zhang, B., Xu, N., Zhou, J., Shi, J. and Diao, Z. (2023). Vision-based navigation and guidance for agricultural autonomous vehicles and robots: a review. *Computers and Electronics in Agriculture* 205, 107584.

Bargoti, S. and Underwood, J. (2017). Deep fruit detection in orchards. In: (vol. 2017). IEEE International Conference on Robotics and Automation (ICRA) (pp. 3626–3633). IEEE.

Bhattarai, U. and Karkee, M. (2022). A weakly-supervised approach for flower/fruit counting in apple orchards. *Computers in Industry* 138, 103635.

Bhattarai, U., Bhusal, S., Majeed, Y. and Karkee, M. (2020). Automatic blossom detection in apple trees using deep learning. *IFAC-PapersOnLine* 53(2), 15810–15815.

Bhattarai, U., Sapkota, R., Mo, C., Whiting, M. D., Zhang, Q. and Karkee, M. (2023a). *Design and Evaluation of Robotic Pollination System*. [Unpublished manuscript].

Bhattarai, U., Zhang, Q. and Karkee, M. (2023b). Design, Integration, and Field Evaluation of a Robotic Blossom Thinning System for Tree Fruit Crops. *arXiv preprint arXiv: 2304.04919*.

Bhusal, S., Bhattarai, U. and Karkee, M. (2022). Trellis wire detection for obstacle avoidance in apple orchards. *IFAC-PapersOnLine* 55(32), 72–77.

Blok, P. M., van Boheemen, K., van Evert, F. K., IJsselmuiden, J. and Kim, G. (2019) Robot navigation in orchards with localization based on Particle filter and Kalman filter. *Computers and Electronics in Agriculture* 157, 261–269.

Botterill, T., Paulin, S., Green, R., Williams, S., Lin, J., Saxton, V., Mills, S., Chen, X. and Corbett-Davies, S. (2017). A robot system for pruning grape vines. *Journal of Field Robotics* 34(6), 1100–1122.

CNH Industrial (n.d.). *Advanced Farming Systems*. [Brochure]. Retrieved January 6, 2023 Available at: https://assets.cnhindustrial.com/caseih/NAFTA/NAFTAASSETS/Products/Advanced-Farming-Systems/Brochures/AFS%20Brochure_02-21_CIH21020101_pages.pdf.

Corbett-Davies, S., Botterill, T., Green, R. and Saxton, V. (2012). An expert system for automatically pruning vines. In: Proceedings of the 27th Conference on Image and Vision Computing New Zealand (pp. 55–60).

Corke, P. (2017). *Robotics, Vision and Control* (2nd edn.). Springer Nature, Cham, Switzerland, ISBN: 978-3-319-54412-0, DOI: 10.1007/978-3-319-54412-0.

Dai, B., He, Y., Gu, F., Yang, L., Han, J. and Xu, W. (2017). A vision-based autonomous aerial spray system for precision agriculture. In: IEEE International Conference on Robotics and Biomimetics (ROBIO) (pp. 507–513). IEEE.

De-An, Z., Jidong, L., Wei, J., Ying, Z. and Yu, C. (2011). Design and control of an apple harvesting robot. *Biosystems Engineering* 110(2), 112–122.

Deckers, T., Schoofs, H. and Verjans, W. (2010). Looking for solutions for chemical fruit thinning on apple. *Acta Horticulturae* XI(884), 237–244.

Dias, P. A., Tabb, A. and Medeiros, H. (2018a). Apple flower detection using deep convolutional networks. *Computers in Industry* 99, 17–28.

Dias, P. A., Tabb, A. and Medeiros, H. (2018b). Multispecies fruit flower detection using a refined semantic segmentation network. *IEEE Robotics and Automation Letters* 3(4), 3003–3010.

Dischinger, L. M., Cravetz, M., Dawes, J., Votzke, C., VanAtter, C., Johnston, M. L., Grimm, C. M. and Davidson, J. R. (2021). Towards intelligent fruit picking with in-hand sensing. In: (vol. 2021). IEEE/RSJ International Conference on Intelligent Robots and Systems (IROS) (pp. 3285–3291). IEEE.

Dos Santos, F. N., Sobreira, H. M. P., Campos, D. F. B., Morais, R., Moreira, A. P. G. M. and Contente, O. M. S. (2015). Towards a reliable monitoring robot for mountain vineyards. In: (vol. 2015). IEEE International Conference on Autonomous Robot Systems and Competitions (pp. 37–43). IEEE.

Ellis, J. D., Evans, J. D. and Pettis, J. (2010). Colony losses, managed colony population decline, and Colony Collapse Disorder in the United States. *Journal of Apicultural Research* 49(1), 134–136.

Freitas, G., Hamner, B., Bergerman, M. and Singh, S. (2012). A practical obstacle detection system for autonomous orchard vehicles. In: (vol. 2012). 2012 IEEE/RSJ International Conference on Intelligent Robots and Systems (pp. 3391–3398). IEEE.

Fu, L., Gao, F., Wu, J., Li, R., Karkee, M. and Zhang, Q. (2020). Application of consumer RGB-D cameras for fruit detection and localization in field: A critical review. *Computers and Electronics in Agriculture* 177, 105687.

Ganesh, P., Volle, K., Burks, T. F. and Mehta, S. S. (2019). Deep orange: mask R-CNN based orange detection and segmentation. *IFAC-PapersOnLine* 52(30), 70–75.

Gan-Mor, S., Clark, R. L. and Upchurch, B. L. (2007). Implement lateral position accuracy under RTK-GPS tractor guidance. *Computers and Electronics in Agriculture* 59(1–2), 31–38.

Gao, F., Fu, L., Zhang, X., Majeed, Y., Li, R., Karkee, M. and Zhang, Q. (2020). Multi-class fruit-on-plant detection for apple in SNAP system using Faster R-CNN. *Computers and Electronics in Agriculture* 176, 105634.

Gené-Mola, J., Sanz-Cortiella, R., Rosell-Polo, J. R., Morros, J. R., Ruiz-Hidalgo, J., Vilaplana, V. and Gregorio, E. (2020). Fruit detection and 3D location using instance segmentation neural networks and structure-from-motion photogrammetry. *Computers and Electronics in Agriculture* 169, 105165.

Gongal, A., Amatya, S., Karkee, M., Zhang, Q. and Lewis, K. (2015). Sensors and systems for fruit detection and localization: a review. *Computers and Electronics in Agriculture* 116, 8–19.

Goodfellow, I., Pouget-Abadie, J., Mirza, M., Xu, B., Warde-Farley, D., Ozair, S., Courville, A. and Bengio, Y. (2020). Generative adversarial networks. *Communications of the ACM* 63(11), 139–144.

GoodFruitGrower (2016). Robotic apple picker trials continue in Washington. *YouTube*. Retrieved April 10, 2023 Available at: https://www.youtube.com/watch?v=mS0coCmXiYU.

Grisetti, G., Stachniss, C. and Burgard, W. (2007). Improved techniques for grid mapping with rao-blackwellized particle filters. *IEEE Transactions on Robotics* 23(1), 34–46.

Hayashi, S., Ganno, K., Ishii, Y. and Tanaka, I. (2002). Robotic harvesting system for eggplants. *Japan Agricultural Research Quarterly: JARQ* 36(3), 163–168.

He, K., Gkioxari, G., Dollár, P. and Girshick, R. (2017). Mask r-cnn. In: Proceedings of the IEEE International Conference on Computer Vision (2961–2969).

Hohimer, C. J., Wang, H., Bhusal, S., Miller, J., Mo, C. and Karkee, M. (2019). Design and field evaluation of a robotic apple harvesting system with a 3D-printed soft-robotic end-effector. *Transactions of the ASABE* 62(2), 405–414.

Huang, Z., Miyauchi, G., Gomez, A. S., Bird, R., Kalsi, A. S., Jansen, C., Liu, Z., Parsons, S. and Sklar, E. (2020). An experiment on human-robot interaction in a simulated agricultural task. Annual Conference Towards Autonomous Robotic Systems (pp. 221–233).

Hussain, M., He, L., Schupp, J. and Heinemann, P. (2022). Green fruit removal dynamics for development of robotic green fruit thinning end-effector. *Journal of the ASABE* 65(4), 779–788.

Ilg, E., Mayer, N., Saikia, T., Keuper, M., Dosovitskiy, A. and Brox, T. (2017). Flownet 2.0: evolution of optical flow estimation with deep networks. In: Proceedings of the IEEE Conference on Computer Vision and Pattern Recognition (pp. 1647–1655).

Isola, P., Zhu, J.-Y., Zhou, T. and Efros, A. A. (2017). Image-to-image translation with conditional adversarial networks. In: Proceedings of the IEEE Conference on Computer Vision and Pattern Recognition (pp. 5967–5976).

Jia, W., Tian, Y., Luo, R., Zhang, Z., Lian, J. and Zheng, Y. (2020). Detection and segmentation of overlapped fruits based on optimized mask R-CNN application in apple harvesting robot. *Computers and Electronics in Agriculture* 172, 105380.

John Deere (n.d.). Guidance. Retrieved January 6, 2023 Available at: https://www.deere.com/en/technology-products/precision-ag-technology/guidance/.

Kamilaris, A. and Prenafeta-Boldú, F. X. (2018). Deep learning in agriculture: A survey. *Computers and Electronics in Agriculture* 147, 70–90.

Kao, W. W. (1991). Integration of GPS and dead-reckoning navigation systems. In: (vol. 2) SAE Technical Paper Series Vehicle Navigation and Information Systems Conference. IEEE.

Karkee, M. and Adhikari, B. (2015). A method for three-dimensional reconstruction of apple trees for automated pruning. *Transactions of the ASABE* 58(3), 565–574.

Karkee, M. and Bhattarai, U. (2022). Applied machine vision technologies in specialty crop production. In: *Sensing, Data Managing, and Control Technologies for Agricultural Systems*. Springer (pp. 41–73).

Karkee, M. and Zhang, Q. (2021). *Fundamentals of Agricultural and Field Robotics*. Springer Nature, Cham, Switzerland, ISBN: 978-3-030-70399-8, DOI: 10.1007/978-3-030-70400-1.

Karkee, M., Silwal, A. and Davidson, J. R. (2018). Mechanical harvest and in-field handling of tree fruit crops. In *Automation in Tree Fruit Production: Principles and Practice*. Wallingford UK: CABI, 179–233.

Karkee, M., Adhikari, B., Amatya, S. and Zhang, Q. (2014). Identification of pruning branches in tall spindle apple trees for automated pruning. *Computers and Electronics in Agriculture* 103, 127–135.

Karpyshev, P., Ilin, V., Kalinov, I., Petrovsky, A. and Tsetserukou, D. (2021). Autonomous mobile robot for apple plant disease detection based on cnn and multi-spectral vision system. In: (vol. 2021). 2021 IEEE/SICE International Symposium on System Integration (SII) (pp. 157–162). IEEE.

Khanal, S. R., Sapkota, R., Ahmed, D., Bhattarai, U. and Karkee, M. (2023). Machine vision system for early-stage apple flowers and flower cluster detection for precision thinning and pollination. IFAC World Congress, Yokohoma, Japan.

Kohlbrecher, S., Meyer, J., Graber, T., Petersen, K., Klingauf, U. and Von Stryk, O. (2014). Hector open source modules for autonomous mapping and navigation with rescue robots. In: *Robot Soccer World Cup*. Springer, Berlin, Heidelberg (pp. 624–631).

Kondo, N., Monta, M. and Fujiura, T. (1996). Fruit harvesting robots in Japan. *Advances in Space Research: The Official Journal of the Committee on Space Research* 18(1–2), 181–184.

Kondo, N., Monta, M. and Noguchi, N. (2011). *Agricultural Robots, Mechanisms and Practice*. Kyoto University Press, Kyoto, ISBN 978-1-920901-83-7.

Landrieu, L. and Simonovsky, M. (2018). Large-scale point cloud semantic segmentation with superpoint graphs. In: Proceedings of the IEEE Conference on Computer Vision and Pattern Recognition (pp. 4558–4567).

Lauri, P.-E. and Corelli Grappadelli, L. (2014). Tree architecture, flowering and fruiting-thoughts on training, pruning and ecophysiology. *Acta Horticulturae* 1058, 291–298.

Lee, C. G. (1982). Robot arm kinematics, dynamics, and control. *Computer* 15(12), 62–80.

Lepej, P. and Rakun, J. (2016). Simultaneous localisation and mapping in a complex field environment. *Biosystems Engineering* 150, 160–169.

Li, G., Suo, R., Zhao, G., Gao, C., Fu, L., Shi, F., Dhupia, J., Li, R. and Cui, Y. (2022). Real-time detection of kiwifruit flower and bud simultaneously in orchard using YOLOv4 for robotic pollination. *Computers and Electronics in Agriculture* 193, 106641.

Li, Z., Luo, Y., Shi, Z., Xie, D. and Li, C. (2020). Design and research of end-effector for naval orange harvesting. *Journal of Mechanical Transmissions* 4, 67–73.

Libby, J. and Kantor, G. (2010). Accurate GPS-free positioning of utility vehicles for specialty agriculture. In: (vol. 2010). Pittsburgh, Pennsylvania, June 20–June 23, 2010 (p. 1). American Society of Agricultural and Biological Engineers.

Liu, S., Wang, X., Li, S., Chen, X. and Zhang, X. (2022). Obstacle avoidance for orchard vehicle trinocular vision system based on coupling of geometric constraint and virtual force field method. *Expert Systems with Applications* 190, 116216.

Liu, W., Anguelov, D., Erhan, D., Szegedy, C., Reed, S., Fu, C. Y. and Berg, A. C. (2016). SSD: single shot multibox detector. In: European Conference on Computer Vision (pp. 21–37). Springer.

Liu, Z., Wu, J., Fu, L., Majeed, Y., Feng, Y., Li, R. and Cui, Y. (2019). Improved kiwifruit detection using pre-trained VGG16 with RGB and NIR information fusion. *IEEE Access* 8, 2327–2336.

Lu, S., Chen, W., Zhang, X. and Karkee, M. (2022). Canopy-attention-YOLOv4-based immature/mature apple fruit detection on dense-foliage tree architectures for early crop load estimation. *Computers and Electronics in Agriculture* 193, 106696.

Lu, X. and Liu, M. (2016). Optimal design and tuning of PID-type interval type-2 fuzzy logic controllers for delta parallel robots. *International Journal of Advanced Robotic Systems* 13(3), 96.

Lyons, D. J., Heinemann, P. H., Schupp, J. R., Baugher, T. A. and Liu, J. (2015). Development of a selective automated blossom thinning system for peaches. *Transactions of the ASABE* 58(6), 1447–1457.

Ma, B., Du, J., Wang, L., Jiang, H. and Zhou, M. (2021). Automatic branch detection of jujube trees based on 3D reconstruction for dormant pruning using the deep learning-based method. *Computers and Electronics in Agriculture* 190, 106484.

Majeed, Y., Zhang, J., Zhang, X., Fu, L., Karkee, M., Zhang, Q. and Whiting, M. D. (2018). Apple tree trunk and branch segmentation for automatic trellis training using convolutional neural network based semantic segmentation. *IFAC-PapersOnLine* 51(17), 75–80.

Majeed, Y., Zhang, J., Zhang, X., Fu, L., Karkee, M., Zhang, Q. and Whiting, M. D. (2020). Deep learning based segmentation for automated training of apple trees on trellis wires. *Computers and Electronics in Agriculture* 170, 105277.

Mu, X., He, L., Heinemann, P., Schupp, J. and Karkee, M. (2023). Mask R-CNN based apple flower detection and king flower identification for precision pollination. *Smart Agricultural Technology* 4, 100151.

Ohi, N., Lassak, K., Watson, R., Strader, J., Du, Y., Yang, C., Hedrick, G., Nguyen, J., Harper, S., Reynolds, D., Kilic, C., Hikes, J., Mills, S., Castle, C., Buzzo, B., Waterland, N., Gross, J., Park, Y., Li, X. and Gu, Y. (2018). Design of an autonomous precision pollination robot. In: (vol. 2018). IEEE/RSJ International Conference on Intelligent Robots and Systems (IROS) (pp. 7711–7718). IEEE.

Redmon, J., Divvala, S., Girshick, R. and Farhadi, A. (2016). You only look once: unified, real-time object detection. In: Proceedings of the IEEE Conference on Computer Vision and Pattern Recognition (pp. 779–788).

Ren, S., He, K., Girshick, R. and Sun, J. (2015). Faster r-cnn: towards real-time object detection with region proposal networks. *Advances in Neural Information Processing Systems* 28.

Robinson, T., Hoying, S., Sazo, M. M., DeMarree, A. and Dominguez, L. (2013). A vision for apple orchard systems of the future. *NY Fruit Q* 21(3), 11–16.

Rovira-Más, F., Zhang, Q. and Hansen, A. C. (2010). *Mechatronics and Intelligent Systems for Off-Road Vehicles*. Springer, London, ISBN: 978-1-84996-468-5, DOI: 10.1007/978-1-84996-468-5.

Sa, I., Ge, Z., Dayoub, F., Upcroft, B., Perez, T. and McCool, C. (2016). Deepfruits: A fruit detection system using deep neural networks. *Sensors* 16(8), 1222.

Sarmento, J., Silva Aguiar, A., Neves dos Santos, F. and Sousa, A. J. (2021). Autonomous robot visual-only guidance in agriculture using vanishing point estimation. In: EPIA Conference on Artificial Intelligence (pp. 3-15). Springer.

Seitz, N., Traynor, K. S., Steinhauer, N., Rennich, K., Wilson, M. E., Ellis, J. D., Rose, R., Tarpy, D. R., Sagili, R. R., Caron, D. M., Delaplane, K. S., Rangel, J., Lee, K., Baylis, K., Wilkes, J. T., Skinner, J. A., Pettis, J. S. and vanEngelsdorp, D. (2015). A national survey of managed honey bee 2014–2015 annual colony losses in the USA. *Journal of Apicultural Research* 54(4), 292–304.

Shi, Y., Zhu, K., Zhai, S., Zhang, D., Liu, L., Zhao, J., Long, Y. and Cui, Y. (2018). Design of an apple-picking end effector. *Strojniski Vestnik/Journal of Mechanical Engineering* 64(4).

Shutske, J. M. (2023). Agricultural automation & autonomy: safety and risk assessment must be at the forefront. *Journal of Agromedicine* 28(1), 5–10.

Shutske, J. M., Sandner, K. J. and Jamieson, Z. (2022). Risk assessment methods for automated agricultural machines: current practice and future needs. In: (vol. 2022). ASABE Annual International Meeting (p. 1). American Society of Agricultural and Biological Engineers.

Silwal, A., Davidson, J. R., Karkee, M., Mo, C., Zhang, Q. and Lewis, K. (2017). Design, integration, and field evaluation of a robotic apple harvester. *Journal of Field Robotics* 34(6), 1140–1159.

Silwal, A., Karkee, M. and Zhang, Q. (2016). A hierarchical approach to apple identification for robotic harvesting. *Transactions of the ASABE* 59(5), 1079–1086.

Silwal, A., Parhar, T., Yandun, F., Baweja, H. and Kantor, G. (2021). A robust illumination-invariant camera system for agricultural applications. In: (vol. 2021). IEEE/RSJ International Conference on Intelligent Robots and Systems (IROS) (pp. 3292–3298). IEEE.

Sivakumar, A. N., Modi, S., Gasparino, M. V., Ellis, C., Velasquez, A. E. B., Chowdhary, G. and Gupta, S. (2021). Learned visual navigation for under-canopy agricultural robots. *arXiv Preprint ArXiv:2107.02792*.

Strader, J., Yang, C., Gu, Y., Nguyen, J., Tatsch, C., Du, Y., Lassak, K., Buzzo, B., Watson, R., Cerbone, H. and Ohi, N. (2019). Flower interaction subsystem for a precision pollination robot. In: (vol. 2019). IEEE/RSJ International Conference on Intelligent Robots and Systems (IROS) (pp. 5534–5541).

Sujiwo, A., Ando, T., Takeuchi, E., Ninomiya, Y. and Edahiro, M. (2016). Monocular vision-based localization using ORB-SLAM with LIDAR-aided mapping in real-world robot challenge. *Journal of Robotics and Mechatronics* 28(4), 479–490.

Tian, Y., Yang, G., Wang, Z., Li, E. and Liang, Z. (2020). Instance segmentation of apple flowers using the improved mask R-CNN model. *Biosystems Engineering* 193, 264–278.

Tian, Y., Yang, G., Wang, Z., Wang, H., Li, E. and Liang, Z. (2019). Apple detection during different growth stages in orchards using the improved YOLO-V3 model. *Computers and Electronics in Agriculture* 157, 417–426.

Underwood, J. P., Jagbrant, G., Nieto, J. and Sukkarieh, S. (2016). Tree centric localisation in almond orchards. *Acta Horticulturae* 1130, 619–624.

Vasconez, J. P., Kantor, G. A. and Cheein, F. A. A. (2019). Human-robot interaction in agriculture: A survey and current challenges. *Biosystems Engineering* 179, 35–48.

Wan, S. and Goudos, S. (2020). Faster R-CNN for multi-class fruit detection using a robotic vision system. *Computer Networks* 168, 107036.

Wan, Z., Sun, Y., Qin, Y., Skorina, E. H., Gasoto, R., Luo, M., Fu, J. and Onal, C. D. (2023). Design, analysis, and real-time simulation of a 3D soft robotic snake. *Soft Robotics* 10(2), 258–268.

Wang, D. and He, D. (2021). Channel pruned YOLO V5s-based deep learning approach for rapid and accurate apple fruitlet detection before fruit thinning. *Biosystems Engineering* 210, 271–281.

Wang, Y. (2010). Research on motion planning and control system of the cucumber picking robot. MS Thesis. Zhejiang University of Technology, Hangzhou, China.

Wang, Z., Underwood, J. and Walsh, K. B. (2018). Machine vision assessment of mango orchard flowering. *Computers and Electronics in Agriculture* 151, 501–511.

Westling, F., Underwood, J. and Bryson, M. (2021). A procedure for automated tree pruning suggestion using LiDAR scans of fruit trees. *Computers and Electronics in Agriculture* 187, 106274.

Whiting, M., Taylor, K. and Das, P. (2022). Supplemental artificial pollination can improve fruit set in tree fruit. *Acta Horticulturae* International Symposium on Integrating Canopy, Rootstock and Environmental Physiology in Orchard Systems XII(1346), 121–128.

Williams, H., Nejati, M., Hussein, S., Penhall, N., Lim, J. Y., Jones, M. H., Bell, J., Ahn, H. S., Bradley, S., Schaare, P., Martinsen, P., Alomar, M., Patel, P., Seabright, M., Duke, M., Scarfe, A. and MacDonald, B. (2020). Autonomous pollination of individual kiwifruit flowers: toward a robotic kiwifruit pollinator. *Journal of Field Robotics* 37(2), 246–262.

Wong, J. Y. (1993). *Theory of Ground Vehicles*. John Wiley & Sons.

Woo, S., Park, J., Lee, J.-Y. and Kweon, I. S. (2018). Cbam: convolutional block attention module. In: Proceedings of the European Conference on Computer Vision (ECCV) (pp. 3–19).

World, B. Agriculture, forestry, and fishing, value added (current US$). Retrieved January 6, 2023 Available at: https://data.worldbank.org/indicator/NV.AGR.TOTL.CD.

World, B. Employment in agriculture (% of total employment) (modeled ILO estimate). Retrieved January 6, 2023 Available at: https://data.worldbank.org/indicator/SL.AGR.EMPL.ZS.

Wu, D., Lv, S., Jiang, M. and Song, H. (2020). Using channel pruning-based YOLO v4 deep learning algorithm for the real-time and accurate detection of apple flowers in natural environments. *Computers and Electronics in Agriculture* 178, 105742.

Wulf, W. A. (2000). Great achievements and grand challenges. *The Bridge* 30(3), 4.

Yang, Z., Ouyang, L., Zhang, Z., Duan, J., Yu, J. and Wang, H. (2022). Visual navigation path extraction of orchard hard pavement based on scanning method and neural network. *Computers and Electronics in Agriculture* 197, 106964.

Ye, Y., Wang, Z., Jones, D., He, L., Taylor, M. E., Hollinger, G. A. and Zhang, Q. (2017). Bin-dog: A robotic platform for bin management in orchards. *Robotics* 6(2), 12.

You, A., Grimm, C. and Davidson, J. R. (2022a). Optical flow-based branch segmentation for complex orchard environments. *2022 IEEE/RSJ International Conference on Intelligent Robots and Systems (IROS)*. (pp. 9180–9186).

You, A., Kolano, H., Parayil, N., Grimm, C. and Davidson, J. R. (2022b). Precision fruit tree pruning using a learned hybrid vision/interaction controller (vol. 2022) International Conference on Robotics and Automation (ICRA) (pp. 2280–2286).

You, A., Parayil, N., Krishna, J. G., Bhattarai, U., Sapkota, R., Ahmed, D., Whiting, M., Karkee, M., Grimm, C. M. and Davidson, J. R. (2022c). An autonomous robot for pruning modern, planar fruit trees. *ArXiv Preprint ArXiv:2206.07201*.

You, A., Sukkar, F., Fitch, R., Karkee, M. and Davidson, J. R. (2020). An efficient planning and control framework for pruning fruit trees. In: (vol. 2020). IEEE Publications IEEE International Conference on Robotics and Automation (ICRA) (pp. 3930–3936).

Yuan, T., Zhang, S., Sheng, X., Wang, D., Gong, Y. and Li, W. (2016). An autonomous pollination robot for hormone treatment of tomato flower in greenhouse. In: (vol. 2016). IEEE Publications 3rd international conference on systems and informatics (ICSAI) (pp. 108–113).

Zahid, A., Mahmud, M. S., He, L., Choi, D., Heinemann, P. and Schupp, J. (2020). Development of an integrated 3R end-effector with a Cartesian manipulator for pruning apple trees. *Computers and Electronics in Agriculture* 179, 105837.

Zahid, A., Mahmud, M. S., He, L., Heinemann, P., Choi, D. and Schupp, J. (2021). Technological advancements towards developing a robotic pruner for apple trees: a review. *Computers and Electronics in Agriculture* 189, 106383.

Zhang, J., Karkee, M., Zhang, Q., Zhang, X., Yaqoob, M., Fu, L. and Wang, S. (2020a). Multi-class object detection using faster R-CNN and estimation of shaking locations for automated shake-and-catch apple harvesting. *Computers and Electronics in Agriculture* 173, 105384.

Zhang, K., Lammers, K., Chu, P., Li, Z. and Lu, R. (2021). System design and control of an apple harvesting robot. *Mechatronics* 79, 102644.

Zhang, X. (2020). Study of canopy-machine interaction in mass mechanical harvest of fresh market apples. PhD Dissertation. Washington State University Press, Pullman, WA.

Zhang, X., He, L., Majeed, Y., Whiting, M. D., Karkee, M. and Zhang, Q. (2018). A precision pruning strategy for improving efficiency of vibratory mechanical harvesting of apples. *Transactions of the ASABE* 61(5), 1565–1576.

Zhang, X., He, L., Zhang, J., Whiting, M. D., Karkee, M. and Zhang, Q. (2020b). Determination of key canopy parameters for mass mechanical apple harvesting using supervised machine learning and principal component analysis (PCA). *Biosystems Engineering* 193, 247–263.

Zhang, X., Karkee, M., Zhang, Q. and Whiting, M. D. (2021). Computer vision-based tree trunk and branch identification and shaking points detection in Dense-Foliage canopy for automated harvesting of apples. *Journal of Field Robotics* 38(3), 476–493.

Zhang, X., Zhang, Q., Karkee, M. and Whiting, M. D. (2021b). Machinery-canopy interactions in tree fruit crops. In: *Fundamentals of Agricultural and Field Robotics*. Springer (pp. 415–442).

Zhao, Y., Gong, L., Liu, C. and Huang, Y. (2016). Dual-arm robot design and testing for harvesting tomato in greenhouse. *IFAC-PapersOnLine* 49(16), 161–165.

Zsombor-Murray, P. J. (2004). *Descriptive Geometric Kinematic Analysis of Clavel's "Delta" Robot*. Centre of Intelligent Machines, McGill University (p. 1).

Chapter 20

Advances in the use of robotics in greenhouse cultivation

Jochen Hemming and Jos Balendonck, Wageningen University & Research, The Netherlands

1 Introduction
2 Key challenges in using robotics in protected cultivation
3 State of the art and challenges: selective harvesting
4 State of the art and challenges: crop management
5 State of the art and challenges: plant propagation, grafting and planting
6 State of the art and challenges: crop scouting/disease and insect detection and control
7 State of the art and challenges: transportation and logistics
8 Conclusion and future research trends
9 Where to look for further information
10 References

1 Introduction

The production of crops and ornamental plants in greenhouses differs from open field production due to its intensive use of resources including space, water, nutrients and energy. Human labor plays a key role in most activities. The layout and structure of greenhouses offer a more defined infrastructure than open field production. Depending on the level of technology, various environmental parameters can be regulated and controlled, ranging from low-tech greenhouses with natural ventilation and simple irrigation systems to high-tech greenhouses equipped with sophisticated heating and cooling installations, supplemental lighting, comprehensive fertigation and precise control over relative humidity and CO_2 levels. Greenhouse cultivation can extend the season in which crops can be grown when compared to open field production. Many crops can be grown and harvested throughout the year. In contrast to most arable crops, greenhouse crops often require selective harvesting over an extended period (e.g. as differing fruits ripen) as well as specific manual tasks such as pruning. The use of automation and robotics in greenhouses is therefore of great interest.

http://dx.doi.org/10.19103/AS.2023.0124.24

Many highly automated systems have already been applied in greenhouse horticulture, starting with more industrial greenhouse production processes. This includes logistics and autonomous transport of plants and harvested products from the greenhouse to sorting and packing facilities, as well as machine vision-based sorting systems for pot-plants and cut flowers (Hemming, 2020). However, most of these systems do not fall under the category of 'robots' and are therefore not further addressed here. For the purposes of this chapter, a robot is defined as an autonomous or semi-autonomous machine capable of sensing its environment, carrying out computations to make decisions and performing actions in the real world. This definition complies with the commonly used sense–plan–act paradigm (Bartneck et al., 2021). In contrast to a machine, a robot exhibits a degree of intelligence, including the ability to solve complex problems, engage in interactions and the ability to learn, all of which it accomplishes through artificial intelligence (AI).

Robots can be used to perform a variety of tasks in greenhouses, including harvesting, plant manipulation, propagation, transplanting, crop scouting, spraying, transport and packaging. Robots have the potential to lower production costs and reduce the need for manual labor. Robots can be constructed in such a way to perform better than humans, with fewer risks in high temperature and high humidity environments. Robots follow precision horticulture techniques in which plants are treated on an individual basis, thereby improving quantity and quality of crop production whilst using resources as efficiently as possible (Henten et al., 2013).

This chapter discusses the role of robotics in greenhouse cultivation. First, the key challenges are introduced. Current technology is then presented, grouped by the main tasks executed by the robots. The selective harvesting of major greenhouse crops (tomato, cucumber, sweet pepper, strawberries and flowers) is explored, followed by a description of crop maintenance operations, including automatic leaf removal. Plant propagation operations, including grafting and autonomous planting of cuttings, are then analyzed. Attention is then given to crop scouting and the detection of disease and insects and control tasks. A section on autonomous transport and logistics in the greenhouse concludes this section, before future trends in greenhouse robotics research are discussed.

2 Key challenges in using robotics in protected cultivation

The use of robotics in protected cultivation is characterized by a number of challenges distinct from robotic applications in other domains. In greenhouses, robotic systems need to be able to cope with extremely high natural variables in both products (size, shape and color) and the environment (location, obstacles, light and temperature). There is, in addition, a large diversity of crops, different varieties within each crop and different cultivation systems. Products are typically delicate and need to be handled and stored with great care to prevent damage and irreversible quality loss. Additionally, plants should remain undamaged to ensure continuous growth and production. As greenhouses are climate controlled and have a complex internal infrastructure, space inside is precious and a high occupancy rate is maintained. The maximum practical number of plants is planted to achieve optimum yield per square meter of greenhouse area. This often results in a very dense plant canopy and thus limited space for machinery and robots to move and operate.

As an example, Fig. 1 illustrates a full-grown crop of sweet yellow pepper. To automate the harvesting of this crop, robots are required to move between the double plant rows, in which ripe fruit appear on the upper half of the plants, more than 2 m above the floor, often (partially) occluded by leaves, stems or other neighboring fruit. Unripe and ripe fruits hang next to each other, as seen in a similar situation with strawberries in Fig. 2. Moreover, the stems of the plants are not stiff but flexible and will easily move when touched. In contrast to many industrial robotic applications in which temperature and humidity levels can be controlled, environmental conditions in the greenhouse can be hostile for mechatronic systems. Depending on the crop, air temperatures can range between +10°C and +40°C and relative humidity levels are often >80%, reaching 100% on a regular basis. Moreover, plants are regularly irrigated and plant protection chemicals are sprayed. Light levels vary depending on outside weather conditions and can transition from direct sunlight to diffuse light within seconds during a partially cloudy day. These variations can be influenced by shading screens and/or artificial illumination.

Figure 1 Dense vegetation of a full-grown greenhouse crop of sweet yellow pepper.

Figure 2 Organifarms' strawberry-harvesting robot, BERRY. Source: Adapted from: Organifarms.

These conditions are very challenging when it comes to imaging sensors. For many crops, ripening is indicated by a change in color, although measuring color in changing light conditions is demanding. In addition to color, 3D data is needed if a robotic arm or gripper is required to interact with an object. Solutions that provide both combined color and high-quality 3D data from a short distance (typically in the range of 0.01–1.00 m) are limited. When interacting with plants, some fruits and leaves can become rotten if they have been affected by pests and diseases or other prior mechanical damage, which can potentially cause contamination to the robot. Some plants, including tomatoes, emit sticky sap that is difficult to clean.

3 State of the art and challenges: selective harvesting

The current use of autonomous robots in commercial greenhouse horticulture remains limited, although research into robotics for crop production in greenhouses is not new. Studies and applications on robotic systems in greenhouse crops started in the 1980s, with pioneering research carried out in Japan, the Netherlands and the US. Tomatoes, cucumbers and strawberries were the first greenhouse crops on which the research focused, with selective harvesting tasks the primary area of investigation. Several existing review articles provide a good overview (Bac et al., 2014; Kootstra et al., 2021; Wang et al., 2022) and the field is rapidly evolving due to better and more affordable hardware and software, as well as the increased need to automate such tasks. This chapter primarily focuses on developments in greenhouse automation and robotics over the past 15 years, from 2008 to 2023. It starts by looking at selective harvesting of key greenhouse crops.

3.1 Tomato

Tomatoes are the most significant crop produced in greenhouses globally, making them an obvious candidate for automation. Tomatoes are cultivated and harvested in many different ways. Traditional bush-type cropping systems are common, as well as the so-called high-wire cropping system in which the main plant vines are guided along a vertical plastic wire toward a horizontal trellis wire in the upper part of the greenhouse. In this system, plants are lowered at regular intervals so that ripe fruit consistently appears in the same horizontal region. Tomatoes can be harvested as single full-size fruits or as single snack or cherry tomatoes (with or without the green crown attached). Increasingly popular with customers are tomatoes on the vine (also called truss tomatoes) in which a cluster of fruits remain attached to the vine after harvesting. The various cropping systems and fruit sizes necessitate the development of different sensing and harvesting mechanisms.

Research on robotic harvesting of tomatoes was first reported in 1984 in Japan (Kawamura et al., 1984) where there was a highly active research and development community. Kondo et al., for example, developed end-effectors for harvesting single tomatoes as well as fruits on the vine (Kondo et al., 2010). Since 2014, there has been an annual national tomato harvesting robot competition organized in Japan to promote agricultural robots and to provide a test environment for tomato harvesting. Performance is improving gradually. In 2020, the winning robot could harvest a tomato within 30 s in full sunlight (Ishii et al., 2021).

A Chinese cherry tomato harvester has been reported with a reported harvest rate of 83% (Qingchun et al., 2018). Each successful harvest required 1.4 attempts on average, and a single successful harvesting cycle took 8 s, excluding the time needed to move the platform. Ling et al. (2019) developed a dual-arm robot using a binocular vision sensor for fruit detection and localization, reporting a fruit picking success rate of 87.5%. However, to date, most developed systems have only been tested under greatly simplified conditions. Examples of such conditions include

- double plant rows in which the need for the robot to grasp the fruit at the back of the plant is avoided;
- leaves that are occluding fruit being removed first;
- manual isolation of ripe fruit touching each other;
- stems of neighboring plants spaced at a greater distance; and

- experiments carried out in controlled light conditions (e.g. at night with artificial illumination or even indoors in a lab environment).

As such, it can be expected that performance will reduce when deploying these systems in a commercial production environment. End-effectors used to detach single tomatoes from the plant have been designed based on different principles, including fingers, rotary claw-types or with a snare or lasso. More recently, research has focused on force-feedback soft grippers in which the fingers are actuated pneumatically and the force applied is controlled by a micro controller (Kultongkham et al., 2021).

The development of selective tomato harvesting robots has shifted from academic research into start-up companies and bigger industrial enterprises. The US-based company AppHarvest (https://www.appharvest.com/agtech/) is working on what it calls a universal harvester robot that in the first instance will be deployed to harvest single cherry tomatoes using a rotary claw-type end-effector. The Israel-based MetoMotion (https://www.metomotion.com) has also joined forces with Dutch horticulture technology supplier Ridder (https://info.ridder.com/robotics). The two companies announced that they have passed the concept and validation phase of their dual robotic arm harvester for tomatoes on the vine, in which the robot is able to harvest on both sides of the greenhouse row. Photos of the robot are shown in Fig. 3. Ripe fruit detection is based on color imaging and deep-learning algorithms. The elongated end-effector cuts and catches the vine in a single operation. The detection system of the robot is self-learning and accumulates new data during operation to further improve its performance.

In Japan, the Panasonic Corporation has introduced a prototype of a harvesting robot for single tomatoes detached from the plant by a lasso principle. It claims it can match the speed of a human worker by harvesting ten tomatoes every minute (Humphries, 2017). In 2020, the global robotic and automation company Denso (Japan) presented a cherry tomato harvesting

Figure 3 Tomato harvesting robot GroW, from Ridder/Metomotion. Source: Adapted from: Ridder-MetoMotion.

Figure 4 Certhon harvest robot for harvesting tomatoes. Source: Adapted from: Certhon/DENSO.

robot called FARO (Fig. 4) for harvesting tomatoes on the vine. Denso and the Dutch company Certhon (https://certhon.com/) are now collaborating to commercialize this robot under the name Certhon harvest robot.

The Japanese start-up Inaho (https://en.inaho.co/ and https://www.youtube.com/watch?v=tcfEtP130fo) is working on a fully automated harvesting robot for single cherry tomatoes using a two-finger gripper. The Dutch start-up SAIA (https://www.saia-agrobotics.com/) is also working on an automatic harvesting solution for tomatoes. Their approach includes a major change in the cropping system as they aim to make the plants mobile and move them to the robot(s) instead of moving the robot to the plants.

3.2 Strawberry

Despite the fact that strawberries are a very delicate and easily damaged fruit, much research to automate its harvesting has been carried out (Arima et al., 2001; Hayashi et al., 2010; Kondo et al., 1998). The reason behind this interest might be that the strawberry plant is small and a single fruit is lightweight, which allows for smaller and therefore cheaper robots (Defterli et al., 2016). As with tomatoes, strawberries turn from green to red when they ripen and this can be detected in a green canopy relatively easily using color cameras. It currently requires a lot of manual work to harvest strawberries. Most robots grip the strawberry by the peduncle (fruit stem), but there are also examples where the fruit is grasped with a suction cup. More recent research is working on sensing principles that are less influenced by light variations and grippers that are able to push aside the surrounding obstacles to pick strawberries located in clusters (Xiong et al., 2020).

Although there is not yet a commercially successful robot for harvesting strawberries, several mature initiatives exist, such as Agrobot in Spain (http://www.agrobot.com), DogTooth (https://dogtooth.tech) in the UK, Traptic in the US (Heater, 2022) and Octinion in Belgium (http://octinion.com). Octinion merged its horticultural robotic activities with Priva under the brand name Octiva in 2022. Octiva is currently selling its strawberry-harvesting robot Rubion to pilot customers. The robot uses a soft touch gripper to grip the fruit body and is able to pick one strawberry every 3 s. The Norwegian/UK company SAGA Robotics is investigating strawberry-picking with its robot platform Thorvald (Saga Robotics, 2022). The German start-up Organifarms (https://www.organifarms.de) launched a harvesting robot for strawberries in 2022 that also handles quality control and packing. The robot is equipped with two RGB-D cameras on the end-effector and two cameras on the robot platform to detect the fruits and their peduncles. It uses deep-learning image analysis for detection and evaluation of ripeness (Fig. 2).

3.3 Cucumber

The concept of a modular autonomous robot for selective harvesting of cucumbers in greenhouses was introduced by researchers at Wageningen University & Research in the Netherlands in 2002 (Henten et al., 2002). Results from field tests involving a prototype were published a year later (Henten et al., 2003). The robot consisted of an autonomous vehicle that could utilize the greenhouse's heating pipes as rails to drive along, a seven degree-of-freedom (DOF) industrial manipulator, an end-effector, two camera vision systems and miscellaneous electronic and pneumatic hardware (Fig. 5). The end-effector gripped the peduncle of the fruit and fixed the fruit body in place by use of a suction cup. A thermal cutting device, based on the principle of an electric surgical knife, was incorporated into the end-effector to detach the fruit from the plant. This method of cutting significantly reduced the risk of disease transmission from one plant to another. To facilitate the detection of green fruit within a green canopy, a custom-made dual-spectral vision system was created that allowed for discrimination between fruit and leaves, making use of the water absorption bands in the visible and near-infrared range. The robot achieved an average harvest success rate of 74% and each successful harvest cycle took 65 s per cucumber (Henten et al., 2003).

The robotic harvesting of cucumbers in the greenhouse remains a dynamic field of interest, on both a scientific and a more commercial level. Scientific articles have been published on the development of cucumber-harvesting robots in China, including crawler-type platforms and robot arms

Figure 5 A robot harvesting cucumbers at Wageningen University & Research. Source: Adapted from: Henten et al. (2003).

with different degrees of freedom (Wang et al., 2022). Researchers at Vineland Research and Innovation Centre in Canada are developing a system that uses a lightweight cobot manipulator (Kinova) on an autonomous navigating platform (Clearpath Husky) with multiple RGB-D cameras (Kuak, 2021). A deep-learning vision system detects fruit, determines its ripeness, and locates it within the robot's workspace. The robot is able to harvest without pausing in its forward movement. Its developers state that the robot is able to pick one cucumber every 15 s with a 90% success rate (Vineland Research & Innovation Centre, 2021). While numerous initiatives are dedicated to the development of cucumber-harvesting robots, many start-ups and commercial companies are understandably reluctant to disclose details while development remains in progress.

3.4 Sweet pepper

In the commonly used high-wire cropping system for tomatoes and cucumbers in greenhouses, the ripe fruit is generally present in a defined horizontal region above the ground due to frequent manual lowering of the plants. The situation with sweet pepper is different as plants continue to grow upwards throughout the growing season and ripe fruit thus need to be harvested higher above the ground later in the season. This must be taken into consideration when developing harvesting robots for this crop. The platform or robotic arm must be able to be positioned over a wide vertical range. The European research projects Clever Robots for Crops (CROPs) (Bac et al., 2017) and SWEEPER, an acronym for SWEEt PEpper Robot (Wageningen University & Research, 2018), worked on the application of sweet pepper harvesting. SWEEPER resulted in the robot prototype shown in Fig. 6. It comprises an end-effector to cut and grasp the pepper; an RGB-D camera to detect and locate the fruit and to estimate

Figure 6 Prototype of the sweet pepper-harvesting robot SWEEPER (left) (photo: Boaz Arad); a sample color image acquired with the camera in the end-effector (right top); and deep-learning -based classification result (right bottom) (Barth et al., 2019).

fruit ripeness; active LED flashlights on the gripper to better detect fruit and obstacles under alternating environmental conditions; an industrial six DOF robot arm to search for, move, and convey fruit; a mobile platform including a scissor lift to position the robot within a large vertical range and to move the robot through the greenhouse; and logistics to convey and store-picked fruits. The system is one of the first to apply deep-learning models to detect fruits as well as obstacles. The average cycle time to harvest a fruit is reported to be 24 s. The harvest success rate was 61% under optimal crop conditions (Arad et al., 2020).

In another project, a greenhouse pepper-harvesting robot called Harvey was developed and evaluated in real-world conditions (Lehnert et al., 2020). This robot achieved a harvesting success rate of 76.5% in a modified crop and 47% in an unmodified crop with a cycle time of 36.9 s, excluding platform navigation. It should be noted that the performances of both SWEEPER and Harvey cannot be compared directly with each other as the robots were tested on different cropping systems. The relatively poor performance of both robots was caused by clutter and occlusion of the target fruits and the dense and complex crop structure. Navigating a robotic arm in that environment with the rather large end-effectors used resulted occasionally in fruit not being reachable as well as limited fruit and plant damage. AGRIST, an agricultural robotics company in Japan, is working on a sweet pepper-harvesting robot with the name 'L'. The robot also uses RGB-D cameras and AI-based image analysis to localize the fruits. A distinguishing feature is that it does not drive on the ground but moves on horizontal wires installed in the greenhouse. Details on the performance of this robot are not yet available (Agrist, 2022).

3.5 Flowers

The ability to selectively harvest cut flowers autonomously has been attempted by researchers for many years. For more than 15 years, various computer vision techniques such as stereo vision, laser triangulation, X-ray imaging and volumetric intersection have been used for robotic harvesting of roses in greenhouses. These sensing systems are designed to locate the stem and trace it down to the optimal cutting position (Noordam et al., 2005). A prototype system for selective harvesting of roses on moveable benches has been developed. Unlike a conventional set-up, in this greenhouse, the plants were moved to the robot instead of the robot moving to the plants (Bergerman et al., 2016). However, this concept never became commercially viable due to its overall technical complexity and high investment costs. More recently, a Dutch consortium of technology providers and growers deployed a mobile robot that assists manual harvesters to pick only ripe roses by highlighting the flower to be picked with a laser (RoboCrops, 2021). The same system is also under development for gerbera flowers. As virtual information is added to the existing real-world environment, this can be considered one of the first examples of applying augmented reality in greenhouse horticulture.

Selective harvesting of gerbera flowers is another highly labor-intensive task that requires specific knowledge in order to assess the maturity of each individual flower. Whilst the first studies on gerbera harvesting utilized classical image analysis approaches to identify and 3D-localize the flowers and pedicels (Rath and Kawollek, 2009), more recent work is applying AI and deep-learning-based methods. One example is the robot under development at Wageningen University & Research shown in Fig. 7. When a stem is detected by the camera system, the arm with the end-effector moves toward the stem and then a part of the end-effector moves downwards. The end-effector stops descending when it meets resistance, a sign that the base of the plant has been reached. At that point, the end-effector cuts the stem (Ruizendaal, 2022).

3.6 Other crops

Research into robotic harvesting has been conducted on several other greenhouse crops, such as eggplants (Hayashi et al., 2002). However, most of these activities have resulted in more fundamental studies and have not yet been followed up by commercial initiatives.

Mushrooms do not need daylight to grow and are not considered a greenhouse crop. However, there are a number of similarities between the cultivation of mushrooms and greenhouse crops. Mushrooms are typically grown in controlled climate conditions. Cultivation is labor-intensive and in most cases a manual selective harvest takes place. Moreover, mushrooms are

Figure 7 Gerbera harvesting robot, full system overview (left) and detail of end-effector (right).

very delicate and easily damaged, hence their inclusion in this chapter. Research into robotic harvesting of mushrooms has been carried out globally. Sensing and gripper developments have been carried out on the popular champignon or button mushroom varieties and as far back as 2001 it was reported that robotic harvesting of high-quality mushrooms for the fresh market was entirely feasible (Reed et al., 2001). A robotic mushroom-picking mechanism was designed by Huang et al., including a picking end-effector based on a bending motion, a four DOF positioning end-effector for moving the picking end-effector, a mushroom stipe trimming end-effector and an electro-pneumatic control system. Laboratory test results indicated that the picking mechanism had the potential to be implemented in automatic mushroom harvesting (Huang et al., 2021). Another example is the initiative by Vineland Research & Innovation Centre in Canada, in which a smart vision system identified mushrooms ready to be harvested. This information was used in a decision support system (Kashkoush and Avigad, 2019; Vineland Research & Innovation Centre, 2020). The EU-funded project SoftGrip (https://www.softgrip-project.eu) works on self-actuating soft robotic grippers for harvesting mushrooms. One of the objectives of this project is to develop grippers with built-in sensing and embodied intelligence and adaptability to object shape for reliable and efficient picking of mushrooms.

4 State of the art and challenges: crop management

Researchers and developers have extended their focus beyond selective harvesting to tackle other labor-intensive crop maintenance operations in greenhouse settings. A very time-consuming task when growing tomato and cucumber crops (the two most important greenhouse vegetable crops) in a high-wire cultivation system is to remove older leaves from the main stem of the plant. Leaves at the lower end of the plants are removed due to their reduced vitality, their negligible contribution to canopy photosynthesis and their heightened susceptibility to fungal diseases. Additionally, leaf removal results in a more open structure of the canopy in which the fruits are clearly visible and accessible, a feature beneficial for automatic harvesting (Arad et al., 2020; van Henten et al., 2005). A functional model of an autonomous robot to remove the leaves from cucumber plants was presented by Henten et al. (2005). Essentially, the end-effector mounted on the robotic arm was positioned at the stem at the lower end of the plants from which leaves had earlier been removed. This part of the stem was easy to detect and to approach. The end-effector then moved upwards along a predefined stem segment while removing the leaves. Leaf petioles were caught by teeth and the presence of leaves was detected by electrodes and a thermal knife was used to cut the leaves off the stem.

In 2019, the start-up SAIA presented the concept of a leaf removal robot for cucumber. This system used a high-flow vacuum tube to suck-in leaves detected by a deep-learning based 3D vision system and a vibrating knife to cut the leaves off the main stem (SAIA, 2019). The Dutch company VDL announced in 2022 the commercial introduction of a modular robot platform named CropTeq for removing leaves in high-wire grown cucumber. Their vision system is also deep-learning based and with their gripper/cutter end-effector they aim for more than 1000 cuts per hour (VDL, 2023). The development of a de-leafing robot for tomato plants has been in progress for many years by the company Priva in partnership with Dutch growers, technology partners and other experts, finally resulting in the robot Kompano (Priva, 2022; Sparks, 2016).The Kompano (Fig. 8) is an autonomous vehicle that utilizes the greenhouse's heating pipes

Figure 8 Tomato de-leafing robot Kompano. Source: Adapted from: Octiva.

to move between the crop rows. It detects the cutting points of the leaves to be removed using a flashlight-supported 3D camera system. A telescopic arm with an end-effector consisting of a gripper and knife is extended from the base of the robot platform to grip and cut each target leaf. The Kompano is currently being extensively piloted by Dutch growers. The Dutch research program FlexCRAFT researches active perception, grippers, world modeling and planning and control. In this research program, the removal of leaves from tomato plants is also being investigated (FlexCRAFT, 2022). The spin-off company Floating Robotics from ETH Zürich in Switzerland is developing multiple robotic systems for greenhouse applications (Floating Robotics, 2023), one of which is robotic de-leafing for tomato. A deep-learning network is used for image analysis. In addition to the network's ability to distinguish objects, the software estimates stem thickness to ensure the main tomato trunk is not unintentionally cut.

For the three major high-wire vegetable crops - tomatoes, cucumber and pepper - a number of crop maintenance and crop manipulation activities require a great deal of manual labor. In addition to leaf removal, these include tying up the plant and pruning. When at the beginning of the growing season, the young plant is put on the substrate slab, the stem(s) of the plant must be connected to the vertical crop wire(s). In practice, a woven plastic wire is used. This wire is knotted to the horizontal trellis wire (often a metal wire) at gutter level at regular intervals. The stem is then twisted around the vertical wire so that the wire spirals around the stem. As the plant grows, this task of twisting the stem around the vertical wire will repeat itself during the growing season. A research paper by Long et al. describes a rather complex but successful device to knot the plastic crop wire to the horizontal trellis wire in the greenhouse (He et al., 2013). A number of technologies have been described in research studies and have been put into effect on the market in the hope that they may partially or fully automate this task. An example is the Ringmaster (Priva, the Netherlands), a gun-like device that can place small copper rings around the stem and the wire to attach the plant to the wire.

For high-wire grown tomatoes and cucumbers, regular lowering of the plant is a labor-intensive operation. The Plant Elevation System (Atgrow, 2022) uses plastic pulleys holding the vertical plastic wires. The pulleys are hung on the horizontal trellis wire in the greenhouse and are connected to each other by a thin air tube. Computer-controlled air pulses are sent through the tubes that activate a gear mechanism. This mechanism ensures that the plant lowers about half a centimeter per pulse. The Israel-based company Arugga AI farming (Arruga AI Farming, 2022b) is working on a greenhouse robot for tomato plant lowering. The system mimics the manual operation of unwinding the vertical crop wire from the plant hook (Arruga AI Farming, 2022a). A comparable development for lowering plants is ongoing at the abovementioned company

Floating Robotics. A custom-made hook for the crop wire allows this robot to lower the plants and move them forward with the same tool used for harvesting and de-leafing (Floating Robotics, 2023). A combination of an automatic knotting technique, a vision system to find plant stems, a device to mechanically tie the plant stem to the crop wire and a guidance system for the top of the plant looks technically feasible. However, to date, no automatic solution has been demonstrated.

Pruning determines plant growth and future development. One action of pruning is to remove excess young fruits or flowers. The delicacy of the job, the speed at which it carried out, the amount of detailed information needed to do it correctly and the small and likely gentle mechanics required to selectively remove flowers and/or fruits make this process very difficult to automate. The authors are not aware of any technology to automate this task.

Planting of young plants in the greenhouse is done once or twice a year over a short period. There are devices nowadays to plant some cut flowers semi-automatically in greenhouses, but due to the short timeframe and the large machinery involved, it is most likely not economically feasible to automate this process in the near future.

5 State of the art and challenges: plant propagation, grafting and planting

Robotic cutting, (trans)planting and grafting machines have also been introduced to horticultural practice. These machines are not generally installed inside the greenhouse production compartments but in the processing or sorting area in an attached building. They often have a fairly industrial layout (e.g. a safe robotic cell inside enclosed cabinets with defined environmental conditions and connections to power and pressured air) but must still be able to address the natural variability and non-uniform characteristics of the products they deal with (Hemming, 2020). A grafting robot based on the cleft grafting method has, for example, been developed to improve the efficiency of grafting Solanaceae vegetables. This robot consists of four parts, including a clamping manipulator and carrying, cutting and feeding mechanisms. These not only enable the robot to conveniently perform clamping, carrying, positioning, cutting, joining and binding rootstock and scion but also improve grafting efficiency. The average success rates for cutting rootstock, cutting scion and conjugation are 98.7%, 99% and 59.6%, respectively. The developed robot was able to perform consistent operations, suggesting potential value for practical applications (Sun et al., 2016).

Grafting robots offered by the company ISOGroep (ISO-Groep, 2022a) have a capacity of up to 1000 grafts per hour. They cut each one at the same angle such that the scion and rootstock fit nicely, which stimulates union. The ability to disinfect with steam reduces the risk of bacterial propagation. The ISO Cutting Planter machine (ISO-Groep, 2022a) automates the manual work of planting cuttings in trays. Self-learning 3D vision software is used, and therefore cutting types of many species of plants can be planted, such as chrysanthemums and lisianthus. The operator places the cuttings on the conveyor in no specific order. The conveyor shakes the cuttings to enable the vision system to spot individual sprigs. Then a robot arm takes one cutting and plants it immediately. The robotic gripper is equipped with a pushing device to make sure the cutting is planted firmly in the root medium. Using this process, every cutting is pushed down to the same depth, giving high planting uniformity as a result. One person can operate up to seven machines.

In terms of planting in the greenhouse, one commercially available solution is the RoBoPlant (ISO-Groep, 2022b). This machine plants fully automatic peat blocks with young plants in chrysanthemum flower production. It automatically takes peat blocks with young plants out of the boxes, separates them and plants them in the chosen pattern into the holes of the hoist netting. In addition to reducing the workload, the fully automatic planting machine delivers a highly uniform plant quality.

6 State of the art and challenges: crop scouting/ disease and insect detection and control

To maintain crop health and productivity, a large amount of manual plant monitoring takes place in the greenhouse. This ranges from crop assessment, monitoring plant development and growth to climate and CO_2 level monitoring, the detection of pests and diseases and monitoring fruit ripeness and quality. Workforce availability and cost issues mean that sampling resolution and rate are often low. Production forecasting is becoming increasingly important in enabling the delivery of the right quantity and quality of product at the right moment.

An intermediate step before enabling full hands-free production in greenhouses with robotic plant manipulation and selective harvesting is to first automate scouting and monitoring activities. Whilst the overall environmental conditions in a modern greenhouse are very well monitored and controlled by climate-control computers that retrieve data from fixed installed sensors in the greenhouse, additional information on local and microclimate can be collected by autonomous roaming vehicles. A Smart Greenhouse Horticulture research project developed a Gerberascout platform developed. The goal of the device, which drives autonomously on the heating pipe rail system in the greenhouse, was to assess the plant load of gerbera plants (number of flowers and flower buds) for yield prediction. Initial results showed that plant load can be predicted with an accuracy of more than 93% using a Faster Region Based

Convolutional Neural Network (F-RCNN) for image analysis (Ruizendaal et al., 2018).

The company Hortikey (Hortikey, 2022) offers a combination of autonomous moving vehicle with multiple layers of stereovision cameras that can automatically count and classify tomatoes on the plant. Together with forecasting software, this results in detailed information on future crop development and harvest. With this information, the grower can optimize the process between supply and demand, and thereby the production chain and customer satisfaction. A photo of this robot, the Plantalyzer, is shown in Fig. 9. Octiva is developing a scouting implement called Curion to monitor the state of strawberry crops and data on ripeness categories (Octiva, 2022). Octiva is also producing the autonomous navigating lighting platform Lumion, which can treat powdery mildew infections on strawberries and other crops with UV-C light.

As indicated earlier, automatic stress and disease detection (e.g powdery mildew) are under development, using imaging sensors (color cameras and/ or multi- or hyperspectral cameras). A number of commercial devices are in an advanced state of development. The company Ecoation (Ecoation, 2022), for example, is already deploying a pest management and yield forecasting platform for pepper and tomato crops in greenhouses. Data is collected by both human observation and sensors such as cameras and microclimate sensors on a mobile platform. AI algorithms are used to forecast pest pressure to assist growers in decision-making.

Figure 9 The Hortikey Plantalyzer with three rows of cameras to count tomatoes and assess their ripeness while still on the plant. Source: Adapted from: HortiKey.

The research community is currently working on hardware and software for the detection of insects and the detection and mapping of insect-pest damage. A robotic detection system has been developed to detect two major threats to bell pepper plants - powdery mildew and tomato-spotted wilt virus. The robot autonomously scans the plant and analyses leaves using RGB and multispectral cameras (Schor et al., 2017). Methods for automatic counting of whiteflies and beneficial insects trapped on yellow sticky traps using deep-learning image analysis methods have been developed (Nieuwenhuizen et al., 2018). For more background on pest and disease detection, see for example the articles by Liu and Wang (2021) and Che'Ya et al. (2022).

Research and development on the use of autonomous drones in greenhouses for the purpose of data collection on the status of the crop or the environmental conditions is also being carried out. However, flying drones in the greenhouse environment can be challenging due to the many obstacles, including infrastructure and the plants themselves. The start-up company PATS drones (https://pats-drones.com/) uses a combination of a stationary camera and computer vision system to detect and track harmful insects (such as specific moths) and a small-sized drone that will autonomously navigate in real-time to the localized insect with the aim of destroying it with its rotor blades. This futuristic application can only be used to control flying insects of a considerable size, but most pests are very small and also tend to stay on the plant.

Trolleys that drive along the heating pipe rail system among the plant rows to spray fungicides and pesticides have been commercialized for many years. To reduce the use of plant protection chemicals and to comply with environmental legislation, there is a need to make these sprayers more intelligent by performing precision spraying of infected plant parts only. In spite of ongoing efforts to achieve this, the sensing capabilities and intelligence of these machines remain limited in most cases and as such they do not fit the definition of robots given at the beginning of this chapter.

7 State of the art and challenges: transportation and logistics

Internal transport within a commercial operation is an important issue for horticultural companies. The faster the product goes from the greenhouse to the sorting and packing facilities, the better for the quality of the product. As well as manually driven trolleys, automatic chain track systems are also used. A forward-moving metallic chain is embedded in the concrete floor of the main corridor of the greenhouse that can pull both full and empty harvesting trolleys to the necessary locations. These systems are primarily mechanical systems that are simple to operate and require little maintenance.

Automated guided vehicles (AGVs) are also increasingly used for internal transportation. The AGV typically uses induction wires placed along the concrete path as the route to follow. More advanced systems include full course and traffic rules management and advanced safety sensors (ultrasonic, lidar, cameras) and can automatically open and close gates. As more automated systems have to function autonomously or work together, the ability to precisely locate a position in the greenhouse is extremely important. Real-time position measurements using Global Navigation Satellite Systems (GNSS) are not always reliable and other positioning methods currently in use give varying results. There are efforts to establish a global standard based on a local, three-dimensional coordinate system for the greenhouse (AVAG, 2022). Some systems use QR codes mounted on greenhouse construction elements for localization, and the use of simultaneous localization and mapping (SLAM)-based techniques have been investigated in recent years (Jiang et al., 2022).

8 Conclusion and future research trends

A number of highly automated operations exist in the horticultural sector. This applies especially to modern types of greenhouses found in Western Europe, North America and some Asian countries. However, this is not yet the case for robots, at least according to the definition given in the introduction to this chapter. The few commercial greenhouse robots available include strawberry-harvesting robots, a leaf-picking robot for tomato (both Octiva) and robots for grafting and automatic planting of cuttings (ISO Group). Despite the great technical progress that has been made over the past few decades, many systems still lack practical robustness, failure control and the ability to fully integrate robots into overall company logistics. The success rate, accuracy and speed of most developed systems are still too low to allow economically viable implementation. However, flawless working robots and high speed are not necessarily prerequisites for economic viability, and robots with lower performance can be of value. This requires a different way of thinking, in terms of partial solutions. Even if the robot does not solve the entire problem, for example a robotic system that is only able to perform pre-harvest tasks, it can still add sufficient value to be profitable. Such achievements could be the starting point for the next innovation cycle.

Addressing the challenges mentioned previously - such as continuously changing conditions (e.g. temperature and light level), high variability in both products (e.g. size, shape and color) and the environment, the delicate nature of the products and hostile environmental conditions - requires advanced sensing, perception and manipulation. Robotic hardware (e.g. sensors, robotic arms and end-effectors) that are fully suited to operating in greenhouse conditions barely exist. Control systems for robots for horticulture need to be highly sensitive and accurate to prevent unwanted damage. Research into all these aspects is still required, including on modular and soft grippers, tactile and imaging sensors, and agile robotic manipulators, and all must fulfill a high ingress protection (IP) class (see also https://www.iec.ch/ip-ratings). Improved low- and high-level control, sensor fusion and AI-based control is needed.

The use of neural networks and deep-learning algorithms for robotic tasks in horticulture has proven to be highly successful over the past few years. The increased availability of rapid and affordable computing hardware for these methods offers new possibilities. Machine learning is frequently used by computer vision algorithms employed in the context of robotics. Techniques that allow a robot to acquire novel skills or adapt to its environment are highly relevant but not widely implemented and should be addressed further in the future. This all involves the complex integration of multiple disciplines and technologies.

Growing systems can be modified to make detection and approach of fruits and flowers easier, thus allowing for simplified design and operation of automated systems (Bergerman et al., 2016). This is confirmed by the fact that the performance of robots greatly improves when the crop is simplified, such as by removing some of the leaves and fruits (Arad et al., 2020; Bac et al., 2017). A systemic approach in which the crop production system is designed together with the robotic system as just one part of its components is a promising route. A future trend is also to develop robots that work on an even smaller spatial resolution. This could be robots that scout through the greenhouse and automatically diagnose pests and diseases, not on a crop but on a plant or plant organ level, and directly take the appropriate action such as spraying locally or releasing beneficial insects, or removing plant parts or taking small tissue samples for further analysis. Such developments will make production in greenhouses more sustainable with fewer inputs and less environmental impact.

Another trend is to apply a modular and configurable design that will keep development costs to a minimum by applying a basic configuration to a range of tasks. Yet another option is to develop multifunctional robots that are capable of doing more than just one task. While it might be difficult to secure the business case for a robot that is only able to perform one specific task in a limited period of the growing season, such as harvesting, there is often the potential to include other tasks. It might be possible to add a sensor to detect stress in the plants or collect other data that can be analyzed. A robot that is carrying out harvesting tasks could also collect information regarding fruit size and the location in which the fruit is located on the plants. This data could be used for yield and crop development forecasts or to detect local spots in the greenhouse that require special attention. The robot might also be used to sort the fruit by size or ripeness during harvesting to deliver a specific quality requested by a customer. Whilst full automation is not yet feasible, human-robot collaboration and 'human in the loop' applications will be important stepping stones toward full automation. This topic has been gaining importance in the scientific literature in recent years (Edan, 2019; Huang et al., 2020; Vasconez et al., 2019). In situations where humans work closely together with robots, issues of user safety and user acceptance must be addressed.

9 Where to look for further information

A comprehensive introduction to robotics can be found in the *Springer Handbook of Robotics* (Bergerman et al., 2016). A dedicated chapter on robotics in agriculture and forestry gives a good introduction to the field. Members of the European robotics topic group 'Agriculture Robots' have recently published an overview paper and technology roadmap on AI, data and robotics in agri-food production (Sander et al., 2021). This paper provides a vision and mission on what robotics should contribute and how to facilitate the transition to a more sustainable agri-food production. The EU-funded network Agrobofood (https://agrobofood.eu/project/) is currently building a European ecosystem of Digital Innovation Hubs for the effective adoption of robotic technologies in the agri-food sector, in which universities, research centers, large- and small-scale industries and other relevant stakeholders are connected.

A number of scientific journals publish original research and review articles on the use of robotics in greenhouses. These include *Computers and Electronics in Agriculture*, *Biosystems Engineering*, *Journal of Field Robotics* and papers from conferences and workshops of the American Society of Agricultural and Biological Engineers (ASABE) and the European Society of Agricultural and Biosystems Engineering (EurAgEng). More specific to horticulture, the International Society for Horticultural Science (ISHS) publishes relevant papers in *Acta Horticulturae*. The International Federation of Automatic Control (IFAC) also regularly organizes conferences on sensing, control and automation technologies for agriculture and horticulture.

Data and statistics on crops grown in greenhouses and their economic value can be retrieved from websites. The Food and Agriculture Organization of the United Nations (FAO) provides free access to food and agriculture data at www.fao.org/faostat/en/#home. Free access to detailed global trade data can be found at https://comtrade.un.org/. More specific to the Netherlands, the agro and food portal website of Wageningen University & Research gives insight into the people, planet and profit performance of the Dutch agricultural and horticultural sector https://www.agrimatie.nl/. Economical statistics can also be retrieved from the Dutch Central Bureau of Statistics (CBS) https://www.cbs.nl/nl-nl/economie/landbouw.

10 References

Agrist. (2022). Agrist company homepage. Retrieved from https://agrist.com/. Accessed 24.5.2022.

Arad, B., Balendonck, J., Barth, R., Ben-Shahar, O., Edan, Y., Hellström, T., Hemming, J., Kurtser, P., Ringdahl, O., Tielen, T. and van Tuijl, B. (2020). Development of a sweet pepper harvesting robot. *Journal of Field Robotics* 37(6), 1027–1039. doi:10.1002/rob.21937. Retrieved from https://onlinelibrary.wiley.com/doi/abs/10.1002/rob.21937.

Arima, S., Kondo, N., Yagi, Y., Monta, M. and Yoshida, Y. (2001). Harvesting robot for strawberry grown on table top culture, 1: Harvesting robot using 5-DOF manipulator [in Japanese with English abstract]. *Shokubutsu Kojo Gakkaishi* 13(3), 159–166. doi:10.2525/jshita.13.159.

Arruga AI Farming. (2022a). Arugga tomato plant-lowering robotic system. Retrieved from https://www.youtube.com/watch?v=BcRSezt5FZw. Accessed 28.8.2022.

Arruga AI Farming. (2022b). Company website. Retrieved from https://www.arugga.com. Accessed 28.8.2022.

Atgrow. (2022). Plant elevation system. Retrieved from http://www.atgrow.nl/. Accessed 12.8.2022.

AVAG. (2022). Hortivation Point global standard for positioning in greenhouses (in Dutch language). Retrieved from https://www.avag.nl/nl/nieuws/mondiale-standaard-voor-plaatsbepaling-in-kassen/. Accessed 30.8.2022.

Bac, C. W., van Henten, E. J., Hemming, J. and Edan, Y. (2014). Harvesting robots for high-value crops: State-of-the-art review and challenges ahead. *Journal of Field Robotics* 31(6), 888–911. doi:10.1002/rob.21525.

Bac, C. W., Hemming, J., Tuijl, B. A. J. V., Barth, R., Wais, E. and Henten, E. J. V. (2017). Performance evaluation of a harvesting robot for sweet pepper. *Journal of Field Robotics* 34(6), 1123–1139. Retrieved from https://edepot.wur.nl/411193.

Barth, R., Ijsselmuiden, J., Hemming, J. and Van Henten, E. J. (2019). Synthetic bootstrapping of convolutional neural networks for semantic plant part segmentation. *Computers and Electronics in Agriculture* 161, 291–304. doi:10.1016/j.compag.2017.11.040. Retrieved from https://www.sciencedirect.com/science/article/pii/S0168169917307664.

Bartneck, C., Lütge, C., Wagner, A. and Welsh, S. (2021). What is AI? In: Bartneck, C., Lütge, C., Wagner, A. and Welsh, S. (Eds.), *An Introduction to Ethics in Robotics and AI* (pp. 5-16). Cham: Springer International Publishing.

Bergerman, M., Billingsley, J., Reid, J. and van Henten, E. (2016). Robotics in agriculture and forestry. In: Siciliano, B. and Khatib, O. (Eds.), *Springer Handbook of Robotics* (pp. 1463-1492). Cham: Springer International Publishing.

Che'Ya, N. N., Mohidem, N. A., Roslin, N. A., Saberioon, M., Tarmidi, M. Z., Arif Shah, J., Fazlil Ilahi, W. F. and Man, N. (2022). Mobile computing for pest and disease management using spectral signature analysis: A review. *Agronomy* 12(4). doi:10.3390/agronomy12040967.

Defterli, S. G., Shi, Y., Xu, Y. and Ehsani, R. (2016). Review of robotic technology for strawberry production. *Applied Engineering in Agriculture* 32(3), 301-318. doi:10.13031/aea.32.11318.

Ecoation. (2022). Ecoation company web site. Retrieved from https://www.ecoation.com/. Accessed 12.8.2022.

Edan, Y. (2019). Human-robot collaboration in agricultural robots. In: Billingsley, J. (Ed.), *Robotics and Automation for Improving Agriculture* (pp. 77–100). Cambridge: Burleigh Dodds Science Publishing Limited.

FlexCRAFT. (2022). Cognitive robotics for flexible agro-food technology. Retrieved from https://flexcraftprogram.com/. Accessed 29.6.2022.

Floating Robotics. (2023). Company hompage. Retrieved from https://www.floatingrobotics.com/. Accessed 26.02.2023.

Hayashi, S., Ganno, K., Ishii, Y. and Tanaka, I. (2002). Robotic harvesting system for eggplants. *Japan Agricultural Research Quarterly: JARQ* 36(3), 163–168. doi:10.6090/jarq.36.163.

Hayashi, S., Shigematsu, K., Yamamoto, S., Kobayashi, K., Kohno, Y., Kamata, J. and Kurita, M. (2010). Evaluation of a strawberry-harvesting robot in a field test. *Biosystems Engineering* 105(2), 160–171. doi:10.1016/j.biosystemseng.2009.09.011. Retrieved from https://www.sciencedirect.com/science/article/pii/S1537511009002797.

He, L., Zhang, Q. and Charvet, H. J. (2013). A knot-tying end-effector for robotic hop twining. *Biosystems Engineering* 114(3), 344–350. doi:10.1016/j.biosystemseng.2012.12.012. Retrieved from https://www.sciencedirect.com/science/article/pii/S1537511012002231.

Heater, B. (2022). Following acquisition by Bowery, Traptic's strawberry-picking robotics pivot to vertical farming. Retrieved from https://techcrunch.com/2022/02/16/following-acquisition-by-bowery-traptics-strawberry-picking-robotics move-into vertical-farming/ accessed 27.3.2022.

Hemming, J. (2020). Current developments in greenhouse robotics and challenges for the future. *Acta Horticulturae* 1296, 975–986. doi:10.17660/ActaHortic.2020.1296.124.

Henten, E. J. V., Hemming, J., Tuijl, B. A. J. V., Kornet, J. G., Meuleman, J., Bontsema, J. and Os, E. A. V. (2002). An autonomous robot for harvesting cucumbers in greenhouses. *Autonomous Robots* 13(3), 241–258. ISSN 0929-5593. Retrieved from http://library.wur.nl/edepot/38467.

Henten, E. J. V., Tuijl, B. A. J. V., Hemming, J., Kornet, J. G., Bontsema, J. and Van Os, E. A. (2003). Field test of an autonomous cucumber picking robot. *Biosystems Engineering* 86(3), 305–313. doi:10.1016/j.biosystemseng.2003.08.002.

Henten, E. J. V., Tuijl, B. A. J. V., Hoogakker, G. J., Weerd, M. J. V. D., Hemming, J., Kornet, J. G. and Bontsema, J. (2005). An autonomous robot for de-leafing cucumber plants in a high-wire cultivation system. *Acta Horticulturae* 691, 877–884.

Henten, E. J. V., Bac, C. W., Hemming, J. and Edan, Y. (2013). Robotics in protected cultivation. 4th IFAC Conference on Modelling and Control in Agriculture, Horticulture and Post HarvestIndustry, Espoo, Finlalnd, 170–177. doi:10.3182/20130828-2-SF-3019.00070.

Hortikey. (2022). Hortikey company homepage. Retrieved from https://hortikey.nl/. Accessed 12.8.2022.

Huang, Z., Gomez, A. S., Bird, R., Kalsi, A. S., Jansen, C., Liu, Z., Miyauchi, G., Parsons, S. and Sklar, E. I. (2020). *Understanding Human Responses to Errors in a Collaborative Human-robot Selective Harvesting Task*. Paper presented at the UKRAS20 Conference: "Robots into the real world". doi:10.31256/Xp9Yb4H.

Huang, M., Jiang, X., He, L., Choi, D., Pecchia, J. and Li, Y. (2021). Development of a robotic harvesting mechanism for button mushrooms. *Transactions of the ASABE* 64(2), 565–575. doi:10.13031/TRANS.14194. Retrieved from http://www.scopus.com/inward/record.url?scp=85105565783&partnerID=8YFLogxK.

Humphries, M. (2017). Panasonic's new robot harvests tomatoes as fast as a human. Retrieved from https://www.pcmag.com/news/panasonics-new-robot-harvests-tomatoes-as-fast-as-a-human. Accessed 17.3.2022.

Ishii, K., Matsuo, T., Takemura, Y., Sonoda, T., Nishida, Y., Yasukawa, S. and Fujinaga, T. (2021). *Tomato-harvesting-robot Competition Towards Smart Agriculture*. Paper presented at the International Conference on Artificial Life and Robotics (ICAROB2021), Japan.

ISO-Groep. (2022a). The ISO cutting planter. Retrieved from https://www.iso-group.nl/en/machines/iso-cutting-planter-2500. Accessed 12.08.2022.

ISO-Groep. (2022b). RoBoPlant. Retrieved from https://www.iso-group.nl/en/machines/roboplant. Accessed 12.08.2022.

Jiang, S., Wang, S., Yi, Z., Zhang, M. and Lv, X. (2022). Autonomous navigation system of greenhouse mobile robot based on 3D lidar and 2D lidar SLAM. *Frontiers in Plant Science* 13, 815218. doi:10.3389/fpls.2022.815218.

Kashkoush, M. and Avigad, G. (2019). *A Decision Support System for Automated Mushroom Harvesting*. Paper presented at the Intelligent Systems and Applications, Cham.

Kawamura, N., Namikawa, K. and Fujiura, T. (1984). *Study on Agricultural Robot (II)*. Paper presented at the Proceedings of the 7th Vehicle Automation Symporium JAACE.

Kondo, N., Monta, M. and Arima, S. (1998). Strawberry harvesting robot on hydroponic system. *IFAC Proceedings Volumes* 31(5), 181–185. doi:10.1016/S1474-6670(17)42119-7. Retrieved from https://www.sciencedirect.com/science/article/pii/S1474667017421197.

Kondo, N., Yata, K., Iida, M., Shiigi, T., Monta, M., Kurita, M. and Omori, H. (2010). Development of an end-effector for a tomato cluster harvesting robot. *Engineering in Agriculture, Environment and Food* 3(1), 20–24. doi:10.1016/S1881-8366(10)80007-2.

Kootstra, G., Wang, X., Blok, P. M., Hemming, J. and van Henten, E. (2021). Selective harvesting robotics: Current research, trends, and future directions. *Current Robotics Reports* 2(1), 95–104. doi:10.1007/s43154-020-00034-1.

Kuak, D. (2021). How soon will there be a robot in your greenhouse? *gpb – Greenhouse Product News*, November 2021 (pp. 26–29).

Kultongkham, A., Kumnon, S., Thintawornkul, T. and Chanthsopeephan, T. (2021). The design of a force feedback soft gripper for tomato harvesting. *Journal of Agricultural Engineering* 52(1), 1090.

Lehnert, C., McCool, C., Sa, I. and Perez, T. (2020). Performance improvements of a sweet pepper harvesting robot in protected cropping environments. *Journal of Field Robotics* 37(7), 1197–1223. doi:10.1002/rob.21973.

Ling, X., Zhao, Y., Gong, L., Liu, C. and Wang, T. (2019). Dual-arm cooperation and implementing for robotic harvesting tomato using binocular vision. *Robotics and Autonomous Systems* 114, 134–143. doi:10.1016/j.robot.2019.01.019. Retrieved from https://www.sciencedirect.com/science/article/pii/S092188901830808X.

Liu, J. and Wang, X. (2021). Plant diseases and pests detection based on deep learning: A review. *Plant Methods* 17(1), 22. doi:10.1186/s13007-021-00722-9. Retrieved from https://doi.org/10.1186/s13007-021-00722-9.

Nieuwenhuizen, A. T., Hemming, J. and Suh, H. K. (2018). *Detection and Classification of Insects on Stick-traps in a Tomato Crop using Faster R-CNN*. Paper presented at the Netherlands Conference on Computer Vision NCCV 2018, Eindhoven, Netherlands.

Noordam, J. C., Hemming, J., Heerde, C. J. E. V., Golbach, F. B. T. F., Soest, R. V. and Wekking, E. (2005). Automated rose cutting in greenhouses with 3D vision and robotics: Analysis of 3D vision techniques for stem detection. *Acta Horticulturae* 691, 885–892. doi:10.17660/ActaHortic.2005.691.110.

Octiva. (2022). Octiva company homepage. Retrieved from https://www.privakompano.com/. Accessed 4.3.2023.

Priva. (2022). Priva Kompano – Automous Deleafing. Retrieved from https://www.privakompano.com/. Accessed 22.2.2022.

Qingchun, F., Zou, W., Fan, P., Zhang, C. and Wang, X. (2018). Design and test of robotic harvesting system for cherry tomato. *International Journal of Agricultural and Biological Engineering* 11(1), 96–100. doi:10.25165/j.ijabe.20181101.2853.

Rath, T. and Kawollek, M. (2009). Robotic harvesting of Gerbera Jamesonii based on detection and three-dimensional modeling of cut flower pedicels. *Computers and Electronics in Agriculture* 66(1), 85–92.

Reed, J. N., Miles, S. J., Butler, J., Baldwin, M. and Noble, R. (2001). Automatic mushroom harvester development. *Journal of Agricultural Engineering Research* 78(1), 15–23. doi:10.1006/jaer.2000.0629. Retrieved from https://www.sciencedirect.com/science/article/pii/S0021863400906292.

RoboCrops. (2021). Studio RoboCrops on tour – GearRover. Retrieved from https://youtu.be/PuN5oVcwZNg. Accessed 24.6.2022.

Ruizendaal, J. (2022). Harvesting robot uses scissors to harvest gerberas. Retrieved from https://www.wur.nl/en/Research-Results/Research-Institutes/plant-research/greenhouse-horticulture/show-greenhouse/Harvesting-robot-uses-scissors-to-harvest-gerberas.htm. Accessed 24.6.2022.

Ruizendaal, J., Barth, R. and Balendonck, J. (2018). *Flower and Bud Detection of Gerbera L. using Faster Region based Convolutional Networks*. Paper presented at the EurAgEng 2018 Wageningen, The Netherlands.

Saga Robotics. (2022). Saga Robotics Thorvald platform. Retrieved from https://sagarobotics.com/thorvald-platform/. Accessed 22.6.2022.

SAIA. (2019). SAIA Agrobotics homepage. Retrieved from http://www.saia-agrobotics.com/. Accessed 01.12.2019.

Sander, S., van Henten, E. J., Lokhorst, C., Pekkeriet, E. J. and Steckel, T. (2021). *European Robotics in Agri-food Production*. Opportunities and Challenges. Working paper.

Schor, N., Berman, S., Dombrovsky, A., Elad, Y., Ignat, T. and Bechar, A. (2017). Development of a robotic detection system for greenhouse pepper plant diseases. *Precision Agriculture* 18(3), 394–409. doi:10.1007/s11119-017-9503-z.

Sparks, B. (2016). Priva's tomato deleafing robot nabs innovation award at GreenTech. Retrieved from https://www.greenhousegrower.com/technology/privas-tomato-deleafing-robot-nabs-innovation-award-at-greentech/. Accessed 25.8.2022.

van Henten, E. J., Van Tuijl, B. A. J., Hoogakker, G.-J., Van Der Weerd, M. J., Hemming, J., Kornet, J. G. and Bontsema, J. (2005). An autonomous robot for de-leafing cucumber plants grown in a high-wire cultivation system. *Acta Horticulturae* 691, 877–884. doi:10.17660/ActaHortic.2005.691.109.

Vasconez, J. P., Kantor, G. A. and Auat Cheein, F. A. (2019). Human-robot interaction in agriculture: A survey and current challenges. *Biosystems Engineering* 179, 35–48. doi:10.1016/j.biosystemseng.2018.12.005. Retrieved from https://www.sciencedirect.com/science/article/pii/S1537511017309625.

VDL. (2023). VDL CropTeq robotics homepage. Retrieved from https://www.vdlcropteqrobotics.com/. Accessed 4.3.2023.

Vineland Research & Innovation Centre. (2020). Automated mushroom harvesting system. Retrieved from https://www.youtube.com/watch?v=xjHM3wDV6Wg. Accessed 29.6.2022.

Vineland Research & Innovation Centre. (2021). Opportunity knocking: Reducing labour costs with automated cucumber harvesting. Retrieved from https://youtu.be/fbJzuClZO14. Accessed 24.6.2022.

Wageningen Univeristy & Research. (2018). SWEEPER, the sweet pepper harvesting robot. Retrieved from https://www.wur.nl/en/project/SWEEPER-the-sweet-pepper-harvesting-robot.htm. Accessed 24.6.2022.

Wang, Z., Xun, Y., Wang, Y. and Yang, Q. (2022). Review of smart robots for fruit and vegetable picking in agriculture. *International Journal of Agricultural and Biological Engineering* 15(1), 33–54. doi:10.25165/j.ijabe.20221501.7232.

Xiong, Y., Ge, Y. Y., Grimstad, L. and From, P. J. (2020). An autonomous strawberry-harvesting robot: Design, development, integration, and field evaluation. *Journal of Field Robotics* 37(2), 202–224. doi:10.1002/rob.21889.

Chapter 21

Advances in the use of robotics in livestock production

Kees Lokhorst, Wageningen Livestock Research, The Netherlands; and Tomas Norton, Katholieke Universiteit Leuven, Belgium

1 Introduction and main drivers for livestock robotics
2 The role of robotics in precision livestock farming
3 Examples of robotic development: poultry
4 Examples of robotic development: pig production
5 Examples of robotic development: dairy
6 Challenges for research
7 Conclusion
8 References

1 Introduction and main drivers for livestock robotics

The use of robotics in livestock production dates back to the 1980s. One of the first robots to be developed was for shearing sheep (Trevelyan, 1989; Trevelyan, 1987). More recent developments in robotics have been made in aquaculture (Antonucci and Costa, 2020; Wu et al., 2022; Yue and Shen, 2022). This chapter presents recent advances in the development and use of robotic solutions in various livestock farming systems, focusing on dairy, pig and poultry production. These three livestock sectors have different characteristics in terms of the general design of the farming and housing systems used. Moreover, there can be significant diversity within each sector, which has both pros and cons when it comes to the use of robotics. These aspects will be discussed in this chapter in terms of their relationship with robotic applications.

This chapter is inspired by two quotes related to the field of robotics: *'Robots create value by performing physical tasks that people cannot, should not, or will not do'* and *'A robot is a programmable machine capable of carrying out a physical task or a complex series of actions automatically'* (Zillner et al., 2020). Given the level of drudgery involved in agricultural tasks, it is easy to see the potential applications of robotics. However, these practical definitions leave room for scientific and practical disputes over what falls under the umbrella of

http://dx.doi.org/10.19103/AS.2023.0124.25

robotics. When it comes to livestock farming, for example, robotics may fall within the field of precision livestock farming (PLF), although recent practical developments in the PLF area lack robotic-based innovations (Banhazi et al., 2022). Given that robotics and PLF are based on sensor data and require a pre-defined 'world' to operate in, Section 2 will attempt to connect robotics to the concept of PLF.

It is our belief that livestock farming systems can be strengthened by integrating robotics into PLF and that robotic-based PLF solutions should be adopted and discussed in a broader context. In this discussion, awareness should be on the main driving factors for innovation. To be able to innovate with cutting-edge technology is essential.

Based on experience, and looking at previous significant innovations, Lokhorst (2018) identified developments in labor, technology, society, market, legislation and science as the key drivers for innovation in the livestock sector. Reduction in labor is directly connected to cost savings. Reduction in labor does not mean that humans will work less, but that efficiency is improved so they can produce more in the available time. This makes living and working more fun and efficient. In this respect, developments in technology can have a significant impact. Developments in information and computer technologies (ICTs) are predicted regularly by Gartner (www.gartner.com) and are used by many organizations for strategic investments. Conolly (2022) indicates that developments in ICT have the capacity to change the agricultural sector. Drones, robots, sensors, 3D printing, the internet of things (IoT), artificial intelligence (AI), virtual reality (VR), augmented reality (AR) and Blockchain will find their way into the smart farming domain. Professionals in the livestock sector must be aware of these emerging technologies and judge them on their usability for these innovations to be successful in the sector.

Wider discussions around outdoor cow grazing, or the size of mega-farms and the use of robotics, to name just two, can also drive innovation. Food producers are continuously developing new products for their customers. Markets are heavily influenced by food trends and transparency in the food chain has become an extremely important societal issue. Legislation can limit undesirable production volumes and emissions of greenhouse gases. In itself, this drives innovation, as people find smarter ways to work within the limits of the legislation. Understanding the science behind how and why processes and behaviors in the food production chain work can also lead to innovation. In conclusion, there are many aspects that stimulate innovation in smart farming.

An awareness of what is happening in society, the market, science and legislation assists in the development of robotics solutions (technological) to make living and working (labor) on livestock farming systems more effective and efficient. The main challenges we face are:

- a lack of sufficient and qualified labor due to farmer ageing, waning interest from younger workers and restrictions on migrant workers;
- scale of economics due to the level of investment needed for environment, welfare, health, biodiversity and quality of products, which has led to fewer farms with a higher number of animals; and
- adoption of green deals and farm-to-fork strategies with more emphasis on climate, circular and biodiversity in the complexity and variety of livestock farming systems.

A recent EU report ('European Robotics in agri-food Production: Opportunities and Challenges', Sander et al., 2021) identified a vision that 'future agri-food production networks will be flexible, responsive, and transparent, providing sufficient high-quality and healthy products and services for everyone at a reasonable cost while preserving resources, biodiversity, climate, environment, and cultural differences'. Its mission is to 'stimulate the development and integration of innovative robotic, AI, and Data solutions that can successfully be used in flexible, responsive, and transparent agri-food production networks'. In this chapter, we will look at several specific challenges in the livestock sector that can be addressed using robotic technology.

2 The role of robotics in precision livestock farming

The concept of PLF is being developed in parallel with that of precision agriculture (PA). PLF aims to '*manage livestock farming by continuous real-time monitoring of health, welfare, production, reproduction and environmental impact*' (Berckmans, 2008, 2017). The International Society for Precision Agriculture (ISPA) defines PA as '*Precision Agriculture is a management strategy that gathers, processes and analyses temporal, spatial and individual data and combines it with other information to support management decisions according to estimated* variability *for improved resource use efficiency, productivity, quality, profitability and sustainability of agricultural production*'.

This definition was formulated and agreed via an open and thorough process and also covers PLF.

In *Introduction to Smart Dairy Farming*, Lokhorst (2018) highlights the main dilemmas and issues involved in the application of precision farming to dairy production. To deal with the inter- and intra-variability of dairy farms, management tools need to be adaptive in order to support daily management of farmers and their advisors/service providers. Connected to the concept of *Livestock Farming With Care* (Scholten et al., 2013) and the central position of the farmer, PLF can be internalized if a farmer is willing to accept the following statement:

> As a FARMER I will guarantee that every SUBJECT OF INTEREST gets the CARE it requires at the right moment, in the right place, and to the right extent, and I am TRANSPARENT about this.

The 'subjects of interest' are the most granular units in the process that can be controlled in farm situations. For dairy farming, this might be the individual cow or calf, as well as parcels of land. For pig and poultry farmers, this might be a group in a pen or barn, or in the case of breeding farms, it could be individual sows. The choice of the subject of interest depends on the farm and the farmer.

To support a farmer in their daily management, the implementation of PLF should align with the farm's organizational structure. When organizing tasks on the farm, farmers have the following options:

- Carrying out the work independently, with the farmer and their family members and friends sharing the workload;
- Employing animal caretakers to help manage daily operations;
- Engaging the services of specific providers and experts (e.g. contractors, veterinarians, hoof trimmers, advisors and so on); and
- Automation through the use of robots.

In practice, a combination of the above will be used, depending on the farmer's preference and ability to organize and coordinate these basic labor resources.

PLF involves continuous measurements with sensors and software applications acting as extra eyes, ears and noses, and models as extra brains to continuously interpret the context and status of the subjects of interest and translate these findings into concrete actions that can be performed by the available labor resources. This is part of a system and control approach. Digital data are key in this respect and need to be transparent.

In practice, robotics and PLF are mostly treated as separate concepts. We are of the belief that both should be integrated and can strengthen one another. Figure 1 shows how this concept might work. In the center, the circle has to come from data, through information and knowledge to data-driven actions. There should always be a process goal that creates awareness of what should be monitored or controlled. Even routine tasks have this at the start. Achieving this integration involves consideration of a number of key concepts.

2.1 Standard operating procedures

The concept of working with standard operating procedures (SOPs) is focused on instructing farm employees. A SOP describes work instructions. For the subject of interest (e.g. a particular activity), who should do what, when, where and how needs to be accurately and exactly described (Wind et al., 2017).

Working with protocols can be beneficial, making farms more competitive and safer. In the case of a sudden reduction in the labor force, tasks can be taken on more easily by someone else. SOP protocols help encourage the farmer to

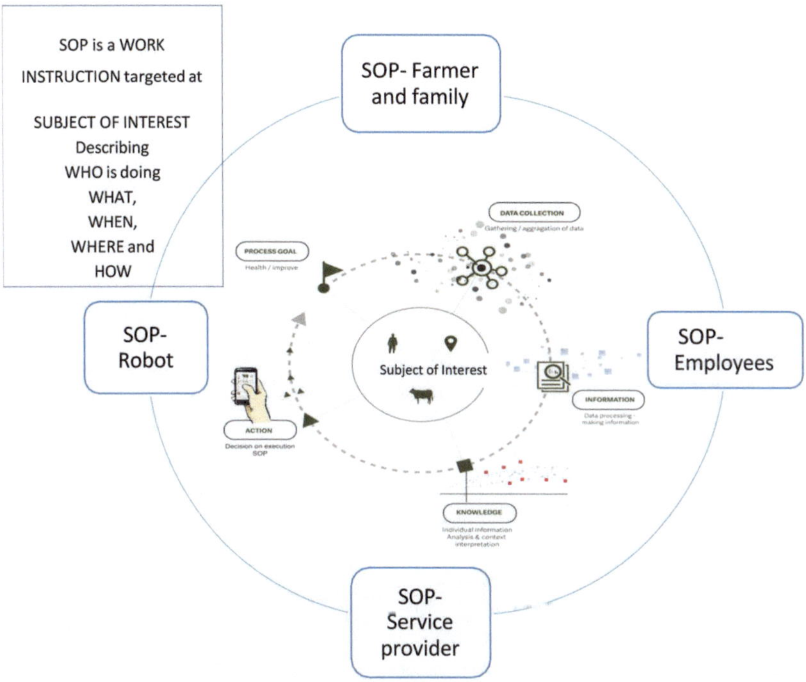

Figure 1 SOPs connecting the cycle from data to action, for different labor resources, including robots.

design farm processes in a more optimal and structural way, in accordance with their expectations and references. SOPs should not just be written instructions but also made available in electronic form. Both humans and machines can be targeted. Naturally, the instructions will differ according to the labor resource, but it is worthwhile to discover how different SOPs can be developed and contribute to greater transparency and interoperability.

While Fig. 1 demonstrates how robotics can perform different parts of the cycle of data collection to action as seen with PLF, it is clear there are core processes in the design, development and maintenance of robotic systems that are not commonly seen in the PLF framework.

Figure 2 adapts the work of Porter and Heppelman (2014) to show the increasing complexity that connects the different aspects that must be managed. With robotics, we are now at Product level 4 (Product system) and the next step will be to reach the 'system of systems' level, involving robots. To be able to reach that, the following aspects are important in the (re)design of robotic systems.

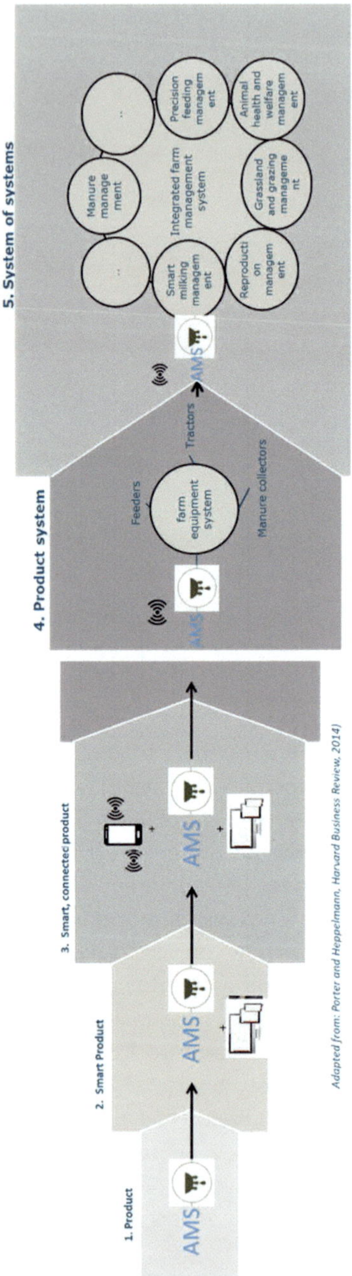

Figure 2 Redefinition of industrial boundaries. Source: Adapted from: Porter and Heppelman (2014).

2.2 Systems engineering

Systems engineering-oriented frameworks are a core component of robotic applications, and livestock robotics should be no different. The starting point is the implementation of methods such as quality functional design (QFD) to reveal the requirements of the system and provide suggestions for services. These can be further analyzed and translated into technical system requirements. *Software and Hardware Architectural Design Processes* should then be employed to ensure the robot complies with all requirements and an *Implementation Process* is needed to produce the specified system units, starting with the most essential sub-systems (e.g. milking, cleaning and monitoring). This is typically followed by an *Integration Process* in which we will integrate the sub-systems and sensing units according to the architecture. Finally, a *Validation Process* guarantees that the services provided by the robot align with what was initially specified by the prospective users. The *Demonstration Process* verifies if the robot will deliver the services in practical application.

2.3 Architecture

Robotic software applications such as robotic operating systems (ROSs) architecture, which is a cross-platform architecture that is required to build a solution on in order to run different robotic operations.

2.4 Systems integration

Achieving interoperability of systems from different original equipment manufacturer is essential to make cost-effective robotic applications. Exploiting methods and tools to implement this will enable slight changes to the specification. Core to this is the definition of standards, in which appropriate normative standards will be identified and considered within the design, implementation and evaluation phases.

2.5 System of systems

Robots can be integrated within the complete ICT system of the overall farming system, utilizing their capacity to interact with a diverse range of devices, from actuators of other control systems to tablets and smartphones. Such integration can enable smooth running of the system and contribute to continual farm process improvement.

2.6 Modeling and knowledge engineering

Information derived from a robotic system can be embellished and integrated with other available information to assist in farm model-based management decision-making. Big Data approaches employ techniques to continuously update the resolution of the models based on captured data. Data from the robotic platforms can support simulation in the decision-making loop, ranging from early warnings of outbreak of diseases to signaling the most appropriate times to carry out an operation on the farm, such as cleaning, animal removal, etc.

2.7 Quality validation, maintenance and documentation

Appropriate, statistically supported procedures are identified and applied to assure quality validation of systems performance and services, coupled with supporting maintenance procedures and documentation.

The following section provides examples of robotic developments in three different livestock sectors to show where robotics have already been successfully applied and their current focus. These developments illustrate the advances in robotic solutions in research and in the market.

3 Examples of robotic development: poultry

Poultry production is a specialized sector that produces approximately 100 million tons of meat (i.e. broiler production) every year globally. Production of broilers and eggs is a low margin business that requires large-scale production, consisting of sheds housing tens of thousands of birds. Modern poultry breeds are advanced to maximize feed conversion efficiency, with the result that chickens are one of the most efficient feed-converting animals in modern agriculture. Typically, they grow from under 50 g to as much as 2.2 kg in a 5- to 6-week cycle. They are, however, very delicate animals and can easily become stressed by their living environment. As such, accurate control of their growing environment is essential.

Poultry farmers determine what management actions to carry out based on their assessment of the living environment and chicken behavior to determine if they are healthy. Correct and timely decisions deliver improved chicken welfare, better use of feed, better health and less waste, resulting in increased income. Poultry farmers therefore depend on a mixture of intuition and experience, supplemented by automatic monitoring and control systems. A fully fledged PLF solution can offer continuous measurements, utilizing data-driven decision-making. The poultry sector has seen remarkable progression in the application of ICT. Use of climate and light control computers is common and there are many options on the market. Research into real-time monitoring of animals has made significant progress and various techniques are being implemented and commercialized.

Sensor systems that continuously and automatically monitor the growth and welfare of birds are now available and provide farmers with support to manage their animals remotely (even at individual animal level) (Sluis et al., 2022). Bird growth is measured by the use of automatic weighing scales and, more recently, by camera technology and sound information (Fontana et al., 2017). Welfare is more complicated and as such is not measured, but indirect parameters such as activity, animal distribution across the floor and drinking and eating behavior can be observed using commercial tools. While this type of ICT application is part of PLF, exploiting robotics to remove simple, monotonous tasks from the farmer while guaranteeing a return on investment can make a significant impact. By so doing, farmers can focus on other aspects of their business, as well as improving standards of care for their animals, leading to better productivity, health and welfare and thus an economically viable business. Through automation, the role of the stock person can change, as opposed to being reduced, as robots will allow time to be used differently. There is no doubt that robotics can be an enabler for a more productive sustainable future for poultry production.

The field of robotics is already infiltrating the poultry sector. One such example in the modern cage egg industry is the use of integrated conveyor belts to collect and transport eggs to the packing room. Conveyor belts eliminate the need to handle the eggs manually, which is especially pertinent in modern sheds with tiered cages. Egg collection via conveyor belts also reduces occupational health and safety (OH&S) risks associated with manual egg collection at height and under low levels of illumination. Other areas of application have occurred in recent years, such as the introduction of autonomous robotic vehicles that move throughout the poultry house, performing a number of tasks such as egg collection from the floor (Vroegindeweij et al., 2018) and litter management (Elijah et al., 2022). Certain key developments can have an immediate impact on the poultry industry in the following ways.

3.1 Dead bird detection

As part of daily inspection rounds, broiler farmers are required to remove dead birds from the buildings, which is not a pleasant task. Moreover, it is fraught with risk of disease control and biosecurity issues as the farmer moves between houses. The potential to automate dead-bird detection and possibly even removal from the building can be seen in the development of robotic technologies (e.g. Chickenboy ('Poultry farming: The robots are coming', 2018) currently on the market. Actual bird removal is undergoing research with many researchers proposing solutions, albeit under controlled conditions.

3.2 Litter quality monitoring

Farmers must continuously monitor litter quality in the house, as moist poultry litter is susceptible to crusting, which prevents moisture from penetrating the litter, making it slippery and uncomfortable for the birds. Poor litter quality is the primary cause of podo-dermatitis outbreaks. Moreover, nipple drinkers regularly block, and the resulting spillages cause significant deterioration in litter quality. By the time it is brought to the attention of the farmer, very often a 'dig-out' of the litter is required. Autonomous litter quality sampling is therefore a highly desirable scenario. Examples of prototypes include the Spoutnic ('Spoutnic, le robot de ponte qui limite les pertes', 2019) and the FloxLitterBox (Elijah et al., 2022).

3.3 Abnormal behavior of broiler chickens

This is an important welfare indicator, currently measured by 'gait scoring' via human observation. While new techniques such as optical flow are being studied by researchers, there is no suitable commercial system to date. The key is that the observations need to happen close to ground level, but accurate assessment of the birds is difficult without removing the birds from the building. The potential for robotics to monitor bird movement has been demonstrated by Demmers et al. (2018) at the Royal Veterinary College, who have developed a device to measure bird behavior.

4 Examples of robotic development: pig production

The application of PLF technologies in the pig sector has been the subject of research for many years. Key elements of PLF include identification of the animal via electronic identification (e-ID) or radio frequency identification (RFID) followed by automatic monitoring of parameters including growth, through the use of weighing platforms and camera technology (also known as visual image analysis – VIA), which exploits the strong correlation between the live weight and the area of pigs in plain view (Schofield et al., 1999). Another example includes a robotic system capable of placing a sensor in contact with any one of a number of pre-determined positions on the body of a loosely constrained live animal (Frost et al., 2000), the use of multiple sensors to monitor pig health and welfare (Wang et al., 2022) and the design of an intelligent monitoring system for pig health (Yanchang et al., 2021). There have also been developments in robotic solutions. Some of the technologies are highlighted next.

4.1 Bedding distribution

As with dairy production, various robotic solutions in the pig sector are utilized to provide the animals with bedding material. This is not a highly sophisticated technology and works without the need for environmental perception to control its actuation (spreading the straw). The greatest level of application of such technology exists in countries such as Denmark where the need for straw

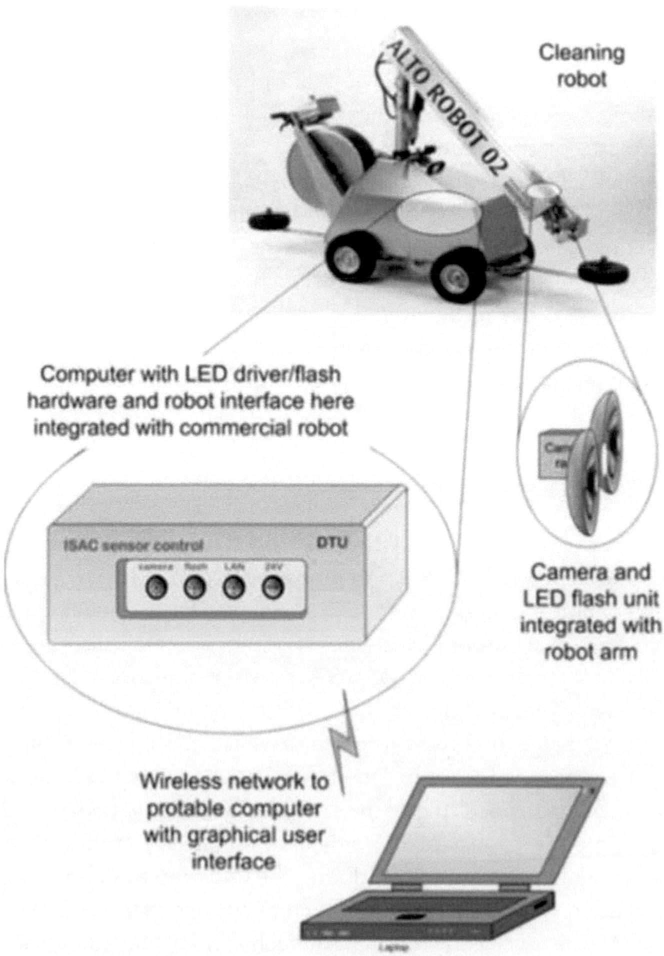

Figure 3 Overview of the novel pig building-cleaning robot by Nil Axel et al. (2005).

(or other fibrous materials) to be available to pigs is set out by legislation. As far back as 2008, Wattagnet.com highlighted a simple robotic solution in which a straw distribution unit is transported throughout a pig house via a fixed rail system installed in the building (https://www.wattagnet.com/articles/902-robot-takes-the-strain-out-of-strawing).

4.2 Cleaning

Another area of interest is barn cleaning via a robot equipped with power washing equipment. A rationale behind using robotic applications for cleaning pig houses is the significant labor required for this repetitive, unpleasant, but highly necessary job (Lofqnvust, 2014). As with the bedding robot, these technologies function quite simply. The level of perception is limited to halting the robot when it gets close to a wall. These robots also emerged in early 2000 and are currently used in large-scale pig production sites. One of the early cleaning systems was announced in 2004 (https://www.pigprogress.net/home/simple-to-operate-but-technically-advanced/). Current building cleaning robots are not efficient in their use of water, leading to a high risk of wastage (Nielsen et al., 2012). This has been reported by some as the main reason these robots are only utilized on 60–70% of pig farms (Gjødesen, 2007).

Nils Axel et al. (2005) carried out research to improve the performance of these devices, in order to develop a 'cleanliness sensor', adding functionality to the motion controller to enable it to determine the level of cleanliness when executing its cleaning plan (Fig. 3). Further development of the water application solution has been carried out by Nielsen et al. (2012).

4.3 Precision feeding

Robotic technologies with more sophisticated functions are also undergoing research. One such application is precision feeding. Providing a diet that meets the necessary nutrient requirements without any spillage or over-consumption is one of the most important elements for economic success, especially considering that feed represents 65–75% of total production costs. Feed costs can be reduced by taking into account the variation in nutrient needs between animals, as opposed to current practice in which the nutrients supplied are based on the average of a group. Due to the variation in nutrient requirements and growth potential, not all pigs exhibit the same response to certain nutrients. The growth potential of each pig is partly determined by factors such as gender and genetics, which do not change during the fattening period but depend to a large extent on varying components such as supplied feed components, the physical and social micro-environment of each animal, their health and animal welfare.

In practice, fattening pigs are fed as a group. With advancements in technology (including electronic animal identification, weight, health status, etc.) it has become possible and affordable to feed group-housed fattening pigs individually. The idea of precision feeding for pigs is not completely new. The benefits of nutrient efficiency are mentioned in the literature (Pomar and Remus, 2019) but to date have only been demonstrated through simulation models, never in practical applications.

Relatively few applications related to the dynamic precision feeding of fattening pigs have been investigated. One of the current concepts consists of feeding meat pigs according to individual lysine requirements (Pomar and Remus, 2019). A mechanistic-empirical model for the real-time estimation of individual lysine requirements was used by Hauschild et al. (2012). After estimating the lysine requirement for maintenance, the optimal lysine concentration of the feed was determined. Pigs received feed through an individual feeding system (Automatic and Intelligent Precision Feeder) that consisted of an Archimedes transport screw mixing up to four different feeds in the desired ratio (Pomar et al., 2011). After the head of the pig appears in the feeding system, the pig is recognized and an amount of feed is delivered that meets the specific lysine requirements of the animal, determined by the mechanistic-empirical model (Hauschild et al., 2012). This method of feeding resulted in a 27% decrease in lysine intake and a 22% and 27% reduction in nitrogen and phosphorous excretion, respectively, compared to three-phase feeding.

In group housing systems for sows, the use of electronic feeders has become a standard practice. The ability to feed each individual sow to her unique needs was recognized two decades ago. Use of sorting systems also depend on the type of housing systems. In group housing systems for fattening pigs and sows, sorting systems can be used.

The research clearly demonstrates the utility of precision feeding techniques by switching from a pre-determined feed composition for an entire group of pigs to the real-time determination of individual needs to achieve a particular goal. Although weight gain and backfat thickness are important parameters in pig production, the pig farmer needs to know how much a finisher produces in net terms.

5 Examples of robotic development: dairy

In this section, we offer insight into the current state of development (and use) in professional managed dairy farms. Milking robots for cows were one of the first agricultural robots to be commercialized and appear to be most developed. Development started more than 40 years ago and a thriving professional market now exists, with robot providers including companies such as Lely, DeLaval, GEA,

Boumatic, Fullwood and Westfalia Surge. Sharipov et al. (2021) provide a historic overview and classification. The main reason for this successful development and its impressive penetration into the market (https://www.researchdive.com/8651/milking-robots-market) is reduction of labor on dairy farms of all sizes, allowing a more flexible lifestyle for farmers (Rodenburg, 2017). Robotic milking is a very good alternative for farms up to 250 cows, but recently robotic milking has also been applied on bigger farms. As current milking robots are reliable, promote excellent cow care and allow farmers to do other farm or off-farm work, the use of robotic milkers will continue to grow on farms in the developed world. The market for robotic milking systems is expected to rise from US$1.25 billion in 2019 to US$2.94 billion by the end of 2027 (https://www.dairyglobal.net/industry-and-markets/smart-farming/robotic-milkers-where-we-are-and-where-were-going/).

5.1 Feeding

The use of robotic solutions for feeding dairy cows depends on the different feed types. Concentrate feeders have been introduced alongside milking robots, both using electronic identification of the cows. Concentrate feeders are used widely in practice. Several attempts have been made to robotize roughage feeding in the barn but robot roughage feeders such as the Mollerup feeding system (Patent # 10 660 307), which feeds individual cows, have not yet succeeded. Only in such experimental stations as the Hokofarm RIC are such methods used. Nevertheless, there have been attempts to robotize the feeding of roughage to production groups in front of the feeding gates. Trioliet and Lely are active in this market and have put various robotic products on the market. Related to this is the rapid development of feed pushers, which have the task of pushing roughage feed towards the feeding fence for the cows to easily eat. By so doing, the idea is that feed intake is stimulated. Several companies deliver such feed pushers and it can be seen as a silent success in the market. Lely introduced the fresh grass concept in 2020 and developed the Lely Exos to harvest fresh grass and bring it directly to the feed alley. This concept needs to be further developed and tested. However, cows themselves are good harvesters of fresh grass.

To support grazing, two robotic developments should be mentioned. One is the development of sorting gates between the barn and the parcels of land to prove that cows are outside in the field for a certain period of time, assuming that they graze. This has become part of certain product and marketing developments. The second tool to support grazing is the development of virtual fences or moving wired fences, the latter of which began more than 20 years ago and is still not implemented in the field (Butler et al., 2006), (nn, 2021), (Comis, 2000; Umstatter et al., 2015). Regarding feed types, we also must consider the upcoming use of feed additives, which improve the health and resilience of the cows as well as reduce emissions such as methane. Current feeding robots dispense the additives either to individuals (using the concept of concentrate feeders) or to groups. The final feeding category in which we are seeing an increase in robotics is the feeding of milk to calves. Many companies (GEA, Forster, Lely, DeLaval) and products have been developed and their use in practice is rising. However, using robotics to feed cows until they give birth to their first calf is not yet fully developed. In addition, the feeding/dispensing of water is not yet supported by robots.

5.2 Monitoring and management

The monitoring of cows has witnessed many developments in recent years. Observing cows can be highly labor-intensive, especially when the group size grows. Farms consist of multiple barns with separate compartments and require round-the-clock observation. The most effective and widely accepted measurements are those made in the milking robot/parlor, consisting of milk quantity, cell count, milk conductivity, fat percentage, protein percentage and timely detection of estrous cycles. Due to improvements in sensor quality, accuracy and longevity, as well as the capability of the IoT, the remote acquisition of activity data has become extremely successful. These technological opportunities have driven the development of a variety of new sensor and monitoring systems to observe characteristics related to cow behavior, health and welfare (Mottram, 2016; Rutten et al., 2013). The website http://www.koesensor.be/ provides an overview of available sensors in the dairy sector, such as measuring the weight and body condition of animals. This is essential at all life stages, but to date has had limited use when integrated with milking robots. Even following decades of experimentation, the use of weighing platforms has yet to reach practical mass adoption. However, most of these products and services still face challenges in the market. A lack of trust by farmers, in addition to a limited proven business model and interoperability, all hinder mass uptake. In the coming years, some of these solutions will be integrated in farming systems, especially those that focus on early disease detection and those capable of measuring and verifying that farm (and cow) level emissions and welfare are managed properly.

A relatively new field is the use of drones to observe and count cows in pasture environments (Yousefi et al., 2022) provide a useful overview of the underlying technology and data analysis.

To guide cows inside the barn, the use of selection gates is widespread. Most are integrated in the design of the milking parlor, their main functions being to separate cows that require extra care or cows that are not allowed to enter the milking robot. The development of smart feeding fences in the barn appears to have stopped. Where used, it is a mechanical solution that works well. Today's robots do not yet support catching and driving specific cows, e.g. to guide a cow to be milked or to be inseminated. Over the last decade, tools have been developed to use location information for the cows. GEA, SmartBow and NEDAP have each introduced their own location awareness systems, but uptake in the market remains low and the task is till carried out by humans.

5.3 Cleaning

Using robots to treat cows is limited to the cleaning of the udder in the milking robots and automatic brushes for grooming. The animals show a genuine appreciation of this treatment, leading to several different companies selling these products. The robots are positioned in specific locations in the barn and the cows visit them voluntarily. Other treatments, such as insemination, hoof trimming, shaving and the delivery of medication are not yet supported by robots. These treatments are rather complex and require specific skills from specialists such as veterinarians, inseminators, hoof trimmers and farmers. Robots and PLF can assist the farmer and others but cannot replace the human factor in these complex processes.

Cleaning of cows takes place when they are treated. Cleaning of the milking robot itself is integrated within the robot's own functions and cleaning of the milking parlor or the milking robot environment is mostly still done by hand. The biggest cleaning activity is carried out in the areas where the cows roam and rest. Manure-scraping belts and specific manure scraping/removing robots have been developed and are used quite a bit in practice. The main producers in the Netherlands are JOZ, GEA and Lely.

Robots for cleaning the lying beds and in confinement barns are not yet well developed or utilized. Various systems are used in practice for the distribution of fresh straw in straw beds. We do not foresee specific solutions for cleaning dairy farms themselves. In terms of cleaning the air, various new products are coming on to the dairy market that are able to 'wash' the air and remove ammonia, for example. Dairy barns are mostly open systems and air handling and washing therefore needs specific design. The Lely Sphere is one such system, but it has yet to find its position in the market.

5.4 Milk processing

Processing of milk takes place mostly at specific processing plants. The storage, transport and sampling of milk are part of a highly professional industry. In addition, there is also on-farm processing of milk. Homemade cheese, butter and yoghurt is produced by the farmer without robotic support. The introduction of the Lely Orbiter in 2020 was an innovative development that shocked the market with its ability to process, pasteurize and package the milk on the farm itself. This is a new concept and requires further development in the coming decades.

6 Challenges for research

6.1 Sectoral challenges

With over 35 years of experience, one of the authors, working at Wageningen Research, has witnessed much scientific interest in robotics, including the development of robotic milking systems and the design of innovative barn types that harnessed the potential to identify individual cows by the use of electronic identification and concrete floors. Animal scientists collaborated with agricultural engineers during this period, contributing to systemic breakthroughs. After a period in which the focus on robotics in livestock systems waned, there was a strategic shift to invest in the Wageningen Agro Food Robotics program (https://www.wur.nl/en/research-results/projects-and-programmes/agro-food-robotics.htm). The various challenges included (1) knowledge development to apply robotization to far-reaching animal monitoring (behavior and environment of individuals in groups that are kept both indoors and outdoors and interact with robots); (2) knowledge development to apply robotization in specific land-based farm systems with grassland that focus on nature inclusivity; and (3) encompassing intelligence (learning and adaptability) of robotic systems for feeding, harvesting (milking/collecting eggs), animal care, cleaning and animal routing, whereby local and remote collaboration between robot(s), humans and animals is optimally utilized.

In 2016, challenges for robotic solutions in the poultry sector were discussed with farmers in the Netherlands. The idea was that robotics could reduce the need for humans to carry out jobs such as litter maintenance, removing dead birds and determining the potential cause of death, collecting eggs outside the laying nests, carrying out health checks for blood lice, selection of non-producing birds, cleaning the barn and performing vaccinations. Considering the latest developments discussed in Section 4 on pig production, the majority of these challenges are still on the table. Reasons for this could be that the large industrial companies are focusing on climate control and egg packaging, whilst poultry farmers remain hesitant, and newcomers in the market with see a highly fragmented poultry sector with small and highly specialized sub-sectors. The number of customers for a product needs to come directly from a global market.

As outlined in Section 4 on pig production, robots in the pig sector are fairly successful in controlling the climate and feeding pigs in groups and

Table 1 Impression of 12 identified use cases and the 11 key challenges mentioned per use case

Use case theme	**Key challenges (technical, ecosystem, business, training and human capital development)**										
Innovative/ disruptively novel agri-food systems enabled by robotics	Global modeling, simulation and benchmarking	Robot-to-X interaction	24/7 level 5 cooperative systems and fleet and swarm management	Ferception in robotics	Multi-dimensional manipulation	Interactive design of trustful, safe, and ethical robotic system	Sustainable pan-European agROBOfood network	Push-to-market for agricultural robots and systems, support, education and training	Specialized robots to be used by seasonal unskilled labor	Infrastructure for practical training with access to robotics	Lifelong learning: connecting people from agri-food with people from robotics and analytics
	X		x	x		X					
Robotics, AI, and data science for breeding	X	X	x	x							
Complex handling and manipulation in primary production	X	X	x								X
Complex handling and manipulation in post-harvest		X	x	x	X						

Realizing full autonomy of already mechanized tasks			x			X	
AI and robotics for livestock farming		X	x			X	
AI and robotics for precision agriculture	X	X	x	x			
Cleaning in agri-food				X	X		
Connectivity, distributed intelligence and pervasive technology		X	x			X	
Logistics and transport	X	X	X			x	x
Ocean farming and agri-food	X		x	x		X	
Food preparation and presentation							

Source: Adapted from: Sander et al. (2021).

sows in individual feeding stations. This market is driven by large multinational companies. It is possible that the development of robotic solutions might be hindered by the current structures used in barn and pens, which make it difficult to design autonomous robots to observe pig behavior, play the role of entertainer, and maintain the environment of the pigs, keeping both the interior and exterior of the buildings in good shape. The cleanliness of the environment in which the pigs live remains an important factor and the challenge will be to develop robotic systems that replace or support this task. New farming system designs might help to stimulate these developments.

In terms of robotic developments in the dairy sector (see Section 5), it is clear that the main industrial drive comes from the companies involved in robotic milking. DeLaval is using its R&D Hamra test farm to develop and test all kinds of new innovations, and Lely, for example, is working on the vision and concept of 'fully' autonomous dairy farms, on which its product portfolio is based. Its main challenge will be to strengthen the already adopted robotic solutions and to further develop and bring to market its newly introduced concepts. An additional challenge will be to allow the different robots to collaborate with each other and with humans. It is expected that animal intrinsic treatments such as insemination, claw trimming, shaving and medical treatments will not and cannot be done by robots in the near future. For the many relative newcomers to the robotic dairy market, it will be quite challenging to develop robotic solutions that fit into the farming systems. Additional attention is required for interoperability. It is expected that most companies will focus on robotic systems for cow and barn observations. Some well-developed companies have created robots for specific tasks and niches. Examples include JOZ for manure handling, Trioliet for roughage feeding and NEDAP for finding cows. The next section looks at the challenges that need to be overcome in terms of practical uptake, as well as the further development and testing of robotic solutions in the dairy sector.

6.2 Overall challenges

It is worthwhile connecting the aforementioned sectoral challenges with those identified in the 'European Robotics in agri-food Production: Opportunities and Challenges' report (Sander et al., 2021). Table 1 gives an overview of the 12 identified use cases and which of the 11 key challenges are mentioned per use case. It can be surmised that all use cases might have a link with livestock production systems, but if we limit ourselves to robotics that support poultry, pig and dairy farmers, the following uses cases can be used to inspire future developments.

6.2.1 Artificial intelligence and robotics for livestock farming

The monitoring, feeding and care of production animals such as cows, pigs, fish and poultry are daily recurring tasks. Smart sensor systems provide the necessary additional 'eyes, noses and ears' to the workers who take care of these animals, which form groups and are housed in general on ever-growing farms. Robotic support can relieve farmers in terms of inspection, sorting and feeding, so they can concentrate their care on the animals that require their support at the right moment, the right place and with the proper care. Core functions include observing animal behavior and physiology of individuals that are part of a group; feeding and treating individuals according to their needs; and harvesting the product from the animals (such as eggs, milk and wool), as well as sorting and regrouping animals. In addition, technology providers and feed-, food-, pharma- and breed companies, as well as contractors, are involved.

6.2.2 Realizing full autonomy of already mechanized tasks

Many agricultural tasks are mechanized, but the machinery still requires operators and major decisions are carried out by human intervention, based on experience rather than on (validated) data. Supervision and monitoring tasks are not yet automated, which is a barrier to a more sustainable method of production. When sensors gather data autonomously, when data-loops are closed and enhanced with automated decision (-support) algorithms, full autonomy can begin. The farmer can override a decision at any time, injecting specific knowledge not visible in the data. With autonomous systems, robots can become smaller (less soil compaction/energy use) and/or limit the use of chemicals, water, manure/nutrients and labor costs. This can help improve biodiversity, allow new farming methods such as intercropping/pixel farming and improve timing and 24/7 capabilities, as well as added crop value on minimal land use. The core function is the integration and demonstration of full autonomy or hybrid solutions in which robots and humans collaborate. Systems should act reliably and comply with safety standards for autonomous robots and provide convenience to the end user.

6.2.3 Cleaning in agri-food

Cleaning is a task that occurs in all stages of the agri-food production system. It is often carried out in harsh environments and at irregular times. The task requires dedication to do it properly, the key being to remove organic contamination from infrastructure and machines to prevent cross-contamination with pathogens. Examples include cleaning of animal boxes

Table 2 Identified key challenges

Targets	EU robotics requirement	Requirements related to livestock robot development
Configurability	Software and hardware reconfiguration considerations, including use of intuitive programing.	Aspects of configurability and re-configurability should be accommodated in the design brief for a robotic platform that is multifunctional and has a modular tooling selection to support farming tasks.
Adaptability	Response to changes in the operating environment, including the ability to self-learn and apply auto-configuration strategies.	Sensor and robotic platform needed to provide data on the operating environment (farm) for operational purposes, proving also the opportunity to facilitate adaptive response functions for the platform itself.
Interaction capability	Interaction with operators, other robots and other systems within a production environment.	Identification of animals with radio frequency or optical identification can be used to develop and exploit identification-assisted robotics that can be applied for 'systems of systems' interaction.
Dependability	Factoring periphery and environmental integration into the reliability and dependability requirement for mean time between failures.	Probability engineering principles and failure modes and effects analysis can feature as part of system design, with the livestock farm providing a platform for modeling robotic-agricultural environment interaction.
Capability	Kinematics and dynamics of manipulators as well as the positioning and navigation of autonomous platforms.	Livestock robotics systems should exploit developments in both radio frequency identification (RFID) and associated optical or electromagnetic sensing for positioning, navigation and mapping where needed.
Manipulation ability	Ability to handle material objects and tools in a manufacturing context. Adaptability and robustness are primary goals, along with the need for accuracy and repeatability.	Intrinsic to the design being proposed is the facility for robust and adaptable handling material objects in the form of animals/animal body parts and tools for performing farm tasks. While accuracies may not equate to levels experienced in manufacturing the same principles of measurement, measurement assurance, repeatability and reproducibility will be applied in measurements undertaken, adding to the precision base for PLF.

(*Continued*)

Table 2 (*Continued*)

Targets	EU robotics requirement	Requirements related to livestock robot development
Perception ability	Suitable choice of sensing modality, efficient signal and data analysis, as well as generating the maximum information output from the data at hand. Guaranteed safe perception is also a key requirement.	Integration of multiple sensor systems and real-time data capture, processing and communication can be seen as an integral part of the system design, with the animal-derived data sets also considered for inclusion in 'Big Data' set developments and system readiness for future 'Big Data' services and Internet-of-Things (IoT) object-connected services.
Decisional autonomy	Primary goal to increase the level of responsibility in the control processes of production systems, in which the resulting autonomy is focused on reducing energy consumption, increasing throughput and providing context-aware task control in interactions with operators.	Livestock robot systems should support self-awareness and enhance decisional functionality and contribute to the knowledge base for integrated support in farm decision-making.
Cognitive ability	Potential is for functions that contribute to a reduction in programing and configuration requirements in deployed systems.	Should be possible to enhance cognitive ability and learn the individual characteristics of animals, while also contributing to the knowledge base for integrated support in agricultural robotic systems.

Source: Adapted from: Sander et al. (2021).

in livestock systems. Stables and protected horticultural environments (e.g. greenhouses, mushroom cells and vertical farms) are cleaned upon completion of a production cycle. Lorries for livestock logistics should be cleaned following completion of delivery. Increased awareness of and demands on food safety have dramatically improved intensive hygiene over the last decade. In practice, this time-consuming task is commonly carried out manually. The chemicals are not user friendly. Another task is cleaning the products themselves. Cleaning can be carried out as a physical intervention to the product itself, such as washing table eggs. Cleaning robots should be able to navigate through constructed environments and detect organic or mineral contamination in moist environments. They should be able to apply effective strategies to remove contamination from infrastructure elements to create low-risk environments for pathogens and for cross-contamination in the production system. Sensor-based evaluation of cleaning effectiveness (detecting organic contamination) is an important part of the solution.

6.2.4 Connectivity, distributed intelligence and pervasive technology

Robotization, AI, IoT, VR, AR, Blockchain, 3D printing, sensors and drones are all disruptive technologies that have the potential to transform agriculture.

All these disruptive technologies come together under the umbrella of cooperative robotic systems with the ability to deal with the variety of small- and large-scale operations in future agricultural production systems. In agriculture, there are a variety of products and production scenarios. For a period, the trend was to increase the scale of operation and therefore the capacity of the machines used in the production process. Robotic systems emerged as an alternative to replace this large machinery. However, advances in robotization and progress in other domains such as IoT and sensors have allowed a pivotal shift from big machinery to swarms of small robots or multi-robot systems that cooperate, are more intelligent and form a pervasive part of the cyber-physical agri-food system. This could create farms in which the social and ethical structures and requirements align more harmoniously, where the economical success of the farm is increasingly correlated to the robotic and AI-driven support provided to the farmer rather than to the size of the farm, as is traditionally the case.

The first core function addressed in this use case is that robot systems should be connected so that they can be remotely contacted, known as localization, i.e. knowing when and where the robot system is deeply integrated into this functionality. The second function in this use case is fleet management. Cooperation between robots and decisions as to where to embed intelligence should become a basic functionality. The third functionality is that the robot systems become much smaller in such a way that they can be integrated into an environment without disrupting the process they support.

6.2.5 Robotics, artificial intelligence and data science for breeding

Creating and introducing new food products based on new or improved breeds of production crops, animals, insects and fish typically takes 5–10 years. The last decade's focus has been on genomics, which has resulted in sophisticated data analytics and automation of genomic laboratory procedures. The focus is now shifting toward gathering and analyzing phenotypic data.

Traditionally this is also based on manual labor, but the variety of circumstances and the phenotypic requirement for data demands advanced methodologies for data gathering and analysis. It is expected that robotics can stimulate the demand for phenotypes in view of challenges such as increased resistance to environmental changes (e.g. increased temperature range based on climate change) or weeds, increased periods of low or high water availability or an increased output or resistance to illness or pests. Automated data and AI-driven decision-making and selection processes are needed to ensure diversity, traceability and transparency (e.g. no genetic modification) of phenotypic data gathering.

Table 1 demonstrates that the technical challenges are generally mentioned in the use case. For livestock robotics, it appears that the key challenge of '24/7 level 5 cooperative systems and fleet and swarm management' in which robot systems should be able to perform tasks 24/7 with a degree of autonomy similar to those currently defined for autonomous cars (SAE Level 5), will be most important, directly followed by 'Robot-to-X interaction' and 'Interactive design of trustful, safe, and ethical robotic system'. These nicely match the described challenges for the livestock sector. It is also important to take note of the ecosystem, business, training and human capital development challenges. Although not mentioned in the use cases, they are a prerequisite for further development and practical implementation of robotics in the livestock sectors.

Table 2 shows the identified key challenges. To deliver robotics that safely and effectively support the monitoring and management of livestock farming processes, research is required to advance capabilities and key technologies relevant to industrial and service robotics. As the system should be a key part of a production process in which it is operating with livestock, it needs to ensure safe, cost-effective performance, reliability and ease of use. To this end, due attention should be paid to the EU Robotics 2020 strategic research agenda and multi-annual roadmap, addressing each of the key targets in the Table 2, in the context of our target application, and the prospects for advancing key robotics technologies in the different fields. While many of the EU Robotics' key ability targets are directed at robotics in manufacturing, they may also apply within the agricultural sector, including PLF.

7 Conclusion

This chapter has demonstrated that robotics is an important topic in livestock management. Aligned with the concept of PLF, it truly offers freedom of choice to farmers. A growing awareness of what has already been achieved and what is currently being utilized demonstrates that robotics is already at the center of modern farming systems. Based on the desire for innovation, we have seen many attempts to (re)design new robotic solutions. These challenges must be met with increased vigor and accompanied by the development of forward-looking and resilient farming systems. Robotics can play a pivotal role in this field.

8 References

Antonucci, F. and Costa, C. (2020). Precision aquaculture: A short review on engineering innovations. *Aquaculture International* 28(1), 41-57.

Banhazi, T., Halas, V. and Maroto-Molina, F. (2022). *Practical Precision Livestock Farming: Hands-On Experiences with PLF Technologies in Commercial and R&D Settings*. Wageningen Academic Publishers. https://doi.org/10.3920/978-90-8686-934-3.

Berckmans, D. (2008). Precision livestock farming (PLF). *Computers and Electronics in Agriculture* 62(1), 1.

Berckmans, D. (2017). General introduction to precision livestock farming. *Animal Frontiers* 7(1), 6-11. https://doi.org/10.2527/af.2017.0102.

Butler, Z., Corke, P., Peterson, R. and Rus, D. (2006). From robots to animals: Virtual fences for controlling cattle. *The International Journal of Robotics Research* 25(5-6), 485-508.

Comis, D. (2000). The cyber cow whisperer and his virtual fence. *Agricultural Research* 48(11), 4-7.

Conolly, A. (2022). *The Future of Agriculture*. Published by Agritechcapital.com.

Demmers, T. G. M., Cao, Y., Gauss, S., Lowe, J. C., Parsons, D. J. and Wathes, C. M. (2018). Neural predictive control of broiler chicken and pig growth. *Biosystems Engineering* 173, 134-142. https://doi.org/10.1016/j.biosystemseng.2018.06.022.

Elijah, A., Nzebo Richard, A., Thomas George, T. and Fumiya, I. (2022). Autonomous detection and sorting of litter using deep learning and soft robotic grippers 9. DOI: 10.3389/frobt.2022.1064853.

Fontana, I., Tullo, E., Carpentier, L., Berckmans, D., Butterworth, A., Vranken, E., Norton, T., Berckmans, D. and Guarino, M. (2017). Sound analysis to model weight of broiler chickens. *Poultry Science* 96(11), 3938-3943. https://doi.org/10.3382/ps/pex215.

Frost, A. R., Tillett, R. D. and Welch, S. K. (2000). The development and evaluation of image analysis procedures for guiding a livestock monitoring sensor placement robot. *Computers and Electronics in Agriculture* 28(3), 229-242.

Lokhorst, C. (2018). An introduction to Smart Dairy Farming. https://doi.org/10.31715/20181.

Mottram, T. (2016). Animal board invited review: Precision livestock farming for dairy cows with a focus on oestrus detection. *Animal* 10(10), 1575-1584.

Nielsen, M., Petersen, A. and Nielsen, K. (2012). Machine vision guided cleaning for autonomous pig sty cleaning. http://www2.atb-potsdam.de/cigr-imageanalysis/images/images12/tabla_137_C0555.pdf.

nn (2021). Virtual fence keeps cattle where you want them. *Farm Industry News*.

Pomar, C. and Remus, A. (2019). Precision pig feeding: A breakthrough toward sustainability. *Animal Frontiers* 9(2), 52-59.

Pomar, C., Hauschild, L., Zhang, G. H., Pomar, J. and Lovatto, P. A. (2011). Precision feeding can significantly reduce feeding cost and nutrient excretion in growing animals. In *Modelling Nutrient Digestion and Utilisation in Farm Animals* (pp. 327-334). DOI: 10.3920/978-90-8686-712-7_36.

Rodenburg, J. (2017). Robotic milking: Technology, farm design, and effects on work flow. *Journal of Dairy Science* 100(9), 7729-7738. https://doi.org/10.3168/jds.2016-11715.

Rutten, N., Velthuis, A., Steeneveld, W. and Hogeveen, H. (2013). Sensor systems for dairy cow health management: A review. *Proceedings of the Precision Dairy Conference and Expo: A Conference on Precision Dairy Technologies*. University of Minnesota. pp. 89–90. https://edepot.wur.nl/430185.

Sander, S., van Henten, E. J., Lokhorst, C., Pekkeriet, E. J. and Steckel, T. (2021). European robotics in agri-food production: Opportunities and challenges. https://doi.org/10.5281/zenodo.4742481.

Scholten, M. C. T., de Boer, I. J. M., Gremmen, B. and Lokhorst, C. (2013). Livestock farming with care: Towards sustainable production of animal-source food. *NJAS – Wageningen Journal of Life Sciences*. http://www.scopus.com/inward/record.url?eid=2-s2.0-84879221855&partnerID=40&md5=62232784f969290112f3b2bd469c2b3f.

Sharipov, D. R., Yakimov, O. A., Gainullina, M. K., Kashaeva, A. R. and Kamaldinov, I. N. (2021). Development of automatic milking systems and their classification. *IOP Conference Series: Earth and Environmental Science* 659(1), 012080. https://doi.org/10.1088/1755-1315/659/1/012080.

Sluis, M. V. D., Rodenburg, T. B. P. D., Ellen, E. D. D. and Haas, Y. D. D. (2022). The chicken and the tag: Automated individual-level activity tracking and the relationships between activity, body weight and leg health in broilers Wageningen University. Wageningen, WorldCat.org.

Spoutnic, le robot de ponte qui limite les pertes (2019). *Les Echos*

Trevelyan, J. P. (1989). Sensing and control for sheep shearing robots. *IEEE Transactions on Robotics and Automation* 5(6), 716–727.

Trevelyan, J. P. (1987). Robots in the shearing shed: Automated shearing of sheep using robots. *Advanced Robotics* 2(1), 3–8.

Umstatter, C., Morgan-Davies, J. and Waterhouse, T. (2015). Cattle responses to a type of virtual fence. *Rangeland Ecology and Management* 68(1), 100–107. https://doi.org/10.1016/j.rama.2014.12.004.

Vroegindeweij, B. A., Blaauw, S. K., Ijsselmuiden, J. M. M. and van Henten, E. J. (2018). Evaluation of the performance of PoultryBot, an autonomous mobile robotic platform for poultry houses. *Biosystems Engineering* 174, 295–315. https://doi.org/10.1016/j.biosystemseng.2018.07.015.

Wang, S., Jiang, H., Qiao, Y., Jiang, S., Lin, H. and Sun, Q. (2022). The research progress of vision-based artificial intelligence in Smart pig farming. *Sensors* 22(17), 6541.

Wind, T., Biewenga, G. and Lokhorst, C. (2017). Deliverable 4.1 standard operating procedures. Copyright © 2016 - 2019 – 4D4F Consortium, all rights reserved. https://www.4d4f.eu/content/standard-operating-procedures.

Wu, Y., Duan, Y., Wei, Y., An, D. and Liu, J. (2022). Application of intelligent and unmanned equipment in aquaculture: A review. *Computers and Electronics in Agriculture* 199, 107201.

Yanchang, L., Haisheng, Z., Zexu, L., Zhixia, Z., Xiangang, Z. and Guohou, L. (2021). Design of intelligent monitoring system for pig healthy breeding based on robot. *Journal of Chinese Agricultural Mechanization* 42(8), 187.

Yue, K. and Shen, Y. (2022). An overview of disruptive technologies for aquaculture. *Aquaculture and Fisheries* 7(2), 111–120.

Yousefi, D. B. M., Rafie, A. S. M., Al-Haddad, S. A. R. and Azrad, S. (2022). A systematic literature review on the use of deep learning in precision livestock detection and localization using unmanned aerial vehicles. *IEEE Access* 10, 80071–80091. https://doi.org/10.1109/ACCESS.2022.3194507.

Zillner, S., Bisset, D., Milano, M., Curry, E., García Robles, A., Hahn, T., Irgens, M., Lafrenz, R., Liepert, B., O'Sullivan, B. and Smeulders, A. (Eds.). (2020). Strategic research, innovation and deployment agenda - AI, data and robotics partnership. Third release. September 2020, Brussels. BDVA, euRobotics, ELLIS, EurAI and CLAIRE.

Index

3D digital model 429

ABAG. *see* Adaptive gain adaptive bias (ABAG)
Accurate diagnosis 266
Ackermann steering mechanism 591, 592
Ackerman steering geometry 153
Action Server 212–214
Active learning 14–16, 22
Active perception 23
Active visual perception 20–21
Adaptable diagnostics 266
Adaptive gain adaptive bias (ABAG) 45
Agreenculture 186
Agreenculture robot 189, 190, 193–194
Agricultural and Veterinary Chemicals Code Act 1994 524
Agricultural machine systems 139
  case study
    basic orchard mowing mission 157–159
    multi-vehicle mission planning, for citrus grove operations 159–163
  guidance and navigation in 141–145
  mission planning for 145–147
  navigation and motion control 153–156
  path planning 147–150
  search and optimization methods
    genetic algorithms 152–153
    heuristic-based algorithms 150–152
    rapidly exploring random trees 152
Agricultural unmanned aerial vehicles, sensing applications
  application in precision agriculture
    abiotic and biotic stress detection 429–430
    growth vigor and yield assessment 440–442
    insects and pathogens 438
    nutrients, precision fertilization 433–434
    phenotyping 438–440
    precision livestock farming (PLF) 442–443
    water, precision irrigation 430–433
    weeds, precision weed management 434–437
  flight planning and pre-processing
    image pre-processing 428–429
    measurements and standardization targets 427
    multiple interacting factors influence 426–427
    radiometric calibration targets 427–428
  future trends 443–444
  overview 415–417
  platforms for precision agriculture
    fixed-wing UAVs 424–426
    multi-rotor models 425–426
    rotary-wing UAVs 424–426
  synopsis 443–444
  unmanned aerial vehicle remote sensing sensors 417–424
Agricultural unmanned ground vehicles (AUGVs) 324
Agri-EPI Centre 515, 517, 524
Agri-food industry (AFI)
  legal frameworks 462
  robotics and automation 461
Agri-food robotics xxiv, 387. *see also individual entries*
  fault detection. *see* Fault detection
  machine learning. *see* Machine learning
  world model. *see* World model
AGRIST 644
AgROBOfood network xxix
AgroIntelli, 186
AGVs. *see* Automated guided vehicles (AGVs)
AgXeed 186
AIC. *see* Akaike Information Criterion (AIC)
Akaike Information Criterion (AIC) 551
AlexNet 109, 538
All-manual harvesting 174
AMS. *see* Automatic milking systems (AMS)
ANNs. *see* Artificial neural networks (ANNs)
Anthropomorphic gripper 401
Anti-resonance frequencies 52
Antiresonance frequency 51

Arable farming worksites 224
Arable fields 224
Area coverage path planning 148
Artificial intelligence (AI) xxiv, xxv, 233-236
  algorithms 83
  edge computing and distributed AI xxv
Artificial lighting 113
Artificial neural networks (ANNs) 572
  based algorithms 92
0A* search algorithm 151
Asterix 578
Augmented reality (AR) glasses 189
AUGVs. *see* Agricultural unmanned ground vehicles (AUGVs)
Automated and semiautomated robots 355
Automated guided vehicles (AGVs) 653
Automatic milking systems (AMS) 223
Automatic weed detection 94
Automating orchard crop management xxiv
Autonomous agricultural robotics 23-25
Autonomous crop machines
  co-robotic arrangement 493-494
  economic analyses 493-494
  economics 489-490
  farm labour 494
  Hand Free Hectare (HFH) 491
  HFH LP model 491-492
  on-site supervision 492-493
  ownership 493
  robotic harvesting 493
  study 489-490
  swarm robot cost 492
Autonomous machine/equipment 481-482
Autonomous navigation 79-80. *see also* Agricultural machine systems
Average reflectance 551
Azure inect ToF camera 600

Bell Laboratories 229
Bernoulli gripper 402-403
Bidirectional reflectance distribution function (BRDF) 55
Big Data approach 668
BIM. *see* Building information model (BIM)
Bin-Dog system 295, 594
Binocular stereo cameras 85
Bitcoin 245
Blockchains 245
BoniRob field robot 116
BRDF. *see* Bidirectional reflectance distribution function (BRDF)
Building information model (BIM) 61
Bush-type cropping systems 639
Butterworth filter 298

CaffeNet 109
Candidatus liberibacter solanacearum 94
Canopy interpretation 598-599
  flower detection 601-602
  fruit detection 602-604
  trellis wire detection 604
  trunk and branch detection 599-600
Canopy positions 202
Case-based reasoning algorithm (CBR) 279
CBAM. *see* Convolutional block attention module (CBAM)
CBR. *see* Case-based reasoning algorithm (CBR)
Challenges of action (CoA)
  dense vegetation environments 390-391
  diversity between different crop variants 389
  diversity within a crop variant (inherent diversity) 389-390
  high diversity in agricultural products 389
  safety/quality requirements in the handling of crops 390
  task complexity 390
Chassis design 329-330
  coupling systems
    draw-bar 331
    non-standard coupling 331
    standard 3-point hitch 331
  design parameter
    conventional 330
    gantry 330
    overrunning 330
    reconfigurable 330
  levels of mobility
    articulated 330
    fixed 330
Circular Hough Transformation 122
Clear diagnosis 266
Closed-loop inverse kinematic (CLIK) method 45
Closest Point algorithm 54
CNNs. *see* Convolutional neural networks (CNNs)
CoA. *see* Challenges of action (CoA)
Coefficient of variation 551
Cognitive ergonomics approaches 171
Coherent point drift (CPD) algorithm 54
Collaboration 172

Collision avoidance faults 264
Colour-based cameras 81
Colour cameras 95
Commercial-grade NIR stereo cameras 110
Commercial greenhouse 114
Commercial guidance system 144
Commercial orchards and vineyards 113
Communication and cloud processing 97
Communication networks 229–231
  Agricultural Industry Electronics Foundation (AEF) 230
  application programming interface (API) 232
  cellular networks 230
  consumer electronics 229
  C-V2X technologies 230
  external third-party services 229
  FIWARE platform 232
  industrial applications 229
  Internet of things 231–233
  Internet of Things–Architecture Reference Model (IoT-ARM) 232
  LinkedData 233
  semantic interoperability 233
  single GNSS positioning 231
  Vehicle-to-Everything (V2X) communication technologies 230
  Virtual Private Networks 230
Complementary grasping technologies 406
Complex traits 534
Composition 64
Computational power 110
Computer vision (CV) 91, 109, 235
Configuration 64
Conventional robots 388
Conventional tractor and sprayer 189
Convolutional block attention module (CBAM) 603
Convolutional neural networks (CNNs) 11, 83, 538, 539, 546, 572, 597, 601
Cooperative robotics 179
CountNet 602
COVID-19 pandemic 185
Crop phenotyping, robotics in
  case study
    abiotic stress 547–549
    biotic stress 549–554
    single view 3D reconstruction 545–546
  four-dimensional plant imaging 543–545
  future perspectives 554–555
  plant phenotyping 533, 534
  robotic manipulators 536
  structural phenotyping 534
  temporal phenotyping 534
  three-dimensional plant imaging
    growth and yield monitoring 542
    phenotypical information extraction 541–542
  two-dimensional plant imaging 537
    biotic and abiotic stress factors 538–539
    fruit phenotyping 540
    phenotypic parameters 539–540
Crop sensors 81
CropTeq 647
CV. *see* Computer vision (CV)
Cyber-physical system (CPS) 138, 188

DAI. *see* Decentralised Artificial Intelligence (DAI)
Data acquisitions 115
Data-based methods xxv, 266
Data fusion approach 88
Data spaces
  automatic and autonomous devices and robots 245
  buying and selling data assets 244
  communication and networking 244
  defined 243
  developments 245–247
  distributed ledger technology (DLT) 245
  FlexiGroBots concept 247–248
  future trends 250–252
  M2M communications 244
  and robots for agriculture tasks 247
  Semantic Web concept 244
  trust 245
DDPG. *see* Deep deterministic policy gradient algorithm (DDPG)
Decentralised Artificial Intelligence (DAI) 237
Deep belief networks 270
Deep deterministic policy gradient algorithm (DDPG) 121
Deepfield 186
Deep learning (DL) xxiv, xxvi, 11–12, 104, 597
  driven methods 538–540
  techniques 83
Deep learning networks 109, 648
Deep neural networks 4, 11
DeepSeg3-Dmaize 542
Degrees of freedom (DoFs) 326, 370–371
DeLaval 680

Delta mechanism 609
Dempster-Shafer (DS) theory 280
DenseNet 603
Department of Mines, Industry Regulation and Safety 524
Depth cameras 5
Design gripper, sorghum breeding
  coupled sliding mechanism 376
  design and evaluation 374
  load cell 376
  one-sided equivalency test 376–377
  operations 374
  plunges 374–376
  rind penetrometer 374
  testing and validation 376
  v-block 376
Designing skills 60
Detailed robot tasks 225
Developmental psychology 25
Digital twin (DT)
  agricultural system 190–193
  applications of 187
  augmented reality (AR) glasses 189
  case study results 214–215
  communication
    farming controller and implement 206
    implement and vehicle via ISOBUS 204–206
  discussion 215-217
  Farming Controller 206–214
  human-robot collaboration (HRC) 189
  implementation 188
  predictive maintenance 188
  production system 187
  ROBS4CROPS 189
  simulation outcomes 189
  smart implements
    for arable farming, in Netherlands 196–197
    canopy density 201-204
    for vineyards in France 197–199
    for vineyards in Greece and orchards in Spain 199-201
  vehicles
    agreenculture robot 193–194
    GNSS 195–196
    retrofitted tractor 194–195
    Robotti 194
  visualization tools 189
Dijkstra's algorithm 150–151
Direct sunlight 113
Discrete control plan 47
Distributed intelligence
  artificial intelligence 233-236
  edge computing and 236-237
  smart agriculture solutions and applications 237-238
Diverse pot-and bin-handling robots 224
3D LiDARs 108
1DOF actuator 49, 50
Domain knowledge 23
Dormant pruning 616
Drones and ground robots 237
DT. *see* Digital twin (DT)
Dual-mode gripper 403
Dutch horticulture technology 640

Economics of agricultural robotics
  adoption history 484–485
    autonomous crop machines 486–487
    robotics in livestock production 485–486
    uncrewed aerial vehicles (UAVs) 487–488
  agricultural technology choices 482–483
  automation 482
  broader implications, farm sector 497–499
    impact of swarm robotics on farm structure 497–498
    small irregularly shaped fields efficiently with swarm robots 498
    swarm robots radically alter farm equipment market structure 498–499
  estimated wheat cost of production in Britain 491
  expected benefits of agricultural robotics 488
    autonomous crop machines 489–494
    milking robots 488–489
  information gap 483
  milestones 485
  objectives 483–484
  overview 481–484
  policy, regulatory and institutional issues 501
    agricultural robotics 502-503
    AI 503
    data privacy and security 503
    guidance on human supervision of autonomous crop machines 501-502

public good benefits 504
public sector research and education 503–504
technical training and retraining 503
theft prevention 503
training required for human supervisors of robots 502
potential robotics, low-and middle-income countries 494–495
UAVs for smallholder farms 495
vision for low-cost autonomous crop machines 495–497
social impact 499–501
entrepreneurial opportunities 500
large-scale commercial arable farming sector 499–500
low-and middle-income countries 500–501
small-and medium-scale arable farms 500
synopsis 504–505
technology adoption 483
technology innovations 482
UK whole farm economic studies 492
US whole farm net returns from autonomous machines 490
Ecorobotix ARA spot sprayer 187
Ecosystem integration 139
EDG. *see* Embedded deformation graph (EDG)
EE. *see* End effector (EE)
EFDI. *see* Extended FMIS Data Interface (EFDI)
Effective learning 26
EfficientNet 109
Efficient oil seed pest management 247
EID. *see* Electronic tags (EID)
EKF. *see* Extended Kalman Filter (EKF)
Electrical faults 264
Electroadhesion soft gripper 398, 406
Electroadhesive grasper 403
Electronic tags (EID) 486
Embedded deformation graph (EDG) 543, 544
End effector (EE) 43
Entropic thresholding 84
Environmental variations 9–10
Environment model (EM) 62
EOLO sprayer 192
Error Estimate (EE) 277
EU. *see* European Union (EU)
EU General Data Protection Regulation (GDPR) 2016 472
EU RHEA project 577
EU Robotics 2020 685
Europe 75
European Telecommunications Standards Institute (ETSI) 229
European Union (EU) xxvii, 75
European Union Joint Research Centre document 515
Extended FMIS Data Interface (EFDI) 240
Extended Kalman Filter (EKF) 268
Exteroceptive sensors 3
Eye-in-hand systems 7

False Isolation Rate (FIR) 276
FarmDroid 186
Farm Management Information Systems (FMISs) 144, 225
Farm tasks 225
Faster-RCNN 109
Fault detection and isolation (FDI) xxv
comparing diagnostic methods 270–272
components 287–291
data-based methods 268–270
deep learning 298, 299
defined 264
hardware redundancy 270
human/robot interaction module
adjustment of sub-methods 286
and criticality assessment 280–282
fault identified by operator 286
human operator 287
proposal for solution 282–285
hybrid diagnostic method 274–278
hybrid diagnostic method and human/robot interaction module 307–314
Kalman filters 295–297
knowledge-based methods 268
model-based diagnostic method 267–268
OMOPAMED2 270
particle filter 300–304
principles of 264–266
Robot Fault Diagnosis Supervision System (S2D2R) 270
robot movement and potential faults 291–295
wheel locking faults 298–306
Federated learning (FL) 237
Field crop cultivation, robots in
applications
harvesting 578

monitoring plant diseases and pests 575-578
transplanting 574
in weed control 575-578
future trends 578-579
overview of 567-569
requirements 569-571
technologies 571-573
FIR. *see* False Isolation Rate (FIR) 5-DOF manipulator 608
FIWARE platform 232
FL. *see* Federated learning (FL)
Flavescence doree disease 576
Fleet tasks 225
Flourish 186
FlowNet2 600
FollowStem task 47
Food and Agriculture Organization (FAO) 355, 515
4-DOF manipulator 607, 608
Four-wheels/more robots 333
Fresh Fruit Robotics 611-613
Fruit classification 109
Fruit tree pruning 616

GAIA-X 246
GAN. *see* Generative adversarial network (GAN)
Gantry systems 7
Generalized Traveling Salesman Problem (GTSP) 161
Generative adversarial network (GAN) 600
Generative learning 17-18
Genetic algorithms xxiv, 153
Genome-wide association studies (GWAS) 539
Geometric maps 42
German start-up Organifarms 642
Global navigation satellite system (GNSS) 41, 80, 195-196, 427, 653
Global System for Mobile Communications standard 230
GLSD. *see* Grapevine leaf stripe disease (GLSD)
GNSS. *see* Global navigation satellite system (GNSS)
GoogleNet 109
GPS fault isolation 296
GPS fault scenario 312
Grafting robot 649
Grain Producers Australia 524
Grapevine leaf stripe disease (GLSD) 576
Graphical model-based methods 268
Grasping techniques in agricultural robots
cutaway of the custom gripper 377
design of robotic manipulators 357
different types of finger designs 366
grasping mechanisms evaluated for stalk grasping 375
gripper design 363
actuation mechanisms 363-364
finger design 364-365
soft grippers 365-367
histogram of penetrometer measurements and difference 378
interpreting the adequacy of dairy farm management from cow milk yields 374
mechanics of grasping 358
dynamics 362-363
kinematics 360-362
problem formulation 358-360
overview 355-356
real-world applications 370
case study, design of a gripper for sorghum breeding 374-378
food handling and transportation 372-373
in-field tasks 370-372
robotic manipulators 356-357
samples of human hand grasping 359
soft gripper actuated with air 364
synopsis 378-379
systems integration 367
motion planning and control 369-370
perception 367-368
three generic gripper configurations for grasping tomatoes 361
visualization of the slip point 377
Greedy Algorithm 151
Greenhouse applications 169
Greenhouse cultivation
crop management 647-649
crop scouting 650-652
future research trends 653-655
grafting and planting 649-650
harvesting 638-639
cucumber 642-643
flowers 645
mushroom 645
strawberry 641-642
sweet pepper 643-644
tomato 639-641
insect detection and control 650-652

?-650

38
tics 652–653

Grid-based representations 149
Ground control points (GCPs) 427–428
Guava harvesting 121
GWAS. *see* Genome-wide association studies (GWAS)

Hand Free Hectare (HFH) 491
'Hands Free Farm' 515, 517, 524
Hardware redundancy 266
Harvesting fruits 369
HASL-based gripper 396–397, 403
HFH. *see* Hand Free Hectare (HFH)
Hidden Markov model (HMM) 543
High-wire cropping system 639
'Hill Control' control system 187
HIS. *see* Hue-saturation-intensity (HIS)
HMM. *see* Hidden Markov model (HMM)
Horticulture 223
Hortikey Plantalyzer 651
HRC. *see* Human-robot collaboration (HRC)
HRI. *see* Human-robot interaction (HRI)
HS camera 550, 551
Hue-saturation-intensity (HIS) 538
Human-Oriented Semi-Autonomy 174
Human-robot collaboration (HRC) 170, 171
  in agriculture xxv
  viz. synergetic systems 171
Human-robot interaction (HRI) 170, 605
Hybrid Diagnostic Method (MHD) xxvi
Hybrid topometric map 44
Hyperspectral (HS) sensor 537

i4Trust 246
iCassava2019 117
ICP. *see* Iterative closest point (ICP) algorithm
ICTs. *see* Information and computer technologies (ICTs)
Identification and removal technology 94
IFAC. *see* International Federation of Automatic Control (IFAC)
Image-acquisition systems 6–7
Image analysis methods. 48
Image augmentation 12–13
ImageNet dataset 118
Image pre-processing
  geometric rectification 428
  orthomosaic, 3D digital model and SfM photogrammetry 429
  radiometric calibration 428–429
Image segmentation algorithms 84, 92
Imaging sensors 4–6
IMU. *see* Inertial measurement unit (IMU)
Incremental learning 14
Inertial measurement unit (IMU) 595
Inexpensive diagnosis 266
Informational filter (IF) 268
Information and computer technologies (ICTs) 662
Intercropping system 406–407
International Data Spaces Association (IDSA) 245
International Federation of Automatic Control (IFAC) 655
International Society for Precision Agriculture (ISPA) 663
Internet of Robotic Things (IoRT) 226, 227, 229
Internet of Things–Architecture Reference Model (IoT-ARM) 232
Internet of Things stations 169
Invariants 64, 65
IR thermal camera 539
iSHARE 246
ISOBUS control technology 324
ISOBUS Task Controller (TC) 239
ISODATA 84
ISPA. *see* International Society for Precision Agriculture (ISPA)
Israel-based company Arugga AI farming 648
Israel-based MetoMotion 640
Iterative closest point (ICP) algorithm 543

Kalman Filter (KF) xxvi, 268
Kiwifruit picking 121
K-means 117
  algorithm 84
  clustering 542
K-NN 117
Knowledge-based methods xxv, 266
KnowRob 62
Kompano robot 647, 648

Labeled property graphs (LPGs) 63
LadyBird dataset 116
Land preparation 170
Laser range finder (LRF) 43, 48
Laser technology 110

LatticeNet 111
Lely Exos 674
Level of automation (LOA) 172
LiDAR. *see* Light detection and ranging (LiDAR)
Light detection and ranging (LiDAR) 43, 48, 91
  devices 82
  scanning 617
  sensors 4–6
  system 600
Linear Matrix Inequalities 188
LinkedData 233
Livestock production, robotics in
  dairy production 673–674
    cleaning 676
    feeding 674–675
    milk processing 676–677
    monitoring and management 675–676
  overview 661–663
  pig production 670–671
    bedding distribution 671–672
    cleaning 672
    precision feeding 672–673
  poultry production 668–669
    broiler chickens abnormal behavior 670
    dead bird detection 669–670
    litter quality monitoring 670
  precision livestock farming 663–664
    architecture 667
    maintenance and documentation 668
    modeling and knowledge engineering 668
    standard operating procedures 664–666
    system of systems 667
    systems engineering 667
    systems integration 667
  research challenges 679–680, 682–683
    agri-food cleaning 681, 683–684
    artificial intelligence and data science 684–685
    artificial intelligence and robotics 681
    autonomous systems 681
    distributed intelligence and pervasive technology 684
  sectoral challenges 677–680
Local planning functions 146
Locomotion principle 326
Logistic regression 551
M2M communication 232
Machine-aided harvesting 174
Machine learning xxiv, 83, 654
  data availability
    available datasets 116–117
    coping, developments in 117–119
  definition 104–105
  environment complexity and variability 112–115
  integration and field-testing use-cases 121–124
  long cycle times 119–121
  for perception 107–108
  perception and planning tasks 103
  for robotic planning 111–112
  robust autonomous or semi-autonomous agricultural operations 103
  sensors for 106–107
  three-dimensional perception data 110–111
  two-dimensional perception data 108–109
Machine learning operations (MLOps) 236
Management zones (MZs) 415
Manned precision agriculture systems 146
Map-based navigation 196
MAS. *see* Multi-agent systems (MAS)
Mask-RCNN 109
Mask Scoring R-CNN 601, 618
MAVLink protocol 249
MDR. *see* Missed Detection Rate (MDR)
Mechanical faults 264
Minimum error thresholding 84
Missed Detection Rate (MDR) 276
Mission control 239
Mission control platforms
  by ISOBUS 239–240
  need for 238–239
  QGroundControl drone control system 240–241
  service framework 241–243
Mission planning 139, 140
MLOps. *see* Machine learning operations (MLOps)
MMAE. *see* Multiple-Model Adaptative Estimation (MMAE)
MobileNets 109, 120
Mobile robot design 326
Mobility configurations
  articulated steering 336
  coordinated steering 335
  differential steering 334–335

independent steering 335–336
skid steering 334
Mobility platforms for agricultural robots
agricultural unmanned ground vehicles, mobility classification 326–328
chassis and implement coupling 329–331
locomotion 328–329
suspension 329
conclusion and future trends 347–348
farmer robot Polibot 342
hybrid and other unconventional unmanned ground vehicles mobility platforms for agriculture applications 346–347
legged mobile robots 344–346
overview 323–326
tracked mobile robots (TMRs) 339–341
case study, polibot tracked robot 341–344
wheeled mobile robots 332–333
case study, Agri.Q–sustainable rover 336–339
classification based on the number of wheels 333–334
mobility configurations 334–336
Model-based methods xxv, 266
Modern robots 3
Moment-preserving algorithm 84
Monitoring mode 310
Monte Carlo approach 195
Motion control systems 156
Multi-agent systems (MAS) 235, 237
Multimodal gripper 405
Multiple-Model Adaptative Estimation (MMAE) 268
Multiplicative faults 265
Multi-robot systems (MRS) management 229
Multispectral cameras 91
Multi-view stereo (MVS) techniques 541, 542, 546
Multi-view visual perception 19–20
MVS. *see* Multi-view stereo (MVS) techniques
MZs. *see* Management zones (MZs)

Naïo 186
Natural grippers 405–406
Natural Language Processing (NLP) 235
Natural pollination 619
Navigation and motion control 141
Navigation faults 264
Near-infrared (NIR) 81
sensor 537, 538
Neural network approach 57
Newtonian (unicycle) kinematic model 291
NIR. *see* Near-infrared (NIR)
Nominal Mode Return Delay (NMRD) 277
Non-complex traits 534
Normalised Difference Vegetation Index (NDVI) 93
North America 76
Norwegian/UK company SAGA Robotics 642

OAK-D Power-over-Ethernet (PoE) cameras 197
Object pose 8
Object variation 9
Occupancy grid map (OGM) 42
Octiva 651
Omnidirectional driving mechanism 592
OODA loop 173
Open-access datasets 115
Open street map (OSM) 61
Orchard operations 588–589
challenges and opportunities 620–623
end-effector technologies
flower thinning end-effectors 614–615
harvesting end-effectors 612–613
pollination end-effectors 615
pruning end-effectors 613–614
human-machine-canopy interaction 605–606
machine vision system 596
canopy interpretation. *see* Canopy interpretation
machine learning tools 596–597
nighttime capability 604–605
obstacle detection 598
orchard vision system 596–597
orchard navigation 592–593
simultaneous localization and mapping 594–595
vision-based guidance 593–594
overview 587–588
pruning 589
robotic manipulation 606
coordinate manipulator 610–612
hybrid manipulator 609–610
parallel manipulator 608–609
serial manipulator 606–608
robotic orchard operations 615–616

robotic pollination 619–620
robotic pruning 616–618
robotic thinning 618–619
robotics for 589–591
wheel-type platforms 591–592
Orchard robotics
ambient light conditions and visibility 79
automation and robotics 76
canopies 80
canopy detection for automated spraying 95–96
consumer demand 76
dairy management robots 79
detecting fruit for harvesting 91–92
detection of weeds 94–95
digitalisation 76
environmental regulations 76
fruit-bearing orchards, bushes and vineyards 76
industrialised agriculture 76
innovation and modernisation 76
labour costs 76
multidimensional sensors for 84–88
pests, diseases and stress, detection of 93–94
plant monitoring 92–93
sensor and data analysis technologies 80–84
sensor fusion for augmented perception 88–91
Oregon State University 600, 617
Otsu method rank 84

PA. *see* Precision agriculture (PA)
Panasonic Corporation 640
Parallel Programming 237
Parameter estimation 267
Partial least squares regression (PLS) 547
Passive visual perception 18–19
Path planning 141, 149
PATS drones 652
PCA. *see* Principal component analysis (PCA)
Perception 3, 108
challenge 387
Phenotypic traits 108
Photochemical Reflectance Index 93
Photonics 83
Photovoltaic (PV) panels 338
Physics-based model 188
Picking robots 80
Pinhole camera model 58
Plant Elevation System 648
Plant monitoring 92–93
Plant phenotyping robotic system 536
PlantVillage 117
PlantVillage dataset 117
PLF. *see* Precision livestock farming (PLF)
PLS. *see* Partial least squares regression (PLS)
PLS-DA model 548
PM. *see* Pruning mechanism (PM)
PointNet 110, 542
PointNet++ 111
Point Quadrant 202
Point-to-point planning 148
Pose 141
Pose sensing 156
Poultry xxix
Power takeoff (PTO) 331
Precision agriculture (PA) 323, 415–417, 663
abiotic and biotic stress detection 429–430
growth vigor and yield assessment 440–442
nutrients
Agricultural Research Organization (ARO) 434
nitrogen, phosphorous and potassium 433
partial least squares regression (PLS-R) model 434
UAV stress detection study 433–434
phenotyping
field-based henotyping (FBP) 439
high-throughput phenotyping platforms (HTPPs) 439
plant breeding 438–439
UAVs images 439–440
precision livestock farming (PLF)
real-time monitoring 442
UAV images 442–443
water
canopy water content (CWC) 430–433
plant growth and development 430
plant temperature 430–431
thermal sensing 431
UAVs images 431
weeds
biotic factors 434–437
precision weed management (PWM) 437
UAVs images 437

Precision livestock farming (PLF) 662–665, 685
Principal component analysis (PCA) 542, 547, 548
Production job outcome 139
Proprioceptive sensors 3
Pruning and harvesting 124
Pruning mechanism (PM) 371–372
PTO. *see* Power takeoff (PTO)

QGroundControl software 241
QR code 47
QTL. *see* Quantitative trait locus (QTL) mapping
Quantitative trait locus (QTL) mapping 540

Random forest models 548, 552
Rao-Blackwell version 268
Rapid diagnosis 266
Rapidly-exploring Random Trees (RRT) xxiv
Raw sensor data 38
R-CNN. *see* Region-based CNN (R-CNN)
Reactive diagnosis 314
Recurrent neural networks (RNNs) 597
Red-green-blue (RGB) cameras 5, 55, 95, 122, 123, 537, 643, 644
Region-based CNN (R-CNN) 539, 601–603
Regulatory frameworks and standards, European Union
  data privacy and data ownership
    The Data Act 473–474
    digital assets 473
    EU, data protection 472
    example of 473
    GDPR aims 472
    high-tech applications 471–472
  European Union directive, liability for defective products
    AI-based system 471
    applications 468
    defectiveness 469
    EU directive 469
    EU framework 470
    EU Product Liability Directive 468–469
    European Commission 470–471
    PLD 469–470
    product safety regulation 468
    tendency 471
  The European Union's Artificial Intelligence Act
    banned 475–476
    EU's AI strategy 474
    high risk 475
    low risk 475
    techniques and approaches 474
  ISO Standards applicable to agribots 465
  overview 461–463
  product safety regulation and standardization
    autonomous mobile robots (AMRs) 467
    CE marking 467–468
    compliance 464–466
    consensus and approved 464
    General Product Safety Directive (GPSD) 463
    harmonized standards 466
    ISO 12100 466–467
    ISO18497 467
    Member States (MS) 463–464
    Rapid Exchange of Information System (RAPEX) 468
    regulations and directives 463–464
    robots 464–465
  synopsis 476–477
Reinforcement learning 105, 111–112
Remaining Useful Life (RUL) 188
Remote sensing 81
Remote sensing (RS) technology 416
Replanning 146
Required Data Storage (RDS) 277
ResNet 109, 603
ResNet18 538
Resonance frequencies 53
Resource 65
Retinex algorithm 54
Retrofitted tractor 194–195
RGB. *see* Red-green-blue (RGB) cameras
RGB-D cameras 108
RNNs. *see* Recurrent neural networks (RNNs)
Robocrop system 187
RoboEarth 61
Robot entrepreneur's Management Information System (RMIS) 225
Robot Fault Diagnosis Supervision System (S2D2R) xxvi
Robotic gripper 650
Robotic hardware 653
Robotic harvesting 114
Robotic manipulators 536
Robotic milking 674
Robotic pruner 605

Robotic pruning xxviii
Robotics 235
Robot model (RM) 62
Robot Operating System (ROS) 206, 288
Robot swarm mission control 238
Robot tasks 225
Robotti 194
ROBS4CROPS xxv, 189
Robust mission management tools 146
Robustness 63
Robust perception 125
RTK-GPS techniques 592

SEARFS. *see* Simulation Environment for Precision Agriculture Tasks using Robot Fleet (SEARFS)
Segmentation 124
Sensor fusion 21
Sensorimotor contingencies 24
Sensor systems 669
Service Framework 226
7-DOF manipulator 607
SfM-photogrammetry 429
Short-wave infrared (IR) regions 81
Side grasping 399
Silent Spring 514
Simulation Environment for Precision Agriculture Tasks using Robot Fleet (SEARFS) 572
Simultaneous localization and mapping (SLAM) 41, 195, 594–595
Situation awareness (SA) 172
6 degrees of freedom (6-DOF) 607, 608, 617
  robotic manipulator 576, 578
Skid-steering faults 315
Skill 64
SLAM. *see* Simultaneous localization and mapping (SLAM)
Small contact forces and singularities 45
Smallholders 224
Smart farming 169
Smart sensor system 681
SMOTE. *see* Synthetic Minority Over-Sampling Technique (SMOTE)
Social and ethical considerations, agricultural robotics
  ethical concerns 516–517
  overview 513–514
  responsible agricultural robotics
    anticipating impacts 514–515, 517
    inclusion to date 521–522
    inclusion with participatory methods 518–521
    reflexivity and responsiveness 523–525
    in robotic futures 522–523
Soft grasping in agriculture
  case study, use of soft robotics in intercropping 406–407
  challenges of action in agricultural robotics 389–391
  designing the ideal versatile gripper 405–406
  developing an assessment matrix for soft grippers 394–396
  different ways of detaching crops from peduncle 393
  how soft robotics can address challenges of action 391–394
  number of soft grippers 392
  overview 387–388
  performance analysis 399
    concave objects 403–405
    flat objects 401–402
    object softness 399–401
    side pose 399
  performance analysis of current soft agri-food grippers 396–398
  performance matrix for a soft robotic gripper 395
  synopsis 408–409
The SoftHandler 406
Soft robotics 388, 409
  in intercropping xxvi
Software-defined Wide Area Network (SD-WAN) 206
Soil sensors 81
Sonar sensors 88
SOP. *see* Standard operating procedures (SOP)
Space exploration 263
Speciality crops 75
Spectral sensing 91
Spectral sensors 81
Spectro-radiometer 551
Spraying 121
Spraying robot 123
Standard deviation 551
Standard operating procedures (SOP) 664–666
Stereo vision 91
Strawberry harvesting 121
  robot 637, 638

system 54
Studying natural grippers 405–406
SugarBeets2016 dataset 116
SugarBeets dataset 118
Summer pruning 616
Supervised learning systems 16–17, 108
Support vector machines (SVMs) 105, 542, 543
Suspension type
active 329
hybrid 329
passive 329
rigid 329
SVMs. *see* Support vector machines (SVMs)
SWEEPER 643, 644
Sweet peppers 114
harvesting 119, 121
Synthetic data 17–18
Synthetic Minority Over-Sampling Technique (SMOTE) 548

Tactile sensors 48
Task 64
Task variation 10
Teleoperation 173
Temporal information 8
Thermal cameras 94
Thermal imaging 82
Think Local Act Personal Partnership 518
3-DOF manipulator 578
Three-dimension (3D) 367
Three-finger gripper with soft palm 406
Three-wheeled robots 333
TIM. *see* Tractor Implement Management (TIM)
Time-of-flight (ToF) 48
cameras 82, 91
Time Per Cycle (TPC) 277
TMA. *see* Tractor and Machinery Association (TMA)
TMRs. *see* Tracked mobile robots (TMRs)
Top-down cropping method 115
Topological map 44
Toroidal hydrostat gripper 405
Tracked mobile robots (TMRs)
example 341
full and half-track 340
limited top speed 339–340
obstacles and uneven and soft terrain 339
polibot tracked robot 341–344
track design configurations 340
vibration 340
Tractor and Machinery Association (TMA) 524
Tractor Implement Management (TIM) 205
Training set 13–14
Transfer learning 12–13
Traveling Salesman Problem (TSP) 161
Tree canopy vegetation 144
TSP. *see* Traveling Salesman Problem (TSP)
TURTLE robot of TechUnited 60
Two-wheeled robots 333

UAVs. *see* Unmanned aerial vehicles (UAVs)
UFO. *see* Upright Fruiting Off-shoot (UFO)
UKF. *see* Unscented (UKF)
Ultrasonic rangefinders 96
U-Net model 540
Unified Robot Description Format (URDF) 572
United Nation's State of Food and Agriculture report 515
University of Auckland 619
Unmanned aerial vehicles (UAVs) xxvii, 6, 96, 323, 416, 576, 577
remote sensing sensors 417
applications 423–424
light detection and ranging (LiDAR) 419–420
passive and active optical sensors 417–418
PA study, suitability 423
soil sensing 418–419
solar-induced fluorescence (SIF) 418
ULS systems 420–423
use of 416–417
Unmanned ground vehicles (UGVs) 6, 323–324
Unscented (UKF) 268
Unsupervised domain adaptation techniques 236
Unsupervised learning 16–17, 22–23
Upright Fruiting Off-shoot (UFO) 616
UR5 608
URDF. *see* Unified Robot Description Format (URDF)
US-based company AppHarvest 640
User interfaces (UI) 176, 226

Vacuum end-effectors 613
Variable rate technology (VRT) 200
VGG 109
VineScout robot 94

Virtual Private Networks 230
Vision-based detection 54
Visual information 7-8
Visualization tools 189
Visual perception, for agricultural robotics
  challenges of incomplete information 10–11
    active visual perception 20–21
    multi-view visual perception 19–20
    passive visual perception 18–19
  challenges of variation 9–10
    active learning 14–16
    composition of training set 13–14
    deep learning 11–12
    incremental learning 14
    synthetic data and generative learning 17–18
    transfer learning and image augmentation 12–13
    unsupervised and self-supervised 16–17
  future research
    active learning and uncertainty 22
    active perception 23
    autonomous agricultural robotics 23-25
    domain knowledge 23
    quality of training set 22
    unsupervised learning 22-23
  image-acquisition systems 6–7
  imaging sensors 4-6
  LiDAR sensors 4
  perception 3
  proprioceptive and exteroceptive sensors 3
  visual information 7–8
Visual sensing 108
V-shaped fruiting wall 588

Washington State University 603, 617
Water-sensitive paper 202
Weather sensors 81
Web of Things 244
Weed detection 109
Weeding and pest control robots 223
WEED-IT spot sprayer 186
WGS84 coordinate system 213
Wheeled autonomous mobile robots 263
Wheeled mobile robots 138
Wheeled vehicles 332-333
Wide-boomed sprayers 137
World Bank 587
World model
  affordances model 39
  defined 38
  features 39
  invariant-based world models 39
    approaches 41
    open-field applications 40
    problem 41
    SLAM 41
  multi-robot cooperation 59-61
  for reasoning 61-63
  sensors, data fusion and path planning communicate 39
  for skill configuration 53-58
  skills and resources 38
  static discrete controllers 38
  task redundancy 48-53
  topological maps 42-47

XGBoost 548

You Only Look Once (YOLO) 109
  algorithm 92
  v3 603
    based model 601
    network 541
  v4
    based model 601
    DL network 121
    network 603
  v5 603
    architecture 539